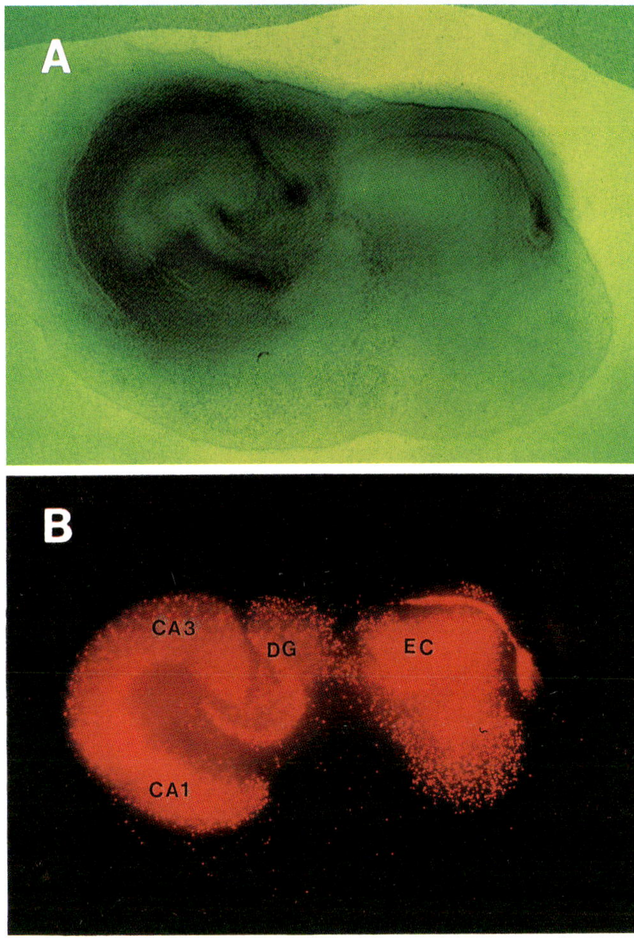

PLATE 1. Photomicrograph of hippocampal explant culture: phase contrast photomicrograph of hippocampal explant culture of 14 d *in vitro* (A). Fluorescent photomicrograph of the same culture 24 h after the 45 min of ischemic insult shows extensive labeling of most of the neuronal layers (B). Subregions are labeled on B: CA1and CA3 region; DG, dentate gyrus; EC, entorhinal cortex.

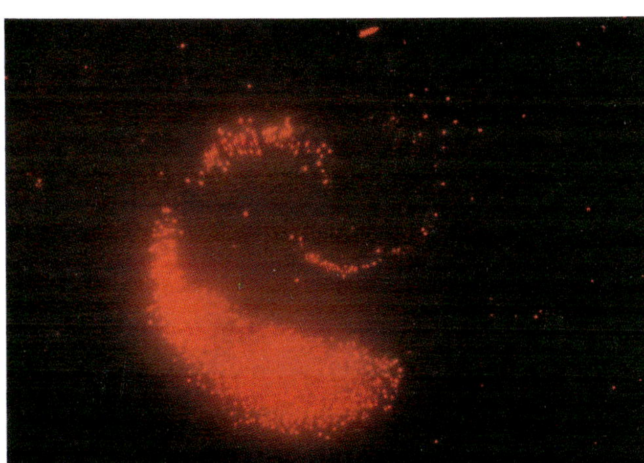

PLATE 2. Selective ischemic injury of hippocampal pyramidal neurons: with ischemia maintained for 30 min, more selective injury in CA1 region of the hippocampus is obtained. Minimal to mild injury is apparent in dentate gyrus and CA3 region.

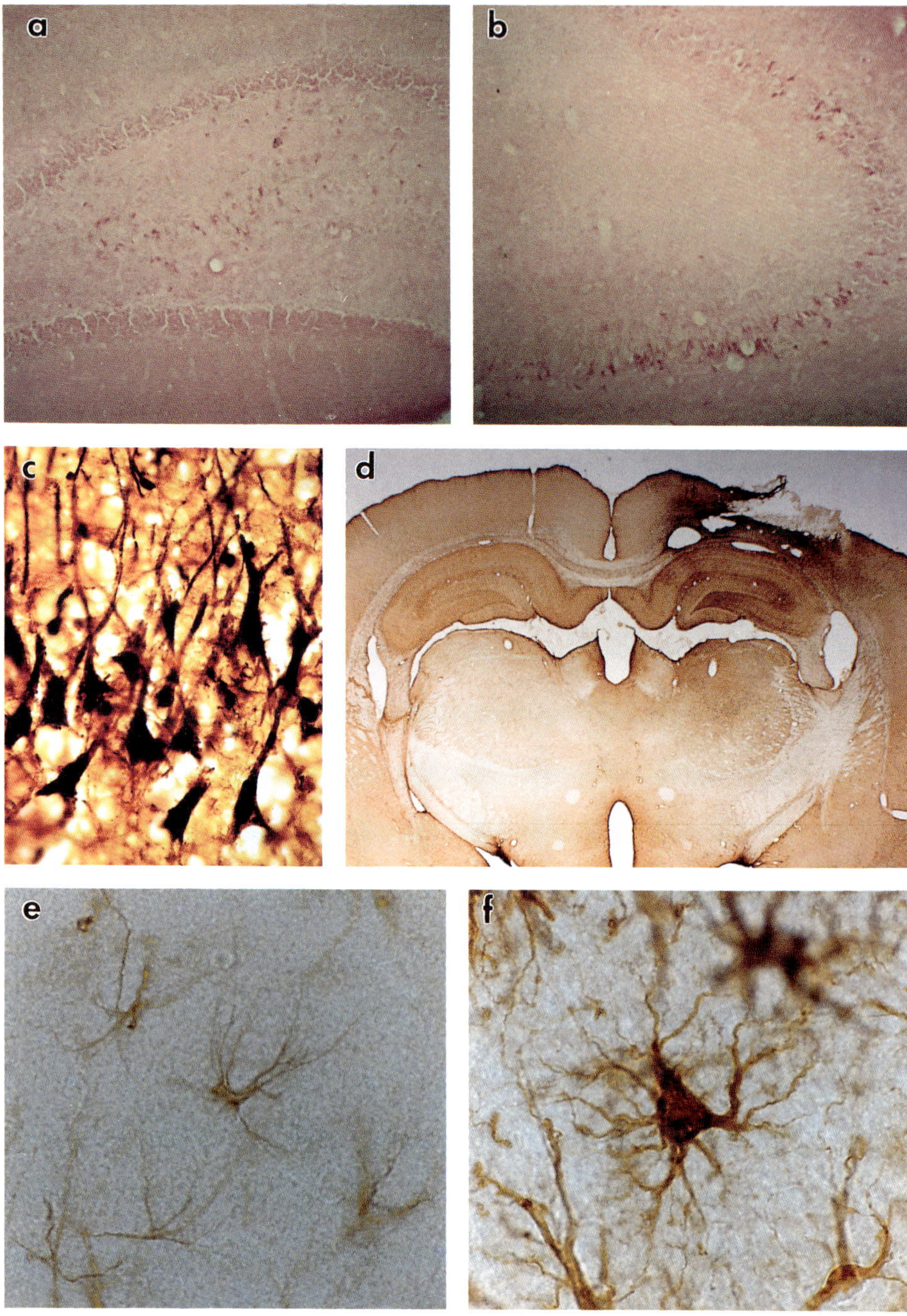

PLATE 3. Histopathology observed in rats sacrificed at acute time points following contusion (3.22 m/s impact velocity; 2 mm cortical compression) of the right sensorimotor cortex. Acid fuchsin staining of brain tissue 6 h after injury reveals numerous acidophilic (pink-red) neurons within the ipsilateral hilar region (a) and CA3 sector (b) of the hippocampus. Silver staining reveals a similar pattern of neuronal damage (argyrophilia; dark black perikarya) in the CA3 sector of the hippocampus at 6 h after contusion (c). GFAP staining illustrates that a robust astrogliosis has occurred within the contused cortex and in the ipsilateral hippocampus and thalamic regions by 8 d after TBI (d). The GFAP-positive cells in the ipsilateral hippocampus (and other ipsilateral structures) exhibit a marked hypertrophy from 24 h (e) to 8 d (f) following injury.

CRC Series in Membrane-Linked Diseases

CENTRAL NERVOUS SYSTEM TRAUMA

Research Techniques

CRC Series in Membrane-Linked Diseases

S. Tsuyoshi Ohnishi and Tomoko Ohnishi, Series Editors
Philadelphia Biomedical Research Institute
and
University of Pennsylvania School of Medicine

Cellular Membrane: A Key to Disease Processes (1993)

Membrane Abnormalities in Sickle Cell Disease and in Other Red Blood Cell Disorders (1994)

Malignant Hyperthermia: A Genetic Membrane Disease (1994)

Central Nervous System Trauma: Research Techniques (1995)

CRC Series in Membrane-Linked Diseases

CENTRAL NERVOUS SYSTEM TRAUMA

Research Techniques

Edited by

S. Tsuyoshi Ohnishi, Ph.D.
Director
Philadelphia Biomedical Research Institute
King of Prussia, Pennsylvania

Tomoko Ohnishi, Ph.D.
Research Professor
Department of Biochemistry and Biophysics
and the Johnson Research Foundation
University of Pennsylvania School of Medicine
Philadelphia, Pennsylvania

CRC Press
Boca Raton New York London Tokyo

Library of Congress Cataloging-in-Publication Data

Central nervous system trauma : research techniques / edited by S. Tsuyoshi Ohnishi, Tomoko Ohnishi.
p. cm.
Includes bibliographical references and index.
ISBN 0-8493-8094-4 (alk. paper)
1. Brain damage—Research—Methodology. 2. Central nervous system—Wounds and injuries—Research—Methodology. I. Ohnishi, S. Tsuyoshi. II. Ohnishi, Tomoko.
[DNLM: 1. Central Nervous System—injuries. 2. Head Injuries. 3. Diagnosis, Laboratory. 4. Technology, Medical. WL 300 C3978 1995]
RC387.5.C46 1995
616.8—dc20
DNLM/DLC
for Library of Congress 94-41529
CIP

International Standard Book Number 0-8493-8094-4
Library of Congress Card Number 94-41529
Printed in the United States of America 1 2 3 4 5 6 7 8 9 0
Printed on acid-free paper

DEDICATION

To those who have suffered from CNS injury,
and who are still in a wheelchair or bedridden.

PREFACE

Let's Work Together!

Dear Karen,
How is your dad doing?
It is a total surprise to hear
That your dad was injured while he was skiing,
And is still on a ventilator.
Since I used to compete in local ski races,
I know how exciting skiing is.
But I also know so well that the enjoyment
Is always a hairline away from a possible accident.
I am praying for his quick recovery.

Dear Mrs. Lewis,
Thank you very much for sending me the information.
I now know precisely what happened.
While your husband was attending
A winter conference in Copper Mountain, Colorado,
Your entire family was enjoying skiing together.
When everyone else turned right to stop at the lodge for lunch,
He attempted to stop by turning toward the left.
Just a few steps from the door,
Was a drainage ditch, which he did not recognize
Because the snowy day gave no shadow to the object.
It caught the tip of his ski. . . .
He had no speed,
But he was thrown forward to the ground on his face!
What kind of karma made him turn left?
Why was the ditch without a warning flag?

Dear Mrs. Lewis,
Thank you very much for sending me a newspaper article.
I am so impressed with his accomplishments.
As a surgeon, and as a missionary,
He had worked for a developing country
By living there with the family.
He is still working for the welfare of these people,
By co-founding a fellowship to send health professionals overseas.
He recently received an award
Which only a handful of people in the world have received
Including Mother Theresa.
How can a man of dedication like him meet such a tragic accident
And injure his spinal cord at the first and second vertebrae?

Dear Dr. Lewis,
Thank you very much for your message.
I am so encouraged by your determination
To challenge this difficulty.
As a doctor, you know exactly what this injury means.

You understand precisely how difficult it is to recover completely.
You were somehow protected!
Your physician friends rushed to your side and applied CPR.
Your children and their spouses prayed all night.
The rescue team worried that you might die on that night,
But with your strong will power,
And with your family's sincere prayer,
You survived!
Even though you are paralyzed from the neck down,
Even though you are using a ventilator,
You have already started
An intensive rehabilitation program!
You have already volunteered to enroll yourself
In a two year double blind study program
To test the efficacy of GM-1, a ganglioside.

Dear Dr. Lewis,
Your undaunted, rock solid spirit reminds me of my friend,
Mr. Kent Waldrep, once a pro-football star in Dallas,
Who injured his spine during a game and became paraplegic.
He determined to be cured.
Moving his hand-driven wheelchair,
He has launched a life-long journey
To establish a project to cure paralyzed patients.
Because of his determination,
Because of his dedicated wife,
Many research projects were supported, and finally,
They built the Center for Nerve Growth and Regeneration
At the University of Texas South Western Medical Center in Dallas.

Dear Dr. Lewis,
In your life you have met with so many difficulties,
But you have won victories.
You are now meeting with the greatest difficulty.
Your mission is to show the world
That you can be cured!
Your challenge will be the source of inspiration,
The goal for our research, and
Wordless encouragement for all!
For us, it is a strange coincidence.
In the summer two years ago,
A neuro-injury conference was held in Copper Mountain
With topics of how to minimize secondary injury,
How to accelerate the recovery, and
How to regenerate the nerve fibers!
Mr. and Mrs. Waldrep were present encouraging us.
Please have confidence, Dr. Lewis!
You will win victory in this battle, too!
Your dedicated wife,
Your determined family,
Who will be with you in rotation,

And all of us, neuroscientists,
Will work toward the goal!

Dear Lewis family,
Let's work together!
We will pray for your success.
We will work hard toward the goal
Until the day arrives
When we can prevent secondary injury,
And have a tool at hand
To regenerate the injured nerve fibers!

Dear Lewis family,
Let's work with hope and confidence!
The day will come
When all wheelchaired patients,
All bed-ridden athletes,
Will stand up, walk, run . . . and ski!

S. Tsuyoshi Ohnishi

THE EDITORS

S. Tsuyoshi Ohnishi, Ph.D., is the Director of the Philadelphia Biomedical Research Institute, King of Prussia, Pennsylvania, and is Visiting Professor of Hahnemann University School of Medicine, Philadelphia, Pennsylvania.

Dr. S. T. Ohnishi graduated from Kyoto University, Kyoto, Japan, in 1954 with a B.S. degree in Physics. He obtained his Master's Degree in Chemistry in 1956 and his Ph.D. from Nagoya University in 1959. He was a Post Doctoral fellow at Nagoya University (Biophysics) from 1960 to 1962, and Visiting Scientist fellow between 1963 and 1965 at the University of Tokyo School of Medicine (Pharmacology). He served as Associate Professor at Waseda University, Tokyo from 1956 to 1967, Visiting Associate Professor at the Johnson Foundation, University of Pennsylvania from 1967 to 1968, Assistant Professor at the Medical College of Pennsylvania from 1969 to 1972, Associate Professor at Hahnemann University School of Medicine from 1973 to 1984, and Research Professor at the same university in 1985. He was Director of the Membrane Research Institute from 1985 to 1989 and Director at the Philadelphia Biomedical Research Institute from 1989 to the present.

Dr. Ohnishi was Visiting Professor at Hokkaido University, Japan in 1979 and Visiting Professor at Mayo Medical School in 1982. He was an Invited Research Scientist at MIT, Boston in 1983.

He is a member of the American Society of Biological Chemistry, the Biophysical Society, the Society of Neuroscience, and the American Hematological Society.

In the early 1960s he pioneered the discovery that a muscle protein actin exists in biological membranes. Working with Dr. Setsuro Ebashi at the University of Tokyo, Dr. Ohnishi developed the murexide method to measure the concentration of calcium ions in physiological media, and demonstrated that the sarcoplasmic reticulum takes up calcium ions with a rate rapid enough to account for the speed of muscle relaxation. He developed a spectrophotometer using a rotating sector-semiconductor switching circuit. The technology has been widely used since. He contributed in the development of the concept of "membrane-acting" antisickling agents. In the early 1980s he demonstrated that irreversibly sickled cells can be formed *in vitro* when the red blood cells from sickle cell anemia patients were exposed to repeated cycles of sickling and unsickling. He also identified the cause of malignant hyperthermia: the abnormally high calcium release from the sarcoplasmic reticulum of genetically susceptible subjects. He has done much work on the spin-trapping of reactive oxygen radicals and nitric oxide radicals in CNS injury.

Dr. S. T. Ohnishi has made over 50 presentations and invited lectures in national and international meetings. He has published 120 papers and is a co-editor of 6 books. He has been the recipient of 20 research grants from the National Institutes of Health, National Science Foundation, and other private foundations. His current research interest is investigating the role of antioxidants in ischemia-reperfusion injury, malaria, cancer, and AIDS. He is also working on three-dimensional EPR imaging of nitric oxide radicals in the brain.

Tomoko Ohnishi, Ph.D., is a Research Professor at the University of Pennsylvania School of Medicine and the Johnson Research Foundation.

Dr. Ohnishi graduated from Kyoto University in 1956 with a B.S. degree in Biochemistry and obtained her Ph.D. in 1962 from Nagoya University. She was a Visiting Scientist fellow between 1961 and 1963 and worked in the Osaka University School of Medicine (Biochemistry). She was Research Associate at Wenner-Gren Institute, University of Stockholm, Sweden from 1965 to 1966, and Research Associate at cornell University from 1966 to 1967. Dr. Tomoko Ohnishi was Visiting Assistant Professor at the Johnson Research Foundation, University of Pennsylvania from 1967 to 1971 and Assistant Professor at the same foundation from 1971 to 1977. She served as Research Associate Professor at the

University of Pennsylvania Department of Biochemistry and Biophysics from 1977 to 1984. She is currently Research Professor of the same department and also of the Johnson Research Foundation.

Dr. Ohnishi received the International Cell Research Award to attend the international conference and workshop in Stockholm in 1965. She was Invited Professor of the Organization of American States Scientific and Technological Program, Buenos Aires in 1977, Invited Scholar of the Tokyo Metropolitan Institute of Medical Science, Tokyo in 1982, and Invited Foreign Scholar of CNRS, Paris from 1983 to 1984.

She is a member of the American Society of Biological Chemistry. She was the Chairperson of the Bioenergetics Subgroup of the Biophysical Society from 1992 to 1995.

She was the first to isolate tightly coupled, intact mitochondria from yeast and demonstrate that they are different from bovine heart mitochondria because they oxidize exogenous NADH in an ADP-controlled manner. She has applied liquid helium temperature EPR to the study of the mitochondrial respiratory chain. She characterized basic thermodynamic and EPR properties of iron-sulfur clusters as well as bound forms of semiquinone species and flavin free radicals. She pioneered the discovery that mitochondrial genetic diseases can be caused by the deficiency of iron-sulfur clusters in the NADH-ubiquinone oxidoreductase. She also developed a paramagnetic probe technique to determine spatial organization of intrinsic redox components within the energy-transducing membranes.

She has given over 20 invited lectures and 50 presentations in national and international meetings. She has published 170 papers and has been a recipient of consecutive grants from the National Institutes of Health and the National Science Foundation. In addition, she recently received an NIH Fogarty International Research Collaboration Award to work with scientists at the Moscow State University.

CONTRIBUTORS

Peter G. Aitken
Department of Cell Biology
Duke University Medical Center
Durham, North Carolina

Minoru Akino
Department of Physiology
The Ohio State University
Columbus, Ohio

John S. Althaus
CNS Diseases Research
The Upjohn Company
Kalamazoo, Michigan

Paula K. Andrus
CNS Diseases Research
The Upjohn Company
Kalamazoo, Michigan

Takao Asano
Department of Neurosurgery
Saitama Medical Center
Kawagoe, Saitama, Japan

Frank C. Barone
Cardiovascular Pharmacology
SmithKline Beecham Pharmaceuticals
King of Prussia, Pennsylvania

Joseph S. Beckman
Department of Anesthesiology
University of Alabama at Birmingham
Birmingham, Alabama

Raul Busto
Cerebral Vascular Disease Research Center
University of Miami
School of Medicine
Miami, Florida

Pak H. Chan
Department of Neurosurgery
University of California
San Francisco, California

Robert K. Clark
Cardiovascular Pharmacology
SmithKline Beecham Pharmaceuticals
King of Prussia, Pennsylvania

Robert S. B. Clark
Departments of Anesthesiology and Critical Care Medicine, and Pediatrics
University of Pittsburgh
Pittsburgh, Pennsylvania

Sien T. Chen
Department of Neurology
Chang Gung Memorial Hospital and Medical School
Taipei, Taiwan

Sylvia F. Chen
Department of Neurosurgery
University of California
San Francisco, California

Andrea M. Corse
Department of Neurology
Johns Hopkins University
Baltimore, Maryland

Shamay Cotev
Department of Anesthesiology
Hebrew University
Hadassah Medical School
Jerusalem, Israel

John P. Crow
Department of Anesthesiology
University of Alabama at Birmingham
Birmingham, Alabama

W. Dalton Dietrich
Cerebral Vascular Disease Research Center
Department of Neurology
University of Miami
School of Medicine
Miami, Florida

C. Edward Dixon
Department of Neurosurgery
University of Texas
Health Science Center at Houston
Houston, Texas

Dennis G. Emery
Department of Zoology and Genetics
Iowa State University
Ames, Iowa

Charles J. Epstein
Department of Pediatrics
University of California
San Francisco, California

Giora Z. Feuerstein
Department of Cardiovascular Pharmacology
SmithKline Beecham Pharmaceuticals
King of Prussia, Pennsylvania

Paul C. Francel
Department of Neurosurgery
University of Virginia
Health Sciences Center
Charlottesville, Virginia

Nancy N. Futrell
Cerebral Vascular Disease Research Center
Department of Neurology
University of Miami
School of Medicine
Miami, Florida
(currently)
Department of Neurology
Medical College of Ohio
Toledo, Ohio

Myron D. Ginsberg
Cerebral Vascular Disease Research Center
University of Miami
School of Medicine
Miami, Florida

Mordecai Y.-T. Globus
Cerebral Vascular Disease Research Center
University of Miami
School of Medicine
Miami, Florida

Guenter W. Gross
Center for Network Neuroscience
Department of Biological Sciences
University of North Texas
Denton, Texas

B. J. Gwag
Department of Neurology
Washington University
School of Medicine
St. Louis, Missouri

Edward D. Hall
Central Nervous System Diseases Research
The Upjohn Company
Kalamazoo, Michigan

Robert J. Hamm
Department of Psychology
Virginia Commonwealth University
Richmond, Virginia

Ronald L. Hayes
Department of Neurosurgery
University of Texas
Health Science Center at Houston
Houston, Texas

Yong Y. He
Department of Neurology
Washington University
School of Medicine
St. Louis, Missouri

Philip J. Horner
Department of Physiology
The Ohio State University
Columbus, Ohio

Chung Y. Hsu
Department of Neurology
Washington University
School of Medicine
St. Louis, Missouri

Z. Y. Hu
Department of Neurology
Washington University
School of Medicine
St. Louis, Missouri

Harry Ischiropoulos
University of Pennsylvania
Institute for Environmental Medicine
Philadelphia, Pennsylvania

John A. Jane
Department of Neurosurgery
University of Virginia
Health Sciences Center
Charlottesville, Virginia

Hideyuki Kamii
Department of Neurosurgery
University of California
San Francisco, California

Hideyaki Kanemitsu
Cerebral Vascular Disease Research Center
Department of Neurology
University of Miami
School of Medicine
Miami, Florida
(currently)
Department of Neurosurgery
Teikyo University
Tokyo, Japan

Resesh D. Kapadia
Physical and Structural Chemistry
SmithKline Beecham Pharmaceuticals
King of Prussia, Pennsylvania

Hiroshi Karibe
Department of Neurological Surgery
School of Medicine
University of California
San Francisco, California

Stephen E. Karpiak
Division of Neuroscience
N.Y.S.P.I.
Department of Psychiatry
Columbia University
College of Physicians and Surgeons
New York, New York

Yoichi Katayama
Department of Neurological Surgery
Nihon University
School of Medicine
Tokyo, Japan

Hiroyuki Kinouchi
Division of Neurosurgery
Institute of Brain Diseases
Tohoku University
School of Medicine
Sendai, Japan

Patrick M. Kochanek
Safar Center for Resuscitation Research, and Departments of Anesthesiology and Critical Care Medicine and Pediatrics
University of Pittsburgh
Pittsburgh, Pennsylvania

Irvin Kron
Department of Surgery
University of Virginia
Health Sciences Center
Charlottesville, Virginia

Danxia Liu
Marine Biomedical Institute and Department of Human Biological Chemistry and Genetics
University of Texas Medical Branch
Galveston, Texas

P. K. Liu
Division of Restorative Neurology
Baylor College of Medicine
Houston, Texas

Jen Hill Lucas
Department of Physiology
The Ohio State University
Columbus, Ohio

Anthony Mancuso
Department of Neurological Surgery
School of Medicine
University of California
San Francisco, California

Carrie G. Markgraf
Cerebral Vascular Disease Research Center
Department of Neurology
University of Miami
School of Medicine
Miami, Florida
(currently)
Marion Merrell Dow Research Institute
Cincinnati, Ohio

Parthena M. Martin
Neurotoxicology Division
U.S. Environmental Protection Agency
Research Triangle Park, North Carolina

Toru Matsui
Department of Neurosurgery
Saitama Medical Center
Kawagoe, Saitama, Japan

Mark P. Mattson
Sanders-Brown Research Center on Aging
Department of Anatomy and Neurobiology
University of Kentucky
Lexington, Kentucky

David J. McAdoo
Marine Biomedical Institute
and Department of Human Biological Chemistry and Genetics
University of Texas Medical Branch
Galveston, Texas

Tracy K. McIntosh
Department of Neurosurgery
University of Pennsylvania School of Medicine
Philadelphia, Pennsylvania

Tatsuo Mima
Department of Cell Biology and Anatomy
Cornell University Medical College
New York, New York
(currently)
Department of Neurosurgery
Kochi Medical School
Kochi, Japan

John H. Mullen, Jr.
Philadelphia Biomedical Research Institute
King of Prussia, Pennsylvania

Hitoshi Nakayama
Cerebral Vascular Disease Research Center
Department of Neurology
University of Miami
School of Medicine
Miami, Florida
(currently)
Department of Neurosurgery
Teikyo University School of Medicine
Mizonakuchi, Hospital
Kawasaki, Japan

Allen H. Nelson
Department of Cardiovascular Pharmacology
SmithKline Beecham Pharmaceuticals
King of Prussia, Pennsylvania

Edwin M. Nemoto
Department of Anesthesiology and Critical Care Medicine
University of Pittsburgh
Pittsburgh, Pennsylvania

James P. O'Callaghan
Neurotoxicology Division
U.S. Environmental Protection Agency
Research Triangle Park, North Carolina

Ryo Ogawa
Department of Anesthesiology
Nippon Medical School
Tokyo, Japan

S. Tsuyoshi Ohnishi
Director
Philadelphia Biomedical Research Institute
King of Prussia, Pennsylvania

Tomoko Ohnishi
Department of Biochemistry and Biophysics
University of Pennsylvania
Philadelphia, Pennsylvania

Richard Paczynski
Department of Neurology
Washington University
School of Medicine
St. Louis, Missouri

Ricardo Prado
Cerebral Vascular Disease Research Center
Department of Neurology
University of Miami
School of Medicine
Miami, Florida

Jeffrey D. Rothstein
Department of Neurology
Johns Hopkins University
Baltimore, Maryland

Kenneth K. Sadanaga
Philadelphia Biomedical Research Institute
King of Prussia, Pennsylvania

Atsuhiro Sakamoto
Department of Anesthesiology
Nippon Medical School
Tokyo, Japan

A. Salminen
Department of Cell Biology
University of Jyvaskyla
Jyvaskyla, Finland

Henry M. Sarau
Pulmonary Pharmacology
SmithKline Beecham Pharmaceuticals
King of Prussia, Pennsylvania

Susanta K. Sarkar
Physical and Structural Chemistry
SmithKline Beecham Pharmaceuticals
King of Prussia, Pennsylvania

Shinya Sato
Department of Neurosurgery
Tokoku University
School of Medicine
Sendai, Japan

Joanne K. Schiding
Department of Anesthesiology
and Critical Care Medicine
University of Pittsburgh
Pittsburgh, Pennsylvania

Dulcie B. Schmidt
Pulmonary Pharmacology
SmithKline Beecham Pharmaceuticals
King of Prussia, Pennsylvania

Yoram Shapira
Division of Anesthesiology
Ben Gurion University of the Negev
Beer Sheva, Israel

Frank R. Sharp
Department of Neurology
University of California
San Francisco, California

Cheolsu Shin
Departments of Neurology
and Pharmacology
Mayo Clinic
Rochester, Minnesota

Esther Shohami
Department of Pharmacology
Hebrew University
Hadassah Medical School
Jerusalem, Israel

Bradford T. Stokes
Department of Physiology
The Ohio State University
Columbus, Ohio

Xiaofeng Sun
Department of Neurology and Pharmacology
Mayo Clinic
Rochester, Minnesota

Richard L. Sutton
Director
Neurosurgical Research
Department of Surgery
Hennepin County Medical Center
and Assistant Professor
Department of Neurosurgery
University of Minnesota
Minneapolis, Minnesota

Teiji Tominaga
Department of Neurosurgery
Tohoku University
School of Medicine
Sendai, Japan

Dan Torbati
Division of Critical Care Medicine
Miami Children's Hospital
Miami, Florida

Maritsa N. Tzimas
Inflammatory Pharmacology
SmithKline Beecham Pharmaceuticals
King of Prussia, Pennsylvania

Robert Vink
Division of Biochemistry
and Human Physiology
James Cook University
Townsville, Queenland, Australia

Philip F. VonVoigtlander
CNS Diseases Research
The Upjohn Company
Kalamazoo, Michigan

Brant D. Watson
Cerebral Vascular Disease Research Center
Department of Neurology
and Department of Biomedical Engineering
University of Miami
School of Medicine
Miami, Florida

Per Wester
Cerebral Vascular Disease Research Center
Department of Neurology
University of Miami
School of Medicine
Miami, Florida
(currently)
Department of Internal Medicine
University of Umea
Umea, Sweden

Philip R. Weinstein
Department of Neurological Surgery
School of Medicine
University of California
San Francisco, California

Robert N. Willette
Department of Cardiovascular Pharmacology
SmithKline Beecham Pharmaceuticals
King of Prussia, Pennsylvania

Hiroshi Yao
Cerebral Vascular Disease Research Center
Department of Neurology
University of Miami
School of Medicine
Miami, Florida
(currently)
2nd Department of Internal Medicine
Kyushi University
Fukuoka, Japan

Yao Zu Ye
Department of Anesthesiology
University of Alabama at Birmingham
Birmingham, Alabama

TABLE OF CONTENTS

I. Introduction

Chapter 1

CNS TRAUMA RESEARCH USING SMALL ANIMALS

S. Tsuyoshi Ohnishi

CNS trauma, which includes stroke, head injury in car accidents, as well as other brain and spinal cord injury, is one of the leading causes of death in developed countries. Although thousands of research papers have been published every year, the mechanisms of these ailments have not been clearly understood, and pharmacologic interventions are yet to be established. There have been a variety of theories regarding the underlying mechanisms, but there is one thing in common—the involvement of the membranes. This is the reason the fourth volume of the series *Membrane-Linked Diseases* is dedicated to the research on CNS trauma.

Many different species have been used as animal models to study the mechanism of CNS trauma. Primates or domestic animals are more suited to the study because the mechanism in these animals may be close to that in human CNS trauma. However, for various reasons, it is becoming progressively more difficult to use these animals for trauma study. Therefore, in this book, animal models were limited to rodents. The species dealt with in this book were primarily the rat and the mouse. However, the gerbil and the rabbit were also included. The former has a simpler cerebral vasculature and has been used for brain ischemia studies; the latter is known to be suitable for spinal cord ischemia studies.

Table 1 depicts what kind of CNS studies were conducted using what kind of models. Most of them use animals; however, certain investigators employ tissue culture models. The

TABLE 1
Rodent Models and Techniques Used in This Book (chapter numbers are shown)

	Animals			
Models/techniques	**Mouse**	**Gerbil**	**Rat**	**Rabbit**
In vitro				
Na-K-ATPase		6		
Mitochondria			3	
Cell culture			3,23,26	
Tissue slice			24,25	
Free radicals			3	
In vivo				
Brain ischemia	10,11,31,32	6	5,7,8,12,27,28, 30,33,34,35,39	
Head injury	14,15		16,17,18	
SAH			13	13
Spinal cord ischemia				19
Spinal cord injury			20,21,29	
Molecular biology	30,31		30	
Free radicals	31,32		33,34,35,36	
EPR/NMR	11		33,34	
Microdialysis			27,28,29	
Histology/staining	11		37,38	
Cognitive function/motor function	14,15		18,21,22	19

0-8493-8094-4/95/$0.00+$.50

numbers in the table correspond to the chapter numbers of this book. Needless to say, this table does not necessarily represent the entire interests of CNS researchers. It happened to be the interests of the authors who contributed to this book. The editor's intention is that the table should help readers not only to find their interested subjects, but also to recognize the current trends in CNS research. This table could also help to stimulate investigators to discover new areas where more active research may be needed.

II. Mechanisms of CNS Injury

Chapter 2

EFFECTS OF HYPOXIA AND HYPEROXIA ON CNS

Dan Torbati

TABLE OF CONTENTS

0-8493-8094-4/95/$0.00+$.50

I. CNS HYPOXIA

A. THE SCOPE

This review attempts to compare cellular reactions and clinical events in the CNS during hypoxia and hyperoxia with the tolerated tensions of oxygen, termed normoxia. The CNS physiologic responses to hypoxia, normobaric hyperoxia (NH), and hyperbaric oxygenation (HBO) from the organelle to the systemic level is examined. Focus is specifically on the "simple" example of acute oxygen deprivation alone, rather than the more complex forms of ischemic hypoxia, chronic hypoxia, or adaptation to high-altitude hypoxia. In addition, the neuronal properties that may produce the well-known differences in sensitivity to hypoxia and hyperoxia between neonate mammals and adults are discussed. The relative contribution of various molecular and cellular pathways that may be involved in the etiology of both hypoxic and hyperoxic neuronal injury are also reevaluated. It is acknowledged that disturbances in cellular energy production, impairment of membranes and cytosolic ionic balance, and neurotransmitter release may play a major role in acute neuronal injury resulting from both hypoxia and hyperoxia. In this scenario, perhaps, oxidative stress is the triggering factor in the pathogenesis of acute hypoxia- and hyperoxia-induced neural injury. In addition, interaction of superoxide with the nitric oxide ($NO^{\cdot}$), and interference with signal-transduction mechanisms, may be the precursor of delayed neurological deficiencies consequent to both hypoxic and hyperoxic insults. Ultimately, it is argued that while using oxygen, specifically HBO, as a "drug" for various illnesses may be beneficial, one must also consider its possible side effects.

B. DOES NORMOXIA EXIST IN REALITY?

The borderline between biologically safe and harmful tensions of oxygen, in terms of cellular functions, is obscure, and therefore tissue hypoxia and hyperoxia may be regarded as a continuum with normoxia (Jones, 1985). Because normal consumption of oxygen can generate reactive oxygen species (ROS), and considering the "free-radical theory of aging" (Harman, 1956, 1988; Mehlhorn and Cole, 1985), it could be argued whether any PO_2 at cellular level, in the long run, is safe enough for biological systems. For our relatively short life span, however, one may neglect the slow effect of ROS in the complex aging process.

Accordingly, we may define normal levels of tissue PO_2 (normoxia) as a range of PO_2 that causes no immediate structural or functional damage at cellular or subcellular levels. Given that ROS are also generated during normal cellular functions (Chance et al., 1979; Freeman and Crapo, 1982), such a convention for normoxia is compatible with the view that some free radicals (i.e., superoxide and nitric oxide) mediate regulatory functions in normal physiological conditions, such as signal transduction (Saran and Bors, 1989; Dawson et al., 1992; Culotta and Koshland, 1992; Halliwell, 1992; Schmidt, 1992).

C. WHEN DOES THE BRAIN BECOME HYPOXIC?

Cerebral hypoxia may result from a variety of pathologic, environmental, and toxicological conditions that occur acutely or chronically as a result of circulatory, respiratory, or metabolic failures originating in the heart, lung, or locally in the brain itself (Figure 1). The effect of hypoxia on the CNS varies with its severity, rate of development, duration, and particularly the stage of brain development (Fazekas et al., 1941; Glass et al., 1944; Blass and Gibson, 1979; Ginsberg, 1979; Bomont et al., 1992). Classification of various types of cerebral hypoxia is necessary because of different etiology and treatment requirements. For example, acute hypoxia affects the CNS differently than chronic hypoxia (Choi, 1990; Monge and Leon-Velarde, 1991; Kuwahira et al., 1993; Haddad and Jiang, 1993). Responses to hypoxia occur at molecular, cellular, organ, and systemic levels. Experimental models of hypoxia have utilized *in vitro* conditions to study hypoxia-induced molecular and cellular events for understanding of the basic mechanisms of neuronal injury. The pathophysiology of cerebral hypoxia, however, has been mostly studied by whole-organ or system models of hypoxia. In this review, the general term "hypoxia" is used to indicate acute reductions in the inspired PO_2, arterial oxygen tension, or brain tissue PO_2, regardless of the cause, form, or experimental modality of hypoxia.

The terms "hypoxia", "O_2 deprivation", and "deficiency" have been used interchangeably to indicate shortage of oxygen at various levels of biological organization (Haddad and Jiang,

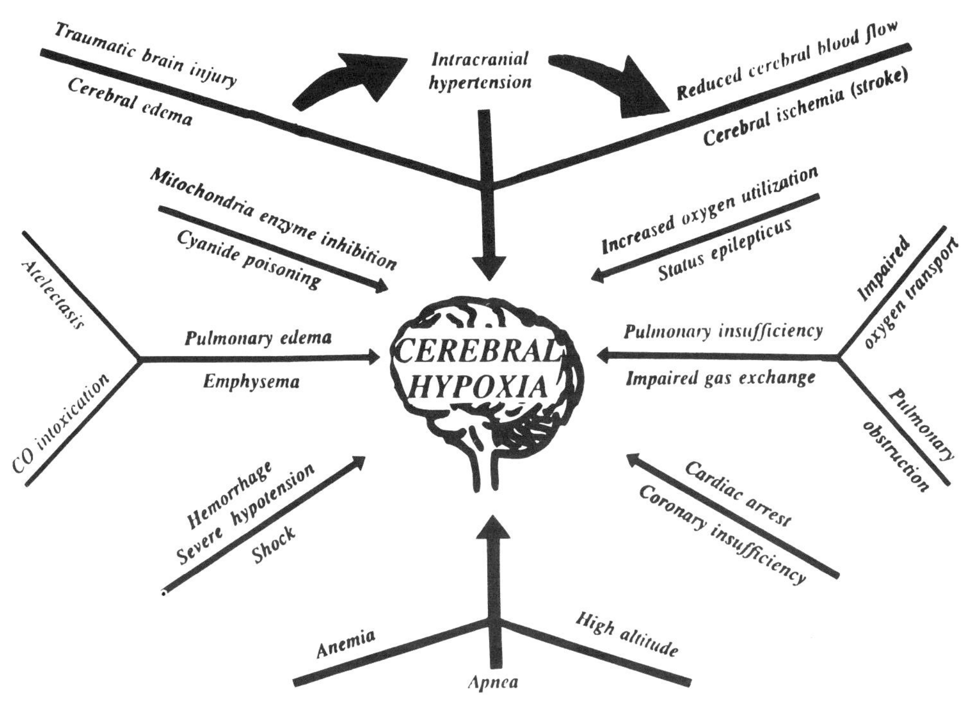

FIGURE 1. Pathophysiologic, toxicological, and environmental conditions that may lead to cerebral hypoxia.

1993). The brain, compared to other organs, is extremely vulnerable to shortage of oxygen because it cannot function without an adequate supply of energy, produced mostly by consumption of oxygen in the mitochondria (Siesjo, 1978). The alternative energy sources in the brain, rather than aerobic oxidation of glucose, are scarce. Hence, at the same level of hypoxia, brain damage may occur before the failure of cardiac or respiratory functions. Cerebral hypoxia could be divided into a physiologic phase, also termed the compensated period, followed by a pathologic phase or uncompensated stage (Saugstad, 1988). In the physiologic phase, hypoxia increases both cerebral blood flow (CBF) and glycolytic activity, thus maintaining a relatively normal level of energy supply for neuronal demands (Cohen et al., 1967; MacMillan and Siesjo, 1972; Norberg and Siesjo, 1975; Pulsinelli and Duffy, 1979; Dahlgren, 1990). In the pathologic phase, energy demands are not met and neuronal injury may ensue. Accordingly, tissue hypoxia can also be defined as "oxygen deficiency resulting in altered or interrupted energy metabolism."

D. DOES THE BRAIN RESPOND SIMILARLY TO ALL TYPES OF HYPOXIA?

Obviously, the common denominator in all kinds of acute hypoxia is a critical reduction in brain tissue oxygenation. However, since the origin and cause of cerebral hypoxia are different, each may induce characteristic pathophysiologic responses, which may influence the neurologic outcome. In general, chemical regulatory mechanisms of CBF attempt to mitigate the harmful effects of hypoxia by a vasodilatory response (Cohen et al., 1967; Johannson and Siesjo, 1975; Siesjo et al., 1979). However, cerebral vasoconstriction may develop on ascent to high altitudes (hypoxia) because of arterial hypocapnia induced by chemoreflex stimulation of respiration (Pierce et al., 1962). Conversely, apnea, which can lead to an anoxic state, induces hypercapnia and cerebral vasodilation (Kety and Schmidt, 1948; Reivich, 1964; Blass and Gibson, 1979). These hypoxic circulatory phenomena create a complex relationship between the degree and origin of hypoxia and the magnitude of the neurological impairment.

E. ISCHEMIA AS A COMPLEX FORM OF HYPOXIA

Clinically, the most common type of cerebral hypoxia develops concurrently with brain ischemia (stroke), consequent to blockade of the regional CBF (rCBF), cerebral hemorrhage, brain trauma, or cardiac arrest (Siesjo, 1981; Choi, 1990). Global or local cerebral ischemia interrupts the supply of oxygen, glucose, and other metabolites. The question is whether the pathophysiologic and biochemical alterations during ischemia can be ascribed to the reduced tissue oxygenation alone or also to interruption in the supply of glucose and removal of waste chemicals. The answer to this question may be sought in studies in which hypoxia alone can be simulated. This can be obtained by *in vitro* studies, simulating hypoxia alone or with ischemia (Schiff and Somjen, 1987; Choi, 1990; Silver and Erecinska, 1990). Logically, since oxygen and glucose are the two most important substrates for brain energy metabolism, the lack of both, plus an impaired waste removal should be more harmful than the absence of either one alone. Whether reduced tissue oxygenation rather than glucose availability is the major factor in neural cell injury is not yet clear.

F. ANOXIA VS. HYPOXIA

Anoxia is the condition in which the supply of oxygen to a tissue is completely cut off, while hypoxia is characterized by decreases in the intracellular concentration of oxygen, leading to a decline in aerobically produced ATP (Gutierrez, 1991). In cellular terms, anoxia is defined as local oxygen tensions that are too low to maintain mitochondrial phosphorylation (MacMillan and Siesjo, 1972). Although anoxia may be simply a severe form of hypoxia, they are distinguished from each other by different pathophysiologic consequences. One major difference is that during anoxia both supply of oxygen and production of energy by

ATP abruptly cease, while during hypoxia oxidative metabolism is impaired but not abolished (Blass and Gibson, 1979). Moreover, the state of anoxia can lead to loss of EEG activity and consciousness within seconds (Sugar and Gerard, 1938). Anoxia (PO_2 = 0 mmHg; *in vitro*) induces strong depolarization of neuronal membranes in less than 3 min, accompanied by an increase in synaptic potentials and marked changes in the extracellular ion concentrations (Hansen, 1985; Raflin et al., 1991; Jiang and Haddad, 1991). These phenomena are considerably delayed during mild or severe hypoxia (Somjen, 1988; Choi, 1990). It appears that anoxia, unlike hypoxia, has no compensated interval, or perhaps this phase is so rapid that the intermediary reactions could not be detectable. This implies that one cannot directly extrapolate results from the study of anoxia alone, or with ischemia to the understanding of hypoxia. For example, K^+ leakage is much faster and higher when nerve cells are deprived of both O_2 and glucose than glucose alone (Jiang and Haddad, 1992).

G. HIGH-ALTITUDE HYPOXIA

This type of hypoxia is similar to that resulting from breathing hypoxic air mixtures at sea level. The usual symptoms of moderate hypoxia in man include headache, dizziness, breathlessness, weakness, nausea, sweating, palpitation, blurring of vision, partial deafness, and sleeplessness. This pattern of symptoms on ascent to high altitudes has been known as mountain sickness (Talbot and Dill, 1936; Kety and Schmidt, 1948). The inspired PO_2 (P_IO_2) at any altitude can be calculated as: barometric pressure minus the sum of (PH_2O + PCO_2 + PN_2). The alveolar PH_2O is independent of altitude and is determined by the body temperature (normal PH_2O = 47 mmHg). Using this formula, theoretically, a person must breathe pure O_2 at an altitude of about 11000 m in order to maintain an alveolar PO_2 equal to that during air breathing at sea level (normoxia). At higher altitudes even 100% oxygen cannot maintain a normoxic alveolar PO_2 because of the reduced barometric pressure. At an approximate altitude of 16000 m, the P_IO_2 is about 40 mmHg. To prevent cerebral hypoxia at such altitudes, both the percentage of O_2 and cabin pressure of spacecrafts or space suits must be increased. The first behavioral manifestations of cerebral hypoxia in human occurs in visual acuity at about a 10% fall in the P_IO_2 relative to that at sea level (Siesjo et al., 1974). This is equivalent to an altitude of 1200 m above sea level. The ability to carry out a complex task is impaired at hypoxic levels of about 25% reduction in the P_IO_2, which is equivalent to breathing air at an altitude of 2500 m above sea level. The short-term memory is impaired by further reducing the P_IO_2 by 35% of that at sea level. Below a PO_2 of 60 to 70% of that at sea level, the individual becomes unconscious (Siesjo et al., 1974). The latter occurs at PO_2 levels that still can support cardiac, renal, and other organs' functions, demonstrating the extreme vulnerability of the brain to hypoxia compared to other organs.

H. TISSUE OXYGEN GRADIENTS

Metazoa exist only because of transport systems that have developed to carry oxygen from the environment to distant cells. They have developed an enormously complex transport system consisting of lungs, heart, and vessels, with brain coordination. The most important component of this system is the hemoglobin. This iron-containing flavoprotein serves both as an oxygen carrier and a temporary storage molecule. Distribution of oxygen in the tissue is affected by capillary density, intercapillary distances, diffusivity (characterizes the intrinsic conductance of the medium for oxygen), and solubility (Bicher et al., 1971; Bruley et al., 1973). Because of a great deal of heterogeneity, it is difficult to isolate the PO_2 of one point of the capillary from other parts (Van Liew and Rodgers, 1975). This heterogeneous pattern of oxygen gradient may also be extended for intracellular diffusion gradients of oxygen, particularly under some O_2- and ATP-limited conditions, namely hypoxia (Jones, 1986).

Krogh and Erlang (1919) developed a single model for the final step in oxygen transport in muscle consisting of a single capillary surrounded by homogeneously metabolizing muscle fibers whose axes were parallel to that of the capillary. In actual tissue, however, there are many capillaries, not all of which are parallel to the consuming cells. Moreover, both capillary blood supply and oxygen consumption are variable, creating a marked heterogeneity (Forster, 1994). In fact, the capillary geometry is so variable that the classic Krogh and Erlang cylinder model of the interaction between the capillary supply of oxygen, its diffusion, and cellular utilization can be used only as a qualitative tool (Honig, 1977). Thus, oxygen is always unevenly distributed within any volume of brain tissue, and may not be considered as a singular or lumped value (Van Liew and Rodgers, 1975). Similarly, at the cellular level no single PO_2 value can represent a hypoxic cell (Jones, 1986). Such a pattern of oxygen distribution is consistent with the view that tissue oxygen concentration is a continuum from hypoxia to hyperoxia (Jones, 1985).

I. WHEN DOES HYPOXIC NEURONAL INJURY OCCUR?

The alveolar PO_2 during air breathing at sea level may be as high as 155 mmHg, but it quickly falls to about 100 mmHg in the larger arteries. Since brain tissue is a major consumer of oxygen, its PO_2 gradient may drop as low as 1 to 3 mmHg (MacMillen and Siesjo, 1971). The question is whether such low tensions of oxygen are adequate to maintain the high energy demand of the brain. Since isolated mitochondria can respire maximally at a PO_2 of 1 to 3 mmHg (Chance et al., 1962), normal levels of oxidative phosphorylation should also take place at such low gradients of tissue PO_2 (MacMillen and Siesjo, 1971). During hypoxia, oxygen distribution in the tissue is shifted to even lower values, and in the extreme, the tissue becomes completely anoxic. The large fall in local oxygen tension is the result of oxygen consumption along the arterial and venous ends of the capillary network. Some cells adjacent to the arterial end of the capillary may have a PO_2 as high as the arterial PO_2, while those at the venous site of the capillary may be close to an anoxic state (Figure 2).

Brain cells are among the most sensitive type of cells to hypoxia and can be irreversibly damaged after only 4 to 6 min of hypoxia (Siesjo, 1981). The factors that constitute an irreversible neuronal injury may include any abnormal reaction that impairs cellular functions (Choi, 1990). Early metabolic derangements may remain within the boundaries of physiologic response if the minimum threshold of oxygen for energy production is maintained. It is only at a point of no return that permanent neuronal damage will occur and lead to neuronal death. Overall, the transition from a reversible stage of hypoxia into an irreversible phase depends on the level of metabolic activity of the cells, as well as the severity and duration of an hypoxic insult (Siesjo, 1981; Schiff and Somjen, 1987; Gutierrez, 1991). In a given species, development of irreversible neuronal damage is also affected by stress reactions, age, and anesthetic agents, and is accelerated by ischemia and/or preexisting trauma (Siesjo, 1978; Torbati et al., 1986; MacNeill and Bryan, 1988; Jiang et al., 1992).

J. CELLULAR SENSING OF HYPOXIA

Cells, presumably, "sense" hypoxia when the energy requirements for membrane ionic balance and membrane functions are not maintained. In the absence of adequate energy, the biosynthesis and release of neurotransmitters may be affected. A specific case, which is compatible with this concept, is the chemotransduction in the glomus cells of the carotid bodies, which are specialized for oxygen sensing. These cells have neural origin and perform complex and important sensory and regulatory functions. They are innervated by nerve fibers running in the sinus nerve, and are located dorsal to the common carotid artery bifurcation (Eyzaguirre and Zapata, 1984; Acker, 1989; Anand and Paintal, 1991; Lahiri, 1991). Glomus cells respond vigorously to changes in arterial PO_2, PCO_2, and pH, both *in vivo* and *in vitro* (Lahiri et al., 1978, 1981; Eyzaguirre et al., 1989). The sensory discharges of the sinus nerve increase in frequency when arterial PO_2 or pH falls, or when PCO_2 increases (hypercapnia).

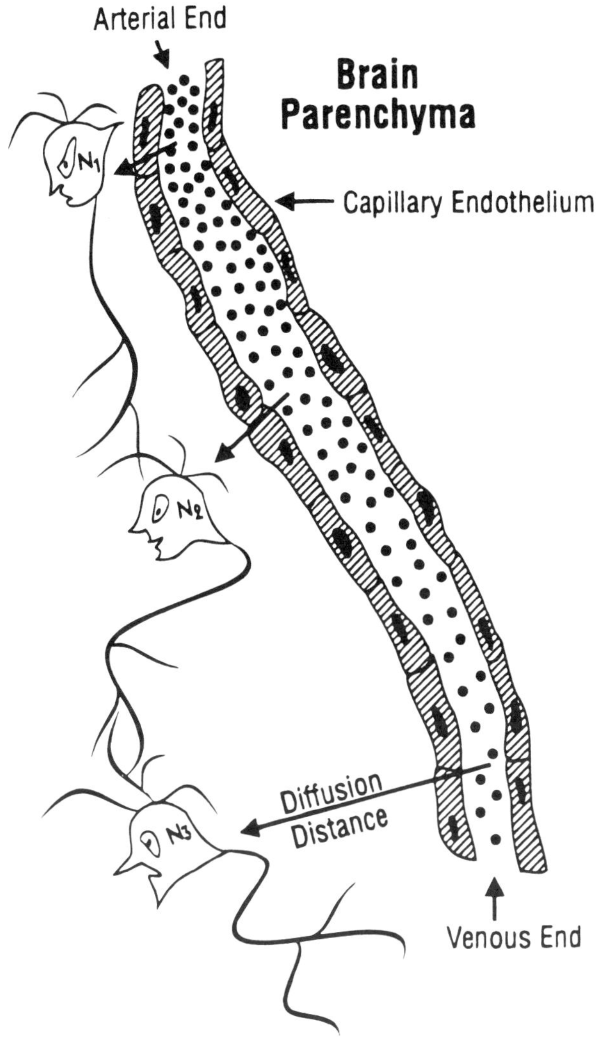

FIGURE 2. A graphic presentation of a capillary oxygen gradient demonstrating the fall in oxygen concentration from the arterial end to the venous end of a capillary, and the effect of diffusion distances on oxygenation (PO_2) of the neuronal elements. Providing that the levels of metabolic activity and blood flow are equal for all three representative neurons, then N_1 is relatively hyperoxic as compared to N_2. Now, assuming that N_2 is receiving an optimal (normoxic) amount of O_2, then N_3 is at a hypoxic spot relatives to both N_1 and N_2. Nevertheless, this is only a snapshot of the local tissue PO_2, which is subject to extreme fluctuations with any alteration in the supply, or demand, of oxygen by circulation, lung, and alterations in the local metabolic activity, respectively. Hyperoxia, without substantial decrease in the capillary blood flow, may upgrade N_3 to a more normoxic status, while N_1 may face toxic O_2 levels (oxidative stress). In a hypoxic situation, on the other hand, N_3 may become completely anoxic if there is no substantial increase in the capillary blood flow or decrease in the energy metabolism.

Conversely, the discharge frequency decreases with hyperoxia and low arterial PCO_2 (hypocapnia). On the hyperoxic side of the oxygen spectrum, it has been demonstrated that the delicate chemoreceptor mechanisms in the carotid body, particularly mitochondrial structure and function, are damaged following prolonged exposures to NH or acute HBO (Lahiri et al., 1987; Torbati et al., 1993). The hyperoxic damage then affects the hypoxic ventilatory drive (Arieli et al., 1988; Liberzon et al., 1989).

K. GLOMUS CELLS VS. NEURONAL CELLS RESPONSE TO HYPOXIA

The existing wealth of the accumulated data on carotid body glomus cells may shed some light on the cellular mechanisms by which unspecialized cells react to hypoxia (S. Lahiri, personal communication). Among several hypotheses trying to describe the molecular basis of the carotid body oxygen-sensing mechanism, the involvement of mitochondrial oxidative phosphorylation, known as metabolic hypothesis, is of special interest (Anichkov and Belenkii, 1963; Mulligan et al., 1981). The details of how changes in glomus cells' energy metabolism lead to the release of neurotransmitters and alter the frequency of action potentials in the sensory fibers are yet speculative (Biscoe and Duchen, 1990, 1990a). However, it is generally agreed that mitochondrial function of the chemoreceptor cell initiates oxygen chemoreception (Eyzaguirre and Zapata, 1984; Eyzaguirre et al., 1989; Acker, 1989). Accordingly, changes in oxidative phosphorylation, required for maintenance of the intracellular high-energy phosphate levels, form the basis of O_2 chemoreception in the glomus cells (Mulligan et al., 1981).

The question is whether oxygen-sensing mechanisms in the carotid body constitute the organized form of cellular response to either hypoxic or hyperoxic insults. Oxygen-sensing mechanisms may not be unique to this type of cellular transducer, and they can utilize general and fundamental biological processes. Thus, it is likely that the glomus cells' ability to respond to hypoxia represents one pole of a spectrum, ranging from normal to pathological (Biscoe and Duchen, 1990). The hypoxic response appears to create a more severe pathologic outcome in neural cells than in the specialized glomus cells. This sensitivity to the destructive effects of hypoxia implies that the major difference between glomus cells and other unspecialized neural cells may be the rate of the response to hypoxia. The glomus cells' responsiveness to hypoxia could be determined by the level of sinus nerve impulse activity. The maximum discharge rate of the sinus nerve occurs at about an arterial PO_2 of 40 mmHg, while the lowest activity appears upon breathing of 100% oxygen (Lahiri et al., 1978, 1981). While the critical threshold for supply of energy in the brain is at an arterial PO_2 of about 20 mmHg (MacMillan and Siesjo, 1971, 1972), the mitochondrial PO_2 sufficient to sustain oxidative metabolic processes may not be more than 1 mmHg of the nearly 160 mmHg inspired (Chance et al., 1962). Thus, carotid body glomus cells are able to detect differences in tissue PO_2 at a mild range of hypoxia, without being permanently damaged, whereas other cells' responsiveness to hypoxia occurs at a severe range with irreversible consequences.

L. BIOCHEMICAL INDICATORS OF BRAIN HYPOXIA

The biochemical study of brain hypoxia *in vivo* is complicated because in addition to neurons there are several types of glia and vascular elements that can contribute to hypoxia-induced biochemical changes. Within the CNS, cells in different locations, and even different types of cells in the same region, vary widely in their susceptibility to hypoxic damage (Donnelly et al., 1992). The CNS sensitivity to hypoxia is also variable at different developmental phases (Haddad and Donnelly, 1990). Such differences are attributed to variations in vascularization and biochemical characteristics, specifically as related to oxidative and glycolytic metabolism (Crain and Nadler, 1988; Ransom et al., 1990; Xia et al., 1992). Clearly, the energy requirements of some white matter structures during hypoxia are higher than gray matter structures (Pulsinelli and Duffy, 1979; Torbati et al., 1986; Ransom et al., 1990). Whether the increased metabolic demand is a characteristic response to white matter hypoxia, or it reflects mainly the energy requirements of axonal or glial elements is not known. At the present time there is no unique or sensitive biochemical index that can differentiate between the compensated and uncompensated stages of cerebral hypoxia. However, a major molecular event that may ultimately become involved in acute neuronal injury is alteration in the mode of energy production from oxidative phosphorylation to glycolysis. This reduces energy production, increases lactate concentration, and decreases pH because of an increased glycolytic flux (Cohen et al., 1967; MacMillan and Siesjo, 1972; Duffy et al., 1972; Norberg and Siesjo, 1975; Gyulai et al., 1987; Gutierrez, 1991). Other reactions that take place during prolonged or intermittent hypoxia also modify enzymatic activities which are closely related to energy transduction (Dagani et al., 1984).

Hypoxia of 11% inspired O_2 is associated with a transient increase in the cerebral glucose utilization rate (MacNeill and Bryan, 1988). Moreover, redistribution of local rates of brain energy metabolism in rats occurs with moderate hypoxia of 7% P_IO_2, which is equivalent to an arterial PO_2 of about 27 mmHg (Miyaoka et al., 1979; Torbati et al., 1986). This level of hypoxia is sufficient to produce behavioral abnormalities without loss of consciousness (Miyaoka et al., 1979). The release of neurotransmitters seems to be among the early neurochemical manifestations of acute cerebral hypoxia (Davis et al., 1979; Goldberg et al., 1988; Choi, 1990a). In fact, measurement of neurotransmitter metabolism may offer a unique advantage for detection of the early biochemical changes during hypoxia, because neurotransmitters are confined to neurons and not found in glia or vascular tissue (Davis et al., 1979). Thus, subtle indices of hypoxia are reflected as impairment in the biosynthesis of various neurotransmitters and amino acids before brain energy metabolism, as a whole, is affected (Duffy et al., 1972; Norberg and Siesjo, 1975). This is known to occur at hypoxic levels above 7% inspired oxygen (Gibson and Duffy, 1981; Miwa et al., 1986; Ninomiya et al., 1988).

The hypoxic-induced neurotransmitter and metabolic derangements may set the stage for development of a variety of abnormal molecular events, including: (a) generation of ROS, increased production of nitric oxide radical ($NO^{\cdot}$), and the release of free fatty acids (Fridovich, 1979; Imaizumi et al., 1984; Jones, 1985; Ueno et al., 1988; Pelligrino et al., 1993); (b) disturbance in cellular homeostasis of Ca^{2+} (McBurney and Neering, 1987; Siesjo and Bengtsson, 1989; Marcoux et al., 1990; Young, 1992); (c) modification of ion channels activity, excitotoxicity, and Ca^{2+} influx (Choi, 1987; Goldberg et al., 1988; Goldberg and Choi, 1990; Anwyl, 1991; Haddad and Jiang, 1993, 1993a); (d) activation of the signal transduction pathways (Friedman and Haddad, 1992; Mozzi and Horrocks, 1992; Qi and Dawson, 1993); (e) triggering of spreading depression (SD)-like depolarization (Leao, 1944; Balestrino and Somjen, 1986; Somjen et al., 1990); and (f) tissue lactacidosis (Siesjo, 1978a, 1988). In the following sections, the possible sequence of these pathophysiologic pathways, and their relative contribution to neuronal cell injury are discussed. The exact role of each of the proposed responses to hypoxia has not yet been firmly established. However, it is recognized that the shortage of energy and its complex relationship to membrane dysfunction are apparently the major events in the development of both acute pathologic responses and delayed neurological deficiencies (Figure 3).

M. HOW CELLULAR ENERGY PRODUCTION IS AFFECTED BY HYPOXIA

A brief account of energy-producing reactions during normoxia may demonstrate the pivotal role of the modified energy metabolism in hypoxic neuronal injury. The overall reaction for energy production in the brain, as in other tissues, consists of glycolytic events, the tricarboxylic acid reactions (Krebs cycle), and the transport of electrons along the respiratory chain (Siesjo, 1978 and 1978b). Glucose, which is the major energy-producing fuel in the CNS, is converted to lactate after a long series of glycolytic reactions during which no oxygen is consumed, and only two molecules of ATP are produced for each molecule of glucose consumed. Glycolysis serves as an important emergency mechanism capable of yielding some energy for short periods when oxygen is scarce or unavailable. A much larger portion of cellular energy, however, is produced during oxidative phosphorylation. This occurs during transfer of electrons from substrates to molecular oxygen through highly organized multienzyme systems, consisting of the Krebs cycle and the chain of mitochondrial electron-carrying enzymes in the inner membrane of the mitochondria. In contrast to glycolysis, the energy stored during oxidative phosphorylation contributes 36 molecules of ATP for

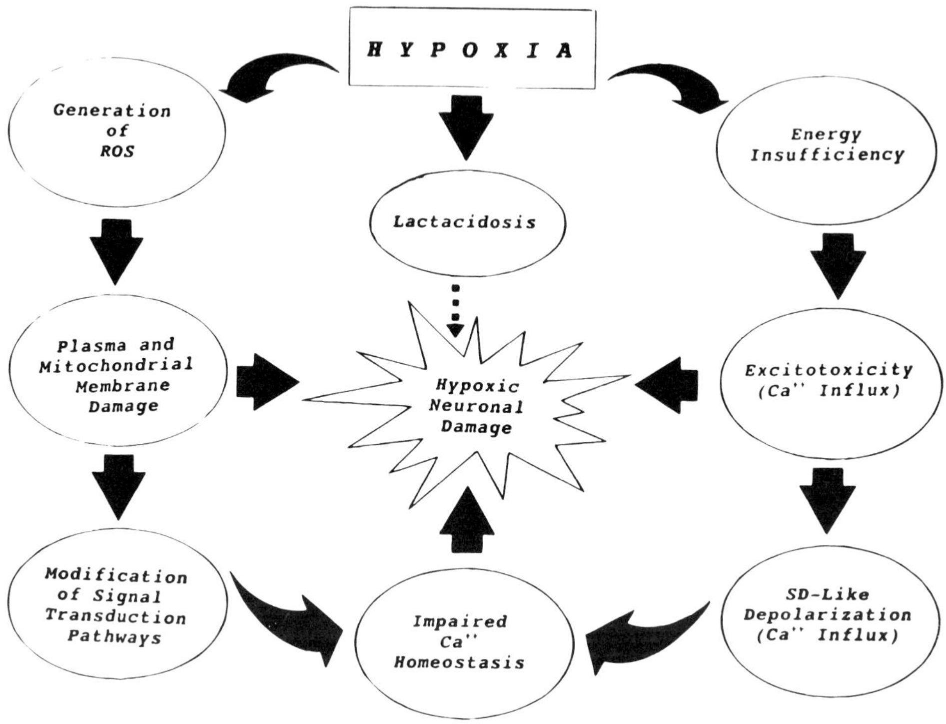

FIGURE 3. Hypothetical links between cerebral hypoxia and various cellular events leading to hypoxic neuronal damage.

each molecule of glucose. Thus, the oxidative pathways of energy production are about 15 times as efficient as the glycolytic reactions in providing ATP (Siesjo, 1978).

A key player in the regulation of mitochondrial electron-carrying enzymes is the cytochrome-c oxidase complex. This complex is the terminal agent in the electron-transport chain and catalyses oxidation of reduced cytochrome-c by molecular oxygen. The rate of reduction in molecular oxygen (mitochondrial respiratory rate) is PO_2 dependent (Wilson et al., 1977, 1977a). With the fall in oxygen tension below 16 mmHg, the mitochondrial respiratory rate begins to decline slightly, but a sharp decline is usually associated with PO_2 below 3 mmHg (Sugano et al., 1974; Wilson et al., 1979). This indicates that a severe hypoxia can reduce mitochondrial respiratory rate, thereby diminishing oxidative phosphorylation. Shortage of energy, in the form of ATP, will then affect the complex cellular economy, particularly those related to the maintenance of plasma and mitochondrial membrane potentials (due to impairment of the ATPase transport functions), and signal-transduction mechanisms (Richter and Frei, 1985; Carafoli, 1987; Erecinska and Silver, 1989; Haddad and Jiang, 1993).

N. OXIDATIVE STRESS-MEDIATED HYPOXIC NEURONAL INJURY

Excessive production of ROS beyond the antioxidant capacity of the cells is defined as oxidative stress. ROS, including superoxide, hydrogen peroxide (H_2O_2), and hydroxyl radical ($HO^{\cdot}$), are formed in biological systems via both enzymatic and nonenzymatic pathways (Chance et al., 1979; Freeman and Crapo, 1982; Fridovich, 1983; Halliwell et al., 1992). Free radicals can modify and damage cellular structures and functions via oxidation of membrane lipids, proteins, and genetic materials (Fridovich, 1978; Cohen, 1985; Richter and Frei, 1985; Grisham and McCord, 1986; Floyd, 1990; Pal Yu, 1994). Paradoxically, lowering of the tissue PO_2 (hypoxia), likewise increased concentration of oxygen (hyperoxia), may enhance the potential of "oxygen poisoning" rather than diminishing it, as speculated in Figure 4.

One possibility for hypoxic-induced generation of ROS is the failure of the electron-carrier system in the mitochondria. In the absence of oxygen as an electron acceptor, a reductive condition is created in the mitochondria, which increases the concentration of potential electron donors. This process results in enhanced formation of superoxide either via direct electron donation to oxygen or by redox cycling (Nohl and Hegner, 1978; Kappus and Sies, 1981; Jones, 1985). Another pathway for hypoxic generation of ROS is the reduction in ATP synthesis and accumulation of AMP during hypoxia; ATP can be degraded to hypoxanthine and xanthine, which are substrates for xanthine oxidase, a superoxide producer enzyme (Saugstad, 1988; Halliwell and Gutteridge, 1985). Alternatively, the release of excitatory amino acids and Ca^{2+} influx during hypoxia increases intracellular accumulation of free Ca^{2+}, which in turn activates phospholipase A_2 (Richter and Frei, 1985; Choi, 1990a; Siesjo, 1990) and a number of intracellular oxidases, particularly, xanthine oxidase (Kinuta et al., 1989). Activation of phospholipase A_2 degrades membrane phospholipids, generating arachidonic acid, which with further metabolism results in the production of ROS (Chan et al., 1985). The release of arachidonic acid has been demonstrated in the rat brain during both moderate and severe hypoxia, and from rat cerebral synaptosomes under *in vitro* hypoxic conditions (Gardiner et al., 1981; Ueno et al., 1988). The contribution of the arachidonic acid metabolism to hypoxic-ischemic cell damage has been downplayed by Rehncrona et al. (1982). These investigators suggest that free-radical damage plays little part, if any, in cell damage produced either with complete or incomplete hypoxic-ischemia. Gilman et al. (1993), however, have demonstrated that oxidative stress exerts an adverse effect on the presynaptic nerve terminals, resulting in excessive release of the excitatory amino acids. Since increased release of free fatty acids, from hydrolysis of membrane phospholipids, can potentiate excitatory synaptic transmission (Choi, 1990a; Siesjo, 1990), further cellular

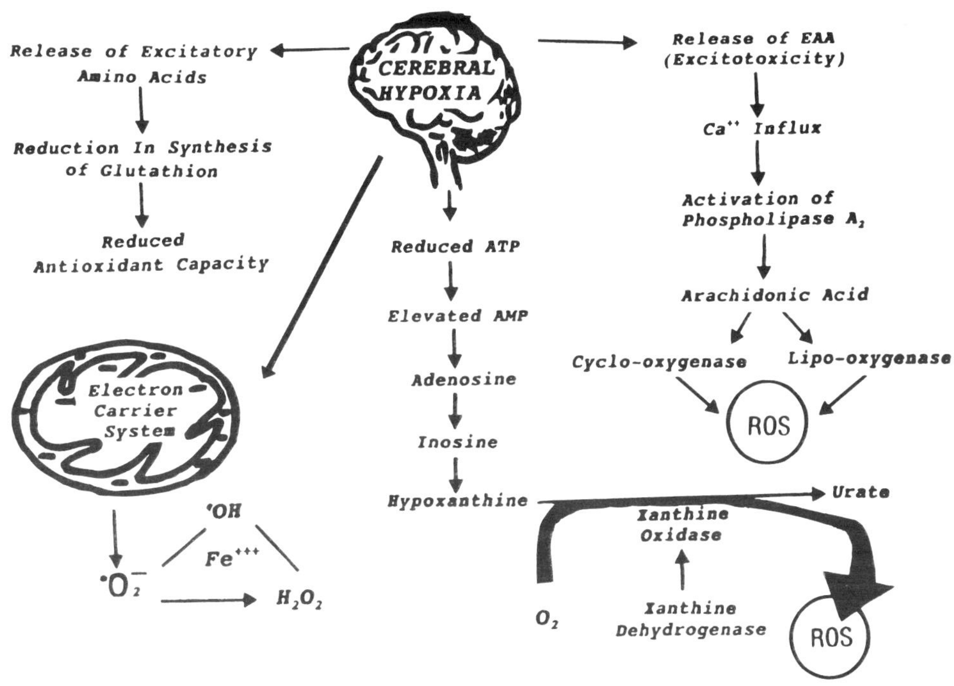

FIGURE 4. Possible pathways for generation of reactive oxygen species (ROS) during development of cerebral hypoxia.

damage may occur because of excitotoxicity processes (see below). Using spin-trapping and electron spin resonance (ESR) techniques, Sakamoto et al. (1991) demonstrated that ROS were produced *in vivo* upon reperfusion of the ischemic rat brain. They suggested that the production of ROS may be related to the subsequent lipid peroxidation of the brain tissue. In a later study, they found that both the production of ROS and the tissue lipid peroxidation were inhibited by the preadministration of a prostaglandin derivative with antioxidant activity (Sakamoto et al., 1991a; Ohnishi et al., 1992). Interestingly, some enzymatic free-radical scavengers, namely, superoxide dismutase and glutathion peroxidase, are known to be inhibited by hypoxia in the kidney (Rosenberg and Paller, 1991). If this happens in the brain, then ROS contribution to hypoxic neuronal damage could be intensified.

O. NITRIC OXIDE RADICAL-MEDIATED HYPOXIC NEURONAL INJURY

Nitric oxide (NO·), a gaseous radical, has been recently recognized as a key physiologic and pathologic mediator in vertebrates (Bredt and Snyder, 1992; Palmer, 1993). Nitric oxide formation was first demonstrated in the vascular endothelial cells, where it acts as a vasodilator agent (Palmer et al., 1987, 1988; Moncada et al., 1989). Subsequently, it was demonstrated that NO· plays a major role in a number of biological systems including, but not limited to, leukocyte functions (tumoricidal and bactericidal effects), inflammation, and immune regulation. Moreover, nitric oxide may act as a neurotransmitter for certain nonadrenergic, noncholinergic neurons in the peripheral nervous system (Moncada et al., 1991; Nathan, 1992; Bredt and Snyder, 1992; Palmer, 1993; Stefanovic-Racic et al., 1993), and has been suspected as a signal-transduction molecule that regulates soluble guanylyl cyclase activity (Schmidt, 1992). NO· production is catalyzed by the enzyme(s) nitric oxide synthetase (NOS), which is also widespread in the brain tissue (Hope et al., 1991). During cerebral hypoxia, NO· may participate in both metabolic changes and in the regulation of CBF (Bredt et al., 1990; Iwamoto et al., 1992). Reduction in oxygen supply may inhibit the NO-generating capacity in the brain (Pelligrino et al., 1993), while increased production of NO·, due to concurrent hypercapnia, may lead to both vasodilation and oxidative damage. The molecular mechanism for NO·-induced neuronal damage during hypoxia may be related to its reaction with hypoxia-induced superoxide radical generation. This leads to production of peroxynitrate, which decomposes into the potent oxidants OH· and NO_2· (Beckman et al., 1990; Hogg et al., 1992; Cazevieille et al., 1993). Another possibility for NO·-induced hypoxic neuronal damage is linked to the observation that NO· donors mimic glutamate neurotoxicity in primary cortical cultures (Dawson et al., 1991). Until recently, definite evidence for this reaction *in vivo* has been lacking (Moncada, 1992). Using spin-trapping and ESR techniques, Tominaga et al. (1993, 1994), and Sato et al. (1993) demonstrated that NO· is produced in the ischemic rat brain. They demonstrated that both global and focal ischemia potentiated the production of NO· *in vivo,* suggesting that this may be a protective reaction by increasing the CBF. Nevertheless, the possibility that the production of NO· may be also related to tissue damage cannot be ruled out (Sato et al., 1994).

P. CALCIUM HOMEOSTASIS AND HYPOXIC NEURONAL INJURY

Calcium has a central role in the regulation of a variety of cell functions, including inhibition or activation of specific enzymes and release of neurotransmitters (Messick and Milde, 1987; Choi, 1988a; Milde, 1989; Siesjo, 1990). The free-Ca^{2+} concentration gradient is 10,000 to 1 between the extracellular fluid and the cytosol (Siegel et al., 1981). Calcium influx into the cell and Ca^{2+} release from mitochondria are spontaneous processes (Fiskum, 1985). The high concentration gradient of Ca^{2+} across the plasma membrane is maintained by active removal from plasma membrane, mitochondria, and endoplasmic reticulum, by utilizing four different energy-dependent transport systems (Siegel et al., 1981). Moreover, the active involvement of Ca^{2+} messenger systems in signal-transduction mechanisms have been firmly established (Rasmussen and Barrett, 1984; Haddad and Jiang, 1993). Extensive oxidative damage to the plasma membrane transport systems or increased permeability of the plasma membrane to Ca^{2+} results in increased intracellular Ca^{2+} and mitochondrial loading (Richter and Frei, 1985; Siesjo, 1988a, 1990; Young, 1992). Excessive intracellular accumulation of Ca^{2+} represents a final common pathway for hypoxic-ischemic injury, known as the Ca^{2+} hypothesis (Siesjo, 1981; Miller, 1987; Haddad and Jiang, 1993).

Because hypoxia causes massive leakage of glutamate and aspartate into the extracellular space, the intracellular concentration of Ca^{2+} may rise (Siesjo, 1988a, 1990; Silver and Erecinska, 1990). This influx of Ca^{2+} is mediated by activation of both *N*-methyl-D-aspartate (NMDA) and non-NMDA receptors. However, NMDA receptors, apparently, are more important in mediating subsequently delayed neuronal disintegration (Choi, 1988, 1988a; Collingridge and Lester, 1989; Michaels and Rothman, 1990). Hypoxia also induces Ca^{2+} release from intracellular stores including mitochondria (Dubinsky and Rothman, 1991). Moreover, the capacity of the intracellular Ca^{2+}-binding proteins is reduced by intracellular acidosis, which develops during hypoxia (Carafoli, 1987). The increased cytosolic Ca^{2+} then activates a variety of intracellular enzymes, including proteases, lipases, and endonucleases, resulting in cellular destruction (Orrenius et al., 1989; Boobis et al., 1989; Arai et al., 1990, 1991). Despite substantial evidence supporting the Ca^{2+} hypothesis in hypoxic neuronal injury, some studies suggest alternative mechanisms to hypoxic-induced cellular damage (Dubinsky and Rothman, 1991). In this regard, Friedman and Haddad (1993) have demonstrated that disintegration of the hippocampal neurons may occur during anoxia *in vitro,* whether or not the intracellular Ca^{2+} increases.

Q. NEUROTRANSMITTERS-MEDIATED HYPOXIC NEURONAL INJURY

Hypoxia can induce the release of neurotransmitters, particularly glutamate. This leads to neuronal depolarization and activation of various ion channels including NMDA receptor-mediated Na^+/Ca^{2+} channels, Na^+ channels, and voltage-dependent Ca^{2+} channels (Choi, 1990). The possibility that the excitatory amino acid, glutamate, is involved in hypoxic/ischemic neuronal damage (the glutamate hypothesis) was indirectly derived from studies by Olney (1969). This investigator demonstrated that excessive administration of glutamine (from 50 to 400 mg/kg) in newborn mice and 500 to 700 mg/kg in adult mice can affect CNS development and produce brain lesions, particularly in the hypothalamus. Consequently, growing mice showed endocrinological deficiencies, marked obesity, and female sterility. Since then, the term "excitotoxicity" has been used as a general property of excitatory amino acids (glutamate and aspartate) to damage neuronal cells. Recently, evidence has been accumulating that extracellular levels of glutamate are increased during hypoxia and that the synaptic release of excitatory amino acids mediates hypoxic/anoxic neuronal cell injury (Rothman, 1984; Choi et al., 1987; Goldberg et al., 1988; Choi, 1988, 1990a). The mechanism by which glutamate is released during hypoxia is not yet understood, but a close relationship between glutamate neurotoxicity and the influx of extracellular Ca^{2+} has been proposed (Siesjo, 1981, 1990; Choi, 1988a, 1990a; Beal, 1992; Haddad and Jiang, 1993) (Figure 5).

Hypoxic-induced excitotoxicity may intensify oxidative stress via two different pathways: (a) accumulation of intracellular Ca^{2+} may lead to formation of ROS via arachidonic acid metabolism (Figure 4) (Chan et al., 1985; Siesjo et al., 1989); and (b) potentiation of the excitatory transmission by free fatty acids (Choi, 1990a; Siesjo, 1990), which can lead to excessive release of glutamate. This may potentially reduce synthesis of glutathion via inhibition of neuronal cysteine uptake (Murphy et al., 1989). Additionally, as mentioned earlier, it is also possible that $NO^{\cdot}$ mediates hypoxic neuronal damage subsequent to stimulation of the excitatory amino acid receptors (Shibuki and Okada, 1991; Nathan, 1992; Lancaster, 1992; Dawson et al., 1992). Other neurotransmitters that may be involved in either hypoxic neuronal damage, or even protection against it, are serotonin and the inhibitory

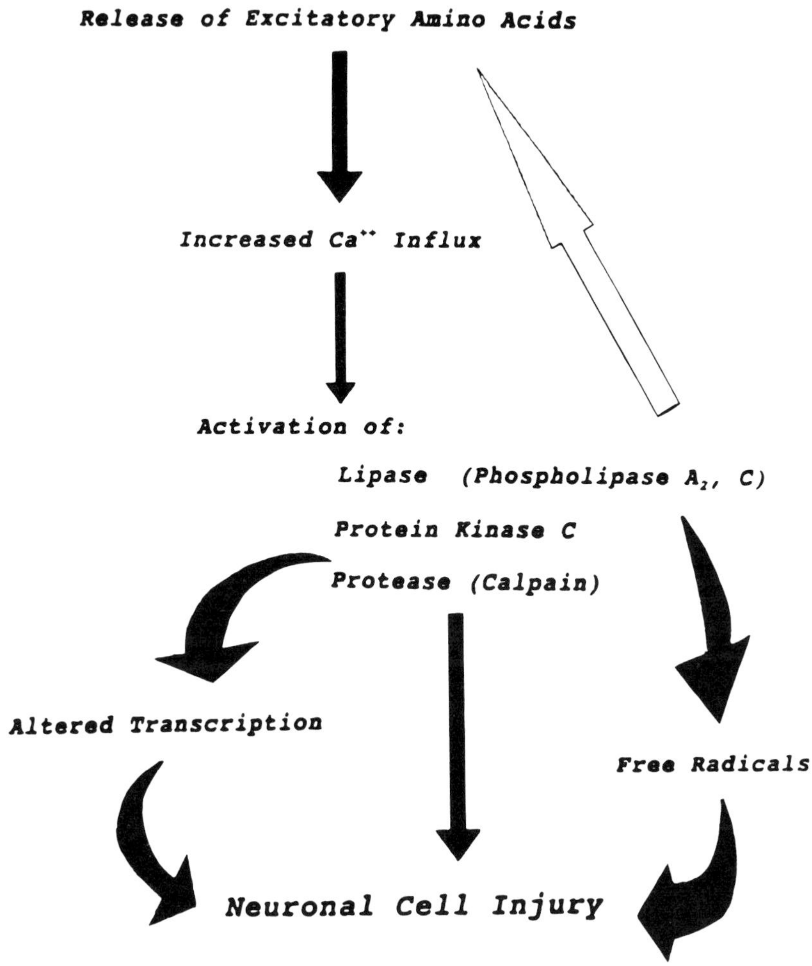

FIGURE 5. Hypothetical links between the release of excitatory amino acids, Ca^{2+} influx, free-radical generation, signal-transduction systems, and hypoxic neuronal injury.

neurotransmitters GABA and glycine, all of which show an increase during hypoxia (Joseph et al., 1992; Young et al., 1992). However, there is no direct evidence implicating serotonin in neurotoxicity during acute hypoxia (Haddad and Jiang, 1993). The increased release of the inhibitory neurotransmitters during hypoxia may activate K^+ and Cl^- channels, thereby hyperpolarizing neurons and depressing their activity. This, in turn, would protect neurons because of decrease in energy consumption and membrane excitability (Haddad and Jiang, 1993).

R. HYPOXIA AND SIGNAL-TRANSDUCTION PATHWAYS

For a stimulus to elicit a cellular response, the effector, (hormone, drug, antigen) must transport its signal across the cell membrane into the cell or nucleus. The mechanisms that allow specific substances to enter the cell in a highly controlled fashion are known as the signal-transduction systems, or pathways. These include ion channels, receptor-mediated second messenger, receptor internalization, and passive diffusion or infiltration (Carcillo, 1988). Ion channels provide a means of communication that allows the cells to respond immediately, as well as chronically, to an external stimulus (Miller, 1987). The three major ion channels are the Na^+, K^+, and Ca^{2+} channels. The receptor-mediated second-messenger systems provide the major mechanisms for rapid signal transduction. In this process, the ligand (specific molecule) binds to the receptor in the plasma membrane, producing an activated receptor complex. The latter stimulates an enzyme that, in turn, produces a second messenger inside the cell. Currently we know about three second-messenger systems, including the cyclic adenosine monophosphate (cAMP), the cyclic guanosine monophosphate (cGMP), and diacyglycerol (DAG) (Schmidt, 1992). The second messengers in these systems activate cellular responses leading to the process of signal transduction. Another signal-transduction mechanism is known as receptor internalization, which is the major route for the transport of extracellular macromolecules into the cells (Stahl and Schwartz, 1986). In this process, a specific receptor recognizes a specific ligand, attaches to it, and begins the signal-transduction process via internalization of the receptor-ligand complex into the cells. The passive diffusion is a type of signal transduction that occurs with the passage of lipophilic agents across the lipid membrane, such as some inhaled anesthetics (Carcillo, 1988).

Observations on the effect of acute or persisting hypoxia on neuronal signal-transduction systems in adult and newborn mammals have been recently reviewed by Haddad and Jiang (1993). There is now ample evidence supporting the effect of hypoxia on various ionic pumps and channels, particularly inhibition of Na^+-K^+-ATPase, reversal of the activity of the Na^+-Ca^{2+} exchange leading to Ca^{2+} accumulation, opening of the ATPase-sensitive K^+ channels due to a decrease in cystosolic ATP, and consequent membrane hyperpolarization. The latter would reduce energy consumption and thus prevent activation of several cation channels. Studies demonstrate that activation of Ca^{2+}-dependent K^+ channels may also account for the anoxia-induced hyperpolarization and decrease in input resistance in hippocampal neurons. This activation can be an adaptive mechanism in some central neurons, limiting depolarization caused by excitotoxicity and Ca^{2+} influx, thus protecting nerve cells against hypoxic damage. Moreover, hypoxia-induced Na^+ channel inhibition, and Cl^- channel activation may lead to the characteristic electrophysiological changes that occur during anoxia (see bibliography in Haddad and Jiang, 1993). Haddad and Jiang (1993a) further studied the mechanisms underlying the depolarization induced by anoxia (PO_2 = 0 mmHg within 20 s) in single neurons from brainstem slices, using the patch clamp technique. They tested (a) whether the increase in extracellular glutamate or decrease in intracellular ATP plays an important role in producing the membrane electrophysiologic phenomena during anoxia, and (b) whether the extracellular Na^+ plays a role in the anoxia-induced depolarization. The study indicates that the anoxia-induced depolarization is not a function of an increase in extracellular glutamate concentration, and it rather depends on Na^+- and ATP-mediated processes. A direct involvement of Ca-dependent K channels in hypoxic-ischemia, and trauma-related injury was suggested by Ohnishi et al. (1988) and by Tominaga et al. (1988). These investigators demonstrated that the recovery of animals after injury was enhanced by the preadministration of chaybdoxin, a specific inhibitor of these channels.

S. SECOND-MESSENGER SYSTEMS AND HYPOXIA

Studies have shown that the concentrations of cAMP and Ca^{2+} increase following hypoxia (Friedman and Haddad, 1993). cAMP, in turn, can activate cAMP-dependent protein kinase (kinase A), which phosphorylates other enzymes or proteins. The increased cytosolic free Ca^{2+} binds to calmodulin, forming a Ca^{2+}-calmodulin complex, which then binds to various proteins, including Ca^{2+}-calmodulin-dependent protein kinase (kinase C), thereby altering their activity (Haddad and Jiang, 1993). It is suggested that the release of Ca^{2+} from intracellular stores during hypoxia is mediated by inositol 1,4,5-triphosphate (IP_3) (Dubinsky and Rothman, 1991). This second messenger is activated during cardiac ischemia and reperfusion (Prasad et al., 1991). A similar event in neuronal elements could trigger a chain response of Ca^{2+}-activated Ca^{2+} increase in the cytosol, potentially leading to irreversible injury (Haddad and Jiang, 1993).

Another example of the effect of hypoxia on signal-transduction mechanisms is incorporation of serin into phosphatidylserine in hypoxic rat brain slices and homogenates (Mozzi and Horrocks, 1992). In this study, hypoxia caused an increase of the incorporation of the base into phosphatidylserine. Serine glycerophospholipids are involved in signal transduction through the membrane by contributing to the activation of the protein kinase C. Therefore, increased synthesis of phosphatidylserine may affect signal-transduction mechanisms and thus participate in the modification of lipid metabolism as observed in hypoxia and/or ischemia (Mozzi and Horrocks, 1992). Qi and Dawson (1993) exposed postnatal rat oligodendrocytes in monolayer culture to 6 h of hypoxia. They demonstrated a dramatic inhibition of the basal protein kinase C-mediated phosphorylation of myelin basic protein. Because hypoxia did not affect protein kinase C activity or its translocation to the membrane fraction, they suggested that the mechanism of inhibition of phosphorylation involves the uncoupling of signal transduction at a site before phospholipase C. Six hours of hypoxia also completely inhibited both muscarinic and α-1 adrenergic receptor-mediated inositol monophosphate release. They hypothesized that hypoxia may specifically target GTP-dependent processes, resulting in metabolic abnormality in lipids, proteins, and signal-transduction cascades of the oligodendrocytes (Qi and Dawson, 1993). As mentioned earlier, some free radicals, particularly, NO·, are present during hypoxic insults and participate in hypoxic neuronal injury (Pelligrino et al., 1993). Because NO· is involved in regulating soluble guanylyl cyclase activity, it may also function as a sensor for intracellular events, such as Ca^{2+} movements (Schmidt, 1992).

T. HYPOXIA-INDUCED GENE EXPRESSION

Because signal-transduction systems are closely related to gene expression (Bishop, 1987), it is possible that hypoxic-induced signal-transduction events may mediate neuronal responses not only through ion channels or Ca^{2+} levels and phosphorylation, but in the long-term through gene expression (Carcillo, 1988). In fact, hypoxia stimulates the expression of genes encoding for several cellular proteins and enzymes and thus may affect protein synthesis by altering translation and posttranslational processes. These effects of hypoxia are seemingly the result of activation of several intracellular second-messenger systems (Haddad and Jiang, 1993). Lorez et al. (1989) determined the levels of nerve growth factor in the cerebral cortex and hippocampus in rats that exhibited hypoxic neuronal cell damage after a bilateral carotid artery occlusion and exposure to 7% O_2. They found increased synthesis of the nerve growth factor that was apparently not related to a retrograde transport by cholinergic neurons. However, expression of a large number of genes is induced by ischemia rather than hypoxia alone (see bibliography in Haddad and Jiang, 1993). Whether hypoxia by itself is the main factor that triggers gene expression, except for c-fos protein, is not known.

U. SPREADING DEPRESSION AND HYPOXIC NEURONAL INJURY

Spreading depression (SD)-like depolarization, a phenomenon first described by Leao (1944), is characterized by a slowly spreading wavefront of suppression of spontaneous and evoked responses, a large negative shift of extracellular electric potential, and almost total loss of membrane potential in neuronal and glial cells. These events are accompanied by a massive loss of cellular K^+ and uptake of Ca^{2+}, Na^+, Cl^-, and water (Balestrino et al., 1988; Somjen et al., 1990). The SD-like depolarization is known to occur in some neuroanatomical structures consequent to anoxia, hypoxia, ischemia, and epileptic insults, and it may play an important role in neuronal damage (Mayevsky and Chance, 1975; Hansen, 1985; Balestrino and Somjen, 1986). This is because the probability of recovery from an acute hypoxic insult is inversely related to the time spent in hypoxic SD. Seemingly, only certain selective ion channels open during SD, and this can occur even in the absence of Ca^{2+} in the medium (Aitken et al., 1991). Nevertheless, excessive uptake of Ca^{2+} during SD may signal the onset of irreversible neuronal injury. In this regard, the duration of SD is a critical factor, and a short-lived hypoxic SD-like depolarization does not necessarily result in brain damage, as does a prolonged SD (Kawasaki et al., 1988; Balestrino et al., 1988). Moreover, in normoxic conditions K^+-induced SD depolarization causes an increase in the cortical oxygen consumption, adequately matching an increase in local blood flow. These effects are quickly over after termination of the wave of SD (Mayevsky and Weiss, 1991). Finally, despite the observation that prolonged SD-like depolarization allows massive influx of Ca^{2+}, it may not be considered as the only mechanism by which neuronal injury could be elicited (Balestrino et al., 1988).

V. HYPOXIC-INDUCED LACTACIDOSIS AND NEURONAL INJURY

Hypoxia-induced neuronal and glial acidification, mediated by activation of the glycolytic pathway, can affect cellular function in several ways (Siesjo, 1988; Nedergaad et al., 1991). These include membrane depolarization (Dean et al., 1989; Lehmenkohler et al., 1989), increase in cytosolic free-Ca^{2+} concentration (Burns et al., 1991), and denaturation of cytosolic and membrane proteins and enzymes (Chesler, 1990). Nevertheless, some studies have actually demonstrated a protective effect of hypoxia/ischemia-induced acidosis, such as selective attenuation of glutamate neurotoxicity in cortical and hippocampal neurons (Giffard et al., 1990; Tombaugh and Sapolski, 1990). Therefore, the question is whether hypoxia-induced tissue lactacidosis by itself can account for neuronal cell injury (Norberg and Siesjo, 1975; Myers, 1979; Rehncrona et al., 1981; Kalimo et al., 1981; Siesjo, 1988). *In vivo* studies demonstrate that delayed neuronal damage is increased in hyperglycemic animals subjected to hypoxic-ischemia, compared to normoglycemia (Myers, 1979, 1979a; Rehncrona et al., 1981). Since glycolysis is activated during hypoxia, the enhanced neurological damage in hyperglycemic animals has been attributed to development of a severe lactic acidosis (Siesjo, 1981, 1988). *In vitro* studies with hippocampal slices, however, have indicated that lactic acid has no detrimental effect on hypoxic neuronal tissue, even at concentrations capable of producing a pH of 5.5 (Schurr et al., 1988). Moreover, hypoglycemia accelerates and hyperglycemia delays the onset of SD-like depolarization, though neuronal damage is worse in hyperglycemia (Hansen, 1985). These findings are in agreement with observations demonstrating protection against hypoxic and anoxic damage by an increased level of glucose in the bathing medium of brain slices (Schurr et al., 1987, 1988; Schurr and Rigor, 1989; Grigg and Anderson, 1989; Roberts and Sick, 1992).

Perhaps, the harmful effect of hyperglycemia *in vivo* may be related to cerebral edema that develops following ischemia/reperfusion, that being the reason it cannot affect *in vitro* preparations (Myers, 1979a). In this case, one should investigate if hypoxia without ischemia can also produce cerebral edema. Norris and Pappius (1970) subjected cats to asphyxia and severe hypoxia with or without hypercapnia. They demonstrated that despite the presence of profound neurological damage, (coma, gross abnormal EEG), severe hypoxia alone or with hypercapnia did not produce cerebral edema either acutely or chronically (up to 8 d). Unlike a great deal of evidence on the harmful effect of hyperglycemia following hypoxic-ischemic insults *in vivo* (Siesjo, 1988; Smith and Siesjo, 1988), there is no information on neurological outcome of hyperglycemia after hypoxic insults alone. Hence, the answer for indirect effect of lactacidosis on hypoxic neuronal injury, *in vivo*, conditions remains obscure.

W. HYPOXIC NEURONAL INJURY AS A MEMBRANE PHENOMENON

The main pathways in the overall response to hypoxic neuronal injury are linked to plasma and mitochondrial membranes and their electrochemical and metabolic functions. The major players are ATP, ion channels and signal-transduction systems, neurotransmitters, excitotoxicity, and Ca^{2+} influx. In such multifactorial phenomena, the contribution of ROS, reduction in energy metabolism (lowering of ATP), and the limits of reversibility of neuronal injury remain obscure. Moreover, the role of gene activation, oxygen sensors, main inhibitory factors, as well as differences between hypoxia, anoxia, and ischemia have not been completely resolved (Haddad and Jiang, 1993). Nevertheless, it appears that the cellular reactions that lead to hypoxic neuronal injury are initiated by hypoxia-induced energy insufficiency and/or by oxidative stress, both of which primarily target plasma and mitochondrial membranes. The shortage of energy during hypoxia does not allow membrane ions and electrical potentials to be sustained. Such uncoupling between membrane functions and energy production eventually leads to high membrane permeability. The early electrochemical manifestations of membrane ion pumps (neuronal depolarization) could be related to a massive ouflux of K^+ from the neurons and an influx of Na^+ into the neurons (Haddad and Jiang 1993, 1993a). Concurrently, perhaps, glutamate-receptor overstimulation produces the electrochemical changes, which are identified as SD-like depolarization. If the K^+ concentration in the extracellular fluid falls below 12 m*M*, changes in membrane potential become large enough to activate voltage-dependent Ca^{2+} channels (Hansen, 1985).

Concurrently, the hypoxia-induced generation of ROS damages specific membrane proteins, particularly, enzymes that contain critical sulfhydryl groups (Fridovich, 1978; Jones, 1985; Davies, 1987; Floyd, 1990). The plasma membrane, in fact, is the primary site of ROS-mediated injury owing to its content of peroxidizable lipids and proteins and to its physical proximity to both intra- and extracellular radical sources (Grisham and McCord, 1986; Scott and Rabito, 1988). Thus, ROS may contribute to disorganization of the plasma and mitochondrial membranes that are unable to maintain ion homeostasis, or regulate energy production and consumption. Consequently, hypoxia-induced excitotoxicity and Ca^{2+} influx, reduction in antioxidant capacity, and, perhaps, intracellular synthesis of $NO^{\cdot}$ may irreversibly damage the membrane and cellular integrity, resulting in cellular death. Intracellularly, the influx of the Ca^{2+} stimulates the activity of various intracellular enzymes, such as Ca^{2+}-dependent phospholipase A_2 or receptor-mediated phospholipase C. These enzymes catalyze hydrolysis of the membrane phospholipids, causing the release of free fatty acids, and further disintegration of the plasma and mitochondrial membranes. The disruption of intracellular Ca^{2+} homeostasis may also stimulate the activities of diacyglycerol lipase, calpain, protein kinase C, guanylate cyclase, nitric oxide synthase, and endonuclease (Siesjo, 1981; Hochachka, 1986; Umemura et al., 1992). These reactions produce a cascade of events, thereby modifying signal transduction to aberrant neuronal signaling and cytosolic chaos.

X. NEURONAL SENSITIVITY TO HYPOXIA AS A SECONDARY INSULT

If arterial PO_2 is in the normal range, the local brain tissue PO_2 will then be determined by parameters of CBF, diffusion, and CMR-O_2 (Duling and Pittman, 1975; Wilson et al., 1979; Duling and Klitzman, 1980). Hypoxia rapidly reduces tissue oxygenation from either a normoxic or hyperoxic state. Upon supplementation of oxygen (reoxygenation), the hypoxic tissue quickly regains adequate oxygenation for oxidative phosphorylation, providing normal CBF and hemodynamic conditions. This concept is the rationale for the therapeutic use of oxygen at high concentrations when tissue oxygenation is impaired (Lambertsen, 1978). Regulation of CBF is governed by various mechanisms, involving metabolic, mechanical (perfusion pressure), and chemical stimuli (O_2 and CO_2), and neurogenic factors (Kuschinsky and Wall, 1978; Siesjo 1978a; Heistad and Contos, 1983; Prough and Rogers, 1989). These reactions are mediated by changes in the synthesis and release of active substances, ions, and free radicals, including prostaglandins, adenosine, angiotensin, K^+, and H^+, as well as $NO^{\cdot}$ (Winn et al., 1981; Seifert et al., 1988; Van Wylen et al., 1989; McBean et al., 1989; Faraci and Heistad, 1992; Szabo et al., 1992; Prado et al., 1993).

In normal conditions, with an intact blood-brain barrier (BBB), there is a tight coupling between regional CMR-O_2 and regional CBF (Siesjo, 1978a; Pickard, 1981). These delicate regulatory mechanisms of CBF and CMR-O_2 can be easily disturbed by trauma (Nilsson et al., 1990; DeWitt et al., 1992; Bouma and Muizelaar, 1992), ischemia (Ginsberg et al., 1978; Welsh et al., 1978; Siesjo et al., 1979; Koch et al., 1984; Grogaard et al., 1980), intracranial hypertension (Eldridge et al., 1991), hemorrhagic hypovolemic hypotension, and systemic hypertension (Miller et al., 1975; Ishige et al., 1988; Szabo et al., 1992). When cerebral hypoxia occurs secondary to these conditions, the basic rules for cerebral tissue oxygenation, as described earlier, may not apply any more. Consequently, reversal of tissue hypoxia to normoxia or hyperoxia (by hyperoxygenation) does not take place completely. This may indicate that the overall cerebrovascular and metabolic responses to a secondary insult by hypoxia are greatly impaired (Lewelt et al., 1982; Feustel and Nelson, 1987). Thus, hypoxia, in face of a widespread edema and ischemia, may lead to a poor neurologic outcome (Marmarou et al., 1984; Ishige et al., 1987; Chan et al., 1990; Gabb et al., 1990).

Y. WHY A DAMAGED BRAIN IS MORE VULNERABLE TO HYPOXIA

Lack of improvement in tissue oxygenation, in face of restoration of systemic hemodynamics and oxygenation, may indicate impaired cerebrovascular responses to chemical or neurogenic stimuli (Saunders et al., 1979; Lewelt et al., 1982; Koch et al., 1984; Ishige et al., 1987). Brain trauma and/or ischemia can also affect the autoregulatory responses to changes in systemic and intracranial pressures (Lewelt et al., 1980; Ishige et al., 1988; Torbati et al., 1989; Bouma and Muizelaar, 1992), resulting in "no reflow" phenomenon (Ginsberg et al., 1978; Kagstrom et al., 1983; Pool et al., 1987; Wisner et al., 1989). The "no reflow" could be also related to generation of free radicals, loss of Ca^{2+} homeostasis, and leukocyte-mediated injury in the endothelium (Grogaard et al., 1989; Hallenbeck and Dutka, 1990). Posttraumatic or postischemic brain oxygenation may be affected because of development of cerebral edema and intracranial pressure, which can reduce both cerebral perfusion pressure and CBF (Poole et al., 1987; Wisner et al., 1989; Hallenbeck and Dutka, 1990). Once brain edema has developed, it results in the release of free radicals, kinins, eicosanoids, and glutamate (Bazan and Rodriguez de Turco, 1980; Whittle et al., 1990). These mediators of edema may then increase capillary hydraulic conductivity and further enhance edema formation (Untenberg et al., 1990).

The basic mechanisms of the cerebrovascular response to hypoxia in an intact brain may provide some explanation for hypersensitivity to hypoxia as a secondary insult. The cerebrovascular response to hypoxia, in an intact brain, is mediated by H^+ (Siesjo and Nilsson, 1971; Johannson and Siesjo, 1975; Shinozuka et al., 1989), adenosine (Winn et al., 1981; Morii et al., 1987), cAMP (Rehncrona et al., 1978; Folbergrova et al., 1981), and K^+ (Dora et al., 1979). Brain injury modifies the regulation of CBF and energy metabolism (Cold, 1990; Nilsson et al., 1990; Katayama et al., 1990; Hovda et al., 1992), increases the level of oxidative stress (Chan and Fishman, 1985; Kontos and Povlishock, 1986; Shohami et al., 1987), and disturbs membrane ion balance, particularly of Ca^{2+} (Siesjo, 1981, 1988a; Young, 1992). Some of these molecular events seem to be common with those suggested for hypoxic injury alone. Therefore, their excessive production or presence can only worsen the outcome of a secondary hypoxic insult. One mechanism that may specifically agonize the effect of posttraumatic hypoxia, or reoxygenation after ischemia, may be enhancement of preexisting oxidative damage in the neurovascular system (Demopoulos et al., 1980; Halliwell and Gutteridge, 1985; Halliwell et al., 1992). Moreover, since both hypoxia (via energy insufficiency) and oxidative stress can impair the ability to maintain Ca^{2+} homeostasis, either can enhance the potential hazard of the other (Jones, 1985). Thus, any type of brain injury (ischemic, trauma, seizures) that has the potential to increase oxidative stress may also enhance the effect of hypoxia, both as primary or secondary insults.

Z. APNEA AS A COMMON BY-PRODUCT OF HEAD TRAUMA

Hypoxia resulting from transient apnea or reduced ventilation is very common in all kinds of brain trauma (Horsly, 1894; Denny-Brown and Russel, 1941; Levine and Becker 1979; Genneralli et al., 1982). Unfortunately, this type of systemic hypoxia usually occurs in the face of other complications, including reduction in CBF and loss of autoregulatory ability, reduction in cardiac output, intracranial hypertension, and systemic hypotension (Crackard et al., 1977; Cold, 1990; Eldridge et al., 1991; Torbati et al., 1992a). Obviously, in such a situation the highest priority should be given to a prompt respiratory and cardiovascular resuscitation. Investigators in the 19th century had already recognized that restoration of normal respiration after brain trauma in humans can profoundly reduce mortality and prevent delayed neurological deficiencies (Polis, 1884; Horsly, 1894). Since this simple "first aid" procedure is not always immediately available at the scene of an accident involving brain trauma, drugs and/or other therapeutic procedures are needed to prevent delayed or secondary hypoxic brain damage.

AA. BRAIN DEVELOPMENT AND RESISTANCE TO HYPOXIA

Hypoxia is a common insult during the prenatal and neonatal period (Vannucci, 1978). Fortunately, fetuses and neonates in mammals display a greater tolerance to both cerebral anoxia and hypoxia than the adult animals (Fazekas et al., 1941; Glass et al., 1944; Duffy et al., 1975). It has been shown that the newborn has a different respiratory responsiveness to hypoxia than the adult (Haddad and Mellins, 1984). At the molecular level, the difference in tolerance to hypoxia has been attributed to differences in membrane properties between immature and adult CNS (Hochachka, 1986). Haddad and Donnelly (1990) studied the neuronal response to hypoxia with maturation, using hypoglossal neurons from adult and neonatal rat from 3 to 28 d old. They found that while all neurons depolarized when exposed to hypoxia of 15 to 20 mmHg O_2 for 5 min, the magnitude of depolarization was three times larger in adult hypoglossal neurons. In these experiments, measurement of K^+ concentrations in the extracellular space revealed a major increase in the extracellular K^+ in the adult neurons, but not in newborn. This indicates that the difference in hypoxic-induced membrane depolarization between the neonate and adult can be attributed, in part, to differences in extracellular K^+ (Haddad and Donnelly, 1990). They have also suggested that the difference in extracellular K^+ may be due, to some extent, to different rates of intracellular ATP depletion and failure of the Na^+-K^+ pump. This hypothesis was later supported by another study, from the same group (Jiang et al., 1992), showing that in the adult rat K^+ is lost during 3 to 4 min anoxia from neurons through sulphonylurea receptor or K-ATP channels in a major way, while K-ATP channels are poorly expressed in the newborn rat.

Xia et al. (1992) compared the role of oxidative and glycolytic pathways in the brainstem of adult and newborn rats during anoxia, showing about a ten-fold increase in the extracellular K^+ in adult, but not in newborn. However, the extracellular K^+ increased remarkably in both adult and newborn brainstem when glycolysis was blocked with iodoacetic acid. These investigators concluded that the newborn rat brain must rely either on higher glycolytic capacity or on a reduction in the metabolic rate during hypoxia. Interestingly, Jiang et al. (1991) have discovered that the level of tissue PO_2 in neonate brain slices was always higher than adult brain slices at comparable depths and ambient O_2 levels of 5, 10, and 21%, and hyperoxia of 600 mmHg in the perfusate. This implies that the severity of cerebral hypoxia could be always less in the neonate than the adults which are subjected to similar experimental conditions. Overall, the inherent properties that may create a relative tolerance to hypoxia in newborn mammals, compared to adults, may be related to following factors: (a) low levels of cerebral circulatory and energy requirements, perhaps with different metabolic pathways (Medina, 1985; Nehlig et al., 1988, 1989; Bomont et al., 1992; Xia et al., 1992); (b) the ability of the developing brain to maintain neuronal ionic gradients, such as K^+ and Na^+ at minute levels of ATP (Mishra and Shankar, 1980; Haddad and Donnelly, 1990; Jiang and Haddad, 1992; Haddad and Jiang, 1993a); (c) much higher levels of tissue PO_2 in neonatal brain, compared with an adult brain at all ambient O_2 concentrations (Jiang et al., 1991); (d) a low concentration of polyunsaturated fatty acids in the brain, which reduces brain vulnerability to lipid peroxidation by oxidative stress (Crawford and Sinclair, 1972); (e) a lower susceptibility of immature neurons to glutamate neurotoxicity (Peterson et al., 1989); and (f) lack of release of excitatory amino acids (Cherici et al., 1991), which can attenuate Ca^{2+} influx, thereby reducing neuronal hypoxic damage (Schurr and Rigor, 1987). Despite the "tolerance" of the developing brain to hypoxia, it may retard the intellectual functions of the brain and/or make it prone to postnatal neurologic disorders (Vannucci, 1990; Jensen et al., 1991).

AB. IS HYPOXIC NEURONAL INJURY PREVENTABLE?

Because of the diversity of the hypoxia-induced metabolic and molecular events, there are no simple solutions for an absolute prevention of hypoxic neuronal injury. Nevertheless, based on the proposed molecular mechanisms of neuronal injury, there are a number of ways that we may maximize cellular protection and promote recovery following an hypoxic event. For example, inhibition of excitatory transmission or Ca^{2+}-channels blockade may ameliorate neuronal hypoxic injury to a great extent, because Ca^{2+} influx can lead to both acute and delayed neuronal destruction (Choi, 1990; Beal, 1992). Moreover, excessive release of glutamate may reduce the synthesis of glutathion (Murphy et al., 1989), thereby reducing the antioxidant capacity of the neuronal elements. Thus, interruption of the excitatory amino acids release lessens the degree of injury inflicted by both Ca^{2+} influx and oxidative stress. Another approach, for slowing the multiple hypoxic reactions in the CNS, is to devise more than a single "magic remedy". A combination of drugs or agents that are engineered to attack more than one target (e.g., being both a Ca^{2+}-channel blocker and an antioxidant) may prove more effective than drugs or procedures that are designed to take care of a single molecular disturbance at a time. Manipulation of the signal-transduction systems may become the future strategy in preventing delayed consequences of hypoxic neuronal injury. Bearing in mind the side effects of any drug or procedure, the best approach seems to be a multidirectional assault, launched simultaneously on several cellular targets. Technically, the repair of hypoxic neuronal damage that has already occurred may be difficult, unless advances in gene therapy, transplantation, and neural plasticity prove otherwise (Backlund, 1987). However, deterioration of hypoxic damage is preventable even after relatively long episodes of hypoxia. At late stages of hypoxia, therapeutic intervention should be directed to prevent intracellular proteolysis, a strategy that has been proven useful with calpain inhibitors (Arai et al., 1990, 1991; Lee et al., 1991).

II. HYPEROXIA AND THE CNS

A. INTRODUCTION

The opposite pole of oxygen domain, termed hyperoxia, affects the CNS in a unique way. Such an "experience" had not been known to terrestrial organisms throughout the natural history of the earth until the discovery of oxygen and its concentrated use by humans. It began when scientists searched for new therapeutic modalities, using oxygen as a drug (Davis and Hunt, 1977; Gilbert, 1981). This, with man's desire to expand his territory into the depths of oceans and distant spaces demanded physiological adjustments to abnormally high hydrostatic pressures of gases and exposure to artificial gas mixtures that contained high concentrations of oxygen. The major physiological barriers of pressure-related maladies were discovered by Paul Bert (1878), a French physiologist of the 19th century. Bert's pioneering studies described the effects of increased barometric pressure (of air and oxygen) on various organs, body parts, and plants. He demonstrated the effect of HBO on intact animals, indicating that the CNS is the primary target of HBO poisoning. The most visible neurological manifestations of HBO-induced oxygen toxicity were identified as convulsions similar to those observed during grand-mal epilepsy (Bert, 1878; Bean, 1945). A major question to be answered here is how the CNS in highly developed mammals can cope with such an extremely unnatural environment, created solely by man's adventurous ingenuity rather than the slow evolutionary progress of nature.

B. WHEN IS A TISSUE HYPEROXYGENATED?

In the section on tissue oxygen gradient, using Jones' concept (1985), it was concluded that tissue oxygen concentration is a continuum from hypoxia to hyperoxia. Nevertheless, in this specific section, discussing the effect of high concentrations of O_2, particularly at high barometric pressure, one needs a more precise definition of tissue or cellular levels of oxygen that could be considered hyperoxia. While there is no numeric value for normoxic tissue PO_2 when breathing air at normobaric pressure, application of NH, HBO, or gas mixtures containing oxygen at high barometric pressures, will dramatically increase the alveolar, arterial, and tissue PO_2. The state of hyperoxia could be then defined as a tissue PO_2 elevated above the maximal levels obtainable during air breathing at sea level (inspired PO_2 of about 160 mmHg). Because of different levels of blood flow and metabolism, this convention also implies that the hyperoxic thresholds can vary in different tissues in a given species, as well as among the different species.

Normally, hyperoxic exposures increase the lung's PO_2 to the greatest extent (Lambertsen, 1978). The next organ to be overoxygenated is the blood, because hemoglobin can already be saturated up to 98% during air breathing. Oxygen breathed at high barometric pressure will be physically dissolved in the liquid fraction of the blood. This extra oxygen content would be readily available for extracapillary diffusion and cellular consumption. Physically, about 2.3 ml O_2 can be dissolved in 100 ml blood for each atmosphere absolute (ATA) of inspired oxygen at the normal body temperature. However, the level of arterial PO_2 does not always reflect tissue PO_2 values, because the capillary density, rates of blood flow, and oxygen consumption vary in different organs, particularly in the brain. This organ, compared to others, has much lower tissue PO_2 because of the high degree of oxygen consumption (Sokoloff, 1976). During HBO, providing a normal gas exchange in the lungs and unchanged CBF and CMR-O_2, the brain parenchyma turns hyperoxic even in locations that previously had a hypoxic/anoxic distribution pattern during hypoxia or even normoxic air breathing (Whalen et al., 1970; Torbati, et al., 1976; Hunt et al., 1978).

C. HOW IS TISSUE HYPEROXIA PRODUCED?

Normobaric hyperoxia and HBO are extensively used for treatment of a variety of diseases in need of improved tissue oxygenation (Jacobson, 1971; Snider and Rinaldo, 1980; Fairley, 1980; Myers and Schnitzer, 1984; Jain, 1994). This is accomplished by using masks, head tents or endotracheal tubes, as well as pressure chambers designed for HBO therapy (Bland, 1980; Fulmer and Snider, 1984). Tissue hyperoxygenation can also be achieved by breathing oxygen at lower barometric pressures (160 mmHg $< PO_2 <$ 760 mmHg), a condition termed hypobaric hyperoxia. These relatively low levels of hyperoxia could be generated during shallow diving with air, during deep diving, using mixtures of oxygen with a variety of noble gases, or during flight when using gas mixtures with high concentrations of oxygen (Strauss, 1979; Thorsen et al., 1993). The percent of oxygen in the gas mixtures used for deep diving is usually lower than 21% (concentration at sea level). But, with increasing depth, the inspired PO_2 will increase to levels of hypobaric hyperoxia (760 mmHg $> PO_2 >$ 160 mmHg). Oxygen at hypobaric pressure is also used for altitude simulations inside closed hypobaric/hyperbaric chambers (Quattrone and Staley, 1966). Increased PO_2, however, does not always result in elevated tissue oxygenation. The conditions in which brain may become hypoxic, even if the individual is subjected to HBO therapy, include impairment of gas exchange or transport of oxygen, substantial reduction in local CBF, or abnormal increase in local oxygen consumption (status epilepticus; Figure 1).

D. ORGANS AND SYSTEMIC SENSITIVITY TO HYPEROXIA

The rate of development of oxygen toxicity in each organ and the time of onset of generalized oxygen poisoning for a given species are different, and are determined by rate of response to both physiologic and toxic effects of hyperoxia. The major intrinsic factors that determine the sensitivity of an organ to hyperoxia include chemical composition, the degree of metabolic activity, and its antioxidant capacity (Lambertsen, 1978; Torbati, 1987). Because these factors vary from organ to organ and are not the same in different species, the rate of development and characteristic signs of oxygen poisoning would also be different. In mammals, the respiratory system and the CNS are among the primary targets of the toxic effects of hyperoxia (Bean, 1945; Clark and Lambertsen, 1971; Balentine, 1982). Application of HBO in medicine imposes a great challenge on regulatory mechanisms of CBF, cerebral energy metabolism, and the neurotransmitter system. Presumably, alterations in these parameters modifies neuronal excitability, leading to preconvulsive electrical and behavioral manifestations of CNS oxygen toxicity.

Because of the toxic effects of oxygen, pressures that are used for HBO therapy, diving, and aerospace activities are usually limited to 1 to 3 ATA, with exposures lasting for a few hours depending on the P_IO_2. The effects of oxygen that can influence the outcome of CNS and systemic oxygen therapy include (a) physical effects, which involve no chemical or other active influence upon cells; (b) physiologic effects, which modify function but are fully reversible on return to normobaric normoxic air breathing; and (c) the toxic effects, which may progress to a lethal degree (Lambertsen, 1978). Hyperoxia can simultaneously affect multiple organ systems (Balentine, 1982). Thus, any complication caused by hyperoxia, even when it does not involve CNS and/or the respiratory system, should be considered as an end point for termination of hyperoxic exposure in medicine, diving, and aerospace activities (Simon and Torbati, 1982).

E. CNS CIRCULATORY, ENERGY METABOLISM, AND ELECTRICAL ACTIVITIES DURING HBO

An immediate physiologic response to hyperoxia is cerebral vasoconstriction. This effect of oxygen could be a transient phenomenon, depending on oxygen pressure (Lambertsen et al., 1955, 1963; Bean et al., 1972; Torbati et al., 1978; Balentine, 1982). At relatively high pressures of oxygen (3 to 5 ATA), the vasoconstrictive response fails before the onset of the neurological signs of oxygen toxicity (Bean et al., 1972; Torbati et al., 1978). The vasoconstrictive effect of oxygen could be related to an early generation of ROS, particularly, superoxide generation, which has a strong vasoconstrictive property (Lefer and Lefer, 1993). Superoxide may also inactivate the nitric oxide radical ($NO^{\cdot}$), which acts as a vasodilatory agent in the brain (Halliwell, 1992; Muhvich et al., 1992; Lefer and Lefer, 1993), thereby increasing the cerebral vascular resistance, reducing CBF. The failure to maintain hyperoxic-induced vasoconstriction may be related to the vasodilatory effect of CO_2, the concentration of which is increased in the blood during HBO. The author also speculates that the late breakdown of the HBO-induced cerebral vasoconstriction (Bean et al., 1972; Torbati et al., 1978) may be the result of an excessive production of $NO^{\cdot}$ in the face of an increased generation of ROS (Figure 6).

During HBO-induced cerebral vasoconstriction, conscious animals do not show any visible neurological manifestation of oxygen toxicity, although large-scale changes in the brain energy metabolism take place during this clinically asymptomatic exposure to HBO. Autora-

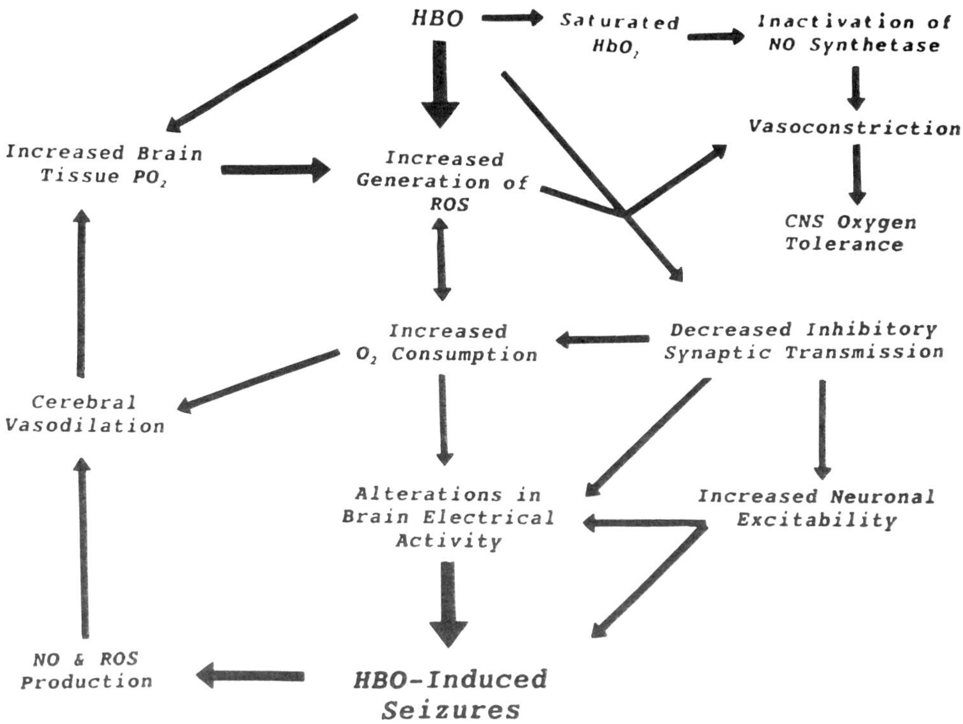

FIGURE 6. Hypothetical pathophysiologic pathways which may lead either to recurrent HBO-induced electrical and clinical seizures, or to CNS tolerance to oxygen toxicity.

diographic studies have demonstrated increases in regional brain glucose utilization rate at a wide range of pressures and duration of exposure to HBO (Torbati et al., 1983, 1984; Torbati and Lambertsen, 1983 and 1985). The distribution pattern of cerebral glucose utilization changes at the onset of HBO-induced electrical or convulsive discharges, when both increases and decreases in regional glucose utilization rate occur simultaneously in different brain structures (Torbati and Lambertsen, 1987). The early circulatory and metabolic events during HBO correspond with increase in EEG slow-wave activity and the appearance of paroxysmal electrical discharges, as demonstrated in conscious rats and rabbits (Rucci et al., 1967; Harel et al., 1969; Torbati et al., 1971, 1972, 1981; Raday et al., 1975). The delayed metabolic changes following HBO-induced seizures could be related to either activation of membrane Na-K-ATPase (Gottlieb et al., 1977; Hemrick and Gottlieb, 1977; Mayevsky, 1983) or its inhibition (Kovachich et al., 1981), depending on duration of exposure, anatomical location, and the level of functional activity (Torbati, 1987).

F. EFFECT OF HBO ON NEUROTRANSMITTERS

Early studies by Wood and Watson (1963), and Wood et al., (1963, 1969, 1971) demonstrated that the effects of HBO on the CNS could occur at the level of synaptic transmission. They found that only a few minutes exposure to HBO in rats was necessary to cause a significant decrease in gamma-aminobutyric acid (GABA) concentration. The decrease in GABA preceded the neurological manifestations of CNS oxygen toxicity as manifested by seizures (Wood and Watson, 1963). Furthermore, rats injected interaperitoneally with various concentrations of GABA were more resistant to CNS oxygen toxicity than control animals, as judged by a decrease in the number of rats convulsed and a decrease in the mortality rate (Wood et al., 1963). Moreover, they found a linear correlation between the relative rate of GABA metabolism in chicks from 2 to 21 d old (evaluated by the ratio of the activity of glutamic acid decarboxylase (GAD), which synthesizes GABA, and GABA-α-oxoglutarate transaminase that degrades GABA) and the susceptibility to HBO-induced seizures (Wood et al., 1971). They also demonstrated a significant decrease in GAD activity in whole-brain and synaptosomes in chicks, rats, and mice that were exposed to 1 to 5 ATA O_2.

The question is whether the changes in GABA metabolism and the onset of convulsions are two independent events, having no cause-and-effect relationship. Wood's observations (1963, 1969, 1971) imply that the increased excitability of the brain to seizures could be modulated by the effects of HBO on regulation, synthesis, and/or release of the inhibitory neurotransmitters. This argument is supported by Wheeler and Blackburn (1983) who measured GABA uptake in rat brain synaptosomes from animals that had been exposed to HBO. They found significant decrease in velocity of GABA uptake in HBO-exposed rats compared to air breathing controls. More recently, Colton and Colton (1986) used the lobster walking-leg neuromuscular preparation to study the effect of oxygen at 1, 1.7, and 5.1 ATA on the inhibitory transmission as manifested by a fall in inhibitory synaptic conductance. Because the lobster walking leg uses GABA as the inhibitory neurotransmitter, and has an accessible inhibitory axon and large post-synaptic fibers, it can provide direct electrophysiological evidence on GABA-mediated inhibitory neurotransmission at the cellular level (Colton and Colton, 1986). They found a similarity in the action of HBO to the action of isoniazid, a known GAD inhibitor in the same lobster-leg preparation, suggesting that one possible site of oxygen action is on GABA production.

G. PRECONVULSIVE SIGNS OF CNS OXYGEN TOXICITY

In humans, in addition to a clinically asymptomatic period during HBO, there are other subtle neurological manifestations of CNS oxygen toxicity, such as nausea, mental confusion, dizziness, and twitching (Lambertsen, 1978; Butler and Thalmann, 1986). It is reasonable to assume that the early metabolic, neurochemical, and electrical events, which are observed in experimental animals, may also constitute the biological base for the preconvulsive neurological signs of oxygen poisoning in humans. Thus, both physiologic and pathologic reactions to HBO are key players in the dramatic appearance of the HBO-induced convulsions, which are the end point of CNS oxygen toxicity (Torbati et al., 1983; Torbati, 1987). The CNS responses to HBO, preceding convulsions, usually do not result in visible neurologic or behavioral signs in animals that also show a variable asymptomatic period (Torbati, 1987). The question is whether the clinically "asymptomatic" HBO exposures in man are safe in terms of delayed neurological damage. Because most HBO therapies usually consist of intermittent and relatively short-term HBO exposures, it is assumed that no long-term neurological effect may develop. This also implies that most, if not all, physiological changes that occur during the visibly asymptomatic period of HBO exposure should be reversible.

H. FREE RADICAL THEORY OF OXYGEN TOXICITY

Earlier studies on cellular mechanisms of oxygen toxicity were focused on inhibition of enzymes activity, particularly SH-containing enzymes and those that are involved in energy production (glycolytic enzymes, tricarboxyllic acid cycle, electron-transport system of mitochondria), enzymes engaged in neurotransmission, and proteolysis (Stadie and Haugaard, 1945; Dickens, 1946; Haugaard, 1968; Wood et al., 1969; Kovachich and Haugaard, 1981; Faiman et al., 1981). More recent studies have emphasized both inhibitory and activating effects of some membrane transport enzymes, specifically, Na-K-ATPase in etiology of CNS oxygen toxicity (Gottlieb et al., 1977; Kovachich et al., 1981). The most likely explanation for the enzyme inhibition theory of oxygen poisoning seems to be generation of ROS due to excessive tissue oxygenation. This "free radical theory of oxygen poisoning" was first suggested by Gerschman et al. (1954a), who proposed that oxygen toxicity and X-irradiation may share a common mechanism. Experimental evidence supporting this theory came from studies demonstrating increased peroxidation of brain lipids following single and short-term exposures to HBO, under both *in vivo* and *in vitro* conditions (Becker and Glavior, 1962; Jerret et al., 1973; Fridovich, 1976; Kovachich and Mishra, 1980; Kovachich and Haugaard, 1981; Dricks and Faiman, 1982; Noda et al., 1983; Jamieson, 1989). Further support for the "free radical theory" came from observations showing that vitamin E, liposomal superoxide dismutase, and catalase can protect against both neurological and pulmonary manifestations of oxygen toxicity (Kann et al., 1964; Zirkle et al., 1965; Turrens et al., 1984; Yusa et al., 1984), while hyperoxia increases production of H_2O_2 in the brain (Yusa et al., 1987).

Direct evidence for free radical generation, before the onset of the neurological signs of oxygen toxicity, was provided by detection of ROS in the rat brain using *in vivo* spin trapping and ESR spectroscopy (Torbati et al., 1992). These observations seem to contradict the study by Jamieson (1991) showing no significant lipid peroxidation (estimated by the thiobarbituric acid reaction substances) in the brain of mice exposed to HBO at several pressures for 30 min. In Jamieson's study, however, varying degrees of convulsive activity occurred during HBO exposures. Interestingly, convulsions diminish the appearance of the spin adduct free radical signals in the ESR spectra from brain lipid extracts of rats that were exposed to 5 ATA O_2 (Torbati et al., 1992). The latter observation, in the author's opinion, may explain the low level of lipid peroxidation in the brain of convulsed mice after HBO exposure, as observed by Jamieson (1991). In conclusion, it is sensible to hypothesize that, likewise in hypoxic neuronal injury, a large-scale production of ROS is responsible for initiating the chain of reactions underlying the mechanism of the hyperoxic neuronal injury.

I. HBO-INDUCED FREE RADICAL GENERATION AS THE PRIMARY CAUSE OF CNS OXYGEN TOXICITY

How production of ROS is increased during hyperoxia is still unknown. Two possibilities are (a) a direct mass effect of oxygen on cellular components (Fridovich, 1976, 1983; Yusa et al., 1987), indicating that free-radical production is proportional to O_2 availability (Stock et al., 1990); and (b) an increased rate of oxygen consumption (Nohl and Hegner, 1978; Freeman and Crapo, 1982; Torbati et al., 1983). Whether a change in energy-producing processes during hyperoxia is involved in ROS production or conversely, changes in energy production are in response to generation of ROS is not known. Lahiri et al. (1987, 1991), and Torbati et al. (1993) have demonstrated that exposure to either NH or HBO can damage carotid body mitochondrial membrane and hypoxic chemotransduction function in the cat. Because the carotid body is part of the neural system, these observations may suggest both a link between oxidative metabolism and glomus cells chemosensory ability, and a strong relationship between changes in energy metabolism and hyperoxic neuronal injury. Since mitochondrial membranes are closely related to oxidative energy metabolism, the ROS-induced enzyme inhibition or membrane derangement can affect cellular energy resources. Mitochondria, by themselves, are a major source of endogenous generation of ROS, which is enhanced by increased levels of cellular PO_2. Although this speculation may explain hyperoxic-induced neuronal damage, alternative mechanisms involving other functionally complex membranes (plasma and endoplasmatic reticulum) may also be involved. In fact, plasma membrane seems to be a more likely site of ROS actions, where the active ion transport and consequently ion balance could be readily affected, leading to altered excitability. The process of repair requires energy that becomes scarce because of mitochondrial oxygen poisoning.

Another potential site for hyperoxic neuronal damage is at the level of genetic material. This is consistent with the hypothesis suggesting a common molecular mechanism for both oxygen poisoning and irradiation injury (Gerschman et al., 1954a). Although the specific effect of hyperoxia on gene expression/regulation, DNA or RNA synthesis has not been investigated, impairment in protein synthesis or induction of antioxidant enzymes may be related to the effect of ROS on either transcription or replication processes. Since generation of ROS has been implicated in the initiation and/or promotion of carcinogenesis (Floyd, 1990), one should not exclude the possibility that excessive and repeated use of HBO may promote carcinogenesis. Regardless of the exact mechanism(s) of the CNS oxygen toxicity, it is reasonable to propose that lipid peroxidation, caused by excessive generation of ROS, disorganizes the delicate structures of the plasma and mitochondrial membranes, thereby impairing cellular function and integrity (Balentine, 1982; Torbati et al., 1992, 1993). In the plasma membrane this can perhaps occur because of HBO-induced inhibition or modulation of Na-K-ATPase activity (Gottlieb et al., 1977; Kovachich et al., 1981; Mishra et al., 1989), as well as alterations in ion channels function, and, perhaps, modification of the signal-transduction pathways. In conclusion, the exact abnormal events during HBO that generate ROS, damage membranes, impair energy production, or modify neurotransmitter functions and neuronal excitability remain to be discovered. However, it may be speculated that HBO-induced oxidative stress may primarily affect mitochondria, plasma membrane, and protein synthesis, thus triggering the chain of molecular events that ultimately lead to the neurological manifestations of oxygen poisoning (Figure 6).

J. SUPEROXIDE VS. NITRIC OXIDE IN CNS OXYGEN TOXICITY

McCall et al. (1989) demonstrated that NO· and O_2^- generated by activated leukocytes could interact in such a way as to cancel out the biological activity of NO·. Moreover, it has been postulated that physiologic interaction between O_2^- and NO· may be involved in tuning of the vascular muscle, and that endothelial cells might produce both free radicals as antagonistic agents (Halliwell, 1989). However, interaction of O_2^- and NO· at physiologic pH can yield a nonradical product, peroxynitrite (Saran et al., 1990; Beckman et al., 1990; Halliwell, 1992). This reaction may damage biological structures and functions because peroxynitrite may be directly cytotoxic, and can oxidize thiol groups or decompose to form ·OH (Beckman, 1991; Radi et al., 1991). Studies suggest that NO· could be involved in the normal functioning of excitatory amino acids such as glutamate, while its excessive generation may damage neuronal elements (Dawson et al., 1991). Furthermore, NO· may share the ability of ROS to activate proteases (Nathan, 1992). Thus, interaction of ROS with NO· seems to be highly relevant to both normal brain functioning and development of the neurodegenerative diseases (Beckman, 1991; Halliwell, 1992). Oury et al. (1992) and Zhang et al. (1993) have suggested that NO· may be an important mediator in CNS oxygen toxicity, while O_2^- mediates inactivation of NO·, preventing neurological and lethal manifestations of oxygen toxicity. This novel hypothesis seems compatible with our pre- and postconvulsive pathophysiologic responses to HBO (Torbati, 1987). Assuming that O_2^- is a vasoconstrictive agent, its excessive production by HBO can maintain cerebral vasoconstriction, thus preventing dangerously elevated tissue PO_2. Presumably, once NO-synthase is activated, and NO· is released, it may act both as a vasodilatory agent (increasing CBF and elevating tissue PO_2) or it can react with O_2^-, leading to formation of peroxynitrite. The latter is by far more cytotoxic than O_2^- alone (Figure 7).

This hypothetical pathway for CNS oxygen toxicity may also explain the postconvulsive attenuation, and disappearance, of the free radical signals detected by spin trapping and ESR spectroscopy during a visibly asymptomatic exposure to HBO in rats (Torbati et al., 1992). Theoretically, repeated HBO-induced convulsions may create a vicious cycle in which NO· production increases CBF and tissue PO_2, and the latter generates more superoxide. These free radicals then can react with each other and produce the peroxynitrite. Since the spin-trapping agents compete with NO· for reaction with superoxide or HO· radical, this might then inhibit spin-adduct production and thereby prevent their accumulation, which leads to an intense ESR signal.

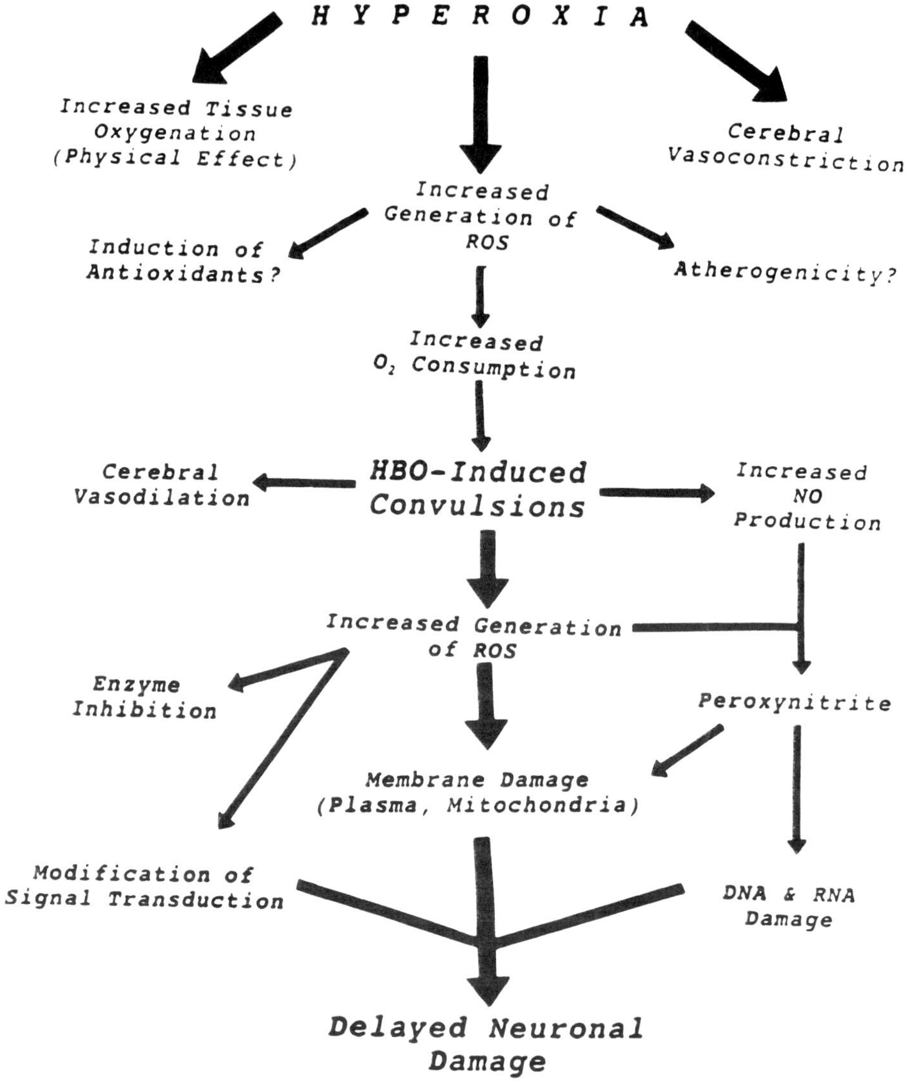

FIGURE 7. Cellular and molecular events that may lead to acute (Paul Bert effect) and/or chronic CNS oxygen toxicity (John Bean effect).

K. HBO-INDUCED NEUROPATHOLOGY VS. HYPOXIC-INDUCED NEURAL INJURY

Histopathological changes have been observed in the rat brain and spinal cord following single or repeated exposures to HBO (Balentine, 1966; Balentine and Gutsche, 1966; Schnakenburg and Nolte, 1973; Balentine, 1974). Studies using electron microscopy have demonstrated morphological alterations in the mitochondria of the proximal tubular epithelium of adult rats, within 1 h after a single exposure to 3 ATA O_2 (Greene et al., 1979). Mitochondrial degeneration has also been observed in rat hepatocytes following 3 h HBO exposure at 3 ATA (Schaffner and Felig, 1965). As mentioned earlier, alterations in mitochondrial ultrastructure and function have been identified in the carotid body glomus cells of the cats that were exposed either to NH or HBO at 5 ATA (Lahiri et al., 1987; Torbati et al., 1993). In these studies, the chemosensory responsiveness to hypoxia was also diminished, corresponding to decreases in the cristated area of mitochondria, which is a possible site of oxygen-sensing mechanism in the glomus cells (Torbati et al., 1993).

Balentine (1966, 1973, 1982), comparing the topography of HBO-induced neural necrosis with necrosis induced by an anoxic-ischemia, concluded that these two insults produced distinctly different neuropathologic patterns. In Balentine's studies, HBO-induced neural damage spares the neocortex, neostriatum, and hippocampus, areas that are the most vulnerable to hypoxic damage. In these studies no neurological defficiencies, except paralysis, were examined before the animals were autopsied for ultrastructural analysis (Balentine, 1982). However, the neuropathologic outcomes of CNS oxygen toxicity may be complicated because of recurrent convulsions (Schnakenburg and Nolte, 1973; Dam, 1992). Single or repeated convulsions may reduce generation of ROS, produced either by HBO (Torbati et al., 1992), brain injury, or ischemia (Anis, 1977; Schallert et al., 1986; Feeney et al., 1987). It has been demonstrated that only a couple of electroconvulsive seizures after unilateral contusion of the sensory motor cortex in rats accelerated the recovery of the motor function ability (Feeney et al., 1987). Conversely, convulsions can also serve as ROS generators (Armstead et al., 1989; Bauknight et al., 1992) and add to HBO-induced neuropathology. Thus, the question whether HBO, which, similar to hypoxia, generates both ROS and affects neuronal energy metabolism, acts on different neuroanatomical structures than hypoxic injury, remains unresolved.

L. CNS TOLERANCE TO VARIOUS LEVELS OF HYPEROXIA

High concentrations of oxygen (40 to 80%) with N_2 are utilized in order to improve arterial PO_2 in patients with respiratory insufficiency (Bland, 1980; Phelps, 1982; Fulmer and Snider, 1984). This "low-pressure oxygen therapy," as opposed to HBO therapy, appears to be neurologically safe in adults, providing that there is no preexisting hypoxic-ischemia or brain damage (see Section I.Y). Moreover, no immediate or delayed neurological effects have been reported for short-term HBO exposures up to 3 ATA, normobaric and hypobaric hyperoxia, or subtoxic levels of oxygen that are actually used for oxygen therapy, deep diving, and aerospace activities. Hence, it has been widely believed that the neurological manifestations of oxygen toxicity may occur only after a relatively short exposure to HBO above 3 ATA (Lambertsen, 1978; Balentine, 1982). However, it has been shown that the potential for electrical seizures and consequent neuronal damage by hyperoxia may also exist after prolonged exposures (8 to 12 h) in conscious rats (Torbati, 1987).

The following section discusses the CNS sensitivity to various hyperoxic levels that are, or may become, applicable to medicine, or aerospace and diving activities.

M. NEURAL SENSITIVITY TO HYPOBARIC HYPEROXIA

Concentrations of oxygen, below the barometric pressure, are used during ascent to high altitudes (mountain climbing and flying in an unpressurized aircraft). These applications of oxygen allow the body to maintain normoxic levels of arterial PO_2 in the face of a reduced barometric pressure that occurs at high altitudes (Blumen et al., 1992). Although percent composition of air at high altitude does not change significantly (about 21%), both the PO_2 and barometric pressure decrease. Despite the fact that the upper limit or permanent living for acclimated man at high altitude is 5300 m above sea level, most mammals, if suddenly exposed to such altitudes, may need immediate supplementation of oxygen to survive. At altitudes about 11000 m, which is equivalent to 160 mmHg barometric pressure, a person must breathe pure oxygen in order to maintain an alveolar PO_2 equal to that during air breathing at sea level. At such altitudes increasing the percentage of O_2 in the breathing air is the only way to prevent hypoxia. This supplementation of oxygen to the breathing air at high altitudes does not pose any threat to either the CNS or other vital organs.

Some aerospace and deep-diving activities may require long periods of oxygen breathing with PO_2 ranging from 160 to 400 mmHg (Welch et al., 1963; Thompson et al., 1970; Webb et al., 1989; Gazenko et al., 1990). These hyperoxic exposures are tolerated for much longer periods than NH, because the danger of pulmonary oxygen toxicity is greatly reduced (Harrison, 1974; Hayatdavoudi et al., 1981). Nevertheless, there is some evidence that prolonged exposures to such subtoxic levels of oxygen may become mildly toxic to the lung tissues. For example, subjects undergoing a 28-d saturation dive with variable hyperoxia (mostly 400 mmHg) showed a significant reduction in the transfer factor for carbon monoxide in their lungs (Thorsen et al., 1993). Oxygen-induced pulmonary damage, in turn, may indirectly affect the CNS oxygenation and function (Welch et al., 1963; Crapo et al., 1986; Harris et al., 1991). Whether any long-term or persisting neurological defficiency can develop at 160 mmHg $< PO_2 <$ 450 mmHg is not known.

N. INTERACTION BETWEEN HYPEROXIA AND INERT GASES AT HIGH PRESSURE

Oxygen, like any other gas, in addition to its distinguished chemical effect imposes also a physical effect, which is related to the degree of hydrostatic pressure, on biological systems. This effect is negligible at the relatively low pressures that are used for either oxygen therapy (1 to 3 ATA O_2) or shallow diving ($<$1 ATA). Oxygen, in combination with other gases such as He and N_2, can produce a relative hyperoxia when used during deep or saturation diving. Nitrogen is highly narcotic at relatively low hydrostatic pressures (4 to 5 ATA), creating a gas-related disease known as nitrogen narcosis. Oxygen molecules also have some narcotic properties, but only at extremely high pressures (Paton, 1967; Bennett and Ackles, 1970). Moreover, the presence of inert gases may enhance superoxide-radical production thereby promoting oxygen toxicity (Thom, 1992). An interesting gas-related disease is known as high-pressure nervous syndrome (HPNS). It can occur at high hydrostatic pressure during deep diving, even with normoxic mixtures of about 160 mmHg inspired PO_2 or less. The neurological manifestations of HPNS, such as tremors and convulsions, can be either enhanced or weakened, depending on interactions among various gases in the breathing mixtures (Bennett and Dosset, 1973; Lambertsen et al., 1977; Halsey, 1982). In gas mixtures used for deep diving, the percentage of oxygen is usually reduced, depending on the maximum depth of diving. For example, a breathing mixture composed of 5% O_2 in 95% He at 100 m of sea water can generate an inspired PO_2 (P_IO_2) of up to 380 mmHg. At 200 m of seawater the same mixture can generate up to 760 mmHg, which is equivalent to NH (1 ATA O_2). In these conditions the interaction of high concentration of oxygen with He or other inert gases may act synergistically to accelerate the onset of the neurological manifestations of the HPNS (Thompson et al., 1970; Brauer and Beaver, 1982). However, the safety limits of P_IO_2 at extremely high hydrostatic pressures are not known, and it is not certain whether a relative hyperoxia during deep diving can actually accelerate the onset of CNS oxygen toxicity (Thompson et al., 1970) or attenuate it (Burns, 1972).

O. NEURONAL SENSITIVITY TO NORMOBARIC HYPEROXIA

It is generally accepted that NH primarily affects the respiratory system rather than the CNS (Bean, 1945; Clark and Lambertsen, 1971; Lambertsen, 1978; Balentine, 1982; Small, 1984). This convention is generally the case for adult mammals, but as we shall see later, it may not be the case for newborn mammals or prematurely born infants who exhibit selective neural vulnerability to NH or even subtoxic levels of oxygen (Gyllensten, 1959; Grave et al., 1972; Sokoloff, 1977; Torbati, 1993a). Physiologic changes in respiratory functions, such as reduction in breathing frequency, occur as early as 18 to 24 h after exposure to NH (Clark and Lambertsen, 1971; Simon and Torbati, 1982; Torbati and Reilly, 1988). Biochemical and morphological changes in the lungs, however, take place after much longer exposures (Schafer and Citoler, 1978; Matalon et al., 1982; Crapo, 1986). For example, 11 d exposure to 80% oxygen is not lethal for rats as a whole, but results in overt cellular damage to lung cells and blood erythrocytes (Webster et al., 1987).

The question is how an apparently asymptomatic exposure to NH is capable of affecting the CNS. The degree of the CNS tolerance to NH in adult mammals, in addition to the level of P_IO_2, also depends on preceding physiologic responses. In man, significant reduction in CBF occurs early after exposure to NH (Kety and Schmidt, 1948; Lambertsen et al., 1963), whereas in the rat 10 to 30 min exposure to 1 ATA O_2 does not produce significant changes in regional CBF (Torbati et al., 1978; Onarheim and Tyssebotn, 1980). Longer exposures (up to 40 h) to 1 ATA O_2 in conscious sheep also do not show changes in cortical, brain stem, and cerebellar blood flow (Matalon et al., 1982). These studies indicate that during asymptomatic exposures to NH, even in the presence of some vasoconstriction (human studies), the brain should be well hyperoxygenated. Longer exposures to NH (40 to 60 h) may lead to significant rise in regional CBF, perhaps because of the onset of pulmonary manifestations of oxygen poisoning and development of hypercapnia (Matalon et al., 1982). Despite the potential for elevated PO_2 in the intact brain by NH it should be noted that application of hyperoxia, particularly following brain ischemia or trauma, may actually enhance vasoconstriction and result in cerebral hypoxia rather than hyperoxia (Torbati and Carey, 1989a; Mickel et al., 1990).

Normobaric hyperoxia also affects brain energy metabolism, as determined by a slight increase in regional glucose utilization, demonstrated only after 6 h of exposure (Torbati et al., 1984). The increased rate of glucose utilization, however, disappears after 24 h of ongoing exposure to NH (Torbati and Reilly, 1988). Interestingly, alterations in brain glucose utilization correspond roughly to the transient changes in oxidative reactions *in vivo,* as observed in the rat brain synaptosomes exposed to NH from 1 to 24 h (Gordon-Majszak and Gajkowska, 1992). The latter investigators have demonstrated that the content of thiobarbituric acid reaction substances (indicating lipid peroxidation) and the level of SH-proteins were already significantly increased after 1 to 2 h exposure to NH. No further increase in the degree of lipid peroxidation occurred at 24 h of exposure, indicating, perhaps, adaptation to oxidative stress. Ahotupa et al. (1992), determined oxidative damage to rat's brain and other organs during NH by utilizing the appearance of conjugated diene double bonds and amount of fluorescent chromolipids. They found over twofold increases in these compounds at 36 and 48 h of exposure, respectively. In these experiments, the activity of superoxide dismutase in the brain initially decreased, followed by slight induction of activity at later time points. Whether changes in the activity of endogenous antioxidants, such as superoxide dismutase, is related to these early alterations in brain glucose utilization, such as activation of the pentose shunt, remains to be investigated.

P. SENSITIVITY OF THE IMMATURE RETINA TO NORMOBARIC HYPEROXIA

Survival of prematurely born infants is often dependent on respiratory support using high concentration of oxygen over a long period of time. Newborn mammals, compared to adults, can survive much longer exposures to NH (Stevens and Autor, 1980; Frank, 1991). Despite generalized resistance to the lethal effects of oxygen poisoning, the retina in newborn mammals, including prematurely born infants, could be severely affected even by ordinarily subtoxic levels of oxygen (Ashton and Cook, 1954; Ashton et al., 1954; Patz, 1965, 1984; Flynn and Tasman, 1992). This sensitivity of the immature retina to NH, known as retinopathy of prematurity (ROP; old terminology: retrolental fibroplasia), may impair vision or cause blindness in prematurely born infants subjected to hyperoxic therapy. Potential for ROP development creates a major neonatal emergency care problem, particularly for babies born with body weight below 1250 g (Flynn et al., 1987; Ben-Sira et al., 1988; Flynn and Tasman, 1992).

Q. ETIOLOGY OF RETINOPATHY OF PREMATURITY

It is generally accepted that ROP is a vascular disease that may be related to hyperoxic-induced vasoconstriction in the immature retinal vessels (Ashton et al., 1954; Ernest and Goldstick, 1984; Flynn et al., 1987). Animal studies suggest a biphasic response of the immature retinal vasculature to high concentrations of oxygen (Ashton and Cook, 1954; Patz, 1965, 1984). Accordingly, an initial vasoconstriction of the major retinal vessels is followed by vasoobliteration of the capillary bed. Consequently, when air breathing resumes (second stage), the retina becomes hypoxic-ischemic. This results in neovascular proliferation, perhaps following secretion of an unknown angiogenic substance (Patz, 1984). The outgrowth of the newly formed vessels into the vitreous then leads to pathological phenomena that constitute various clinical and experimental stages of ROP (Ben-Sira et al., 1988; Phelps, 1992; Chang-Ling et al., 1992).

R. BRAIN DEVELOPMENT AND NORMOBARIC HYPEROXIA

Clinical studies indicate that a larger proportion of children with ROP show mental retardation than would be expected in either sighted or those blinded by other conditions (Williams, 1958; Parmelee et al., 1959; Genn and Silverman, 1964). Animal studies also support the possibility that hyperoxic exposures may affect brain development (Gyllensten, 1959, 1959a, 1959b; Grave et al., 1972; Sokoloff, 1977; Hannah and Hannah, 1980). The question is whether brain damage and ROP share a common etiology. Gyllensten (1959, 1959a) observed a 20% reduction in vascular mass in the visual and motor cortices of neonatal mice exposed to 0.9 to 1 ATA O_2 for 5 to 30 d. The decrease in blood vessels was apparent at day 5 and became maximal at day 20. This hyperoxic-induced reduction in vascular mass occurred during a period in which there is normally a proliferation of capillaries, a condition similar to postnatal vascularization in the immature retina. Moreover, a vasoproliferation phenomenon in the visual and motor cortices, similar to that observed during ROP development, occurs in mice subjected to hyperoxia at days 1 to 5 and 11 to 15 after birth (Gyllensten, 1959b). In fact, obliteration of the developing brain vessels by hyperoxia has been also observed in frontal cortices of neonatal hamsters (Hannah and Hannah, 1980). Additionally, significant reduction on cellular growth has been found in the cerebral cortex of mice exposed to 0.9 to 1 ATA for 20 d postnatal period (Gyllensten, 1959a). Further evidence, supporting hyperoxic-induced retardation of brain development, are the observations showing that neonatal rats exposed to 0.7 to 0.8 ATA O_2 for the first 9 d after birth had reduced levels of total RNA, DNA, and protein content of the brain (Grave et al, 1972). The average brain weight in these rats was significantly less than controls (570 mg vs. 605 mg). None of these variables, however, was associated with either nutritional status or total body weight. These studies, therefore, support the hypothesis that, in addition to generation of ROP, hyperoxia in newborn mammals may result in brain damage or retardation (Grave et al., 1972; Sokoloff, 1977; Torbati et al., 1993a).

The hyperoxic-induced alterations in morphological and biochemical parameters of the brain during development are in contrast with the relative resistance of newborn mammals, as a whole, to oxygen poisoning. The generalized oxygen poisoning, however, is measured in terms of pulmonary dysfunction, seizures, and mortality rather than delayed and specific neurological defficiencies (Balentine, 1973; Stevens and Autor, 1980; Frank, 1991; Torbati et al., 1993a). Theoretically, the higher sensitivity of immature neural system to NH indicates a different physiologic response to oxygen than those that occur in the adult CNS. Besides the unique neuronal membrane biochemical characteristics in newborns, which were discussed in Section I.AA, the unique CNS response to hyperoxia in newborn mammals could also be attributed to an enhanced cerebral vasoconstriction, and perhaps extreme reduction in ventilation (Torbati et al., 1993a). In conclusion, the potential for neuropathological changes in newborns following hyperoxic exposures may necessitate restriction of oxygen administration in the prematurely born infants.

S. LONG-TERM EFFECTS OF HYPERBARIC OXYGENATION

The questions here are (a) are there any long term effects after a single exposure to HBO? and (b) do repeated exposures lead to accumulated CNS damage? The neuropathologic findings, in adult and newborn mammals exposed either once or repeatedly to extreme pressures of 4 to 5 ATA O_2, indicate that they can cause brain lesions, even after a single exposure (Balentine, 1966, 1973, 1974, 1982; Schnakenburg and Nolte, 1973). However, no systematic research has been conducted on the CNS or peripheral effects of low-dose, short, intermittent exposures (2 to 3 ATA O_2; 1 to 2 h), which are traditionally used for HBO therapy. One reason for this may be that it is a very rare problem to which attention has not yet been drawn. Nevertheless, since generation of ROS precedes the onset of the acute neurological signs of CNS oxygen toxicity (Torbati et al., 1992), possible accumulated effect of ROS on carcinogenesis or neural degeneration should be investigated. Enriched oxygen mixtures are repeatedly used under high pressure during shallow and deep saturation diving. The rate of cancer production in divers, as opposed to the general population, has not been investigated. However, chromosome aberrations have been shown in 3.9% lymphocyte cultures of 153 commercial divers, an unusually higher number when compared to 127 nondivers occupied on oil rigs or industrial sites (Fox et al., 1984).

The author also draws attention to the long-term immunological consequences of repeated O_2 exposures, or prolonged and repeated hypoxic episodes, particularly in hyperlipidemic subjects. While the role of tissue hypoxia on atherosclerosis has been recognized (Barker et al., 1993), no information on possible atherogenic effect of prolonged or intermittent NH or HBO exposures are available. Because it has been demonstrated that low-density lipoprotein (LDL) particles must undergo oxidative modification before they can give rise to foam cells, which are the key component of the fatty streak lesion of atherosclerosis (Palinski et al., 1989), the author hypothesizes that repeated HBO exposures, or hypoxemia, can promote atherogenicity. This hypothesis is indirectly supported by the theories indicating enhanced hyperoxia- and hypoxia-induced generation of ROS, as described in previous sections. Whether free radicals generated during hyperoxia or hypoxia can accelerate oxidation of LDL and convert it to an immunogenic compound warrants investigation. Such oxidative reactions may, in turn, increase the uptake of the oxidized LDL by macrophage scavenger receptors leading to formation of foam cells (Ross, 1993).

T. MODIFICATION OF TOLERANCE TO CNS OXYGEN TOXICITY

The onset of the neurological signs of oxygen poisoning could be either delayed or accelerated by a variety of environmental factors including alterations in external temperature, inspired PCO_2 (P_ICO_2), diet, illumination, noise, and other stressful conditions (Campbell, 1938; D'Aoust, 1971; Berry et al., 1977; Nagai et al., 1977; Clark, 1981; Bitterman et al., 1986; Jamieson and Carmody, 1989). Inherent factors, such as sex or endocrinological status and, particularly, the age of an individual, can greatly modify the sensitivity to CNS oxygen toxicity (Gerschman et al., 1954; Cross and Houlihan, 1969; Sperling, 1970; Torbati et al., 1971, 1972, 1993a; Balentine, 1973). Also, cold-blooded animals are relatively resistant to the CNS effects of HBO, probably because they have a lower cerebral metabolic rate for oxygen than warm-blooded animals (Baeyens et al., 1973). However, the sensitivity to CNS oxygen poisoning in mammals is apparently not related to body size, despite the higher cerebral metabolic rates (Arieli, 1988).

A practical way to extend tolerance to CNS oxygen toxicity is simply achieved by intermittent exposures. This modality is based on the assumption that the passage from the physiologic or asymptomatic stage into a pathologic or irreversible stage of HBO occurs gradually and is both time and pressure dependent (Bean, 1945; Lambertsen, 1978; Simon and Torbati, 1982). The major endogenous factor that reduces sensitivity to CNS oxygen toxicity, besides antioxidants, is cerebral vasoconstrictive response to hyperoxia (Kety and Schmidt, 1948; Lambertsen et al., 1955, 1963; Bean et al., 1972; Torbati et al., 1978). As long as this physiologic defense mechanism persists, the collapse of the asymptomatic stage will be delayed and neuronal damage may not occur. Whether, at the same time, the vascular layers, particularly the endothelium tissue, are protected against cellular oxygen poisoning is not known. However, with continuous exposure to HBO, a series of pathological events will eventually erupt in other tissues. The latter will occur earlier with the reversal of the cerebral vasoconstriction and subsequent development of the CNS oxygen toxicity. To break the gradually increasing toxic effects of oxygen in various organs, investigators have tried to alternate HBO exposures with air exposure intervals. Such intermittent exposures have significantly extended the total time of exposure to various pressures of oxygen without creating obvious signs of CNS or pulmonary oxygen poisoning (Widell et al., 1974; Hendricks et al., 1977; Clark and Lambertsen, 1984; Torbati et al., 1984a). However, it should be emphasized that the extension of tolerance to CNS oxygen toxicity may not entirely prevent oxygen poisoning of the lungs or other peripheral organs from occurring.

U. BRAIN DEVELOPMENT AND HBO

As in the case of NH, newborn mammals exposed to acute HBO are able to survive much longer than adults (Sperling, 1970; Balentine, 1973; Torbati et al., 1993a). Sperling (1970) was the first investigator to demonstrate that newborn mice exposed to 4.4 ATA O_2 were markedly less sensitive to hyperoxia than adults, as determined by survival time, onset of convulsions, and pulmonary pathology. Balentine (1973, 1982) studied the delayed neuropathologic effects of HBO in newborn rats, comparing the ultrastructural changes with those observed in adult rats (Balentine 1966, 1974; Balentine and Gutsche, 1966). In Balentine's studies, newborn rats were intermittently exposed to 5 ATA O_2 for 1 h daily from the day of birth until death. The young adult rats were three to four months old and were treated daily with the same dose and duration of oxygen. The mean onset of newborn mortality was 23.5 ± 2.4 (mean $\pm$ SD) exposures, compared to only 4.1 ± 2.3 exposures in adult rats (Balentine, 1973). Despite much longer exposure time in newborns, unlike adults, they did not develop spinal cord necrosis, but had forebrain and brainstem lesions similar to adults (Balentine, 1966, 1974; Balentine and Gutsche, 1966).

Recently, we have demonstrated that 4- to 5-d-old rats were able to survive up to 8 h at the extreme pressure of 5 ATA O_2, even with hypercapnia at 90 to 190 mmHg P_ICO_2 (Torbati et al., 1993a, 1995). In adult mammals, on the other hand, the onset of HBO-induced convulsions and survival time are significantly accelerated by a slight hypercapnia. This effect is primarily due to the early breakdown of the HBO-induced cerebral vasoconstriction (Lambertsen et al., 1955; Bean et al., 1972; Clark, 1981). In our study with newborn rats, the survival time became significantly reduced by both mild hypercapnia (P_ICO_2 = 60 and 90 mmHg) and extreme hypercapnia (P_ICO_2 = 380 mmHg) at 5 ATA O_2, while moderate hypercapnia at P_ICO_2 of 140 and 190 mmHg did not alter the CNS sensitivity to HBO. However, newborn rats that survived the 8-h exposure to 5 ATA O_2 developed extreme reductions in respiratory frequency (down to 97 to 98%) during HBO exposure. The same rats, however, died within a few hours after returning to air because of severe pulmonary damage and consequent respiratory insufficiency (Torbati et al., 1993a, 1995). We hypothesized that pre- and post-HBO events in newborn mammals, fetuses, or prematurely born infants subjected to hyperoxic therapy (NH and HBO) may set the stage for the development of some delayed neurological disorders, including ROP and brain retardation. Accordingly, the damaged suffered by the brain or retina, either during or after hyperoxic exposures, may paradoxically be the consequence of oxygen "tolerance" in newborn mammals (Figure 8).

V. RETINOPATHY OF PREMATURITY AND HBO

Theoretically, it is important to know whether ROP is uniquely produced following prolonged NH therapy or can it also be initiated by short-term exposures to HBO. Ricci et al. (1987, 1990; Ricci and Calegero, 1988; Ricci, 1990) attempted to test this hypothesis but failed to produce ROP in newborn rats exposed to 1.8 ATA O_2. They suggested that constriction of choriocapillaries and retinal vessels by HBO may have protected the retina, thus preventing the development of ROP. Our hypothesis, as illustrated in Figure 8, predicted that ROP perhaps could be developed at oxygen pressures higher than 1.8 ATA, providing that ROS generation and retinal/vascular damage occur either directly by HBO, or via hypoxic-ischemia and recirculation upon return to air (Torbati et al., 1993a). This hypothesis is now supported by our results in newborn rats, demonstrating increased retinal vascular proliferation six to eight weeks after exposure to 5 ATA O_2 for 5 h (Torbati et al., 1995a).

W. HBO THERAPY

The philosophical and moral point view of Priestly, the discoverer of oxygen, was that "the air which nature has provided for us is as good as we deserve" (Gilbert, 1981). The 19th and 20th century biologist ignored "nature morality" and tried to apply artificial mixtures

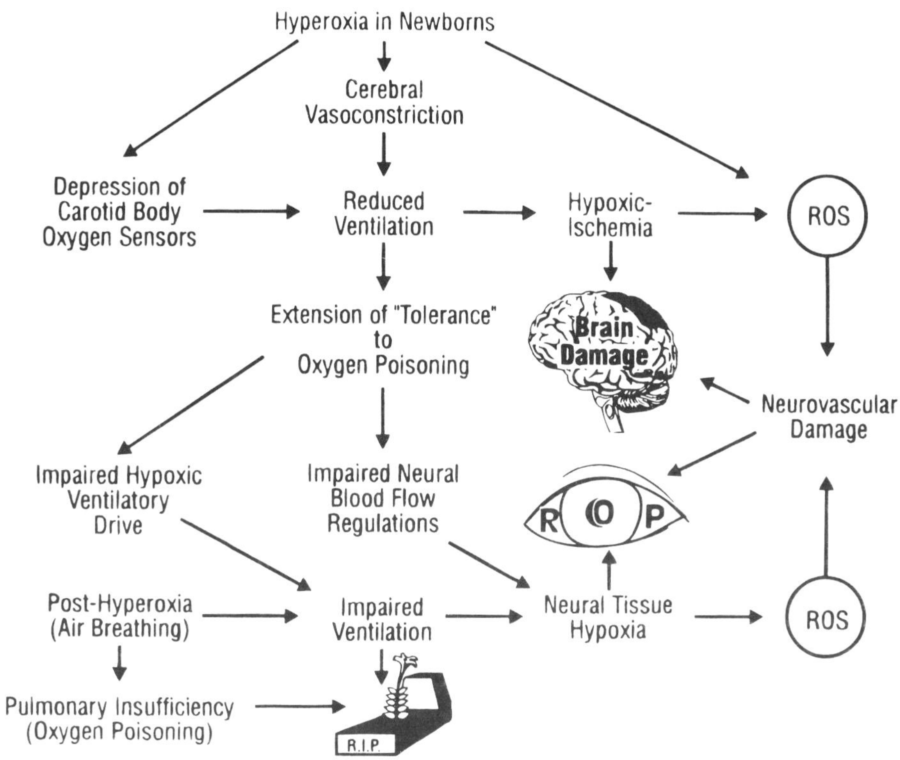

FIGURE 8. CNS and retinal responses to normobaric and/or hyperbaric hyperoxia that may explain the apparent tolerance to CNS oxygen toxicity in newborn mammals, compared to adults. The same physiologic responses, however, may also initiate the development of chronic neural diseases, such as retinopathy of prematurity and/or brain retardation in newborn mammals. (Reprinted from Torbati, D. et al., *Free Radic. Biol. Med.*, 13, 101, 1993. With permission from Elsevier Sci. Ltd, Pergamon imprint, Kidlington, UK).

of gases containing far more oxygen than any creature should deserve. Thus, the clinical applications of NH and HBO have become an important therapeutic modality in modern medicine. Because HBO therapy has never been proven to be completely effective for a variety of diseases for which its use has been advocated, its critics have called it "a treatment in search of a disease" (Heimbach, 1993). For safety reasons, rather than a scientific rationale, HBO therapy is usually practiced as intermittent, short-term, 100% oxygen breathing at pressures greater than 1 ATA, but rarely exceeding 3 ATA (Davis and Hunt, 1977; Jain, 1994). Although the release of any new drug requires extensive testing in animals and randomized controlled clinical trials to prove both efficacy and safety for each of the indications claimed, such rigorous trials, so far, have not been conducted with HBO therapy for any of its claimed indications (Heimbach, 1993). Obviously, oxygen at high concentrations or high pressures is acting as a drug, and, therefore, the same pharmacodynamic and pharmacokinetic rules that are ordinarily followed for the investigation of any xenobiotic drug must also be applied to high-dose oxygen therapy.

The general factors that can influence oxygen therapy include, (a) physical effects, which involve no chemical influence upon cells; (b) physiological effects, which modify function but are fully reversible on return to normal air breathing; and (c) the toxic effect, which may progress to a lethal degree (Lambertsen, 1978). These factors are subject to alteration following trauma, ischemia, and local or general respiratory or circulatory failure because of disruption of the circulatory and metabolic regulatory mechanisms (Ishige et al., 1988; Torbati et al., 1989; Hallenbeck and Dutka, 1990). In addition to being costly, HBO therapy may cause a few acute and chronic side effects. Practically, the incidence of seizures in patients undergoing repeated HBO therapy (1:4690 compressions to 2 to 2.8 ATA O_2 for 30 to 90 min sessions) is very low (Hart et al., 1979). At these relatively low pressures of oxygen, pulmonary oxygen poisoning may still constitute a limiting factor. However, the rate of development of pulmonary oxygen toxicity during continuous exposure to 2 ATA O_2, as examined by reduction in vital capacity, is very slow (Clark and Lambertsen, 1971a). Furthermore, as mentioned earlier, the pulmonary tolerance to HBO could be extended in man by intermittent exposures (Widell et al., 1977; Hendricks et al., 1977).

The rationale for using HBO as treatment for certain neurological disorders is to improve brain tissue oxygenation until adequate respiration and circulation are restored. If this goal is overruled by potentially harmful effects of HBO, its use must be limited and alternative therapeutical solutions should be sought. However, in some conditions, such as CNS decompression sickness, CO toxicity, and infections that are unaccessible to antibiotics (Hart, 1974; Thom and Thom, 1989; Thom, 1990, 1993), supplementation of oxygen is lifesaving. In such cases, one may accept a higher degree of risk (in using HBO) than cases in which traditional and low-cost therapies are available.

III. EPILOGUE

It appears that most metazoa on our planet have evolved in an oxidizing atmosphere, enriched with a free radical agent known as oxygen. Although this gas is essential to life, in too great or too little amounts it becomes life threatening. Balancing these two extremes will ensure our survival, at least for a while. We may not always be in control of a hypoxic environment that has been suddenly imposed on our vital organs. What we can do is use oxygen intelligently, with the same caution exercised with other drugs that are prescribed for individual cases. Thus, this artificially created hyperoxic environment does not damage that which it is supposed to improve.

ACKNOWLEDGMENT

The author wishes to thank Dr. Anabel Aron, M.D., Tel-Hashomer Sheba Medical Center, Ramat-Gan, Israel, for her useful comments on the manuscript.

REFERENCES

Acker, H. (1989) PO_2 chemoreception in arterial chemoreceptors. *Annu. Rev. Physiol.* 51, 835–844.

Ahotupa, M., Mantyla, E., Peltola, V., Puntala, A., and Toivonen, H. (1992) Pro-oxidant effects of normobaric hyperoxia in rat tissues. *Acta Physiol. Scand.* 145, 151–157.

Aitken, P. G., Jing, J., Young, J., and Somjen, G. G. (1991) Ion channel involvement in hypoxia-induced spreading depression in hippocampal slices. *Brain Res.* 541, 7–11.

Anand, A. and Paintal, A. S. (1991) Oxygen sensing by arterial chemoreceptors, in *Response and adaptation to hypoxia organ to organelle,* eds., Lahiri, S., Cherniack, N. S., and Fitzgerald, R. S. (Am. Physiol. Soc., Oxford University Press, New York), pp. 81–94.

Anichkov, S. V. and Belenkii, M. L., eds. (1963) *Pharmacology of the carotid body chemoreceptors.* (Pergamon, Oxford), 225 p.

Anis, G. (1977) Parkinson's disease, depression and electroconvulsive therapy, a review and case study. *Am. J. Psychiatry.* 134, 191–195.

Anwyl, R. (1991) Modulation of vertebrate neuronal calcium channels by transmitters. *Brain Res. Rev.* 16, 265–281.

Arai, A., Kessler, M., Lee, K., and Lynch, G. (1990) Calpain inhibitors improve the recovery of synaptic transmission from hypoxia in hippocampal slices. *Brain Res.* 532, 63–68.

Arai, A., Vanderklisk, P., Kessler, M., Lee, K., and Lynch G. (1991) A brief period of hypoxia causes proteolysis of cytoskeletal proteins in hippocampal slices. *Brain Res.* 555, 276–280.

Arieli, R., Kerem, D., and Melamed Y. (1988) Hyperbaric exposure affects the ventilatory response to hypoxia in awake rats. *J. Appl. Physiol.* 64, 181–186.

Arieli, R. (1988) Oxygen toxicity is not related to mammalian body size. *Comp. Biochem. Physiol.* 91, 221–223.

Armstead, W. M., Mirro, R., Liffler, C. W., and Busija, D. W. (1989) Cerebral superoxide anion generation during seizure in newborn pigs. *J. Cereb. Blood Flow Metab.* 9, 175–179.

Ashton, N., Ward, B., and Serpell, G. (1954) Effect of oxygen on developing retinal vessels with particular reference to the problem of retrolental fibroplasia. *Br. J. Ophthalmol.* 38, 397–432.

Ashton, N. and Cook, C. (1954) Direct observation of the effect of oxygen on developing vessels. *Br. J. Ophthalmol.* 38, 433–440.

Backlund, E.-O. (1987) Transplantation to the brain—a new therapeutic principle or useless venture? *Acta Neurochir. Suppl.* 41, 46–50.

Balentine, J. D. (1966) Pathologic effects of exposure to high oxygen tension. A review. *New Engl. J. Med.* 275, 1038–1040.

Balentine, J. D. and Gutsche, B. B. (1966) Central nervous system lesions in rats exposed to oxygen at high pressure. *Am. J. Pathol.* 48, 107–127.

Balentine, J. D. (1973) Pathologic observations in newborn and adult rats fatally exposed to hyperbaric oxygen, in *Proc. 5th Int. Hyperbaric Cong.*, eds., Trapp, W. G., Banister, E. W., Davidson, A. J., and Trapp, P. A. (Simon Fraser Univ., British Columbia, Canada), pp. 102–108.

Balentine, J. D. (1974) Ultrastructureal pathology of hyperbaric oxygenation in the central nervous system. Observations in anterior horn gray matter. *Lab. Invest.* 31, 580–592.

Balentine, J. D. (1982) The nervous system, in *Pathology of oxygen toxicity,* ed., Balentine, J. D. (Academic Press, New York), pp. 1172–1214.

Balestrino, M. and Somjen, G. G. (1986) Chlorpromazine protects brain tissue in hypoxia by delaying spreading depression-mediated calcium influx. *Brain Res.* 385, 219–226.

Balestrino, M., Aitken, P. G., Jones, L. S., and Somjen, G. G. (1988) The role of spreading depression-like hypoxic depolarization in irreversible neuron damage and its prevention. *Adv. Behav. Biol.* 35, 291–302.

Baeyens, D. A., Hoffert, J. R., and Fromm, P. O. (1973) A comparative study of oxygen toxicity in the retina, brain and liver of the teleost, amphibian and mammal. *Comp. Biochem. Physiol.* 45A, 925–932.

Barker, S. G. E., Baskerville, P. A., and Martin, J. F. (1993) The arterial adventitia: a role for vasa vasorum and hypoxia in atherosclerosis? *Atheroscler. Rev.* 25, 201–208.

Bauknight, G. G., Jr., Wei, E., and Kontos, H. A. (1992) Superoxide production in experimental seizures in cats. *Stroke,* 23, 1512–1514.

Bazan, N. G. and Rodriguez de Turco, E. B. (1980) Membrane lipids in the pathogenesis of brain edema: phospholipids and arachidonic acid, the earliest membrane components changed at the onset of ischemia. *Adv. Neurol.* 28, 197–205.

Beal, M. F. (1992) Mechanisms of excitotoxicity in neurologic diseases. *FASEB J.* 6, 3338–3344.

Bean, J. W. (1945) Effects of oxygen at high pressure. *Physiol. Rev.* 25, 1–145.

Bean, J. W., Lingnell, J., and Burgess, D. W. (1972) Cerebral O_2, CO_2 regional cerebral vascular control and hyperbaric oxygenation. *J. Appl. Physiol.* 32, 650–657.

Becker, N. H. and Glavior, J. F. (1962) Effect of oxygen-rich atmosphere on cerebral lipid peroxides. *Aerosp. Med.* 33, 985–987.

Beckman, J. S., Beckman, T. W., Chen, J., Marshal, P. A., and Freeman, B. A. (1990) Apparent hydroxyl radical production by peroxynitrite: implications for endothelial injury from nitric oxide and superoxide. *Proc. Natl. Acad. Sci. U.S.A.* 87, 1620–1624.

Beckman, J. S. (1991) The double-edged role of nitric oxide in brain function and superoxide-mediated injury. *J. Dev. Physiol.* 15, 53–59.

Bennett, P. B. and Ackles, K. N. (1970) The narcotic effects of hyperbaric oxygen, in *Proc. 4th Int. Cong. Hyperbaric Med.*, eds., Wada, J. and Iwa, T. (Igaku Shoin, Ltd., Tokyo), pp. 74–79.

Bennett, P. B. and Dossett, A. N. (1973) EEG activity of rats compressed by inert gases to 700 feet and oxygen-helium to 4000 feet. *Aerosp. Med.* 44, 239–244.

Ben-Sira, I., Nissenkorn, I., and Kremer, I. (1988) Retinopathy of prematurity. *Surv. Ophthalmol.* 33, 1–6.

Berry, S., Fitch, J. W., and Schatte, C. L. (1977) Influence of sex and age on the susceptibility of mice to oxygen poisoning. *Aviat. Space Environ. Med.* 48, 37–39.

Bert, P. (1878) *Barometric pressure: researches in experimental physiology, Paris.* English translation: Hitchcock, M. A. and Hitchcock, F. A. (College Book Co., Columbus, OH), 1943; Published by Undersea Biomed. Soc., Bethesda, 1978.

Bicher, H. I., Bruley, D. F., Kinsely, M. H., and Reneau, D. D. (1971) The effect of microcirculation changes on brain tissue oxygenation. *J. Physiol.* 217, 689–707.

Biscoe, T. J. and Duchen, M. R. (1990) Cellular basis of transduction in carotid chemoreceptors. *Am. J. Physiol.* 258, L271–L278.

Biscoe, T. J. and Duchen, M. R. (1990a) Monitoring PO_2 by the carotid chemoreceptors. *NIPS* 5, 229–233.

Bishop, J. M. (1987) The molecular genetics of cancer. *Science* 235, 305–311.

Bitterman, N., Melamed, Y., and Perlman, I. (1986) CNS oxygen toxicity in the rat: role of ambient illumination. *Undersea Biomed. Res.* 13, 19–25.

Bland, R. D. (1980) Special consideration on oxygen therapy for infants and children. *Am. Rev. Respir. Dis.* 122, 45–54.

Blass, J. P. and Gibson, G. E. (1979) Consequences of mild, graded hypoxia. *Adv. Neurol.* 26, 229–250.

Blumen, I. J., Abernetly, M. K., and Dunne, J. M. (1992) Flight physiology: clinical consideration. *Crit. Care Clin.* 8, 597–618.

Bomont, L., Bilger, A., Boyet, S., Vert, P., and Nehlig, A. (1992) Acute hypoxia induces specific changes in local cerebral glucose utilization at different postnatal ages in the rat. *Develop. Brain Res.* 66, 33–45.

Boobis, A. R., Fawthrop, D. J., and Davis, D. S. (1989) Mechanisms of cell death. *TIPS* 10, 275–280.

Bouma, G. J. and Muizelaar, J. P. (1992) Cerebral blood flow, cerebral blood volume, and cerebrovascular reactivity after severe head injury. *J. Neurotrauma* 9, Suppl. 1, S333–S348.

Brauer, R. W. and Beaver, R. W. (1982) Synergism of hyperoxia and high helium pressures in the causation of convulsions. *J. Appl. Physiol. Respir. Environ. Exer. Physiol.* 53, 192–202.

Bredt, D. S., Hwong, P. M., and Snyder, S. H. (1990) Localization of nitric oxide synthase indicating a neural role for nitric oxide. *Nature*, 347, 768–770.

Bredt, D. S. and Snyder, S. H. (1992) Nitric oxide, a novel neuronal messenger. *Neuron* 8, 3–11.

Bruley, D. F., Reneau, D. D., Bicher, H. I., and Knisely, M. H. (1973) Modeling cerebral tissue oxygenation with autoregulation, in *Oxygen supply: theoretical and practical aspects of oxygen supply and microcirculation of tissue*, eds., Kessler, M., Bruley, F., Clark, L. C., et al. (Univ. Park Press, Baltimore), pp. 32–3.

Burns, J. D. (1972) Concentration-dependent attenuation of hyperbaric oxygen toxicity. *Aerosp. Med.* 43, 989–992.

Burns, K. D., Homma, T., Breyer, M. D., and Harris, R. C. (1991) Cytosolic acidification simulates a calcium influx that activates Na^+-H^+ exchange in LLC-RK. *Am. J. Physiol.* 261, F617–F625.

Butler, F. K. Jr. and Thalmann, E. D. (1986) Central nervous system oxygen toxicity in closed-circuit scuba divers. *Undersea Biomed. Res.* 13, 193–223.

Campbell, J. A. (1938) Effects of oxygen pressure as influenced by external temperatures, hormones and drugs. *J. Physiol. (London)* 92, 29–31.

Carafoli, E. (1987) Intracellular calcium homeostasis. *Annu. Rev. Biochem.* 56, 395–433.

Carcillo, J. (1988) Signal transduction: how the cells process information. *Crit. Care Clin.* 4, 679–693.

Cazevieille, C., Muller, A., Meynier, F., and Bonne, C. (1993) Superoxide and nitric oxide cooperation in hypoxia/reoxygenation-induced neuron injury. *Free Radic. Biol. Med.* 14, 389–395.

Chan, P. H., Fishman, R. A., Longar, S., Chen, S., and Yu, A. (1985) Cellular and molecular effects of polyunsaturated fatty acids in brain ischemia and injury. *Prog. Brain Res.* 63, 227–235.

Chan, P. H. and Fishman, R. A. (1985) Free fatty acids, oxygen free radicals, and membrane alterations in brain ischemia and injury, in *Cerebrovascular diseases*, eds., Plum, F. and Pulsinelli, W. (Raven Press, New York), pp. 161–168.

Chan, P. H., Fishman, R. A., Wesley, M. A., and Longar, S. (1990) Pathogenesis of vasogenic edema in focal cerebral ischemia. Role of superoxide radicals. *Adv. Neurol.* 52, 177–183.

Chance, B., Cohen, P., Jobsis, F., and Schoener, B. (1962) Intracellular oxidation-reduction states *in vivo. Science* 137, 499–508.

Chance, B., Sies, H., and Boveris, H. (1979) Hydroperoxide metabolism in mammalian organs. *Physiol. Rev.* 59, 527–605.

Chang-Ling, T., Tout, J., Hollander, H., and Stone, J. (1992) Vascular change and their mechanisms in the feline model of retinopathy of prematurity. *Invest. Ophthalmol. Vis. Sci.* 33, 2128–2147.

Cherici, G., Alesiani, M., Pellegrini-Giampietro, D. E., and Moroni, F. (1991) Ischemia does not induce the release of excitotoxic amino acids from the hippocampus of newborn rats. *Dev. Brain Res.* 60, 235–240.

Chesler, M. (1990) The regulation and modulation of pH in the nervous system. *Prog. Neurobiol.* 34, 401–427.

Choi, D. W. (1987) Ionic dependence of glutamate neurotoxicity. *J. Neurosci.* 7, 369–379.

Choi, D. W., Maulucci-Gedde, M., and Kriegstein, A. R. (1987) Glutamate neurotoxicity in cortical cell culture. *J. Neurosci.* 7, 375–368.

Choi, D. W. (1988) Glutamate neurotoxicity and diseases of the nervous system. *Neuron* 1, 623–634.

Choi, D. W. (1988a) Calcium-mediated neurotoxicity: relationship to specific channel types and role in ischemic damage. *Trends Neurosci.* 11, 465–469.

Choi, D. W. (1990) Cerebral hypoxia: some new approaches and unanswered questions. *J. Neurosci.* 10, 2493–2501.

Choi, D. W. (1990a) The role of glutamate neurotoxicity in hypoxic-ischemic neuronal death. *Annu. Rev. Neurosci.* 13, 171–182.

Clark, J. M. and Lambertsen, C. J. (1971) Rate of development of pulmonary O_2 toxicity in man during O_2 breathing at 2 ATA. *J. Appl. Physiol.* 30, 739–752.

Clark, J. M. and Lambertsen, C. J. (1971a) Pulmonary oxygen toxicity: a review. *Pharmacol. Rev.* 23, 37–133.

Clark, J. M. and Lambertsen, C. J. (1984) Effects of inspired oxygen pressure on the nature and degree of oxygen tolerance modifications, in *Proc. 8th Symp. Underwater Physiol.*, eds., Bachrach, A. J. and Matzen, M. M. (Undersea Med. Soc., Bethesda), pp. 31–41.

Clark, J. M. (1981) Effects of acute and chronic hypercapnia on oxygen tolerance in rats. *J. Appl. Physiol. Respir. Environ. Excercise Physiol.* 50, 1036–1044.

Cohen, P. G., Alexander, S. C., Smith, T. C., Reivich, M., and Wollman, H. (1967) Effects of hypoxia and normocarbia on cerebral blood flow and metabolism in conscious man. *J. Appl. Physiol.* 23, 183–189.

Cohen, G. (1985) Oxidative stress in the nervous system, in *Methods in enzymology Vol. 186, Oxygen radicals in biological systems, Part B. Oxygen radicals and antioxidants*, ed., Packer, L. and Glazer, A. N. (Academic Press, London), pp. 383–402.

Cold, G. E. (1990) Cerebral blood flow in acute head injury. The regulation of CBF and metabolism during acute phase of head injury, and its significance for therapy. *Acta Neurochir. Suppl.* 49, 1–64.

Collingridge, G. L. and Lester, R. A. J. (1989) Excitatory amino acid receptors in the vertebrate central nervous system. *Pharmacol. Rev.* 40, 143–210.

Colton, C. A. and Colton, J. S. (1986) The action of oxygen and oxygen at high pressure on inhibitory transmission. *Brain Res.* 364, 151–158.

Crain, B. J. and Nadler, V. J. (1988) Selective neuronal cell death after transient forebrain ischemia in the mongolian gerbil. *Adv. Behav. Biol.* 35, 35–47.

Crakard, H. A., Brown, F. D., Calica, A. B., Johns, L. M., and Mullan, S. (1977) Physiological consequences of experimental missile injury and use of data analysis to predict survival. *J. Neurosurg.* 46, 784–794.

Crapo, J. D. (1986) Morphological changes in pulmonary oxygen toxicity. *Annu. Rev. Physiol.* 48, 721–731.

Crawford, M. A. and Sinclair, A. J. (1972) Nutritional influences in the evolution of mammalian brain, in *Lipids, malnutrition, and the developing brain*, Ciba Found. Symp. (Associated Sci. Publisher), pp. 267–292.

Cross, M. H. and Houlihan, R. T. (1969) Sympathoadrenomedullary response of the rat to high oxygen pressure. *J. Appl. Physiol.* 27, 523–527.

Culotta, E. and Koshland, D. E., Jr. (1992) Molecule of the year. *Science*, 258, 1862–1865.

Dagani, F., Marzatico, F., Curti, D., Zanada, F., and Benzi, G. (1984) Effect of prolonged and intermittent hypoxia on some cerebral enzymatic activities related to energy transduction. *J. Cereb. Blood Flow Metab.* 4, 615–624.

Dahlgren, N. (1990) Local cerebral blood flow in spontaneously breathing rats subjected to graded isobaric hypoxia. *Acta Anesthesiol. Scand.* 34, 463–467.

Dam, M. (1992) Quantitative neuropathology in epilepsy. *Acta Neurol. Scand. Suppl.* 137, 51–54.

D' Aoust, B. G. (1971) Natural resistance to oxygen poisoning, in *Underwater physiology IV*, ed., Lambertsen, C. J. (Academic Press New York), pp. 24–34.

Davies, K. J. A. (1987) Protein damage and degradation by oxygen radicals. *J. Biol. Chem.* 262, 9895–9901.

Davis, J. C. and Hunt, T. K. (1977) Preface, in *Hyperbaric oxygen therapy,* eds., Davis, J. C. and Hunt, T. K. (Undersea Biomed. Soc., Bethesda). pp. XI.

Davis, J. N., Giron, L. T., Jr., Stanton, E., and Maury, W. (1979) The effect of hypoxia on brain neurotransmitters synthesis. *Adv. Neurol.* 26, 219–228.

Dawson, V. L., Dawson, T. M., London, E. D., Bredt, D. S., and Snyder, S. H. (1991) Nitric oxide mediates glutamate neurotoxicity in primary cortical cultures. *Proc. Natl. Acad. Sci. U.S.A.,* 88, 6368–6371.

Dawson, T. M., Dawson, V. L., and Snyder, S. H. (1992) A novel neuronal messenger molecule in brain: the free radical, nitric oxide. *Ann. Neurol.* 32, 297–311.

Dean, J. B., Lawing, W. L., and Millhorn, D. E. (1989) CO_2 decreases membrane conductance and depolarizes neurons in the nucleus tractus solitarii. *Exp. Brain Res.* 76, 656–661.

Demopoulos, H. B., Flamm, E. S., Pietronigro, D. D., and Seligman, M. L. (1980) The free-radical pathology and the microcirculation in the major central nervous system disorders. *Acta Physiol. Scand.* 492, 91–119.

Denny-Brown, D. and Russell, W. R. (1941) Experimental cerebral concussion. *Brain* 64, 93–164.

DeWitt, D. S., Prough, D. S., Taylor, C. L., and Whitley, J. M. (1992) Reduced cerebral blood flow, oxygen delivery, and electroencephalographic activity after traumatic brain injury and mild hemorrhage in cats. *J. Neurosurg.* 76, 812–821.

Dikens, F. (1946) The toxic effect of oxygen on brain metabolism and on tissue enzymes, I. *Brain Metab. Biochem. J.* 40, 145–171.

Donnelly, D. F., Jiang, C., and Haddad, G. G. (1992) Comparative responses of brainstem and hippocampal neurons to O_2 deprivation: *in vitro* intracellular studies. *Am. J. Physiol.* 262, L549–L554.

Dora, E., Zeuthen, T., Silver, I. A., Chance, B., and Kovach, A. G. B. (1979) Effect of arterial hypoxia on the cerebrocortical redox state, vascular volume, oxygen tension, electrical activity and potassium ion concentration. *Acta Physiol. Acad. Sci. Hung.* 54, 319–331.

Dricks, R. C. and Faiman, M. D. (1982) Free-radical formation and lipid peroxidation in rat and mouse cerebral cortex slices exposed to high oxygen pressure. *Brain Res.* 248, 355–360.

Dubinsky, J. M. and Rothman, S. M. (1991) Intracellular calcium concentrations during "chemical hypoxia" and excitotoxic neuronal injury. *J. Neurosci.* 11, 2545–2551.

Duffy, T. E., Nelson, S. R., and Lowry, O. H. (1972) Cerebral carbohydrate metabolism during acute hypoxia and recovery. *J. Neurochem.* 19, 959–977.

Duffy, T. E., Kohle, S. J., and Vannucci, R. C. (1975) Carbohydrate and energy metabolism in perinatal rat brain: relation to survival in anoxia. *J. Neurochem.* 24, 271–276.

Duling, B. R. and Pittman, R. N. (1975) Oxygen tension: dependent or independent variable in local control of blood flow? *Fed. Proc.* 34, 2012–2019.

Duling, B. R. and Klitzman, B. (1980) Local control of microvascular function: role in tissue oxygen supply. *Annu. Rev. Physiol.* 42, 373–382.

Eldridge, P. R., Hope, D. T., Yeoman, P. M., Farquhar, I. K., Mitchell, M., et al. (1991) Somatosensory evoked potentials in intracranial hypertension: analysis of the effects of hypoxia. *J. Neurosurg.* 75, 108–114.

Erecinska, M. and Silver, I. A. (1989) ATP and brain function. *J. Cereb. Blood Flow Metab.* 9, 2–19.

Ernest, J. T. and Goldstick, T. K. (1984) Retinal oxygen tension and oxygen reactivity in kittens. *Invest. Ophthalmol. Vis. Sci.* 25, 1129–1134.

Eyzaguirre, C. and Zapata, P. (1984) Prespectives in carotid body research. *J. Appl. Physiol. Respir. Environ. Exercise Physiol.* 57, 931–957.

Eyzaguirre, C., Monti-Bloch, L., Margarita, B., Hayashida, Y., and Woodbury, J. W. (1989) Changes in glomus cells membrane properties in response to stimulants and depressants of carotid nerve discharge. *Brain Res.* 477, 265–279.

Faiman, M. D., Nolan, R. J., Dodd, D. E., Waechter, J. M., Dricks, R. C., et al. (1981) Mechanism(s) of central oxygen toxicity: a re-evaluation, in *Underwater Physiology VII.* eds., Bachrach, A. J. and Matzen M. M. (Undersea Med. Soc., Bethesda), pp. 25–36.

Fairley, H. B. (1980) Oxygen therapy for surgical patients. *Am. Rev. Respir. Dis.* 122, 37–44.

Faraci, F. M. and Heistad, D. D. (1992) Endothelium-derived relaxing factor inhibits constrictor responses of large cerebral arteries to serotonin. *J. Cereb. Blood Flow Metab.* 12, 500–506.

Fazekas, J. F., Alexander, F. A. D., and Himwich, H. E. (1941) Tolerance of newborn to anoxia. *Am. J. Physiol.* 134, 281–287.

Feeney, D. M., Bailey, B. Y., Bayeson, M. G., Hovda, D. A., and Sutton, R. L. (1987) The effect of seizures on recovery of function following cortical contusion in the rat. *Brain Injury* 1, 27–32.

Feustel, P. J. and Nelson, L. R. (1987) Cortical blood flow regulation during hypoxemia in experimental head injury. *J. Surg. Res.* 43, 86–93.

Fiskum, G. (1985) Mitochondrial damage during cerebral hypoxia. *Ann. Emerg. Med.* 114, 810–815.

Floyd, R. A. (1990) Role of oxygen free radicals in carcinogenesis and brain ischemia. *FASEB J.* 4, 2587–2597.

Flynn, J. T., Buncalari, E., Bachynski, B. N., et al. (1987) Retinopathy of prematurity: diagnosis, severity, and natural history. *Ophthalmology* 94, 620–629.

Flynn, J. T. and Tasman, W. (1992) Retinopathy of prematurity: unfinished agenda, in: *Retinopathy of prematurity; a clinician's guide,* eds., Flynn, J. T. and Tasman, W. (Springer-Verlag, New York), pp. 1–5.

Folbergrova, J., Nilsson, B., Sakabe, T., and Siesjo, B. K. (1981) The influence of hypoxia on the concentrations of cyclic nucleotides in the rat brain. *J. Neurochem.* 36, 1670–1674.

Forster II, R. E. (1994) A book review on respiratory physiology. *Science* 263, 998–999.

Fox, D. P., Robertson, F. W., Brown, T., Whitehead, A. R., and Douglas, J. D. W. (1984) Chromosome aberrations in divers. *Undersea Biomed. Res.* 11, 193–204.

Frank, L. (1991) Developmental aspects of experimental pulmonary oxygen toxicity. *Free Radic. Biol. Med.* 11, 463–494.

Freeman, B. A. and Crapo, J. D. (1982) Biology of diseases: free radicals and tissue injury. *Lab. Invest.* 47, 412–426.

Friedman, J. and Haddad, G. G. (1992) Major differences in Ca^{++} response to anoxia between neonatal and adult rat CA_1 neurons; role of Ca^{++}_o and Na^+_o. *J. Neurosci.* 13, 63–72.

Fridovich, I. (1976) Oxygen radicals, hydrogen peroxide, and oxygen toxicity, in *Free radicals in biology,* ed., Pryor, W. A. (Academic Press, New York), Vol. 1, pp. 239–277.

Fridovich, I. (1978) The biology of oxygen radicals. *Science* 201, 875–880.

Fridovich, I. (1979) Hypoxia and oxygen toxicity. *Adv. Neurol.* 26, 255–259.

Fridovich, I. (1983) Superoxide radical: an endogenous toxicant. *Am. Rev. Pharmacol. Toxicol.* 23, 239–257.

Fulmer, J. and Snider, G. L. (1984) ACCP Conference on oxygen therapy. *Chest* 86, 234–256.

Gabb, M. R., Poch, B., and Heller, V. (1990) Oxygen tension, oxygen metabolism, and microcirculation in vasogenic brain edema. *Adv. Neurol.* 52, 247–256.

Gardiner, M., Nilsson, B., Rehncrona, S., and Siesjo, B. K. (1981) Free fatty acids in the rat brain in the moderate and severe hypoxia. *J. Neurochem.* 36, 1500–1505.

Gazenko, O. G., Grigor'ev, A. I., Melenshko, G. I., and Shepelev, E. (1990) Habitability and biological life support systems for man. *Kosm. Biol. Avikosm. Med.* 24, 12–17.

Genn, M. M. and Silverman, W. A. (1964) The mental development of ex-premature children with retrolental fibroplasia. *J. Nerv. Ment. Dis.* 138, 79–86.

Gennarelli, T. A., Segawa, H., Wald, U., Czerniki, Z., Marsh, K., and Thomson, C. (1982) Physiological response to angular acceleration of the head, in *Head injury: basic and clinical aspects,* eds., Grossman, R. G. and Gildenberg, P. L. (Raven Press, New York), pp. 129–140.

Gerschman, R., Gilbert, D. L., Nye, S. W., Nady, P. W., and Fenn, W. O. (1954) Role of adrenalectomy and adrenocortical hormones in oxygen poisoning. *Am. J. Physiol.* 178, 346–350.

Gerschman, R., Gilbert, D. L., Nye, S. W., Dwyer, P., and Fenn, W. O. (1954a) Oxygen poisoning and X-irradiation: a mechanism in common. *Science* 119, 623–626.

Gibson, G. E. and Duffy, T. E. (1981) Impaired synthesis of acetylcholine by mild hypoxic hypoxia or nitrous oxide. *J. Neurochem.* 36, 28–33.

Giffard, R. G., Monyer, H., Christine, C. W., and Choi, D. W. (1990) Selective vulnerability of cultured corticalglia to injury by extracellular acidosis. *Brain Res.* 530, 138–141.

Gilbert, D. L. (1981) Prespective on the history of oxygen and life, in *Oxygen and living processes: an interdisciplinary approach,* ed., Gilbert, D. L. (Springer-Verlag, New York), pp. 1–43.

Gilman, S. C., Bonner, M. J., and Pellmar, T. C. (1993) Effect of oxidative stress on excitatory amino acid release by cerebral cortical synaptosomes. *Free Radic. Biol. Med.* 15, 671–675.

Ginsberg, M. D., Budd, W. W., and Welsh, F. A. (1978) Diffuse cerebral ischemia in the cat. I. Local blood flow during severe ischemia and recirculation. *Ann. Neurol.* 3, 482–492.

Ginsberg, M. D. (1979) Delayed neurological deterioration following hypoxia, *Adv. Neurol.* 26, 21–44.

Glass, H. G., Snyder, F. F., and Webster, E. (1944) The rate of decline in resistance to anoxia of rabbits, dogs and guinea pigs from the onset of viability to adult life. *Am. J. Physiol.* 140, 609–615.

Goldberg, M. P., Monyer, H., and Choi, D. W. (1988) Hypoxic neuronal injury *in vitro* depends on extracellular glutamate. *Neurosci. Lett.* 94, 52–57.

Goldberg, M. P. and Choi, D. W. (1990) Intracellular free calcium increases in cultured cortical neurons deprived of oxygen and glucose. *Stroke* 21 (Suppl. III), 75–77.

Gordon-Majszak, W. and Gajkowska, B. (1992) Biochemical and morphological changes in rat brain synaptosomes after exposure to normobaric hyperoxia *in vivo. Exp. Pathol.* 44, 96–101.

Gottlieb, S. F., Schmidt, P. L., Stewart, R. J., and Bendixon, G. (1977) The *in vivo* effect of oxygen on Na-K-ATPase: the beginning of a new cellular theory of oxygen toxicity, in *Proc. 6th Int. Cong. Hyperbaric Med.,* ed., Smith G. (Aberdeen Univ. Press, Scotland), pp. 8–12.

Grave, G. D., Kennedy, C., and Sokoloff, L. (1972) Impairment of growth and development of rats brain by hyperoxia at atmospheric pressure. *J. Neurochem.* 19, 187–194.

Greene, W. B., Balentine, J. D., and Hennigar, G. R. (1979) Selective mitochondrial degeneration in renal tubules following hyperbaric oxygen pressure. *Am. J. Pathol.* 96, 737–752.

Grigg, J. J. and Anderson, E. G. (1989) Glucose and sulfonylureas modify different phases of the membrane potential change during hypoxia in rat hippocampal slices. *Brain Res.* 489, 302–310.

Grisham, M. B. and McCord, J. M. (1986) Chemistry and cytotoxicity of reactive oxygen metabolites, in *Physiology of oxygen radicals*, eds., Taylor, A. E., Matalon, S., and Ward, P. A. (Williams and Wilkins, Baltimore), pp. 1–18.

Grogaard, B., Schurer, L., Gerdin, B., and Arfors, K. E. (1989) Delayed hypoperfusion after incomplete forebrain ischemia in the rat. The role of polymorphonuclear leukocytes. *J. Cereb. Blood Flow Metab.* 9, 500–505.

Gutierrez, G. (1991) Cellular energy metabolism during hypoxia. *Crit. Care Med.* 19, 619–626.

Gyllensten, L. (1959) Influence of oxygen exposure on the postnatal vascularization of the cerebral cortex in mice. *Acta Morph. Neerl. Scand.* 2, 289–310.

Gyllensten, L. (1959a) Influence of oxygen exposure on the differentiation of the cerebral cortex in growing mice. *Acta Morph. Neerl. Scand.* 2, 311–330.

Gyllensten, L. (1959b) Postnatal development of visual cortex in darkness (mice). *Acta Morph. Neerl. Scand.* 2, 331–345.

Gyulai, L., Schnall, M., McLaughlin, A. C., Leigh, J. S., Jr., and Chance, B. (1987) Simultaneous ^{31}P- and ^{1}H nuclear magnetic resonance studies of hypoxia and ischemia in the cat brain. *J. Cereb. Blood Flow Metab.* 7, 543–551.

Haddad, G. G. and Mellins, R. B. (1984) Hypoxia and respiratory control in early life. *Ann. Rev. Physiol.* 46, 629–643.

Haddad, G. G. and Donnelly, D. F. (1990) O_2 deprivation induces a major depolarization in brainstem neurons in the adult but not in the neonate rat. *J. Physiol.* 429, 411–428.

Haddad, G. G. and Jiang, C. (1993) O_2 deprivation in the central nervous system: on mechanisms of neuronal responses, differential sensitivity and injury. *Prog. Neurobiol.* 40, 277–318.

Haddad, G. G. and Jiang, C. (1993a) Mechanism of anoxia-induced depolarization in brainstem neurons: *in vitro* current and voltage clamp studies in the adult rat. *Brain Res.* 625, 261–268.

Hallenbeck, J. M. and Dutka, A. J. (1990) Background review and current concepts of reperfusion injury. *Arch. Neurol.* 47, 1245–1254.

Halliwell, B. (1989) Superoxide, iron, vascular endothelium and reperfusion injury. *Free Radic. Biol. Res. Commun.* 5, 315–318.

Halliwell, B. (1992) Reactive oxygen species and the central nervous system. *J. Neurochem.* 59, 1609–1623.

Halliwell, B. and Gutteridge, J. M. C. (1985) Oxygen radicals and the nervous system. *Trends Neurosci.* 8, 22–26.

Halliwell, B., Gutteridge, J. M. C., and Cross, C. E. (1992) Free radicals, antioxidants, and human disease: where are we now? *J. Lab. Clin. Med.* 119, 598–620.

Halsey, M. J. (1982) Effects of high pressure on the central nervous system. *Physiol. Rev.* 62, 1341–1377.

Hannah, R. S. and Hannah, K. J. (1980) Hyperoxia: effects on the vascularization of the developing central nervous system. *Acta Neuropathol.* 51, 141.

Hansen, A. J. (1985) Effect of anoxia on ion distribution in the brain. *Physiol. Rev.* 65, 101–148.

Harel, D., Kerem, D., and Lavy, S. (1969) The influence of high oxygen pressure on the electrical activity of the brain. *EEG Clin. Neurophysiol.* 26, 310–317.

Harman, D. (1956) Aging: a theory based on free-radical and radiation chemistry. *J. Gerontol.* 11, 198–200.

Harman, D. (1988) Free radicals in aging. *Mol. Cell. Biochem.* 84, 155–161.

Harris, J. B., Chang, L-Y., and Crapo, J. D. (1991) Rat lung alveolar type I epithelial cell injury and response to hyperoxia. *Am. J. Respir. Cell Mol. Biol.* 4, 115–125.

Harrison, G. A. (1974) Ultrastructural response of rat lung to 90 days' exposure to oxygen at 450 mmHg. *Aeros. Med.* 45, 1041–1045.

Hart, G. B. (1974) Treatment of decompression illness and air embolism with hyperbaric oxygen. *Aerosp. Med.* 45, 1190–1193.

Hart, G. B., Strauss, M. B., and Peirce, E. C. II (1979), in *Proc. 6th Int. Cong. Hyperbaric Med.*, eds., Smith, G. (Aberdeen Univ. Press, Scotland), pp. 47–52.

Haugaard, N. (1968) Cellular mechanisms of oxygen toxicity. *Physiol. Rev.* 48, 311–373.

Hayatdavoudi, G., O'Neil, J. J., Barry, B. E., Freeman, B. A., and Crapo, J. D. (1981) Pulmonary injury in rats following continuous exposure to 60% O_2 for 7 days. *J. Appl. Physiol. Respir. Environ. Exercise Physiol.* 51, 1220–1231.

Heimbach, D. (1993) Editorial response: use of hyperbaric oxygen. *Clin. Infect. Dis.* 17, 239–240.

Heistad, D. D. and Kontos, H. A. (1983) Cerebral circulation, in *Handbook of physiology. The cardiovascular system,* Sec. 2, Vol. 3, Part 1 (Am. Physiol. Soc., Bethesda), pp. 137–182.

Hemrick, S. K. and Gottlieb, S. F. (1977) Effect of increased pressure of oxygen, nitrogen, and helium on activity of Na-K-ATPase of beef brain. *Aviat. Space Environ. Med.* 48, 40–43.

Hendricks, P. L., Hall, D. A., Hunter, W. L., Jr., and Haley, P. J. (1977) Extension of pulmonary O_2 tolerance in man at 2 ATA by intermittent O_2 exposure. *J. Appl. Physiol. Respir. Environ. Exercise Physiol.* 42, 593–599.

Hochachka, P. W. (1986) Defense strategies against hypoxia and hypothermia. *Science* 231, 234–241.

Hogg, N., Darley-Usmar, V. M., Wilson, M. T., and Moncada, S. (1992) Production of hydroxyl radicals from the simultaneous generation of superoxide and nitric oxide. *Biochem. J.* 281, 419–424.

Honig, C. R. (1977) Hypoxia in skeletal muscle at rest and during the transition to steady work. *Microcirc. Res.* 13, 377–398.
Hope, B. T., Michael, G. J., Knigge, K. M., and Vincent, S. R. (1991) Neuronal NADPH diaphorase is a nitric oxide synthetase. *Proc. Natl. Acad. Sci. U.S.A.* 88, 2811–2814.
Horsley, V. (1894) The destructive effects of small projectiles. *Nature* 50, 104–108.
Hovda, D. A., Becker, D. P., and Katayama, Y. (1992) Secondary injury and acidosis. *J. Neurotrauma* 9, Suppl. 1, S47–S60.
Hunt, R. C., Blackburn, J. G., Ogilvie, R. W., and Balentine, J. D. (1978) Oxygen tension in the globus pallidus and neostriatum of unanesthetized rats during exposure to hyperbaric oxygen. *Exp. Neurol.* 62, 698–707.
Imaizumi, S., Kayama, T., and Suzuki, J. (1984) Chemiluminescence in hypoxic brain—the first report. Correlation between energy metabolism and free-radical reaction. *Stroke* 15, 1061–1065.
Ishige, N., Pitts, L. H., Berry, I., Carlson, S. G., Nishimura, M. C., et al. (1987) The effect of hypoxia on traumatic head injury in rats: alterations in neurologic function, brain edema, and cerebral blood flow. *J. Cereb. Blood Flow Metab.* 7, 759–758.
Ishige, N., Pitts, L. H., Berry, I., Nishimura, M. C., and James, T. L. (1988) The effects of hypovolemic hypotension on high-energy phosphate metabolism of traumatized brain in rats. *J. Neurosurg.* 68, 129, 136.
Iwamoto, J., Yang S-P., Yoshinaga, M., Kransney, E., and Kransney J. (1992) N-Nitro-L-arginine influences cerebral metabolism in awake sheep. *J. Appl. Physiol.* 73, 2233–2240.
Jacobson, E. (1971) The theory and indications of hyperbaric oxygen, a review. *Laval Med.* 42, 291–300.
Jain, K. K., ed. (1994) *Textbook of hyperbaric medicine.* 2nd ed. (Hogrefe and Huber, Kirkland, WA).
Jamieson, D. D. (1989) Oxygen toxicity and reactive oxygen metabolites in mammals. *Free Radic. Biol. Med.* 17, 87–108.
Jamieson, D. D. (1991) Lipid peroxidation in brain and lungs from mice exposed to hyperoxia. *Biochem. Pharmacol.* 41, 749–756.
Jamieson, D. D. and Carmody, J. (1989) Low temperature worsen mammalian oxygen toxicity. *Aviat. Space Environ. Med.,* 60, 639–643.
Jensen, F. E., Applegate, C. D., Holtzman, D., Belin, T. R., and Burchfiel, J. L. (1991) Epileptogenic effect of hypoxia in the immature rodent brain. *Ann. Neurol.* 29, 629–637.
Jerrett, S. A., Jefferson, D., and Mengel, C. E. (1973) Siezures, H_2O_2 formation and lipid peroxides in brain during exposure to oxygen under high pressure. *Aerosp. Med.* 44, 40–44.
Jiang, C. and Haddad, G. G. (1991) Effect of anoxia on intracellular and extracellular potassium activity in hypoglossal neurons *in vitro*. *J. Neurophysiol.* 66, 103–111.
Jiang, J. and Haddad G. G. (1992) Effect of differential responses of neocortical neurons to glucose and O_2 deprivation in the human and rat. *J. Neurophysiol.* 68, 2165–2175.
Jiang, C., Agulian, S., and Haddad, G. G. (1991) O_2 tension in adult and neonatal brain slices under several experimental conditions. *Brain Res.* 568, 159–164.
Jiang, C., Xia, Y., and Haddad, G. G. (1992) Role of ATP sensitive K^+ channels during anoxia: major differences between rat (newborn and adult) and turtle neurons. *J. Physiol.* 448, 599–612.
Johannson, H. and Siesjo, B. K. (1975) Cerebral blood flow and oxygen consumption in the rat in hypoxic hypoxia. *Acta Physiol. Scand.* 93, 269–276.
Jones, D. P. (1985) The role of oxygen concentration in oxidative stress: hypoxic and hyperoxic models, in *Methods in enzymology,* ed., Parker, L. and Glazor, A. N. (Academic Press, London), Vol. 186: Part B. pp. 151–195.
Jones, D. P. (1986) Intracellular diffusion gradients of O_2 and ATP. *Am. J. Physiol.* 250, C663–C675.
Joseph, R., Tsering, C., Grunfeld, S., and Welch, K. M. A. (1992) Serotonin may have neurotoxic properties. *Neurosci. Lett.* 136, 15–18.
Kagstrom, E., Smith, M-L., and Siesjo, B. K. (1983) Local cerebral blood flow in the recovery period following complete cerebral ischemia in the rat. *J. Cereb. Blood Flow Metab.* 3, 170–182.
Kalimo, H., Rehncrona, S., Soderfeldt, B., et al. (1981) Brain lactic acidosis and ischemic cell damage. 2. Histopathology. *J. Cereb. Blood Flow Metab.* 3, 346–353.
Kann, H. E., Jr., Mengel, C. E., Smith, W., and Horton, B. (1964) Oxygen toxicity and vitamine E. *Aerosp. Med.* 35, 840–844.
Kappus, H. and Sies, H. (1981) Toxic drug effects associated with oxygen metabolism: redox cycling and lipid peroxidation. *Experientia* 37, 1233–1241.
Katayama, Y., Becker, D. P., Tamura, T., and Hovda, D. A. (1990) Massive increase in extracellular potassium and the indiscriminate release of glutamate following concussive brain injury. *J. Neurosurg.* 73, 889–900.
Kawasaki, K., Czeh, G., and Somjen, G. G. (1988) Does Leao's spreading depression cause irreversible brain damage? *Adv. Behav. Biol.* 35, 305–306.
Kety, S. S. and Schmidt, C. F. (1948) The effect of altered arterial tensions of CO_2 and O_2 on cerebral blood flow and cerebral oxygen consumption of normal young men. *J. Clin. Invest.* 27, 484–492.
Kinuta, Y., Kimura, M., Itokava, Y., Ishkawa, M., and Kikuchi, H. (1989) Changes in xanthine oxidase in ischemic rat brain. *J. Neurosurg.* 71, 417–420.

Koch, K. A., Jackson, D. L., Schmiedl, M., and Rosenblatt, J. I. (1984) Total cerebral ischemia: effects of alterations in arterial PCO_2 on cerebral microcirculation. *J. Cereb. Blood Flow Metab.* 4, 343–349.

Kontos, H. A. and Povlishock, J. T. (1986) Oxygen radicals in brain injury. *Cent. Nerv. Syst. Trauma* 3, 257–263.

Kovachich, G. B., and Mishra, O. P. (1980) Lipid peroxidation in rat brain cortical slices as measured by the thiobarbituric acid test. *J. Neurochem.* 35, 1449–1452.

Kovachich, G. B., Mishra, O. P., and Clark, J. M. (1981) Depression of cortical Na-K-ATPase activity in rats exposed to oxygen at 4 ATA O_2 *Brain Res.* 206, 229–231.

Kovachich, G. B. and Haugaard, N. (1981) Biochemical aspects of oxygen toxicity in the metazoa, in *Oxygen and living processes, an interdisciplinary approach,* ed., Gilbert, D. L. (Springer-Verlag, New York), pp. 210–234.

Krogh, A. (1919) The supply of oxygen to the tissues and the regulation of the capillary circulation. *J. Physiol.* 52, 457–474.

Kuschinsky, W. and Wahl, M. (1978). Local chemical and neurogenic regulation of cerebral vascular resistance. *Physiol. Rev.* 58, 656–689.

Kuwahira, I., Heisler, N., Piiper, J., and Gonzalez, C. N. (1993) Effect of chronic hypoxia on hemodynamics, organ blood flow and O_2-supply in rats. *Respir. Physiol.* 92, 227–238.

Lahiri, S., Mokashi, A., Delaney, R. G., and Fishman, A. P. (1978) Arterial PO_2 and PCO_2 stimulus threshold for carotid chemoreceptors and breathing. *Respir. Physiol.* 34, 359–375.

Lahiri, S., Mokashi, A., Mulligan, E., and Nishino, T. (1981) Comparison of aortic and carotid chemoreceptor responses to hypercapnia and hypoxia. *J. Appl. Physiol. Respir. Environ. Exercise Physiol.* 51, 55–61.

Lahiri, S., Mulligan, E., Andronikou, S., Shirahata, M., and Mokashi, A. (1987) Carotid body chemosensory function in prolonged normobaric hyperoxia in the cat. *J. Appl. Physiol.* 65, 1924–1931.

Lahiri, S. (1991) Oxygen biology of peripheral chemoreceptors, in *Responses and adaptation to hypoxia organ to organelle,* eds., Lahiri, S., Cherniack, N. S., and Fitzgerald, R. S. (Am. Physiol. Soc., New York), pp. 95–106.

Lambertsen, C. J., Ewing, J. H., Kough, R. H., Gould, R., and Stroud, M. W., III. (1955) Oxygen toxicity. Arterial and internal jugular blood gas composition in man during inhalation of air, 100% O_2 and 2% CO_2 at 3.5 atmospheres ambient pressure. *J. Appl. Physiol.* 8, 255–263.

Lambertsen, C. J., Kough, R. E., Cooper, D. Y., Emmel, H. H., Loeschcke, C. F., and Schmidt, C. F. (1963) Oxygen toxicity, effects in man of oxygen inhalation at 1 and 3.5 atmospheres upon blood gas transport, cerebral circulation and cerebral metabolism. *J. Appl. Physiol.* 5, 471–484.

Lambertsen, C. J., Gelfand, R., Peterson, R., Strauss, R., Wright, W. B., et al. (1977) Human tolerance to He, Ne, and N_2 at respiratory gas densities equivalent to He-O_2 breathing at 1200, 3000, 4000 and 5000 feet of sea water. *Aviat. Space Environ. Med.* 48, 843.

Lambertsen, C. J. (1978) Effects of hyperoxia on organs and their tissues, in *Extrapulmonary manifestations of respiratory disease,* ed., Robin, E. D. (Marcel Dekker, New York), pp. 239–303.

Lancaster, J. R., Jr. (1992) Nitric oxide in cells. *Am. Sci.* 80, 248–259.

Leao, A. A. P. (1944) Spreading depression of activity in the cerebral cortex. *J. Neurophysiol.* 7, 359–390.

Lee, K. S., Frank, S., Vanderklish, P., Arai, A., and Lynch, G. (1991) Inhibition of proteolysis protects hippocampal neurons from ischemia. *Proc. Natl. Acad. Sci. U.S.A.* 88, 7233–7237.

Lefer, A. M. and Lefer, D. J. (1993) Pharmacology of the endothelium in ischemia-reperfusion and circulatory shock. *Annu. Rev. Pharmacol. Toxicol.* 33, 71–90.

Lemenkohler, A., Bingman, D., and Speckman, E. J. (1989) Neuronal and glial responses to hypoxia and hypercapnia. *Biomed. Biochim. Acta* 48, 155–160.

Levine, J. E. and Becker, D. P. (1979) Reversal of incipient brain death from head-injury apnea at the scene of accidents. *New Engl. J. Med.* 301, 109.

Lewelt, W., Jenkins, L. W., and Miller, J. D. (1980). Autoregulation of cerebral blood flow after experimental fluid percussion injury of the brain. *J. Neurosurg.* 53, 500–501.

Lewelt, W., Jenkins, L. W., and Miller J. D. (1982) Effects of experimental fluid percussion injury of the brain on cerebrovascular reactivity to hypoxia and hypercapnia. *J. Neurosurg.* 56, 332–338.

Liberzon, I., Arieli, R., and Kerem, D. (1989) Attenuation of hypoxic ventilation by hyperbaric O_2: effects of pressure and exposure time. *J. Appl. Physiol.* 66, 851–856.

Lorez, H., Keller, F., Ruess, G., and Otten, U. (1989) Nerve growth factor increases in adult rat brain after hypoxic injury. *Neurosci. Lett.* 98, 339–344.

MacMillan, B. and Siesjo, B. K. (1971) Critical oxygen tensions in the brain. *Acta Physiol. Scand.* 82, 412–414.

MacMillan, B. and Siesjo, B. K. (1972) Brain energy metabolism in hypoxemia. *Scand. J. Clin. Lab. Invest.* 30, 127–136.

MacNeil, C. and Bryan, R. M., Jr. (1988) Regional cerebral glucose utilization transiently increases during mild hypoxia. *J. Cereb. Blood Flow Metab.* 8, 495–501.

Marcoux, F. W., Probert, A. W., and Weber, M. L. (1990) Hypoxic neuronal injury in tissue culture is associated with delayed calcium accumulation. *Stroke* 21 (Suppl. III), 71–74.

Marmarou, A., Nakamura, T., Sakamoto, H., Jenkins, L. W., and Becker, D. P. (1984) Development of brain edema following fluid percussion injury. in *Brain edema,* eds., Inaba, Y., Klatzo, I., and Spatz, M. (Springer, Berlin), pp. 88–91.

Matalon, S., Nesarajah, M. S., and Farhi, L. E. (1982) Pulmonary and circulatory changes in conscious sheep exposed to 100% O_2 at 1 ATA. *J. Appl. Physiol. Respir. Environ. Exercise Physiol.* 53, 110–116.

Mayevsky, A. and Chance, B. (1975) Metabolic responses of the awake cerebral cortex to anoxia, hypoxia, spreading depression and epileptiform activity. *Brain Res.* 98, 149–165.

Mayevsky, A. (1983) Multiparameter monitoring of the awake brain under hyperbaric oxygenation. *J. Appl. Physiol.* 54, 740–748.

Mayevsky, A. and Weiss, H. R. (1991) Cerebral blood flow and oxygen consumption in cortical spreading depression. *J. Cereb. Blood Flow Metab.* 11, 829–836.

McBean, D. E., Rome, J., and Harper, A. (1989) A cerebral vasoconstrictive effect of some adenosine analogues. A study of adenosine analogues on local cerebral blood flow and glucose utilization in the rat. *J. Cereb. Blood Flow Metab.* 9, 548–555.

McBurney, R. N. and Neering, R. I. (1987) Neuronal calcium homeostasis. *Trends Neurosci.* 10, 164–169.

McCall, T. B., Boughton-Smith, N. K., Palmer, R. M. J., Whittle, B. J. R., and Moncada, S. (1989) Synthesis of nitric oxide from L-arginine by neutrophils: release and interaction with superoxide anion. *Biochem. J.* 261, 293–296.

Medina, J. M. (1985) The role of lactate as an energy substrate for the brain during the early neonatal period. *Biol. Neonate* 48, 237–244.

Mehlhorn, R. J. and Cole, G. (1985) The free-radical theory of aging: a ciritical review. *Adv. Free Rad. Biol. Med.* 1, 165–223.

Messick, J. M., Jr. and Milde, L. M. (1987) Brain protection. *Adv. Anesthesiol.* 4, 47–88.

Michaels, R. L. and Rothman, S. M. (1990) Glutamate neurotoxicity *in vitro:* antagonist pharmacology and intracellular calcium concentrations. *J. Neurosci.* 10, 283–292.

Mickel, H. S., Kempski, O., Fenerstein, G., Parisi, J. E., and Webster, H. D. (1990) Prominant white matter lesions develop in mongolian gerbils treated with 100% O_2 normobaric oxygen after global brain ischemia. *Acta Neuropathol.* 79, 465–472.

Milde, L. N. (1989) Pathophysiology of ischemic brain injury. *Crit. Care Clin.* 5, 729–753.

Miller, J. D., Garibi, J., North, J. B., and Teasdale, G. M. (1975) Effects of increased arterial pressure on blood flow in the damaged brain. *J. Neurol. Neurosurg. Psychiat.* 38, 657–665.

Miller, R. J. (1987) Calcium channels and neuronal function. *Science* 230, 186–188.

Mishra, O. P. and Shankar, R. (1980) Na-K-ATPase in developing brain during undernutrition. *Nutr. Metab.* 24, 114–121.

Mishra, O. P., Delivoria-Papadopoulos, M., Cahillance, G., and Wagerle, L. C. (1989) Lipid peroxidation as the mechanism of modification of the affinity of Na-K-ATPase active site for ATP, K^+, Na^+ and strophanthidine *in vitro. Neurochem. Res.* 14, 845–851.

Miwa, S., Fujiwara, M., Inoue, M., et al. (1986) Effects of hypoxia on the activities of nonadrenergic and dopaminergic neurons in the rat brain. *J. Neurochem.* 47, 63–69.

Miyaoka, M., Shinohara, M., Kennedy, C., and Sokoloff, L. (1979) Alterations in local cerebral glucose utilization (LCGU), in rat brain during hypoxemia. *Ann. Neurol.* 6, 151–154.

Moncada, S., Palmer, R. M. J., and Higgs, E. A. (1989) Biosynthesis of nitric oxide from L-arginine: a pathway for the regulation of cell function and communication. *Biochem. Pharmacol.* 38, 1709–1715.

Moncada, S., Palmer, R. M. J., and Higgs, E. A. (1991) Nitric oxide: physiology, pathophysiology and pharmacology. *Pharmacol. Rev.* 43, 109–142.

Moncada, S. (1992) Nitric oxide gas: mediator, modulator, and pathophysiologic entity. *J. Lab. Clin. Med.* 120, 187–191.

Monge, C. and Leon-Velarde, F. (1991) Physiological adaptation to high altitude: oxygen transport in mammals and birds. *Physiol. Rev.* 71, 1135–1172.

Morii, S., Ngai, A. C., Ko, K. R., and Winn, H. R. (1987) Role of adenosine in regulation of cerebral blood flow: effects of theophylline during normoxia and hypoxia. *Am. J. Physiol.* 253, H165–H175.

Mozzi, R. and Horrocks, L. A. (1992) Serine incorporation into phosphatidylserine in hypoxic rat brain cortex. *Mol. Chem. Neuropathol.* 17, 183–188.

Muhvich, K. H., Piano, M. R., Myers, R. A. M., Ferguson, J. L., and Marzella, L. (1992) Hyperbaric oxygen decreases blood flows in normal and septic rats. *Undersea Biomed. Res.* 19, 31–40.

Mulligan, E., Lahiri, S., and Story, B. T. (1981) Carotid body O_2 chemoreception and mitochondrial oxidative phosphorylation. *J. Appl. Physiol. Respir. Environ. Exercise Physiol.* 51, 435–446.

Murphy, T. H., Miyamoto, M., Sastre, A., Schnara, R. L., and Coyle, J. T. (1989) Glutamate toxicity in a neuronal cell line involves inhibition of cyctine transport leading to oxidative stress. *Neuron* 2, 1547–1558.

Myers, R. E. (1979) A unitary theory of causation of anoxic brain pathology. *Adv. Neurol.* 26, 195–213.

Myers, R. E. (1979a) Lactic acid accumulation as a cause of brain edema and cerebral necrosis resulting from oxygen deprevation, in: *Advances in perinatal neurology*, eds., Korobkin, R. and Guilleminault, G. (Spectrum, New York), pp. 85–114.

Myers, R. O. M. and Schnitzer, B. M. (1984) Hyperbaric oxygen use update. *Postgrad. Med.* 76, 83–95.

Nagai, S. H., Levy, A., Finck, A. D., Yang, J. C., and Spector, S. (1977) Central nervous system toxicity of hyperbaric oxygen. Effects of light, norepinephrine depletion and beta-adrenergic blockade. *Neuropharmacology* 16, 675–679.

Nathan, C. (1992) Nitric oxide as a secretory product of mammalian cells. *FASEB J.* 6, 3051–3064.

Nedergaad, M., Goldman, N. A., Desal, S., and Pulsinelli, W. A. (1991) Acid-induced death in neurons and glia. *J. Neurosci.* 11, 2489–2497.

Nehling, A., Pereira de Vasconcelos, A., and Boyet, S. (1988) Quantitative autoradiographic measurement of local cerebral glucose utilization in freely moving rats during postnatal development. *Neuroscience* 8, 2321–2333.

Nehling, A., Pereira de Vasconcelos, A., and Boyet, S. (1989) Postnatal changes in local cerebral blood flow measured by the quantitative autoradiographic ^{14}C-iodoantipyrine technique in freely moving rats. *J. Cereb. Blood Flow Metab.* 9, 579–588.

Nilsson, P., Hillered, L., Ponten, U., and Ungerstendt, U. (1990) Changes in cortical extracellular levels of energy-related metabolites and amino acids following concussive brain injury in rats. *J. Cereb. Blood Flow Metab.* 10, 631–637.

Ninomiya, H., Taniguchi, T., Kameyama, M., and Fujiwara, M. (1988) Increased binding of [^3H] Flunitrazepam in the rat brain under hypoxia. *J. Neurochem.* 51, 1111–1117.

Nishizuka, Y. (1986) Studies and perspectives of protein kinase C. *Science* 233, 305–312.

Noda, Y., McGeer, P. L., and McGeer, G. G. (1983) Lipid peroxide distribution in brain and the effect of hyperbaric oxygen. *J. Neurochem.* 40, 1329–1332.

Nohl, H. and Hegner, D. (1978) Do mitochondria produce oxygen radicals *in vivo? Eur. J. Biochem.* 82, 563–567.

Norberg, K. and Siesjo, B. K. (1975) Cerebral metabolism in hypoxic hypoxia. I. Pattern of activation of glycolysis: a re-evaluation. *Brain Res.* 86, 31–44.

Norris, J. W. and Pappius, H. M. (1970) Cerebral water and electrolytes, effect of asphyxia, hypoxia and hypercapnia. *Arch. Neurol.* 23, 248–258.

Ohnishi, S. T., Barr, J. K., Katagi, C., and Sadanaga, K. K. (1988) Charybdotoxin improves motor recovery of the rat after spinal cord injury. *Pharmacol. Biochem. Behav.* 31, 187–191.

Ohnishi, S. T., Sakamoto, A., Ohnishi, T., and Ogawa, R. (1992) Inhibition of lipid peroxidation by prostaglandin oligomeric derivatives. *Prostag. Leukotr. Esstl. Fat. Acids* 45, 217–221.

Olney, J. M. (1969) Brain lesion, obesity and other disturbances in mice treated with monosodium glutamate. *Science* 164, 719–721.

Onarheim, J. and Tyssebotn, I. (1980) Effect of high ambient pressure and oxygen tension on organ blood flow in the anesthetized rat. *Undersea Biomed. Res.* 7, 47–60.

Orrenius, S., McConkey, D. J., Bellomo, G., and Nicotera, P. (1989) Role of Ca^{++} in toxic cell killing. *TIPS* 10, 281–285.

Oury, T. D., Ho, Y-S., Piantadosi, C. A., and Crapo, J. D. (1992) Extracellular superoxide dismutase, nitric oxide, and central nervous system O_2 toxicity. *Proc. Natl. Acad. Sci. U.S.A.* 89, 9715–9719.

Pal Yu, B. (1994) Cellular defenses against damage from reactive oxygen species. *Physiol. Rev.* 74, 139–162.

Palinsky, W., Rosenthal, M. E., Yla-Herthuala, S., Gurtner, G. F., Sofer, S. S., et al. (1989) Low-density lipoprotein undergoes oxidative modification *in vivo. Proc. Natl. Acad. Sci. U.S.A.* 86, 1372–1376.

Palmer, R. M. J., Ferrige, A. G., and Moncada, S. (1987) Nitric oxide release accounts for the biological activity of endothelium-derived relaxing factor. *Nature* 327, 524–526.

Palmer, R. M. J., Ashton, D. S., and Moncada, S. (1988) Vascular endothelial synthesize nitric oxide from L-arginine. *Nature* 333, 664–666.

Palmer, R. M. J. (1993) The discovery of nitric oxide in the vessel wall: a unifying concept in the pathogenesis of sepsis. *Arch. Surg.* 128, 396–401.

Parmelee, A. H., Fiske, C. E., and Wright, R. H. (1959) The development of 10 children with blindness as a result of retrolental fibroplasia. *AMA J. Dis. Child.* 98, 74–96.

Paton, W. D. M. (1967) Experiments on the convulsant and anaesthetic effects of oxygen. *Br. J. Pharmacol. Chemother.* 29, 350–366.

Patz, A. (1965) The effect of oxygen on immature retinal vessels. *Invest. Ophthalmol.* 4, 988–999.

Patz, A. (1984) Current concepts of the effect of oxygen on the developing retina. *Curr. Eye Res.* 13, 159–163.

Pelligrino, D. A., Koenig, H. M., and Albrecht, R. F. (1993) Nitric oxide synthesis and regional cerebral blood flow responses to hypercapnia and hypoxia in the rat. *J. Cereb. Blood Flow Metab.* 13, 80–87.

Peterson, C., Neal, J. H., and Cotman, C. W. (1989) Development of N-methyl-D-aspartate excitotoxicity in cultured hippocampal neurons. *Dev. Brain Res.* 48, 187–195.

Phelps, D. L. (1982) Neonatal oxygen toxicity—is it preventable? *Pediatr. Clin. N. Am.* 29, 1233–1239.

Phelps, D. L. (1992) New horizons in clinical retinopathy of prematurity research, in *Retinopathy of prematurity: a clinicians' guide*, eds., Flynn, J. T. and Tasman, W. (Springer-Verlag, New York), pp. 125–128.

Pickard, J. D. (1981) Role of prostaglandins and arachidonic acid derivatives in the coupling of cerebral blood flow to cerebral metabolism. *J. Cereb. Blood Flow Metab.* 1, 361–384.

Pierce, E. C., Jr., Lambertsen, C. J., Strong, M. G., Alexander, S. C., and Steele, D. (1962) Blood PCO_2 and brain oxygenation at reduced ambient pressure. *J. Appl. Physiol.* 17, 899–908.

Polis, A. (1884) Recherches experimentales sur la commutione cerebrale. *Rev. Chir.* 273, 645–655.

Poole, G. V., Jr., Prough, D. S., Johnson, J. C., Stullken, E. H., Stump, D. A., and Howard, G. (1987) Effects of resuscitation from hemorrhagic shock on cerebral hemodynamics in the presence of an intracranial mass. *J. Trauma* 27, 18–23.

Prado, R., Watson, B. D., and Wester, P. (1993). Effects of nitric oxide synthase inhibition on cerebral blood flow following bilateral carotid artery occlusion and recirculation in the rats. *J. Cereb. Blood Flow Metab.* 13, 720–723.

Prasad, M. R., Popescu, L. M., Moraru, I. I., Lui, X. K., Maity, S., et al. (1991) Role of phospholipases A_2 and C in myocardial ischemia reperfusion injury. *Am. J. Physiol.* 260, H877–H883.

Prough, D. S. and Rogers, A. T. (1989) Physiology and pharmacology of cerebral blood flow and metabolism. *Crit. Care Clin.* 5, 713–727.

Pulsinelli, W. A. and Duffy, T. E. (1979) Local cerebral glucose metabolism during controlled hypoxemia. *Science* 204, 626–629.

Qi, Y. and Dawson, G. (1993) Effects of hypoxia on oligodendrocytes signal transduction. *J. Neurochem.* 61, 1097–1104.

Quattrone, P. D. and Staley, R. W. (1966) Exposure system for small animals at atmospheric and reduced pressure. *J. Appl. Physiol.* 21, 741–744.

Raday, N., Conforti, N., Harel, D., and Lavy, S. (1975) Analysis of preseizure electrocorticographic changes in rats during hyperbaric oxygenation. *Exp. Neurol.* 46, 9–19.

Radi, R., Beckman, J. S., Bush, K. M., and Freeman, B. A. (1991) Peroxynitrite oxidation of sulfhydryls. The cytotoxic potential of superoxide and nitric oxide. *J. Biol. Chem.* 266, 4244–4250.

Raflin, C. N., Harrison, M., Sick, T. J., and Rosenthal, M. (1991) EEG suppression and anoxic depolarization: influence on cerebral oxygenation during ischemia. *J. Cereb. Blood Flow Metab.* 11, 407–415.

Ransom, B. R., Stys, P. K., and Waxman, S. G. (1990) The pathophysiology of anoxic injury in central nervous system white matter. *Stroke* 21 (Suppl. III), III52–III57.

Rasmussen, H. and Barrett, P. Q. (1984) Calcium messenger system: an integrated view. *Physiol. Rev.* 64, 938–984.

Rehncrona, S., Siesjo, B. K., and Westerberg, E. (1978) Adenosine and cyclic AMP in cerebral cortex of rats in hypoxia, status epilepticus and hypercapnia. *Acta Physiol. Scand.* 104, 453–463.

Rehncrona, S., Rosen, I., and Siesjo, B. K. (1981) Brain lactic acidosis and ischemic cell damage. 1. Biochemistry and neurophysiology. *J. Cereb. Blood Flow Metab.* 1, 297–311.

Rehncrona, S., Westerberg, E., Akesson, B., and Siesjo, B. K. (1982) Brain cortical fatty acids and phospholipids during and following complete and severe incomplete ischemia. *J. Neurochem.* 38, 84–93.

Reivich, M. (1964) Arterial PCO_2 and cerebral hemodynamics. *Am. J. Physiol.* 206, 25–35.

Ricci, B. (1987) Effects of hyperbaric, normobaric and hypobaric oxygen supplementation on retinal vessels in newborn rats: a preliminary study. *Exp. Eye Res.* 44, 459–464.

Ricci, B. and Calegero, G. (1988) Oxygen-induced retinopathy in newborn rats: effects of prolonged normobaric and hyperbaric oxygen supplementation. *Pediatrics* 82, 193–198.

Ricci, B. (1990) Oxygen-induced retinopathy in the rat model. *Doc. Ophthalmol.* 74, 171–177.

Richter, C. and Frei, B. (1985) Ca^{++} movements induced by hydroperoxides in mitochondria, in *Methods in enzymology, Vol. 186, Oxygen radicals in biological systems, Part B. Oxygen radicals and antioxidants*, eds., Packer, L. and Glazer, A. N. (Academic Press, London), pp. 221–241.

Roberts, E. L. and Sick, T. J. (1992) Glucose enhances recovery of potassium ion homeostasis and synaptic excitability after anoxia in hippocampal slices. *Brain Res.* 570, 225–230.

Rosenberg, M. E. and Paller, M. S. (1991) Differential gene expression in the recovery from ischemia renal injury. *Kidney Int.* 39, 1156–1161.

Ross, R. (1993) The pathogenesis of atherosclerosis: a perspective for the 1990s. *Nature* 362, 801–809.

Rothman, S. (1984) Synaptic release of excitatory amino acid neurotransmitter mediates anoxic neuronal death. *J. Neurosci.* 4, 1884–1891.

Rucci, F. S., Giretti, M. L., and Laroca, M. (1967) Changes in electrical activity of the cerebral cortex and some subcortical ceters in hyperbaric oxygen. *EEG Clin. Neurophysiol.* 22, 231–238.

Sakamoto, A., Ohnishi, S. T., Ohnisih, T., and Ogawa, R. (1991) Relationship between free-radical production and lipid peroxidation during ischemia-reperfusion injury in the rat brain. *Brain Res.* 554, 186–192.

Sakamoto, A., Ohnishi, S. T., Ohnishi, T., and Gawa, R. (1991a) Protective effect of a new antioxidant on the rat brain exposed to ischemia-reperfusion injury: inhibition of free-radical formation and lipid peroxidation. *Free Radic. Biol. Med.* 11, 385–389.

Saran, M. and Bors, W. (1989) Oxygen radicals acting as chemical messengers: a hypothesis. *Free Radic. Res. Commun.* 3–6, 213–220.

Saran, M., Mitchel, C., and Bors, W. (1990) Reaction of NO with O_2^-. Implications for the action of endothelium-derived relaxing factor. *Free Radic. Res. Commun.* 10, 221–226.

Sato, S., Tominaga, T., Ohnishi, T., and Ohnishi, S. T. (1993) Spin trapping study of nitric oxide formation during bilateral carotid occlusion in the rat. *Biochim. Biophys. Acta.* 1181, 195–197.

Sato, S., Tominaga, T., Ohnishi, T., and Ohnishi, S. T. (1994) Electron paramagnetic resonance study of nitric oxide production during brain focal ischemia and reperfusion in the rat. *Brain Res.* 674, 91–96.

Saugstad, O. D. (1988) Hypoxanthine as an indicator of hypoxia: its role in health and disease through free radical production. *Pediatr. Res.* 23, 143–150.

Saunders, M. L., Miller, J. D., Stablein, D., and Allen, G. (1979) The effects of graded experimental trauma on cerebral blood flow and responsiveness to CO_2. *J. Neurosurg.* 51, 18–26.

Schaffner, F. and Felig, P. (1965) Changes in hepatic structure in rats produced by breathing pure oxygen. *J. Cell Biol.* 27, 505–517.

Schafer, G. and Citoler, P. (1978) Blood gas tension and development of lung damage in mice exposed to oxygen at 1 ATA. *Aviat. Space Environ. Med.* 49, 476–479.

Schallert, T., Hernandez, T. D., and Barth, T. M. (1986) Recovery of function after brain damage: severe and chronic disruption by diazepam. *Brain Res.* 379, 104–111.

Schiff, S. J. and Somjen, G. G. (1987) The effect of graded hypoxia on the hippocampal slice: an *in vitro* model of the ischemic penumbera. *Stroke,* 18, 30–37.

Schmidt, H. H. W. (1992) NO·,CO and ·OH endogenous soluble guanylyl cyclase-activating factors. *FEBS* 307, 102–107.

Schnakenburg, K. V. and Nolte, H. (1973) Results of electron microscopic studies in the rat brain under oxygen at high pressure. *Aeros. Med.* 44, 259–264.

Schurr, A., West, C., Reid, K. H., et al. (1987) Increased glucose improves recovery of neuronal function after cerebral hypoxia *in vitro*. *Brain Res.* 421, 135–139.

Schurr, A., Dong, W.-Q., Reid, K. H., et al. (1988) Lactic acidosis and recovery of neuronal function following cerebral hypoxia *in vitro*. *Brain Res.* 438, 311–314.

Schurr, A. and Rigor, B. M. (1987) The mechanism of neuronal resistance and adaptation to hypoxia. *FEBS Lett.* 224, 4–8.

Schurr, A. and Rigor, B. M. (1989) Cerebral ischemia revisited: new insights as revealed using *in vitro* brain slice preparations. *Experientia* 45, 684–695.

Scott, J. A. and Rabito, C. A. (1988) Oxygen radicals and plasma membrane potential. *Free Radic. Biol. Med.* 5, 237–246.

Seifert, V., Stolke, D., Kunz, U., and Resch, K. (1988) Influence of blood volume on cerebrospinal fluid levels of arachidonic acid metabolites after subarachnoid hemorrhage: experimental study on the pathogenesis of cerebral vasospasm. *Neurosurgery* 23, 313–321.

Shibuki, K. and Okada, D. (1991) Endogenous nitric oxide release required for long-term synaptic depression in the cerebellum. *Nature (London)* 349, 326–328.

Shinozuka, T., Nemoto, E. M., and Winter, P. M. (1989) Mechanisms of cerebrovascular O_2 sensitivity from hyperoxia to moderate hypoxia in the rat. *J. Cereb. Blood Flow Metab.* 9, 187–195.

Shohami, E., Shapira, Y., Sidi, A., and Cotev, S. (1987) Head injury induces increased prostaglandin synthesis in rat brain. *J. Cereb. Blood Flow Metab.* 7, 58–63.

Siegel, G. J., Stahl, W. L., and Swanson, P. D. (1981) Ion transport, in *Basic Neurochemistry,* eds., Siegel, G. J., Albers, R. W., Katzman, R., and Agranoff, B. W. (Little Brown and Co., Boston) 2nd ed., pp. 107–143.

Siesjo, B. K. and Nilsson, L. (1971) The influence of arterial hypoxemia upon labile phosphates and upon extracellular and intracellular lactate and pyruvate concentration in the rat brain. *Scand. J. Clin. Lab. Invest.* 27, 83–96.

Siesjo, B. K., Johannson, H., Ljunggren, B., and Norberg, K. (1974) Brain dysfunction in cerebral hypoxia and ischemia, in *Brain dysfunction in metabolic disorders,* ed., Plum, F. (Raven Press, New York), pp. 75–112.

Siesjo, B. K. (1978) Metabolism of substrates in the brain, in *Brain energy metabolism,* ed., Siesjo, B. K. (Wiley & Sons, New York), pp. 151–209.

Siesjo, B. K. (1978a) Circulation and oxygen consumption in the brain, in *Brain energy metabolism,* ed., Siesjo, B. K. (Wiley & Sons, New York), pp. 56–100.

Siesjo, B. K. (1978b) Cellular work in the brain, in *Brain energy metabolism,* ed., Siesjo, B. K. (Wiley & Sons, New York), pp. 29–55.

Siesjo, B. K., Berntman, L., and Rehncrona, S. (1979) Effect of hypoxia on blood flow and metabolic flux in the brain. Adv. Neurol. 26, 267–283.

Siesjo, B. K. (1981) Cell damage in the brain: a speculative synthesis. *J. Cereb. Blood Flow Metab.* 1, 155–185.

Siesjo, B. K. (1988) Acidosis and ischemic brain damage. *Neurochem. Pathol.* 9, 31–88.

Siesjo, B. K. (1988a) Historical overview: calcium, ischemia and death of brain cells. *Ann. N.Y. Acad. Sci.* 522, 638–661.

Siesjo, B. K., Agardh, C. D., and Bengtsson, F. (1989) Free radicals and brain damage. *Cerebrovasc. Brain Metab. Rev.* 1, 165–211.

Siesjo, B. K. and Bengtsson, F. (1989) Calcium fluxes, calcium antagonists and calcium-related pathology in brain ischemia, hypoglycemia, and spreading depression: a unifying hypothesis. *J. Cereb. Blood Flow Metab.* 9, 127–140.

Siesjo, B. K. (1990) Calcium, excitotoxines and brain damage. *NIPS* 5, 120–125.

Silver, I. A. and Erecinska, M. (1990) Intracellular and extracellular changes of [Ca^{++}] in hypoxia and ischemia in rat brain *in vivo*. *J. Gen. Physiol.* 95, 837–866.

Simon, A. J. and Torbati, D. (1982) Effect of hyperbaric oxygen on heart, brain and lung functions in rat. *Undersea Biomed. Res.* 9, 263–275.

Small, A. (1984) New prespectives on hyperoxic pulmonary toxicity—a review. *Undersea Biomed. Res.* 11, 1–24.

Smith, M.-L. and Siesjo, B. K. (1988) Acidosis related brain damage and delayed events. *Adv. Behav. Biol.* 35, 57–72.

Snider, G. L. and Rinaldo, J. E. (1980) Oxygen therapy in medical patients hospitalized outside the intensive care unit. *Am. Rev. Respir. Dis.* 122, 29–36.

Sokoloff, L. (1976) Circulation and energy metabolism of the brain, in *Basic neurochemistry*, eds., Siegel, G. L., Albers, R. W., Katzman, R., and Agranoff, B. W. (Little Brown Co., Boston), 2nd ed., pp. 388–413.

Sokoloff, L. (1977) Toxic effect of elevated oxygen tension on brain maturation, in *Brain, fetal and infant*, ed., Berenberg, S. R. (Martinus Nijhoff Med. Div, The Hague), pp. 178–183.

Somjen, G. G. (1988) Basic mechanisms in cerebral hypoxia and stroke. *Adv. Behav. Biol.* 35, 447–466.

Somjen, G. G., Aitken, P. G., Balestrino, M. Herreras, O., and Kawasaki, K. (1990) Spreading depression-like depolarization and selective vulnerability of neurons. A brief review. *Stroke* 21 (Suppl. III), 179–183.

Sperling, D. R. (1970) Hyperbaric oxygen toxicity in newborn and developing mice. *J. Appl. Physiol.* 29, 472–474.

Stadie, M. E. and Haugaard, N. (1945) Oxygen poisoning. V. The effect of high oxygen pressure upon enzymes succinate dehydrogenase and cytochrome oxidase. *J. Biol. Chem.* 161, 153–174.

Stahl, P. and Schwartz, A. L. (1986) Receptor mediated endocytosis. *J. Clin. Invest.* 77, 657–662.

Stefanovic-Racic, M., Stadler, J., and Evans, C. H. (1993) Nitric oxide and arthritis. *Arthritis Rheum.* 36, 1036–1044.

Stevens, J. B. and Autor, A. P. (1980) Proposed mechanism for neonatal rat tolerance to normobaric hyperoxia. *Fed. Proc.* 39, 3138–3143.

Stock, M. K., Silvernail, K. K., and Metcalfe, J. (1990) Prenatal oxidative stress. I. Malondialdehyde in hypoxic and hyperoxic chick embryos. *Free Radic. Biol. Med.* 8, 313–318.

Strauss, R. H. (1979) Diving medicine. *Am. Rev. Respir. Dis.* 119, 1001–1023.

Sugano, T., Oshino, N., and Chance, C. (1974) Mitochondrial functions under hypoxic conditions: the steady states of cytochrome c reduction and of energy metabolism. *Biochem. Biophys. Acta* 347, 340–358.

Sugar, O. and Gerard, R. W. (1938) Anoxia and brain potentials. *J. Neurophysiol.* 1, 558–572.

Szabo, C., Csaki, C., Benyo, Z., Reivich, M., and Kovach, A. G. B. (1992) Role of the L-arginine-nitric oxide pathway in the changes in cerebrovascular reactivity following hemorrhagic hypotension and retransfusion. *Circ. Shock* 37, 307–316.

Talbot, J. H. and Dill, D. B. (1936) Clinical observations at high altitude; observations on six healthy persons living at 17,500 feet and a report of one case of chronic mountain sickness. *Am. Med. Sci.* 192, 626–639.

Thom, S. R. (1989) Carbon monoxide poisoning: a review of epidemiology, pathophysiology, clinical findings, and treatment options including hyperbaric oxygen therapy. *Clin. Toxicol.* 27, 141–156.

Thom, S. R. (1990) Antagonism of carbon monoxide-mediated brain lipid peroxidation by hyperbaric oxygen. *Toxicol. Appl. Pharmacol.* 105, 340–344.

Thom, S. R. (1992) Inert gas enhancement of superoxide radical formation. *Arch. Biochem. Biophys.* 295, 391–396.

Thom, S. R. (1993) Editorial response: a role for hyperbaric oxygen in clostridial myonecrosis. *Clin. Infect. Dis.* 17, 238.

Thompson, R. E., Neilson, T. W., and Ackers, T. K. (1970) Synergistic oxygen-inert gas interactions in laboratory rats in a hyperbaric environment. *Aeros. Med.* 41, 1388–1392.

Thorsen, E., Segadal, K., Reel, J. W., Elliott, C., Gulsick, A., and Hjelle, J. O. (1993) Contribution of hyperoxia to reduced pulmonary function after deep saturation dives. *J. Appl. Physiol.* 75, 657–662.

Tombaugh, G. C. and Sapolsky, R. M. (1990) Mild acidosis protects hippocampal neurons from injury induced by oxygen and glucose deprivation. *Brain Res.* 506, 343–345.

Tominaga, T., Katagi, K., and Ohnishi, S. T. (1988) Is Ca^{++}-activated K efflux involved in the formation of ischemic brain edema? *Brain Res.* 460, 376–378.

Tominaga, T., Sato, S., Ohnishi, T., and Ohnishi, S.T. (1993) Potentiation of nitric oxid formation following bilateral carotid occlusion and focal cerebral ischemia in the rat: *in vivo* detection of the nitric oxide radical by electron paramagnetic resonance spin trapping. *Brain Res.* 614, 342–346.

Tominaga, T., Sato, S., Ohnishi, T. and Ohnishi, S. T. (1994) Electron paramagnetic resonance detection of nitric oxide produced during focal ischemia in the rat. *J. Cereb. Blood Flow Metab.* 14, 715–722.

Torbati, D., Harel, D., and Lavy S. (1971) Influence of adrenalectomy on electrical activity of the brain under high oxygen pressure. *Aerosp. Med.* 42, 658–660.

Torbati, D., Harel, D., and Lavy, S. (1972) The time of appearance of paroxismal electrical discharges in normal and hypophysectomized rats under hyperbaric conditions. *Isr. J. Med. Sci.* 8, 39–42.

Torbati, D., Parolla, D. and Lavy S. (1976) Changes in electrical activity and PO_2 of rats brain under high oxygen pressure. *Exp. Neurol.* 50, 439–447.

Torbati, D., Parolla, D., and Lavy, S. (1978) Blood flow in rat brain during exposure to high oxygen pressure. *Aviat. Space Environ. Med.* 49, 963–967.

Torbati, D., Simon, A. J., and Ranade, A. (1981) Frequency analysis of EEG in rats during the preconvulsive period of oxygen poisoning. *Aviat. Space Environ. Med.* 52, 598–603.

Torbati, D., Greenberg, J., and Lambertsen, C. J. (1983) Correlation of brain glucose utilization and cortical electrical activity during development of brain oxygen toxicity. *Brain Res.* 262, 267–273.

Torbati, D. and Lambertsen, C. J. (1983) Regional cerebral metabolic rate for glucose during oxygen-induced convulsions. *Brain Res.* 279, 382–386.

Torbati, D., Lambertsen, C. J., and Greenberg, J. (1984) Regional cerebral glucose utilization rates in rats during asymptomatic period of exposure to 1, 2, and 3 atmospheres absolute oxygen. *Neuroscience* 11, 947–950.

Torbati, D., Lo, P., and Lambertsen, C. J. (1984a) Local cerebral glucose utilization rate following intermittent exposures to 2 atmospheres absolute oxygen. *Neurosci. Lett.* 50, 79–84.

Torbati, D. and Lambertsen, C. J. (1985) The relationship between cortical electrical activity and regional cerebral glucose metabolic rate in rats exposed to 3 atmospheres absolute oxygen. *Brain Res.* 344, 186–190.

Torbati, D., Lambertsen, C. J., and Greenberg, J. (1986) Regional cerebral glucose metabolic rate during 30 minutes hypoxia of 7% oxygen in adult rats. *Neurosci. Lett.* 65, 253–258.

Torbati, D. and Lambertsen, C. J. (1987) Effects of hyperbaric oxygenation on regional cerebral glucose utilization rate in awake rat. A review, in *Proc. 8th Int. Cong. Hyperbaric Med.*, eds., Kindwall, G., Hart, G., and Jacobson, J. (Best Publishing Co., San Pedro, CA), pp. 48–56.

Torbati, D. (1987) Oxygen and brain physiologic functions: a review, in *Undersea and hyperbaric physiology IX*, eds., Bove, A. A., Bachrach, A. J., and Greenbaum, A. J. (Undersea and Hyperbaric Med. Soc., Bethesda), pp. 659–690.

Torbati, D. and Reilly, K. J. (1988) Effect of prolonged normobaric hyperoxia on regional metabolic rate for glucose in conscious rats. *Brain. Res.* 459, 187–191.

Torbati, D., Farrell, J. B., Davidson, J. F., and Carey, M. E. (1989) Effect of hemorrhagic hypotension on CBF and intracranial pressure after brain missile wounding in anesthetized paralyzed cats, in *ICP VII*, eds., Hoff, J. T. and Betz, A. L. (Springer-Verlag, New York), pp. 657–659.

Torbati, D. and Carey, M. E. (1989a) Effect of normobaric hyperoxia on regional cerebral blood flow before and after brain missile wounding in anesthetized cats. *Undersea Biomed. Res. Suppl.* 16, 78.

Torbati, D., Church, D. F., Keller, J. M., and Pryor, W. A. (1992) Free-radical generation in the brain precedes hyperbaric oxygen-induced convulsions. *Free Radic. Biol. Med.* 13, 101–106.

Torbati, D., Jacks, A. F., Carey, M. E., Davidson, J. F., and Farrell, J. B. (1992a) Cerebral cardiovascular and respiratory variables after an experimental brain missile wound. *J. Neurotrauma* 9, Suppl. S143–S155.

Torbati, D., Sherpa, A. K., Lahiri, S., Mokashi, A., Albertine, K. H., and DiGiulio, C. (1993). Hyperbaric oxygenation alters carotid body ultrastructure and function. *Respir. Physiol.* 92, 183–196.

Torbati, D., Wafapoor, H., and Peyman, G. A. (1993a) Hyperbaric oxygen tolerance in newborn mammals; hypothesis on mechanisms and outcome. *Free Radic. Biol. Med.* 13, 101–106.

Torbati, D., Peyman, G. A., Rodriguez, A. J., Wafapoor, H., and Navarro, G. C. (1995) Modulation of sensitivity to hyperbaric oxygenation by CO_2 in newborn rats. *Undersea and Hyperbaric Med.*, in press.

Torbati, D. Peyman, G. A. Wafapoor, H., Shaibani, S. B., and Khoobehi, B. (1995a) Experimental retinopathy by hyperbaric oxygenation. *Undersea and Hyperbaric Med.*, 22, 31–39.

Turrens, J. F., Crapo, J. D., and Freeman, B. A. (1984) Protection against oxygen toxicity by intravenous injection of liposome-entrapped catalase and superoxide dismutase. *J. Clin. Invest.* 73, 87–95.

Ueno, K., Tanaka, Y., and Ando, S. (1988) Release of free fatty acids from rat cerebral synaptosomes under *in vitro* hypoxia. *Neurosci. Lett.* 84, 297–301.

Umemura, A., Mabe, H., Nagai, H., and Sugino, F. (1992) Action of phospholipases A_2 and C on free fatty acid release during complete ischemia in rat neocortex. *J. Neurosurg.* 76, 648–651.

Unterberg, A., Polk, T., Ellis, E., and Marmarou, A. (1990) Enhancement of infusion-induced brain edema by mediator compounds. *Adv. Neurol.* 52, 355–358.

Van Liew, H. D. and Rodgers, B. (1975) Variability of partial pressure of CO_2 and O_2 in resting skeletal muscles. *Microvasc. Res.* 10, 127–138.

Vannucci, R. C. (1978) Neurologic aspects of perinatal asphyxia. *Pediatr. Ann.* 7, 15–30.

Vannucci, R. C. (1990) Experimental biology of cerebral hypoxia-ischemia: Relation to perinatal brain damage. *Pediatr. Res.* 27, 317–326.

Van Wylen, D. G., Park, S., Rubio, R., and Berne, R. M. (1989) The effect of local infusion of adenosine and adenosine analogues on local cerebral blood flow. *J. Cereb. Blood Flow Metab.* 9, 556–562.

Webb, J. T., Olsn, R. M., Krutz, R. W., Jr., Dixon, G., and Barnicott, P. T. (1989) Human tolerance to 100% oxygen at 9.5 psi during five daily simulated 8-h EVA exposures. *Aviat. Space Environ. Med.* 60, 415–421.

Webster, N. R., Toothill, C., and Cowen, P. N. (1987) Tissue responses to hyperoxia: biochemistry and pathology. *Br. J. Anaesth.* 59, 760–771.

Welch, B. E., Morgan, T. E., Jr., and Clamann, H. G. (1963) Time concentration effects in relation to oxygen toxicity in man. *Fed. Proc.* 122, 1053–1055.

Welsh, F. A., Ginsberg, M. D., Reider, W., and Budd, W. W. (1978) Diffuse cerebral ischemia in the cat. II. Regional metabolites during severe ischemia and recirculation. *Ann. Neurol.* 3, 493–501.

Whalen, W. J., Ganfield, R., and Nair, P. (1970) Effects of breathing O_2 or O_2 + CO_2 and the injection of neurohumors on the PO_2 of cat cerebral cortex. *Stroke* 1, 194–200.

Wheeler, D. D., Blackburn, J. G., and Tai, J. J. (1983) Effect of hyperbaric oxygenation on GABA transport in rat brain synaptosomes. *J. Neurol. Transm.* 57, 167–185.

Whittle, I. R., Piper, I. R., and Miller, J. D. (1990) The contribution of secondary mediators to the etiology and pathophysiology of brain. oedema: studies using feline infusion oedema model. *Acta Neurochir. Suppl.* 51, 71–73.

Widell, P. J., Bennett, P. B., Kivlin, P., and Gray, W. (1974) Pulmonary oxygen toxicity in man at 2 ATA with intermittent air breathing. *Aerosp. Med.* 45, 407–410.

Williams, C. E. (1958) Retrolental fibroplasia in association with mental disease. *Br. J. Ophthalmol.* 42, 549–557.

Wilson, D. F., Owen, C. S., and Holian, A. (1977) Control of mitochondrial respiration, a quantitative evaluation of the roles of cytochrome-c and oxygen. *Arch. Biochem. Biophys.* 182, 749–762.

Wilson, D. F., Erecinska, N., Drown, C., and Silver, I. A. (1977a) Effect of oxygen tension on cellular energetics. *Am. J. Physiol.* 233, C135–C140.

Wilson, D. F., Erecinska, M., Drown, C., and Silver, I. A. (1979) The oxygen dependence of cellular energy metabolism. *Arch. Biochem. Biophys.* 195, 485–493.

Winn, H. R., Rubio, R., and Berne, R. M. (1981) Brain adenosine concentration during hypoxia in rats. *Am. J. Physiol.* 241, H235–H242.

Wisner, D., Busche, F., Sturm, J., Gabb, M., and Meyer, H. (1989) Traumatic shock and head injury: effects of fluid resuscitation on the brain. *J. Surg. Res.* 46, 49–59.

Wood, J. D. and Watson, W. J. (1963) Gamma-aminobutyric acid levels in the brain of rats exposed to oxygen at high pressures. *Can. J. Biochem. Physiol.* 41, 1907–1913.

Wood, J. D., Watson, W. J., and Clydesdale, F. M. (1963) Gamma-aminobutyric acid and oxygen poisoning. *J. Neurochem.* 10, 625–633.

Wood, J. D., Radomski, M. W., and Watson, W. J. (1971) A study of possible biochemical mechanisms involved in hyperbaric oxygen-induced changes in cerebral gamma-aminobutyric acid levels and accompanying seizures. *Can. J. Biochem.* 49, 543–547.

Wood, J. D., Watson, W. J., and Murray, G. W. (1969) Correlation between decreases in brain GABA levels and susceptibility to convulsions induced by hyperbaric oxygen. *J. Neurochem.* 16, 281–287.

Xia, Y., Jiang, C., and Haddad, G. G. (1992) Oxidative and glycolytic pathways in rat (newborn and adult) and turtle brain: role during anoxia. *Am. J. Physiol.* 262, (Regulat. Integrat. Comp. Physiol. 31): R595–R603.

Young, W. (1992) Role of calcium in central nervous system injuries. *J. Neurotrauma* 9, Suppl. 1, S9–S25.

Young, R. S. K., During, M. J., Donnelly, D. F., Aquila, W. J., Perry, V. L., and Haddad, G. G. (1992) Effect of anoxia on excitatory amino acids in brain slices of rats and turtles: *in vitro* microdialysis. *Am. J. Physiol.* 264, R716–R719.

Yusa, T., Crapo, J. D., and Freeman, B. A. (1984) Liposome-mediated augmentation of brain SOD and catalase inhibits CNS O_2 toxicity. *J. Appl. Physiol. Respir. Environ. Physiol.* 57, 1674–1681.

Yusa, T., Beckman, J. S., Crapo, J. D., and Freeman, B. A. (1987) Hyperoxia increases H_2-O_2 production by brain *in vivo*. *J. Appl. Physiol.* 63, 353–358.

Zhang, J., Su, Y., Oury, T. D., and Piantadosi, C. A. (1993) Cerebral amino acids, norepinephrine and nitric oxide metabolism in CNS oxygen toxicity. *Brain Res.* 606, 56–62.

Zirkle, L. G., Jr., Mengel, C. E., Horton, B. B., and Duffy, E. J. (1965) Studies on oxygen toxicity in the central nervous system. *Aerosp. Med.* 36, 1027–1032.

Chapter 3

ROLE OF CALCIUM IN CNS TRAUMA

Mark P. Mattson

TABLE OF CONTENTS

I. INTRODUCTION

It is beyond the scope of this chapter to review the evidence supporting the involvement of calcium in the neuronal damage that results from traumatic injury to the CNS. Recent excellent reviews include those of Waxman et al. (1991), Young (1992), and Siesjo (1993). Here the focus is on emerging concepts of the role of calcium regulation and free radical metabolism in neuronal injury and neuroprotection with an emphasis on work from this laboratory. The spinal cord is not considered here, with the presumption that similar mechanisms are operative in the brain and spinal cord. Because the preponderance of data indicates that mechanisms of traumatic CNS injury are similar to those in ischemic brain injury, this chapter draws on the larger body of data from ischemia research. Finally, it is assumed that the neuronal cell injury in CNS trauma involves an excitotoxic component. To aid the would-be CNS trauma researcher, several specific examples of technical approaches to the study of neural injury are presented.

II. EVENTS INITIATED BY CNS TRAUMA THAT LEAD TO DISRUPTION OF CALCIUM HOMEOSTASIS

Metabolic activity is rapidly depressed (minutes) following CNS injury with concomitant edema and hemorrhage. Reduced energy availability to neurons results in ATP depletion,

0-8493-8094-4/95/$0.00+$.50

which compromises function of the plasma membrane and endoplasmic reticulum Ca^{2+}-ATPases (Figure 1). Injured cells release K^+ resulting in membrane depolarization and Ca^{2+} influx through voltage-dependent channels. Elevation of intracellular free-calcium levels ($[Ca^{2+}]_i$) stimulates glutamate release, which activates *N*-methyl-D-aspartate (NMDA) and α-amino-3-hydroxy-5-methyl-4-isoxazole-propionate (AMPA) receptors leading to further elevation of $[Ca^{2+}]_i$ (Faden et al., 1989; Bullock and Fujisawa, 1992; Katayama et al., 1990; Nilsson et al., 1990; Hayes et al., 1992). Therefore, neurons that express high levels of excitatory amino acid receptors are particularly vulnerable. Intracellular calcium overload

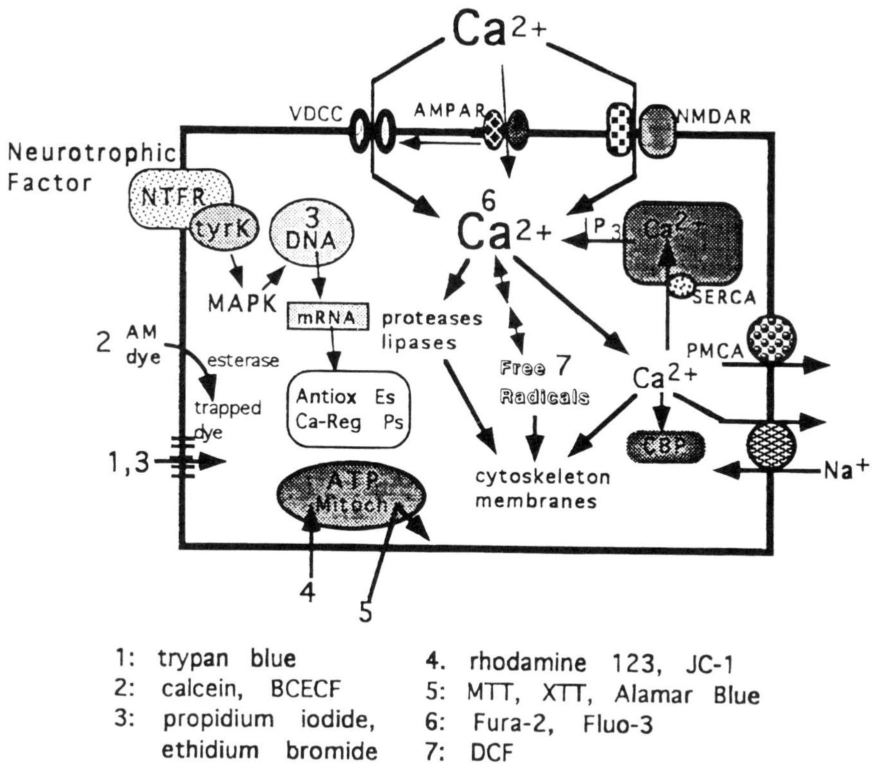

FIGURE 1. Approaches for examining the roles of calcium and free radicals in traumatic injury to CNS neurons. Activation of glutamate receptors results in Ca^{2+} influx through *N*-methyl-D-aspartate receptors (NMDAR) and voltage-dependent Ca^{2+} channels (VDCC) secondary to membrane depolarization mediated predominantly by non-NMDA ionotropic (AMPA) receptors. Elevation of cytosolic free-Ca^{2+} levels is normally transient due to the activity of: the plasma membrane Ca^{2+}-ATPase (PMCA); plasma membrane Na^+-Ca^{2+} exchanger, endoplasmic reticulum Ca^{2+}-ATPase (SERCA); and Ca^{2+}-binding proteins (CBP). Overactivation of glutamate receptors results in sustained elevations of $[Ca^{2+}]_i$, which can directly activate proteases and can promote the formation of free radicals by activating lipases. Both Ca^{2+} and free radicals cause damage to structural proteins, enzymes, and membranes, which can lead to loss of ion homeostasis and cell degeneration. Neurotrophic factors bind to receptors (NTFR) that have tyrosine kinase activity (tyrK). Activation of NTFR activates MAP kinases (MAPK), which affect the activity of transcription factors involved in the regulation of genes encoding proteins involved in calcium and free-radical metabolism. Probes for assessing cell viability include: trypan blue, propidium iodide, and ethidium bromide, which are only taken up in cells with a damaged plasma membrane; the acetyoxymethylester (AM) forms of calcein and BCECF, which will only accumulate in healthy cells with esterase activity; rhodamine 123 and JC-1, whose uptake into mitochondria is dependent on transmembrane potential; MTT and XTT, which are reduced in mitochondria and form a precipitate; and alamar blue, which is reduced in mitochondria and released into the culture medium. Intracellular free-calcium levels can be assessed using the Ca^{2+} indicator dyes fura-2 and fluo-3. Levels of cellular peroxides can be assessed using 2,7-dichlorofluorescin (DCF). (Modified from Mattson, *M.P., Methods Cell. Biol.*, 46, 187–216.)

causes neuronal death by multiple mechanisms, the most prominent being promotion of free radical production and activation of proteases.

Calcium can promote the production of free radicals in several ways. Calcium activates phospholipase A_2 resulting in the accumulation of arachidonic acid, and subsequent oxidation by cyclooxygenase and lipoxygenase (Verity, 1993). Cyclooxygenase inhibitors can improve neurological recovery following CNS injury (Faden and Salzman, 1992) and the lipoxygenase inhibitor nordihydroguaiaretic acid can protect CNS neurons against excitotoxicity (Rothman et al., 1993). The conversion of xanthine dehydrogenase to xanthine oxidase by Ca^{2+}-activated proteolysis and subsequent catalysis of the oxidation of hypoxanthine or xanthine to uric acid releases superoxide radical (McCord, 1985). Stimulation of nitric oxide synthase by Ca^{2+}-calmodulin results in nitric oxide production and nitric oxide reacts with superoxide anion leading to the formation of peroxynitrite, a highly destructive radical (Lipton et al., 1993).

Intervention studies have clearly shown that both calcium and free radicals are causally involved in the neuronal injury and death resulting from traumatic (metabolic/excitotoxic) injury to the CNS. For example, NMDA and AMPA receptor antagonists and Ca^{2+} channel blockers reduced neuronal injury (for review, see Choi, 1990; Faden and Salzman, 1992). High doses of magnesium significantly improved the neurologic outcome following experimental traumatic brain injury (McIntosh et al., 1989). Anticonvulsants (e.g., phenytoin and carbamazepine) protected cultured hippocampal neurons against excitotoxic insults (Mattson and Kater, 1989). The severity of CNS trauma can be reduced by antioxidants including vitamin E and 21-aminosteroids (Clifton et al., 1989; Anderson et al., 1991; Hall et al., 1992). An iron chelator (iron promotes free radical production) was effective in enhancing neurologic outcome following CNS injury (Panter et al., 1992). Ischemic brain injury in rodents was reduced by the spin-trapping compound *N-tert*-butyl-α-phenylnitrone (Carney et al., 1991) and by polyethylene glycol-conjugated superoxide dismutase (He et al., 1993).

III. MECHANISMS OF CNS TRAUMA AND CHRONIC NEURODEGENERATIVE DISORDERS

CNS trauma is most commonly considered as an isolated incident of considerable severity that results in the rapid destruction of neural tissue. Less appreciated is traumatic injury to the CNS involving minor "incidents" that occur over periods of years and decades. The incidents may involve conventional trauma (e.g., a blow to the head) or metabolic incidents (e.g., transient ischemic episodes or hypoglycemia as in insulin-dependent diabetics). Alzheimer's disease is characterized by the presence of neurofibrillary tangles and senile plaques in the brain that are particularly prominent in regions involved in the processes of learning and memory (e.g., entorhinal cortex, hippocampus, and association cortices). Neurofibrillary tangles consist of abnormal accumulations of cytoskeletal filaments comprised of the microtubule-associated protein tau, and senile plaques that consist of aggregated amyloid β-peptide (Aβ) (for review, see Selkoe, 1991). The source of the 40 to 42 amino acid peptide Aβ is the β-amyloid precursor protein (βAPP), a large (695 to 770 amino acid) transmembrane protein. An enzymatic cleavage in the middle of Aβ precludes deposition of Aβ, but an alternative processing pathway results in release of soluble Aβ, which has the potential to aggregate (Selkoe, 1993). Increasing evidence suggests a role for perturbed metabolism of βAPP and loss of neuronal calcium homeostasis in the pathogenesis of Alzheimer's disease (for review see Mattson et al., 1993a). Aβ can be directly neurotoxic and can render neurons vulnerable to excitoxic and metabolic insults. The mechanism of Aβ toxicity involves disruption of calcium homeostasis resulting in elevated rest and enhanced sensitivity to excitation (Mattson et al., 1992, 1993b). Aggregation of Aβ is required for the calcium-destabilizing and neurotoxic activities (Pike et al., 1993), and the plasma membrane appears to be the site of Aβ action since that is where the Aβ accumulates (Mattson et al., 1993b). We recently discovered that when Aβ is placed in solution it fragments and forms free radical peptides, which may promote cross-linking (aggregation) of the peptide (Hensley et al., 1994). Radicals formed during the fragmentation/aggregation process can damage cellular enzymes including plasma membrane proteins involved in maintenance of calcium homeostasis (Hensley et al., 1994; R. J. Mark and M. P. Mattson, unpublished data). Aβ toxicity can be attenuated by antioxidants (Behl et al., 1992; Goodman and Mattson, 1994) and growth factors that stabilize calcium homeostasis (Mattson et al., 1993b).

There is a greater incidence of Alzheimer's disease in individuals such as boxers with a known history of head trauma. Roberts et al. (1991; Gentleman et al., 1993) reported a strong correlation between history of head trauma and the number of Aβ deposits and neurofibrillary tangles in the brain at autopsy. Presumably, the traumatic incidents caused deposition of Aβ, which then contributed to the neurofibrillary degeneration. In fact, increasing data indicate that the mechanism of neuronal injury in Alzheimer's disease is similar to that in traumatic CNS injury involving metabolic/excitotoxic compromise and mediated by calcium and free radicals. For example, glutamate can induce antigenic, biochemical, and ultrastructural changes in cultured rat and human CNS neurons similar to those seen in the neurofibrillary tangles of Alzheimer's disease (Mattson, 1990; Mattson et al., 1991a; Sautierre et al., 1992). Similar neurofibrillary tangle-like alterations can be induced *in vivo* by injection of kainic acid into the hippocampus and these alterations are exacerbated by glucocorticoids, which impair energy availability to neurons (Elliott et al., 1993). In addition, energy availability to brain cells may be reduced in Alzheimer's disease (Hoyer et al., 1988) and recent findings indicate increased levels of protein oxidation (increased free radicals) in the brain in Alzheimer's disease (Smith et al., 1991).

IV. ENDOGENOUS NEUROPROTECTION FACTORS: ROLES IN REGULATING CALCIUM AND FREE RADICAL METABOLISM

Neurons possess very sophisticated systems for regulating levels of calcium (for review, see Tsien and Tsien, 1990) and free radicals (for review, see Jesberger and Richardson, 1991). These various systems are subject to regulation by neuroprotection factors (NPFs). The rest $[Ca^{2+}]_i$ in CNS neurons is maintained in the range of 75 to 150 n*M* by activity of the plasma membrane and endoplasmic reticulum Ca^{2+}-ATPases. Following an excitatory stimulus cytoplasmic free-calcium levels are rapidly reduced by activity of the plasma membrane Na^+-Ca^{2+} exchanger and by sequestration by calcium-binding proteins. Free radicals are constantly being produced in the body during oxidation-reduction reactions. Cells have evolved mechanisms to protect themselves from free radicals and antioxidant enzymes and free radical scavengers are employed for this purpose. Enzymes that suppress free radicals include superoxide dismutase, catalase, glutathione reductase, and glutathione peroxidase. Free radical scavengers include vitamin E (α-tocopherol), glutathione, and beta carotene. In addition, proteins that bind transition metals such as iron (e.g., ferritin) play an important role in protecting cells against metal-catalyzed oxidation.

Data is rapidly accumulating to support the hypothesis that two general mechanisms of NPF action are stabilization of neuronal calcium homeostasis and stimulation of antioxidant systems (for review, see Mattson et al., 1993c). Basic fibroblast growth factor (bFGF) can protect cultured hippocampal, cortical, septal, and striatal neurons against glutamate toxicity and glucose deprivation-induced injury (Mattson et al., 1989a; Cheng and Mattson, 1991; Freese et al., 1992). Ischemic and excitoxin-induced injury *in vivo* is also attenuated by bFGF (Nozaki et al., 1993). Neurotrophins including nerve growth factor (NGF), brain-

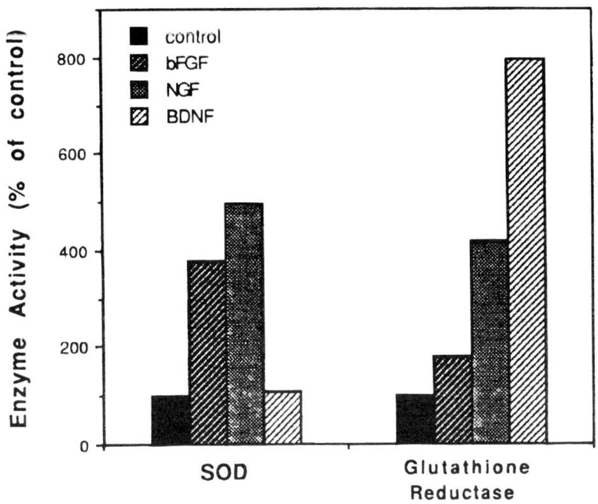

FIGURE 2. Neuroprotective factors increase levels of antioxidant enzyme activity in rat neocortical cell cultures. Dissociated cell cultures of embryonic rat cerebral cortex (7 d in culture; for culture methods, see Cheng and Mattson, 1991) were exposed for 24 h to bFGF (10 ng/ml), NGF (10 ng/ml), or BDNF (100 ng/ml). Activities of superoxide dismutase (SOD) and glutathione reductase in cell homogenates were determined. Note that bFGF and NGF greatly increased SOD activity and all three neurotrophic factors increased glutathione reductase levels to varying degrees.

derived neurotrophic factor (BDNF), and neurotrophin-3 (NT-3) protected cultured hippocampal neurons against glucose deprivation-induced injury (Cheng and Mattson, 1991, 1994). NGF was effective in reducing neuronal damage in animal models of ischemic brain injury (Shigeno et al., 1991) and excitotoxicity (Frim et al., 1993). NGF, bFGF, and insulin-like growth factors (IGFs) protected cultured rat hippocampal neurons against iron-induced (free-radical-mediated) injury (Zhang et al., 1993). BDNF protected mesencephalic neurons against MPTP toxicity in a rodent model of Parkinson's disease (Spina et al., 1992). Traumatic injury to neurons is also allayed by growth factors. For example, NGF (Hefti et al., 1989) and bFGF (Anderson et al., 1988) prevent the degeneration of axotomized septal cholinergic neurons in rats.

Specific mechanisms whereby NPFs protect against metabolic/traumatic injury are beginning to be elucidated (Table 1). Increased expression of calcium-binding proteins is stimulated by several neuroprotective factors including NGF, bFGF, NT-3, and tumor necrosis factors (TNF; Masiakowski and Shooter, 1988; Collazo et al., 1992; Cheng et al., 1994a). Calbindin, a 28-kDa calcium-binding protein induced by bFGF, NT-3, and TNF is believed to confer resistance of neurons to excitotoxicity (Mattson et al., 1991b; Scharfman and Schwartzkroin, 1989). NPFs may also affect the expression of excitatory amino acid receptors as we recently demonstrated in hippocampal cell cultures where bFGF reduced expression of an NMDA receptor protein (Mattson et al., 1993d). Many NPFs are now known to increase the expression of antioxidant enzymes. For example, Nistico et al. (1992) showed that NGF can increase levels of superoxide dismutase and glutathione peroxidase in aging rats. NGF also increased levels of glutathione in PC12 cell line (Pan and Perez-Polo, 1993). We have found that NGF, BDNF, and bFGF can increase superoxide dismutase and glutathione reductase levels to varying degrees in embryonic rat neocortical cell cultures (Figure 2).

The kinds of effects of NPFs on calcium homeostasis and free radical metabolism just described strongly suggest that these and other factors will prove effective in reducing neuronal injury in various animal models of traumatic injury to the CNS. Moreover, a more detailed understanding of the signal-transduction mechanisms of these NPFs should allow

TABLE 1
Examples of Effects of Neurotrophic Factors on the Expression of Proteins Involved in Calcium Homeostasis and Free-Radical Metabolism

Trophic factor	Protein	Effect	Ref.
Basic FGF	Calbindin	Increase	Collazo et al., 1992
	NMDARP-71	Reduce	Mattson et al., 1993
	SOD	Increase	Figure 2
	Glutathione R	Increase	Figure 2
NGF	CBPs	Increase	Masiakowski and Shooter, 1988
	SOD	Increase	Nistico et al., 1991
	Glutathione	Increase	Pan and Perez-Polo, 1993
	Glutathione R	Increase	Figure 2
BDNF	Glutathione R	Increase	Figure 2
NT-3	Calbindin	Increase	Collazo et al., 1992
TNFs	Calbindin	Increase	Cheng et al., 1994a

Note: See Table 2 for abbreviations of trophic factors. NMDARP-71, 71 kDa NMDA receptor protein; SOD, superoxide dismutase; Glutathione R, glutathione reductase; CBPs, calcium-binding proteins.

development of novel therapeutic strategies for traumatic CNS injury (Table 2). A recent example comes from studies in which it was found that staurosporine and K-252 compounds, low-molecular weight agents of bacterial origin, can protect CNS neurons against metabolic/excitotoxic insults (Roberts-Lewis and Siman, 1993; Cheng et al., 1994b). Although these compounds were originally identified as protein kinase inhibitors, we have found that their neuroprotective action results from their ability (at low concentrations, picomolar) to activate tyrosine kinase signal-transduction cascades (Cheng et al., 1994b). Staurosporine and the K-252 compounds induced tryosine phosphorylation of multiple proteins, and the neuroprotective action of these compounds was blocked by the tryosine kinase inhibitor genistein indicating that tyrosine phoshorylation was required for their actions. The latter action of these compounds appears to mimic the action of neurotrophic factors at receptor tyrosine kinases (Schlessinger and Ulrich, 1992). The cell culture studies showed that staurosporine and K-252 compounds stabilized $[Ca^{2+}]_i$, effectively attenuating glucose deprivation-induced elevation of $[Ca^{2+}]_i$ (Cheng et al., 1994b). The latter findings are consistent with the observa-

TABLE 2
Examples of Agents Shown to be Effective in Protecting CNS Neurons against Calcium- and Free-Radical-Mediated Injury

Target	Action	Examples
Glutamate receptor	Antagonists	MK-801, CNQX
VDCC	Antagonists	Nifedipine, omega conotoxin
GABAR (anticonvulsants)	Agonists	Valproic acid, phenytoin carbamazepine
Acetylcholine receptors	Antagonists	Scopalamine, atropine
Potassium channels	Openers	Diazoxide, pinacidil, cromakalim
Free radicals	Antioxidants	Vitamin E, spin traps
Neurotrophic factors	Stabilize $[Ca^{2+}]_i$ and free radicals	bFGF, NGF, BDNF, NT-3 IGFs, CNTF, TNFs
Alkaloids	Growth factor mimicks	Staurosporine, K-252 compounds
Arachidonic acid products	Enzyme inhibitors	NDGA, indomethacin

Note: VDCC, voltage-dependent calcium channels; GABAR, GABA receptor; MK-801, dizocilpine maleate; CNQX, 6-cyano-7-nitroquinoxaline-2,3-dione; bFGF, basic fibroblast growth factor; NGF, nerve growth factor; BDNF, brain-derived neurotrophic factor; NT-3, neurotrophin-3; IGFs, insulin-like growth factors; CNTF, ciliary neurotrophic factor; TNFs, tumor necrosis factors; NDGA, nordihydroguaiaretic acid.

tion that K-252a reduces kainic acid-induced spectrin proteolysis (a Ca^{2+}-mediated process) in the hippocampus *in vivo* (Roberts-Lewis and Siman, 1993).

V. TECHNICAL APPROACHES TO ELUCIDATING MECHANISMS OF TRAUMATIC CNS INJURY

The methods that follow have been used extensively in the author's laboratory and further details and examples of data obtained using these methods can be found in original articles (Mattson et al., 1988; 1989b; 1993e; Goodman and Mattson, 1994) and a recent methods paper (Mattson et al., 1994).

A. MEASUREMENT OF INTRACELLULAR CALCIUM LEVELS

Fluorescent calcium indicator dyes are a valuable, and in many cases required, tool for studying the role of calcium in neuronal injury. Fura-2 is the most commonly used dye used for calcium imaging in cell culture studies (Grynkiewicz et al., 1985; Malgaroli et al., 1987). Cells are grown on glass (we use commercially prepared glass-bottom 35-mm dishes; MatTek, Inc.). The dye is introduced into the cells by incubating for 20 to 40 min in the presence of 5 to 10 μM of the acetoxymethyl ester (membrane-permeant) form of fura-2 (Molecular Probes, Inc.). Cells are washed with fresh medium and allowed to incubate for at least 30 min to allow complete hydrolysis of the AM ester. Immediately prior to imaging, the culture medium is replaced with Hank's Balanced Saline Solution containing 20 m*M* glucose and 10 m*M* Hepes (pH 7.2). An increasing number of calcium-imaging systems are commercially available. We are currently using a Zeiss/Attofluor system, which includes a Zeiss Axiovert 100 inverted microscope, an intensified CCD camera, and a lever-action filter changer. We typically use a 40× oil immersion objective. The $[Ca^{2+}]_i$ is determined from the ratio of the emitted fluorescence acquired using two different excitation wavelengths (340 and 380 nm). The system is calibrated according to the procedure of Grynkiewicz et al. (1985) using the formula $[Ca^{2+}]_i = K_d ((R - R_{min}) / (R_{max} - R)(F_o/F_s))$ where R is the experimental ratio obtained, R_{max} is the ratio obtained from a solution of fura-2 containing calcium, R_{min} is the ratio obtained from a solution of fura-2 lacking calcium, F_o is the fluorescent emission detected from 380 nm excitation of the calcium-free solution, F_s is the fluorescent emission detected from 380 nm excitation of the calcium-containing solution, and K_d is the dissociation constant of fura-2 for Ca_{2+} (typically, 224 to 228 n*M*). An additional dye frequently used is fluo-3, which we employ in confocal laser scanning microscopy studies. This approach is particularly useful for subcellular localization of $[Ca^{2+}]_i$ and comparison to localization of organelles (identified with organelle-specific dyes) or specific proteins (localized by immunofluorescence). In addition, several investigators have used fluo-3 for relative $[Ca^{2+}]_i$ measurements in multiwell culture plates with the fluorescent plate reader described below.

B. MANIPULATION OF INTRACELLULAR CALCIUM LEVELS

There are a variety of pharmacological manipulations that can be employed to evaluate the involvement of Ca^{2+} in neuronal injury paradigms. By varying the concentration of extracellular Ca^{2+} it can be determined whether Ca^{2+} influx is required for neurotoxicity. This approach may, however, not be useful when studying relatively slow forms of neuronal degeneration that occur over periods of many hours to days (e.g., Aβ toxicity) because chronic removal of extracellular Ca^{2+} itself can be deterimental to cells. Inorganic calcium-channel blockers (e.g., Co^{2+}), organic calcium-channel blockers (e.g., nimodipine), and glutamate receptor antagonists (e.g., APV and CNQX) are also useful in evaluating the role of calcium influx in neurotoxicity. Inhibitors of calcium release from endoplasmic reticulum stores such as dantrolene and ruthenium red can be employed to establish the contribution of such calcium stores to neuronal injury (Frandsen and Schousboe, 1991). In order to establish whether elevation of $[Ca^{2+}]_i$ is sufficient to account for the neuronal injury, elevation of $[Ca^{2+}]_i$ can be induced by calcium ionophores such as A23187 and ionomycin. Selective release of Ca^{2+} from endoplasmic reticulum can be induced with thapsigargin, an inhibitor of the endoplasmic reticulum calcium-ATPase (Thastrup et al., 1990). The plasma membrane Na^+-Ca^{2+} exchanger can be inhibited by substitution of extracellular Na^+ with *N*-methyl-D-glucamine; this manipulation elevates $[Ca^{2+}]_i$ (Mattson et al., 1989b).

C. MEASUREMENT OF FREE-RADICAL LEVELS

Two methods we routinely employ to quantify levels of reactive oxygen species generated in cultured neural cells are fluorescence imaging of the oxidation-sensitive compound 2,7-dichlorofluorescin (DCF; Goodman and Mattson, 1994) and measurement of hydroxyl radical release into culture medium using the salicylate trapping method (Zhang et al., 1993). Cultured neurons are incubated in the presence of 1 to 5 μg/ml DCF, which is taken up by neurons and, when exposed to peroxides, is converted to the 2,7 DCF, which is fluorescent. Cells can be imaged using a conventional fluorescence microscope or a confocal laser scanning microscope. Alternatively, DCF fluorescence can be quantified in cells grown in multiwell plates using a fluorescence plate reader. We use a Cytofluor 2350 (Millipore Corp.) for this purpose (see also Kane et al., 1993). A caveat in the use of DCF is that exposure to ultraviolet light can induce photooxidation resulting in increased cell fluorescence. Imaging of oxidative processes in neurons can provide an important complement to calcium-imaging studies and, because the excitation and emission optima for DCF and some Ca^{2+} indicator dyes (e.g., fura-2) differ considerably, it is possible to image both $[Ca^{2+}]_i$ and $[\text{free radicals}]_i$ simultaneously. Another method for quantifying free (hydroxyl) radical production is the salicylate hydroxylation assay (Zhang et al., 1993) in which levels of 2,5- and 2,3-DHBA, and salicylate are determined using reverse phase C_{18} HPLC with electrochemical detection. DHBA levels (which increase directly with increasing hydroxyl radical levels) are expressed as a ratio to salicylate recovered.

D. ASSESSMENT OF MITOCHONDRIAL FUNCTION

Mitochondrial impairment is an important step in the neurodegenerative process resulting from traumatic CNS injury. Mitochondria are the organelles where ATP is produced, and are also a major site of free radical production. Mitochondria that are functioning properly maintain a membrane potential across their inner membrane. The fluorescent dye rhodamine 123 accumulates in mitochondria by a mechanism dependent upon the transmembrane potential of the inner membrane (Johnson et al., 1981). Rhodamine 123 can therefore be used as a sensitive indicator of mitochondrial health (Mattson et al., 1993e). Cultured CNS neurons are incubated for 10 min in the presence of 5 μ*M* rhodamine 123 and are then washed three times in HBSS, and are imaged within 1 h. For comparison of relative levels of fluorescence per cell, images must be acquired using a constant level of excitation and a constant photodetector gain. Average pixel intensity per cell can be determined using the software available with most commercially available imaging systems.

E. MEASUREMENT OF CELLULAR ATP LEVELS

Traumatic injury to the CNS can lead to energy failure, which may contribute to loss of calcium homeostasis and free radical production (Mattson et al., 1993e). In experimental paradigms of CNS injury, it is therefore informative to monitor levels of cellular ATP. We have used a commercially available colorimetric assay (Sigma kit no. 366) similar to that of Adams (1963) to quantify levels of ATP in cultured neocortical cells (Mattson et al., 1993e). Cells in 60-mm dishes (approximately 1 million cells per dish) are rinsed twice with cold PBS and then scraped into 0.5 ml of a solution consisting of a 1:1 mixture of 12% TCA and 0.2 *M* sodium citrate (pH 7.0). The lysate is centrifuged at 400 rpm for 10 min

(4°C) and a 0.4-ml aliquot of the supernatant is added to a solution containing 0.15 mg of NADH-disodium salt, 0.5 ml of PGA-buffered solution (18 m*M* 3-phosphoglyceric acid), and 0.75 ml water. Following recording of the initial absorbance of the solution at 340 nm, 20 μl of GAPD/PDK enzyme mixture (800 U/ml glyceraldehyde-3-phosphate dehydrogenase and 450 U/ml 3-phosphoglyceric phosphokinase) is added to the cuvette and the reduction in absorbance is determined. ATP levels are calculated based upon a millimolar absorptivity of NADH of 6.22.

F. CALCIUM-INDUCED CYTOSKELETAL ALTERATIONS

Neuronal injury resulting from CNS trauma and metabolic/excitotoxic insults induces a number of alterations in the neuronal cytoskeleton, many of which are induced by calcium and precede neuronal death. The cytoskeletal alterations include depolymerization of tubulin and calcium-mediated proteolysis of specific cytoskeletal components including spectrin and the microtubule-associated protein MAP-2 (Siman and Noszek, 1988; Siman, 1992). In addition, excitotoxic insults elicit antigenic changes in tau (Mattson, 1990). Elevation of $[Ca^{2+}]_i$ is causally involved in alterations to each of the above-mentioned cytoskeletal components. Some of the cytoskeletal alterations can be readily identified using antibodies that recognize specific calcium-induced changes. For example, activation of glutamate receptors induces epitopes in the microtubule-associated protein tau similar to those seen in degenerating neurons in Alzheimer's disease (Mattson, 1990; Elliott et al., 1993). Calcium influx also causes proteolysis of MAP-2, which manifests as reduced immunoreactivity following traumatic injury to CNS neurons.

VI. NEUROPROTECTIVE STRATEGIES BASED ON STABILIZATION OF $[Ca^{2+}]_i$

The fact that neurons possess a variety of mechanisms to regulate $[Ca^{2+}]_i$ provides many opportunities for pharmacological intervention in the traumatized CNS. Table 2 contains several examples of pharmacological manipulations that target specific $[Ca^{2+}]_i$-regulating systems, and that have proven effective in reducing excitotoxic/metabolic injury *in vitro* and/or *in vivo*. Antagonists of NMDA and non-NMDA glutamate receptors are effective in attenuating ischemic neuronal injury (Choi, 1990; Meldrum, 1990). Inhibitors of other excitatory transmitter receptors may also reduce traumatic neuronal injury as indicated by data showing that muscarinic receptor antagonists can attenuate excitotoxicity *in vitro* (Mattson, 1989) and traumatic brain injury in rats (Hayes et al., 1992). Ca^{2+}-channel antagonists have also proven effective paradigms of traumatic and ischemic brain injury (Weiss et al., 1990; Young, 1992). Anticonvulsants such as phenytoin and carbamazepine control seizure activity in epilepsy by reducing membrane excitability. These compounds have also proven effective in attenuating excitotoxic neuronal injury (Mattson and Kater, 1989). Another class of agents that reduces membrane excitability is the K^+-channel openers, which can reduce excitotoxicity in cultured hippocampal neurons (Abele and Miller, 1990) by suppressing glutamate-induced elevation of $[Ca^{2+}]_i$ (Figures 3 and 4).

Neuronal injury (and calcium influx) result in free radical production, which contributes to cell death. Antioxidants such as vitamin E can attenuate traumatic and excitotoxic neuronal injury (Dykens et al., 1987; Goodman and Mattson, 1994). As discussed earlier, a variety of neurotrophic factors have been identified that can protect neurons against calcium- and free radical-mediated injury (Mattson et al., 1993e). If the problem of delivery to the brain is surmounted, these NPFs have great potential for reducing neuronal injury and promoting recovery in CNS trauma. Low-molecular weight lipophilic compounds that mimick the actions of NPFs also hold considerable promise as neuroprotective agents (Cheng et al., 1994). Contributing to $[Ca^{2+}]$-induced neuronal injury are free radicals generated during the

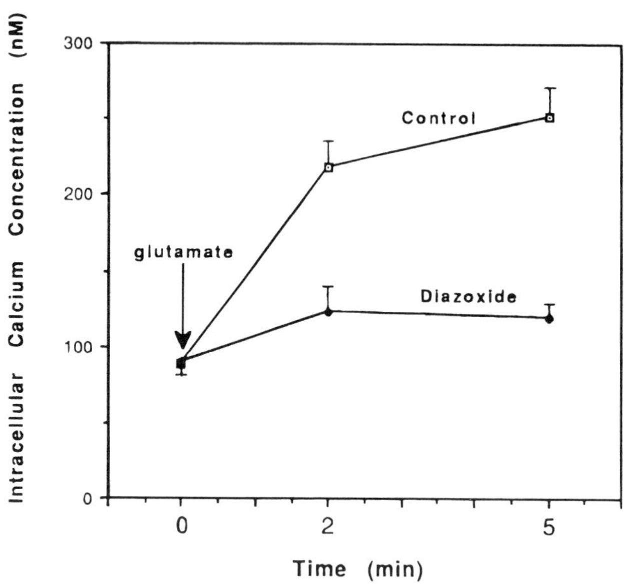

FIGURE 3. The K^+-channel opener diazoxide suppresses glutamate-induced elevation of neuronal $[Ca^{2+}]_i$. The $[Ca^{2+}]_i$ was measured (using the calcium indicator dye fura-2) prior to, and 2 and 5 min following, exposure to 50 μM glutamate. One culture was exposed to 1 mM diazoxide 5 min prior to exposure to glutamate, while a control culture was left untreated. Values represent the mean and SEM of determinations made in 22 (diazoxide) or 24 (control) neurons.

process of inositol phospholipid and arachidonic acid metabolism. Compounds that block key enzymes in these cascades (e.g., cyclooxygenase and 5-lipoxygenase) have proven effective in cell culture and/or *in vivo* studies (e.g., Hall, 1985; Rothman et al., 1993). Additional targets of potential therapeutic value are the plasma membrane and endoplasmic reticulum Ca^{2+}-ATPases, and the Na^+-Ca^{2+} exchanger. Another approach involves activating other second-messenger systems that may serve a neuroprotective function. For example, some data suggest that AMP (Mattson and Kater, 1988) and GMP pathways (Garthwaite and Garthwaite, 1988; Barger et al., 1995) may protect neurons against excitotoxic and/or metabolic insults.

VII. SUMMARY

Trauma to neural tissue initiates a cascade of events that result in prolonged elevation of intracellular free-calcium levels ($[Ca^{2+}]_i$), which stimulates the generation of free radicals. Free radicals, in turn, damage $[Ca^{2+}]_i$-regulating proteins resulting in further dysregulation of calcium homeostasis. Calcium damages proteins by activating proteases and lipases. Ultimately, membrane integrity is compromised by free-radical attack resulting in a collapse of ion gradients and cell death. The CNS utilizes several mechanisms to protect neurons from traumatic injury and to promote repair. Such neuroprotective mechanisms include signals released from injured cells that activate signal-transduction cascades that stabilize $[Ca^{2+}]_i$ and suppress free radicals. Neurotrophins, basic fibroblast growth factor, insulin-like growth factors, and secreted forms of β-amyloid precursor protein (βAPP) are examples of endogenous neuroprotection factors (NPFs) that are mobilized in response to CNS trauma. These NPFs utilize tyrosine kinase signaling cascades that influence the expression of $[Ca^{2+}]_i$-regulating proteins and antioxidant enzymes. Acute trauma to the CNS can initiate events that lead to a protracted neurodegenerative process. For example, head trauma is a major

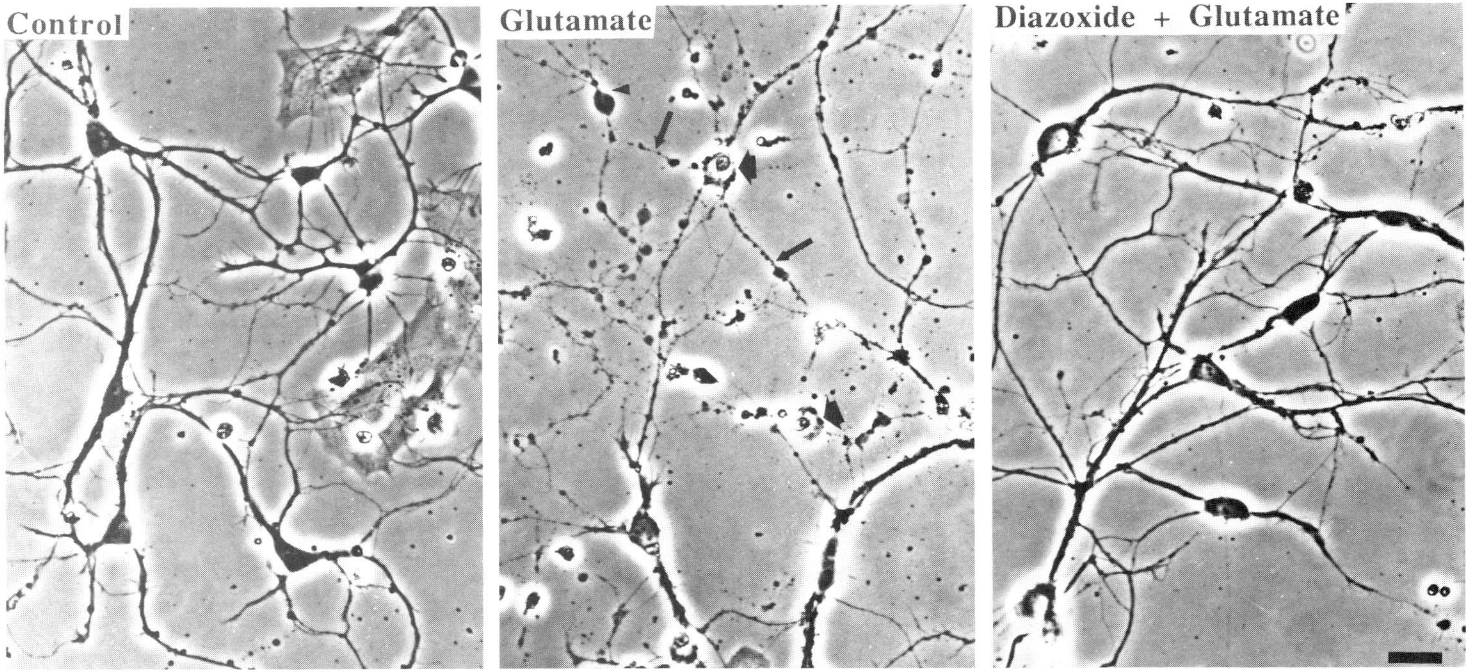

FIGURE 4. The K^+-channel opener diazoxide protects cultured rat hippocampal neurons against excitotoxicity. Phase-contrast of cultured hippocampal neurons. Left, cells in an untreated control culture. Middle, cells exposed to 200 μM glutamate for 4 h; note severe damage to neurites (e.g., arrows) and cell bodies (e.g., arrowheads). Right, cells pretreated with 1 mM diazoxide for 15 min and then exposed to 200 μM glutamate for 4 h. Note damage is much less severe than in culture exposed to glutamate alone. Scale bar, 10 μm.

risk factor for Alzheimer's disease. Head trauma results in a rapid increase in expression of βAPP and a subsequent deposition of amyloid β-peptide (Aβ), which aggregates into the plaques that are a histological hallmark of Alzheimer's disease. Recent studies have shown that Aβ disrupts neuronal $[Ca^{2+}]_i$ homeostasis and renders neurons vulnerable to excitotoxicy, metabolic compromise, and free radical attack. The increased understanding of the role of calcium and free radicals in CNS trauma has identified a variety of cellular targets for therapeutic intervention. These include: glutamate receptors; receptors for other transmitters; ion channels; NPF receptors and downstream kinases; enzymes involved in inositol phospholipid metabolism; and antioxidant systems.

ACKNOWLEDGMENTS

The author thanks S. W. Barger, B. Cheng, R. J. Mark, J. O'Keefe, and V. L. Smith-Swintosky for their contributions to the research in his laboratory, which was supported by grants from the NIH (NS29001 and NS30583), the Alzheimer's Association, and the Metropolitan Life Foundation.

REFERENCES

Anderson, K. J., Dam, D., Lee, S., and Cotman, C. W. (1988) Basic fibroblast growth factor prevents death of lesioned cholinergic neurons *in vivo*. *Nature* 332, 360–361.

Anderson, D. K., Hall, E. D., Braughler, J. M., and Means, E. D. (1991) Effect of delayed administration of U74006F (tirilazad mesylate) on recovery of locomotor function after experimental spinal cord injury. *J. Neurotrauma* 8, 187–192.

Barger, S. W., Fiscus, R. R., Ruth, P., Hofmann, F., and Mattson, M. P. (1994) Role of cyclic GMP in the regulation of neuronal calcium and survival by secreted forms of β amyloid precursor. *J. Neurochem.* 64: 2087–2096.

Behl, C., Davis, J., Cole, G. M., and Schubert, D. (1992) Vitamin E protects nerve cells from amyloid β protein toxicity. *Biochem. Biophys. Res. Commun.* 186, 944–950.

Bullock, K. R. and Fujisawa, H. (1992) The role of glutamate antagonists for the treatment of CNS injury. *J. Neurotrauma* 9, S443–S462.

Carney, J. M., Starke-Reed, P. E., Oliver, C. N., et al. (1991) Reversal of age-related changes in brain protein oxidation, decrease in enzyme activity, and loss in temporal and spatial memory by chronic administration of the spin trapping compound *N*-tert-butyl-alpha-phenylnitrone. *Proc. Natl. Acad. Sci. U.S.A.* 88, 3633–3636.

Cheng, B. and Mattson, M. P. (1991) NGF and bFGF protect rat hippocampal and human cortical neurons against hypoglycemic damage by stabilizing calcium homeostasis. *Neuron* 7, 1031–1041.

Cheng, B. and Mattson, M. P. (1993) NT-3 and BDNF protect hippocampal, septal and cortical neurons against metabolic compromise. *Brain Res.* 640, 56–57.

Cheng, B., Christakos, S., and Mattson, M. P. (1994a) Tumor necrosis factors protect neuron against excitotoxic/metabolic insults and promote maintenance of calcium homeostasis. *Neuron* 12, 139–153.

Cheng, B., Barger, S. W., and Mattson, M. P. (1994b) Staurosporine, K-252a and K-252b stabilize calcium homeostasis and promote survival of CNS neurons in the absence of glucose. *J. Neurochem.* 62, 1319–1329.

Choi, D. (1990) Methods for antagonizing glutamate neurotoxicity. *Cerebrovasc. Brain Metab. Rev.* 2, 105–147.

Clifton, G. L., Lyeth, B. G., Jenkins, J. W., Taft, W. C., Delorenzo, R. J., and Hayes, R. L. (1989) Effect of D α-tocopheryl succinate and polyethylene glycol on performance tests after fluid percussion brain injury. *J. Neurotrauma* 6, 71–81.

Collazo, D., Takahashi, H., and McKay, R. D. G. (1992) Cellular targets and trophic functions of neurotrophin-3 in the developing rat hippocampus. *Neuron* 9, 643–656.

Dykens, J. A., Stern, A., and Trenkner, E. (1987) Mechanism of kainate toxicity to cerebellar neurons in vitro is analagous to reperfusion tissue injury. *J. Neurochem.* 49, 1222–1228.

Elliott, E., Mattson, M. P., Vanderklish, P., Lynch, G., Chang, I., and Sapolsky, R. M. (1993) Corticosterone exacerbates kainate-induced alterations in hippocampal tau immunoreactivity and spectrin proteolysis *in vivo*. *J. Neurochem.* 61, 57–67.

Faden, A. I. and Salzman, S. (1992) Pharmacological strategies in CNS trauma. *Trends Pharmacol. Sci.* 13, 29–35.

Faden, A. I., Demediuk, P., Panter, S. S., and Vink, R. (1989) The role of excitatory amino acids and NMDA receptors in traumatic brain injury. *Science* 244, 798–800.
Frandsen, A. A. and Schousboe, A. (1991) Dantrolene prevents glutamate cytotoxicity and Ca^{2+} release from intracellular stores in cultured cerebral cortical neurons. *J. Neurochem.* 56, 1075–1088.
Freese, A., Finklestein, S. P., and DiFiglia, M. (1992) Basic fibroblast growth factor protects striatal neurons *in vitro* from NMDA receptor-mediated excitotoxicity. *Brain Res.* 575, 351–355.
Frim, D. M., Short, M. P., Rosenberg, W. S., Simpson, J. R., Breakfield, X. O., and Isacson, O. (1993) Nerve growth factor-secreting fibroblasts protect only locally against excitotoxic lesions in the rat striatum. *J. Neurosurg.* 78, 267–273.
Garthwaite, G. and Garthwaite, J. (1988) Cyclic GMP and cell death in rat cerebellar slices. *Neuroscience* 26, 321–326.
Gentleman, S. M., Graham, D. I., and Roberts, G. W. (1993) Molecular pathology of head trauma: altered βAPP metabolism and the aetiology of Alzheimer's disease. *Prog. Brain Res.* 96, 237–246.
Goodman, Y. and Mattson, M. P. (1994) Secreted forms of β-amyloid precursor protein protect hippocampal neurons against amyloid β-peptide-induced and oxidative injury. *Exp. Neurol.* 128, 1–12.
Grynkiewicz, G., Poenie, M., and Tsien, R. Y. (1985) A new generation of calcium indicators with greatly improved flourescence properties. *J. Biol. Chem.* 260, 3440–3450.
Hall, E. D. (1985) Beneficial effects of acute intravenous ibuprofen on neurologic recovery of head-injured mice: comparison of cyclooxygenase inhibition with inhibition of thromobaxane A2 synthetase or 5-lipoxygenase. *J. Neurotrauma* 2, 75–83.
Hall, E. D., Yonkers, P. A., Andrus, P. K., Cox, J. W., and Anderson, D. K. (1992) Biochemistry and pharmacology of lipid antioxidants in acute brain and spinal cord injury. *J. Neurotrauma* 9, S425–S442.
Hayes, R. L., Jenkins, L. W., and Lyeth, B. G. (1992) Neurotransmitter-mediated mechanisms of traumatic brain injury: acetylcholine and excitatory amino acids. *J. Neurotrauma* 9, S173–S187.
He, Y., Hsu, C. Y., Ezrin, A. M., and Miller, M. S. (1993) Polyethylene glycol-conjugated superoxide dismutase in focal cerebral ischemia-reperfusion. *Am. J. Physiol.* 265, H252–256.
Hefti, F., Hartikka, J., and Knusel, B. (1989) Function of neurotrophic factors in the adult and aging brain and their possible use in the treatment of neurodegenerative diseases. *Neurobiol. Aging* 10, 515–533.
Hensley, K., Carney, J., Aksenova, M., Harris, M., Wu, J. F., and Butterfield, D. A. (1994) A new model for β-amyloid aggregation and neurotoxicity based on the free-radical generating capacity of the peptide. *Proc. Natl. Acad. Sci. U.S.A.* 91, 3270–3274.
Hoyer, S., Oesterreich, K., and Wagner, O. (1988) Glucose metabolism as the site of the primary abnormality in early-onset dementia of Alzheimer type? *J. Neurol.* 235, 143–148.
Jesberger, J. A. and Richardson, J. S. (1991) Oxygen free radicals and brain dysfunction. *Int. J. Neurosci.* 57, 1–7.
Johnson, L. V., Walsh, M. L., Bokus, B. J., and Chen, L. B. (1981) Monitoring relative mitochondrial transmembrane potential in living cells by fluorescence microscopy. *J. Cell Biol.* 88, 526–532.
Kane, D. J., Sarafian, T. A., et al. (1993) Bcl-2 inhibition of neural death: decreased generation of reactive oxygen species. *Science* 262, 1274–1277.
Katayama, Y., Becker, D. P., Tamura, T., and Hovida, D. A. (1990) Massive increases in extracellular potassium and the indiscriminate release of glutamate following concussive brain injury. *J. Neurosurgery* 73, 889–900.
Lipton, S. A. et al. (1993) A redox-based mechanism for the neuroprotective and neurodestructive effects of nitric oxide and related nitroso-compounds. *Nature* 364, 626–632.
Malgaroli, A., Milani, D., Meldolesi, J., and Pozzan, T. (1987) Fura-2 measurement of cytosolic free Ca^{2+} in monolayers and suspensions of various types of animal cells. *J. Cell. Biol.* 105, 2145–2155.
Masiakowski, P. and Shooter, E. M. (1988) Nerve growth factor induces the genes for two proteins related to family of calcium-binding proteins in PC12 cells. *Proc. Natl. Acad. Sci. U.S.A.* 85, 1277–1281.
Mattson, M. P. and Kater, S. B. (1988) Intracellular messengers in the generation and degeneration of hippocampal neuroarchitecture. *J. Neurosci. Res.* 21, 447–464.
Mattson, M. P., Dou, P., and Kater, S. B. (1988) Outgrowth-regulating actions of glutamate in isolated hippocampal pyramidal neurons. *J. Neurosci.* 8, 2087–2100.
Mattson, M. P. (1989) Acetylcholine potentiates glutamate-induced neurodegeneration in cultured hippocampal neurons. *Brain Res.* 497, 402–406.
Mattson, M. P. and Kater, S. B. (1989) Excitatory and inhibitory neurotransmitters in the generation and degeneration of hippocampal neuroarchitecture. *Brain Res.* 478, 337–348.
Mattson, M. P., Murrain, M., Guthrie, P. B., and Kater, S. B. (1989a) Fibroblast growth factor and glutamate: opposing roles in the generation and degeneration of hippocampal neuroarchitecture. *J. Neurosci.* 9, 3728–3740.
Mattson, M. P., Guthrie, P. B., and Kater, S. B. (1989b) A role for Na^+-dependent Ca^{2+} extrusion in protection against neuronal excitotoxicity. *FASEB J.* 3, 2519–2526.
Mattson, M. P. (1990) Antigenic changes similar to those seen in neurofibrillary tangles are elicited by glutamate and calcium influx in cultured hippocampal neurons. *Neuron* 4, 105–117.
Mattson, M. P., Engle, M. G., and Rychlik, B. (1991a) Effects of elevated intracellular calcium levels on the cytoskeleton and tau in cultured human cerebral cortical neurons. *Mol. Chem. Neuropath.* 15, 117–142.

Mattson, M. P., Rychlik, B., Chu, C., and Christakos, S. (1991b) Evidence for calcium-reducing and excitoprotective roles for the calcium-binding protein (calbindin-D28k) in cultured hippocampal neurons. *Neuron* 6, 41–51.

Mattson, M. P., Cheng, B., Davis, D., Bryant, K., Lieberburg, I., and Rydel, R. E. (1992) β-Amyloid peptides destabilize calcium homeostasis and render human cortical neurons vulnerable to excitotoxicity. *J. Neurosci.* 12, 376–389.

Mattson, M. P., Barger, S. W., Cheng, B., Lieberburg, I., Smith-Swintosky, V. L., and Rydel, R. E. (1993a) β-Amyloid precursor protein metabolites and loss of neuronal Ca^{2+} homeostasis in Alzheimer's disease. *Trends Neurosci.* 16, 409–415.

Mattson, M. P., Tomaselli, K., and Rydel, R. E. (1993b) Calcium-destabilizing and neurodegenerative effects of aggregated β-amyloid peptide are attenuated by basic FGF. *Brain Res.* 621, 35–49.

Mattson, M. P., Cheng, B., and Smith-Swintosky, V. L. (1993c) Neurotrophic factor-mediated protection from excitotoxicity and disturbances in calcium and free radical metabolism. Semin. *Neuroscience* 5, 295–307.

Mattson, M. P., Kumar, K. N., Cheng, B., Wang, H., and Michaelis, E. K. (1993d) Basic FGF regulates the expression of a functional 71 kDa NMDA receptor protein that mediates calcium influx and neurotoxicity in cultured hippocampal neurons. *J. Neurosci.* 13, 4575–4588.

Mattson, M. P., Zhang, Y., and Bose, S. (1993e) Growth factors prevent mitochondrial dysfunction, loss of calcium homeostasis and cell injury, but not ATP depletion in hippocampal neurons. *Exp. Neurol.* 121, 1–13.

Mattson, M. P., Barger, S. W., Begley, J. G., and Mark, R. J. (1994) Calcium, free radicals, and excitotoxic neuronal death in primary cell culture. *Methods Cell Biol.* 46, 187–216.

McCord, J. (1985) Oxygen-derived free radicals in postischemic tissue injury. *N. Engl. J. Med.* 312, 159–163.

McIntosh, T. K., Vink, R., Yamakami, I., and Faden, A. I. (1989) Magnesium protects against neurological deficit after brain injury. *Brain Res.* 482, 252–260.

Meldrum, B. (1990) Protection against ischaemic neuronal damage by drugs acting on excitatory neurotransmission. *Cerebrovasc. Brain Metab.* 2, 27–57.

Nilsson, P., Hillered, L., Ponten, V., and Ungerstedt, V. (1990) Changes in cortical extracellular levels of energy-related metabolites and amino acids following concussive brain injury in rats. *J. Cereb. Blood Flow Metab.* 10, 631–637.

Nistico, G., Ciriolo, M. R., Fiskin, D., Iannone, M., Demartino, A., and Rotilio, G. (1992) NGF restores decrease in catalase activity and increases superoxide dismutase and glutathione peroxidase activity in the brain of aged rats. *Free Rad. Biol. Med.* 12, 177–181.

Nozaki, K., Finklestein, S. P., and Beal, M. F. (1993) Basic fibroblast growth factor protects against hypoxia/ischemia and NMDA neurotoxicity in neonatal rats. *J. Cereb. Blood Flow Metab.* 13, 221–228.

Pan, Z. and Perez-Polo, J. R. (1993) Role of nerve growth factor in oxidant homeostasis: glutathione metabolism. *J. Neurochem.* 61, 1713–1721.

Panter, S. C., Braughler, J. M., and Hall, E. (1992) Dextran-coupled deferoxamine improves outcome in a murine model of head injury. *J. Neurotrauma* 9, 47–53.

Pike, C., Burdick, D., Walencewicz, A., Glabe, C., and Cotman, C. W. (1993) Neurodegeneration induced by β-amyloid peptides *in vitro:* the role of peptide assembly state. *J Neurosci.* 13, 1676–1686.

Roberts, G. W., Gentleman, S. M., Lynch, A., and Graham, D. I. (1991) βA4 Amyloid protein deposition in brain after head injury. *Lancet* 338, 1422–1423.

Roberts-Lewis, J. M. and Siman, R. (1993) Spectrin proteolysis in the hippocampus: a biochemical marker for neuronal injury and neuroprotection. *Ann. N.Y. Acad. Sci.* 679, 78–86.

Rothman, S. M., Yamada, K. A., and Lancaster, N. (1993) Nordihydroguaiaretic acid attenuates NMDA neurotoxicity-action beyond the receptor. *Neuropharmacology* 32, 1279–1288.

Sautiere, P. E., Sindou, P., Couratier, P., Hugon, J., Wattez, A., and Delacourte, A. (1992) Tau antigenic changes induced by glutamate in rat primary culture model: a biochemical approach. *Neurosci. Lett.* 140, 206–210.

Scharfman, H. E. and Schwartzkroin, P. A. (1989) Protection of dentate hilar cells from prolonged stimulation by intracellular calcium chelation. *Science* 246, 257–260.

Schlessinger, J. and Ulrich, A. (1992) Growth factor signaling by receptor tyrosine kinases. *Neuron* 9, 383–391.

Selkoe, D. J. (1991) The molecular pathology of Alzheimer's disease. *Neuron* 6, 487–498.

Selkoe, D. J. (1993) Physiological production of the β-amyloid protein and the mechanism of Alzheimer's disease. *Trends Neurosci.* 16, 403–409.

Shigeno, T., Mima, T., Takakura, K., Graham D. I., Kato, G., Hashimoto, Y., and Furukawa, S. (1991) Amelioration of delayed neuronal death in the hippocampus by nerve growth factor. *J. Neurosci.* 11, 2914–2919.

Siesjo, B. K. (1993) Basic mechanisms of traumatic brain damage. *Ann. Emerg. Med.* 22, 959–969.

Siman, R. and Noszek, J. C. (1988) Excitatory amino acids activate calpain I and induce structural protein breakdown *in vivo. Neuron* 1, 279–287.

Siman, R. (1992) Proteolytic mechanism for the neurodegeneration of Alzheimer's disease. *Ann. N.Y. Acad. Sci.* 674, 193–202.

Smith, C. D., Carney, J. M., Starke-Reed, P. E., et al. (1991) Excess brain protein oxidation and enzyme dysfunction in normal aging and in Alzheimer disease. *Proc. Natl. Acad. Sci. U.S.A.* 88, 10540–10543.

Spina, M. B., Squinto, S. P., Miller, J., Lindsay, R. M., and Hyman, C. (1992) Brain-derived neurotrophic factor protects dopamine neurons against 6-hydroxydopamine and *N*-methyl-4-phenylpyridinium ion toxicity: involvement of glutathione system. *J. Neurochem.* 59, 99–106.

Thastrup, O., Cullen, P. J., Drobak, B. K., Hanley, M. R., and Dawson, A. P. (1990) Thapsigargin, a tumor promoter, discharges intracellular Ca^{2+} stores by specific inhibition of the endoplasmic reticulum Ca^{2+}-ATPase. *Proc. Natl. Acad. Sci. U.S.A.* 87, 2466–2470.

Tsien, R. W. and Tsien, R. Y. (1990) Calcium channels, stores, and oscillations. *Annu. Rev. Cell Biol.* 6, 715–60.

Verity, M. A. (1993) Mechanisms of phospholipase A2 activation and neuronal injury. *Ann. N.Y. Acad. Sci.* 679, 110–120.

Waxman, S. G., Ransom, B. R., and Stys, P. K. (1991) Non-synaptic mechanisms of Ca^{2+} mediated injury in CNS white matter. *Trends Neurosci.* 14, 461–468.

Weiss, J. H., Hartley, D. M., Koh, J., and Choi, D. W. (1990) The calcium channel blocker nifedipine attenuates slow excitatory amino acid neurotoxicity. *Science* 247, 1474–1477.

Young, W. (1992) Role of calcium in central nervous system injuries. *J. Neurotrauma* 9, S9–S25.

Zhang, Y., Tatsuno, T., Carney, J. M., and Mattson, M. P. (1993) Basic FGF, NGF, and IGF-II protect hippocampal and cortical neurons against iron-induced degeneration. *J. Cereb. Blood Flow Metab.* 13, 378–388.

Chapter 4

ROLE OF MAGNESIUM IN CNS TRAUMA

Robert Vink and Tracy K. McIntosh

TABLE OF CONTENTS

ABSTRACT

Traumatic injury to the central nervous system may result in irreversible tissue damage via a number of mechanisms. From a therapeutic point of view, it is the secondary mechanisms occurring between hours to days after the traumatic event that permits development of an appropriate intervention that may either prevent or attenuate the associated mortality or morbidity. A number of factors have been proposed as being integral in this secondary injury process. These include, amongst others, excitatory amino acids, opioid peptides, free fatty acids, oxygen free radicals, bioenergetic failure, cation changes, and lactic acidosis. Recent evidence suggests that the magnesium cation plays a central role in the pathophysiology of traumatic CNS injury, and that such a role is integrally related to membrane structure and function. This chapter reviews the evidence that magnesium is involved in the injury process, and proposes a mechanism by which the cation can interact with a number of the proposed secondary injury pathways.

I. INTRODUCTION

Traumatic injury to the central nervous system (CNS) is the major cause of death in individuals under 44 years of age in developed countries.[1] Survivors are often left with permanent neurologic deficits that ultimately cost every community billions of dollars in terms of loss of productivity and costs for treatment and rehabilitation.[2] Recent evidence suggests that much of the irreversible tissue damage following CNS trauma is the result of secondary injury factors initiated by the traumatic event, but occurring hours to days later.[3]

0-8493-8094-4/95/$0.00+$.50

These secondary mechanisms are basically of a physiologic and biochemical nature, and a variety of factors have been proposed including, among others, release of excitatory amino acids, opioid peptides, free fatty acids, free radicals, bioenergetic dysfunction, reduced blood flow, and ion changes.[4–8] Despite this extensive list of possible injury factors, a treatment for neurotrauma has not been readily forthcoming. This is largely because of two reasons. First, most techniques have been invasive, and the correlation between a factor and neurologic outcome has been difficult to establish. Second, because of technical reasons, most studies have tended to focus upon a single injury factor with little consideration being given to how that factor interacts with other factors in the development of irreversible tissue injury.

Nuclear magnetic resonance spectroscopy (MRS) has only recently been applied to the study of neurotrauma. In addition to being a nondestructive and noninvasive technique, MRS permits the simultaneous evaluation of a number of injury factors and their relationship to eventual neurologic outcome in the same animal.[9] The advantage of such a technique (apart from its obvious noninvasive nature) is that the interrelationship between several factors can be continuously monitored over time, with selective variation of single factors by pharmacologic means permitting observation of effects on other factors. One such factor that has been seen to play an integral role in the injury process despite a variety of "specific" interventions is the magnesium ion.[8,10] Little attention has been paid to the role of magnesium in metabolic regulation simply because no effective means were available to measure the free concentration of the ion *in vivo*. However, because of the ability of MRS to do precisely that, it has been observed that this divalent cation undergoes significant alterations in concentration that are correlated to posttraumatic outcome.[11] Furthermore, these changes in concentration have been proposed to be initiated by alterations in membrane structure.[12]

II. DETERMINATION OF FREE MAGNESIUM CONCENTRATION

A number of studies have attempted to determine free magnesium concentration in CNS tissue using a variety of methods including microelectrodes, metallochromic indicators, null point methods, enzyme equilibrium determinations, and MRS (for review, see Reference 13). While these methods all have their advantages, only MRS is truly noninvasive and can be used to repeatedly determine free concentration over time. Nonetheless, controversy exists with respect to the variety of free magnesium estimations the technique has produced. These differences can be largely accounted for by methodological differences in the use of MRS, particularly, the manner in which magnesium has been calculated from the data. The methodological issues surrounding the use of MRS in neurotrauma have been reviewed elsewhere,[9] and is beyond the scope of this review. We will therefore concentrate on issues relating to the determination of free magnesium by phosphorus MRS.

The typical phosphorus MRS spectrum of rat brain is shown in Figure 1. A number of peaks can be clearly identified, and include phosphomonoesters, inorganic phosphate, phosphodiesters, phosphocreatine, and the three phosphates of ATP. The chemical shift difference between the α and β peaks of ATP is dependent on ion concentration, particularly H^+ and Mg^{2+} concentration. Between pH 6.6 and pH 8.0, the chemical shift difference of these two peaks is primarily dependent on the magnesium concentration and not pH (Figure 2). Therefore, the chemical shift difference between the two ATP peaks will provide a determination of the concentration of magnesium-bound ATP relative to free ATP according to the equation:[14]

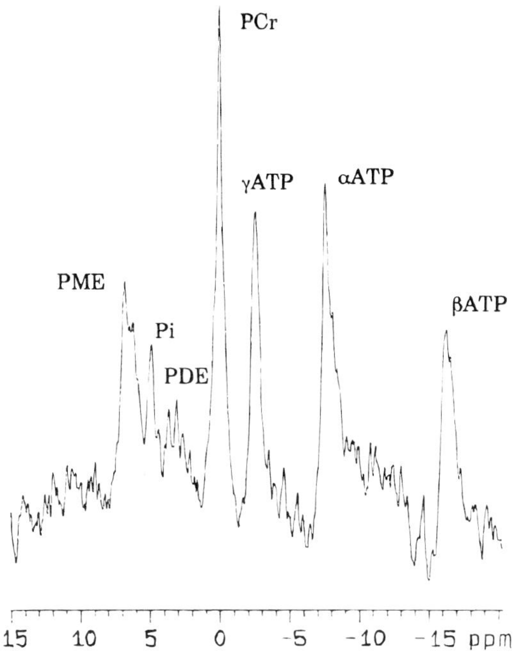

FIGURE 1. Typical phosphorus magnetic resonance spectrum of rat brain obtained *in vivo* at 7.0 tesla over 10 min. Peaks are identified as phosphomonoesters (PME), inorganic phosphate (Pi), phosphodiesters (PDE), phosphocreatine (PCr), and adenosine triphoshate (ATP).

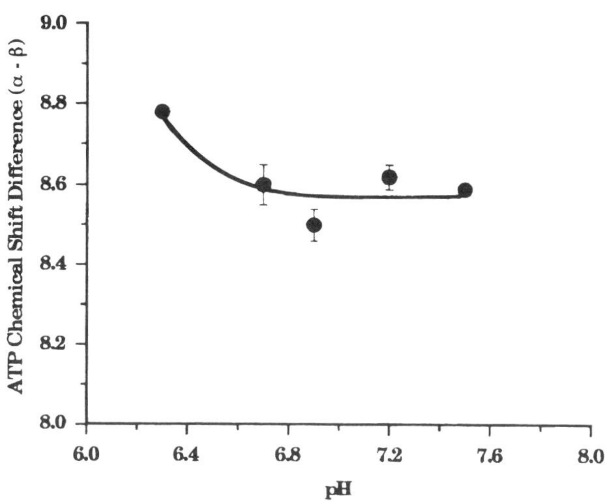

FIGURE 2. Dependence of the MgATP chemical shift on pH. Solution ATP concentration (5 m*M*), free magnesium concentration (0.6 m*M*), and ionic strength (150 m*M*) were set to reflect *in vivo* conditions.

$$\phi = \frac{[\mathrm{ATP}]_{\mathrm{free}}}{[\mathrm{ATP}]_{\mathrm{total}}} = \frac{\delta_{\alpha\beta} - \delta_{\alpha\beta\ \mathrm{Mg\text{-}ATP}}}{\delta_{\alpha\beta\ \mathrm{ATP}} - \delta_{\alpha\beta\ \mathrm{Mg\text{-}ATP}}}$$

where $\delta_{\alpha\beta}$ is the chemical shift difference between the α and β ATP peaks in the acquired spectrum, and $\delta_{\alpha\beta\ \mathrm{Mg\text{-}ATP}}$ and $\delta_{\alpha\beta\ \mathrm{ATP}}$ are the chemical shift difference between the α and β

ATP peaks in a magnesium-saturated and a magnesium-free solution of of ATP, respectively. With a knowledge of the Mg-ATP binding constant, free concentration can then be determined by the equation:

$$Mg_f = K_d(\phi^{-1} - 1)$$

There are two primary sources of error in this calculation. First, the end point titrations for the magnesium-saturated and -free solutions of ATP will significantly affect the calculated free-to-bound-ATP ratio. We have determined the mean value of these two points to be 8.35 and 10.82 under conditions that simulate the *in vivo* condition. Second, the value of the dissociation constant for Mg-ATP has been a source of controversy.[14] Although the value used does not affect the changes that may occur, it does affect the absolute free concentration determined. Recent evidence suggests that the K_d is 50 μ*M*.[15] This value is pH dependent, and can be corrected according to the method described by Bock et al.,[16] although this correction does not significantly affect the calculated free-magnesium concentration, at least within the physiological pH range.

When taking the above into account, free magnesium concentration in rat brain is typically between 0.5 and 1.0 m*M*, with variation within this range being dependent upon external factors such as magnesium content of the diet and type of anesthesia used in the MRS study. This value is in excellent agreement with determinations made using a variety of alternative techniques.[13]

III. MAGNESIUM FOLLOWING BRAIN TRAUMA

A number of MRS studies have reported free magnesium concentration in brain, with the mean value for neurotrauma studies being 0.8 ± 0.1 m*M*.[8] Following traumatic brain injury, free magnesium levels declined by approximately 60% (Figure 3). No recovery in magnesium concentration was observed in the immediate 6-h period posttrauma, except in animals that received a low injury level. These animals recorded a minimum free magnesium concentration at 3 h posttrauma, with a return to preinjury magnesium levels by 4 h postinjury. Similar results have been obtained in studies of spinal trauma,[17] where free magnesium declined from 0.8 to 0.3 m*M* by 3 h posttrauma.

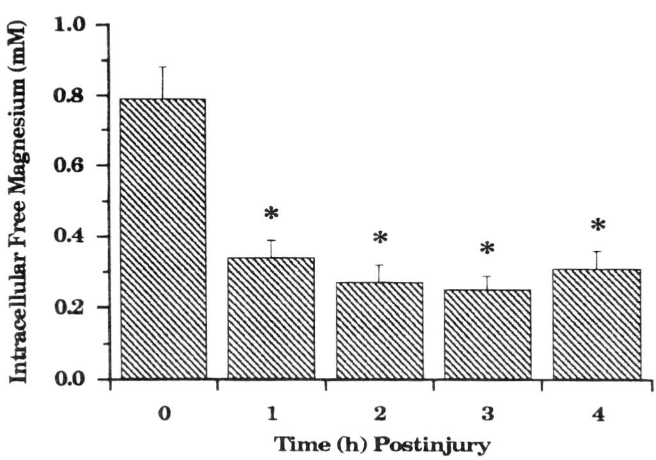

FIGURE 3. Free magnesium concentration in rat brain prior to and following low and moderate fluid percussion-induced traumatic injury. Adapted from Reference 11.

There were also associated declines in total magnesium concentration following both brain and spinal cord trauma, as determined by atomic absorption spectrophotometry.[11,17] Declines in total magnesium were on the order of 10 to 20%, with the declines being localized to the injured tissue (Figure 4). The reduction in total tissue magnesium at the injury site was sustained for up to 3 d posttrauma, with recovery to preinjury levels by 7 d posttrauma. Furthermore, the degree in total tissue magnesium decline was correlated to outcome. The relative changes in free vs. total magnesium concentration have been postulated to reflect the differences in their respective concentrations, with small changes in the total pool being "amplified" in the smaller free pool.[18]

The role of magnesium in trauma was further pursued by McIntosh and colleagues,[19,20] who examined the effects of magnesium depletion and treatment on outcome. These studies reported that dietary reduction of brain free magnesium concentration led to exacerbated injury following moderate traumatic brain injury. Conversely, increasing the availability of magnesium prior to injury afforded neuroprotection following trauma. This neuroprotective action of magnesium was still apparent even when the drug was administered 30 min after the traumatic event. Interestingly, administration of Mg-ATP following trauma worsened neurologic outcome,[20] in contrast to the clear beneficial effects of this treatment in peripheral organ ischemia.[21] The authors have rationalized this in terms of the hypotensive effects of Mg-ATP administration on a brain that has through injury lost autoregulation ability. Any consequent reduction of flow through cerebral vasodilation would therefore be deleterious.[20]

IV. BIOCHEMICAL AND PHYSIOLOGIC EFFECTS OF REDUCED MAGNESIUM

Magnesium has a number of important biochemical and physiologic properties that may influence the development of irreversible tissue injury after neurotrauma. With respect to metabolism, the K_m for magnesium for over 300 enzymes is approximately 1 m*M*.[22] Therefore, any significant alteration of free magnesium concentration from this optimum is likely to exert considerable metabolic control. Included in this list of enzymes are those involved in glycolysis, Krebs cycle activity, and ion homeostasis.[22] Furthermore, reductions in magnesium to less than 0.25 m*M* will significantly depress DNA transcription, RNA aggregation, and protein synthesis.[23] Magnesium is also a natural antagonist of calcium channels, and any

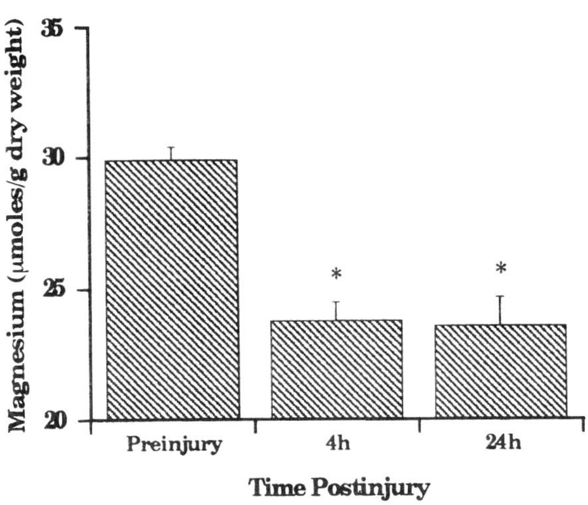

FIGURE 4. Total brain tissue magnesium prior to and following moderate fluid percussion-induced traumatic brain injury.

decline in magnesium concentration is likely to lead to significant increases in calcium concentration. Such an effect is thought to account for the increased cerebral vasospastic response associated with magnesium deficiency.[24] This may account for the localized reductions in blood flow known to occur following CNS trauma.[25]

The role of magnesium in the regulation of NMDA channel activity has been widely recognized,[26] and an increasing recognition of its role in the regulation of the inositol triphosphate pathway is becoming apparent.[27] Similarly, the demonstration that magnesium can change under physiologic conditions has led to a reevaluation of bioenergetic consequences of traumatic injury.[28] If changes in free magnesium concentration are not accounted for, there would seem to be no significant alteration in posttraumatic indexes of bioenergetic state including ADP concentration and cytosolic phosphorylation potential. However, when magnesium declines are taken into account, cytosolic phosphorylation potential decreases by approximately 40% after trauma, a reduction that would significantly affect a number of processes including the Na^+-K^+-ATPase function, and free fatty acid reacylation.

Finally, a decrease in magnesium concentration is known to destabilize the cell membrane, with a resultant increase in membrane fluidity and permeability.[29] One can speculate that this increased fluidity may be a result of the reduced reacylation of free fatty acids indicated above. Under normal conditions, a balance is maintained between phospholipid degradation and synthesis that keeps levels of free fatty acids and lysophospholipids low. Under pathophysiologic conditions, the rate of degradation exceeds the rate of synthesis.[30] The rate of synthesis is dependent on the reacylation of free fatty acids, a step that may be inhibited by magnesium decline. Such inhibition would result in elevated levels of free fatty acids, eicosanoids, and free radicals, all of which have been reported to increase following CNS trauma.

V. MECHANISM OF MAGNESIUM DECLINE

A change in the bound-to-free-ATP ratio would signify that Mg-ATP is losing its bound cation to another ligand with a stronger binding affinity.[31] Because the cell membrane has been proposed to be the buffer for changes in free magnesium concentration,[18] it is reasonable to propose that changes in membrane structure may be associated with alterations in the magnesium distribution. Accordingly, a series of experiments were conducted investigating the role of the membrane in alterations in free magnesium concentration.[12] Phospholipids were extracted from rat brain and reconstituted by sonication into aqueous vesicles containing physiologic levels of Mg-ATP. These suspensions were then incubated at 37°C with a variety of phospholipases either individually or in combination. The author reports that only addition of phospholipase C induced a decline in free magnesium concentration that was similar to the decline noted following traumatic brain injury. Furthermore, the decline was only observed when phospholipase C was the only enzyme added to the suspension, suggesting that *in vivo*, phospholipase C would have to be preferentially activated following a traumatic event.

There is a variety of evidence obtained *in vivo* following traumatic brain injury that supports this mechanism of magnesium decline. Phospholipase C is activated following trauma with maximum activity being recorded at 10 min posttrauma.[32] The mechanism of this activation was not known until it was reported that glutamate can interact with what has now been termed a metabotropic receptor and directly activate phospholipase C.[33] Thus, there is the distinct possibility that phospholipase C can be preferentially activated very early following a traumatic event. Further support for this hypothesis came when the kinetics of glutamate release following trauma was established.[4] Maximum levels of glutamate and aspartate have been shown to occur between 10 and 20 min after a traumatic event. Thus, glutamate is maximal at the same time after injury as phospholipase C activity achieves its maximum. Finally, the kinetics of magnesium decline following trauma indicates that free magnesium levels have commenced decreasing by 15 min after the injury, and have reached their nadir by 30 to 60 min after trauma.[11]

Having proposed a mechanism for free magnesium declines, it remained to rationalize the total tissue declines. Recent evidence from clinical head injury has shown that serum magnesium levels increase after trauma,[34] and the authors propose that the origin of this magnesium is the injured brain tissue. This would only be possible if there was sufficient blood flow in the injured regions to remove the magnesium, presumably bound to membrane-derived lipids from cells that have deteriorated. Blood flow determinations in experimental trauma have shown that while there are reductions in flow, there is still approximately 50% flow in injured tissue.[25] This is well above levels required to adequately perfuse the tissue. It has not been determined following trauma whether the increased serum magnesium is bound to lipids. Studies of alcohol withdrawal have, however, demonstrated that tissue reductions in total magnesium content are correlated with increased serum magnesium concentration and increased serum lipids.[35] When we consider that ethanol upregulates phospholipase C activity,[36] it is reasonable to suggest that as in alcohol withdrawal, posttraumatic increases in phospholipase C activity results in both decreased brain free-magnesium concentration and decreased total brain tissue magnesium concentration.

VI. PROPOSED INJURY PATHWAY

Summarized in Figure 5 is a proposed injury pathway based on the magnesium findings described above. Following the traumatic event, a number of factors are released to initiate the secondary injury process. These factors include excitatory and inhibitory neurotransmitters and neuromodulators. For the sake of clarity, we shall concentrate on the excitatory amino acid glutamate, although it is important to stress that all the neurotransmitters play an interactive role in initiating the response.[37] Once released, glutamate can interact with a number of membrane receptors including the ionotropic NMDA receptor, and the phospholipase C-linked metabotropic receptor, among others.[33,37] The NMDA receptor, and its role in ischemic brain damage, has been well characterized.[38–42] These studies have demonstrated that the opening of the channel is both magnesium and membrane potential dependent. Hence, the dotted lines in Figure 5 represent this "feedback" regulation. Opening of the channel results in the influx of extracellular calcium, overload of the intracellular calcium buffering mechanisms, and an associated increase in intracellular free calcium. Such an increase activates a variety of enzymes and processes including the phospholipases, protein kinase C, and the formation of nitric oxide, all playing a role in what has been termed the calcium cascade.[6,40] An important consideration in the activation of the NMDA pathway is that the calcium-buffering mechanisms must be overcome before the calcium cascade can be initiated.[40] This is most likely to occur under conditions of energy failure.[41,42]

The alternative metabotropic pathway can be activated under both energy replete and energy depleted conditions. Glutamate interacts directly with a phospholipase C-linked receptor that initiates the inositol triphosphate second-messenger pathway.[33] At the same time, activation of phospholipase C will cause membrane breakdown and free fatty acid accumulation, leading to an immediate decline in magnesium concentration.[12] The decline in free magnesium and the increased inositol triphosphate concentration lead to a release of calcium from intracellular stores, and activation of the calcium cascade.[33] It is important to note that the initial activation of phospholipase C causes the decrease in magnesium concentration, and that such activation is independent of energy state.

VII. CONCLUSION

The characterization of the mechanisms involved with secondary injury following CNS trauma will permit the development of pharmacotherapies that will prevent, or at least

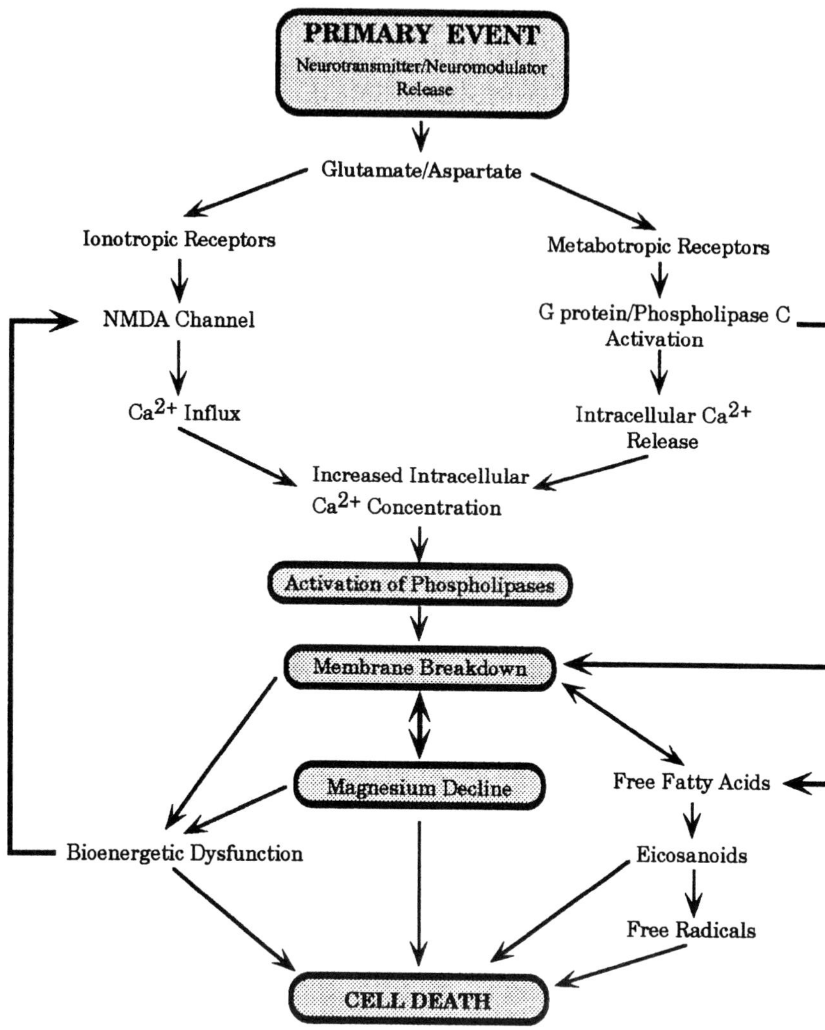

FIGURE 5. Schematic illustration summarizing mechanisms of injury and the interrelationship between magnesium and a number of secondary injury factors.

attenuate, the mortality and morbidity associated with such an event. It has become clear that the events occurring at the membrane level are fundamental to this understanding. While membrane-bound receptors and enzymes are integrally involved in the injury response, it is becoming increasingly apparent that membrane structure and function are what ultimately determine the survivability of tissue. This concept is certainly not new, as virtually all injury pathways that have been proposed for CNS trauma in some way or another incorporate membranes and their catabolic by-products in the injury process.[3,6,7,30] This review has attempted to link a variety of secondary injury factors into a common pathway. Central to this scheme is magnesium, which, although not necessarily a cause of the injury process, is without doubt a marker of tissue injury and recoverability. It remains to be determined whether it is the membrane-induced changes in ion homeostasis that is the critical step in the development of irreversible tissue injury in the CNS.

ACKNOWLEDGMENT

RV is supported in part by the Australian National Health and Medical Research Council. TKM is supported in part by the U.S. National Institutes of Health (RO1-NS26818).

REFERENCES

1. Goldstein, M. (1990) Traumatic brain injury: a silent epidemic. *Ann. Neurol.* 27, 327.
2. Max, W., MacKenzie, E. J., and Rice, D. P. (1991) Head injuries: costs and consequences. *J. Head Trauma Rehabil.* 6, 76–91.
3. Cooper, P. R. (1985) Delayed brain injury: secondary insults, in *Central Nervous System Status Report* (Becker, D. P. and Povlishock, J. T., eds.) pp. 217–228, William Byrd Press, Bethesda.
4. Faden, A. I., Demediuk, P., Panter, S., and Vink, R. (1989) Excitatory amino acids, N-methyl-D-aspartate receptors and traumatic brain injury. *Science* 244, 798–800.
5. Vink, R., Portoghese, T. K., and Faden, A. I. (1991) κ-opioid antagonist improves cellular bioenergetics and recovery after traumatic brain injury. *Am. J. Physiol.* 261, R1527–R1532.
6. Young, W. (1992) Role of calcium in central nervous system injuries, in *Central Nervous System Trauma Status Report 1992* (Jane, J. A., Anderson, D. K., Torner, J. C., and Young, W., eds.) pp. 9–25, Mary Ann Liebert, New York.
7. Siesjo, B. K. (1988) Historical overview: calcium, ischemia and death of brain cells. *Ann. N.Y. Acad. Sci.* 522, 638–661.
8. Vink, R., McIntosh, T. K., and Faden, A. I. (1991) Mg^{2+} in neurotrauma: its role and therapeutic implications, in Mg^{2+} and *Excitable Membranes* (Strata, P. and Carbone, E., eds.) pp. 125–145, Springer-Verlag, Berlin.
9. Vink, R. (1994) Metabolism, in *The Neurobiology of Central Nervous System Trauma* (Salzman, S. K. and Faden, A. I., eds.) pp. 68–78, Oxford University Press, Oxford.
10. Vink, R. and McIntosh, T. K. (1990) Pharmacological and physiological effects of magnesium on experimental traumatic brain injury. *Magnesium Res.* 3, 163–169.
11. Vink, R., McIntosh, T. K., Demediuk, P., Weiner, M. W., and Faden, A. I. (1988) Decline in intracellular free magnesium concentration is associated with irreversible tissue injury following brain trauma. *J. Biol. Chem.* 263, 757–761.
12. Vink, R. (1989) Phospholipase C activity reduces free-magnesium concentration. *Biochem. Biophys. Res. Commun.* 165, 913–918.
13. Alvarez-Leefmans, F. J., Giraldez, F., and Gamino, S. M. (1987) Intracellular free magnesium in excitable cells: its measurement and its biologic significance. *Can. J. Physiol. Pharm.* 65, 915–925.
14. Gupta, R. K., Gupta, P., Yushok, W. D., and Rose, Z. B. (1983) On the noninvasive measurement of intracellular free magnesium by ^{31}P NMR spectroscopy. *Physiol. Chem. Phys. Med.* 15, 265–280.
15. Garfinkel, L. and Garfinkel, D. (1984) Calculation of free-Mg^{2+} concentration in adenosine 5′-triphosphate containing solutions *in vitro* and *in vivo*. *Biochemistry* 23, 3547–3552.
16. Bock, J. L., Wenz, B., and Gupta, R. K. (1985) Changes in intracellular Mg adenosine triphosphate and ionized Mg^{2+} during blood storage: detection by ^{31}P nuclear magnetic resonance spectroscopy. *Blood* 65, 1526–1530.
17. Vink, R., Yum, S. W., Lemke, M., Demediuk, P., and Faden, A. I. (1989) Traumatic spinal cord injury in rabbits decreases intracellular free nagnesium concentration as measured by ^{31}P MRS. *Brain Res.* 490, 144–147.
18. Corkey, B. E., Duszynski, J., Rich, T. L., Matschinsky, B., and Williamson, J. R. (1986) Regulation of free and bound magnesium in rat hepatocytes and isolated mitochondria. *J. Biol. Chem.* 261, 2567–2574.
19. McIntosh, T. K., Faden, A. I., Yamakami, I., and Vink, R. (1988) Magnesium deficiency exacerbates and pretreatment improves outcome following traumatic brain injury in rats: ^{31}P magnetic resonance and behavioral studies. *J. Neurotrauma* 5, 17–31.
20. McIntosh, T. K., Vink, R., Yamakami, I., and Faden, A. I. (1989) Magnesium protects against neurological deficit after brain injury. *Brain Res.* 482, 252–260.
21. Chaudry, I. H. (1983) Cellular mechanisms in shock and ischemia and their correction. *Am. J. Physiol.* 245, R117–R134.
22. Ebel, H. and Gunther, T. (1980) Magnesium metabolism: a review. *J. Clin. Chem. Clin. Biochem.* 18, 257–270.
23. Rubin, H. (1976) Magnesium deprivation reproduces the coordinate effects of serum removal or cortisol addition on transport and metabolism in chick embryo fibroblasts. *J. Cell. Physiol.* 89, 613–626.

24. Altura, B. M., Altura, B. T., and Gebrewold, A. (1983) Alcohol-induced spasms of cerebral blood vessels: relation to cerebrovascular accidents and sudden death. *Science* 220, 331–333.
25. Yamakami, I. and McIntosh, T. K. (1989) Effects of traumatic brain injury on regional cerebral blood flow in rats as measured with radiolabeled microspheres. *J. Cereb. Blood Flow Metab.* 9, 117–124.
26. Meyer, M. L., Westbrook, G. L., and Guthrie, P. B. (1984) Voltage-dependent block by Mg^{2+} of NMDA receptors in spinal cord neurones. *Nature* 309, 261–263.
27. Nicoletti, F., Wroblewski, J. T., and Costa, E. (1987) Magnesium ions inhibit the stimulation of inositol phospholipid hydrolysis by endogenous excitatory amino acids in primary cultures of cerebellar granule cells. *J. Neurochem.* 48, 771–777.
28. Vink, R., Faden, A. I., and McIntosh, T. K. (1988) Changes in cellular bioenergetic state following graded traumatic brain injury in rats: determination by phosphorus-31 magnetic resonance spectroscopy. *J. Neurotrauma* 5, 105–119.
29. Bara, M. and Guiet-Bara, A. (1984) Potassium, magnesium and cell membranes. *Magnesium* 3, 212–225.
30. Wieloch, T. and Siesjo, B. K. (1982) Ischemic brain injury: the importance of calcium, lipolytic activities, and free fatty acids. *Pathol. Biol.* 30, 269–277.
31. London, R. E. (1991) Methods for measurement of intracellular magnesium: NMR and fluorescence. *Annu. Rev. Physiol.* 53, 251–258.
32. Wei, E. P., Lamb, R. G., and Kontos, H. A. (1982) Increased phospholipase C activity after experimental brain injury. *J. Neurosurg.* 56, 695–698.
33. Rescasens, M., Mayat, E., and Guiramand, J. (1991) Excitatory amino acid receptors and phosphoinositide breakdown: facts and perspectives. *Curr. Asp. Neurosci.* 3, 104–175.
34. Calaviere, F., Sciarra, M., Crea, M. A., Rossi, M., and Proieeti, R. (1985) Variazioni del magnesio sierico ed urinario in pazienti traumatizzati cranici. *Recenti. Prog. Med.* 76, 563–566.
35. Flink, E. B. (1986) Magnesium deficiency in alcoholism. *Alcohol. Clin. Exp. Res.* 10, 590–594.
36. Hoek, J. E., Thomas, A. P., Rooney, T. A., Higashi, K., and Rubin, E. (1992) Ethanol and signal transduction in the liver. *FASEB J.* 6, 2386–2396.
37. Nicoll, R. A., Malenka, R. C., and Kauer, J. A. (1990) Functional comparison of neurotransmitter receptor subtypes in mammalian central nervous system. *Physiol. Rev.* 70, 513–565.
38. Gill, R., Foster, A. C., and Woodruff, G. N. (1988) MK801 is neuroprotective in gerbils when administered during the post-ischemic period. *Neuroscience* 25, 847–855.
39. Choi, D. (1987) Ionic dependence of glutamate neurotoxicity. *J. Neurosci.* 7, 369–379.
40. Mayer, M. L. and Miller, R. J. (1990) Excitatory amino acid receptors, second messengers and regulation of intracellular calcium in mammalian neurons. *Trends Pharm. Sci.* 11, 254–260.
41. Henneberry, R. C., Novelli, A., Vigano, M. A., Reilly, J. A., Cox, J., and Lysko, P. G. (1989) Energy-related neurotoxicity at the NMDA receptor: a possible role in Alzheimer's disease and related disorders. *Prog. Clin. Biol. Res.* 317, 143–156.
42. Levy, D. I., Sucher, N. J., and Lipton, S. A. (1990) Redox modulation of NMDA receptor-mediated toxicity in mammalian central neurons. *Neurosci. Lett.* 110, 291–296.

Chapter 5

ION MOVEMENTS AND EDEMA FORMATION IN CNS INJURY

Teiji Tominaga and S. Tsuyoshi Ohnishi

TABLE OF CONTENTS

I. INTRODUCTION

Brain edema is frequently encountered in various central nervous system (CNS) diseases including stroke, trauma, inflammation, and tumors. In addition to neuronal damage due to initial disease, associated brain edema itself may cause serious neurological deficits and, in some patients, may become life-threatening.[11] Therefore, the pathogenesis of brain edema has been intensely investigated using experimental and clinical materials. The mechanism underlying edema formation is different among the causative diseases. Ischemic injury is one of the major causes of edema formation.

Ischemic brain edema is generally classified into cytotoxic edema and vasogenic edema based on histological changes.[13,14] Cytotoxic brain edema (or ionic brain edema) occurs in the early phase of brain ischemia when the blood-brain barrier (BBB) is still intact and characterized by drastic ion movements between intracellular and extracellular fluids (Figure 1). Swelling of neuronal and glial cells is the histological feature at this stage. Consequently, an opening of the BBB leads to vasogenic brain edema (Figure 2). In this stage, extravasation of macromolecules such as albumin creates an osmotic gradient facilitating water movement from intravascular to extravascular spaces. However, the concept of vasogenic brain edema is changing with new insights. Ion movements, particularly sodium entry, are currently considered more important in the development of vasogenic brain edema.[6,17,18]

We report herein a temporal profile of ischemic brain edema caused by focal cerebral ischemia in rat, with detailed methodological description. Current opinions on the pathophysiology of ischemic brain edema are briefly discussed. Further, an involvement of

0-8493-8094-4/95/$0.00+$.50

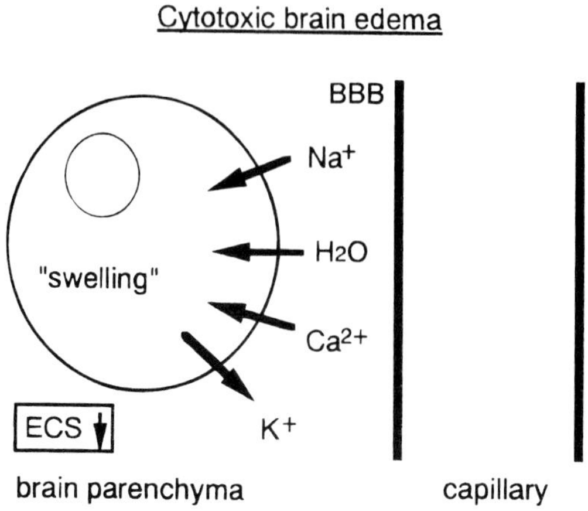

FIGURE 1. Ion movements in cytotoxic brain edema (ECS; extracellular space). (From Tominaga, T. et al., *Brain Res.*, 460, 376–378, 1988. With permission.)

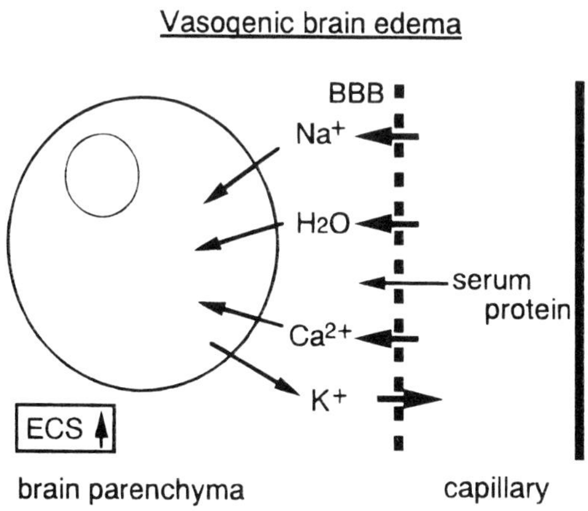

FIGURE 2. Ion movements in vasogenic brain edema (ECS; extracellular space). (From Tominaga, T. et al., *Brain Res.*, 460, 376–378, 1988. With permission.)

Ca^{2+}-activated potassium efflux was studied by examining the effect of charybdotoxin, an inhibitor of this channel, on ischemic brain edema.

II. FOCAL CEREBRAL ISCHEMIA MODEL OF RAT

Since the 1970s, various ischemia models have been used for analysis of ion and water changes. Focal ischemia models of rat are most extensively used for this purpose. Since the report by Tamura et al. in 1981, the middle cerebral artery (MCA) occlusion model of rat has been modified variably and developed to obtain reproducible ischemic lesion.[28] To ascertain the temporal profile of ischemic brain edema and ion movements, we have chosen a rat focal ischemia model that was originally reported by Chen et al., because of good reproducibility and low mortality rate.[4] Rat (male Sprague-Dawley rat, or Long Evans rat) was anesthetized with halothane in a mixture of nitrous oxide and oxygen via fitting facial mask. Systemic arterial blood pressure was monitored by a tail-cuff plethysmographic blood pressure meter. Otherwise, a polyethylene tube (PE 50) was inserted into the femoral artery and the blood pressure was monitored directly using a pressure transducer. Anesthesia was induced with halothane (3 to 4%) and maintained with 1 to 1.5% of halothane. After midline cervical incision, the bilateral common carotid arteries (CCAs) were gently exposed using a surgical microscope. Care was taken to avoid injury to the vagal nerve. The unilateral CCA was ligated by 4-0 silk suture at two places. Then an ipsilateral subtemporal craniotomy was performed with a dental drill at just medial to zygomatic arch. It is desirable to irrigate with saline solution to avoid heat injury during the drilling procedure. Different from the MCA occlusion model originally developed by Tamura et al.,[28] this model does not require proximal (to entorhinal fissure) occlusion of MCA, and thus the zygomatic arch can be left intact. The dura mater was opened by a 27-gauge needle and distal MCA was exposed at just proximal to the bifurcation of the MCA. Taking care to minimize cortical damage, the distal MCA was occluded by 10-0 suture (nylon monofilament; Ethicon). Immediately after the MCA occlusion, the contralateral CCA was occluded by aneurysm clip. The bone window was covered with a small piece of gel-foam (which was soaked in an antibiotic solution) and the wound was closed. The rat was allowed to recover from anesthesia. One hour after the onset of the ischemia, the rat was hand-held and the aneurysm clip released. Histologically, massive cortical infarction was found in the ipsilateral cerebral cortex. In some animals, scattered neuronal necrosis was also seen in the contralateral cerebral cortex. Spontaneous movements were reduced on the day of the surgery. Thereafter, it is difficult to distinguish the ischemic rat from normal rat, unless motor or cognitive tests are performed.[29] Mortality rate is less than 5% in this model.

III. FOCAL FREEZE INJURY MODEL OF RAT BRAIN

Focal freeze injury is a model of vasogenic brain edema. The operative procedures for focal freeze injury are rather simple compared to those of ischemic models. Rat was anesthetized with pentobarbital or inhalation anesthetics, and the head of the rat was fixed in a stereotactic frame. Xylocaine was adequately infiltrated under the scalp at the dorsal aspect of skull. A midline scalp incision was made and the dorsal surface of the skull was exposed. The unilateral coronal and lambda sutures and the sagittal suture are visible at this point. The size and the temperature of a metal probe (usually lead) and the corresponding size of bone window depends on the magnitude of the freezing insult required. A bone window is made by a dental drill with continuous irrigation of saline solution to prevent heat injury. (Care should be taken to avoid dural injury due to drilling procedures.) Then a pre-cooled metal probe is placed on the dura mater for a designated period. As cooling materials, aceton and isopentan, which are precooled to a designated temperature, and liquid nitrogen are popular. To produce extensive freezing lesion involving caudoputamen, a metal tube filled with liquid nitrogen can be applied to the temporal skull without craniotomy. The duration of metal placement on the dura mater again depends on the magnitude of insult required. One minute of contact with the metal probe is usually enough time to produce cortical freezing injury. After the procedure, the bone window is covered with a small piece of gel-foam and the wound is closed. We used a metal probe (5 mm diameter) that was precooled in liquid nitrogen. The probe was placed on the dura mater exposed between the coronal and lambda sutures for one minute. This procedure resulted in rapid histological changes, subsequent brain edema, and DNA injury.[31]

IV. QUANTIFICATION OF ISCHEMIC BRAIN EDEMA

There are several methods to evaluate ischemic brain edema. Histological evaluation provides qualitative observations of ischemic brain edema. Magnetic resonance imaging (MRI) also enables visualization of brain edema *in vivo*. Increase of water content results in prolongation of ^{1}H relaxation time and can be delineated as a high signal intensity area in T2-weighted imaging.

To directly quantitate ischemic brain edema, wet-dry and specific gravity methods have been used extensively. The experimental procedures of the wet-dry method and subsequent measurement of ion contents are indicated in Figure 3. The wet-dry method was first reported by Stewart-Wallace to measure the water content of human brain.[27] The wet weight of rapidly dissected brain sample is immediately measured. Alternatively, the brain sample can be frozen, cut into coronal sections, and a small portion punched out.[23,33] The wet brain sample is usually dried in an oven heated to 95 to 105°C for 5 to 7 d, or freeze-dried for several days. Then the dry weight of the sample is remeasured and the water content is calculated as follows:

$$\text{water content (\%)} = \frac{\text{wet weight} - \text{dry weight}}{\text{wet weight}} \times 100$$

The wet-dry method allows subsequent measurement of ion contents from the dried brain sample. However, when the sample volume is too small, measured values vary to a greater degree. This means that regional analysis of water content is more difficult in small experimental animals such as the gerbil and mouse. To obtain reliable data, rapid dissection of the brain sample is necessary. Further, the duration from the sacrifice of animal to the measurement

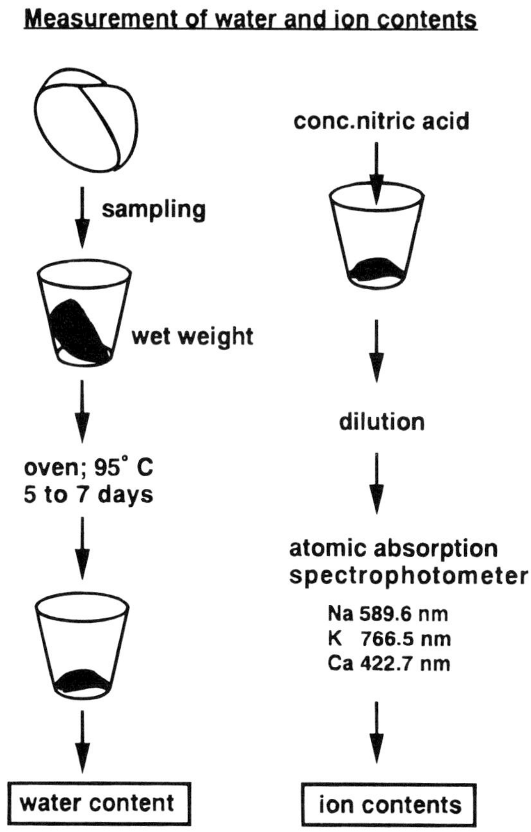

FIGURE 3. Experimental procedures for measurement of water and ion contents. Wet-dry method and an atomic absorption spectrophotometer are used for measurement of water and ion contents, respectively. (Tominaga, T. and Ohnishi, S. T., *Stroke*, 20, 513–518, 1989. With permission. © Am. Heart Assoc.)

of the wet weight of the brain sample should be constant. Humidity and temperature of the experimental room may also affect the reproducibility of the data. The reported values of water content of normal gerbil, rat, cat, and rabbit brain range from 76 to 79% by this method.[6,10,24,29,33] Water content is higher in gray matter than in white matter. In baboon, the water content of gray matter is around 80% and that of white matter around 67%.[32,34] Ion content such as sodium, potassium, and calcium of the dried brain sample can be measured by atomic absorption spectrophotometry. The dried brain sample is digested in concentrated nitric acid and diluted appropriately with distilled deionized water. The concentration of sodium, potassium, and calcium is measured at 589.6, 766.5, and 422.7 nm, respectively.[29] Standard and sample concentrations should be adjusted to fall within the linear range of the instrument. For accurate measurement, ionic contamination from glassware and tips has to be avoided.

The specific gravimetric method was first reported by Nelson et al. in 1971.[20] A gravity column, prepared by mixing of bromobenzene and kerosene, is used for determination of the specific gravity of the brain sample. In this column, gravity continuously increases in the vertical direction. A small brain sample is dropped into this column and its gravity is determined using potassium sulfate (K_2SO_4) as a standard. Higher water content in the brain sample results in a higher gravity value. Since it is possible to determine the gravity of a small brain sample from 1 to over 30 mg, regional differences of brain edema can be detected even in small animals by this method. However, when the brain sample is dropped into the column, it may not stop at one point, but instead fluctuate around its gravity level. This renders accurate measurement of specific gravity difficult, particularly when the sample volume is small. Therefore, the duration from the drop of the sample up to the reading of the gravity value should be constant (usually 1 to 2 min).

V. TEMPORAL PROFILE OF ISCHEMIC BRAIN EDEMA

We measured water and ion contents in the ischemic and contralateral hemispheres of rat exposed to focal ischemia. The wet-dry method and an atomic absorption spectrophotometer (Hitachi; Model Z-8000) were used for measurement. Normal cerebral hemisphere water content was 79.17 $\pm$ 0.03%, and the normal concentrations of sodium, potassium, and calcium were 259 $\pm$ 6, 464 $\pm$ 9, and 5.9 $\pm$ 0.2 μmol/g dry weight, respectively. Figure 4 shows the time courses of water content and ion concentrations of sham-operated and ischemic rats. The water content significantly increased 1 d after the ischemia, reached maximum at 3 d, and declined thereafter. The increase of sodium and the loss of potassium were correlated well with the profile of water accumulation. The amount of sodium gain exceeded that of potassium loss. Calcium was also significantly increased from 1 d after the ischemia. These significant changes resolved two weeks after the ischemia, except the calcium changes. Persistent accumulation of calcium is because of its accumulation in necrotic tissue.

The time course of ischemic brain edema varies depending on the type of ischemic insult; focal or forebrain ischemia, and permanent or transient ischemia. In focal cerebral ischemia, formation of ischemic brain edema is distinct among ischemic core, border (penumbra), and peripheral zones; most prominent changes are found in ischemic core zone.[18,33] In transient forebrain ischemia, brain edema does not occur during the ischemic period when the ischemia is complete.[10,34] While the ischemia is incomplete, water content increases even during ischemia. Reperfusion following substantial ischemic insult accelerates edema formation.[10,34] According to the experiments using ion-sensitive microelectrodes, changes of extracellular calcium and potassium concentrations begin only a few minutes after the onset of ischemia.[2,7] Subsequent accumulation of water becomes significant from several hours after ischemia by the wet-dry method.[33] T2-weighted and diffusion imagings of MRI detect this water accumulation earlier, possibly within 30 min after ischemia.

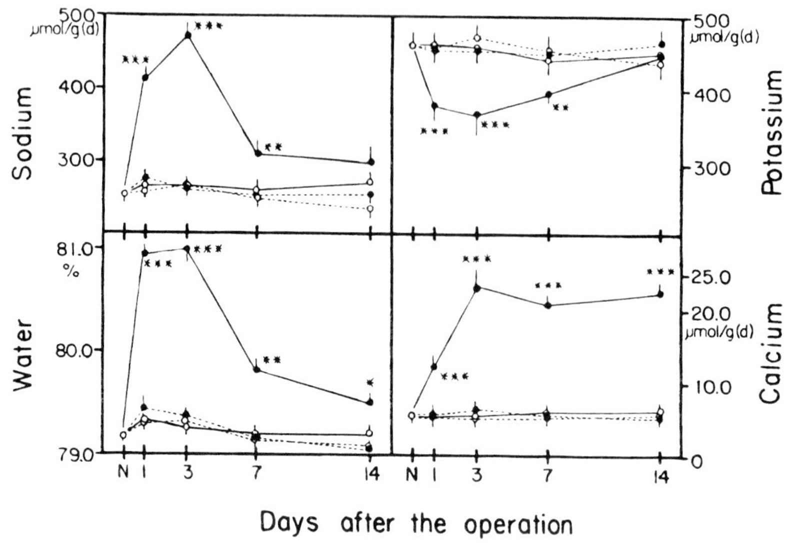

FIGURE 4. Time courses of sodium, potassium, water, and calcium contents in the rat brain exposed to focal ischemia. The values were expressed as mean ± SEM. (—) ischcemic rat; (--) sham-operated rat; (●) ipsilateral hemisphere; (○) contralateral hemisphere. Difference between ischemic and sham-operated ipsilateral hemispheres: *$p < 0.05$, **$p < 0.01$, ***$p < 0.001$. (From Tominaga, T. and Ohnishi, S. T., *Stroke*, 20, 513–518, 1989. With permission.)

VI. THE MECHANISM OF EDEMA FORMATION

In a classical hypothesis, hydrostatic and osmotic pressures of brain parenchymal tissue and blood have a principal role in the development of brain edema. The hydrostatic or osmotic pressure gradient works as a drive force from intravascular fluid to brain parenchymal tissue.[8,9,15] Recent investigations, however, focused more on the role of ion movements and tried to elucidate the mechanism of water accumulation from ion movement. Accumulating evidence indicates that the sodium gain in ischemic tissue may account for water entry.[6,18,33] Gotoh et al. and Young et al. reported that the net ion shift of sodium and potassium correlated well with the changes of water content.[6,33] The increase of water can be calculated as follows:

$$\Delta H_2O \times 145.4 = (\Delta Na - \Delta K) + 10.5$$

It is generally accepted that the increase in sodium exceeds the decrease in potassium in the early stage of ischemic brain edema. As a mechanism of sodium entry from blood, Na-K-ATPase present in the capillary endothelial cells may be activated in the early phase of ischemia by an unknown mechanism.[12,16] The capillary endothelial cells are known to be rather resistant to ischemic insult. Activation of this enzyme mediates entry of sodium from blood and loss of potassium from brain parenchyma, while Na-K-ATPase present in the neuronal and glial cells may be inactivated by rapid energy failure. The net effect of these Na-K-ATPase activities is the accumulation of sodium and loss of potassium in brain parenchyma. However, detailed activity of Na-K-ATPase is still to be determined and further study is necessary to clarify the exact mechanism of sodium and water accumulations.

It has been generally considered that serum proteins penetrating the BBB create an osmotic gradient enabling water movement in the development of vasogenic brain edema. Recent evidence, however, suggests that albumin may play only a minor or no role in edema formation.[18] Indeed, prolonged administration of albumin resulted in reduction of ischemic brain edema.[17] Additionally, even if the BBB is opened by intracarotid injection of high-osmotic solution, vasogenic brain edema does not occur.[22] Instead of serum proteins, ion shift is again thought to be involved in the late phase of edema formation.

Other than ion shifts, several substances have been assumed to be related to ischemic brain edema, such as free radicals, eicosanoids, arachidonic acid, etc.[1,3] These substances are thought to be involved in edema formation mainly through an increase of permeability of the capillary.

VII. Ca^{2+}-ACTIVATED K CHANNELS

Unlike sodium and calcium, potassium has received little attention in the pathophysiology of ischemic brain edema. It has been considered only as a counterpart of sodium transport across membranes. Potassium is important as well in maintaining membrane excitation and, thus, neuronal functions in brain. Potassium channels have been categorized according to biophysical and pharmacological properties. Voltage-dependent K channels comprise a family of proteins, namely, Shal, Shaw, Shab, and Shaker (see Reference 25). These channels exist in various organisms, from flies to humans, and are involved in forming action potential in all excitable cells. Ca^{2+}-activated K channels also represent an additional family of proteins. This family of ligand-gated channels are divided into three subtypes: large conductance (maxi-K), intermediate conductance, and small conductance.[5] Maxi-K channels are present in a variety of excitable and nonexcitable membranes, including neurons, and are involved in a variety of cellular processes such as neuroendocrine, excitation-contraction coupling in muscle, and fluid secretion. Charybdotoxin (CTX) is a minor peptidyl component of the venom of the scorpion *Leiurus quinquestriatus,* and inhibits maxi-K in a freely reversible manner. Originally CTX was extracted from this scorpion venom as a 37 amino acid peptide, while at present, synthesized products are available commercially.[19,26] CTX concentration is not related to the blocked time of maxi-K, while the mean duration of bursts of channel activity is inversely proportional to the CTX concentration.[5] CTX is a potent inhibitor of maxi-K, but its effect is not selective to maxi-K. CTX also inhibits voltage-dependent K channels in T-lymphocytes and neuronal tissue, and intermediate and small-conductance K channels (for further details of Ca^{2+}-activated K channel and CTX, see Reference 5).

VIII. THE EFFECT OF CHARYBDOTOXIN, AN INHIBITOR OF CA^{2+}-ACTIVATED K CHANNELS, ON ISCHEMIC BRAIN EDEMA

To study whether Ca^{2+}-activated K channels are involved in the formation of ischemic brain edema, we examined the effect of CTX on ischemic brain edema using the rat focal ischemia model. CTX was purified from the venom of the scorpion *Leiurus quinquestriatus* (Latoxan; Rosans, France) according to the procedures of Smith et al.[26] CTX in saline solution was administered intravenously via the jugular vein 20 to 30 min prior to the onset of ischemia. Three days after the ischemia, the rat was sacrificed and water, sodium, potassium, and calcium contents of the cerebral hemispheres were measured according to the procedures mentioned earlier.

The increase of water content was significantly reduced by CTX at the concentration of 0.15 mg/kg (Figure 5).[30] The smaller dose of 0.05 mg/kg did not affect the water accumulation, while the larger dose of 0.45 mg/kg significantly deteriorated (Figure 6). Therefore, the dose around 0.15 mg/kg of CTX may be optimal to obtain protective effect against ischemic brain edema. Concurrently, ion movements were also mitigated by 0.15 mg/kg of CTX. In addition to the loss of potassium, accumulations of sodium and calcium were also reduced. This is in accordance with experimental evidence that CTX has protective effect against cellular

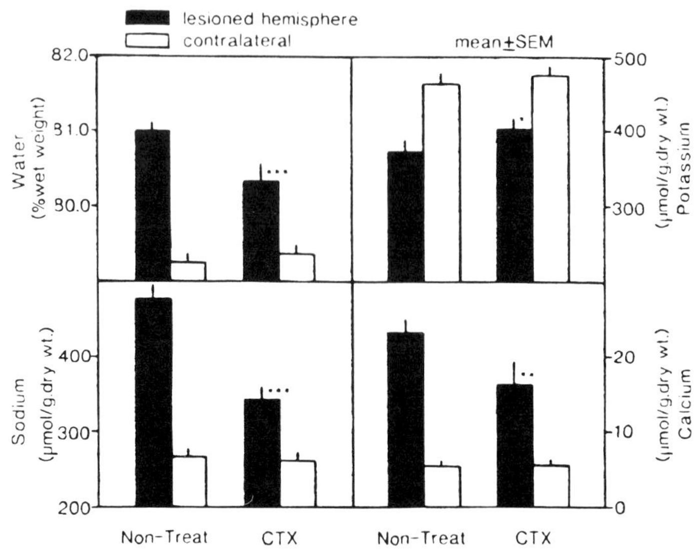

FIGURE 5. The contents of water, potassium, sodium, and calcium in ischemic lesioned and contralateral hemispheres. The data were obtained 3 d after focal ischemia. The number of nontreated and CTX-treated animals were 10 and 11, respectively. CTX (0.15 mg/kg) was injected intravenously 20 to 30 min prior to the ischemic insult. The values were expressed as mean ± SEM. $^*p < 0.05$, $^{**}p < 0.01$, $^{***}p < 0.001$ (Student's *t*-test). (From Tominaga, T. et al., *Brain Res.*, 460, 376–378, 1988. With permission.)

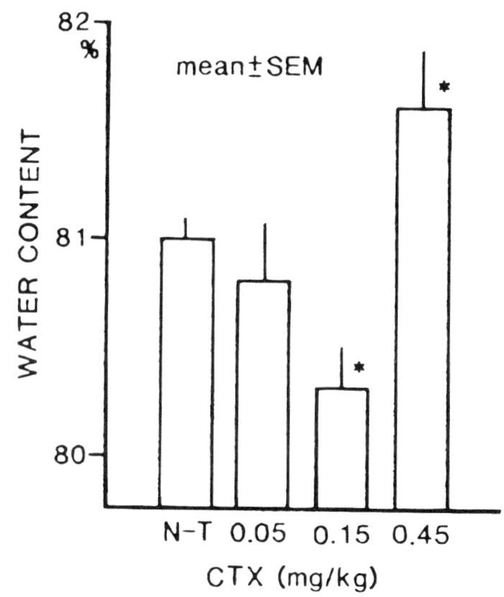

FIGURE 6. The water contents (% wet weight) of the ischemic lesioned hemispheres obtained from various concentrations of CTX treatment. The values were expressed as mean ± SEM. *Significantly different from the nontreated (N-T) group. (From Tominaga, T. et al., *Brain Res.*, 460, 376–378, 1988. With permission.)

injury associated with drastic ion movements. We previously found that CTX prevents the sickling phenomenon of red blood cells closely associated with potassium loss. Further, CTX also mitigated ionic shifts caused by spinal cord injury that are very similar to those caused by ischemic brain injury.[21] More detailed study is necessary to clarify the mechanism of the beneficial effect of CTX. However, prevention of potassium loss by CTX may provide new insight into brain pathology closely associated with ionic shifts and a novel therapeutic avenue in its treatment.

REFERENCES

1. Asano, T., Gotoh, O., Koide, T., and Takakura, K. Ischemic brain edema following occlusion of the middle cerebral artery in the rat. II. Alteration of the eicosanoid synthesis profile of brain microvessels. *Stroke* 16: 110–113, 1985.
2. Branston, N. M., Strong, A. J., and Symon, L. Extracellular potassium activity, evoked potential and tissue blood flow: relationships during progressive ischaemia in baboon cerebral cortex. *J. Neurol. Sci.* 32: 305–321, 1977.
3. Chan, P. H., Schmidley, J. W., Fishman, R. A., and Longar, S. M. Brain injury, edema, and vascular permeability changes induced by oxygen-derived free radicals. *Neurology* 34: 315–320, 1984.
4. Chen, S. T., Hsu, C. Y., Hogan, E. L., Maricq, H., and Balentine, J. D. A model of focal ischemia stroke in the rat: reproducible extensive cortical infarction. *Stroke* 17: 738–743, 1986.
5. Garcia, M. L., Carcia-Calvo, M., Knaus, H.-G., and Kaczorowski, G. J. High-conductance Ca^{2+}-activated K^+ channels: pharmacology and molecular characterization. *Semin. Neurosci.* 5: 107–115, 1993.
6. Gotoh, O., Asano, T., Tamura, A., and Takakura, K. Ischemic brain edema following occlusion of the middle cerebral artery in the rat. I. The time course of the brain water, sodium, and potassium contents and blood-brain barrier permeability to 125I-albumin. *Stroke* 16: 101–109, 1985.
7. Harris, R. J., Symon, L., Branston, N. M., and Bayhan, M. Changes in extracellular calcium activity in cerebral ischemia. *J. Cereb. Blood Flow Metab.* 1: 203–209, 1981.
8. Hatashita, S. and Hoff, J. T. Role of hydrostatic pressure gradient in the formation of early ischemic brain edema. *J. Cereb. Blood Flow Metab.* 6:, 546–552, 1986.
9. Hatashita, S., Hoff, J. T., and Salamat, S. M. Ischemic brain edema and the osmotic gradient between blood and brain. *J. Cereb. Blood Flow Metab.* 8: 552–559, 1988.
10. Hossmann, K.-A., Sasaki, S., and Zimmermann, V. Cation activities in reversible ischemia of the cat brain. *Stroke* 8: 71–76, 1977.
11. Katzman, R., Clasen, R., Klazto, I., Meyer, J. S., Pappius, H. M., and Waltz, A. G. Report of the joint committee for stroke resources. IV. Brain edema in stroke. *Stroke* 8: 512–540, 1977.
12. Koide, T., Asano, T., Matsushita, H., and Takakura, K. Enhancement of ATPase activity by a lipid peroxide of arachidonic acid in rat brain microvessels. *J. Neurochem.* 46: 235–242, 1986.
13. Klatzo, I. Neuropathological aspects of brain edema. *J. Neuropathol. Exp. Neurol.* 26: 1–14, 1967.
14. Klatzo, I. Pathophysiological aspects of brain edema. *Acta Neuropathol.* 72: 236–239, 1987.
15. Kogure, K., Busto, R., and Scheinberg, P. The role of hydrostatic pressure in ischemic brain edema. *Ann. Neurol.* 9: 273–282, 1982.
16. Lo, W. D., Betz, A. L., Schielke, G. P., et al. Transport of sodium from blood to brain in ischemic brain edema. *Stroke* 18: 150–157, 1987.
17. Matsui, T., Sinyama, H., and Asano, T. Beneficial effect of prolonged administration of albumin on ischemic cerebral edema and infarction after occlusion of middle cerebral artery in rats. *Neurosurgery* 33: 293–300, 1993.
18. Menzies S. A., Betz, L., and Hoff, J. T. Contributions of ions and albumin to the formation and resolution of ischemic brain edema. *J. Neurosurg.* 78: 257–266, 1993.
19. Miller, C., Moczydlowski, E., Latorre, R., and Phillips, M. Charybdotoxin, a protein inhibitor of single Ca^{2+}-activated K^+ channels from mammalian skeletal muscle. *Nature (London)* 313: 316–318, 1985.
20. Nelson, S. R., Mantz, M. L., and Maxwell, J. A. Use of specific gravity in the measurement of cerebral edema. *J. Appl. Physiol.* 30: 268–271, 1971.
21. Ohnishi, S. T., Barr, J. K., and Katagi, C. Charybdotoxin improves motor recovery of the rat after spinal cord injury. *Pharmacol. Biochem. Behav.* 31: 187–191, 1988.
22. Rappaport, S. I., Matthews, K., and Thompson, H. K. Absence of brain edema after reversible osmotic opening of blood-brain-barrier. In: Pappius, H. M. and Feindel, W. (eds.), *Dynamics of Brain Edema*, Springer-Verlag, Berlin, 1976, pp. 18–22.
23. Rappaport, Z. H., Young, W., and Flamm, E. S. Regional brain calcium changes in the rat middle cerebral artery occlusion model of ischemia. *Stroke* 18: 760–764, 1987.
24. Schuier, F. J. and Hossmann, K.-A. Experimental brain infarcts in cats. II. Ischemic brain edema. *Stroke* 11: 593–601, 1980.
25. Salkoff, L., Baker, A., Butler, M., and Covarrubias, M. An essential 'set' of K^+ channels conserved in flies, mice and humans. *Trend Neurosci.* 15: 161–166, 1992.
26. Smith, C., Phillips, M., and Miller, C. Purification of charybdotoxin, a specific inhibitor of the high-conductance Ca^{2+}-activated K^+-channel. *J. Biol. Chem.* 262: 14607–14613, 1986.
27. Stewart-Wallace, A. M. A biochemical study of cerebral tissue, and of the changes in cerebral oedema. *Brain* 62: 426–438, 1939.

28. Tamura, A., Graham, D. I., McCulloch, J., and Teasdale, G. M. Focal cerebral ischemia in the rat: I. Description of technique and early neuropathological consequences following middle cerebral artery occlusion. *J. Cereb. Blood Flow Metab.* 1: 53–60, 1981.
29. Tominaga, T. and Ohnishi, S. T. Interrelationship of brain edema, motor deficits and memory impairment in the rat exposed to focal ischemia. *Stroke* 20: 513–518, 1989.
30. Tominaga, T., Katagi, H., and Ohnishi, S. T. Is Ca^{2+}-activated potassium efflux involved in the formation of ischemic brain edema? *Brain Res.* 460: 376–378, 1988.
31. Tominaga, T., Kure, S., and Yoshimoto, T. DNA fragmentation in focal cortical freeze injury of rats. *Neurosci. Lett.* 139: 265–268, 1992.
32. Watanabe, O., West, C. R., and Bremer, A. M. Experimental regional cerebral ischemia in the middle cerebral artery territory in primates. Part 2: Effects on brain water and electrolytes in the early phase of MCA stroke. *Stroke* 8: 71–76, 1977.
33. Young, W., Pappaport, Z. H., Chalif, D. J., and Flamm, E. S. Regional brain sodium, potassium, and water changes in the rat middle cerebral artery occlusion model of ischemia. *Stroke* 18: 751–759, 1987.
34. Zimmerman, V. and Hossmann, K.-A. Resuscitation of the monkey brain after one hour's complete ischemia. II. Brain water and electrolytes. *Brain Res.* 85: 1–12, 1975.

III. Surgical Injury Models

A. Brain Study

B. Head Injury Models

C. Spinal Cord Study

A. Brain Study

Chapter 6

A GLOBAL ISCHEMIA MODEL: THE USE OF Na,K-ATPase ANALYSIS

Stephen E. Karpiak

TABLE OF CONTENTS

I. INTRODUCTION

Several animal models of CNS ischemia are available for study by neurobiologists (for review, see Karpiak et al., 1989). Assessment of resources and research goals is essential before employing any of the studied and accepted animal models. Using any animal model of CNS ischemia requires a period of time where considerable experience with the surgical and animal care procedures is gained. This period of time is essential in order to achieve a level of adequacy as well as a working understanding of the model and its variables. This is imperative in order to execute the model in such a way as to achieve the levels of reproducibility that are mandatory for studies of CNS ischemic injury processes. Achieving reproducibility is most critical when studies are undertaken that assess the efficacy parameters of neuroprotective compounds. The researcher should be forewarned not to view any of the animal CNS ischemic models as being *readily* executed. Even the "simplest" models, such as those that induce global ischemia by carotid artery occlusion, are fraught with intrinsic and often intractable variables. These variables reach maximal levels in focal ischemia models, such as the occlusion of the middle cerebral artery. The variables are a function of differences in anatomy (especially vasculature) as well as often obscure (not obvious) individual and environmental (laboratory) settings. These variables must be controlled and/or well understood before the experimenter can validly assess any proposed hypothesis regarding CNS ischemia and the efficacy of neuroprotective compounds that are hypothesized to ameliorate such neural damage.

The basis for all animal models of CNS ischemia injury is to cause neural tissue damage by blocking (partially, temporarily, or completely) blood flow to either discrete or more global CNS regions. The neurobiologist should be aware that while some of these models

0-8493-8094-4/95/$0.00+$.50

approximate the human condition, they do not mimic the human pathophysiology. The animals used in most CNS ischemia models are young adults in which are absent the classic underlying risk factors that are concomitant with CNS ischemic occurrences in the human population, namely, hypertension and cerebral vascular pathologies. Extrapolation from the model to the human condition should be undertaken with considerable caution (Hernandez et al., 1994).

Last, after choosing the appropriate model and establishing its reproducibility in the laboratory, the variable(s) to be monitored must also be chosen with caution. Functional (behavioral and electrophysiological) measures, as well as classic edema, biochemical (e.g., ion transmitter levels), and histological assessments are characteristic of the independent variables used in the study of CNS ischemia. In all cases these assessments are subject to the extraordinarily dynamic elements that comprise a CNS ischemic episode and injury cascade. These critical elements include the time after the insult and the array of differences between the primary and periischemic tissue areas, as well as the differences in ischemic injury processes that occur in white vs. gray matter. Ischemic injury is a process, not a static event. This dynamism complicates the investigative process, making it susceptible to imprecise interpretation and conclusions.

II. THE GLOBAL ISCHEMIA MODEL

Animal models for CNS ischemic injury can be classified into three categories, each of which is dependent upon the size and location of the ischemic tissue damage: (a) entire cerebral hemisphere(s) (i.e., global ischemia); (b) focal regions of the brain; and (c) multiple loci in the brain (multi infarct dementia). In this chapter the global ischemia model will be described and discussed. The attentive choice of sensitive and meaningful assessments (independent variables) is essential in the study of ischemia injury and recovery (supra). Our laboratories have included the use of the assessment of membrane ATPase activity as a marker for assessing ischemic injury processes and recovery. The use and significance of membrane Na-K-ATPase analyses will be described and discussed in the context of our research on the neuroprotective agent GM1 ganglioside (monosialoganglioside) (for review, see Karpiak et al., 1990).

A. BILATERAL CAROTID OCCLUSION

The reduction of blood supply into the two cerebral hemispheres is characterized by massive cerebral infarction and edema formation. The ease (i.e., simplicity) in inducing that type of global ischemia is the attraction of this bilateral carotid occlusion model. Its primary use is for studies of massive irreversible ischemia that results in death. The bilateral occlusion model was one of the first techniques used in the rat to study ischemia wherein both arterial as well as venous blood supply is interrupted using extrinsic transcutaneous compression of the neck with an inflatable collar (Kowada et al., 1968). In the human condition, global ischemia is typically a consequence of myocardial infarction, hypertension, or ventricular fibrillation. The levels of achievable relevance using this model approach is problematic since the end point is death. In the rat this ischemic episode is achieved by isolation of the common carotid artery through an incision on the ventral surface of the neck. The salivary glands are moved laterally, and the carotid sheath exposed. Both the vagus and sympathetic nerves are separated from the common carotid artery, which is then permanently ligated. In the rat, with permanent bilateral ligations of the common carotid artery, there results a better than 45% mortality 12 h after occlusion. The typical variables include temperature, sex, animal strain, age, and specifics of the procedure. Death results from brain swelling (edema) and focal infarcts. The ability to use numerous animals coupled with the dependent measure mortality has made the use of this global ischemia model attractive. Unlike focal ischemia models that involve surgical opening of the calvarium and intracranial manipulations (leading to alterations in blood brain barrier integrities), the global model avoids these variables, making it more akin to the human pathology. The use of temporary bilateral occlusion of the carotids has been used as well. This approach usually results in focal (hippocampal) infarcts. This approach must assume the added complication of damage due to reperfusion, thereby making the model one wherein two types of tissue damage occur in the same loci.

B. UNILATERAL HEMISPHERE ISCHEMIA

Unilateral hemispheric global ischemia can be accomplished in the Mongolian gerbil because most (not all) have an incomplete circle of Willis, wherein there is little or no communicating vascular system between the hemispheres (Levine et al., 1966, 1969). Following either a unilateral partial or complete ligation of the common carotid artery, ischemia is largely localized to the ipsilateral cerebral hemisphere in the majority of the animals (45 to 50%). This is advantageous, because the animal *can* be used (and often *must* be used when assaying ischemic tissue taken from the surviving animals only) as its own control wherein comparisons can be made between the *ligated and nonligated* hemispheres. In the gerbils that survive the permanent unilateral common carotid artery occlusion there can be expansive variability in the extent, *if any,* of the resulting cerebral injury. The injury is not exclusively ischemic, and, since most studies do not assay tissue from those animals that die, the population sample is *biased.* Therefore, analyses that are confined to only surviving animals can present methodological problems when analyzing certain injury (morphological and biochemical) parameters. These difficulties can be modified if a model utilizing temporary ligation is employed.

The gerbil is not particularly amenable to quantitative functional (neurological and behavioral) testing. After unilateral ligation some gerbils present with hyperactivity as well as periods of rolling, jumping, or running that last as long as 5 min. Mildly affected animals show transient contralateral hemiparesis and ipsilateral circling (Kitagawa et al., 1989). While the gerbil is not ideally suited for neurological testing (supra) some motor deficits can be observed and a large number of gerbils (35%) will develop either focal or generalized seizures (Karpiak et al., 1989).

When global ischemia is induced by unilateral ligation the locus of the primary infarct is initially in the posterior portion of the hemisphere. Anterior areas become ischemic with time, indicative of the spread of the ischemic and nonischemic zones of tissue damage (including periinfarct loci). Although the model implies that there is no vascular circulation to the ischemic hemisphere, there *is* cross-circulation from the intact carotid to the anterior portion of the hemisphere (the anterior cerebral arteries join and enter the interhemispheric region as a single vessel); (Kitagawa et al., 1989). Therefore, the primary infarct (following unilateral common carotid artery occlusion) is induced due to the absence of posterior communicating arteries, and the global aspect of the infarct is a function of the spread of damage *(not specifically ischemic)* into anterior hemispheric regions that *do* have blood supply. This anterior damage occurs as a result of the direct compression (not lack of vasculature) by the substantially enlarged posterior ischemic hemisphere. Accordingly, there is considerable evidence of partial blood flow to the infarcted cerebral hemisphere. Since the CNS tissue damage is not exclusively ischemic, discussion of this model must take into account the heterogeneity of the type of damage. As compared to focal models of ischemia, the global model produces more varied types of tissue injury and injury process.

At 48 h after the unilateral occlusion and the induction of global ischemia, approximately 50% of all animals will be dead. The remaining 50% will exhibit a range of ischemic damage from severe to none. This variability is considerable and must be taken into consideration for any study or pharmacological assessment. The surviving animals do have varying degrees of intrahemispheric communicating arteries. This variance, taken together with the fact that the damage that occurs as a result of the initial induction of the global ischemia is not exclusively ischemic (supra), should alert the experimenter to the need to interpret the results with care. This caution is essential, particularly, in studies whose purpose is to assess and better understand ischemic injury processes exclusively (lack of blood flow). While there are procedures (Kitagawa et al., 1989) that can be used to identify and predict which gerbils are stroke-prone, such methodologies are cumbersome and mitigate against the facility of the model.

The variation in the vasculature of the gerbil makes it necessary for any researcher to seriously consider the need to employ large groups of animals in certain experimental paradigms. The typical n = 10 is generally not adequate for any valid interpretation of results or hypothesis acceptance/rejection. This is particularly true in studies where the only independent variable is mortality. For example, it is *highly probable* to assign ten gerbils to one treatment category in a study, and only one or two of those gerbils might have incomplete posterior communicating arteries. The final results in our laboratories that established the efficacy of GM1 ganglioside to reduce mortality after global ischemia in the gerbil (mortality rates were decreased to 25%) were evident and statistically significant ($p < 0.001$) when 50 to 100 animals were assigned to each of the experimental and control groups (Karpiak et al., 1987) (Figure 1).

C. THE METHOD

Both male and female Mongolian gerbils weighing 60 to 80 g have been used to execute this model. In our laboratories gerbils are anesthetized (IM) with a mixture of (7:3 by volume) of Vetalar (ketamine hydrochloride, 87.5 mg/kg) and Rompun (xylazine hydrochloride, 7.5 mg/kg). The procedure is done on a Harvard Apparatus operating table and the animals are kept in a cage resting on a heating pad kept at the same ambient temperature (37°C) as that of the operating table.

Temperature regulation (during surgery and recovery as well as in the animal care/housing environment) is essential since the animals become hypothermic after the occlusion. This hypothermia variable is critical and must be controlled in order to achieve the reproducibility necessary for the model to be applicable for any systematic study. An incision is made at the ventral surface of the neck, the salivary glands are moved laterally, and the right carotid sheath is exposed. Both the vagus and sympathetic nerves are separated (blunt dissection) from the right common carotid artery, which is then permanently ligated using O-gauge silk

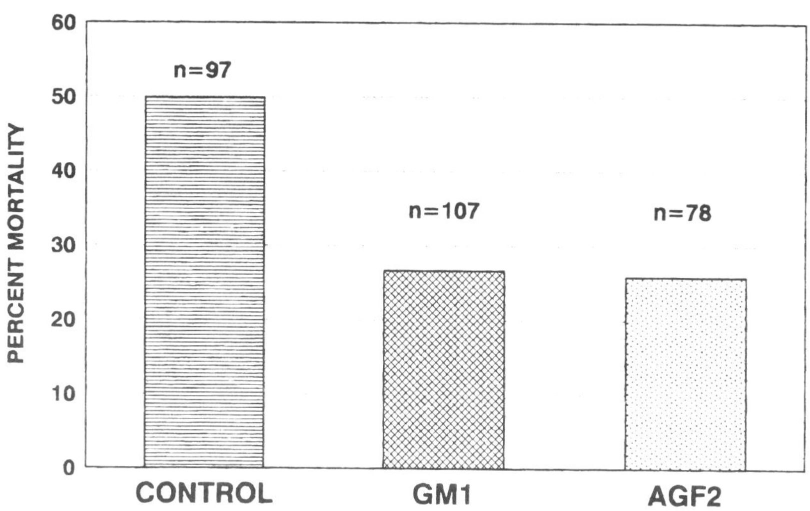

FIGURE 1. Mortality rate in three experimental groups at 48 h after induction of global ischemia by permanent unilateral carotid ligation.

thread (Karpiak et al., 1987). There are several surgical procedural variations for executing this model (Hattakeyama et al., 1988; Kato et al., 1989). The important factor is consistency in the model execution.

III. ASSAYING FOR MEMBRANE ATPase

Analyses of membrane ATPase activity has been used as a primary ***index*** in our laboratories since the activity of these enzymes provides information about the functional status of the plasma membrane as well as its structural integrity. The integrity of the membrane can be interpreted by the activity of the ATPase enzymes since that activity is dependent upon the membrane content and order of membrane lipids and proteins. Further analyses of the integrity of the plasma membrane can be achieved by assays of fatty acid levels (membrane content) as well as membrane fluidity parameters (Mahadik et al., 1993).

A. ATPase AFTER ISCHEMIA: GANGLIOSIDE EFFECTS

In order to best illustrate the suitability and advantages of ATPase analyses for studies of ischemic injury processes, and mechanisms by which neuroprotective agents are effective, our laboratory's data and experiences using ATPase analyses and the study of the neuroprotective compound GM1 ganglioside are examined and discussed (Karpiak et al., 1986, 1987; Mahadik et al., 1989, 1992; Mazzari et al., 1990).

We found that at 48 h after permanent unilateral ligation of the common carotid artery in the gerbil, saline-injected gerbils had a 52% mortality (Figure 1). GM1- and AGF2-injected gerbils showed almost identical mortalities of 27 and 26%, respectively. AGF2 is an internal ester GM1 ganglioside derivative (Karpiak et al., 1987). The ganglioside treatments reduced mortality from 52 to 26% ($p < 0.001$). By comparing levels of membrane Na-K-ATPase activity in the occluded vs. nonoccluded side of the brain in saline-injected controls, we found significant ($p < 0.01$) decreases in enzyme activity (Figure 2). Specifically, we found a 19.8% decrease in the cortex and a 34.7% decrease in the hippocampus. There were no significant differences (losses) in enzyme activity between the occluded and nonoccluded sides of the brain in either the GM1- or AGF2-treated rats. It is important to note that since

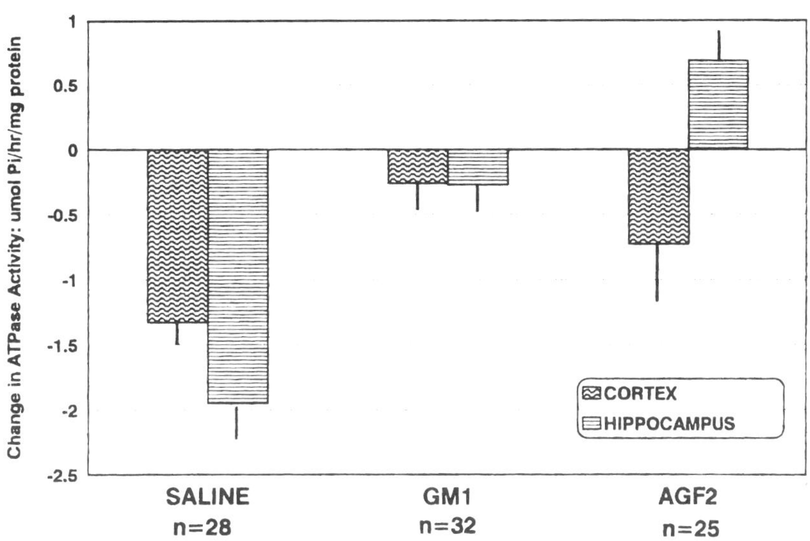

FIGURE 2. The differences (nonoccluded vs. occluded hemisphere) in Na-K-ATPase activity for each of three experimental groups. The increase in Na-K-ATPase activity in the hippocampus in the AGF2 group likely represents an increase in binding capacity (kinetics) for substrate.

we were unable to conduct any analyses on tissue from the dead animals, all comparisons were intraanimal (comparing the occluded vs. nonoccluded hemisphere). If it were possible to include tissue from those animals that had died, the difference in loss of activity would be markedly greater, and the degree of neuroprotection by GM1 or AGF2 also greater. These changes in Na-K-ATPase after ischemia and their protection with GM1 treatment have also been seen when using analyses for Mg^2-ATPase (Mahadik et al., 1992).

Analyses of membrane ATPase activity in our studies (Karpiak et. al, 1991) using a focal model of ischemia (middle cerebral artery occlusion) indicate that results occur similar to those seen with the global ischemia model. Mahadik et al. (1992) reported that losses of Na-K-ATPase after focal ischemia are reduced by GM1 in the primary infarct area, and are almost *normal* in periischemic regions. These studies also demonstrated that there is a time-dependent change (days 1 to 14) in the levels of the α^+ form of Na-K-ATPase (largely neuronal) and in the levels of α^- (largely nonneuronal—glial) membrane Na^+-K^+-ATPase. Identical Ca-ATPase (Ca^{2+} pump) patterns in losses after ischemia are reduced with GM1 treatment. This protection of the Ca-ATPase activity is directly related to the fact that high Ca^{2+} levels that occur after ischemic injury are reduced following GM1 treatment (Karpiak et al., 1991, 1994; Manev et al., 1990a,b; Mazzari et al., 1990).

In the focal ischemia studies, no changes (losses) were observed in enzyme ATPase activities in either primary or periischemic areas 24 h after ischemia. This data suggests that within the ischemic area sampled the enzymes remain *functional* within 24 h after the ischemic insult (the plasma membrane retains sufficient structural integrity to maintain the activity of these enzymes). In the primary infarct area this *functionality* is not indicative of any *irreversibility* of cell damage, but is likely indicative of a state of irreversibility in periischemic areas where some cerebral blood flow remains. *This conclusion can be applied to the global ischemia model as well, where posterior infarct areas are likely not reversible, but anterior injury processes are reversible and amenable to rescue by neuroprotective agent treatment.* In a prior study the changes in levels of tissue Na^+ (increased), K^+ (decreased), and edema (increased), considered to be a result of a loss of Na^+-K^+-ATPase activity due to a loss of membrane structure, were found to be maximal at 24 h after ischemia. This indicates that these ionic changes and edema formation are a function of other cellular processes that do not exclusively involve Na^+-K^+-ATPase activity. One such mechanism can be the decrease in ATP levels since ion transport across the plasma membrane is energy dependent.

The largest reductions in ATPase enzyme activity occurred in both ischemic and periischemic areas at 48 to 72 h after the ischemic insult. This result indicates maximal structural and functional membrane alterations. Fourteen days after ischemia, the ATPase enzyme activities in the periischemic areas return to control levels. This recovery of enzyme activity probably indicates that the membrane changes that occur in these periischemic areas are largely reversible. The primary and periischemic areas differ in the degree of loss (70 to 75% vs. 42 to 56%, respectively) of enzyme activity (indicative of the extent of membrane injury) and the degree of recovery (marginal vs. complete).

The persistent loss of ATPase [both isozymes − and +] in the primary ischemic area at two weeks after the insult indicates the irreversible failure of plasma membrane function of both neuronal and glial cells. A slight increase in α^- ATPase (22%) activity in the primary area at two weeks after ischemia supports the conclusion that nonneuronal (glial) cell proliferation is occurring. While our laboratory's studies of the distribution of α^+ and α^- is similar to previous reports (60 and 40%, respectively) and is consistent with the distribution of neuronal and nonneuronal mass in cortex, morphological analysis using antibodies specific to these isozymes as well as high- and low-affinity ouabain binding confirm the differential effects of ischemic injury on these cells.

The loss in N^+-K^+-ATPase activities observed following ischemia are thought to result from dysregulation of membrane lipid metabolism rather than losses of enzyme protein. This is supported by the less than 10% loss in membrane protein in the ischemic vs. nonischemic primary infarct areas and no losses in membrane protein in periischemic areas at 3 d after ischemia (Mahadik et al., 1992). In addition, similar to other membrane enzyme proteins, the half-life of ATPase is long (>15 d). It is unlikely that losses in enzyme activity seen at 3 d are a result of losses in membrane protein. Further, in the periischemic areas, after 5 d the activities of the ATPase isozymes show a significant recovery, reaching control levels at 14 d. This recovery implies that the plasma membrane disorganization that occurs during the acute stages following ischemia is reversible. We reported similar results following global cerebral ischemia in gerbils.

The kinetic analyses of the changes (losses and protection by GM1) indicate that the losses are not due to protein losses, but rather to changes in activity, indicative of membrane disorganization (Mahadik et al., 1989). Similarly, the decrease in ATPase losses when animals are treated with GM1 are not the result of increases in enzyme, but rather the preservation of enzyme activity (binding capacity) though the maintenance of membrane integrity. This interpretation is significant, not only in defining the mechanism by which GM1 is an effective neuroprotective agent, but also in illustrating the *power* of using ATPase activity as an analytic tool.

The mechanisms associated with the loss of the ATPase enzymes in ischemic injury are several. These (supra) ATPase analyses were provocative and led to studies that attempted to focus on the possible mechanism underlying ischemic injury processes as well as the mechanisms by which GM1 was effective. Following ischemic injury, membrane phospholipids and total fatty acids are reduced in the gerbil model of global ischemia and in the model of cortical focal ischemia, thereby leading to altered membrane fluidity (Mahadik et. al., 1989, 1993; Karpiak et al., 1990). This is paralleled by increased levels of free fatty acids, particularly arachidonic acid, which has been shown to inhibit lipid-dependent membrane enzymes such as Na^+-K^+-ATPase, and which can induce the production of oxygen radicals and associated membrane damage (Mahadik et al., 1993). Free-radical-induced lipid peroxidation has been shown to inhibit Na^+-K^+-ATPase activity.

It is evident from the ATPase results discussed here that considerable irreversible neuronal injury occurs in the primary *and* periischemic areas, and takes place between 2 and 3 d after ischemia, while the center of the primary infarct is irreversibly damaged within hours. There are several plausible explanations for this delayed ischemic neuronal plasma membrane damage. Initial microenvironment changes involve ionic imbalances with associated increased Ca^{2+} influx. Release of glutamate as a result of ischemia can also induce further Ca^{2+} influx. This combined increased Ca^{2+} influx and the reduction in efflux as a result of Ca^{2+} pump (Ca^{2+}-ATPase) failure due to a loss of membrane structure can lead to increased Ca^{2+} loading, which has been observed in our focal ischemia model after 72 h of ischemia (Karpiak et al., 1991). It has been proposed that the Ca^{2+} influx-dependent activation of phospholipases leads to release of membrane phospholipid fatty acids, which results in loss of membrane structural integrity. The biochemical events responsible for the loss of membrane integrity must occur early in the injury cascade, setting the stage for subsequent irreversible plasma membrane damage.

B. ATPase CHANGES AND THE GM1 MECHANISM IN REDUCING ISCHEMIC INJURY

The data discussed here supports our hypothesis that the neuroprotective effects of GM1 result in a maintenance of plasma membrane mechanisms that are responsible for the efflux of Ca^{2+} after ischemic injury (functional ATPase enzyme). This reduction in injury due to neurotoxic Ca^{2+} increases is likely the result of maintained efflux processes. The reduced losses of Na-K-ATPase and Ca-ATPase in the GM1-treated animals support this conclusion (Karpiak et al., 1994). We have concluded that GM1 does not affect the ATPase enzyme directly, but maintains the activity of the ATPase enzymes through the maintenance of the structure of the plasma membrane, since the activity of the ATPase enzymes is fully dependent upon the integrity (lipid environment) of the plasma membrane. The data supports our previous research reports, which hypothesized that the protective effects of GM1 result from its maintenance of plasma membrane structure and function. While cells can sustain (buffer) high Ca^{2+} levels (due to increased influx precipitated by ischemia), the viability of the cell is dependent on its ability to maintain the Ca^{2+} efflux. This Ca^{2+} reduction (efflux) is primarily achieved through maintenance of ATPase activity (for example Ca-ATPase, the Ca^{2+} pump). Our studies have shown that plasma membrane ATPase activity is maintained in injured CNS tissue taken from animals treated with GM1. The reduced loss of plasma membrane ATPase *activities* (Ca- and Na-K-) in GM1-treated animals reflects a protection and maintenance of plasma membrane structure. ATPase activity is critically dependent upon the structure of the membrane (its protein/lipid content and topography). If GM1 maintains plasma membrane integrity, then a functional Ca-ATPase enzyme would be able to restore the calcium ion gradient required to achieve recovery (the transition from irreversible to reversible injury).

Our hypothesis that GM1 maintains plasma membrane integrity after injury is further supported by our studies that indicate that in CNS tissue taken from GM1-treated ischemia animals there are increased levels of membrane fatty acid levels. These increased levels may be indicative of increased membrane metabolism that could contribute to a reversal of the magnitude of membrane failure that occurs after ischemic injury.

C. THE METHOD FOR Na-K-ATPase ANALYSES

Na-K-ATPase is assayed in a total plasma membrane preparation using procedures modified from published reports (Enseleit et al., 1984; Mahadik et al., 1989, 1992). Occluded animals are anesthetized (ether), perfused with phosphate-buffered saline, and their left and right parietal cortices and hippocampus dissected out (tissue dissected is subject to experiment design). A membrane fraction is prepared by homogenizing specimens in ice-cold buffer (40 m*M* in Tris-HCl pH 7.5, 10 m*M* $MgCl_2$, 10 m*M* EDTA), and centrifuging (30 min, 100,000 *g*). The resulting pellet is suspended in the homogenizing buffer. Each 500 ml of reaction mixture contains 40 m*M* Tris-HCl at pH 7.5, 150 m*M* NaCl, 40 m*M* KCl, 10 m*M* $MgCl_2$, m*M* EDTA, 2 m*M* EGTA, 4 m*M* ATP, and 100 to 200 μg of membrane protein. The amount of inorganic phosphate released in 30 min at 37°C is determined by the method of Eibl and Lands (1969). Na-K-ATPase activity is calculated by the difference between the activities in the absence and presence of 5 m*M* ouabain.

REFERENCES

Eibl, H. and Lands, W. A new sensitive determination of phosphate, *Anal. Biochem.* 30: 51–57, 1969.

Enseleit, W. H., Domer, F. R., Jarrott, D. M., and Baricos, W. H. Cerebral phospholipid content and Na,K-ATPase activity during ischemia and postischemic perfusion in the Mongolian gerbil. *J. Neurochem.* 43:, 320–327, 1984.

Hattakeyama, T., Matsumoto, M., Brengman, J., and Yanagihara, T. Immunohistochemical investigation of ischemic and postischemic damage after bilateral carotid occlusion in gerbils. *Stroke* 19: 1526–1534, 1988.

Hernandez, N., MacDonall, J. S., Stier, C. T., Belmonte, A., Fernandez, R., and Karpiak, S.E. GM1 ganglioside treatment of spontaneously hypertensive stroke prone rats. *Exp. Neurol.* 126: 95–100, 1994.

Karpiak, S., Li, Y. S., and Mahadik, S. P. GM1 and AGF2 ganglioside reduce mortality from ischemia: protection of membrane integrity. *Clin. Neuropharmacol.* 9: 338–340, 1986.

Karpiak, S. E., Li, Y. S., and Mahadik, S. P. Gangliosides (GM1 and AGF2) reduce mortality due to ischemia. *Stroke* 18: 184–187, 1987.

Karpiak, S. E., Tagliavia, A., and Wakade, M. Animal models for testing drugs in ischemia. *Annu. Rev. Pharmacol. Toxicol.* 29: 403–414, 1989a.

Karpiak, S. E., Hungund, B. L., Wakade, C. G., Bharucha, V., and Mahadik, S.P. GM1 ganglioside protection of plasma membrane structure and function in CNS ischemia. *J. Neurotrauma* 5: 251–252, 1989b.

Karpiak, S. E. and Mahadik, S. P. Gangliosides and ischemic injury. *CRC Crit. Rev. Neurobiol.* 5: 221–237, 1990.

Karpiak, S. E. Wakade, C. G., Bharucha, V. and Mahadik, S. P. GM1 ganglioside reduction of focal ischemia injury: edema, ATPase, K^+, Na^+, Ca^{2+} analyses. *J. Neurosci. Res.* 30: 512–520, 1991.

Kato, H., Kogure, K., and Nakano, S. Neuronal damage following repeated brief ischemia in the gerbil. *Brain Res.* 479: 366–370, 1989.

Kitagawa, K., Matsumoto, M., Handa, N., Fukunaga, R., Ueda, H., Isaka, Y., Kimura, K., and Kamada, T. Prediction of stroke-prone gerbils and their cerebral circulation. *Brain Res.* 479: 263–269, 1989.

Kowada, M., Ames, A., and Manjo, G. Cerebral ischemia. I. An improved experimental method for study. *J. Neurosurg.* 28: 130–157, 1968.

Levine, S. and Payan, H. Effects of ischemia and other procedures on the brain and retina of the gerbil. *Exp. Neurol.* 16: 255–262, 1966.

Levine, S. and Sohn, D. Cerebral ischemia in infant and adult gerbils. Relation to the incomplete circle of Willis. *Arch. Pathol.* 87: 315–317, 1969.

Lowry, O. H., Rosebrough, N. J., Farr A. L., and Randall R. J. Protein measurement with the Folin phenol reagent. *J. Biol. Chem.* 193: 265–275, 1951.

Mahadik, S. P., Hawver, D., Li, Y. S., and Karpiak, S. E. Mechanism by which GM1 protects membrane function in ischemic tissue. *J. Neurochem.* 48: S45c, 1987.

Mahadik, S. P., Hawver, D., Hungund, B. L., Li, Y. S., and Karpiak, S. E. GM1 protects plasma membrane structure and function following global ischemia: changes in membrane fatty acids and ATPase. *J. Neurosci. Res.* 24: 402–412, 1989.

Mahadik, S. P., Bharucha V., Stadlin A., Ortiz A., and Karpiak S. E. Loss and recovery of two isozymes of Na,K-ATPase (α+ and α) in cortical focal ischemia: GM1 ganglioside protects plasma membrane structure and function. *J. Neurosci. Res.* 32: 209–220, 1992.

Mahadik, S. P., Makar, T. K., Murthy, J. N., Ortiz, A., Wakade, C. G., and Karpiak, S. E. Temporal changes in levels of superoxide dismutase, glutathione peroxidase and catalase in primary and peri-ischemic tissue: monosialoganglioside (GM1) treatment effects. *Mol. Chem. Neuropathol.* 18: 1–14, 1993a.

Mahadik, S. P., Hungund, B. L., Makar, T. K., Gokhale, V. S., Ortiz, A., and Karpiak, S. E. Time dependent alterations in membrane fatty acid levels in primary and peri-ischemic areas following cortical focal ischemia in rat: ganglioside GM1 treatment. *Neurochem. Int.* 23: 163–172, 1993b.

Mazzari, S., Karpiak, S. E., Lipartiti, Seren, S., Lazzaro, A., Rubini, R., Koga, T., Fadda, E., Toffano, G., and Leon, A. Monosialoganglioside effects on excitatory amino acid-related neurotoxicity and cerebral ischemia. In *Neurotoxicity of Excitatory Amino Acids*, Guidotti, A., Ed., Raven Press, New York, 281–291, 1990.

Manev, H., Favaron, M., De-Erausquin, G. A., Guidotti, A., Brooker, G., and Costa, E. Destabilization of ionized Ca^{2+} homeostasis in excitatory amino acid neurotoxicity: antagonism by glycosphingolipids. *Cell Biol. Inter-Rep.*, 14(1): 3–14, 1990a.

Manev, H., Favaron, M., Vicini, S., Guidotti, A., and Costa, E. Glutamate-induced neuronal death in primary cultures of cerebellar granular cells: protection by synthetic derivatives of endogenous sphingolipids. *J. Pharmacol. Exp. Ther.* 252: 419–427, 1990b.

Chapter 7

FOREBRAIN ISCHEMIA IN THE RAT: THE FOUR-VESSEL OCCLUSION MODEL

Tatsuo Mima

TABLE OF CONTENTS

1. HISTORICAL BACKGROUND

Specific regions of the brain, particularly the hippocampal CA1 sector, have been reported to be selectively vulnerable to the ischemic insult (Ito et al. 1975, Kirino 1982, Pulsinelli et al. 1982a). Since these reports, a number of researchers, using various animal models of transient forebrain ischemia, have been trying to answer the questions: (a) why is the hippocampal pyramidal neuron (CA_1 sector) selectively vulnerable? and (b) why is the appearance of cell death (necrosis) delayed as much as two days following ischemic insult? Much of the driving force behind this research stems from the idea that neuronal death may be treatable if the mechanisms underlying cell death are understood and appropriate treatment is applied immediately after transient ischemia occurs.

0-8493-8094-4/95/$0.00+$.50

Before delayed neuronal death of the hippocampus was described in rodent models (Ito et al. 1975, Kirino 1982, Pulsinelli et al. 1982a), it was already well known that patients resuscitated from cardiac arrest often suffer from memory impairment. Pathological surveys later confirmed that even a short cessation of cerebral circulation causes damage to the hippocampus, which is important for memory function (Brierley and Graham 1984, Petito et al. 1987).

The cause of delayed neuronal death has been postulated to be due to neuronal hyperexcitability initiated by glutamate and subsequent intracellular calcium influx (Nadler et al. 1978, Siesjö 1981, Gill et al. 1987, Barnes 1988, Choi and Rothmann 1990). The hypothesis of hyperexcitability, however, can only explain why the hippocampus is vulnerable to ischemia but not why the appearance of neuronal necrosis is delayed.

Recently, in terms of the latter question, more attention has been paid to the evidence that protein synthesis in the hippocampal CA1 sector continues to be inhibited after transient ischemia (Dienel et al. 1986, Thilman et al. 1986) and that the ectopic induction of heat shock proteins (Kitagawa et al. 1990, Kirino et al. 1991) and intraventricularly administered neurotrophic factors (Shigeno et al. 1991) can prevent ischemic neuronal death.

II. DIVERSITY OF ANIMAL MODELS OF TRANSIENT FOREBRAIN ISCHEMIA

Three rodent models are most frequently used: (a) the gerbil model with occlusion of bilateral carotid arteries (Kirino 1982); (b) the rat model with occlusion of both bilateral vertebral arteries and bilateral carotid arteries, in other words, the four-vessel occlusion (4-VO) model (Pulsinelli and Brierley 1979, Pulsinelli and Duffy 1983, Pulsinelli and Buchan 1988); and (c) the rat model with occlusion of bilateral carotid arteries combined with systemic hypotension, in other words, the two-vessel occlusion model (Smith et al. 1984).

The blood flow to the brain derives from both carotid arteries and both vertebral arteries, except the anterior spinal artery and collateral circulations of the neck muscles. Since the gerbil has an absence of the posterior communicating artery, which connects the anterior and posterior circulation of the brain, occlusion of both carotid arteries is enough for forebrain ischemia in the gerbil. On the other hand, the rat model requires combination with either the occlusion of bilateral vertebral arteries or systemic hypotension.

Recently the cardiac arrest model has been reevaluated as an appropriate model because it simulates the clinical features of global brain ischemia resulting from cardiac arrest (Kawai et al. 1992).

Each model has advantages and disadvantages and it is difficult to say which is the best (Ginsberg and Busto 1989). Here, the author describes the 4-VO model, with which he has gained extensive experience by working for two years (1990 to 1992) with Dr. W. A. Pulsinelli who was most involved in creating and modifying the 4-VO model.

III. TECHNICAL ASPECTS

A. OUTLINE OF 4-VO

The procedure of the 4-VO model consists of two parts: permanent surgical occlusion of the vertebral arteries (4-VO surgery), followed one day later by ischemic insult (4-VO ischemia).

4-VO ischemia is achieved by transient occlusion of the bilateral carotid arteries under no anesthesia. The criteria for sufficient severity of brain ischemia is determined by the neurological finding that the rat loses bilateral righting reflex, i.e., it is unable to stand up on either side during ischemia. When the neurological criteria is met, cerebral blood flow (CBF) in the forebrain of the rats decreases below 3% of normal CBF (Pulsinelli et al. 1992b).

However, hindbrain (diencephalon, cerebellum, and brain stem) becomes only moderately ischemic (Pulsinelli et al. 1992b). This condition preserves spontaneous respiratory function and a respirator is not required. The 4-VO model is described as a forebrain ischemia model as well as a global cerebral ischemia model.

B. SELECTED STRAINS

It is important to use rats of selected strains for consistent results of 4-VO ischemia. Pulsinelli and his co-workers have been using Wistar male rats from Hilltop Farms (Scottsdale, PA). After rats arrive at the research institute, they are moved to a special animal room and kept there for several days before 4-VO surgery. Rats subjected to 4-VO ischemia occasionally show moderate or almost no damage to the hippocampus, even though the rats have met the criteria of losing bilateral righting reflex (Table 1). Therefore, each experimental protocol must have its own ischemic control using the same batch of rats. One should not use former experimental data as controls for new experiments.

C. WEIGHT RANGE OF RATS

The appropriate weight range of rats is 200 to 300 g (preferentially 230 to 250 g). Bigger rats (>300 g) are not only hard to hold while giving the ischemic insult without anesthesia, but also less often meet the neurological criteria for successful 4-VO ischemia, probably due to more developed collateral circulation. Smaller rats (<200 g) tend to die of respiratory failure at the time of 4-VO ischemia.

D. SURGICAL PROCEDURE

1. Key Points of 4-VO Surgery

For successful 4-VO surgery, i.e., for the operated rat to meet the neurological criteria during 4-VO ischemia the next day, there are two key points in the surgical procedure: (a) to not damage the vagal nerve when separating it from the common carotid artery, and (b) to cauterize bilateral vertebral arteries completely.

Vagal nerve damage may cause respiratory trouble (usually acute pulmonary edema) and failure to completely interrupt blood supply from the vertebral artery never gives rise to forebrain ischemia.

TABLE 1
Diversity of 4-VO Ischemic Control

	Frequency: number of hemispheres (%)					
	Exp.1	**Exp.2**	**Exp.3**	**Exp.4**	**Exp.5**	**Exp.6**
Neuronal damage Grade **3** (severe)	9 (75)	10 (71)	12 (100)	3 (87)	7 (58)	8 (57)
2.5	1 (8)	2 (14)	—	1 (7)	3 (25)	1 (7)
2	—	—	—	4 (29)	—	1 (7)
1.5	1 (8)	—	—	1 (7)	1 (8)	2 (14)
1	1 (8)	—	—	—	1 (8)	2 (14)
0.5	—	2 (14)	—	—	—	—
0 (normal)	—	—	—	—	—	—
Total number	12	14	12	14	12	14
Average	2.7	2.6	3.0	2.6	2.6	2.4
Standard dev.	0.7	0.9	0.0	0.5	0.7	0.8

The diversity of the hippocampal damage caused by 15 min (Experiments 1 to 3) or 10 min (Experiments 4 to 6) of 4-VO ischemia. Experiments 1 to 6 are the ischemic control conducted in different experimental paradigms. The severity of the damage is graded as a scale of 0 to 3; 0 = normal, 1 = mild, 2 = moderate, and 3 = severe damage. Details of the grading is described in the text. Data of 'Average' and 'Standard dev.' indicates the average and standard deviation of the grading of hippocampal CA_1 damage in each experiment. Note that two grades of 0.5 (almost no damage) out of 14 hemispheres occurred in Experiment 2.

2. Anesthesia

Rats are anesthetized with an N_2/O_2/halothane mixture (70:28:2%). Usage of barbiturate is avoided, although its neuroprotective effect in 4-VO ischemia was not observed (Pulsinelli et al. 1979). The depth of anesthesia should be carefully adjusted to avoid asphyxia and cardiac arrest. Indications that anesthesia may be too deep include: (a) weak pulsation and pale color of the carotid arteries when separating them from the vagal nerve, and (b) sudden slowing of bleeding when dissecting the neck muscles from the first vertebra. Anesthesia problems often occur at the late stage of surgery when the vertebral artery is to be cauterized. At that time, halothane anesthesia tends to become deeper and rotating the head positioning for surgery, as will be described later, disturbs respiration.

3. Clasps around Carotid Arteries

In the supine position, a small pillow is placed behind the neck for better visualization of carotid arteries. A 2-cm horizontal incision is made at a level just rostral to episternum. Under a dissection microscope, the carotid artery is meticulously separated from a vagal nerve. The trick to not damaging the vagal nerve is to slide two forceps away from each other in a parallel direction (Figure 1). A silastic tube (Dow Corning No. 602-155: 0.025 in. ID × 0.047 in. OD, Michigan) is loosely placed around the common carotid artery and threaded through a button. The tube is knotted at an appropriate length for occluding the artery (Figure 2). The button is made of teflon and its holes should be large enough for the silastic tube to smoothly pass through.

4. A Suture to Interrupt Collateral Circulation of the Neck Muscles

The 4-VO model was later modified by adding the procedure of tightening the neck muscles, which decreases collateral circulation from the neck muscles to the brain (Pulsinelli and Duffy 1983, Pulsinelli and Buchan 1988). This modification increases the successful rate, which meets the 4-VO ischemic criteria, from about 60 to more than 90%. A 15-cm length of 0-gauge silk (basically a thick and strong string) is placed dorsal to the external jugular veins, common carotid arteries, trachea, and esophagus, but ventral to the musclus sternocleidomastoideus and the paravertebral muscles (Figure 3). For easier placing of 0-gauge silk, penetrate a needle of lumbar puncture (18-gauge) through the above-mentioned anatomical layer and then insert the silk into the needle. Finally, pull out the needle leaving the string in the animal. The silk suture exits the skin on both sides, and the two ends are secured loosely to the nape of the rat's neck with adhesive tape (Figure 4). At 4-VO ischemia, this suture is to be slowly tightened to encircle and occlude the collateral blood flow of the neck muscles immediately after occlusion of the common carotid arteries. The cervical skin

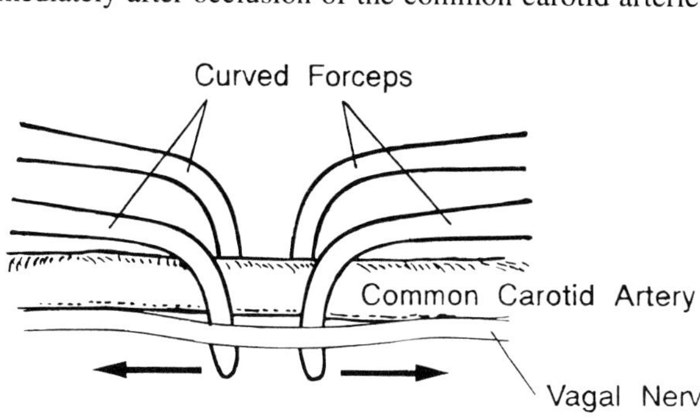

FIGURE 1. Separation of the vagal nerve from the carotid artery. Pull two curved forceps distally at the same time to stretch the artery/nerve complex. Repeat this gentle procedure to separate deeper and more distal regions. The nerve is least damaged as long as forceps are used only in parallel directions.

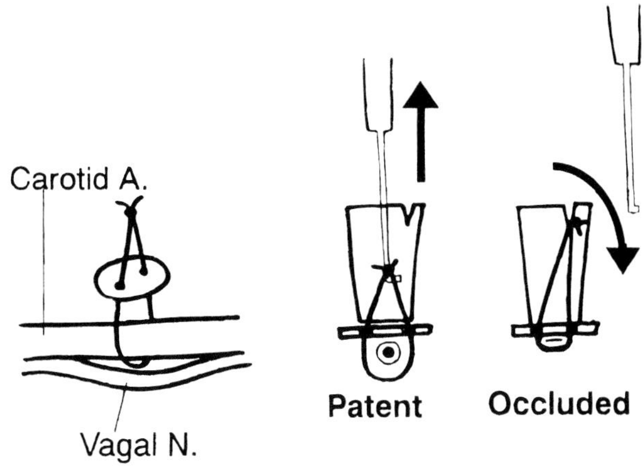

FIGURE 2. Occlusion of the carotid artery with a clasp. The columnar tube is made by cutting off the top and the bottom of a small Eppendorf tube. A V-shaped incision is made at the top for the knot of the clasp to be latched.

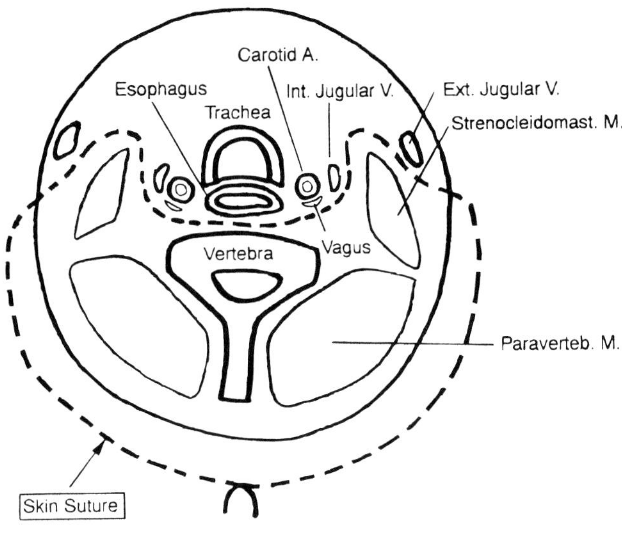

FIGURE 3. Anatomical diagram for placing a suture to tighten the neck muscles and block collateral circulation. The suture is penetrated ventral to vertebra, muscles sternocleidomastoideus, and paravertebral muscles, but dorsal to the carotid arteries, the vagal nerves, esophagus, trachea, and internal and external jugular veins.

incision is closed with a surgical clip so that the wound may be reopened quickly at the manipulation of 4-VO ischemia under no anesthesia.

5. Positioning for Better Visualization of Alar Foramen

Turn the rat to the prone position with its head fixed with stereotactic ear bars. Tilt the head downward at approximately 30° to the horizontal, although this positioning may disturb respiration. Gently stretch the cervical spine by placing tension on the rat's tail with a rubber band anchored to a table. This procedure rotates the first cervical vertebra and provides a better view of the alar foramen. Make a 1-cm dorsal midline incision between the occipital bone and the first vertebra. By dissecting bilateral paravertebral muscles to the lateral side, fully expose the lamina of the first vertebra. The alar foramen is located at each lateral end (Figure 5). A microscope with direct lightning of the surgical field is necessary to carry out the fine surgery of cauterizing the vertebral arteries. A tiny complex of a vessel and nerve

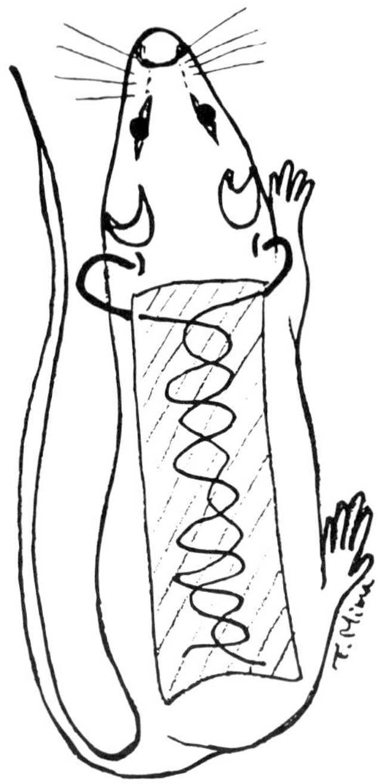

FIGURE 4. The suture outside of the rat needs to be kept long enough. With sticky packaging tape the bundle of the suture is adhered to the back so that the rat cannot scratch it.

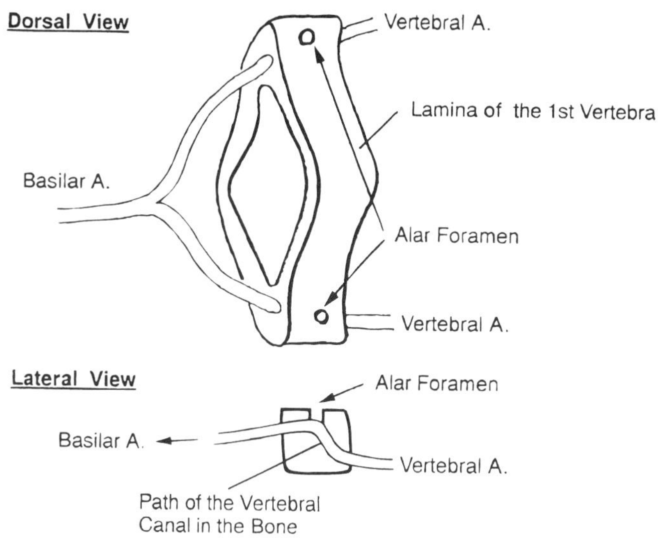

FIGURE 5. The anatomical relationship of the alar foramen and the vertebral artery running through the bony canal of the first vertebra. When the electrocautery needle is inserted vertically down into the alar foramen, it reaches the vertebral artery.

enters into the alar foramen. Sometimes the alar foramen has two holes. Choose the caudal hole for cauterizing the vertebral arteries. When a small (<1 mm) electrocautery needle (Bowie Monopolar Electrocautery, Cincinnati, OH) is inserted vertically down into the alar foramen, it reaches the vertebral artery (Green 1935) (Figure 5). Electrocauterize the vertebral artery with application of current to the needle tip. At this time, overt bleeding from the vertebral artery often occurs. With a suction pipette, make a clear surgical field and continue to cauterize the vertebral artery. However, avoid electrocauterization longer than 2 s, because heat causes spinal cord damage resulting in respiration difficulty during 4-VO ischemia. By doing intermittent electrocauterization, you can safely complete cauterization and stop bleeding. To confirm perfect occlusion of the vertebral arteries, screw a 20-gauge needle into the alar foramen following cauterization. If bleeding occurs, cauterize it again. This confirmation increases the odds of successful 4-VO ischemia. The dorsal skin incision is closed with 2-0 silk (a surgical clip is not needed).

6. Fasting Rats Overnight

Each rat is caged separately. After surgery the animals are fasted overnight but allowed free access to water. Rats without fasting frequently died after 4-VO ischemia; their blood glucose levels after ischemia were found to be extremely high (Pulsinelli, personal communication). Overnight-fasted rats reveal well-controlled blood glucose levels before and after ischemia (Pulsinelli et al. 1982b, Pulsinelli and Duffy 1983). Not only mortality, but also the severity of hippocampal CA1 damage is aggravated by moderate hyperglycemia (Pulsinelli et al. 1982c).

E. PROCEDURE OF PROCEEDING THE ISCHEMIC INSULT

1. Occlusion of Bilateral Carotid Arteries

Before occluding the carotid arteries, 0.1 mg/kg of lidocaine is injected subcutaneously around the cervical wound. One person holds the unanesthetized rat, and another opens the wound with a surgical-clip remover to occlude both carotid arteries. The common carotid artery is occluded transiently by lifting the knot of the silastic tube with a hook and placing it in the notch on a small plastic centrifuge tube made for this purpose (Figure 2).

2. Righting Reflex

After occluding the common carotid arteries, some rats immediately fall down and some do not. When a rat cannot stand up on either side even to pain such as tapping a tail, the neurological finding is called loss of the righting reflex.

As previously described, the criteria for successful 4-VO ischemia is complete and lasting loss of righting reflex during the entire period of ischemic insult. Even if a rat stands for just a few seconds, it may not be used. Such a rat never shows consistent hippocampal damage.

3. Tightening the Suture

If a rat still has the righting reflex, start tightening the neck suture. By doing so, most rats will lose their righting reflex. You should tighten the suture gradually to the optimal strength by observing the respiratory condition. It may take less than 30 s until a rat loses its righting reflex. Then, start the counting time of the duration of ischemia.

It is very difficult to know how strongly to tighten the neck suture. If you tighten the suture maximally, the majority of rats will die due to severe brainstem ischemia causing respiratory failure and/or pulmonary edema. On the other hand, if you do not tighten the suture enough, rats will stand up before reperfusion of the carotid arteries. The optimal window of the appropriate tightness is very narrow and this is an art one must learn for mastering the 4-VO model.

Even though a rat has lost the righting reflex immediately after ischemia, it is better to tighten a neck suture moderately to assure successful 4-VO ischemia. Without further tightening, sometimes a rat may regain its righting reflex because of rapidly developed collateral circulation.

If a rat loses spontaneous respiration, loosen the neck suture a little and assist breathing by blowing intermittently at the rat's mouth through a tube (Figure 6). Most rats will resume spontaneous respiration.

Interestingly, some rats, which appear to suffer from extremely severe ischemia, move their limbs very fast as if they were running (in Pulsinelli's laboratory, it was called ischemic run). The neck suture of those rats should be carefully tightened to very mild strength. They tend to show pulmonary edema discharging pinkish foamy fluid from their nose. If this happens, no treatments rescue the animal.

4. Recirculation

Loosen and cut off the clasps around the carotid arteries when you want to resume cerebral blood flow. The teflon buttons may be reused for the next 4-VO surgery if they are kept in ethanol.

Originally, Pulsinelli's laboratory used ischemic durations of 10, 20, and 30 min (Pulsinelli and Brierley 1979, Pulsinelli et al. 1982). Recently, 10- or 15-min occlusions were used most often as the standard. Just before removing the clasps, check the righting reflex by tapping the tail. Again, exclude all the rats that stand up from experimental data.

5. Monitoring of Body Temperature

Body temperature in transient forebrain ischemia is one of the most important factors that determine the fate of hippocampal neuronal necrosis. A slight decrease of body temperature during ischemia or even after recirculation rescues neuronal death. (Boris-Möller et al. 1989, Bust et al. 1989, Buchan and Pulsinelli 1990a, Kuroiwa et al. 1990, Welsh and Harris 1991). Therefore, maintaining normal body temperature is crucial for research on delayed neuronal death. Since starting the 4-VO model, Pulsinelli and his co-workers have strictly maintained body temperature at more than 37.0°C during ischemia and the early period (approximately 2 h) after reperfusion, using a rectal thermistor (Yellow Springs Inst., Yellow Springs, OH) connected to an overhead heating lamp. The lamp automatically turns on only when rectal temperature drops below 37°C. Almost all rats recover thermal homeostasis after they wake up completely and begin to move around. Body temperature remains above 37°C up to 5 d,

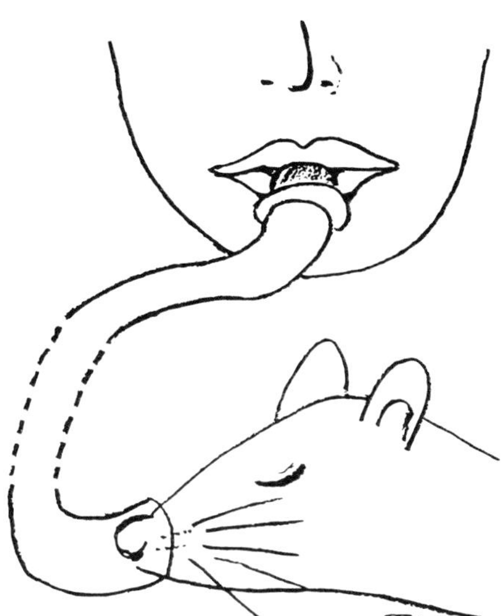

FIGURE 6. If a rat loses spontaneous respiration, loosen the neck suture a little and assist breathing by blowing intermittently at a rat's mouth with a tube. Most rats will regain spontaneous respiration.

at which time the animals are sacrificed (Figure 7). Monitoring of the brain temperature is not always necessary, because the brain and rectal temperature are kept equal before ischemia and after recirculation unless the head skin or skull is open to the air. Therefore, as long as body temperature is monitored, brain temperature is kept almost the same among different experimental groups.

F. FIXATION

Recently, Pulsinelli's laboratory fixed the rats 4 d after 10- or 15-min 4-VO ischemia. Brains are fixed with FAM (40% formaldehyde, glacial acetic acid, methanol, 1:1:8) infused transcardially with a pressure of 130 mmHg for 5 min. The animal is decapitated and the head is immersed in the fixative for more than 24 h. The skull is removed meticulously from the brain. The coronally sliced brain containing hippocampus is placed in methanol for 24 h and embedded in paraffin. The paraffin section is then stained with hematoxylin-eosin.

G. GRADING FOR HIPPOCAMPAL CA1 DAMAGE

As an indicator of the severity of hippocampal CA1 damage, Pulsinelli and his co-workers initiated a grading system (Pulsinelli et al. 1992a). The grading scale has varied slightly (Pulsinelli et al. 1992a, Buchan and Pulsinelli 1990b). The author used a scale of 0 to 3 with 0: normal hippocampus; 1: about 25% of CA1 neurons damaged; 2: about 50% damaged; 3: more than 75% damaged. If it is hard to grade an exact scale, a scale such as 1.5 for between 1 and 2 may be used. The hippocampus of each hemisphere is graded separately. As mentioned before, even rats that met the criteria occasionally show almost no CA1 damage. This problem also occurs in all the other animal models of forebrain ischemia. Table 1 shows the diversity of hippocampal damage caused by 15 or 10 min of 4-VO ischemia. The average varies between 2.4 and 3.0 (Table 1).

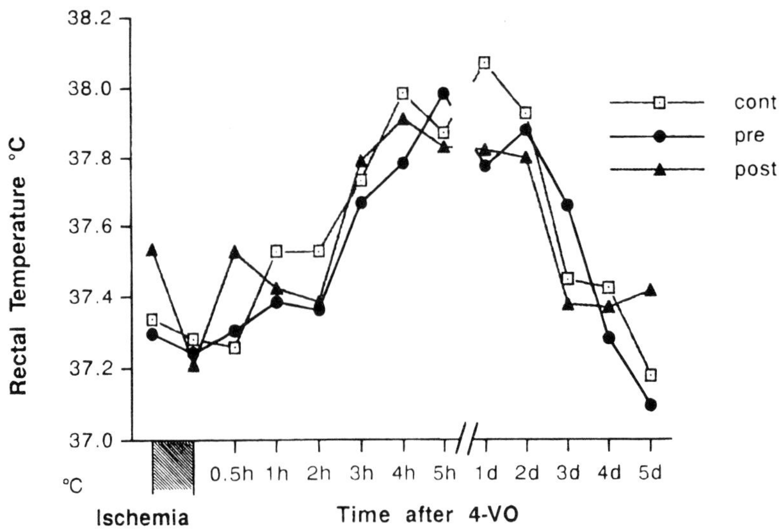

FIGURE 7. Chronological change of the rectal temperature when rats were subjected to 4-VO ischemia. In this experiment, 2.5 mg/kg cycloheximide was injected subcutaneously before ischemia (pre) or after recirculation (post). Rats injected with saline were used as control (cont). Cycloheximide is known to decrease body temperature significantly (Kiessling et al. 1991); however, the rectal temperature stayed higher than 37.0°C during the survival period as long as body temperature was maintained between 37.0°C and 38.0°C for 2 h after ischemic insult. Although cycloheximide was reported to rescue the ischemic hippocampal damage (Goto et al. 1990), the finding was not consistently reproduced (Mima in preparation). The mean value was determined by three to six animals. The standard deviation at each average point is within 0.4°C.

IV. CONCLUDING COMMENTS

The 4-VO model is a practically convenient model once you understand its key points and acquire the standard technique. The advantage of the 4-VO model is that one can confirm the severity of ischemia in each rat by their neurological symptoms without using any monitoring such as electroencephalogram (EEG) or cerebral blood flow (CBF). Therefore, one can produce experimental rats much faster. Some investigators have been reluctant to use the 4-VO model because of its technical difficulty. The author hopes the details described in this paper will be helpful for those who are involved in the research of cerebral ischemia.

ACKNOWLEDGMENT

The author thanks Dr. William A. Pulsinelli for teaching him the 4-VO model. The author also thanks Drs. Donald A. Siegel and Kathryn Ko for suggestions on this paper. The surgical part of the 4-VO model described here is slightly modified from Pulsinelli's procedures.

REFERENCES

Barnes, D. (1988) NMDA receptors trigger excitement. *Science* 239: 254–256.

Boris-Möller, F., Smith, M.-L., Siesjö, B. (1989) Effect of hypothermia on ischemic brain damage: a comparison between preischemic and postischemic cooling. *Neurosci. Res. Commun.* 5: 87–94.

Brierley, J., Graham, D. (1984) Hypoxia and vascular disorders of the central nervous system. Edited by J. Adams, J. Corsellis, and L. Duchen. 125–207 *Greenfield's Neuropathology.* London: Edward Arnold.

Buchan, A., Pulsinelli, W. (1990a) Hypothermia but not the *N*-methyl-D-aspartate antagonist, MK-801, attenuates neuronal damage in gerbils subjected to transient global ischemia. *J. Neurosci.* 10: 311–316.

Buchan, A., Pulsinelli, W. (1990b) Septo-hippocampal deafferentiation protects CA1 neurons against ischemic injury. *Brain Res.* 512: 7–14.

Bust, R., Dietrich, W., Globus, M., Ginsberg, M. (1989) Postischemic moderate hypothermia inhibits CA1 hippocampal ischemic neuronal injury. *Neurosci. Lett.* 101: 299–304.

Choi, D., Rothmann, S. (1990) The role of glutamate neurotoxicity in hypoxic-ischemic neuronal death. *Annu. Rev. Neurosci.* 13: 171–182.

Dienel, G., Kiessling, M., Jacewicz, M., Pulsinelli, W. (1986) Synthesis of heat shock proteins in rat brain cortex after transient ischemia. *J. Cereb. Blood Flow Metab.* 6: 505–510.

Gill, R., Foster, A., Woodruff, G. (1987) Systemic administration of MK-801 protects against ischemia-induced hippocampal neurodegeneration in the gerbil. *J. Neurosci.* 7: 3343–3349.

Ginsberg, M. D., Busto, R. (1989) Rodent models of cerebral ischemia. *Stroke.* 20: 1627–1642.

Green, E. (1935) Anatomy of the rat. *Trans. Am. Philos. Soc.*

Ito, U., Spatz, M., Walker, J. J., Klatzo, I. (1975) Experimental cerebral ischemia in Mongolian gerbils. I. Light microscopic observations. *Acta Neuropathol. (Berl)* 32: 209–223.

Kawai, K., Nitecka, L., Ruetzler, C. A., Nagashima, G., Joó, F., Mies, G., Nowak, J. T. S., Saito, N., Lohr, J. M., Klatzo, I. (1992) Global cerebral ischemia associated with cardiac arrest in the rat. I. Dynamic of early neuronal changes. *J. Cereb. Blood Flow Metab.* 12: 238–249.

Kiessling, M., Xie, Y., Ullrich, B., Thilmann, R. (1991) Are the neuroprotective effects of the protein synthesis inhibitor cycloheximide due to prevention of apoptosis? *J. Cereb. Blood Flow Metab.* 11 Suppl: S357.

Kirino, T. (1982) Delayed neuronal death in the gerbil hippocampus following ischemia. *Brain Res.* 239: 57–69.

Kirino, T., Tsujita, Y., Tamura, A. (1991) Induced tolerance to ischemia in gerbil hippocampal neurons. *J. Cereb. Blood Flow Metab.* 11: 299–307.

Kitagawa, K., Matsumoto, M., Tagaya, M., Hata, R., Ueda, H., Niinobe, M., Handa, N., Fukunaga, R., Kimura, K., Mikoshiba, K., Kamada, T. (1990) 'Ischemic tolerance' phenomenon found in the brain. *Brain Res.* 528: 21–24.

Kuroiwa, T., Bonnekoh, P., Hossmann, K.-A. (1990) Prevention of postischemic hyperthermia prevents ischemic injury of CA1 neurons in gerbils. *J. Cereb. Blood Flow Metab.* 10: 550–556.

Mima, T. (in preparation) Protein synthesis inhibitors (cycloheximide, anisomycin, or actinomycin-D) fail to ameliorate ischemic hippocampal damage in the rat.

Nadler, J., Perry, B., Cotman, C. (1978) Intraventricular kainic acid preferentially destroys hippocampal pyramidal cells. *Nature* 271: 676–677.

Petito, C., Feldmann, E., Pulsinelli, W., Plum, F. (1987) Delayed hippocampal damage in humans following cardiorespiratory arrest. *Neurology* 37: 1281–1286.

Pulsinelli, W., Brierley, J. (1979) A new model of bilateral hemispheric ischemia in the unanesthetized rat. *Stroke* 10: 267–272.

Pulsinelli, W., Rawlinson, D., Plum, F., Brierley, J. (1979) Barbiturate exacerbation of ischemic brain damage following bilateral hemispheric ischemia in the rat. Trans. Am. Neurol. Assoc. 104: 1–4.

Pulsinelli, W., Brierley, J., Plum, F. (1982a) Temporal profile of neuronal damage in a model of transient forebrain ischemia. *Ann. Neurol.* 11: 491–498.

Pulsinelli, W. A., Levy, D. E., Duffy, T. E. (1982b) Regional cerebral blood flow and glucose metabolism following transient forebrain ischemia. *Ann. Neurol.* 11: 499–509.

Pulsinelli, W., Waldman, S., Rawlinson, D., Plum, F. (1982c) Moderate hyperglycemia augments ischemic brain damage: a neuropathologic study in the rat. *Neurology* 32: 1239–1246.

Pulsinelli, W., Duffy, T. (1983) Regional energy balance in rat brain after transient forebrain ischemia. *J. Neurochem.* 40: 1500–1503.

Pulsinelli, W., Buchan, A. (1988) The 4-vessel occlusion rat model: method for complete occlusion of the vertebral arteries and control of the collateral circulation. *Stroke* 19: 913–914.

Shigeno, T., Mima, T., Takakura, K., Graham, D. I., Kato, G., Hashimoto, Y., Furukawa, S. (1991) Amelioration of delayed neuronal death in the hippocampus by nerve growth factor. *J. Neurosci.* 11: 2914–2919.

Siesjö, B. (1981) Cell damage in the brain: a speculative synthesis. *J. Cereb. Blood Flow Metab.* 1: 155–185.

Smith, M.-L., Bendek, G., Dahlgren, N., Rosén, I., Wieloch, T., Siesjö, B. K. (1984) Models for studying long-term recovery following forebrain ischemia in the rat. 2. A 2-vessel occlusion model. *Acta. Neurol. Scand.* 69: 385–401.

Thilman, R., Xie, Y., Kleinhaus, P., Kiessling, M. (1986) Persistent inhibition of protein synthesis precedes delayed neuronal death in postischemic gerbil hippocampus. *Acta Neuropathol. (Berl)* 71: 88–93.

Welsh, F., Harris, V. (1991) Postischemic hypothermia fails to reduce ischemic injury in gerbil hippocampus. *J. Cereb. Blood Flow Metab.* 11: 617–620.

Chapter 8

A FOCAL CEREBRAL ISCHEMIA MODEL DESIGNED FOR BASIC RESEARCH AND EXPERIMENTAL INTERVENTIONS

Richard Paczynski, Yong Y. He, Sien T. Chen, Chung Y. Hsu

TABLE OF CONTENTS

I. INTRODUCTION

Recent advances in the understanding of the pathophysiology of ischemic brain injury have led to the development of a number of therapeutic regimens to protect the brain from ischemic insult (Hossmann, 1988; Paczynski et al., 1994). Unfortunately, therapeutic successes in animal models over the past two decades have generally not been extrapolated to the clinical care of patients with ischemic stroke (Pulsinelli and Buchan, 1989). Lack of success in clinical trials has raised concern about the relevance of animal stroke models in the development of effective therapeutics for stroke patients (Wiebers et al., 1990). However, multiple factors impairing clinical trials such as delays in the initiation of therapy, etiologic heterogeneity, nonuniform distribution of stroke severity, and diversity of the often poorly validated outcome measures employed may explain this lack of success more readily than deficiencies of animal models. A number of recent editorials have critically addressed the strengths and pitfalls of animal stroke models (Graham, 1988; Molinari, 1988; Wiebers et al., 1990; Zivin and Grotta, 1990; Milliken, 1991; Hsu, 1993).

To enhance the quality of animal studies in the development of new stroke therapies, we have designed a model that is suitable for rigorous preclinical trial protocols. This model is characterized by a consistent large lesion in the middle cerebral artery distribution, which is the most common location of cerebral infarction in man (Kase, 1989). The model has

0-8493-8094-4/95/$0.00+$.50

been extensively validated in our laboratory over the past decade, and has been employed in a number of studies designed to investigate the pathophysiologic consequences of focal cerebral ischemia or molecular mechanisms of ischemic brain injury. In this chapter we shall review the basic surgical techniques of this model. Relevant pathological changes that formed the foundation for basic research and preclinical trials will also be reviewed. Molecular biological experiments using this stroke model can be found in another chapter in this monograph.

II. SURGICAL PROCEDURES

The focal cerebral ischemia-reperfusion model has been detailed in previous publications (Chen et al., 1986; Liu et al., 1989; Yip et al., 1991; Lin et al., 1993; Lai et al., 1993). Briefly, Long-Evans male rats (body weight: 250 to 350 g) (Charles Rivers, Wilmington, DE) were used for these studies. Housing and anesthesia concurred with guidelines established by the institutional animal welfare committee, in accordance with the *PHS Guide for the Care and Use of Laboratory Animals,* the USDA regulations, and the AVMA Panel on Euthanasia guidelines. Rats were allowed free access to water and rat chow (Wayne, Chicago, IL) until surgery. In a subset of rats, food but not water was witheld for 24 h (fasted group). Surgery leading to focal cerebral ischemia was conducted under ketamine (100 mg/kg, IP) and xylazine (5 mg/kg, IM) anesthesia. The rectal temperature was monitored and maintained at 37.0 ± 0.5°C via an electronic temperature controller (Versa-therm 2156, Cole-Parmer, Chicago, IL) linked to a heating lamp. Physiologic parameters including blood glucose, arterial blood pressure, and pulse rate were monitored as described previously and animals falling outside of acceptable ranges were excluded from analysis. Animals generally tolerated ketamine/xylazine anesthesia well and there was a very low incidence of significant intraoperative systemic deterioration detected by our monitoring methods (Liu et al., 1989; Yip et al., 1991).

The right middle cerebral artery (MCA) was exposed as previously described (Chen et al., 1986; Yip et al., 1991) using microsurgical techniques. Following a 2-cm vertical skin incision midway between the right eye and ear, the temporalis muscle was split and a 2-mm elliptical burr hole was drilled at the junction of the zygomatic arch and squamous bone. Care was taken to avoid thermal injury to the brain surface by intermittently applying physiologic saline to the drill bit. The meninges were carefully incised and reflected away from the MCA as it passes near the rhinal fissure. An angulated needle attached to a 10-0 suture was passed beneath the MCA immediately proximal to its first major frontal and parietal branches and ligated with a single slip knot. Complete interruption of blood flow at the level of the ligature was confirmed under an operating microscope. Notably, residual blood flow could be observed penetrating the exposed cortex of the MCA distribution in a retrograde fashion via collateral channels, but was typically sluggish and difficult to visualize. Both common carotid arteries (CCAs) were then isolated and clamped using nontraumatic aneurysm clips, carefully avoiding damage to the other contents of the carotid sheath. After the predetermined duration of ischemia (45, 60, or 90 min), either the left CCA clip was removed (partial reperfusion), or both CCA clips and the right MCA ligature were removed (full reperfusion). Restoration of blood flow through the relevant arteries was observed directly under the microscope. All of the rats were kept in air-ventilated incubators with temperature maintained at 24 ± 0.5°C until righting reflexes had returned defining the end of the experiment. Free access to food and water was allowed after recovery from anesthesia.

III. CEREBRAL BLOOD FLOW STUDIES

To delineate anatomically the zone of severe ischemia in this model, we determined relative brain surface blood flow and regional cerebral blood flow (rCBF) using laser Doppler flowmetry (LDF) and 14C-iodoantipyrine autoradiography (14C-IAPA) (4-*N*-methyl-14C, 55 mCi/mmol, American Radiolabeled Chemical Inc., St. Louis, MO), respectively. The methods of Sakurada et al., 1978, and Otsuka et al., 1991 were used. After ligation of the MCA alone, the right MCA and the ipsilateral CCA, or the right MCA and both CCAs, LDF revealed that blood flow to the cortical surface within the MCA territory was reduced by 62, 48, and 18% of baseline, respectively (Chen et al., 1986). Similarly, 14C-IAPA demonstrated blood flow reduction by 88 to 92% restricted only to the cerebral cortex irrigated by the right MCA during triple vessel arterial occlusion.

In other brain regions rCBF showed no change (e.g., the hippocampi) or was only slightly reduced (left MCA cortex or both the right and left basal ganglia) during triple vessel occlusion. Over the time periods studied, rCBF to these mildly ischemic brain regions was well above the threshold for maintaining cellular function and viability, based on information from similar studies in small animals (Heiss, Rosner, 1983.) rCBF to the right MCA cortex was quickly restored to near preischemic levels following reperfusion after 30 to 90 min of severe ischemia. These observations underscore the critical importance of systematic near elimination of collateral flow to the MCA territory during precisely timed periods of triple vessel arterial occlusion in this model. Other investigators have reported considerable variability of infarct volume following MCA occlusion with lesser degrees of collateral flow reduction in normotensive rats (Brint et al., 1988).

IV. PATHOLOGICAL CHANGES

Pathological changes correlate well with the aforementioned region of severe ischemia and are confined to the right MCA territory in this model. With the exception of small foci of pial disruption at the surgery site, the pathological change is characterized by coagulation necrosis with sharply demarcated margins seen on traditional histological preparations (Chen et al., 1986). Within the infarcted zone, the cellular necrosis appeared to proceed in various stages. The affected neurons exhibited eosinophilia, pyknosis, karyorrhexis, and karyolysis. In addition to traditional histological stains such as hematoxylin and eosin, triphenyltetrazolium chloride (TTC; Sigma Chemical, St. Louis, MO) staining—which reflects the integrity of mitochondrial enzymes—was also found to be a fast and efficient method of delineating infarcted regions. TTC staining has been used extensively in morphometric analysis of infarct volume in this and other models (Liszczak et al., 1984). Increasing infiltration of the infarcted region by polymorphonuclear neutrophils within the first 24 h was followed by the appearance of macrophages. Edema confined to the cerebral neocortex is reflected in increases in tissue water content that progress with a characteristic time course to reach maximum values proportional to the duration of severe ischemic insult at 24 to 36 h (Chen et al., 1986; Lin et al., 1993). All of these changes are confined to the neocortex with the rare occurrence of edema and disintegration of a few nerve fibers in the underlying white matter. There is essentially complete sparing of the subcortical nuclei, probably due to the distal site of MCA occlusion beyond the origins of the lateral lenticulostriate perforating vessels.

Using standardized methods for measuring the volume of infarcted tissue, expressed in absolute units or as a percentage of the volume of the right hemisphere, described in detail elsewhere (Chen et al., 1986; Lin et al., 1993), we are able to precisely quantify the impact of different magnitudes of ischemic insult or therapeutic interventions in an objective fashion.

V. MORTALITY/FUNCTIONAL OUTCOME

How to measure therapeutic outcome is highly controversial in both clinical and preclinical trials (Molinari, 1988; Wiebers et al., 1990). Using our model of transient severe focal ischemia followed by either partial or complete reperfusion, overall mortality is very low (less than 5% in over 2000 rats). It is notable that there are no grossly apparent behavioral or motoric abnormalities in the recovered animals. Rats return rapidly to *ad libitum* conditions. In a variant of our model wherein three vessel occlusion is maintained indefinitely (permanent, severe ischemia), mortality within the first 48 h is very high (approximately 60%) (Chen et al., 1986), and animals exhibit grossly abnormal behavior featuring hypokinesia, bradykinesia, and frequent limbic-type convulsions. Other than circling behavior away from or toward the side of the lesion, there are no evident focal abnormalities. Severe cerebral edema, prominently involving the right hemisphere, is consistently observed (Chen et al., 1986).

Grading of neurological behavior or deficit is difficult in small animals (Waltz, 1979). Sophisticated behavioral assessment methods are being developed (Schallert et al., 1990) and should be further explored for their utility in animal stroke preclinical trials. It should be noted, however, that no matter how sophisticated an animal behavior paradigm can be, grading of neurological deficit in animals is likely to be more difficult than it is in humans. The functional anatomy of the motor system of the rat, in particular, is such that gradable focal abnormalities will be limited or absent even after extensive lateralized injuries to the cerebral hemispheres. Scoring functional performance and neurological deficit, which is far from ideal (Molinari, 1988) is currently the mainstay of outcome measurement in human stroke trials. The scoring systems in human trials in the past were often based on arbitrary grading (Hsu et al., 1987, 1988; American Nimodipine Study Group, 1992). They are pseudo-quantitative and subject to interpersonal biases. The application of more objective measures such as brain anatomic and functional imaging has been urged (Molinari, 1988), and have just begun to be incorporated into clinical trial design (Kushner et al., 1990).

A different view emphasizes the need to show "clinical improvement" in animal trials (Wiebers et al., 1990). This proposal is based on the argument that disparity in outcome, as has been noted recurrently between animal and human stroke therapy trials, is primarily caused by disparity of the outcome measures used in animals and humans. Only functional assessment is currently relied upon in human trials. Wiebers et al., therefore, propose the adoption of similar functional outcome assessments in animal preclinical trials. We hold the view that while meaningful measures of functional outcome should be the benchmark of any important advance in clinical stroke therapy, objective and quantitative means to measure ischemic brain injury, including but not limited to anatomic imaging and histologic analysis, are essential. Refinements of method and substantiation of correlations between behavioral change and structural and biochemical pathology should be encouraged. We should not retreat to the use of less desirable outcome measures in animals because of the lack of ideal ones in human trials. The validity of animal behavior as the sole or major parameter in assessing treatment outcome remains to be established. Its utility, at the present time, is to complement other modalities that are more objective and quantitative. These considerations seem particularly relevant at this early stage of exploration of multiple promising prospects for acute stroke therapy.

Part of the discrepancy in the outcomes between animal and human trials may be due to the fact that animal trial protocols are not designed to follow the rigorous criteria demanded by the FDA in human trials. An important focus of our future effort, therefore, is the improvement in the quality of preclinical trials and development of objective outcome measures in clinical trials (Hsu et al., 1993).

VI. GENERAL CONSIDERATIONS OF DESIGN

A valid stroke model should show that the extent of ischemic brain injury is dependent on the magnitude of ischemic insult. Thus, an ischemia-injury curve can be constructed as shown in Figure 1. Establishing an ischemia-injury curve is essential for three reasons. First, the mildest magnitude of ischemic insult should cause very little or no ischemic brain injury.

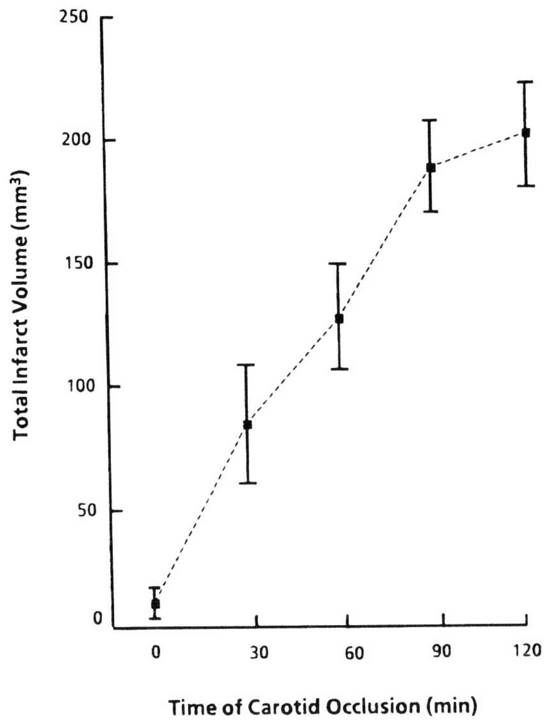

FIGURE 1. The relationship between infarct volumes and duration of ischemia. (From Liu, T. H. et al., *Am. J. Physiol.*, 256, H589–593, 1989. With permission.)

Thus, the investigators should be able to assess the morbidity and mortality caused by anesthesia, surgery, and related injuries but not ischemia per se. If morbidity and mortality is significant even with the mildest degree of cerebral ischemia, the usefulness of such a model in preclinical trials is questionable. Second, validity of the outcome measures intended for ischemic brain injury can be verified. A valid end point measure should demonstrate that the extent of brain injury is dependent on the magnitude of ischemic insult, a construct which encompasses both depth and duration of ischemia (Heiss and Rosner, 1983) as shown in Figure 1. Investigators are in a better position to study neuroprotective agents if they can first demonstrate in their model that outcome can be significantly improved when the magnitude of the ischemic insult is reduced. Third, an ischemia-injury curve allows the investigators to have a broader view of the relationship between magnitude of ischemic insult and the extent of injury. It is desirable that therapeutic interventions be studied within the appropriate range of magnitude of ischemic insult. Mild ischemic insult may cause small infarct size leaving little room for detecting a therapeutic effect. Severe insult, on the other hand, may result in maximum injury that may not be amenable to any therapeutic intervention. As shown in Figure 2, different conclusions could be drawn depending on the severity of ischemia that was selected to study the effects of plasma glucose concentration on brain infarction. Ideally, neuroprotective agents should be tested over a range of different magnitudes of ischemic insult, as has been done (Zivin et al., 1985). The goal of achieving reduction of injury is more realistic using an ischemia-injury curve: the sample size and power, which are essential for a valid trial by FDA criteria, can be estimated prior to the trials using the the derived information. An example is shown in Table 1.

Stroke models allowing construction of an ischemia-injury curve are temporary ligation (followed by complete or partial reperfusion) and graded embolism models. In the former, the location of maximum insult is consistent and the duration of ischemia can be controlled

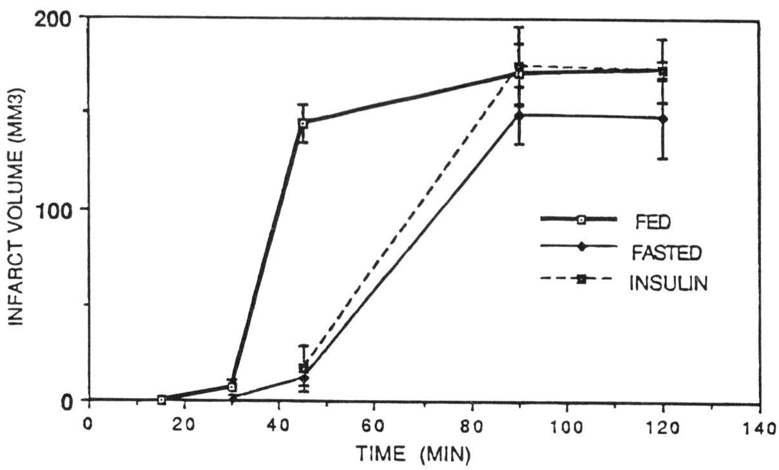

FIGURE 2. The effects of therapeutic interventions (fasting and insulin treatment) on the relationship between infarct volumes and duration of ischemia. Rats were fed until ischemic insult (FED); fasted for 24 h prior to surgery (FASTED); or fed and treated with insulin to normalize blood glucose levels (INSULIN). Note the shift of the curves to the right in the FED and INSULIN groups reflecting interventions that were effective in reducing the infarct volumes only when the duration of ischemia was appropriate. (From Yip, P. K. et al., *Neurology* 41, 899–905, 1991. With permission.)

TABLE 1
Sample Size Estimation Based on Infarct Size Shown in Figure 1

Thapeutic effect (reduction in infarct volume)	$1 - \beta = 0.8$		$1 - \beta = 0.9$	
	$\alpha = 0.05$	$\alpha = 0.01$	$\alpha = 0.05$	$\alpha = 0.01$
20 mm^3	99	147	132	187
30 mm^3	44	65	59	83
40 mm^3	25	37	33	47
50 mm^3	16	24	22	30
60 mm^3	11	17	15	21

From Hsu, C. Y. et al, in *Emerging Strategies in Neuroprotection,* Maranges, P. and Lal, H., Eds., Birkhauser Boston, Inc., Cambridge, MA, 1991. With permission.

as the major independent variable (Liu et al., 1989; Yip et al., in press). In the latter, the "dose" of embolic particles delivered to a given brain region is the variable that is related to the extent of ischemic injury (Zivin et al., 1985). The chief disadvantage of graded embolism is lack of control over the anatomic sites of blood flow disruption, but the need for intracranial surgery is obviated. Rat stroke models entailing permanent occlusion of arteries supplying the brain are commonly used in testing neuroprotective agents (for examples, see Nakayama et al., 1988; Jacewicz et al., 1990). However, permanent occlusion models, based on the assumption that graded ischemic zones exist following permanent arterial occlusions, do not allow for the development of ischemia-injury curves unless carefully controlled studies correlating rCBF with induced pathology are performed concurrently with therapeutic interventions. Estimation of the sample size and power can be done in the permanent occlusion models based on the standard error of infarct size after fixed magnitude of ischemic insult (Brint et al., 1988). The investigators, however, may not know whether a negative treatment outcome is caused by choosing an inappropriate severity of injury as shown in Figure 2. While all of these approaches share some relevance in light of known clinical phenomenology, from the vantage point of statistical analysis and study power the advantages of the temporary ligation and graded embolism models are clear. Less clear are issues pertinent to reperfusion injury in comparing and contrasting these different models. The significance of reperfusion in causing additional ischemia or other forms of secondary injury have not been fully established in either animals or humans.

VII. VARIABLES INHERENT TO ANIMAL MODELS

Each animal stroke model has its inherent variables that may affect the trial outcome. The need to systematically assess these variables in animal models is highlighted by a National Institute of Heart, Lung and Blood-sponsored study entitled "Animal Models for Protecting Ischemic Myocardium," or AMPIM (Reimer et al., 1985). This study was undertaken to understand the major variables that determine infarct size in canine models of myocardial ischemia and to develop well-characterized animal models for more consistent therapeutic outcome in preclinical trials. Results from AMPIM indicate that most of the variation in infarct size in placebo-treated dogs could be related to interindividual differences in anatomical area at risk, efficiency of collateral blood flow, and the rate-pressure product, an index of myocardial metabolic demand. The degree of variability of infarct sizes affect the probability of detecting therapeutic effect of pharmacologic interventions. For instance, an intervention that limits the infarct size by 10% of area at risk, and that takes into account the known variables affecting ischemic injury besides drug treatment, would require 15 dogs each in the placebo and treatment groups to have a 50% probability of detecting the treatment effect using multivariate analysis (Reimer et al., 1985). The number of animals required to detect a greater therapeutic effect is expected to be smaller and vice versa. Without such a systematic analysis of variables, inadequate sample size may be employed and result in inconclusive or faulty outcomes.

Investigators conducting trials should be familiar with variables inherent to their animal models and design protocols accordingly to avoid or minimize the compounding effects of these variables on the trial results. Some of the variables, including the effects of surgery and anesthesia, have been alluded to earlier. Others will be described in this section. General physiological parameters including arterial blood pressure, heart rate, systemic pH and arterial oxygen content may be important determinants of clincial outcome following focal cerebral ischemia (Hossmann, 1988; Paczynski et al., 1994). Effects of therapeutic agents on these parameters should be investigated and are usually carried out in animal trials that have appeared in the literature in recent years. More recently, the profound effects of plasma glucose (Myers and Yamaguchi, 1977) and brain temperature (Busto et al., 1987; Dietrich et al., 1990) on ischemic brain injury were demonstrated. An example of glucose effect on ischemic brain injury is illustrated in Table 2. Ketamine is generally considered to be a neuroprotective agent because of its effect of blocking CNS NMDA receptors. Paradoxically,

TABLE 2
Effect of Plasma Glucose (PG) on Infarct Volume in Insulin-Treated Rats[a]

	Glucose (mg/dL)					
Intraischemic PG range	Pre-ischemia	Intra-ischemia	2hr post-ischemia	3hr post-ischemia	4hr post-ischemia	Infarct volume (mm^3)
(A) $<$60 mg/dL	55 ± 4	50 ± 3	74 ± 7	77 ± 8	110 ± 8	99 ± 26
(B) 60 − 180 mg/dL	79 ± 9	118 ± 11	152 ± 27	149 ± 20	124 ± 13	9 ± 6
(C) $>$180 mg/dL	150 ± 10	238 ± 12	200 ± 36	152 ± 35	150 ± 35	112 ± 36

Note: Insulin-treated rats were divided into 3 groups according to intraischemic PG. Data shown are mean ± SEM. All rats were subjected to ischemia in the right middle cerebral artery territory for 45 min. (Yip, P. K. et al., *Neurology* 41, 899–905, 1991. With permission.)

we noted that animals under isoflurane anesthesia had significantly smaller infarct size than those given ketamine/xylazine anesthesia. One explanation for this paradoxical finding is that ketamine/xylazine anesthesia causes a hyperglycemic response during and after ischemic insult (Hsu et al., unpublished observation). The unexpected hyperglycemic effect of ketamine/xylazine possibly outweighs the potential protective effect of ketamine. Similarly, in a protocol testing the hypothesis that altering brain lipid composition would limit the production of injurious prostaglandin species during ischemia, we found that animals given 20% of total daily caloric intake as marine fish oil developed significantly larger cortical infarctions than those given regular rat chow (Lai et al., 1993). Again, the adverse effect of fish oil was likely related to the greater hyperglycemic response in rats with than without fish oil supplement. These two examples illustrate that altering certain variables may affect the apparent outcome. Another well-publicized example is the hypoythermic effect of MK801 in the global ischemia model. The protective effect of MK801 was no longer detectable when the same body temperature was maintained between control and MK801-treated groups (Buchan and Pulsinelli, 1990). Variations in brain temperature as little as 1 to 2°C can have a profound impact on outcome from otherwise identical ischemic insults (Dietrich et al., 1990). Close attention to key physiologic variables may avoid spurious results caused by unanticipated consequences of therapeutic interventions.

VIII. VARIABLES RELATED TO HUMAN FACTORS

Human factors have rarely been addressed in animal studies. The introduction of human biases in clinical trials are well recognized and can be reduced or eliminated with rigorously designed protocols. Little attention has been devoted to the potential human biases in preclinical stroke trials in published reports. Surgery and preoperative and postoperative care may affect the outcome. While blinding of the investigators who measure the outcome has been enforced in most animal studies, blinding of the surgeon and other researchers who may consciously or subconsciously affect the outcome during surgery and preoperative and postoperative care of the animals have rarely been adopted in preclinical trials. The strict blinding criteria adopted by the FDA for clinical trials can be applied to preclinical trial protocols to eliminate most of the variables caused by human factors. The suggestions presented in the next section contain several features that may minimize human biases that can distort treatment outcome.

IX. EXPERIMENTAL PROTOCOLS FOR PRECLINICAL TRIALS

Partly based on the recent series of provocative editorials (Graham, 1988; Molinari, 1988; Wiebers et al., 1990; Zivin and Grotta, 1990) and partly based on our experience in recent preclinical (Liu et al., 1989; Yip et al., 1991) and clinical trials (Hsu et al., 1987, 1988), we share the sobering view that confirming the therapeutic efficacy of a given neuroprotective regimen is far more difficult in stroke patients than animals. The diversity of underlying pathology, demographic variables, concurrent general medical conditions, and, perhaps most significantly, the unavoidable delay in initiating therapy in stroke patients undoubtedly contribute to the low success rate in clinical trials of "proven" neuroprotective therapies from valid animal pretrials. Because of the substantial investment of human and financial resources in any stroke trial, the decision leading to human studies should be based in part on unequivocal results derived from sound animal trial protocols. We believe that the disparity between animal and human study results can be reduced by bringing the standards of animal studies closer to those enforced in human studies. Attention to the variables inherent in stroke models and human biases introduced in preclinical trials is essential for detecting unequivocal therapeutic effects upon which the decision to launch a clinical trial can be based. Lack of attention to these factors may lead to false-positive outcomes in the animal studies and subsequent waste of vast financial resources and human effort in futile clinical trials. We propose the following criteria be considered in the design of experimental animal protocols intended to test the therapeutic efficacy of neuroprotective regimens:

1. Pretrial assessment of the known variables that may alter the therapeutic outcome should be performed.
2. Pretrial estimation of the sample size and power should be performed.
3. Outcome measurements that are as objective and as quantitative as possible should be utilized.
4. A placebo-controlled design should be used. The pharmacological and physical properties of the placebo should be as close to the drug as possible with the exception that the active ingredient is absent. Controls should be generated concurrently with therapeutic groups to avoid the possible impact of investigator "learning effect" or other unrecognized environmental variables.
5. Blinding of all of the investigators whose knowledge of treatment code may conceivably affect the outcome should be enforced. These investigators include those who perform surgery to create ischemic insult, provide preoperative and postoperative care, and assess the outcome.
6. Morbidity and mortality should be minimized but included in the final analysis of the outcome. The statistical analysis should encompass the intention-to-treat method to avoid bias that may potentially develop because of exclusion of animals in the final analysis from either placebo or treatment group.
7. A neuroprotective agent should be tested over a range of doses to construct a dose-response curve.
8. The efficacy of the promising neuroprotective agent should be verified in other stroke models and in other species to ensure a particular neuroprotective agent is not model- or species-specific.

ACKNOWLEDGMENT

Results from our lab described in this paper were supported by NIH-NINDS grants NS 25545 and NS 28995 and the Vivian L. Smith Foundation for Restorative Neurology and National Stroke Association. We are grateful to the following investigators who participated in some of the works that are cited here: Drs. T. H. Liu, P. K. Yip, M. Miller, J. Beckman, J. Xu, E. L. Hogan, Grace Sun, and Leonard Miller.

REFERENCES

Abramson, N. S., Safar, P., Detre, K. M., Kelsey, S. F., Monroe. J,, Reinmuth, O. M., Snyder, J. V. (1986): Brain resuscitation clinical trial I study group: randomized clinical study of theopental loading in comatose survivors of cardiac arrest. *N Engl J Med* 314: 397–403.

The American Nimodipine Study Group (1992): Clinical trial of nimodipine in acute ischemic stroke. *Stroke* 23: 3–8.

Asano, T., Gotoh, O., Koide, T., Takadura, K. (1985): Ischemic brain edema following occlusion of the middle-cerebral artery in the rat. II. Alteration of the eicosanoid synthesis profile of brain microvessels. *Stroke* 16: 110.

Boxer, P. A., Cordon, J. J., Mann, M. E., Rodolosi, L. C., Vartanian, M. G., Rock, D. M., Talor, C. P., Marcoux, F. W. (1990): Comparison of phenytoin with noncompetitive N-methyl-D-aspartate antagonists in a model of focal brain ischemia in rat. *Stroke* 21(Suppl III):III 47–51.

Brint, S., Jacewicz, M., Kiessling, M., Tanabe, J., Pulsinelli, W. (1988): Focal brain ischemia in the rat: methods for reproducible neocortical infarction using tandem occlusion of the distal middle cerebral artery and ipsilateral common carotid arteries. *J Cereb Blood Flow Metab* 8: 474–485.

Buchan, A., Pulsinelli, W. A. (1990): Hypothermia but not the N-methyl-D-aspartate antagonist, MK-801, attenuates neuronal damage in gerbils subjected to transient global ischemia. *Neuroscience* 10(1): 311–6.

Busto, R., Dietrich, W. D., Globus, M. Y., Valdes, I., Scheinberg, P., Ginsberg, M. D. (1987): Small differences in intraischemic brain temperature critically determine the extent of ischemic neuronal injury. *J Cereb Blood Flow Metab* 7(6): 729–736.

Chen, S. T., Hsu, C. Y., Hogan, E. L., Macriq, H., Balentine, J. D. (1986): A model of focal ischemic stroke in the rat: reproducible extensive cortical infarction. *Stroke* 17: 738–743.

Dietrich, W. D., Busto, R., Halley, M., Valdes, I. (1990): The importance of brain temperature in alterations of the blood-brain barrier following cerebral ischemia. *J Neuropathol Exp Neurol* 49: 486–497.

Ginsburg, M. D., Castella, Y., Dietrich, W. D., Watson, B. D., Busto, R. (1989): Acute thrombotic infarction suppresses metabolic activation of ipsilateral somatosensory cortex: evidence for functional diaschisis. *J Cereb Blood Flow Metab* 9: 329–337.

Globus, M. Y., Busto, R., Dietrich, W. D., Martinez, E., Valdes, I., Ginsberg, M. D. (1988): Effect of ischemia on the *in vivo* release of striatal dopamine, glutamate, and gamma-amniobutyric acid studied by intracerebral microdialysis. *J Neurochem* 51: 1455–1460.

Gotoh, O., Asano, T., Koide, T., Takakura, K. (1985): Ischemic brain edema following occlusion of the middle cerebral artery in the rat. I. The time courses of the brain water, sodium and potassium contents and blood-brain barrier permeability to ^{125}I-Albumin. *Stroke* 16: 101–110.

Graham, D. I. (1988): Focal cerebral infarction. *J Cereb Blood Flow Metab* 8: 769–773.

Heiss, W.-D., Rosner, G. (1983). Functional recovery of cortical neurons as related to degree and duration of ischemia. *Ann Neurol* 14: 294–301.

Hossmann, K.-A. (1988): Pathophysiology of cerebral infarction. In: Viken, P. J., Bruyn, G. W., Klawans, H. L., eds: *Handbook of Clinical Neurology*. New York, Elsevier Scientific Publishing Co. 53: 107–154.

Hsu, C. Y., Faught, R. E., Jr., Furlan, B. M., Coull, B. M., Huang, D. C., Hogan, E. L., Linet, O. I., Yatus, F. M. for Prostacyclin Study Group (1987): Intravenous prostacyclin in acute nonhemorrhagic stroke: a pacebo-controlled double-blind trial. *Stroke* 18: 352–358.

Hsu, C. Y., Norris, J. W., Hogan, E. L., Bladin, P., Dinsdale, H. B., Yatsu, F. M., Earnest, M. P., Scheinberg, P., Caplan, L. R., Karp, H. R., Swanson, P. D., Feldman, R. G., Cohen, M. M., Mayman, C. I., Cobert, B., Savitsky, J. P. for Pentoxifylline Study Group (1988): Pentoxifylline in acute nonhemorrhagic stroke: a randomized, placebo-controlled double-blind trial. *Stroke* 19: 716–722.

Hsu, C. Y., He, Y. Y., Linl, T. N., Wu, G., Marangos, P. (1991): Stroke models for preclinical trials of neuroprotective agents In: Marangos, P., Lai, H. (eds), *Emerging Strategies in Neuroprotection,* Birkhauser Boston Inc., Cambridge, MA., pp 44–56.

Hsu, C. Y. (1993): Criteria for valid preclinical trials using animal stroke models. *Stroke* 24: 117–121.

Jacewicz, M., Brint, S., Tanabe, J., Pulsinelli W. A. (1990): Continuous nimodipine treatment attenuates cortical infarction in rat subjected to 24 hours of focal cerebral ischemia. *J Cereb Blood Flow Metab* 10: 89–96.

Jensen, M. L., Auer, R. N., (1988): The NMDA antagonist ketamine fails to protect against ischaemic neuronal necrosis in the rat. *Br J Anaesth* 61: 206–212.

Karpiak, S. E., Tagliavia, A., Wakade, C. G. (1989): Animal models for the study of drugs in ischemic stroke. *Annu Rev Pharmacol Toxicol* 29: 403–414.

Kase, C. (1988): Middle cerebral artery syndromes. In: Viken P. J., Bruyn, G. W., Klawans, H. L., eds: *Handbook of Clinical Neurology*. New York, Elsevier Scientific Publishing Co. 53.

Kushner, M., Nencini, P., Reivich, M., Rango, M., Jamieson, D., Fazekas, F., Zimmerman, R., Chawluk, J., Alavi, A., Alves, W. (1990): Relation of hyperglycemia early in ischemic brain infarction to cerebral anatomy, metabolism, and clinical outcome. *Ann Neurol* 28:129–135.

Levy, D. E., Brierly, J. B. (1979): Delayed pentobarbital administration limits ischemic brain damage in gerbils. *Ann Neurol* 5: 59–67.

Liszczak, T. M., Hedley-White, E. T., Adams, J. F., Han, D. H., Kolluri, V. S., Vacanti, F. X., Heros, R. C., Zervas, N. T. (1984): Limitations of tetrazolium salts in delineating infarcted brain. *Acta Neuropathol (Berl)* 65: 150–157.

Liu, T. H., Beckman, J. S., Freeman, B. A., Hogan, E. L., Hsu, C. Y. (1989): Polyethelene glycol-conjugated superoxide dismutase and catalase reduce ischemic brain injury. *Am J Physiol* 256, H589–593.

Marcoux, F. W., Goodrich, J. E., Dominick, M. A. (1988): Ketamine prevents ischemic neuronal injury. *Brain Res* 452: 329–335.

Michenfelder, J. D., Milde, J. H., (1975): Influence of anesthetics on metabolics, functional and pathological responses to regional cerebral ischemia. *Stroke* 6: 405–412.

Molinari, G. F. (1986): Experimental models of ischemic stroke. In: Barnett H. J. M., Mohr J. P., Stein B. M., Yatsu F. M., eds: *Stroke, Pathophysiology, Diagnosis, and Management.* New York, Churchill Livingstone.

Molinari, G. F. (1988): Why model strokes? *Stroke* 19: 1197–1203.

Myers, R. E., Yamaguchi, S., (1977): Nervous system effects of cardiac arrest in monkeys. *Arch Neurol* 34:65–73.

Nakayama, H., Ginsberg, M. D., Dietrich, W. D. (1988): (S)-emopamil, a novel calcium channel blocker and serotonin S2 antagonist, markedly reduces infarct size following middle cerebral artery occlusion in the rat. *Neurology* 38: 1667–1673.

Otsuka, T., Wei, L., Acuff, V. R., Shimizu, A., Pettigrew, K. D., Patlak, C. S., Fenstermacher, J. D. (1991): Variation in local cerebral blood flow response to high-dose pentobarbital sodium in the rat. *Am J Physiol* 261: H110–20, Jul

Paczynski, R. P., Diringer, M. N., Hsu, C. Y. (1994): Pathophysiology of ischemic injury. In: Fisher M, ed: *Stroke Therapy.* Boston, Butterworth-Heinemann; 2. 1994 in press.

Pulsinelli, W. A., Brierly, J. B. (1979): A new model of bilateral hemispherical ischemia in the unanesthetized rat. *Stroke* 10: 267–272.

Pulsinelli, W. A., Buchan, A. (1989): The utility of animal ischemia models in predicting pharmacotherapeutic response in the clinical setting. In: Ginsberg, M. D., Dietrich, W. D., eds: *Cerebrovascular Diseases.* New York, Raven Press; 87–91.

Reimer, K. A., Jennings, R. B., Cobb, F. R., Murdock, R. H., Greenfield, J. C. Jr, Becker, L. C., Bulkley, B. H., Hutchins, G. M., Schwartz, R. P. Jr, Bailey, K. R., Passamani, E. R. (1985): Animal models for protecting ischemic myocardium: results of the NHLBI Cooperative Study. *Circ Res* 56: 651–665.

Sakurada, O., Kennedy, C., Jehle, J., Brown, J. D., Carbin, G. L., Sokoloff, L. (1978): Measurement of local cerebral blood flow with iodo [14C] antipyrine. *Am J Physiol* 234: H59–66.

Schallert, T., Jones, T. A., Linder, M. D. (1990): Multilevel transneuronal degeneration after brain damage: behavioral events and effects of anticonvulsant gamma-amionobutyric acid-related drugs. *Stroke* 21(Suppl III):III 143–146.

Waltz, A. G. (1979): Clinical relevance of models of cerebral ischemia. *Stroke* 10: 211–218.

Weiss, J., Goldberg, M. P., Choi, D. W. (1986): Ketamine protects cultured neocortical neurons from hypoxic injury. *Brain Res* 380: 186–192.

Wiebers, B. O., Adams, H. P. Jr, Whisnant, J. P. (1990): Animal models of stroke: are they relevant to human disease? *Stroke* 21: 1–5.

Yip, P. K., He, Y. Y., Hsu, C. Y., Garg, N., Marangos, P., Hogan, E. L. (1991): Effect of plasma glucose on infarct size in focal cerebral ischemia-reperfusion. *Neurology* 41: 899–905.

Zivin, J. A., Grotta, J. C. (1990): Animal stroke models. They are relevant to human disease. *Stroke* 21: 981–983.

Chapter 9

REVERSIBLE FOCAL CEREBRAL ISCHEMIA IN RATS: MIDDLE CEREBRAL ARTERY OCCLUSION BY INTRALUMINAL SUTURE

Hiroshi Karibe, Anthony Mancuso, and Philip R. Weinstein

TABLE OF CONTENTS

I. INTRODUCTION

Despite advances in understanding the pathogenesis of vascular diseases, stroke remains a major cause of morbidity and mortality. Although the mechanisms of ischemic brain damage are not fully understood, extensive metabolic events at the cell membrane level may play an important role in ischemic injury. Intracellular calcium influx and ion pump dysfunction induced by energy failure may lead to degradation of macromolecules and loss of membrane and cytoskeletal integrity; ion homeostasis may be lost, and metabolic acidosis may occur. The extent of metabolic changes and the severity of histopathological damage depend on the severity and duration of blood flow reduction.[1]

In the clinical setting of focal ischemia, blood flow is often restored after spontaneous thrombolysis or after temporary clipping during cerebrovascular surgery. Restoration of blood flow reverses cerebral ischemia and modifies postischemic events. After brief ischemia, reperfusion reduces ischemic tissue injury compared with permanent ischemia;[2] however, after prolonged ischemia (3 to 6 h), reperfusion itself may worsen the outcome by inducing blood-brain barrier disruption, brain edema, and postischemic blood flow disturbances.[3] These results suggest that a reversible focal ischemia model may be useful for studying the pathophysiology and treatment of focal cerebral ischemia.

In rats, experimental focal ischemia simulating clinical stroke has been produced by direct microsurgical occlusion of the middle cerebral artery (MCA).[4] These techniques are complex and invasive, requiring craniectomy, durotomy, and arachnoid incision, and allow cerebrospinal fluid drainage, air influx, local temperature change, and altered intracranial pressure. Most techniques do not allow reperfusion. In this chapter, we describe a rat model of reversible MCA occlusion by a suture inserted into the internal carotid artery (ICA).[5–8] We also discuss the advantages and limitations of the technique, and present an example of its application in an *in vivo* experiment using diffusion-weighted magnetic resonance imaging to investigate early membrane disturbances in the evolution of infarction.

0-8493-8094-4/95/$0.00+$.50

II. ANIMAL MODEL

A. SURGICAL PREPARATION

The basic surgical procedure consists of two parts: (a) reducing collateral blood flow to the MCA territory via the circle of Willis and via leptomeningeal anastomoses, and (b) blocking blood flow into the MCA with a 3-0 monofilament nylon suture (Dermalon, Davis and Geck, Inc., Manati, PR). To avoid perforation of the vessel wall, the tip of the suture is rounded by heating. After adequate training and practice, a surgeon with microsurgical experience in rats can perform the surgery in about 30 min.

In our laboratory, adult male Sprague-Dawley rats weighing 280 to 320 g are anesthetized with 5% isoflurane in an induction chamber; an endotracheal tube is inserted, and the rat is mechanically ventilated. Anesthesia is maintained with 1.5% isoflurane in a mixture of 70% N_2O and 30% O_2. Under an operating microscope, the right common carotid artery (CCA) is exposed by longitudinal midline incision from the submandibular triangle to the supraclavicular notch (Figure 1). To expose the right carotid arteries, a self-retaining retractor is positioned between the digastric and sternomastoid muscle, and the omohyoid muscle is retracted medially (Figure 2a).

To reduce collateral blood flow via leptomeningeal anastomoses, extracranial branches of both the ICA and the external carotid artery (ECA) are occluded. The proximal branches of the ECA (the superior thyroid and occipital arteries) are electrocoagulated and transected close to their origin. The ECA is then dissected and isolated by electrocoagulating and transecting the distal branches of the ECA (the lingual and maxillary arteries) (Figure 2b). The ICA is isolated and carefully separated from the adjacent vagus nerve. To expose the pterygopalatine artery, the only branch of the extracranial ICA, the digastric muscle is retracted laterally to separate the ICA from the hypoglossal nerve. The exposed pterygopalatine artery is ligated with a 5-0 silk suture close to its origin (Figure 2c). At this point, the ICA is the only remaining extracranial branch of the CCA.

Next, two 5-0 silk sutures are tied loosely around the mobilized ECA, one at its origin and the other close to the distal terminal branches. A microvascular clip is placed across the ECA close to its origin, and a 3-0 monofilament suture, its tip rounded by heating, is introduced into the ECA lumen through a puncture at the distal end of the ECA stump. The silk suture encircling the ECA stump is tightened around the intraluminal nylon suture to

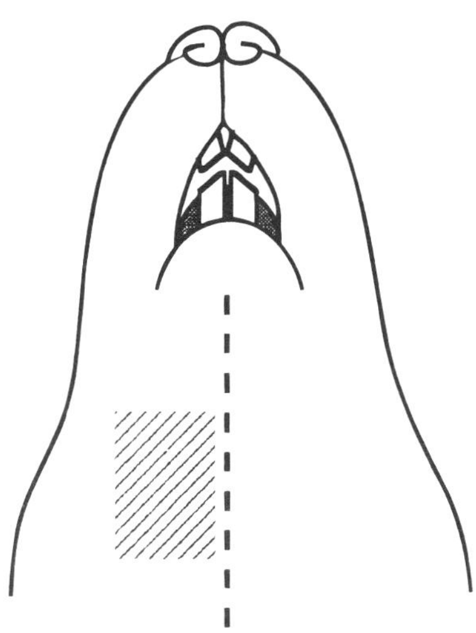

FIGURE 1. Diagram showing the initial cervical skin incision (dotted line) and location of surgical field (shaded area) shown in Figures 2 and 3.

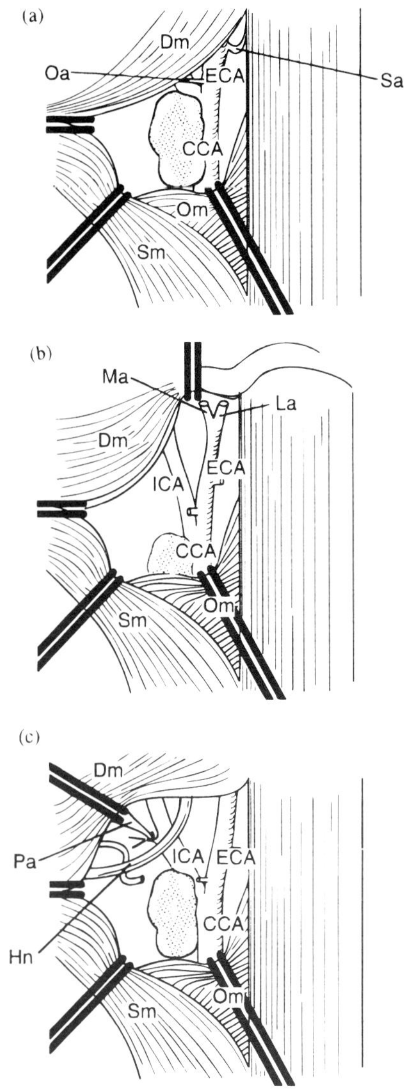

FIGURE 2. Diagram of right carotid arteries. ECA, external carotid artery; CCA, common carotid artery; ICA, internal carotid artery; Dm, digastric muscle; Oa, occipital artery; Sa, superior thyroid artery; Om, omohyoid muscle; Sm, sternomastoid muscle; Pa, pterygopalatine artery; Hn, hypoglossal nerve; Ma, maxillary artery; La, lingual artery.

prevent bleeding (Figure 3a), and the microvascular clip is removed. The intraluminal suture is advanced into the lumen of the CCA, and the loosely tied silk suture at the ECA origin is tightened (Figure 3b). The ECA stump is flipped to introduce the tip of the suture into the ICA (Figure 3c). The suture is gently advanced for a length of 22 mm beyond the bifurcation of the CCA to occlude the origin of MCA and to block collateral circulation via the circle of Willis from both the anterior and posterior communicating arteries. In brains removed at this point in the procedure, the suture tip is found within the proximal segment of the anterior cerebral artery, just beyond the ICA bifurcation. Reperfusion is easily initiated by withdrawing the intraluminal suture from the ICA.

B. ADVANTAGES AND LIMITATIONS

This model has advantages over intracranial microsurgical techniques. No retraction of brain is required; intracranial pressure is not altered; and reperfusion is easily initiated by

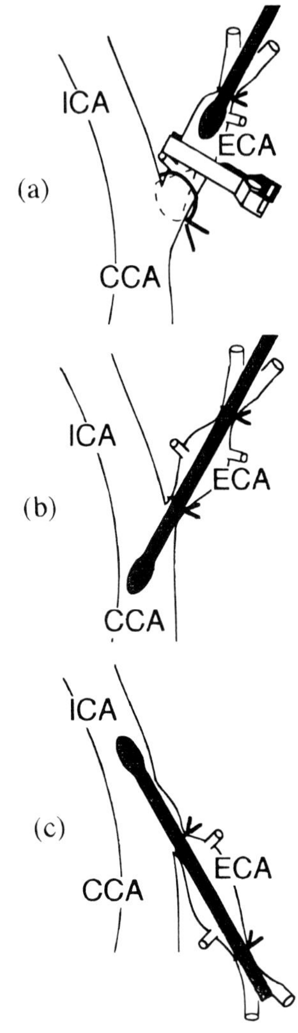

FIGURE 3. Insertion of the intraluminal suture from the stump of the external carotid artery (ECA) into the lumen of the internal carotid artery (ICA). CCA, common carotid artery.

withdrawing the intraluminal suture. Surgical mortality is rare. Infarct volume is reproducible after either permanent or temporary MCA occlusion.[3,6,9] Changes in cerebral blood flow are also reproducible.[3]

This model was refined in Sprague-Dawley rats weighing 280 to 320 g. Extensive pilot studies of various suture materials, lengths, and thicknesses, as well as different sizes of rat revealed that the most reliable infarct volumes are obtained with the protocol described above. Complications may occur if the vagus is mishandled or if the trachea is retracted. If the intraluminal suture is advanced crudely or too far, subarachnoid hemorrhage will occur. Therefore, some practice is required in suture advancement and withdrawal. Reperfusion after more than 3 h of MCA occlusion may cause hemispheric infarction and death from herniation.

C. *IN VIVO* EXPERIMENT WITH THIS MODEL

Recently, magnetic resonance imaging (MRI) has provided a new approach for studying the pathophysiology of ischemia.[7] Diffusion-weighted MRI is a method for detecting alterations in tissue water diffusion (or mobility) early after the onset of ischemia. The mechanisms leading to hyperintensity on MRI images after ischemia are unknown. However, it has been suggested that these changes are related to the occurrence of intracellular edema due to energy depletion and subsequent membrane Na^+-K^+ pump failure.

The MCA occlusion method described above is well suited for MRI studies because it uses only nonmagnetic materials to induce ischemia/reperfusion and creates no cranial window through which air may enter that will cause magnetic susceptibility artifacts (i.e., contaminate the images). In this model, brief temporary ischemia (<90 min) produces smaller infarct size than permanent occlusion; however, infarct size after prolonged temporary ischemia (≥120 min) is not different from that of permanent ischemia. Furthermore, recent results from our laboratory have shown that even delayed induction of hypothermia can protect against ischemic injury, but only if it is initiated within 1 h.[9] These results suggest that the reversibility of tissue damage after MCA occlusion may decrease with time after the onset of MCA occlusion.

The use of the suture occlusion model is illustrated in the following experiment, in which the early evolution of infarction is monitored with diffusion-weighted MRI after MCA occlusion in rats.

Male Sprague-Dawley rats weighing 280 to 320 g were anesthetized, intubated, and subjected to MCA occlusion, as described above. Blood pressure was monitored throughout the experiment; the rats were mechanically ventilated, and respiratory adjustments were made to maintain normal arterial blood gas levels. Rectal temperature was maintained at 37°C.

Diffusion-weighted MR images were obtained with a 7-T imager (Nalorac Cryogenics Corp., Martinez, CA) before and for up to 2 h after MCA occlusion. A coronal image was acquired at the level of the bregma using a conventional spin-echo sequence (TE = 65 ms, TR = 1500 ms, number of acquisitions per phase encoding = 1, diffusion gradient strength = 6.4 gauss/cm; b = 1310 s/mm^2, field of view = 4 cm, slice thickness = 3 mm).

The signal intensity began to increase in the MCA territory immediately after the onset of ischemia. The hyperintense area increased in both size and intensity during the first 2 h after the onset of ischemia (Figure 4). Autoradiographic studies of the rate of glucose metabolism in rat brain after MCA occlusion have shown that the size of the severely damaged area increases with time.[10] Our findings in this illustrative experiment provide further evidence of this phenomenon.

III. SUMMARY

Rat models of cerebral ischemia have the advantages of low cost, relative homogeneity within strains, similarity of cerebrovascular anatomy and physiology to higher species, small brain size that is well suited to fixation procedures for biochemical analysis, and acceptability from both ecological and ethical perspectives.[4] In contrast to direct microsurgical occlusion of the MCA, the rat model of focal ischemia described in this chapter does not require craniectomy and, therefore, does not alter intracranial pressure, does not require direct incision of the brain, and permits reperfusion by withdrawal of the intraluminal suture, while still allowing monitoring of cerebral blood flow and assessment of therapeutic interventions. In addition, because it uses no magnetic material, this model is suitable for acute stroke experiments with MRI and MR spectroscopy. This reversible MCA occlusion model in rats may be useful for studying the pathophysiology and treatment of ischemia and reperfusion injury.

ACKNOWLEDGMENTS

The authors thank Greg Zarow, M.Sc., for assistance with computer graphics, Cheryl Christensen for manuscript preparation, and Stephen Ordway for editorial assistance.

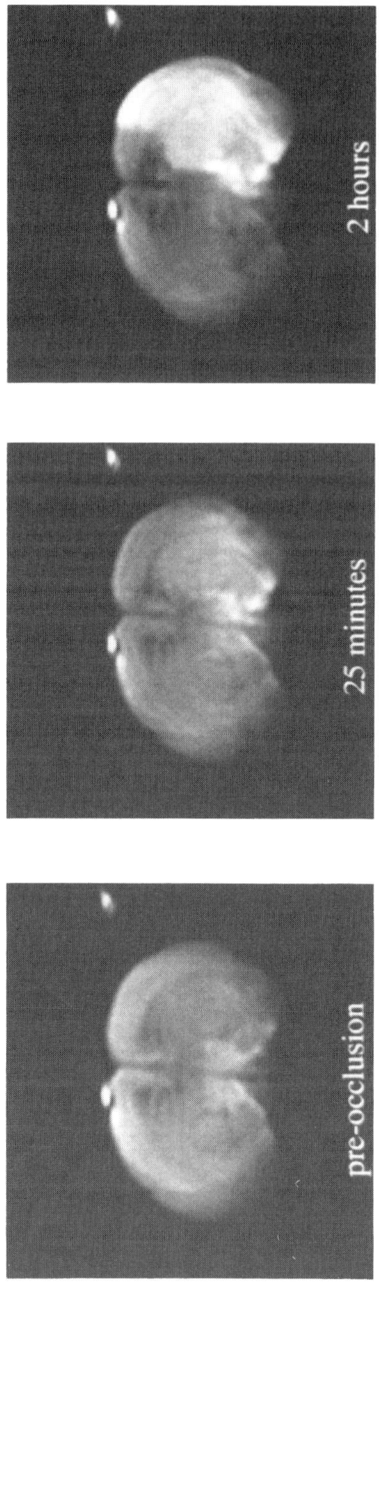

FIGURE 4. Diffusion-weighted MRI obtained in rats before (left) and 25 min (middle) and 2 h (right) after the onset of middle certebral artery occlusion. Hyperintensity in the ischemic hemisphere increased and spread during the first 2 h of ischemia.

REFERENCES

1. Siesjö, B. K. (1992) Pathophysiology and treatment of focal cerebral ischemia. Part I: Pathophysiology. *J. Neurosurg.* 77, 169–184.
2. Memezawa, H., Smith, M. L., and Siesjö, B. K. (1992) Penumbral tissues salvaged by reperfusion following middle cerebral artery occlusion in rats. *Stroke* 23, 552–559.
3. Karibe, H., Zarow, G. J., Graham, S. H., and Weinstein, P. R. (1994) Mild intraischemic hypothermia reduces postischemic hyperperfusion, delayed postischemic hypoperfusion, blood-brain barrier disruption, brain edema, and neuronal damage volume after temporary focal ischemia in rats. *J. Cereb. Blood Flow Metab.* 14, 620–627.
4. Ginsberg, M. D. and Busto, R. (1989) Rodent models of cerebral ischemia. *Stroke* 20, 1627–1642.
5. Koizumi, J., Yoshida, Y., Nakazawa, T., and Ooneda, G. (1986) Experimental studies of ischemic brain edema. 1. A new experimental model of cerebral embolism in rats in which recirculation can be introduced in the ischemic area. *Jpn. J. Stroke* 8, 1–8.
6. Longa, E. Z., Weinstein, P. R., Carlson, S., and Cummins, R. (1989) Reversible middle cerebral artery occlusion without craniectomy in rats. *Stroke* 20, 84–91.
7. Mosley, M. E., Cohen, Y., Mintorovitch, J., Chileuitt, L., Shimizu, H., Kucharczyk, J., Wendland, M. F., and Weinstein, P. R. (1990) Early detection of regional cerebral ischemia in cats: comparison of diffusion- and T2-weighted MRI and spectroscopy. *Magn. Reson. Med.* 14, 330–346.
8. Nagasawa, H. and Kogure, K. (1989) Correlation between cerebral blood flow and histologic changes in a new rat model of middle cerebral artery occlusion. *Stroke* 20, 1037–1043.
9. Karibe, H., Chen, J., Zarow, G. J., Graham, S. H., and Weinstein, P. R. (1994) Delayed induction of mild hypothermia to reduce infarct volume after temporary middle cerebral artery occlusion in rats. *J. Neurosurg.* 80, 112–119.
10. Shiraishi, K., Sharp, F. R., and Simon, R. P. (1989) Sequential metabolic changes in rat brain following middle cerebral artery occlusion: a 2-deoxyglucose study. *J. Cereb. Blood Flow Metab.* 9, 765–773.

Chapter 10

A MODEL OF TRANSIENT FOCAL CEREBRAL ISCHEMIA IN THE MOUSE

Hideyuki Kamii, Hiroyuki Kinouchi, Frank R. Sharp, and Pak H. Chan

TABLE OF CONTENTS

I. INTRODUCTION

The pathophysiology of cerebral ischemia has been investigated in various species including rats, gerbils, cats, and dogs. On the other hand, using mice for stroke study has been limited because of technical difficulties in making a reliable ischemic model, despite the relatively low animal cost, which is important for large-scale studies. Recent improvements in instruments and equipment for the surgical procedure have overcome the difficulties, and some studies on cerebral ischemia have been estimated in a mouse ischemic model (Backhauß et al., 1992; Barone et al., 1993; Bielenberg and Burkhardt, 1990; Birch et al., 1991; Gotti et al., 1990; Himori et al., 1990; Nowicki et al., 1991). However, no transient focal ischemic models were employed in these reports, despite the importance of reperfusion injury in clinical aspects. We have recently developed a model of transient focal cerebral ischemia in mice by modifying procedures used in rats (Longa et al., 1989; Nagasawa and Kogure, 1989). This model is less invasive than others, since craniectomy is not necessary. Furthermore, both transient and permanent focal ischemia can be obtained without any technical difficulties. With the recent successful development of transgenic mice, studies on stroke using a mouse ischemic model became more prevalent (Kinouchi et al., 1991; MacMillan et al., 1993; Yang et al., 1994). This model may be used to elucidate molecular mechanisms underlying the pathogenesis of ischemic brain injury.

II. PROCEDURES FOR FOCAL CEREBRAL ISCHEMIA

Male CD-1 mice (30 to 35 g) were anesthetized with chloral hydrate (350 mg/kg IP) and xylazine (4 mg/kg IP). The rectal temperature of the animals was maintained at 37 ± 0.5°C with a heating pad. The left femoral artery was cannulated with a PE-10 catheter for measurement of mean arterial blood pressure, PaO_2, $PaCO_2$, and pH.

0-8493-8094-4/95/$0.00+$.50

After a midline incision of the neck skin, the left common carotid artery (CCA) was exposed with careful conservation of the vagus nerve, and then the external carotid artery (ECA), superior thyroid artery (STA), and occipital artery (OA) were also isolated. The internal carotid artery (ICA) and pterygopalatine artery (PPA) were seen under the OA (Figure 1).

The STA and OA were electrocoagulated and divided. After a microsurgical clip (Scoville) was placed across the ECA origin, the distal end of the ECA was ligated with 6-0 silk suture and cut. A 6-0 silk suture was tied loosely around the ECA stump. Then, a 2-cm length of 5-0 monofilament nylon suture, blunted at the tip by heat, was introduced into the ECA stump and the silk suture around it was tightened (Figure 2A). After removing the microsurgical clip, the nylon suture was advanced into the ICA through the ECA. Care was taken to avoid introducing the suture into the PPA (Figure 2B).

The suture was gently advanced until resistance was felt. At this point, the tip of the suture was in the proximal segment of the anterior cerebral artery (ACA), where the ACA

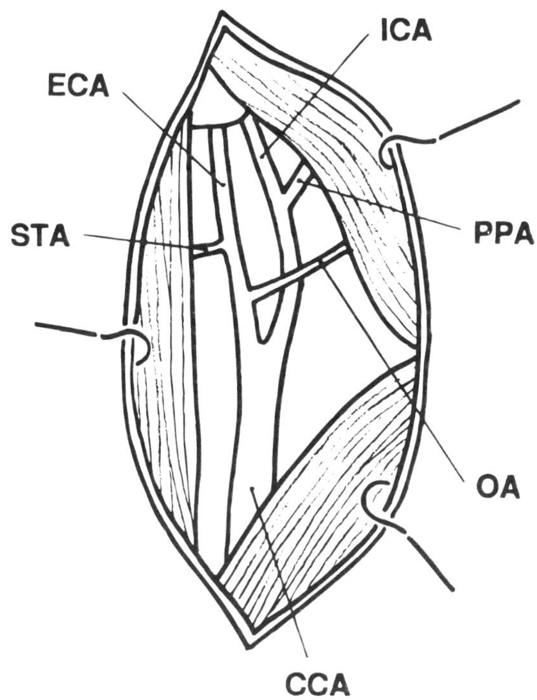

FIGURE 1. Diagram of arterial anatomy in the neck. After a midline incision of the neck skin, the left common carotid artery (CCA) was exposed, and then the external carotid artery (ECA), internal carotid artery (ICA), and their branches were also isolated. PPA, pterygopalatine artery; OA, occipital artery; STA, superior thyroid artery.

TABLE 1
Physiological Parameters during and after Ischemia

Physiological parameter	Before occlusion	10 min after occlusion	10 min after reperfusion
MABP (mmHg)	85.8 ± 10.9	92.2 ± 13.4	91.0 ± 8.2
PaO_2 (mmHg)	93.1 ± 4.1	95.4 ± 3.1	97.5 ± 5.6
$PaCO_2$ (mmHg)	50.6 ± 3.5	49.9 ± 4.2	49.0 ± 3.8
pH	7.33 ± 0.03	7.34 ± 0.02	7.33 ± 0.03

Note: Values are means ± SD (n = 6).

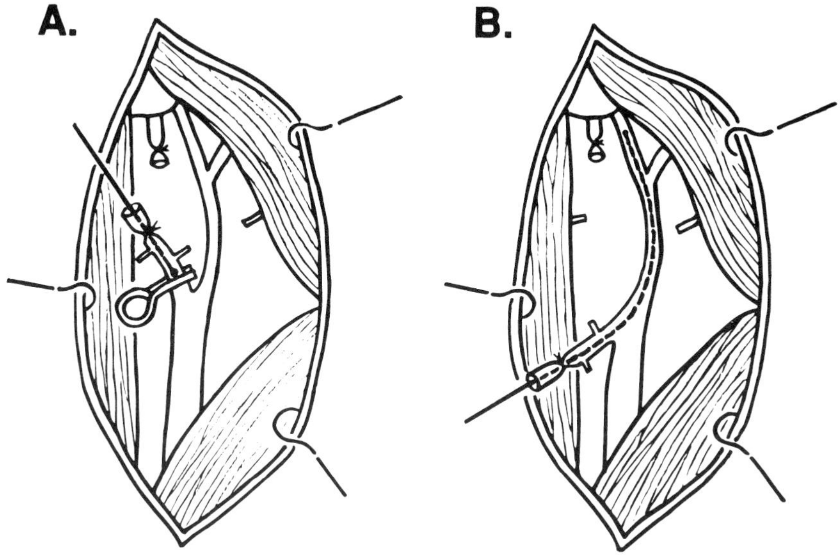

FIGURE 2. Schematic representations showing procedures to introduce the nylon suture into the internal carotid artery (ICA) through the external carotid artery (ECA) stump. After placing a microsurgical clip across the ECA origin, the ECA was divided and the nylon suture was introduced into the ECA stump with a silk suture tightened around it (A). Following removal of the microsurgical clip, the nylon suture was advanced from the ECA to ICA (B).

branches out to the anterior communicating artery (AComA), and the origin of the middle cerebral artery (MCA) was occluded. The distance from the bifurcation of the CCA to the tip of the suture measured approximately 11 mm (Figure 3).

To restore the blood flow of the MCA, the suture was pulled back into the ECA. After cutting the nylon suture protruding over the ECA stump, the incision was closed.

III. CORRELATION BETWEEN ISCHEMIC PERIOD AND BRAIN INJURY

To study correlation between duration periods of the focal cerebral ischemia and the brain injury, the mouse brains were histopathologically examined. At 24 h after 10, 30, and 60 min of transient focal cerebral ischemia, or following 24 h of permanent ischemia, the mice were decapitated and the brains were rapidly removed and frozen in powdered dry ice. Brain sections (20 μm) were cut at 500 μm intervals on a cryostat at −20°C, collected onto slides, and stained with cresyl violet.

Representative photographs of the brain sections were shown in Figure 4. No neuronal necrosis was seen following a 10-min ischemia. A 30-min ischemia produced infarcted areas in the lateral part of the caudate putamen without apparent infarct in the cortex, while a 60-min ischemia did in both the caudate putamen and the dorsolateral cortex. A 24-h permanent ischemia induced extensive infarcted areas including the medial part of the caudate putamen and the temporal cortex with prominent brain edema. However, no neuronal damage was seen in the hippocampus and the frontal cortex, which are out of the MCA distribution.

Figure 5 shows correlation between periods of the ischemia and the infarct volume. The infarcted areas of the brain section were quantitated by an image-analysis system (Swanson et al., 1990), and the infarct volume was calculated by taking the sum of the infarcted area of different brain sections times the interval between the sections (500 μm). These results have demonstrated that the infarct volume depends on the severity of the ischemic insult in our model.

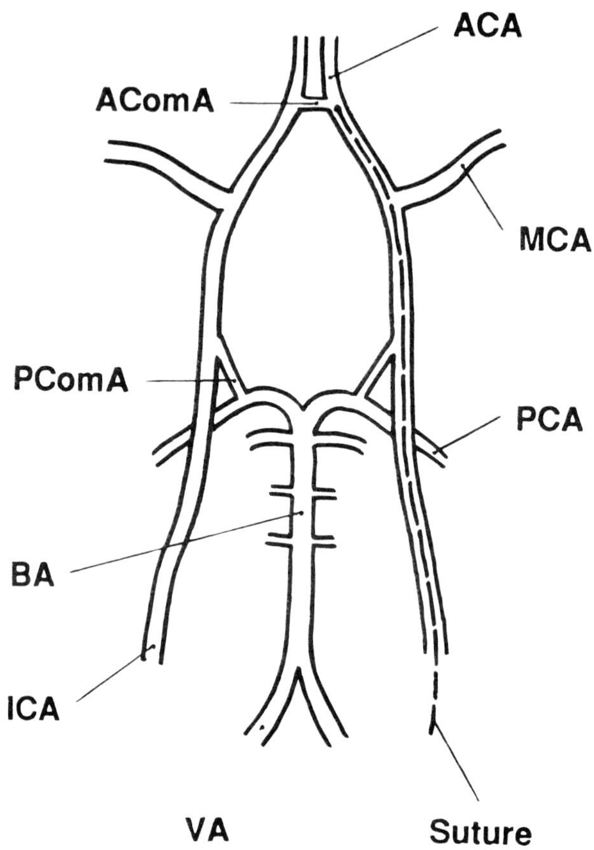

FIGURE 3. Schematic representation showing position of the tip of the nylon suture during ischemia. The origin of the middle cerebral artery (MCA) was occluded by the suture. ACA, anterior cerebral artery; PCA, posterior cerebral artery; AComA, anterior communicating artery; PComA, posterior communicating artery; BA, basilar artery; ICA, internal carotid artery; VA, vertebral artery.

IV. EXPRESSION OF hsp 70 AND c-fos mRNA AFTER ISCHEMIA

A heat shock gene, hsp 70, and a protooncogene, c-fos, are activated by a variety of brain injuries including cerebral ischemia/reperfusion (Kogure and Kato, 1993; Nowak et al., 1990; Sharp et al., 1993). Induction of hsp70 and c-fos mRNA appears to provide useful markers for neurons under various forms of stress and pathological conditions (Gonzalez et al., 1989; Hughes et al., 1993; Sharp et al., 1991).

Following 10 min of the focal cerebral ischemia, the mice were allowed to recover for up to 24 h. At each time point, the mice were decapitated and frozen brain sections (20 μm) were made, and then they were processed for *in situ* hybridization using the method of Schalling et al. (1988) with minor modifications. The sections were hybridized at 37°C for 18 h with 1×10^6 cpm of a ^{35}S-labeled hsp70 or c-fos oligonucleotide probe. After hybridization, the sections were rinsed in 1 × SSC (150 m*M* sodium chloride, 15 m*M* sodium citrate, pH 7.4) at 52°C for 60 min with several changes of 1 × SSC, dehydrated, and exposed to Kodak SB-5 film for a week.

All mice survived the 10 min of transient MCA occlusion. Physiological paremeters during and after ischemia were shown in Table 1. The distribution of hsp70 and c-fos mRNA in the postischemic mouse brains was assessed at caudate and dorsal hippocampal levels (Figure

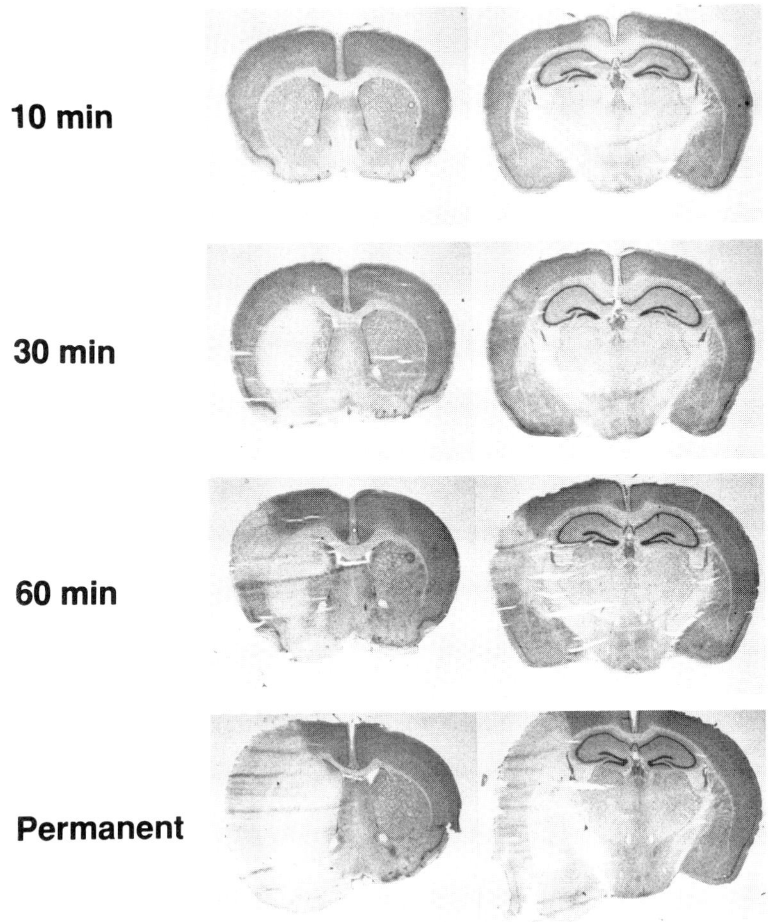

FIGURE 4. Representative photographs of the brain sections at the caudate (left column) and hippocampal (right column) levels following ischemia. At 24 h after 10, 30, 60 min of transient focal cerebral ischemia (10 min, 30 min, 60 min in the figure, respectively), or following 24 h of permanent ischemia (Permanent in the figure), the mice were decapitated and the brain sections were cut and stained with cresyl violet. A 30-min ischemia produced infarcted areas in the lateral part of the caudate putamen, while a 60-min ischemia did in both the caudate putamen and the dorsolateral cortex. A 24-h permanent ischemia induced extensive infarcted areas with prominent brain edema.

6). Expression of hsp70 mRNA was seen in the brain region corresponding to the MCA distribution except for the hippocampus, while c-fos mRNA was induced throughout the hemisphere over the MCA distribution. These findings are consistent with experimental evidences in rats (Kinouchi et al., 1993; Welsh et al., 1992).

V. CONCLUDING REMARKS

With the development of molecular approaches to elucidate mechanisms in the pathogenesis of cerebral ischemia, mouse ischemic models are supposed to become more prevalent. Compared to other focal ischemic models in mice, our model is less invasive and easier to make because of the simple procedures without craniectomy. This ischemic model is useful in various investigations on cerebral ischemia.

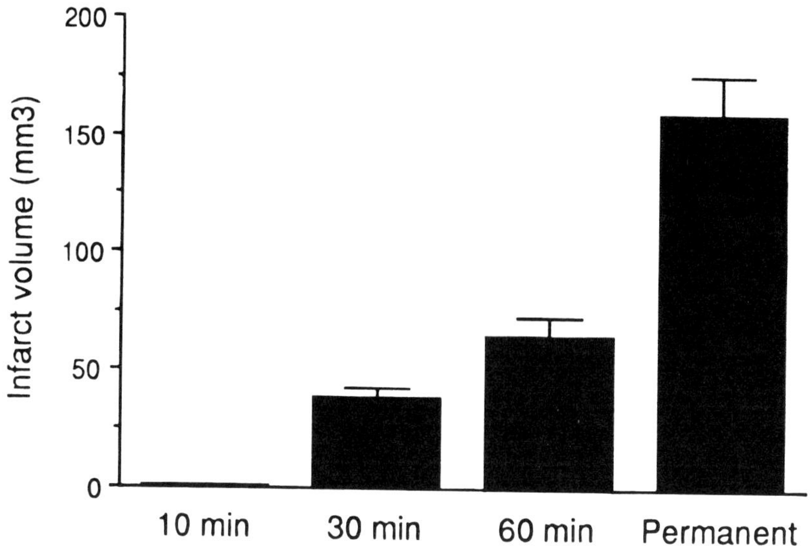

FIGURE 5. Correlation between periods of the ischemia and the infarct volume demonstrating that the infarct volume depends on the severity of the ischemic insult (n = 4 per group, mean ± SEM).

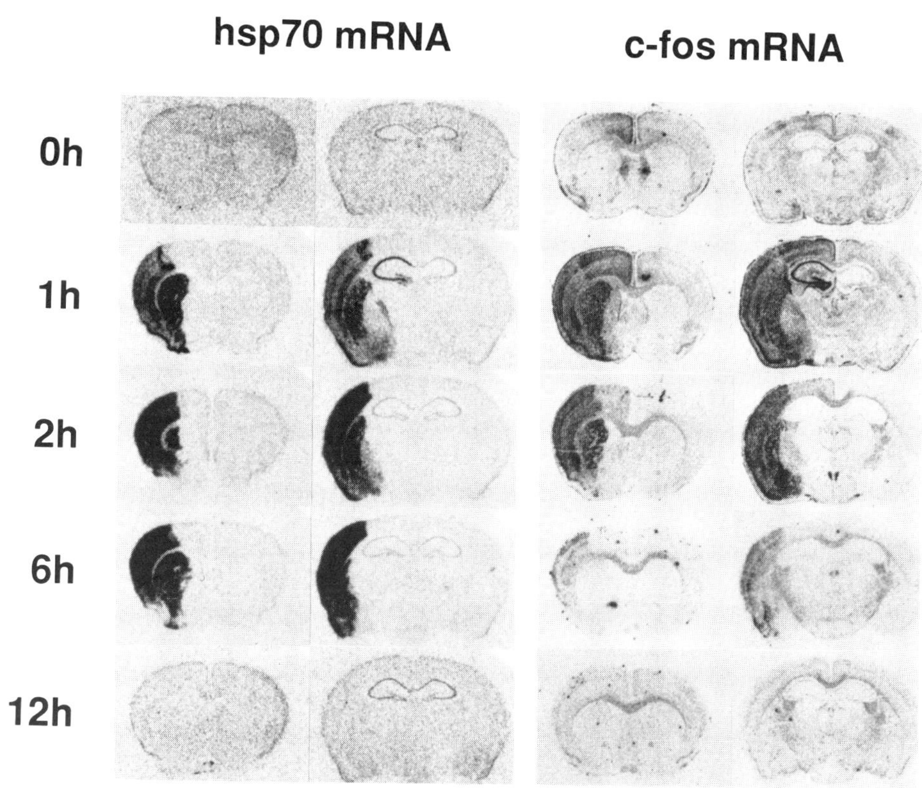

FIGURE 6. Distribution of hsp70 (left columns) and c-fos (right columns) mRNA at the caudate and hippocampal levels following 10 min of the focal cerebral ischemia. Expression of hsp70 mRNA was seen in the brain regions corresponding to the MCA distribution except for the hippocampus, while c-fos mRNA was induced throughout the hemisphere over the MCA distribution.

VI. SUMMARY

A model of transient focal cerebral ischemia in the mice is described. Male CD-1 mice (30 to 35 g) were anesthetized with chloral hydrate (350 mg/kg IP) and xylazine (4 mg/kg IP). The left common carotid artery was exposed, and the external carotid artery and its branches were isolated and coagulated. A 5-0 monofilament nylon suture, blunted at the tip, was introduced into the internal carotid artery through the external carotid artery stump up to the anterior cerebral artery. After various periods of keeping the nylon suture there, blood flow was restored by removing it. This procedure causes ischemic insult in the region corresponding to the middle cerebral artery distribution. Our focal ischemic model is reproducible and can produce infarction depending on duration periods of the ischemic insult.

ACKNOWLEDGMENTS

The authors thank Jari Koistinaho, Matt Morton, and Mary Pat Hill for their technical assistance. This work was supported by NIH grants NS-25372, NS-14543, AG-08938, NS-28167, and the Merit Review Program of the Department of Veterans Affairs.

REFERENCES

Backhauß, C., Karkoutly, C., Welsch, M., Krieglstein, J. (1992) A mouse model of focal cerebral ischemia for screening neuroprotective drug effects. *J Pharmacol Methods* 27: 27–32.

Barone, F. C., Knudsen, D. J., Nelson, A. H., Feuerstein, G. Z., Willette, R. N. (1993) Mouse strain differences in susceptibility to cerebral ischemia are related to cerebral vascular anatomy. *J Cereb Blood Flow Metab* 13: 683–692.

Bielenberg, G. W., Burkhardt, M. (1990) 5-Hydroxytryptamine agonists. A new therapeutic principle for stroke treatment. *Stroke* 21: IV161–IV163.

Birch, P. J., Rogers, H., Hayes, A. G., Hayward, N. J., Tyers, M. B., Scopes, D. I. C., Naylor, A., Judd, D. B. (1991) Neuroprotective actions of GR89696, a highly potent and selective K-opioid receptor agonist. *Br J Pharmacol* 103: 1819–1823.

Gonzalez, M. F., Shiraishi, K., Hisanaga, K., Sagar, M. F., Mandabach, M., Sharp, F. R. (1989) Heat shock proteins as markers of neural injury. *Mol Brain Res* 6: 93–100.

Gotti, B., Benavides, J., MacKenzie, E. T., Scatton, B. (1990) The pharmacotherapy of focal cortical ischemia in the mouse. *Brain Res* 522: 290–307.

Himori, N., Watanabe, H., Akaike, N., Kurosawa, M., Itoh, J., Tanaka, T. (1990) Cerebral ischemia model with conscious mice: involvment of NMDA receptor activation and derangement of learning and memory ability. *J Pharmacol Methods* 23: 311–327.

Hughes, P., Dragunow, M., Beilharz, E., Lawlor, P., Gluckman, P. (1993) MK801 induces immediate-early gene proteins and BDNF mRNA in rat cerebrocortical neurons. *Neuroreport* 4: 183–186.

Kinouchi, H., Epstein, C. J., Mizui, T., Carlson, E., Chen, S. F., Chan, P. H. (1991) Attenuation of focal cerebral ischemic injury in transgenic mice overexpressing CuZn superoxide dismutase. *Proc Natl Acad Sci USA* 88: 11158–11162.

Kinouchi, H., Sharp, F. R., Hill, M. P., Koistinaho, J., Hicks, K., Sagar, S. M., Chan, P. H. (1993) Induction of 70-kDa heat shock protein and hsp70 mRNA following transient focal cerebral ischemia in the rat. *J Cereb Blood Flow Metab* 13: 105–115.

Kogure, K., Kato, H. (1993) Altered gene expression in cerebral ischemia. *Stroke* 24: 2121–2127.

Longa, E. Z., Weinstein, P. R., Carlson, S., Cummins, R. (1989) Reversible middle cerebral artery occlusion without craniectomy in rats. *Stroke* 20: 84–91.

MacMillan, V., Judge, D., Wiseman, A., Settles, D., Swain, J., Davis, J. (1993) Mice expressing a bovine basic fibroblast growth factor transgene in the brain show increased resistance to hypoxic-ischemic cerebral damage. *Stroke* 24: 1735–1739.

Nagasawa, H., Kogure, K. (1989) Correlation between cerebral blood flow and histrogic changes in a new rat model of middle cerebral artery occlusion. *Stroke* 20: 1037–1043.

Nowak, T. S. Jr., Ikeda, J., Nakajima, T. (1990) 70-kDa heat shock protein and c-fos gene expression after transient ischemia. *Stroke* 21 (Suppl III): 107–111.

Nowicki, J. P., Duval, D., Poignet, H., Scatton, B. (1991) Nitric oxide mediates neuronal death after focal cerebral ischemia in the mouse. *Eur J Pharmacol* 204: 339–340.

Schalling, M., Dagerlind, A., Brene, S., Hallma, H., Djurfeldt, M., Persson, H., Terenius, L., Goldstein, M., Schlesinger, D., Hokfelt, T. (1988) Coexistence and gene expression of phenylethanolamine *N*-methyltransferase, and neuropeptide tyrosine in the rat and bovine adrenal gland: effects of reserpine. *Proc Natl Acad Sci USA* 85: 8306–8310.

Sharp, F. R., Jasper, P., Hall, J., Noble, L., Sagar, S. M. (1991) MK-801 and ketamine induce heat shock protein HSP72 in injured neurons in posterior cingulate and retrosplenial cortex. *Ann Neurol* 30: 801–809.

Sharp, F. R., Kinouchi, H., Koistinaho, J., Chan, P. H., Sagar, S. M. (1993) HSP70 heat shock gene regulation during ischemia. *Stroke* 24 (Suppl I): 72–75.

Swanson, R. A., Morton, M. T., Wu, G. T., Davidson C., Sharp, F. R. (1990) A semiautomated method for measuring brain infarct volume. *J Cereb Blood Flow Metab* 10: 290–293.

Welsh, F. A., Moyer, D. J., Harris, V. A. (1992) Regional expression of heat shock protein-70 mRNA and c-fos mRNA following focal ischemia in rat brain. *J Cereb Blood Flow Metab* 12: 204–212.

Yang, G., Chan, P. H., Chen, J., Carlson, E., Chen, S. F., Weinstein, P., Epstein, C. J., Kamii, H. (1994) Human copper-zinc superoxide dismutase transgenic mice are highly resistant to reperfusion injury after focal cerebral ischemia. *Stroke* 25: 165–170.

Chapter 11

MURINE MODELS OF CEREBRAL ISCHEMIA

F.C. Barone, R.N. Willette, R.K. Clark, A.H. Nelson, G.Z. Feuerstein, H.M. Sarau, M.N. Tzimas, D.B. Schmidt, R.D. Kapadia, and S.K. Sarkar

TABLE OF CONTENTS

I. INTRODUCTION

Models of cerebral ischemia in a small and well-established laboratory animal such as the mouse can be very useful. Mice are well controlled genetically to facilitate studies using different strains that might exhibit changes that can be associated with altered physiological or anatomical characteristics. Also, high productivity or throughput can be expected at a low cost (i.e., economically the animals are cheap, and they do not require much anesthesia or large amounts of drugs). Several models of global cerebral ischemia in the mouse have been described including hypoxia (Koida et al., 1989), unilateral common carotid artery occlusion (CCAO) combined with systemic hypoxia (Welsh et al., 1983) and decapitation (Berga et al., 1986; Yamamoto et al., 1987). Bilateral CCAO has been used previously for mouse cerebral ischemia with demonstrated increased mortality, decreased brain metabolism and learning deficits (Himori et al., 1990; 1991; Takeo et al., 1988; Tomida et al., 1987).

Pharmacotherapeutic approaches to the treatment of cerebral ischemia have been evaluated in various mouse strains with results similar to those in rats (Backhauß et al., 1992; Bielenberg and Burkhardt, 1990; Birch et al., 1991; Gotti et al., 1990; Himori et al., 1990). Specifically, NMDA receptor-channel blockers (Backhauß et al., 1992; Gotti et al., 1990; Himori et al., 1990), atypical noncompetitive NMDA receptor antagonists (Gotti et al., 1990), calcium antagonists (Backhauß et al., 1992), κ-opioid receptor agonists (Birch et al., 1991), 5-hydroxytryptamine agonists (Bielenberg et al., 1990; Backhauß et al., 1992), nitric oxide

0-8493-8094-4/95/$0.00+$.50

synthase inhibition (Nowicki et al., 1991), and free-radical scavengers (Backhauß et al., 1992) have demonstrated neuroprotective efficacy in mouse models of cerebral ischemia.

In the present series of experiments we characterized the sensitivity of different mouse strains to cerebral ischemia, which provides a comparison to the effects observed in other species such as the rat and gerbil. In addition, we have evaluated the effects of murine focal ischemia in mice on biochemical indexes of inflammatory cell infiltration, and the ability to visualize mouse focal ischemic damage using MRI. These new data also provided a comparison to that previously demonstrated in the rat. In general, these mouse models of ischemia are readily accessible and provide high-throughput systems directly relevant and comparable to other rodent models that are useful in the cerebral ischemia research laboratory.

II. FOCAL ISCHEMIA SURGERIES AND HISTOLOGY

Methods are as described previously (Barone et al., 1991c; 1993b). Three mouse strains (BDF, 25 to 32g; CFW, 24 to 31g; and BALB/C; 22 to 28g) were obtained from Charles River Labs., Inc. (Wilmington, MA). All mice used were mature males and were given free access to food and water prior to surgery. Animals were anesthetized with chloral hydrate (500 mg/kg, IP; Sigma Chemical Co.). Hand-held surgical techniques were carried out similar to those described previously (Backhauß et al., 1992; Welsh et al., 1987) with body temperature maintained at 37 ± 1°C using a heating lamp and water-heated pad (American Pharmaseal Co., Valencia, CA) during recovery from anesthesia immediately after surgery. Briefly, the right temporal-parietal region of the head was shaved and a 1-cm incision was made between the orbit and ear. Under a dissecting microscope, an incision was made dividing the temporal muscle and the right lateral aspect of the skull was exposed by reflecting the temporal muscle surrounding soft tissue. A small burr hole (1 mm) was made, using a high-speed microdrill (Fine Science Tools, Inc., Foster City, CA), through the outer surface of the semitranslucent skull just over the visibly identified middle cerebral artery at the level of the inferior cerebral vein as previously described in the rat (Barone et al., 1992a). Saline was applied to the area throughout the procedure to prevent heat injury. The inner layer of the skull was removed with fine forceps and the right middle cerebral artery (MCA) was occluded (MCA occlusion, MCAO) by electrocoagulation, using a small vessel cauterizer (Fine Science Tools Inc.) directly through the dura without damaging the brain surface. If the brain surface was visibly damaged or if the MCA bled due to incomplete artery occlusion/coagulation, the animal was euthanized and not used. The muscle and soft tissue were replaced and the skin was closed using 5-0 silk suture. In addition, some animals from each strain were prepared with right common carotid artery occlusion (CCAO) alone or in combination with right MCAO (MCAO + CCAO). Under these conditions, a 1-cm incision was made and the common carotid artery was exposed and dissected free from the vagus and cervical sympathetic nerves under a dissecting microscope. Two 5-0 silk suture ligatures were made and tied securely 3 mm apart, and the artery was completely transected using scissors. Again, the muscle and soft tissue were replaced and the skin was closed using 5-0 silk suture. Sham-operated mice from each strain received similar surgeries but without occlusion of the MCA or CCA. All surgical areas were shaved, cleaned, and treated with betadine prior to and after surgery. Animal survival was monitored for the next 24 h.

At 24 h postsurgery, mice were euthanized by rapid cervical dislocation and the brain removed. Five coronal forebrain slices (1.0 mm thick) were made beginning at the level of the olfactory bulb to just rostral to the cortical-cerebellar junction using a mouse brain slicer prepared in our laboratory. Forebrain slices were immersed immediately in a 1% solution of triphenyltetrazolium chloride (TTC; Sigma Chemical Co.) in phosphate buffer at 37°C for 20 to 30 min (Bederson et al., 1986). Stained tissues then were fixed in 10% phosphate-buffered formalin (Lerner Laboratories, Pittsburgh, PA). The two sides of each TTC-stained section were photographed in color using a Polaroid camera (Polaroid Corp.). These photographs were analyzed for quantification of ischemic damage using an image-analysis system (Amersham RAS 3000; Loats Associates, Inc., Westminster, MD) as described previously for rat focal ischemia (Barone et al., 1992a; 1992b). Briefly, morphological changes following surgery were evaluated for the forebrain (total of nine planar surfaces from each side of the five 1-mm-thick forebrain sections) for each mouse. These planar image surfaces (from photographs) were digitized by the image-analysis system for planimetry determinations of brain swelling and infarction produced by focal ischemia. Individual hemisphere and infarction areas (mm^2) were measured for each forebrain section surface. Hemispheric swelling representing tissue edema (Barone et al., 1992a), was expressed as the percent increase in size of the ipsilateral (i.e., surgery side) hemisphere over the contralateral (normal) hemisphere. Hemispheric infarct size was expressed as the percent infarcted tissue in reference to the contralateral (normal) hemisphere. These parameters were determined for each slice to evaluate the forebrain profile of damage, and for total forebrain changes by using the sum of all individual section data in these calculations. In addition, forebrain profiles of infarct areas were determined and total infarct volumes were calculated from the infarct areas of all forebrain sections.

Focal ischemia in the mouse produced significant cerebral tissue injury. Figure 1 illustrates typical TTC-stained brain sections from BALB/C mice 24 h following sham, MCAO, or MCAO + CCAO surgeries. Significant hemispheric swelling and infarction were observed

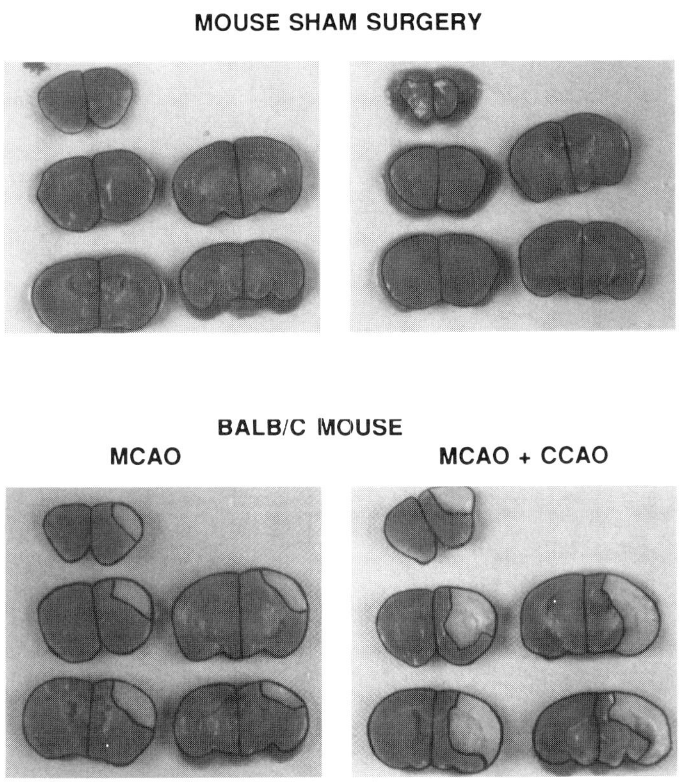

FIGURE 1. Representative photographs of TTC-stained brain sections from BALB/C mice used in the analysis of mouse focal ischemia. Top: All brain sections (photograph of one side of section on left and other side of section on right) illustrating that no cerebral infarction is observed following sham surgery. Bottom: Left side illustrates typical degree of infarction following right MCAO (alone). Right side illustrates increase in degree of infarction following right MCAO + CCAO (photograph of same side of brain sections as for MCAO alone).

following ischemia in all strains studied. Sham surgery did not produce significant hemispheric swelling or any degree of infarction in any strain studied. Figure 2 illustrates the total hemispheric swelling, total percent hemispheric infarction and infarct volume observed in the three mouse strains following sham, MCAO, or MCAO + CCAO surgeries. Hemispheric swelling was not different in the three mouse strains for sham, MCAO, or MCAO + CCAO. Significantly different degrees of total swelling were produced by the different ischemic insults, and the order of increasing effect was generally sham < MCAO < MCAO + CCAO ($p < 0.05$). This relationship was also true for percent hemispheric infarction and infarct volume. However, unlike hemispheric swelling, significant differences in the degree of infarction (hemispheric and volume) between the different strains (BALB/C > CFW > BDF) were observed for both MCAO and MCAO + CCAO surgeries ($p < 0.05$). These differences are clearly illustrated as the forebrain profiles for these measures in Figure 3. Again, no obvious differences between strains were observed for hemispheric swelling (i.e., index of edema; Barone et al., 1992) across forebrain sections. However, strain differences in infarct profiles were noted following MCAO, and especially following MCAO + CCAO (i.e., again BALB/C > CFW > BDF). Although increased edema has been shown to account for overestimation of infarct size (Brint et al., 1988; Swanson et al., 1990), it was unlikely that swelling contributed to the obvious differences in focal infarction between strains, since relatively similar degrees of hemispheric swelling were exhibited by the mouse strains following focal ischemia.

Figure 4 illustrates mortality differences between these mouse strains over the 24 h following the different surgical treatments. No deaths were observed following sham surgery or MCAO treatment in any strain or for any surgical treatment in BDF mice. However, a significant number of deaths were observed following CCAO (alone) and MCAO + CCAO in CFW and BALB/C mice ($p < 0.05$). In addition, multifocal infarctions were clearly produced in the ipsilateral hemisphere following CCAO alone in 36% of BALB/C mice but not at all in the other two strains (data not shown). Typically, following MCAO + CCAO surgery in BALB/C mice, animals were physically twisted or actually rotated in a direction contralateral to the damaged hemisphere. In BALB/C mice, the degree of cerebral ischemic damage was evaluated over 5 d following MCAO. Table 1 lists these data for different groups of mice sacrificed from days 1 to 5 postsurgery. All measures were stable over 4 d. Infarction significantly decreased at day 5, apparently associated with cavitation of the cortex similar to that described previously in the rat (Clark et al., 1993).

III. BLOOD PRESSURE, GASES, AND CELL PROFILES

In order to determine if strain differences in blood pressure and/or gases differed between strains in a manner that might contribute to strain ischemic sensitivity, mice from each strain were anesthetized with chloral hydrate as described for focal ischemia surgery. The left carotid artery was cannulated and connected to a pressure transducer (Stathem; P10EZ) connected to a calibrated polygraph (Beckman; R511A). Mean arterial blood pressure (MABP) was recorded over 3 to 6 min. A heparinized blood sample (0.2 ml) was then obtained from the cannulated carotid artery and blood PO_2, PCO_2, and pH were measured using an acid-base laboratory (Radiometry; ABL330).

Also to determine if strain differences in blood profiles might contribute to strain ischemic sensitivity, mice from each strain were euthanized by CO_2 inhalation and 0.3 ml of blood was collected by cardiac puncture and placed immediately in a microtainer with EDTA (Becton Dickenson; #5961). Blood cell types and complete blood profiles were measured/determined using a hematology system (Technicon).

Mean arterial blood pressure was not significantly different between the three mouse strains (Table 2). Also, blood gases and pH were not different between strains with one

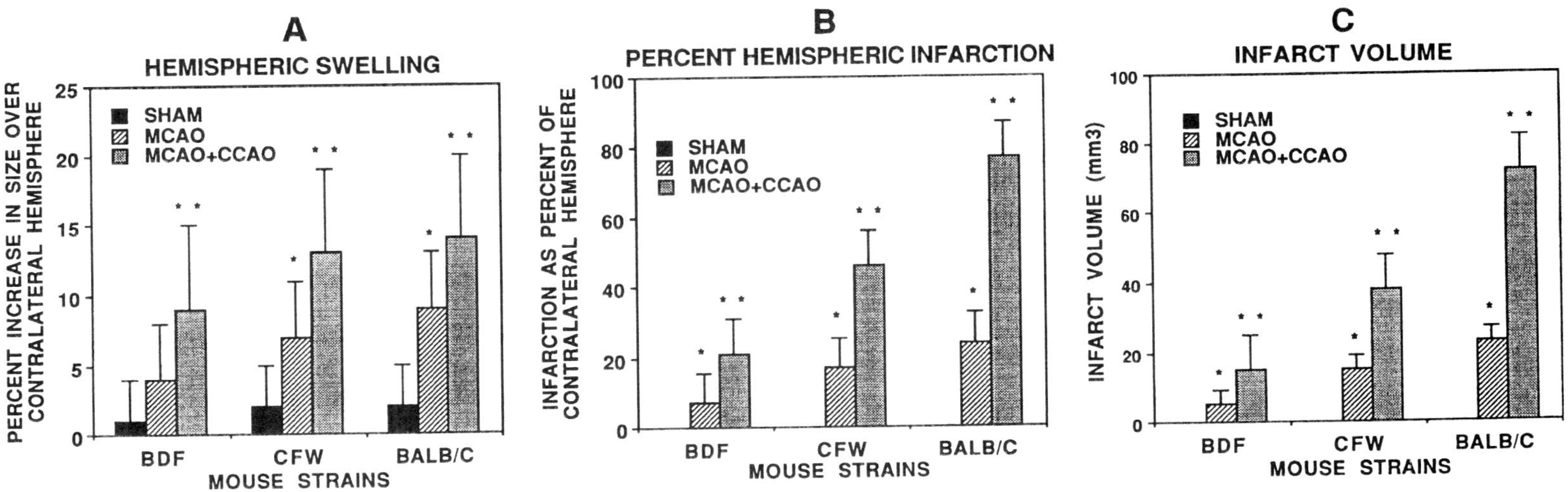

FIGURE 2. Effects of sham surgery, MCAO surgery, and MCAO + CCAO surgery on total focal ischemic damage (i.e., A, hemispheric swelling; B, hemispheric infraction; and C, infarct volume, in BDF (N = 9 to 12), CFW (N = 7 to 8) and BALB/C (N = 10 to 12) mice. Infarcts (percent hemispheric and volume) but not hemispheric swelling following MCAO and MCAO + CCAO surgeries were significantly different between strains (BALB/C > CFW > BDF; $p < 0.05$; ANOVA). For intrastrain differences in the figure: *$p < 0.05$ different from sham, and **$p < 0.05$ different from sham and MCAO; ANOVA with Tukey follow-up test. Means ± standard deviations are presented. (From Barone, F.C. et al., *J. Cereb. Blood Flow Metab.*, 13, 683, 1993. With permission.)

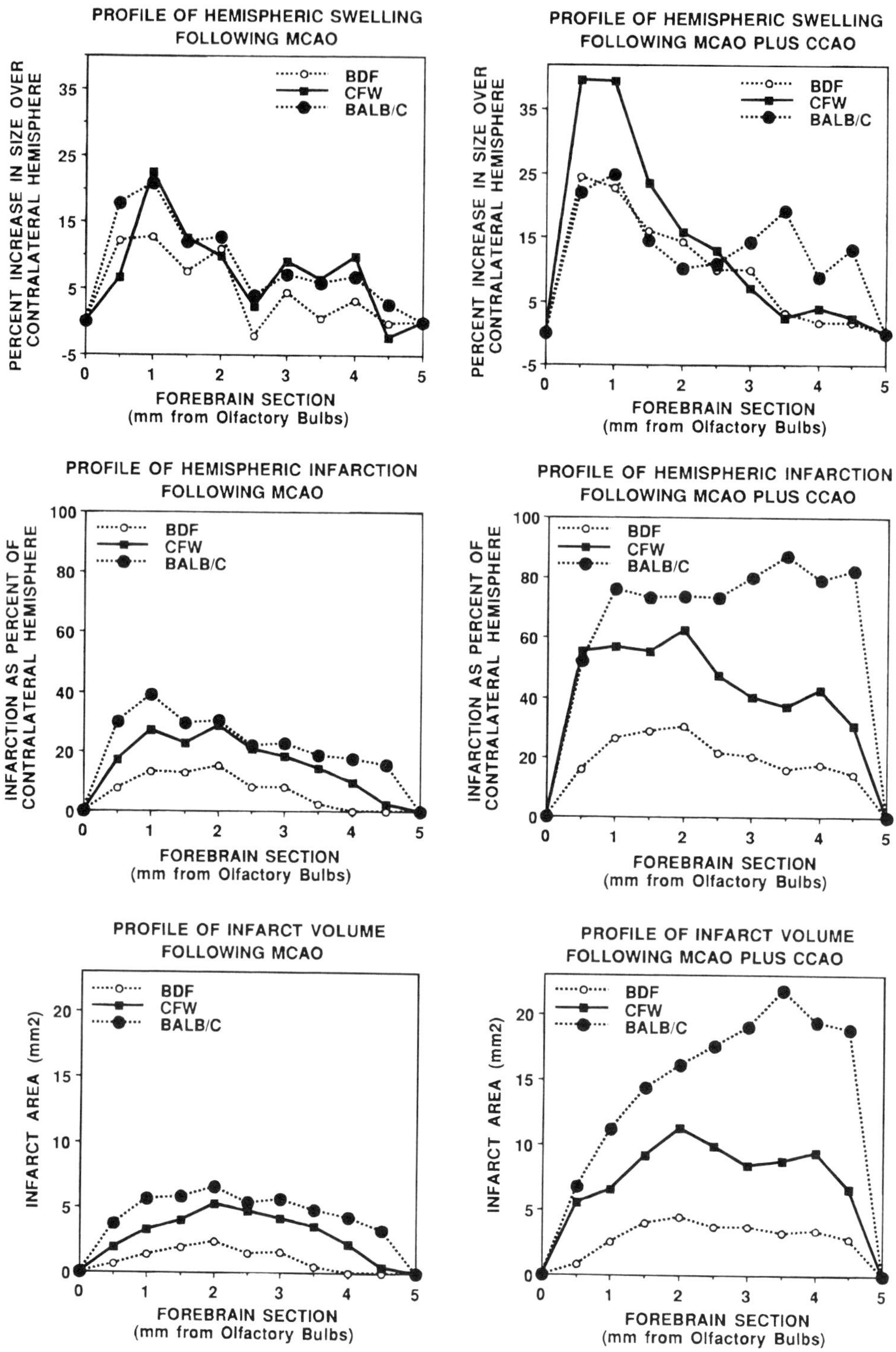

FIGURE 3. Forebrain profiles of hemispheric swelling (top), hemispheric infarction (middle) and infarct volume (bottom) following MCAO and MCAO + CCAO surgeries in BDF (N = 9 to 12), CFW (N = 7 to 8) and BALB/C (N = 10 to 12) mice. Means are presented but standard deviations were omitted for clarity. (From Barone, F. C. et al., *J. Cereb. Blood Flow Metab.*, 13, 683, 1993. With permission.)

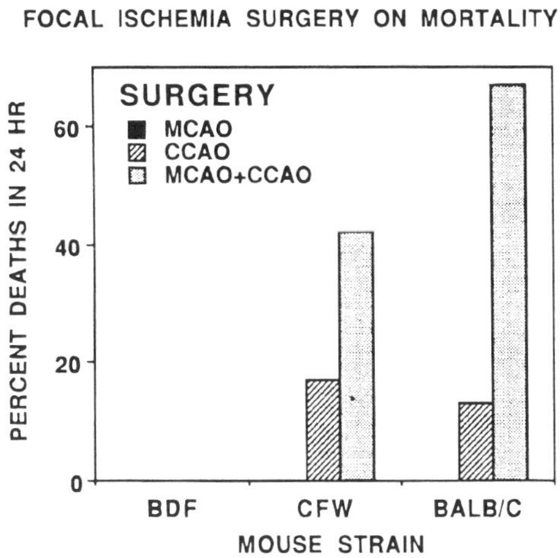

FIGURE 4. Focal ischemia surgery (MCAO + CCAO) produced significant deaths in CFW and BALB/C but not in BDF mice ($*p < 0.05$; χ^2 test). In addition, multifocal infarctions (not shown) were produced following CCAO alone in 36% of BALB/C mice but not at all in the other two strains. (BDF, N = 6 to 12; CFW, N = 6 to 19; BALB/C, N = 12 to 16). (From Barone, F. C. et al., *J. Cereb. Blood Flow Metab.*, 13, 683, 1993. With permission.)

TABLE 1
Focal Ischemic Tissue Damage over Five Days Following MCAO in BALB/C Mice

	Measures		
Day	Hemispheric swelling (%)	Hemispheric infarct (%)	Infarct volume (mm^3)
1	9.5 ± 0.9	18.9 ± 1.9	13.6 ± 1.4
2	9.7 ± 2.2	18.7 ± 0.9	13.9 ± 0.6
3	10.1 ± 1.3	19.4 ± 0.8	14.0 ± 0.6
4	7.6 ± 0.7	17.6 ± 0.9	13.6 ± 0.9
5	6.2 ± 1.4	12.7 ± 0.8*	9.0 ± 0.6*

Note: N = 6 in each group; means ± standard deviations are presented.
* $p < 0.05$; different from day 1; ANOVA with Dunnett follow-up test.

TABLE 2
Blood Pressure, Gases, and pH for Three Mouse Strains

		Mouse Strain		
Variable	N	BDF	CFW	BALB/C
MABP (mmHg)	5–8	65 ± 5	77 ± 13	70 ± 18
PCO_2 (mmHg)	10	44 ± 3*	36 ± 6	36 ± 6
PO_2 (mmHg)	10	93 ± 16	101 ± 19	88 ± 22
pH	10	7.2 ± 0.1	7.2 ± 0.1	7.2 ± 0.1

Note: MABP = mean arterial blood pressure; means ± standard deviations are presented.
* $p < 0.05$; different from other mouse strains; ANOVA with Tukey follow-up test.

exception. The PCO_2 was significantly higher in BDF than the other two strains (Table 1). However, all mouse strains were in the normal range (Loeb and Quimby, 1989), and although statistically significant this one difference cannot explain the strain differences in ischemic sensitivity.

Blood profiles (Table 3) were relatively similar for the three strains and also within normal ranges. Although there were noted statistical differences between strains for neutrophils, eosinophils, and lymphocytes (Table 3), all values were in the normal range (Foster et al., 1983) and again, although statistically significant these differences were not related to the different ischemic sensitivities of the three strains.

IV. FUNCTIONAL CEREBROVASCULAR ANATOMY

Major cerebral arteries were examined in BALB/C, BDF, and CFW mice. Each mouse was anesthetized (as above) and a complete sternotomy was performed to expose the heart. An incision was made in the right atrium and the animal was perfused with 10 ml of 0.9% NaCl via the left ventricle. Following exsanguination, 5 ml of a 10% suspension of carbon black in 0.9% NaCl (Aldrich Chemical Co., Milwaukee, WI; particle size 1 to 2 μm) was administered via the left ventricle to enhance visualization of the cerebral arteries. In some BALB/C mice, the common carotid arteries were ligated bilaterally prior to perfusion with carbon black. Following perfusion, the animals were decapitated and the brains were carefully removed into buffered formalin for 24 h prior to examination. The circle of Willis was examined and photographed under a dissection microscope at 8.4×. Posterior communicating arteries were identified as anastomoses connecting the hindbrain and forebrain circulations (i.e., the basilar and internal carotid arteries).

Differences in the cerebral vascular anatomy were identified between the three mouse strains. The intracardiac administration of carbon black revealed that the presence of patent posterior communicating arteries were significantly decreased in BALB/C compared to BDF mice ($p < 0.05$), while CFW mice had an intermediate degree of patency. Figure 5 illustrates the different distributions of patent posterior communicating arteries in the three mouse strains and a photomicrograph depicting the complete absence of posterior communicating arteries often observed in BALB/C mice. In addition, the carbon-black perfusion technique was utilized to determine the integrity of BALB/C forebrain perfusion during bilateral CCAO

TABLE 3
Blood Profile for Three Mouse Strains

Variable (units)	BDF (N = 19)	CFW (N = 16)	BALB/C (N = 18)
RBC ($\times 10^6/\mu l$)	10.2 ± 0.4	9.8 ± 0.4	10.4 ± 0.4
Platelets ($\times 10^5/\mu l$)	1.2 ± 0.4	1.2 ± 0.4	1.2 ± 0.4
WBC ($\times 10^3/\mu l$)	7.9 ± 1.8	6.2 ± 2.0	7.8 ± 2.1
Neutrophils ($\times 10^3/\mu l$)	0.6 ± 0.4*	1.9 ± 1.2*	1.1 ± 0.4
Eosinophils ($\times 10^3/\mu l$)	0.23 ± 0.09*	0.16 ± 0.08	0.11 ± 0.04
Basophils ($\times 10^3/\mu l$)	0.03 ± 0.04	0.03 ± 0.04	0.04 ± 0.04
Monocytes ($\times 10^3/\mu l$)	0.24 ± 0.04	0.20 ± 0.08	0.19 ± 0.08
Lymphocytes ($\times 10^3/\mu l$)	6.7 ± 1.8	4.0 ± 1.2*	6.3 ± 2.1
Hemoglobin (mg %)	15.1 ± 0.4	15.3 ± 0.4	15.8 ± 0.4
Hematocrit (%)	53.1 ± 1.3	52.5 ± 2.0	53.0 ± 1.7
Mean cell volume (fl)	52.1 ± 0.9	53.4 ± 2.4	50.9 ± 0.8
Mean corpuscular hemoglobin (pg)	14.8 ± 0.4	15.6 ± 0.4	15.1 ± 0.4

Note: RBC = red blood cells; WBC = white blood cells; means ± standard deviations are presented.
* $p < 0.05$; different from other mouse strains; ANOVA with Tukey follow-up test.

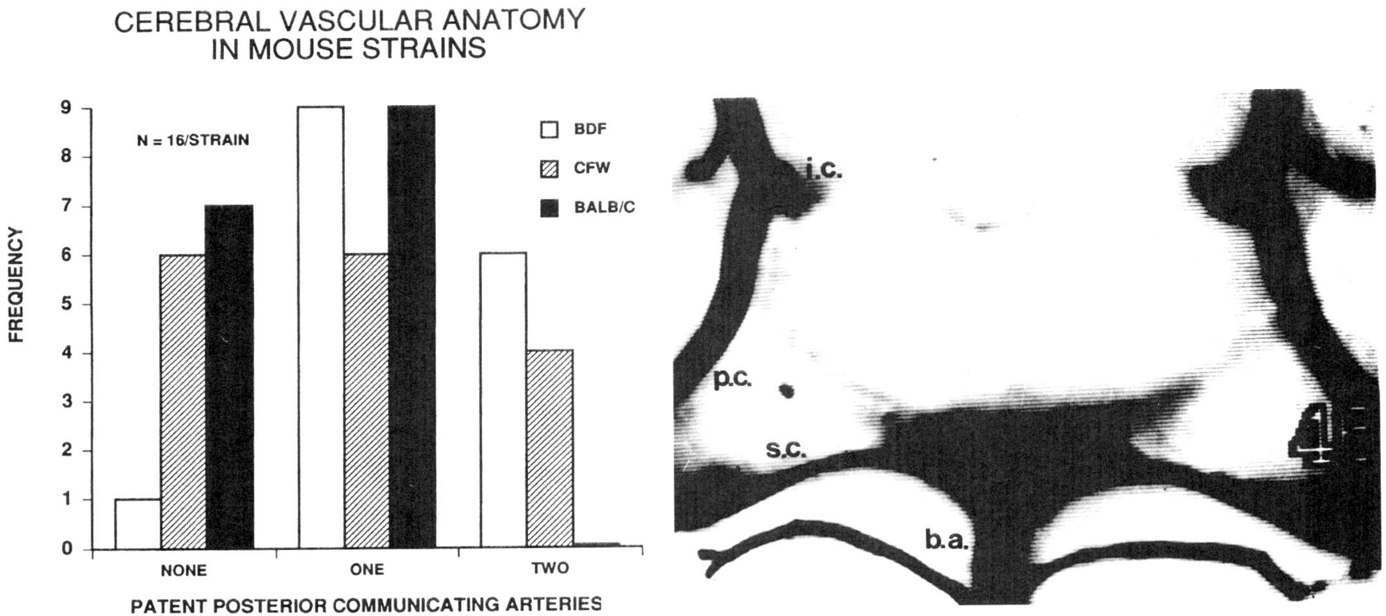

FIGURE 5. The intracardiac administration of carbon black was used to examine the vascular anatomy of the circle of Willis in all three strains of mice (left). Patent posterior communicating arteries (i.e., and anastomosis between p.c. and s.c.) were significantly decreased in BALB/C compared to BDF mice ($p<0.05$; χ^2 test). Abbreviations in the representative (70×) photomicrograph (right) include: b.a., basilar artery; i.c., internal carotid artery; p.c., posterior cerebral artery; s.c., superior cerebellar artery. Note the absence of both posterior communicating arteries in this BALB/C mouse. (From Barone, F.C. et al., *J. Cereb. Blood Flow Metab.*, 13, 683, 1993. With permission.)

(BCCO) in an effort to identify the potential utility of BALB/C mice as a model of transient forebrain (i.e., global) ischemia. Indeed, BCCO eliminated the forebrain perfusion of carbon black and only the posterior cerebral circulation was perfused adequately under these conditions (Figure 6).

V. TRANSIENT FOREBRAIN ISCHEMIA

In BALB/C mice surgical anesthesia was induced with 3% isoflurane and maintained with 1.5% isoflurane in oxygen delivered by means of an open face mask. Each animal was then secured in a supine position on a thermally controlled heating pad to maintain rectal temperature at 37 to 38°C. A ventral midline cervical incision was made to expose and isolate the CCAs. In five BALB/C mice, the effects of BCCO on the EEG and cortical microvascular perfusion (CP) were determined as described previously (Willette et al., 1990). Briefly, each mouse was anesthetized with choral hydrate (as above) and silk suture (4-0) snares were placed around each carotid artery. The mouse was then secured in a stereotaxic instrument (DKI, Tujunga, CA) in a flat skull position. The scalp overlying the left and right parietal bone was reflected and two screw-type electrodes were secured over the left parietal cortex for recording the EEG. In the right parietal bone, a cortical window (3 mm diameter) was prepared with the dura intact. A laser Doppler flowmetry needle probe (ALF, Transonic

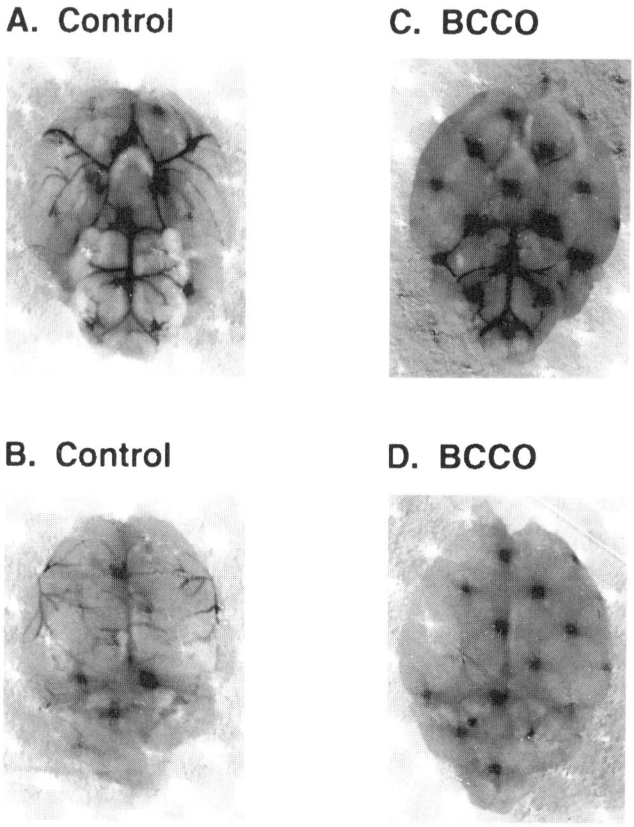

FIGURE 6. The effects of BCCO on the cerebral distribution of carbon black was examined grossly in BALB/C mice. Carbon black filled the major arteries in the basilar, carotid, and forebrain territories following intracardiac administration in control mice (A and B). In contrast, only the posterior cerebral circulation was perfused adequately during BCCO (C and D). (From Barone, F. C. et al., *J. Cereb. Blood Flow Metab.*, 13, 683, 1993. With permission.)

Instruments, Ithaca, NY) was placed with a micromanipulator within the window at the cortical surface. Changes in CP and the EEG induced by tightening (5 min) and loosening the common carotid snares were monitored continuously. EEG power during these studies was determined as described previously (Willette et al., 1992). Briefly, the EEG signal was digitized and the total EEG power spectrum was computed between 1 and 16 Hz from a fast Fourier transform obtained over 10 s epochs (Laboratory Technologies Corp., Wilmington, MA). We had hypothesized that BCCO in the BALB/C mouse would be useful as a model of transient forebrain ischemia. Our results substantiated this hypothesis and recordings from a representative animal are depicted in Figure 7. BCCO produced the expected transient forebrain ischemia profile of a flat EEG and persistent reduction (>85%) in cortical perfusion. In addition, upon reperfusion of the carotid arteries, cortical perfusion recovered in 1 to 2 min. A later recovery of EEG was observed along with a delayed cortical hypoperfusion. The results for a group of BALB/C mice are listed in Table 4.

In order to follow up on these results, we evaluated delayed cell death in susceptible CA_1 neurons of the hippocampus. The CCAs were occluded bilaterally with microaneurysm clips (85 g of pressure). Following the period of occlusion (10, 30, or 60 min), the clips were removed and the incision was sutured closed. With the exception of applying the microaneurysm clips, the same procedure was performed in sham-operated animals. For comparative purposes, in some animals 5-0 silk suture was used to permanently ligate the CCAs bilaterally. All mice were observed for 7 d. On day 7, surviving animals were anesthetized (as above) and perfused via the left ventricle with 0.9% NaCl followed by buffered formalin. The brains were removed and prepared for routine histologic sectioning of paraffin-embedded tissue and mounted and stained with hematoxylin and eosin (Luna, 1968). One 6-μm-thick section taken from the central hippocampus at the level of the lateral thalamic and medial Habenular nuclei (Pellegrino et al., 1979) was examined for each mouse. The percent abnormal cells in the entire CA_1 region of the dorsal hippocampus was quantitated bilaterally and averaged. The duration of BCCO in BALB/C mice was directly related to the degree of delayed neuronal death in the vulnerable hippocampal CA_1 neurons 7 d after ischemia. Figure 8 illustrates the percentage of ischemic/dead CA_1 neurons in the dorsal hippocampus in individual animals and the group data following sham surgery or 10 or 30 min of BCCO. A significant increase in CA_1 ischemic neuronal cell death was observed 7 d following 30 min of BCCO ($p < 0.05$).

The duration of BCCO also was directly related to increased animal mortality. Figure 9 illustrates these effects in different groups of BALB/C mice receiving different durations of BCCO over a 7-d period. Following permanent BCCO, BALB/C mice survived for less than 24 h ($p < 0.05$). BCCO for 60 min produced death in 60% of the animals by the fifth day ($p < 0.05$). BCCO for 10 and 30 min produced death in only 10% of the animals. All sham-operated animals survived the procedure for the duration of observation. It is noteworthy that following 30 min or more of BCCO, autonomic effects were identified, including piloerection, observed in about 80%, and ptosis in one or both eyes, observed in about 50% of the ischemic animals. None of these changes were observed in sham-operated BALB/C mice.

VI. LEUKOCYTES IN FOCAL ISCHEMIC TISSUE

TTC-stained sections from MCAO + CCAO CFW mice used in the planimetric analysis were prepared for histological evaluation. Following fixation (24 h), formalin was replaced with 70% ethanol. Processing for paraffin embedding was performed through routine procedures (Luna, 1968). Briefly, tissues were dehydrated through graded ethanol, cleared in zylenes and infiltrated with Paraplast embedding medium (Monoject Scientific, St. Louis, MO). Sections (6 μm) were cut on a Reichert Model 820 microtome and following drying,

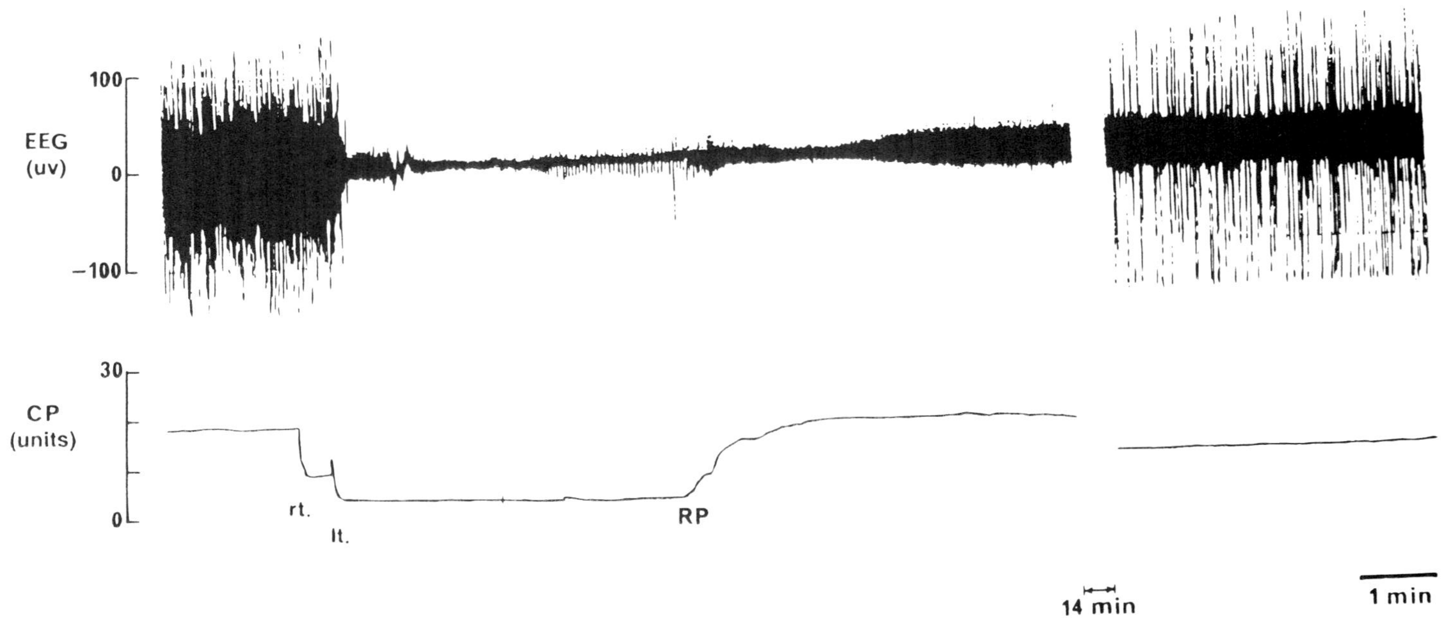

FIGURE 7. Representative effects of transient forebrain ischemia (i.e., BCCO: rt./lt.) and reperfusion (RP) on cortical microvascular perfusion (CP) and EEG in BALB/C mice. (From Barone, F.C. et al., *J. Cereb. Blood Flow Metab.*, 13, 683, 1993. With permission.)

TABLE 4
Changes in Cortical Microvascular Perfusion and EEG Power during BCCO and Reperfusion*

Measure	BCCO	Reperfusion
Cortical perfusion	−86 ± 7%**	5 ± 37%
EEG power	−93 ± 7%**	−62 ± 37%**

* Data were obtained during 5 min of BCCO and following 5 min of reperfusion in five BALB/C mice. Means ± standard deviations are presented.
** $p < 0.05$ different from zero; *t*-test.

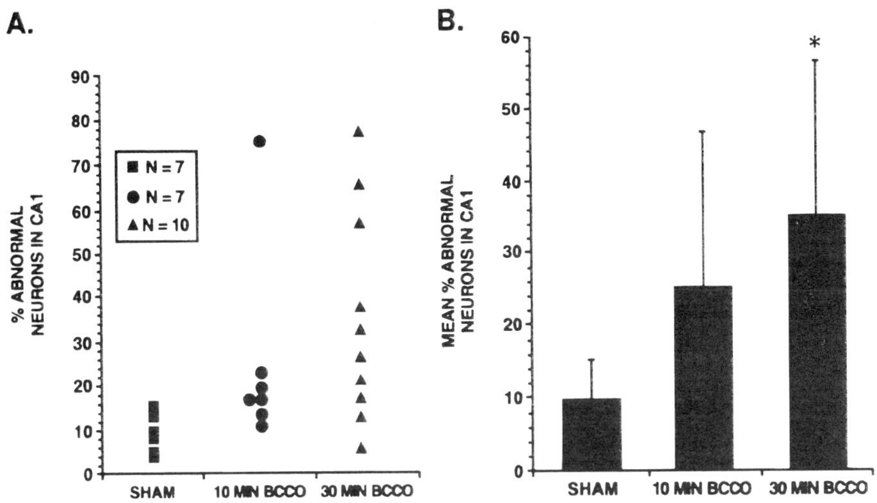

FIGURE 8. BALB/C mouse BCCO produced an ischemia-time related neuronal degeneration in the CA_1 region of the dorsal hippocampus after 7 d (*$p < 0.05$; ANOVA). A: Individual data points. B: Summarized data. Means ± standard deviations are presented. (From Barone, F. C. et al., *J. Cereb. Blood Flow Metab.*, 13, 683, 1993. With permission.)

the sections were deparaffinized and redehydrated through graded ethanol then stained with hematoxylin and eosin (H&E). Leukocytes that accumulated and infiltrated into the tissue were identified and photographed on an Olympus BH2 microscope. Leukocytes were not observed in tissue sections of the surgical hemisphere 24 h after sham surgery or in the control hemisphere at 1 or 5 d postsurgery. Significant neutrophil accumulation was observed in blood vessels and adjacent to blood vessels following 1 d of focal ischemia in the ischemic region. Dramatic infiltration of neutrophils and monocyte/macrophages was observed in the ischemic tissue after 5 d of focal ischemia. Figure 10 illustrates these results, which are similar to that described previously in the rat (Barone et al., 1993b; Clark et al., 1993).

Another group of CFW mice were exposed to sham surgery or MCAO + CCAO for 1 or 5 d, and then euthanized. Brains were removed and dissected into two hemispheres and rapidly frozen and stored at −80°C. The methods used to quantitate neutrophil infiltration into ischemic brain were as described previously (Barone et al., 1991b; 1992b; 1993b). Briefly, the frozen brain hemispheres (surgery and control) were weighed rapidly, homogenized by Polytron (5-s burst at medium intensity) in 4 ml 10 m*M* Tris-HCl (pH 7.4, 4°C) and then centrifuged at 30,000 *g* for 30 min (4°C). The supernatant was discarded and the pellet was resuspended in 4 ml 10 m*M* Tris-HCl, and 1 ml was removed and centrifuged for the myeloperoxidase (MPO) assay while the remaining 3 ml were used for the [^{3}H] leukotriene

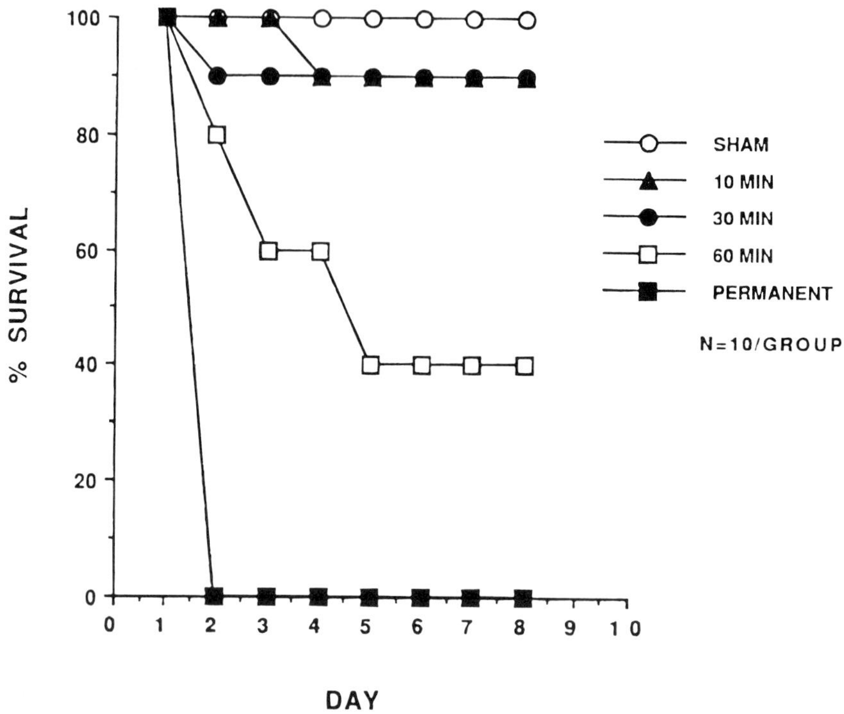

FIGURE 9. The effect of BCCO (on day 1) on survival, over a 7-d period, was examined in BALB/C mice. The duration of BCCO was directly related to survival, with significantly more deaths occurring following 60 min or permanent BCCO ($p < 0.05$ χ^2 test).

B_4 (LTB_4)—binding assay (see below). The supernatant of the MPO aliquots was decanted and the pellet was extracted by suspension in 0.5% hexadecyltrimethyl-ammonium bromide (HTAB; Sigma Chemical Co.) in 50 m*M* potassium phosphate buffer (pH 6.0, 25°C) for 2 min at a tissue wet weight-to-volume ratio of 1:10. Samples were put through three freeze-and-thaw cycles with sonications (10 s, 25°C) between cycles. After the last sonication, samples were incubated at 4°C for 20 min and centrifuged at 12,500 *g* (15 min, 4°C). Supernatant MPO was assayed as described by Bradley et al. (1982). The rate at which a colored product formed during the MPO-dependent reaction with *o*-dianisidine (0.167 mg/ml; Sigma Chemical Co.) was measured (i.e., the rate of change in absorbance at 460 nm was monitored over 2 min) using a Beckman DU-7 spectrophotometer with kinetic analysis capability. One unit of MPO was defined as that amount that degraded 1 μmol of peroxide per minute at 25°C. MPO activity (units) for each hemisphere was normalized per gram wet weight tissue.

The procedure used to quantitate LTB_4 receptor binding from the same mouse hemisphere was as described previously (Sarau et al., 1989; Winkler et al., 1988; Barone et al., 1992b; 1993a). Briefly, equilibrium [^{3}H] LTB_4 binding was achieved with optimum incubation conditions (30 min at 23°C) using ligand concentrations appropriate to establish maximum specific binding. This was accomplished with the remaining 3-ml aliquots (75% of tissue sample) of homogenized segments that were centrifuged at 30,000 *g* (30 min, 4°C). Pellets were washed twice with 10 ml of 50 m*M* Tris-HCl (pH 7.4, 4°C) and resuspended in 50 m*M* Tris-HCl at a protein concentration of 2 mg/ml. Assay tubes contained 200 μg of protein per tube and 1 n*M* [^{3}H] LTB_4 (175 to 200 Ci/mmol; New England Nuclear) and were incubated for 30 min at 23°C with or without the addition of 1 μ*M* unlabeled LTB_4 to

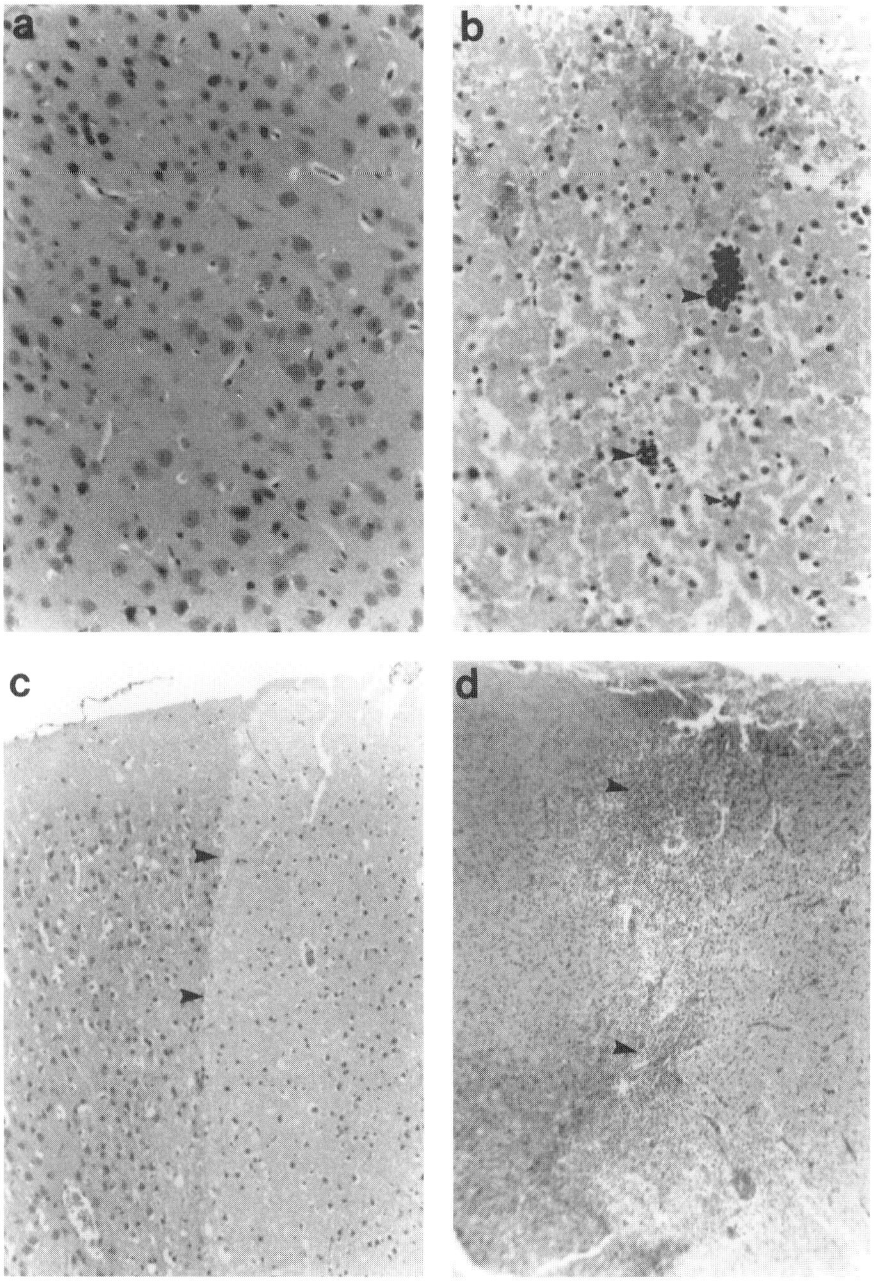

FIGURE 10. A: Control (nonischemic) cortex 1-d postfocal ischemia. B: Ischemic coretex 1-d postfocal ischemia. Arrows depict large number of neutrophils within and adjacent to blood vessels. C: Ischemic (right) and nonischemic (left) cortex 1 d following focal ischemia. Arrows indicate junction of two areas. D: Same as C but 5 d following focal ischemia. Note the larger increase in leukocyte infiltration into ischemic/infarcted tissue from 1 (C) to 5 (D) d after focal stroke.

determine non-specific and total binding, respectively. The reaction was terminated and the bound labeled material was separated from the unbound by vacuum filtration through Whatman GF/C filters and filters washed with 25 m*M* Tris-HCl, pH 7.4. Filters were added to 10 ml of Readi-Protein+ (Beckman Instruments) and radioactivity was quantified by liquid scintillation spectrometry. Specific binding was calculated as the difference between total and nonspecific binding and was expressed as fmol/mg protein. Protein was determined by the Biorad assay using bovine serum albumin as the standard.

Table 5 shows the effects of focal ischemia on MPO activity. Sham surgery (24 h) produced no change, but 1 d of focal ischemia produced a significant increase, and 5 d of focal ischemia produced a dramatic increase in MPO activity. A similar profile of effects was observed for LTB_4-specific binding following focal ischemia (Table 6). These biochemical data correlate with that obtained using histological methodology, and are similar to that observed in rat focal ischemia where these methods, used as an index of leukocyte infiltration have been validated (Barone et al., 1991b; 1992b; 1993a).

VII. MAGNETIC RESONANCE IMAGING OF FOCAL ISCHEMIC DAMAGE

Mice were anesthetized using a continuous flow of 2% isoflurane delivered to a mask that fit over the animal's face/nose during the imaging experiment. Magnetic resonance imaging (MRI) experiments were carried out on a 9.4-T Bruker AMX spectrometer with microimaging accessories as described previously (Barone et al., 1991c; Kapadia et al., 1992; Sarkar et al., 1992). Two-dimensional images were acquired using a spin echo pulse sequence with a TR of 1 s and a TE of 30 ms with 500 μm slice thickness. A 256 × 256 data matrix was used resulting in resolution of 100 × 100 μm. A 2D transverse section through a normal mouse brain is shown in Figure 11. The image exhibits excellent contrast and allows visualization of anatomical details of the brain, as identified in the figure. For example, internal and external capsules, fornix, the third ventricle, and the mammilothalamic tract are observed at this resolution level.

TABLE 5
MPO Activity following Mouse Focal Ischemia

	MPO Activity (units/g wet weight)		
Hemisphere	Sham	1-d Ischemia	5-d Ischemia
Surgery	0.011 ± 0.006	0.034 ± 0.009*	3.034 ± 1.923*
Control	0.011 ± 0.003	0.009 ± 0.003	0.050 ± 0.029

Note: N = 5 in each group; means ± standard deviations are presented.
* $p < 0.05$ vs. control hemisphere or sham surgery; ANOVA with Tukey follow-up test.

TABLE 6
LTB_4 Receptor Binding following Mouse Focal Ischemia

	LTB_4-specific binding (fmol/mg protein)		
Hemisphere	Sham	1-d Ischemia	5-d Ischemia
Surgery	3.9 ± 0.4	5.4 ± 0.3*	11.6 ± 3.3*
Control	3.7 ± 0.4	3.7 ± 0.3	4.5 ± 0.4

Note: N = 5 in each group; means ± standard deviations are presented.
* $p < 0.05$ vs. control hemisphere or sham surgery; ANOVA with Tukey follow-up test.

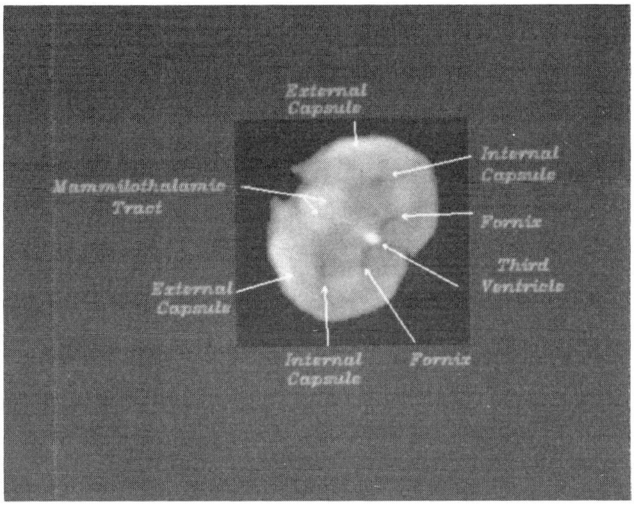

FIGURE 11. 9.4-T magnetic resonance image of a transverse section of a mouse brain obtained using a spin echo sequence with a TR of 3 s and a TE of 32 ms. A slice thickness of 500 μm was used with in-plane resolution of 100 × 100 μm. Resolution is adequate to identify several obvious brain structures.

After initial images were taken, the same BALB/C mouse underwent microsurgery to occlude the MCA using the procedure as described earlier. Images were collected 24 h after the surgery and the infarcted region was clearly identified in the image as shown in Figure 12. The extent of hemispheric infarction in animals ranged from 20 to 30%, and the MRI results paralleled those obtained from histological analyses. Sham surgery did not produce any changes in MR images nor did it show histologically detectable cerebral tissue damage as we have shown previously in the rat (Barone et al., 1991a). In order to better quantify the infarction, we generated 3D rendered volume images by interpolating 2D slices to obtain isotropic voxels of dimension 100 × 100 × 100 μm. A 3D rendered volume image of the mouse brain with infarction is shown in Figure 13. These procedures enhanced the visualization of infarcted tissue in reference to the rest of the brain and helped significantly in quantifying the volume of the infarcted region.

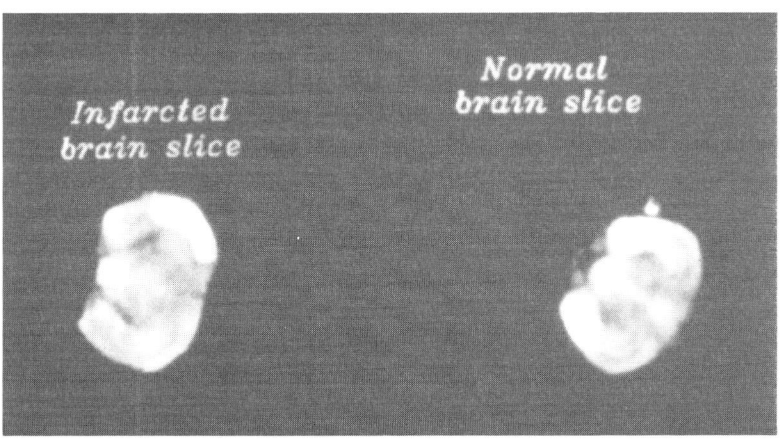

FIGURE 12. 9.4 T-magnetic resonance images of a mouse brain before and 24 h after MCAO. A slice thickness of 500 μm was used with in-plane resolution of 100 × 100 μm. The cortical infarction following focal ischemia can clearly be seen.

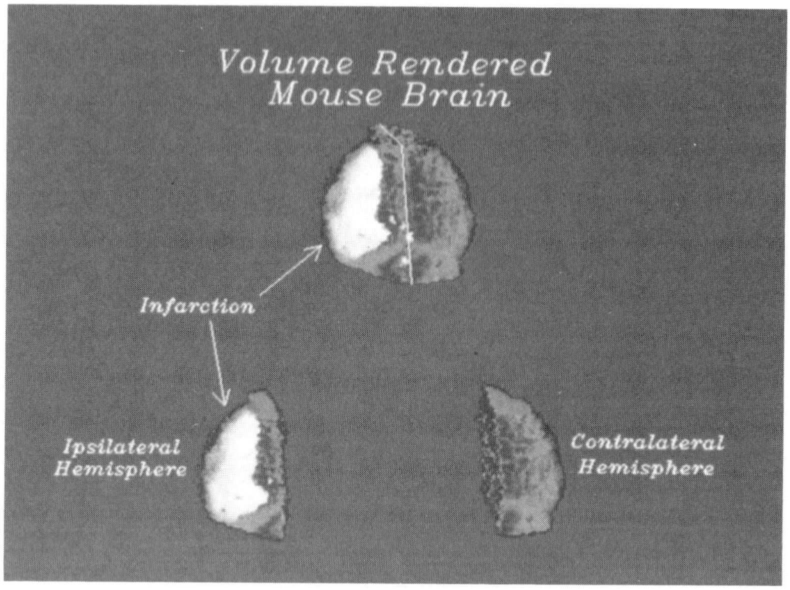

FIGURE 13. A 3D rendered volume image of infarcted mouse brain.

VIII. DISCUSSION

Evidence in the rat has accumulated that suggests that strain is an important determinant in the susceptibility to global and focal cerebral ischemia (Barone et al., 1992a; Globus and Busto, 1989). However, no systematic evaluation of mouse strain differences in cerebral ischemia sensitivity has been carried out to help optimize the use of mouse models of cerebral ischemia. The present series of experiments characterized ischemic tissue damage and mortality in different mouse strains following focal ischemia. The effects of different degrees of focal ischemia were evaluated, and gross cerebrovascular differences between strains were studied in order to help explain the differing ischemic sensitivities between strains. In addition, the utility of BCCO (i.e., global ischemia) in the mouse was evaluated as a potential model, in terms of mortality and susceptible neuronal (i.e., hippocampal CA_1) degeneration.

The data indicate an increased degree of cerebral damage following focal ischemia in BALB/C compared to BDF and CFW mice. CFW mice are also more susceptible to the effects of focal ischemia than BDF mice. Since blood pressure, gases, and cell profiles were relatively similar for the mouse strains, the differing strain susceptibilities to ischemia can be explained on the basis of strain differences in cerebral circulation that appear to increase forebrain risk of ischemic damage. Clearly, differences in susceptibility to focal ischemic damage are demonstrated here in standard, commercially available mouse strains.

Genetically, BDF mice are a hybrid strain and are generated on demand from female C57 black and male DBA_2 (dilute Brown Aglutti) mice. CFW mice are an outbred strain that are also not genetically pure with a controlled origin from inbreeding Swiss mice (cesarean-derived in 1974). In contrast, BALB/C are a very highly controlled inbred strain, originating from albino mice in 1932. Charles River Laboratories (Wilmington, MA) received the BALB/C strain from NIH in 1954 and cesarean-derived them in 1955 (Charles River Laboratories, Inc. 1988; 1992; U.S. Department of Health and Human Services, 1982). The degree of genetic control in the BALB/C strain appear to contribute to the consistent sensitivity to cerebral ischemia via the more compromised cerebral vascular arrangement. Other genetic factors, i.e., genetic hypertension, are important in the increased susceptibility to rat cerebral ischemia and has been discussed recently (Barone et al., 1992a).

Other references have been made to mouse strain differences in sensitivity to cerebral ischemia. The most dramatic recent demonstration is the decreased infarction and edema following MCAO in transgenic mice bearing the human superoxide dismutase gene compared to nontransgenic littermates (Kinouchi et al., 1991). In addition, differences between two other mouse strains, the ICR and ddY strains (Shizuoko Experimental Animal Laboratory, Japan), have also been reported. The ddY strain appears to be more sensitive than the ICR strain to BCCO, exhibiting increased mortality and amnesia, and decreased brain metabolic activity (Himori et al., 1990; Kojima and Kaneto, 1989; Muguruma et al., 1991). Perhaps differences in cerebral circulation can explain the differing susceptibility in these strains as well.

We evaluated BCCO in BALB/C mice based on the discovery that the expected posterior communicating arteries, which in most other species connect the carotid and vertebrobasilar arterial system, were absent or not functional in this mouse strain. This lack of posterior communicating arteries has been identified in gerbils (Levine and Sohn, 1969; Levy and Brierley, 1974), and because of this species abnormality, BCCO in gerbils produces an almost complete and reversible elimination of forebrain cerebral blood flow (Crockard et al., 1980), increased mortality (Kahn, 1972; Tomida et al., 1987), and delayed cell death of susceptible hippocampal CA_1 pyramidal neurons (Kirino, 1982). The present experiments not only demonstrate global ischemia and associated increased mortality in the BALB/C mouse, but also demonstrate delayed neuronal death associated with BCCO, suggesting this mouse strain might be useful as a model of global cerebral ischemia.

Previously, we described in detail the inflammatory cell response that occurs in rat focal ischemia (Clark et al., 1993) and the use of MPO activity and LTB_4 receptor binding to quantitate this cell infiltration (Barone et al., 1991b; 1992b; 1993a). The present studies demonstrate that cell infiltration with an apparent similar time course occurs in mouse focal ischemia. As discussed previously for the rat (Barone et al., 1992b), the mouse might also serve as an ideal model to evaluate pharmacologic intervention to reduce the leukocyte number or their adhesion to blood vessels, which can contribute to the degree of ischemic injury. The plethora of research tools (i.e., antibodies, cDNA probes, etc.) available for the mouse makes additional basic mechanistic research in these murine models of cerebral ischemia highly desirable. Finally, the present data indicate that 3D MRI/microscopy can monitor changes in cerebral tissue following focal ischemia. This approach to studying cerebral ischemic events and their consequences in mouse models, with the very high resolution available under these conditions, provides an excellent experimental system to study pathology and therapy in stroke.

ACKNOWLEDGMENTS

The authors wish to thank Dorothy J. Knudsen for expert technical assistance and Rose Ann Pimpinella for careful preparation of this manuscript.

Animals were housed and cared for in accordance with *Guide for the Care and Use of Laboratory Animals* (DHEW) (DHHS) Publication NO. (NIH) 85-23, revised 1985, Office of Science and Health Reports, DRR/NIH, Bethesda, MD 20205. Procedures using lab animals were approved by the Institutional Animal Care and Use Committee of SmithKline Beecham Pharmaceuticals.

REFERENCES

Backhauß, C., Karkoutly, C., Welsch, M., Krieglstein, J. (1992) A mouse model of focal cerebral ischemia for screening neuroprotective drug effects. *J Pharmacol Meth* 27: 27–32.

Barone, F. C., Clark, R. K., Feuerstein, G., Lenkinski, R. E., Sarkar, S. K. (1991a) Quantitative comparison of magnetic resonance imaging (MRI) and histologic analyses of focal ischemic damage in the rat. *Brain Res Bull* 26: 285–291.

Barone, F. C., Hillegass, L. M., Price, W. J., White, R. F., Feuerstein, G. Z., Sarau, H. M., Clark, R. K., Griswold, D. E. (1991b) Polymorphonuclear leukocyte infiltration into cerebral focal ischemic tissue: myeloperoxidase activity assay and histologic verification. *J Neurosci Res* 29: 336–348.

Barone, F. C., Hillegass, L. M., Tzimas, M. N., Schmidt, D. B., White, R. F., Price, W. J., Feuerstein, G. Z., Clark, R. K., Griswold, D. E., Sarau, H. M. (1993a) Leukocytes and leukotriene B_4 (LTB_4) receptors in stroke. *Soc Neurosci Abstr* 19: 1057.

Barone, F. C., Knudsen, D. J., Nelson, A. H., Feuerstein, G. Z., Tzimas, M. N., Schmidt, D. B., Lee, E. V., Clark, R. K., Sarau, H. M. (1991c) Middle cerebral artery occlusion (MCAO) in the mouse. *Soc Neurosci Abstr* 17: 1094.

Barone, F. C., Knudsen, D. J., Nelson, A. H., Feuerstein, G. Z., Willette, R. N. (1993b) Mouse strain differences in susceptibility to cerebral ischemia are related to cerebral vascular anatomy. *J Cereb Blood Flow Metab* 13: 683–692.

Barone, F. C., Price, W. J., White, R. F., Willette, R. N., Feuerstein, G. Z. (1992a) Genetic hypertension and increased susceptibility to cerebral ischemia. *Neurosci Biobehav Rev* 16: 219–233.

Barone, F. C., Schmidt, D. B., Hillegass, L. M., Price, W. J., White, R. F., Feuerstein, G. Z., Clark, R. K., Griswold, D. E., Sarau, H. M. (1992b) Reperfusion increases neutrophil and LTB_4 receptor binding in focal ischemia. *Stroke* 23: 1337–1348.

Bederson, J. B., Pitts, L. H., German, S. M., Nishimura, M. C., David, R. L., Bartkowski, H. M. (1986) Evaluation of 2,3,5-tetraphenyltetrazolium chloride as a strain for detection and quantification of experimental cerebral infarction in rats. *Stroke* 17: 1304–1308.

Berga, P., Beckett, P. R., Roberts, D. J., Llemas, J., Massingham, R. (1986) Synergistic interactions between piracetam and dihydroergocristine in some animal models of cerebral hypoxia and ischemia. *Arzneimittelforschung* 39: 1314–1320.

Bielenberg, G. W., Burkhardt, M. (1990) 5-Hydroxytryptamine agonists. A new therapeutic principle for stroke treatment. *Stroke* 21:, IV161–IV163.

Birch, P. J., Rogers, H., Hayes, A. G., Hayward, N. J., Tyers, M. B., Scopes, D. I. C., Naylor, A., Judd, D. B. (1991) Neuroprotective actions of GR89696, a highly potent and selective K-opioid receptor agonist. *Br J Pharmacol* 103: 1819–1823.

Brint, S., Jacewicz, M., Kiessling, M., Tanabe, J., Pulsinelli, W. (1988) Focal brain ischemia in the rat: methods for reproducible neocortical infarction using tandem occlusion of the distal middle cerebral and ipsilateral common carotid arteries. *J Cereb Blood Flow Metab* 8: 474–485.

Charles River Laboratories, Inc. (1988) Strain designations and origins of Charles River VAF/PLUS™ and COBS® Animals. Charles River Labs: Wilmington, MA.

Charles River Laboratories, Inc. (1992) Product Catalog. Charles River Labs: Wilmington, MA.

Clark, R. K., Lee, E. V., Fish, C. J., White, R. F., Price, W. J., Jonak, Z. L., Feuerstein, G. Z., Barone, F. C. (1993) Progression of cerebral changes following middle cerebral artery occlusion in the rat: a quantitate planimetric, histologic and immunohistochemical study. *Brain Res Bull* 31: 565–572.

Crockard, A., Iannotti, F., Hunstock, A. T., Smith, R. D., Harris, R. J., Symon, L. (1980) Cerebral blood flow and edema following carotid occlusion in the gerbil. *Stroke* 11: 494–498.

Foster, H. L., Small, J. D., Fox, J. G. (1983) *The Mouse in Biomedical Research, Volume III, Normotensive Biology, Immunology and Husbandry.* Academic Press: New York.

Ginsberg, M. D., Busto, R. (1989) Rodent models of cerebral ischemia. *Stroke* 20: 1627–1642.

Gotti, B., Benavides, J., MacKenzi, E. T., Scatton, B. (1990) The pharmacotherapy of focal cortical ischemia in the mouse. *Brain Res* 522: 290–307.

Himori, N., Watanabe, H., Akaike, N., Kurosawa, M., Itoh, J., Tanaka, T. (1990) Cerebral ischemia model with conscious mice: involvement of NMDA receptor activation and derangement of learning and memory ability. *J Pharmacol Methods* 23: 311–327.

Himori, N., Moreau, J. L., Martino, J. R. (1991) Cerebral ischemia decreases the behavioral effects and mortality rate elicited by activation of NMDA receptors in mice. *Neuropharmacology* 30: 1179–1186.

Kahn, K. (1972) The natural course of experimental cerebral infarction in the gerbil. *Neurology* 22: 510–515.

Kapadia, R. D., Barone, F. C., Knudsen, D. J., Feuerstein, G., Sarkar, S. K. (1992) Cerebral tissue damage in mouse model of focal ischemic stroke: a 3D MR microscopy study. *Soc Mag Res Med* 1: 1806.

Kinouchi, H., Epstein, C. J., Mizui, T., Carlson, E., Chen, S. F., Chan, P. H. (1991) Attenuation of focal cerebral ischemic injury in transgenic mice overexpressing CuZn superoxide dismutase. *Proc Natl Acad Sci (USA)* 88: 11158–11162.

Kirino, T. (1982) Delayed neuronal death in the gerbil hippocampus following ischemia. *Brain Res* 239: 57–69.

Koida, M., Nakamuto, H., Yasuda, K., Muguruma, K., Hiramatsu, Y., Ogawa, Y., Kato, Y. (1989) Carbon monoxide (CO)-induced hypoxia in mice: evaluation as an experimental model of cerebral ischemia for drug screening. *Jpn J Pharmacol* 51: 273–278.

Kojima, M., Kaneto, H. (1989) Preparation of cerebral ischemia-induced amnesic model in mice and ameliorative effect of several compounds on the model. *Nippon Yakurigaku Zasshi (Japan)* 94: 223–228.

Levine, S., Sohn, D. (1969) Cerebral ischemia in infant and adult gerbils: relation to an incomplete circle of Willis. *Arch Pathol* 87: 315–317.

Levy, D. E., Brierley, J. B. (1974) Communications between vertebro-basilar and carotid arterial circulations in the gerbil. *Exp Neurol* 45: 503–508.

Loeb, W. F., Quimby, F. W. (1989) *The Clinical Chemistry of Laboratory Animals.* Pergamon Press: New York.

Luna, L. G. (1968) *Manual of Histologic Staining Methods of the Armed Forces Institute of Pathology.* McGraw-Hill: New York.

Muguruma, K., Nakamuta, H., Tsuji, M., Yanagita, K., Koida, M., Maeda, H., Manno, K., Narama, I., Ogawa, Y., Hiramatsu, Y. (1991) Biochemical and histopathological studies on a mouse brain ischemic model induced by bilateral carotid artery occlusion: comparison with the carbon monoxide inhalation method and other ischemic or anoxic models. *Nippon Yakurigaku Zasshi (Japan)* 98: 23–29.

Nowicki, J. P., Duval, D., Poignet, H., Scatton, B. (1991) Nitric oxide mediates neuronal death after focal cerebral ischemia in the mouse. *Eur J Pharmacol* 204: 339–340.

Pellegrino, L. J., Pellegrino, A. S., Cushman, A. J. (1979) *A Stereotaxic Atlas of the Rat Brain.* Plenum Press: New York.

Sarkar, S. K., Kapadia, R. D., Knudsen, D. J., Barone, F. C. (1992) 3D MR microscopy of cerebral tissue damage in mouse model of focal ischemic stroke. *International Stroke Society Second World Congress of Stroke* S93: P0–C41.

Seigel, S. (1956) *Nonparametric Statistics for the Behavioral Sciences.* McGraw-Hill: New York.

Swanson, R. A., Morton, M. T., Tsao-Wu, G., Savalos, R. A., Davidson, C., Sharpe, F. R. (1990) A semiautomated method for measuring brain infarct volume. *J Cereb Blood Flow Metab* 10: 290–293.

Takeo, S., Tanonaka, K., Hirano, T., Miyake, K., Okamoto, J. (1988) Cerebroprotective action of naftidrofuryl oxalate. I. Prolongation of survival time and protection of cerebral energy metabolism in bilateral carotid artery-ligated mice. *Nippon Yakurigaku Zasshi (Japan)* 91: 267–273.

Tomida, S., Nowak, Jr. T. S., Vass, K., Lohr, J. M., Klatzo, I. (1987) Experimental model for repetitive ischemic attacks in the gerbil: the cumulative effect of repeated ischemic insults. *J Cereb Blood Flow Metab* 7: 773–782.

U.S. Department of Health and Human Sciences (1982) The NIH 1980 Catalog Publication Number 83-606, NIH: Bethesda, MD.

Welsh, F. A., Sakamoto, T., McKee, A. E., Sims, R. E. (1987) Effects of lactatacidosis on pyridine nucleotide stability during ischemia in mouse brain. *J Neurochem* 49: 846–851.

Willette, R. N., Sauermelch, C. F., Ezekiel, M., Feuerstein, G., Ohlstein, E. H. (1990) Effect of endothelin on cortical microvascular perfusion in rats. *Stroke* 21: 451–458.

Willette, R. N., Sauermelch, C. F., Rycyna, R., Sarkar, S., Feuerstein, G. Z., Nichols, A. J., Ohlstein, E. H. (1992) Antithrombotic effects of a platelet fibrinogen receptor antagonist in a canine model of carotid artery thrombosis. *Stroke* 23: 703–711.

Yamamoto, M., Shimizu, M., Sakamoto, N., Ohtomo, H., Kogure, K. (1987) Protective effects of indioxagine hydrochloride on cerebral ischemia in animals. *Arch Int Pharmacodyn Ther* 290: 16–24.

Chapter 12

CONCEPTS AND TECHNIQUES OF EXPERIMENTAL STROKE INDUCED BY CEREBROVASCULAR PHOTOTHROMBOSIS

Brant D. Watson, W. Dalton Dietrich, Ricardo Prado, Hitoshi Nakayama, Hideyaki Kanemitsu, Nancy N. Futrell, Hiroshi Yao, Carrie G. Markgraf, and Per Wester

TABLE OF CONTENTS

I. INTRODUCTION

For many years, the bulk of modern experimental stroke research has been conducted with animal models of mechanically induced cerebral ischemia. The four-vessel model of reversible global ischemia, induced by common carotid artery ligature and subsequent release, is one such widely employed model.[1] Many variants of this approach have been developed and tested subsequently.[2] Owing to the absence of thrombotic processes and their associated potent secretions in these models, however, experimental therapies tested in the context of cerebral ischemia per se may be inadequate when tested in the context of thrombotic stroke (cerebral ischemia induced by thrombosis). With this in mind, in 1983 we began developing a model of thrombotic stroke in rat.[3] The basis of this model is endothelial damage instigated by an intravascular photochemical reaction, which stimulates a thrombotic response in cerebral vessels. A photosensitizing dye is injected intravenously, and light of a wavelength (from a laser) or wavelength band (from an optically filtered arc lamp) appropriate for dye absorption is used to irradiate either isolated cerebral vessels singly, or a region of the cerebral surface containing many vessels. With this approach, several of the common mechanisms of stroke induction can be expressed, including the production of occlusive arterial thrombi designed

0-8493-8094-4/95/$0.00+$.50

to be susceptible or resistant to enzymatic lysis, cerebral embolus formation and enlodgement, and cerebral hemorrhage.

II. PHOTOCHEMICAL PATHWAYS: PHOTOREACTIONS OF TYPE I AND TYPE II

Because the efficiency of thrombus production and its composition are dependent on the photochemical behavior of the sensitizer dye *in vivo,* the mechanisms of photochemical sensitization must be discussed.[4,5] Endothelial damage is mediated by an electronically excited, chemically reactive form of a sensitizer dye molecule following photon absorption by its inert precursor. This initial transition from the dye molecule ground state to the first excited singlet state is followed by radiationless transition to the relatively long-lived (metastable) triplet state manifold. From this point, the two types of recognized photochemical pathway likely determine the nature of the resultant endothelial damage. In Type I photochemistry, the lowest dye triplet state (T) abstracts a hydrogen atom or electron from a susceptible substrate molecule (RH), forming a triplet radical (TH•) or radical ion ($T^{\bar{\bullet}}$) along with a substrate free radical (R•) or radical ion ($RH^{\dot{+}}$). Such substrate radicals often combine with oxygen to form peroxy radicals (ROO•). A peroxy radical is capable of repeating the hydrogen abstraction process, resulting in a hydroperoxide molecule (ROOH) and another substrate radical, which can reinitiate the peroxidation cycle and thereby propagate the chain process among membrane constituents such as unsaturated fatty acids and proteins.

On the other hand, the Type II photochemical mechanism is characterized by direct reaction of the lowest triplet state with molecular oxygen. Most commonly, electronic state energy is transferred from the lowest dye triplet state to molecular oxygen (also a triplet). This results in recovery of the ground-state dye molecule and also creates an excited, singlet-state species of molecular oxygen. Singlet molecular oxygen is not a free radical (there is no charge localization due to unpaired electrons, so it will not initiate a chain process), but it can directly peroxidize certain cell membrane constituents such as unsaturated fatty acids[6,7] and also proteins, especially those containing the amino acids histidine, methionine, cysteine, and tryptophan.[5] Alternatively,[4] the dye triplet state may photoreduce (yield an electron to) oxygen to yield super-oxide anion (O_2^-) or at low pH, its conjugate acid, the perhydroxyl radical ($HO\bullet_2$). *In vivo,* this likely initiates a cascade of active-oxygen reactions mediated by hydrogen peroxide and its metal ion-mediated scission product, the hydroxyl radical (•OH). Either the hydroxyl or perhydroxyl radical can abstract hydrogen atoms from susceptible substrates, again yielding substrate radicals (R•), which can initiate free radical reactions among susceptible molecules leading to their peroxidation, as shown above.

According to these definitions, chain peroxidation of substrate molecules can be induced acutely by photoreactions of Type I and secondarily by those of Type II, while Type II (singlet oxygen) reactions can also generate peroxidation acutely without involvement of chain processes. It is thus possible that either reaction pathway might stimulate similar effects *in vivo,* although modified by such factors as proximity of dye to substrate, reaction compartment (cytosol or membrane), susceptibility of substrate, and location and concentration of reaction quenchers or propagators. However, we have observed quite dissimilar responses of blood vessels (described later in historical sequence) to photosensitizers of Types I and II. Although this suggests that these reaction pathways may be fortuitously simplified for certain dyes *in vivo,* a more prosaic explanation is likely: if Type I processes are defined as producing free radicals acutely, thereby including superoxide anion production,[5] then the Type II pathway is purely singlet oxygen-mediated (thus eliminating secondary production of peroxides by free radicals).

III. VASCULAR EFFECTS OF TYPE II PHOTOCHEMISTRY *IN VIVO*

In 1977 Rosenblum and El-Sabban showed that small, isolated pial vessels in mouse could be occluded with platelet aggregates by means of a fluorescein dye-mediated photochemical reaction.[8] Owing to the high blood concentrations (about 8 m*M*) necessary to facilitate thrombus formation,[8] the intensity of the irradiating blue light was attenuated severely and its photochemical effectiveness was localized to facilitating thrombosis of isolated vessels on the cortical surface. The stimulus for occlusion was said to be endothelial thermal damage induced by the bound, light-absorbing dye. However, because fluorescein emits absorbed energy as fluorescent light with at least 95% efficiency, we suspected that fluorescein was instead photosensitizing the platelet response to endothelial damage through a singlet oxygen mechanism—although with very low efficiency.[9] We thus conjectured that the related dyes rose bengal (disodium tetrachloro-tetraiodofluorescein) or erythrosin B (disodium tetraiodofluorescein) would induce a much more efficient platelet response, because following photon absorption these dyes exhibit the highest known (>70%) quantum efficiencies for singlet oxygen production.[9] (In later work from a different group, the remainder of the yield was attributed to superoxide anion).[10] In 1983 Herrmann offered support for this notion by showing that fluorescein-photoinduced endothelial damage leading to platelet aggregation could be suppressed by beta-carotene, an efficient quencher of singlet oxygen.[11]

By using the much more efficient Type II dye rose bengal, three immediate improvements were anticipated: (a) more intense platelet response due to enhancement in endothelial damage efficiency—despite (b) much less dye dosage required for thrombus induction, and consequently (c) increased range in tissue of the irradiating light, thus enhancing its photochemical effectiveness at depth. This last feature led us to attempt thrombosis of the rat cortical vasculature indirectly by irradiating it through the intact, translucent skull surface. The light source was an arc lamp, optically filtered to emit a wavelength band encompassing 562 nm, the wavelength of maximum absorption for protein- or membrane-bound rose bengal (Figure 1).[12] By leaving the skull intact, optimal rat survival was also anticipated. Subsequently, all of these features were observed, yielding the desired result of cortical thrombotic stroke. A complete description of the technical aspects of this model is given in Method 2, with the animals prepared for surgery according to Method 1 (Appendix A). In the following, we will continue to presume that the photochemical effects of rose bengal on blood vessels are initiated by a singlet oxygen mechanism.

Small cortical vessels (diameter $< 50\ \mu m$) exposed to rose bengal-mediated singlet oxygen attack are easily occluded by pseudopodic platelet agglutinates[13–16] concentrated mainly at the pial surface;[17] fibrin is essentially absent. Platelet adhesion and degranulation are readily observed by transmission electron microscopy.[14,15] We have designated this process of photochemically induced platelet aggregation by the term "photothrombosis", to distinguish it from the fibrin-containing thrombi formed by the coagulation cascade. By means of a patency test such as carbon black infusion, a surface view reveals the typical cortical lesion as pale orange in color owing to a multitude of occluded capillaries, interspersed with red-streaked pial vessels containing trapped or clotted blood.[13] The highly specific platelet response is apparently stimulated by endothelial discontinuities such as luminal surface microruptures.[13] These defects facilitate vasogenic edema, easily observed microscopically as perivascular swelling leading to parenchymal displacement and luminal distortion,[16] and macroscopically as greatly increased content of tissue water.[12,18–20] The edema is sufficiently severe to occlude the deeper cortical vasculature by mechanical compression, thereby enhancing the depth and volume of the resultant brain lesion. Interestingly, the platelet response can be eliminated by maintaining the irradiated brain tissue below 34°C.[16] Despite the lack of occlusive platelet aggregates, an ischemic lesion still results owing to vasogenic edema alone, facilitated by

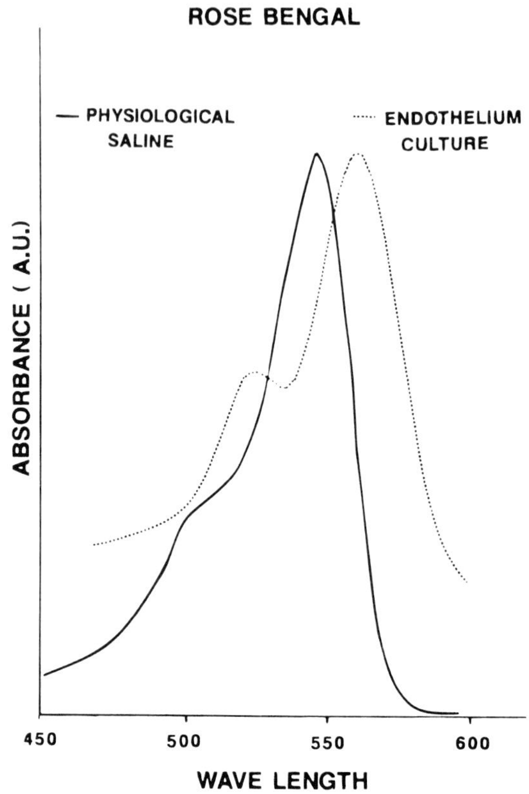

FIGURE 1. Absorption spectra of rose bengal in 0.9% saline solution (black line, maximum at 548 nm) and incubated with human umbilical vein endothelial cells in monolayer culture (dotted line, maximum at 562 nm); in the latter case, the rose bengal solution was replaced with clear saline solution, leaving the membrane-bound rose bengal.

the same photosensitized endothelial breaches. The fundamental manifestation of singlet oxygen-induced damage to endothelium is thus vasogenic edema, not platelet agglutination. The resultant cortical lesion is free of clotted blood and thus is quite pale compared to the lesion obtained at normal temperature.

An early and continuing criticism of the photochemical cortical model of thrombotic stroke is based on the fact that smaller-diameter vessels are occluded preferentially, instead of conduit arteries. It is quite common to observe patent arteries or their branches coursing through an irradiated tissue zone, which will irrevocably die owing to either the direct (photothrombotic) or indirect (mechanically compressive) occlusion of smaller, higher-generation vessels. Selective small-vessel thrombosis allows the depth and ultimate infarct volume of the cortical lesion to be varied consistently: reproducibility appears to be an intrinsic property of this type of lesion, so long as the volume density of microvasculature is constant in the subject animal group. However, the development of a potentially salvageable penumbral region[21] analogous to that associated with arterial occlusion is negated, because tissue bordering the lesion is consumed by vasogenic edema.[18,20] Conversely, a decrease in cortical lesion uniformity may occur if conduit arteries traversing the irradiated field to outlying nonirradiated regions become occluded also (thereby mimicking thrombotic stroke in a distal region). We have, in fact, observed such an inconsistency in a similar model of photothrombotic spinal cord injury, and corrected it by decreasing the irradiation intensity and the dose of injected rose bengal, i.e., by reducing the reaction efficiency.[22]

In order to develop a more clinically relevant stroke model, in 1986 we began investigation of cerebral arterial occlusion in rose bengal-injected rats by means of focused laser irradiation (described in Method 3 of Appendix A). Initially, the middle cerebral artery (MCA) was exposed and irradiated with 25 mW of 514.5-nm argon laser light focused on the proximal trunk. An occlusive thrombus, visible as a colorless discontinuity in flow, developed within 3 min of irradiation.[17] The thrombus again consisted entirely of agglutinated platelets,[23] indicating that the platelet response to singlet oxygen-induced endothelial damage was preserved for arteries (of diameter ca. 200 μm). A very significant concomitant effect was that of vasoconstriction, as indicated by longitudinal corrugations in the luminal surface.[23,24] Both responses are observed consistently during arterial photothrombosis, but we have not determined whether the vasoconstriction precedes or is triggered by the platelet response. In this connection, ritanserin (a 5-HT_2 receptor antagonist) was found recently to inhibit the vasoconstrictive response but not to affect platelet agglutination, except to require enhancement or prolongation of the photothrombotic stimulus in order to occlude the nonconstricted, relatively larger arterial diameter (T. Back, private communication). To gain further information on the mechanism of vasoconstriction, we exposed incubated cultured endothelial cell monolayers to rose bengal photochemistry at similar irradiation intensities (16 W/cm^2). Typically, a cell layer exhibited little change when irradiated with 514.5-nm argon laser light in the presence of rose bengal and serum albumin, but with albumin decreased, rose bengal permeated the cells and photochemically induced a retractive separation of the irradiated monolayer from the surrounding unirradiated cell layer, a response putatively analogous to the constriction observed *in vivo*.[25]

Recently we have been investigating methods to produce a more clearly defined penumbra in the MCA territory. Laser-driven photochemical occlusion of the MCA and its branches is much easier and less traumatic than standard electrocautery or ligature methods, but subject to the same uncertainties in infarct or penumbral volume if the location of the photochemical or mechanical occlusion is similar. Because variable retrograde flow through collateral channels is likely responsible for this inconsistency, it is necessary to irradiate at arterial locations deemed most likely to inhibit such flow. We have irradiated distal MCA branches in this fashion, together with common carotid ligation in order to improve lesion uniformity;[26] however, this task is much complicated by the remarkable variability in the distal MCA distribution. To establish a criterion for irradiation, we classified the MCA distal field by dint of extensive observation into eight categories (Figure 2), along with three locations at which irradiation is to be performed. Sequential irradiation of these locations with 20 mW of laser power at 562 nm yielded a cortical infarct volume of 130 ± 40 mm^3 in Sprague-Dawley rats, but only of 48 ± 27 mm^3 in Wistar rats, which are suspected to be more extensively collateralized.[27] On the other hand, zones of incomplete infarction characterized by selective neuronal necrosis were observed for the first time in a rat model of focal brain ischemia; these zones were found to be nearly twice greater in volume in the Wistar group compared to the Sprague-Dawleys. Subsequently, we modified our beam handling optics to produce three beams simultaneously (cf. Method 3 in Appendix A), and with this system and also with sequential irradiation repeated the above experiment in Sprague-Dawley rats while testing the effect of MK801 on infarct volume. No mitigative effect of MK801 was seen, but also no difference in infarct volume was detected between the sequential and simultaneous methods of irradiation. However, in absolute terms the infarct volumes in these new series were about half[28] those observed previously[27] under the same experimental conditions. This time-variable infarct volume effect among different groups of rats of the same strain made cerebrally ischemic by mechanical means[26] is thus confirmed for the photothrombotic case.

We are currently testing the hypothesis that multiple distal occlusions suppress the collateral response sufficiently to eliminate concurrent common carotid artery ligation, thus potentially

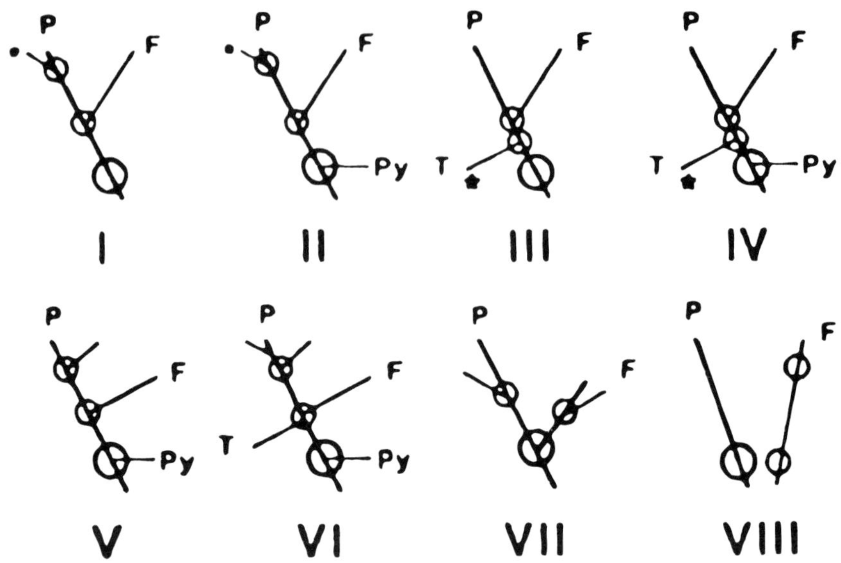

FIGURE 2. Classification of branching patterns in the distal-field distribution of the MCA in Sprague-Dawley rats (from H.Y.). P = parietal, F = frontal, P = pyriform, T = temporal branch. Dots indicate equivalence in function to the temporal branch, but originating distal to the frontal branch; Asterisks indicate that the temporal branch originates proximal to the frontal branch.

improving the clinical simulation. In spontaneously hypertensive rats (SHR) injected with 20 mg/kg rose bengal and irradiated thusly, infarct volume (mean $\pm$ SD) in the distal region at three days was 37.8 $\pm$ 9.8 mm^3 (n = 6, unpublished data). The coefficient of variation (SD/mean) is about 25%, similar to that obtained in SHR for the average infarct volume resulting from concurrent ligation of the distal MCA and common carotid arteries, even though this infarct is much larger (cf. Reference 26 and Model 1 in Reference 29; in both cases, standard deviations must be derived from tabulated standard errors). With regard to detecting drug treatment effects, we tentatively conclude that the infarct derived from the three-beam method without carotid ligation is statistically equivalent to that obtained with the standard ligation method, and the compromised region presents the further advantage of much enhanced accessibility to drug infusions owing to a patent circle of Willis.

In 1986 we extended the rose bengal method to include occlusion of larger arteries such as the common carotid and femoral.[23] With the rat placed on its back, these arteries were irradiated under saline solution, which filled the tissue cavity surrounding the exposed artery. In initial efforts, the beam spot size resulting from focusing with a 39-cm f.l. spherical lens was smaller than the arterial diameter. Mural thrombosis could be obtained, but not occlusion, even with 2 W of 514.5-nm argon laser irradiation in conjunction with a rose bengal dose of 40 mg/kg. However, after the focal spot was enlarged by means of a 61-cm f.l. lens to include the arterial diameter, occlusion could then be observed within about 6 min at intensities of ca. 130 W/cm^2 (1 W average power). An optimal focusing criterion was derived that demonstrated the necessity of including the entire arterial diameter within the irradiating beam profile.[23] The resultant occlusive thrombus consisted of islands of platelets interspersed among erythrocytes in a featureless mileau taken to be coagulum (heat-denatured protein). The next question became whether an occlusion consisting solely of platelets concomitant with vasoconstriction could be generated in arteries larger than the MCA. This was answered positively (cf. Figure 3) by irradiation with similar intensity of the right common carotid artery of 40 mg/kg rose bengal-injected rats with an optimally focused argon-pumped dye laser beam operating at 562 nm. We also found a nonlinear enhancement of occlusion

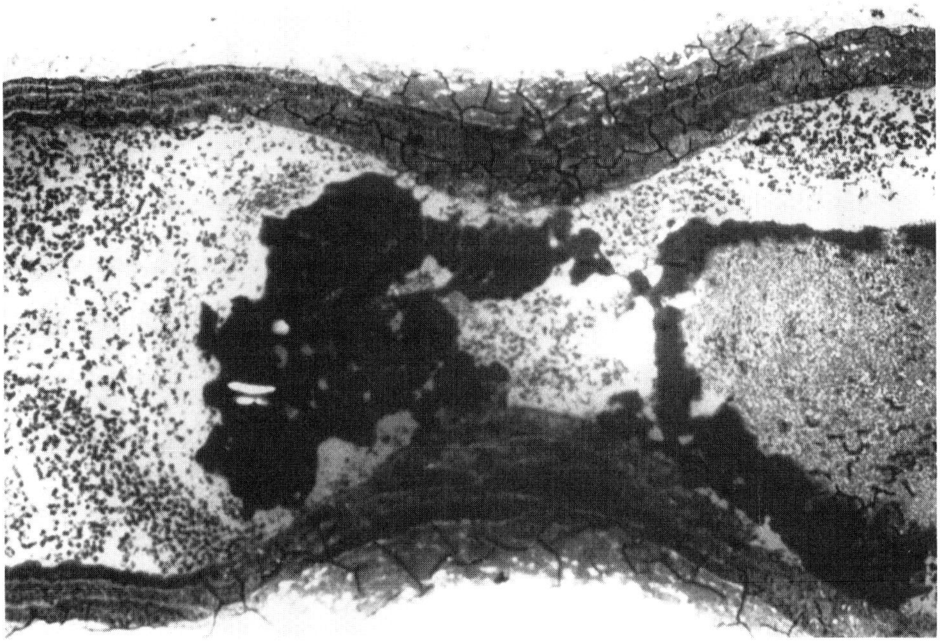

FIGURE 3. Toluidine blue-stained longitudinal section of right common carotid artery of a Wistar rat injected with 40 mg/kg rose bengal and irradiated with 1W of focused 562-nm argon-pumped dye laser light (intensity of 127 W/cm^2) until occlusion at t = 11.4 s. The irradiated segment was prepared by suspension in Millonig's fixative. Note the extreme vasoconstriction; the initial outer diameter of the artery was 1.0 mm. The occlusion at the epicenter appears to be composed solely of platelets; erythrocytes are seen peripheral to the epicenter, but no coagulum is present. Animals irradiated in the absence of rose bengal displayed no luminal alterations.

efficiency with respect to 514-nm irradiation. Although the absorbance at 562 nm is 2.6 times that at 514 nm (cf. Figure 1), the time to reach occlusion ($t_{562,40}$ = 20 ± 17 s, n = 5) was reduced by a factor of 12 ($t_{514,40}$ = 287 ± 50 s, n = 3). This nonlinear relationship is consistent with the chain reaction character of the platelet agglutination response. Interestingly, an occlusive platelet response still occurred with the rose bengal dose reduced to 4 mg/kg, albeit more slowly ($t_{562,4}$ = 160 ± 78 s, n = 6). We now use the 562-nm dye laser for all arterial irradiations for which rose bengal is the chosen sensitizer.

In earlier work with 514.5-nm irradiation we found that by reducing the average irradiation intensity by 2.5 times, a nonocclusive, platelet-rich thrombus (similar to the MCA thrombus) could be maintained in the common carotid artery lumen.[30] The obvious consequence of maintaining an equilibrated thrombus *in vivo* is the production of platelet emboli, resulting in the accumulation of scattered, small cerebral infarcts.[30,31] Emboli can also be produced with the photosensitizer dye Photofrin II (a hematoporphyrin ether composite) with dye or helium-neon laser excitation at 632 nm.[32,33] The resulting carotid thrombus examined at 0.5 to 24 h consists of aggregated platelets, erythrocytes (some hemolyzed), and occasional leukocytes in a putatively fibrin-stabilized matrix.[33] Owing to the presence of fibrin, these primary thrombi provide an alternative embolus structure of complementary clinical relevance (with the possible exception of the hemolyzed erythrocytes). Emboli located in distal cerebral arterial branches at 4 h consisted almost entirely of platelets in various stages of degranulation, but fibrin was not observed explicitly.[33] The Photofrin II method is also reported to require much less laser irradiation intensity (by a factor of 100) than the rose bengal technique, and thus circumvents any increases in blood temperature. In a Photofrin II-based photochemical cortical model of hemorrhagic stroke in cat, fibrin-stabilized microthrombi of similar composition also containing hemolyzed erythrocytes were observed in brain capillaries.[34]

The issue of temperature is important because thrombus composition is likely affected by temperature increases. For example, by means of an implanted thermocouple the temperature immediately distal to the mixed carotid thrombus being formed by the 514.5-nm argon beam/rose bengal interaction increased to ca. 84°C just before the point of occlusion; this quite likely generated the coagulum observed.[33] In contrast, under 2.5 times less irradiation intensity, the nonocclusive carotid platelet thrombus was produced at a blood temperature of 39°C without intermixed coagulum;[30] in this case temperature did not alter the structure expected (but compare with Reference 32). In an improvement of this technique, the 562-nm dye laser is presently operated at a constant 325 mW (intensity of 23 W/cm^2) to produce an equilibrated thrombus subtending about 50% of the carotid lumen, concomitant with a temperature increase of 0.5°C.[35] Recently, however, rose bengal has been used in conjunction with a spectrally filtered xenon arc lamp to produce occlusions of the femoral and middle cerebral arteries.[36–38] These thrombi unexpectedly are composed of erythrocytes and platelets interspersed in a fibrin network similar to that reported above in the case of Photofrin II and also to thrombi produced (with difficulty) by photocoagulation.[39,40] Inasmuch as we observed the evolution of a rose bengal-mediated thrombus from pure platelet to mixed character under elevated temperatures, we suspect that the nonoptimal irradiating wavelengths may similarly affect the structure of these arc lamp-induced thrombi. Remarkably, the arc lamp irradiation intensity was not mentioned in the original method,[36] thus necessitating our conjecture. Arterial vasoconstriction, so prominently seen in our work, was not reported either. The process generating such a platelet-rich, yet fibrin-stabilized thrombus with rose bengal may not be readily interpretable by purely photochemical mechanisms described here.

It is clear that problems in experimental stroke research can be attacked in unique fashion with a laser as the photochemical irradiation source. Another example is the photothrombotic ring model of cortical infarction. The 514.5-nm argon laser beam is focused into a fused silica optical fiber at a 10.8° angle of incidence; the beam emerges with radial symmetry, i.e., in the shape of a thin annulus (ring) when projected onto the exposed skull. Figure 4 shows by carbon black infusion the pattern of vascular patency at 1 h following irradiation with the ring beam, with erythrosin B as the photosensitizer dye. (As explained earlier, erythrosin B produces singlet oxygen with quantum efficiency similar to rose bengal, but for 514-nm irradiation its tissue absorption maximum of 537 nm is matched more optimally than the 562-nm peak of rose bengal.) In further work,[41] we found that the ring lesion continues to evolve centripetally, rendering the ring interior at 24 h entirely ischemic. Until then, however, the surrounding, thickening ring is penetrated by pial arteries, which can be utilized to deliver pharmacological treatment to the ring interior. This, therefore, can be viewed as a model of stroke-in-evolution in which infarction inevitably proceeds to completion in a predesignated tissue volume. Tissue damage in the ring interior may evolve through stages similar to those assumed to occur in a fatally compromised penumbral region (the region of metabolically imbalanced tissue surrounding an ischemic core), the difference being that the compromised tissue lies within rather than perifocal to the acutely ischemic locus. Finally, the 514.5-nm argon beam/erythrosin B interaction has been utilized to generate, by tightly focusing the beam transverse to the exposed spinal column, a model of rat spinal cord infarction with cystic cavity formation;[42,43] the methodology has been described completely.[22]

IV. VASCULAR EFFECTS OF TYPE I PHOTOCHEMISTRY *IN VIVO*

Having established the rose bengal/xenon arc lamp technique of cortical lesioning, we were curious to see what would happen if we utilized a Type I photosensitizer dye. According

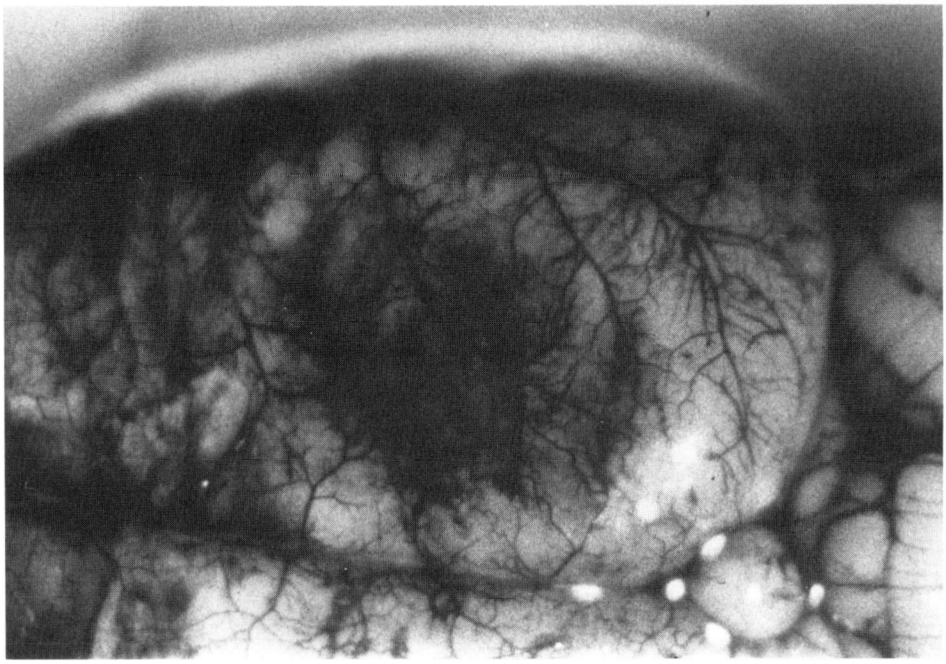

FIGURE 4. Carbon black-perfused brain from a Wistar rat injected with 17 mg/kg erythrosin B, irradiated for 2 min with 65 mW of 514.5-nm light (intensity 0.92 W/cm^2) from an argon laser ring beam, and sampled at 1 h (from P.W.). Note the patency of pial vessels penetrating the ring into the interior (center). Such patency persists for up to 24 h, while occlusion of their higher-order branches proceeds presumably by mechanical compression.

to previous studies,[7] a readily available candidate was flavin mononucleotide (FMN); it requires light in the 450-nm range for its excitation. Upon filtering the arc lamp for this transmission band (±30 nm) and irradiating (5 min, 200 mW/cm^2) the exposed skull of rats injected with 37 mg/kg FMN, we found on carbon black infusion that the cortical surface was covered with small hemorrhages arising from the venous circulation after the formation of venous thrombi (unpublished observation). The same effect was noted in a cortical ring lesion produced by a 457.9-nm argon laser beam in an FMN-injected rat, using the fiber optic system described above. This suggests that the Type I process is complementary to that of Type II. In an irradiated tissue region, the Type I process preferentially occludes small vascular channels on the venous side, while Type II (nonhemorrhagic) photochemistry apparently selects channels on the arterial side.

This concept of complementarity apparently extends to arterial irradiation as well, except that it is manifested in the character of the resultant thrombi. In preliminary studies (1988) we observed that irradiation of the proximal MCA of rats injected with 37 mg/kg FMN and irradiated with a 457.9-nm argon laser beam induced arterial dilation, in contrast to rose bengal-induced vasoconstriction (cf. Method 3 in Appendix A) together with a red thrombus (at the point of irradiation) stabilized distally by a white thrombus (cf. Figure 5). We conjectured that the instability of the slowly forming red thrombus was due to the incapacity of the hypothesized fibrin formed by Type I photochemistry to resist the arterial flow, while at higher powers the occlusion was formed and sealed distally with a platelet-stabilized thrombus. Subsequent staining of the red thrombus for Factor VIII and fibrinogen was negative, consistent with fibrin formation via the coagulation cascade (A.J.-H. Huang and S.C.-G. Tseng, unpublished observations). The white thrombus, associated with a

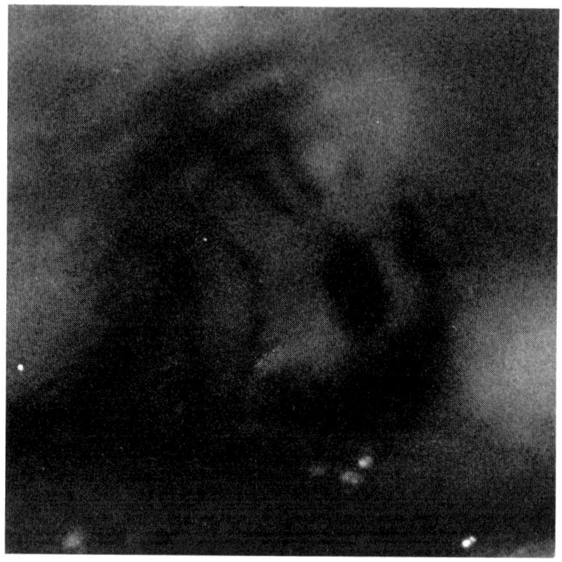

FIGURE 5. Thrombus (at lower center) in the proximal MCA of a Wistar rat injected with 37 mg/kg FMN as photosensitizer and irradiated with 457.9-nm argon laser light. Note the vasodilation in the region of the red thrombus. This contrasts with rose bengal-induced vasoconstriction (not visible) at the distal end (center) in a segment containing a white thrombus occluding the common mouth of three distal branches in parallel. During irradiation at low powers (up to 10 mW) the reddish-colored thrombotic deposit accumulated slowly before embolizing sporadically. Increasing the incident power to 25 mW produced the occlusive proximal red thrombus and a distal white thrombus.

vasoconstricted segment, resembled those produced by Type II photochemistry in its gross appearance and stability.

In our next series, we purposely avoided the common expedient of carotid artery occlusion inasmuch as this would unnaturally suppress the collateral activation response to MCA occlusion. We instead followed the work of Bederson et al.[44] who found that MCA coagulation followed by transection (at one point proximal to the olfactory tract and another just distal to the inferior cerebral vein) would yield consistent MCA territory infarcts; these points of occlusion suggested the optimum placement of FMN-induced occlusive thrombi along the MCA trunk. The desired result is occlusion of an MCA segment much longer than the sum of the occlusion thicknesses. We set out to produce MCA territory infarction using Sprague-Dawley rats irradiated under normotensive conditions (ca. 100 mmHg) as adjusted by the concentration of halothane anesthetic. Irradiation was carried out with 457.9-nm argon laser light in conjunction with intravenously injected FMN and bixin at respective blood concentrations of 1.34 to 2.00 m*M* (37 to 55 mg/kg) and 13.4 μM. Bixin, a reputed quencher of singlet oxygen but not of superoxide anion,[45] was used in order to inhibit formation of the distal white thrombus by presumably a singlet oxygen-mediated process (FMN does not express a pure Type I pathway). The MCA trunk was irradiated with three beams simultaneously, placed to simulate the arrangement of Bederson et al.[44] (cf. Figure 6).

During irradiation, we often observed blood suddenly appearing in the MCA trunk segments between the forming thrombi.[46] The rate of this retrograde blood infiltration and the volume of blood accumulated were variable. The retrograde infiltration would usually ebb (in the anterograde direction) only to return later. This influx and efflux of intrasegmental blood would continue during irradiation until the three-occlusion system stabilized, as signaled by the absence of retrograde infiltration but usually with some trapped blood between the

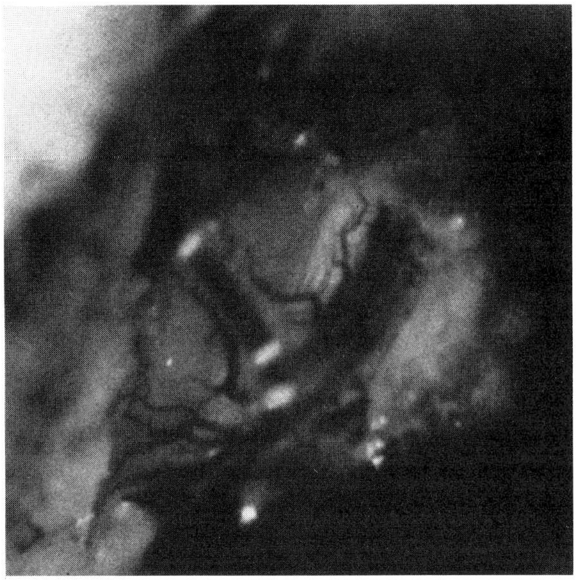

FIGURE 6. Irradiation of the MCA trunk with the sharply focused three-beam system (cf. Method 3 and Figure 9). The primary beam is located proximal to the rhinal branch so as to subtend the lenticulostriates, with the two first-order beams located at the mouth of the rhinal branch and just proximal to the inferior cerebral vein. The array of thrombi usually formed within 15 to 25 min of irradiation with 10 mW of 457.9-nm argon laser light in conjunction with 37 mg/kg of FMN. The segments being irradiated fluoresce bright yellow, but quickly fade as the thrombi form. Reinjection and reirradiation are often necessary.

thrombi. The vascular conduits for the infiltrating blood were not visible but are thought to be located abluminally to the MCA trunk. Subsequently, continuous monitoring of blood pressure during the entire experiment showed that the retrograde blood infiltration was correlated with 5 to 10 mmHg surges in blood pressure. If the blood pressure did not exceed 100 mmHg, the three-clot system would stabilize and persist for at least 3 h. Within times of 40 min to 3 h after irradiation, however, retrograde infiltration often recurred, and the occlusions began to erode and the MCA segment recanalize. Recanalization nearly always occurred from the distal direction. If the two distal thrombi dissolved, the proximal thrombus did not survive either. Branches from the main MCA trunk, such as the rhinal branch, also would recanalize spontaneously.

Transmission electron microscopic analysis was undertaken for selected initial conditions. For thrombi formed during mild hypotension (ca. 70 mmHg) and without bixin injection, the most notable result was the extensive accumulation of aggregated platelets. Thrombi formed with bixin present appeared colorless under either mild hypotension or normotension (ca. 100 mmHg). Although erythrocytes were seen, the most striking difference between the FMN and rose bengal methods was the pervasive accumulation of an unusual amorphous material onto the endothelial surface (Figure 7). Luminal accumulation of a quite similar material was observed also during endothelial photosensitization in small vessels of hamster cheek pouch by fluorescein isothiocyanate-dextran; nafazatrom inhibited its formation.[47] Also noted in our work was the presence of multiple short strands between platelets and/or erythrocytes (Figure 8). We propose that the short-stranded material is likely to be newly formed fibrin, but cannot offer a common mechanism to explain the formation of the amorphous intraluminal material (apparently a Type I photochemical response) by both FMN and conjugated fluorescein. No obvious structural differences between thrombi formed at low (70 mmHg) or higher (100 mmHg) blood pressures could be seen.

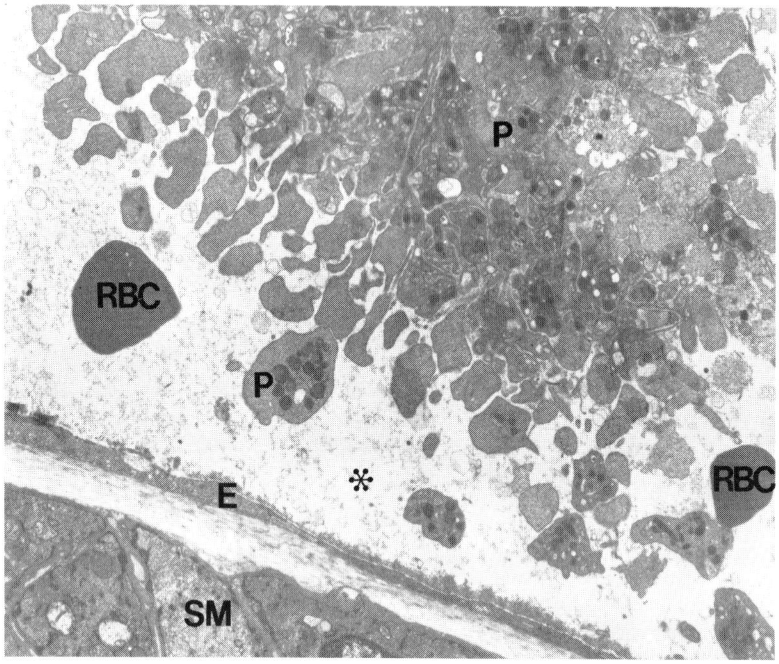

FIGURE 7. Transmission electron micrograph of proximal MCA thrombus formed in a Fischer 344 rat injected with 37 mg/kg FMN and 66 μM bixin (in blood), and irradiated with the zero-order beam (of the three-beam system) for 20 min with 11 mW of tightly focused 457.9-nm argon laser light. The systolic blood pressure surged by 5 to 10 mmHg occasionally, but generally remained at 80 mmHg. Note the amorphous material (*) in the lumen, as well as its apparent accumulation on the endothelial surface (E). Some red blood cells (RBC) are dispersed throughout the massive platelet (P) thrombus, accounting for the mottled color of the thrombus as viewed through the operating microscope.

Under the above irradiation conditions, histologically observed infarct size at 3 d was small and located primarily in the striatum, but correlated with time of occlusion.[46] We then attempted to obtain more stable occlusions by defocusing the beams and positioning them directly at the mouths of the main MCA branches, in order to prevent retrograde infiltration of blood into the unoccluded MCA branches, despite the fact that the latter thrombi were much more difficult to dissolve by human tissue-type plasminogen activator (rt-PA, see Section V). This did result in enhanced occlusion stability but infarct size was inconsistent. At blood pressures exceeding 115 mmHg, no relationship between blood pressure and infarct size could be determined. Thrombolysis by administration of rt-PA (see Section V) by external carotid catheter was attempted, and was eventually successful, but only over prolonged times (ca. 1 h) and with ca. 500 μg, an amount sufficient to induce a lytic state. The rt-PA infiltrated retrogradely into the clot array, just as described above (unpublished observations), and the clots lysed in retrograde sequence.

We concluded that the origin of clot and infarct instability, namely, the apparent retrograde pulsatile infiltration of blood, arose due to the combined effects of blood pressure and collateralization of flow (facilitated by avoidance of carotid artery ligation) into the affected brain region. Collateralization is thought universally to confound consistent progression of a compromised brain region to infarction, which we have certainly observed here. However, our experimental approach allows direct observation and characterization of thrombotic

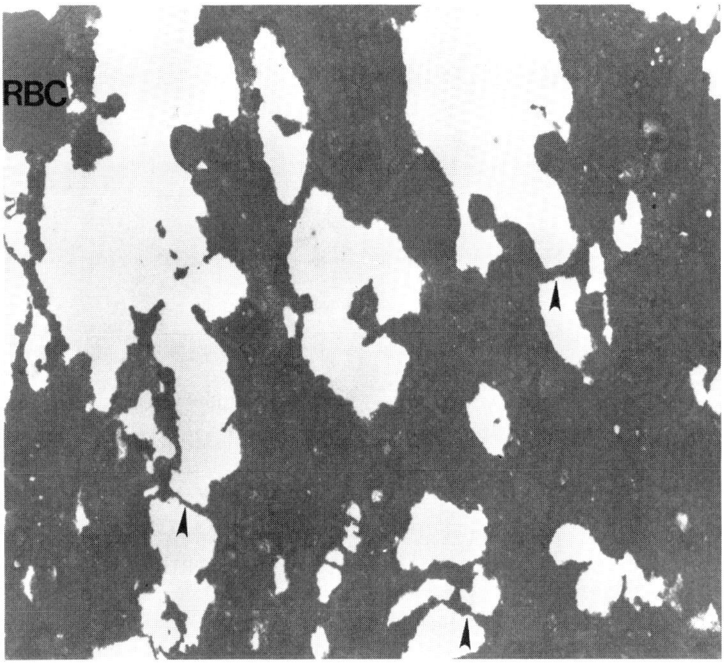

FIGURE 8. Transmission electron micrograph of thrombus formed distal to the rhinal branch in a Fischer 344 rat injected and irradiated as in Figure 7 except that bixin was excluded, and the irradiating beam was first-order with a power of 4.5 mW. Note again the occasional erythrocyte (RBC) associated with a mass of aggregated platelets connected in part by short strands of material suggestive of fibrin (arrowheads).

stroke-in-evolution in response to endothelial injury (e.g., alterations in cerebral blood flow) and also its possible mitigation by release of vasoactive agents.

V. LYTIC PROPERTIES OF TYPE I AND TYPE II PHOTOTHROMBI

Certainly a desirable property of a thrombotic stroke model is reversibility. Efforts to recanalize rose bengal-mediated photothrombotic (Type II) occlusions in the MCA with rt-PA (Genentech, Inc.) were begun in 1984 before the thrombus microstructure was known, and were successful only occasionally over many trials.[17] This stability was attributed subsequently to the morphologically determined absence of fibrin in such thrombi.[13,15–17] This was consistent histologically with fluorescent antibody analysis conducted in thin sections of the irradiated artery, which revealed intense staining for Factor VIII and fibrinogen (A.J.-H. Huang and S.C.-G. Tseng, unpublished observations). This suggested that the intercalated platelet mass was stabilized by thrombospondin-stabilized cross-links between GPIIb-IIIa platelet surface proteins.[48] We have since verified the long-term stability *in vivo* of such thrombi (up to 16 weeks in rabbit corneal neovasculature).[49] Evidently, if Type II photothrombosis proceeds to occlusion in a vascular network (during areal irradiation of a bulk tissue) or in a specific vessel such as a large conduit artery (during focused laser irradiation), the occlusion(s) will be permanent under most irradiation conditions.[50]

Lysis of these thrombi clearly requires an unconventional approach—and in 1987 we became aware of such a possibility. We learned that a rare leech, *Haementeria ghilianii,* may be able to secure a blood meal even from an occluded vascular network, suggesting the presence of a powerful, nonspecific thrombolytic agent. The putative active enzyme, hementin[51] (Biopharm, England), was obtained from the supernatant of an unpurified macerate of the leech salivary gland suspended in 10 m*M* Ca^{2+}. Lysis of rose bengal-mediated MCA photothrombi was facilitated in ca. 20 min (n = 4) with 5 U hementin activity equivalents injected via an external carotid artery catheter. The recanalized irradiated segments appeared through an operating microscope to be completely cleared of thrombus after ca. 1 h, and more surprising, even resumed their preirradiation diameters with no suggestion that endothelial defects had been present (i.e., the photodamaged luminal surface appeared now to be nonthrombogenic). These thrombi disappeared gradually at apparently a uniform rate. By contrast, the few platelet thrombi that were lysed by rt-PA only partially disappeared, incompletely breaking up by means of sporadic showers of microemboli. The lumen remained constricted, and reocclusion was very common.[17] Hementin has since become available as the purified enzyme, but unfortunately in this form exhibits no lytic activity whatsoever for platelet thrombi formed *in vivo* (unpublished observations), although (impertinently) it is able to dissolve fibrin clots *in vitro* (R.T. Sawyer, private communication). Because the platelet thrombi produced by primary rose bengal photochemistry appear to possess similar or identical structural linkages to the thrombi responsible for reoccluding arteries previously cleared by rt-PA infusion,[52] it would seem imperative to recover the previously observed hementin activity against such thrombi so that clinical application can be envisioned. This is the only enzyme known to exhibit activity against these formidable platelet occlusions after their formation. Hementin also dissolved FMN-induced MCA thrombi in two rats. Thus, hementin (with its putative cofactors) is unique in its thrombolytic versatility.

We also investigated the lytic activity of human rt-PA against a single FMN-photoinduced Type I thrombus (which apparently contains fibrin; cf. above). The distal MCA trunk of six normotensive rats injected with 37 mg/kg FMN was irradiated with 1 mW of 457.9-nm argon laser light focused tightly in the transverse geometry (cf. Figure 9). The occlusion typically formed within several minutes, but usually recanalized once spontaneously, thus requiring an additional FMN bolus of 18 mg/kg and reirradiation. Reperfusion with rt-PA, infused retrograde into the external carotid artery, occurred in all rats within 30 min at doses of 0.45 to 0.9 mg/kg. However, the arterial lumens still contained thrombotic material, perhaps composed of platelet masses that would not respond to the rt-PA. Improvement of recanalization in this technique would seem to require either a dye expressing more Type I character, or a more efficient thrombolytic agent. The latter possibility has been realized in rt-PA by changing only one amino acid (Asn-Gln at position 117). The much longer-lived and more potent product, SUN9216, has been used to lyse thrombi formed by the rose bengal/arc lamp irradiation method (described earlier) in large arteries such as the femoral, as well as the MCA, with an efficiency of about 30%.[38] In conjunction with a thromboxane receptor antagonist, vapiprost, this percentage was improved to about 60%. Given the strong possibility that the composition of these thrombi is unnecessarily complicated by heat during their formation, the performance of SUN9216 is exceptional. We conjecture that FMN-induced thrombi should be quite susceptible to this agent.

An alternate method was developed for inducing recanalization of occlusions formed by irradiation with a 543.5-nm He-Ne laser rated at 1 mW.[24] The beam was focused tightly onto the MCA as a thin (40 μm) line transverse to the arterial axis. Irradiation was performed concurrently with separate rose bengal injections, leading to the formation of two "soft" thrombi that were stable for times up to 3 h. Recanalization was initiated by slitting the dura at the irradiation site and applying nimodipine solution; the ensuing vasodilation enabled blood to penetrate the thrombi perimetrically, from which point they gradually degraded. Apparently, Type II photothrombi develop in stages, with structural integrity increasing as the irradiation dose accumulates, but to reach the threshold of permanence a minimum irradiation intensity is required.

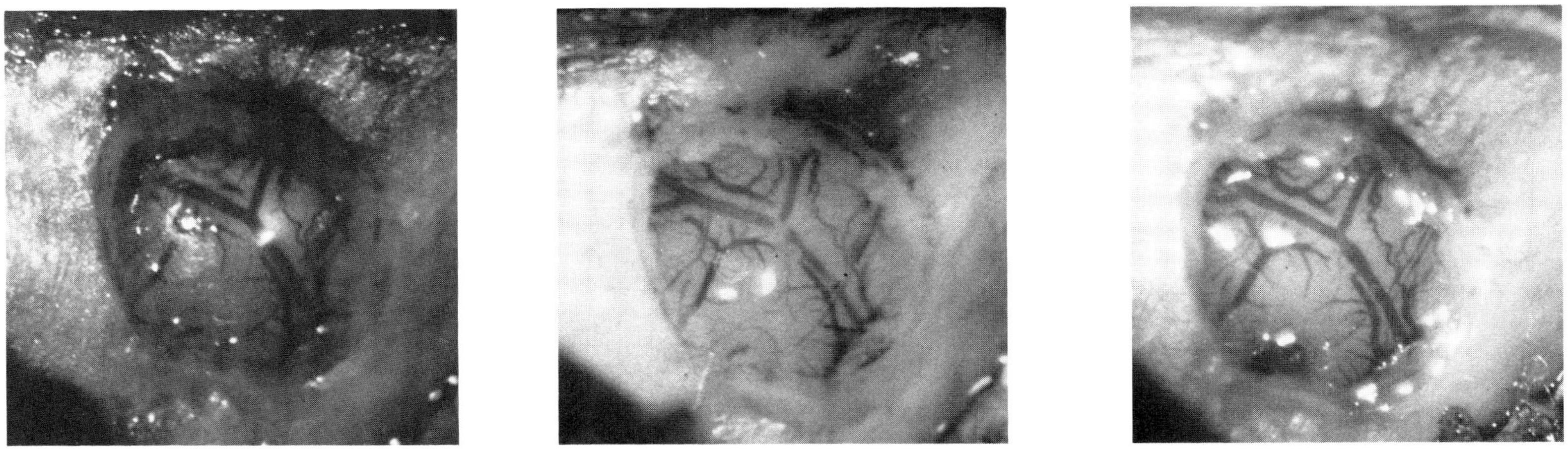

FIGURE 9. Left: Distal MCA trunk of a Wistar rat injected with FMN and irradiated with 1 mW of 457.9-nm argon laser light, focused sharply by means of crossed cylindrical lenses (cf. Method 3 and Figure 6). Center: Mottled (intermixed red and white foci) thrombus resulting after 2 min irradiation with 42 mg/kg FMN, spontaneous reperfusion, and reirradiation for 2 min with an additional 21 mg/kg FMN to achieve stability. Right: Reperfusion effected at 1 h by bolus injections of rt-PA (total dose 0.82 mg/kg) beginning at 15 min. (From C.G. Markgraf.)

VI. SUMMARY OF PHOTOCHEMICAL PRINCIPLES IN VASCULATURE

Most models of central nervous system injury based on a photochemical approach to vascular occlusion predominantly utilize (or claim to utilize) the Type II (singlet oxygen) photochemical pathway. Type II photosensitized damage to endothelium is effectively induced by endothelium-bound rose bengal or erythrosin B because it is mediated by metastable electronic-state transitions. The electronic energy of these transitions is sequestered in the form of membrane-reactive singlet molecular oxygen, and is thus transferred directly to endothelial components. Evidently, a reversal of role complementarity occurs in which the endothelium becomes thrombogenic but the platelets become nonthrombogenic.[53] Therefore, up to the point of occlusion, thrombus formation requires the incorporation of fresh, photochemically undamaged platelets derived from existing blood flow. These thrombi are very stable against spontaneous lysis and are also resistant to tissue-type plasminogen activator, owing to their apparent lack of fibrin substrate. Cortical infarction follows naturally from this occlusive process.

On the other hand, Type I photothrombosis deserves further study owing to its evident tendency to selectively occlude small vessels on the venous side (upon areal irradiation), leading to cortical surface petechiae and a hemorrhagic lesion. This result is reminiscent of that obtained with hematoporphyrin derivative (or Photofrin II) as the photosensitizer in conjunction with red light irradiation.[34] With regard to arterial occlusion, the composition of Type I thrombi is quite different from the platelet thrombi of Type II, inasmuch as fibrin apparently is present along with enmeshed erythrocytes as well as aggregated platelets. These thrombi are similar in composition, except for lack of coagulum, to those produced by photocoagulation. Photocoagulation to the point of occlusion is extremely inefficient compared to photothrombosis of either type, however, because the density of endothelial damage sites that can be induced by heat transfer from a light-absorbing dye is much diluted by the flowing blood, and the quanta of transition (phonons) are much less energetic than electronic-state transitions.

The observations we have delineated with respect to thrombus formation and stability evidently offer operational criteria for identifying the predominant photochemical pathway of a given photosensitizer molecule *in vivo.* Reports of photoinduced arterial occlusion with other dyes said to produce singlet oxygen, such as Photofrin II and chloroaluminum sulfonated phthalocyanine, are very scarce, perhaps because their main intended use is the destruction of tumor microvasculature. Nonetheless, if permanent occlusion of a directly accessible artery is the desired end point, the thrombus must reflect a minimum of Type I character. Type I-photoinduced thrombi are likely to be unstable *in vivo,* as illustrated in the case of FMN photosensitization: as discussed earlier, the structure of microthrombi photoinduced by Photofrin II resembles the structure of arterial thrombi produced by FMN. Therefore, evidence of singlet oxygen generation *in vitro* is probably not sufficient to expect effective Type II photosensitization *in vivo.* For occlusion of arteries at depths in tissue that preclude direct irradiation (e.g., in the therapy of arteriovenous malformations), dyes approaching the performance of rose bengal or erythrosin B, but absorbing at tissue-penetrating wavelengths in the 830-nm region, would have to be synthesized. Alternatively, the luminal surface of arteries that could be penetrated by a catheter containing an optical fiber may be irradiated directly by a ring beam; as long as blood can flow past the dye-photosensitized luminal circumference, a platelet mass will accumulate circumferentially to the point of occlusion.

VII. SIGNIFICANCE OF PHOTOTHROMBOTIC STROKE MODELS

Our original motivation for developing photochemically induced models of thrombotic stroke was driven by the conviction that the effects of stroke on compromised brain tissue are different from and more severe than those due to cerebral ischemia. We have indeed demonstrated this by characterizing and comparing cerebral reperfusion injury as induced by either mechanical or photothrombotic occlusion of the MCA and its recirculation at 1 h; the severity of neuronal and microvascular changes in the latter case were considered irreversible in contrast to the former.[54] Although blood-brain barrier (BBB) integrity is compromised directly in the cortical photothrombotic model of Method 2, it also occurs in cerebral locations downstream from MCA photothrombosis as well.[24,55,56] In fact, blood-borne factors produced during common carotid artery photothrombosis, and which are evidently responsible for BBB leakage,[56] are even transferable via an aliquot of thrombogenically activated blood to acceptor rats, with the same BBB consequence.[57] Such acute phenomena are outside the spectrum of events evinced by mechanical occlusion models, and can be expected to affect pharmacologic evaluations fundamentally.

Owing to its reproducibility, the cortical photothrombotic model has been subjected to studies from several research groups.[12,18–20,34,58–65] With respect to infarct size reduction, this model has been found resistant to standard antiischemic drug therapies owing to the florid and acute production of vasogenic edema arising from small cortical vessels. Although early vasogenic edema of such severity is regarded as unrepresentative of clinical stroke, discovering useful antidotes to it must be regarded as useful, at least with regard to determining the most efficient therapeutic pathways. Recently, reductions in infarct volume of ca. 50% in this model have been reported based on adminstration of the NMDA receptor antagonists memantine[58] and nitroglycerin,[59] complementing an assessment of ca. 40% infarct size reduction with the calcium channel antagonist flunarizine.[60] This same cortical model was shown earlier to produce effects in the brain remote from the infarct, such as depression of flow,[61] and more recently, multiple episodes of spreading depression,[62] and diminution of pulse-pair inhibition.[63] As these works illustrate, with this model the lesion may be placed at any desired cortical location, an advantage we originally exploited to infarct specifically the rat barrel-field region as the basis for metabolic[64] and subsequent behavioral[65] studies.

Photothrombotic models of MCA territory infarction, however, remain subject to the same vagaries of infarct volume reproducibility as standard mechanical models of focal ischemia. On the other hand, neither reproducibility per se, nor the achievement of large infarct volume, is necessarily desirable if pharmacologic efforts to decrease infarct volume are defeated at the outset by maximizing the suppression of collateral flow. Achieving a smaller, reproducible infarct that is nonetheless accessible to collateral flow seems a worthwhile goal, but the difficulties and paradoxes inherent in such an effort based on mechanical models of focal infarction are well known. We propose that the photothrombotic approach based on three-point occlusion of the distal MCA field may prove useful in resolving this difficulty. Our preliminary data indicates that the lesion in SHR rats, although small, can be localized reproducibly with an acceptable coefficient of variation, while being surrounded on all sides by potential channels for retrograde perfusion.

As we have shown, largely owing to their influence on the BBB, the photothrombotic models present much more dynamically evolving insults to the brain compared to mechanical models. This can be appreciated directly through the phenomena evoked by photoembolic stroke models: the detection in downstream blood of serotonin released by platelets aggregating in the carotid artery,[66] and its suppression of blood flow in remote cortical regions mitigated by ketanserin.[67] Proliferation of intimal smooth muscle cells has also been observed at one month following common carotid artery photothrombosis.[31] Finally, recent work has revealed that platelet emboli produced by this technique are mainly transient, as shown by ^{111}In labeling,[68] but leave behind widespread neuronal and BBB changes.[69] We conclude that models of photothrombotic stroke afford fundamentally useful counterparts to traditional models of cerebral ischemia, not only in terms of vascular events and their downstream consequences, but also as setting more relevant and higher standards for judging the success of pharmacologic treatments.

ACKNOWLEDGMENTS

This work was supported by U.S. Public Health Service grants NS23244, NS27127, and NS5820. Brant Watson is the recipient of a Jacob Javits Neuroscience Investigator Award from NINDS (1992–1999).

APPENDIX A. METHODS

METHOD 1: PREPARATION OF RATS FOR PHOTOCHEMICALLY INDUCED THROMBOSIS

Male rats weighing 250 to 350 g are usually used in these studies. The animal is first anesthetized with 4% halothane and a mixture of oxygen and nitrous oxide (30/70) delivered into a closed jar. Anesthesia is then maintained with halothane 1.0 to 1.5% during surgical preparation. Following the surgical procedures, described below, anesthesia is maintained with 0.5% halothane. Femoral venous and arterial catheters (PE-50) are inserted for fluid administration and monitoring of physiological parameters. Endotracheal intubation is performed and then the animal is given 6 mg/kg pancuronium bromide intraperitoneally. Mechanical ventilation is initiated by means of a Stoelting small-animal respirator. In order to avoid atelectasis, a slightly lower than normal respiratory rate (60/min) and higher than normal tidal volume are selected. Rectal temperature is measured with a thermocouple connected to a temperature-regulating device (Bioanalytical Systems, Inc., West Lafayette, IN), which actuates a heating pad beneath the animal. Rectal temperature is maintained at 37 $\pm$ 0.5°C. Following this, the animal is placed prone in a stereotaxic frame, and the head is stabilized by means of ear bars.

Arterial blood samples are drawn at frequent intervals and are assayed for PCO_2 and PO_2 in a Radiometer acid-base analyzer (ABL-30, Copenhagen). Readings are corrected automatically for any deviation of body temperature from 37.0°C. Unless otherwise called for in specific protocols, arterial PCO_2 is maintained at 35 to 40 mmHg. Arterial PO_2 is maintained at or above 110 mmHg. Arterial blood pressure is monitored via an arterial catheter connected to a pressure transducer (Statham) and a polygraphic recording system.

METHOD 2: PHOTOTHROMBOTIC PRODUCTION OF CORTICAL NECROSIS VIA SMALL-VESSEL OCCLUSION

The scalp is incised in the midline and retracted laterally to expose the the underlying fascia and skull. The fascia is incised in the midline and hemostasis achieved with electrocautery. The cortical region to be irradiated (by light transmitted through the translucent skull) is demarcated on the skull surface by an aperture made of brass shim stock. In our original work, we used a 4 $\times$ 6 mm aperture mounted over the left parietal convexity.[3,13] Transmission of green light through this 0.5-mm-thick skull region was estimated to be 70%. The rat's head is then positioned for irradiation with an optically filtered xenon arc lamp, following injection of the photosensitive dye. Rose bengal, erythrosin B, and FMN are known to photochemically facilitate a cortical lesion. The dye originally used was rose bengal (mol wt 1018), available in 97.4% purity (Aldrich Chemical Co., Milwaukee, WI). When dissolved in 0.9% saline solution at 7.5 mg/ml it is stable for months without refrigeration while stored in the dark. A body dose of 10 mg/kg is administered by injecting a solution volume equal in microliters to 1.33 times the rat weight, resulting in a blood concentration of 173 μM (rat blood volume is assumed to be 5.7 ml/100 g.[70] Owing to liver uptake, however, rose bengal is quickly removed from blood with a half-life about 2 min for body doses up to about 40 mg/kg; at higher body doses the half-life increases due to hepatic saturation.[71] Over a period of 90 s for a rose bengal dose of up to 20 mg/kg and 30 s for any dose of erythrosin B the dye should be injected via the catheterized femoral vein by an infusion pump; fluctuations in blood pressure seen occasionally during manual infusion are prevented by infusing at a constant rate. Fluctuations in blood pressure, however, are not observed with erythrosin B. One minute after infusion, irradiation is begun. Because the irradiation intensity and duration are fundamental in obtaining a suitable cortical lesion, these aspects require extensive discussion (cf. the following paragraphs). Before irradiation, a drop of mineral oil is applied to the apertured skull area in order to smooth the surface optically. During irradiation the skull emits a bright orange fluorescence, as viewed through either a 562-nm rejection band interference filter (Omega Optical, Burlington, VT) or special laser goggles (Type 52 from Uvex Winter Optical, Inc., Smithfield, RI). The fluorescence arises chiefly from rose bengal bound to circulating plasma proteins and formed blood elements. The small portion bound to endothelium, from which thrombosis is derived photochemically, does not fluoresce visibly by comparison even after formation of the colorless thrombi.

To induce cortical photothrombosis with a xenon arc lamp, the lamp must first be optically filtered. Ordinarily the ultraviolet and infrared regions are first rejected by a hot mirror. The remaining white light spectrum is further modified by a high-quality interference filter to transmit a spectral band that lies within the *in vivo* absorption spectrum of the chosen photoactive dye. However, the transmitted spectral band still will not be absorbed with photochemically optimum efficiency (except at the peak), so a proportion thereof will be converted to heat in the irradiated tissue. In compensation, isolation of a narrow spectral band bracketing the peak of the dye absorption spectrum is seldom helpful owing to the difficulty of obtaining sufficient intensity. On the other hand, some groups have claimed that the apparatus may be improved by eliminating the visible interference filter, preferring instead to irradiate with the cold white light beam emitted from the optical fiber without actually measuring the power of this beam.[12,18,19,60] This is mystifying because much of the unfiltered white light will serve only to heat the tissue rather than be used for photothrombosis. A useful white light beam may not be cold because intensities of a few watts per square centimeter are obtainable even from an optically filtered xenon arc system (cf. below) and are easily sufficient to heat brain tissue 15 to 20°C beyond its normal temperature. Thus, possible heat-mediated factors may be involved unnecessarily in lesion development and overwhelm the contribution of photothrombosis. We therefore cannot agree that elimination of spectral filtering constitutes an improvement. For this reason, and also in the interest of lesion reproducibility, we strongly recommend frequent measurement of beam intensity in physical units (W/cm^2) with an NBS-calibrated power meter.

Our original irradiation apparatus (ILC Technology, Sunnyvale, CA) features a xenon arc at the focus of a parabolic reflector, thus facilitating efficient collection and partial collimation of the emitted light. The modern equivalent is the Model 302. This compact air-cooled unit contains a hot mirror and enough internal space to emplace a transmission interference filter and even a beam chopper wheel, if necessary. The minimum duty cycle for the beam chopper cannot be less than 33%; however, owing to space limitation the 1-in.-diameter beam from the xenon lamp cannot be subtended more efficiently. Output power from the Model 302 is continuously tunable as desired, but the filters are unavoidably exposed to the full lamp intensity. In our experience, colored glass filters are unable to withstand such an intensity and will crack, but 1-in.-diameter interference filters (562 nm center wavelength, 60 nm bandpass from Omega Optical, Burlington, VT) appear resistant to degradation and thus transmit a consistent spectral band. An internal lens focuses the light into a 1-m liquid light guide (analogous to a thick optical fiber; Oriel, Inc., Stratford, CT), which is mounted in a cylindrical alumimum block. It may be necessary to change slightly the position of the block so as to maximize beam uptake by the light guide at the focal point. Beam power is measured with an NBS-calibrated pyroelectric meter (Model 365, Scientech, Boulder, CO). This optically filtered system produces a bright yellow beam with a maximum power of 235 mW from a 3-mm-diameter light guide, and 340 mW from a 5-mm-diameter guide. The respective average intensities at the output faces of the 3- and 5-mm light guides, however, are 3.32 and 1.73 W/cm^2. The emitted intensity diminishes rapidly with distance because the beam fans out conically at a half angle of 30°. For applications requiring maximum intensity, it is therefore desirable to select the smaller-diameter light guide.

To begin an experimental series, a particular cortical region is selected and the temperature during irradiation measured in a few animals with a thermocouple implanted just under the skull and over the region of interest. Generally, it is desirable to utilize the maximum available intensity in order to minimize the duration of irradiation, but the ancillary heat must be removed in order to avoid expression of nonphysiological events. For this purpose, we have found that miniature piezoelectric fans (Model LP24HT, Piezo Systems, Cambridge, MA) are quite satisfactory. These devices can be positioned within 7 mm of the irradiated surface without interfering with the incident beam. At this distance, one fan operated at full power and at a 45° angle of incidence can neutralize the heating of the cortex from an incident yellow beam, with intensity of 600 mW/cm^2 emitted from the light guide positioned for vertical incidence. If even higher intensities are desired, another fan can be placed on the other side of the head. Once a temperature-neutralized irradiation intensity has been found, a dye dose may be selected. Rats injected with 10 mg/kg rose bengal and irradiated for 2 mins with 600 mW/cm^2 of filtered yellow light at the parietal convexity should yield a lesion extending through the cortical thickness, as determined by carbon black infusion 1 h later. The lesion thickness can be moderated by changing the irradiation time or dye dose, but it is still sufficiently reproducible that a survey series establishing the proper irradiation conditions will likely consist of six to eight animals or fewer. Surprisingly, the apertured irradiating beam is transferred through the skull to the underlying cortex with excellent fidelity; apparently, the light that is scattered is not sufficiently intense to generate thrombi outside the aperture image. This is even true for very small apertures of diameters of 1 to 2 mm,[63] or for apertures overlapping thicker bone such as the parietal ridge overlying the barrel-field region.

Of course, a laser beam may be used to induce a cortical lesion as well.[12,19] Blood-brain barrier breakdown and edema are early and prominent consequences, just as in arc lamp-based systems, but owing to the ease of focusing laser light, care must be taken to avoid temperature effects. For example, irradiation of the skull surface with intensities of 0.64 or 3.2 W/cm^2 (References 12 and 19, respectively) without temperature neutralization will quite likely induce epiphenomena due to excess heat absorption. Even at 200 mW/cm^2 we have found it necessary to cool the irradiated skull surface with a fan. Also, monochromatic irradiation near the 562-nm absorption maximum of rose bengal would be expected to induce a lesion more efficiently than arc lamp irradiation, and this has in fact been demonstrated.[12] In another technical variation, we explored whether a lesion could be produced by irradiating the skull at a 13° angle of incidence with a 514.5-nm argon beam focused in the horizontal plane with a cylindrical lens. The idea was to allow the beam to photothrombose only vessels on the cortical surface, beneath the skull. This apparently did occur: by carbon black infusion the resultant lesions were very uniform in occlusion density, and were colored grossly reddish throughout the cortical thickness with more consistency than the arc lamp-induced lesion (unpublished observations).

METHOD 3: STROKE INDUCED BY PHOTOTHROMBOTIC OCCLUSION OF THE MCA

Distal MCA Occlusion

The distal MCA is exposed by first making a vertical skin incision between the lateral canthus of the eye and the ear. The skin is retracted with silk sutures and the underlying temporal muscle is incised and retracted ventrally. With the aid of a Zeiss dissecting microscope, the right distal MCA is exposed by means of a small craniotomy (2 mm) in the frontal-squamosal bone ventral to the coronal suture and above the zygomatic arch. For a period of 30 min to 1 h before laser irradiation, the animal is monitored physiologically until stabilized, and all wounds are infiltrated with 1% xylocaine.

Proximal MCA Occlusion

The right proximal MCA is exposed via a modified subtemporal approach. A skin incision is made between the eye and ear, the zygomatic arch is removed, and the temporal and masseter muscle flaps are retracted downward together with the mandibular bone. A small portion of the jaw is removed but the mandibular joint is kept intact so as not to disturb normal feeding. A 2-mm craniotomy is then made in the inferotemporal fossa rostral to the foramen ovale. Physiological monitoring and wound dressing is administered as described above.

Single-Beam MCA Photothrombosis of Type II

Thrombotic stroke is induced by a laser-driven, photochemically mediated intravascular reaction in the distal MCA. The beam of an argon laser-pumped dye laser is operated at 562 nm, and initially at low power (less than 1 mW) for alignment purposes (cf. Appendix B for descriptions of it and alternative systems). The beam is focused on the MCA in several ways, depending on the result desired. If thrombotic stroke is to be induced by the simplest means possible, i.e., irradiation of a single MCA segment or sequential irradiation of more than one segment, the beam is focused through a 25-cm-f.l. BK7 antireflection-coated glass lens and positioned with a micrometer-controlled mirror. The focused beam diameter should overlap the target vessel segment somewhat (ca. 6%) for maximum occlusion-forming efficiency.[23] Next, rose bengal, 15 mg/ml in 0.9% saline, is injected via a femoral vein to a body concentration of 20 mg/kg (1.33 ml/kg). Irradiation is then conducted at a power level of 20 mW until occlusion results, usually within 2 min. Thrombosis as observed through a 562-nm interference rejection filter (Omega Optical, Burlington, VT) is characterized initially by intense orange fluorescence due to rose bengal. Typically, the fluorescence decreases during irradiation, yielding a white thrombus consisting entirely of aggregated platelets. This thrombus does not contain fibrin, and therefore is resistant to lysis by exogenous or endogenous tissue plasminogen activator.

Multiple Beam Irradiation System

For experiments requiring reversal of the thrombotic stroke, it is important to induce thrombi that are thinner than the segment diameter and accordingly, more susceptible to exogenous enzymatic lysis, yet stable against the arterial pressure gradient. These requirements led us to develop a special optical system that allows three MCA segments to be irradiated simultaneously. Blood flow is stopped with the minimum total occlusion length, inasmuch as the occlusions reinforce each other during their formation. For this purpose the dye laser beam is directed through a 200 line/in. Ronchi ruling (Rolyn Optics, Covina, CA). The beam diffracts vertically into three components, with the power of the zero-order beam a factor of 2.4 greater than the two first-order beams. The first-order beams are vertically displaced ± 11 mm by a pair of prisms (Figure 10, far right). (The spacing of the displacement prisms is contingent on spatial separation of the diffracted beams. In practice, the Ronchi rule is placed in the near-field region near the laser output, and the diffracted beam travels at least 1.5 m to reach the displacement prisms placed in the far radiation field.) The beam then traverses a 30-cm-f.l. cylindrical lens that controls the width of the beams as they are focused onto the MCA and overlap its diameter. The beams are next focused by another cylindrical lens (25 cm f.l.) oriented perpendicularly to the first lens; this lens defines the thickness (typically less than 100 μm) of each beam on the MCA. The focused rectangular image of the zero-order (fundamental) beam is then rotated by a dove prism (Figure 10,

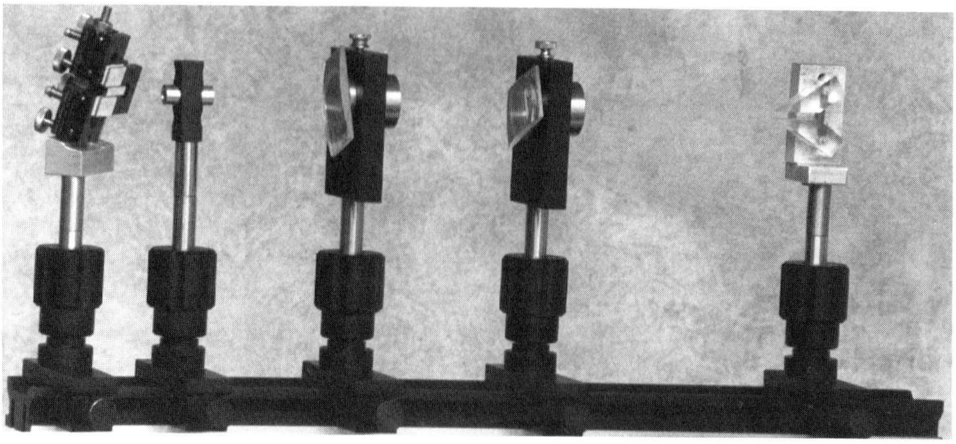

FIGURE 10. Apparatus for separating, routing, and focusing three independently positionable laser beams (cf. Method 3 for detailed description).

second from left) to a right angle with respect to the MCA (cf. Figure 6). Finally, a triplet of precision screw-controlled mirrors (Figure 10, far left) directs the three-beam array independently onto the MCA. In practice, the positions of the components on the optical rail vary slightly from one experiment to the next. Each beam can be positioned independently according to the specific locations desired for thrombus formation, but the zero-order beam is always placed proximally owing to its higher intensity. A common variant, useful for ensuring permanent occlusion,[27,28] replaces the cylindrical lenses with the 25-cm spherical lens, so that the resultant occlusions are at least as thick as the segment diameter.

MCA Photothrombosis of Type I

Flavin mononucleotide (FMN), a natural dye presumed to be a Type I (free radical) generator, is used as the photosensitizer. FMN is photoexcited by the argon laser operating at 457.9 nm. To putatively enhance the free radical pathway, norbixin, a singlet oxygen (but not superoxide radical) quencher, is injected to a blood concentration of 13.4 μ*M*, followed by FMN at 1.34 m*M* (37 mg/kg). FMN is dissolved in hot saline solution to 28 mg/ml, and then cooled to room temperature, while bixin is dispersed ultrasonically into rat plasma (200 μg/ml) where it binds to plasma proteins. Irradiation is carried out with the three-beam optical system described above. As shown in Figure 6, the three-beam array induces the yellow fluorescence of FMN during the irradiation (viewed through a 457-nm rejection filter). At input powers ranging from 10 to 20 mW and for blood pressures below 100 mmHg, three reddish thrombi somewhat mottled in appearance form in about 10 min with the beams focused sharply as in Figure 6. FMN reinjection and reirradiation are sometimes required to obtain stability. At systolic blood pressures exceeding 100 mmHg, higher input power levels are often required to achieve occlusion, in conjunction with defocusing the beams on the MCA trunk to make them thicker. In these thrombi, fibrin content is implied by negative Factor VIII and fibrinogen immunofluorescence, as well as the presence of a short-stranded material linking the surfaces of adjacent blood cells (Figure 8).

MCA Recanalization

For lysis of FMN-mediated thrombi, rt-PA is injected via the carotid catheter in a 200 μg bolus (200 μl). If the thrombi are thin, recanalization should be observed within 10 min. However, as long as retrograde infusion of blood via collateral channels exists, the FMN-mediated array of thrombi is unstable. Increasing the beam thickness does yield much more stable thrombi but they are also much harder to lyse; a systemic lytic state results from the increased levels of rt-PA.

APPENDIX B. PHYSICAL CHARACTERISTICS OF LASER IRRADIATION SYSTEMS

We have been using a 4-W argon ion laser (Model Innova 70-4, Coherent, Inc., Palo Alto, CA) as our irradiation source, in conjunction with a Coherent Model CR599 wavelength-tunable dye laser. The argon laser, operating on all its discrete transitions simultaneously, usually is used to energize the dye laser. The dye laser is loaded with either rhodamine 110 dye (Lambda Physik, Inc., Acton, MA) or rhodamine 560 dye (Exciton, Inc., Dayton, OH) dissolved in ethlyene glycol/methanol; the dye laser is tunable over the wavelength range 539 to 606 nm. The argon-pumped dye laser can produce at least 1 W at 562 nm (the wavelength of maximal rose bengal absorption in tissue) when excited with 6 W of argon laser power. The argon laser can be prism-tuned to produce laser action at several transitions between 457.9 and 514.5 nm, and at 528.7 nm with special mirrors. FMN can be excited with any of the transitions between 457.9 (most efficient) and 488.1 nm, while erythrosin B can be excited by 514.5 or 528.7 nm; the latter wavelength is very close to the 537–nm absorption maximum of erythrosin B in tissue. An alternative source, a tunable argon-krypton laser (Innova Spectrum from Coherent, Inc.), will excite any of these dyes, as well as Photofrin II, efficiently *in vivo.* In particular, the krypton transition at 530.9 nm, rated at 200 mW, will excite erythrosin B quite efficiently, as will the 568.2-nm krypton line, rated at 150 mW, for excitation of rose bengal. The 543.5-nm helium-neon laser will excite either rose bengal or erythrosin B, but its very low power (less than 4 mW) requires very tight beam focusing to reach sufficient intensities,[24] which limits its general use.

Before irradiation is begun, the argon laser and then the dye laser must be tuned to their maximum power levels with the aid of an external power/intensity meter. This will ensure stability of beam power and beam pointing during the irradiation procedure, especially important when coupling the beam to an optical fiber. With pretuning optimization, we have observed the dye laser output power level to be stable to within 1%.

REFERENCES

1. Pulsinelli, W. and Brierley, J. B. (1979) A new model of bilateral hemispheric ischemia in the unanesthetized rat. *Stroke* 10, 267–272.
2. Ginsberg, M. D. and Busto, R. (1989) Progress review: rodent models of cerebral ischemia. *Stroke* 20, 1627–1642.
3. Watson, B. D., Dietrich, W. D., Busto, R., and Ginsberg, M. D. (1983) Photochemically induced thrombotic stroke in rat brain. *Ann. Neurol.* 14, 126.
4. Spikes, J. D. (1989) Photosensitization. In *The Science of Photobiology,* Smith, K. C., ed. (Plenum Press, New York), pp. 79–109.
5. Foote, C. S. (1976) Photosensitized oxidation and singlet oxygen: consequences in biological systems. In *Free Radicals in Biology,* Pryor, W. A., ed. (Academic Press, New York), Vol. 2, pp. 85–134.
6. Chacon, J. N., McLearie, J., and Sinclair, R. S. (1988) Singlet oxygen yields and radical contributions in the dye-sensitized photo-oxidation in methanol of esters of polyunsaturated fatty acids (oleic, linoleic, linolenic and arachidonic). *Photochem. Photobiol.* 47, 647–656.
7. Chan, H. W.-S. (1977) Photo-sensitized oxidation of unsaturated fatty acid methyl esters. The identification of different pathways. *J. Am. Oil Chem. Soc.* 54, 100–104.
8. Rosenblum, W. I. and El-Sabban, F. (1977) Platelet aggregation in the cerebral microcirculation. Effect of aspirin and other agents. *Circ. Res.* 40, 320–328.

9. Gandin, E., Lion, Y., and Van de Vorst, A. (1983) Quantum yield of singlet oxygen production by xanthene derivatives. *Photochem. Photobiol.* 37, 271–278.
10. Plato, P. C. C. and Rodgers, M. A. J. (1987) Laser flash photokinetic studies of rose bengal sensitized photodynamic interactions of nucleotides and DNA. *Photochem. Photobiol.* 45, 79–86.
11. Herrmann, K. S. (1983) Platelet aggregation in the hamster cheek pouch by a photochemical approach with excited fluorescein isothiocyanate-dextran. *Microvasc. Res.* 26, 238–249.
12. Boquillon, M., Boquillon, J. P., and Bralet, J. (1992) Photochemically induced, graded cerebral infarction in the mouse by laser irradiation. Evolution of brain edema. *J. Pharmacol. Meth.* 27, 1–6.
13. Watson, B. D., Dietrich, W. D., Busto, R., Wachtel, M. S., and Ginsberg, M. D. (1985) Induction of reproducible brain infarction by photochemically initiated thrombosis. *Ann. Neurol.* 17, 497–504.
14. Watson, B. D., Dietrich, W. D., Prado, R., and Green B. A. (1988) Photochemically induced vascular thrombosis (photothrombosis). *N.A.T.O. ASI Ser.* H22, 507–524.
15. Dietrich, W. D., Ginsberg, M. D., Busto, R., Watson, B. D., and Yoshida, S. (1987) Vascular aspects and hemodynamic consequences of central nervous system injury. *Cent. Nerv. Syst. Trauma* 3, 265–280.
16. Dietrich, W. D., Watson, B. D., Busto, R., Bethea, J. R., Scheinberg, P., and Ginsberg, M. D. (1987) Photochemically induced cerebral infarction: 1. Early microvascular alterations. *Acta Neuropath (Berl.)* 72, 315–325.
17. Watson, B. D., Prado, R., Dietrich, W. D., Busto, R., Scheinberg, P., and Ginsberg, M. D. (1987) Mitigation of evolving cortical infarction in rats by recombinant tissue plasminogen activator following photochemically induced thrombosis. In *Cerebrovascular Diseases—Fifteenth Research (Princeton) Conference*, Raichle, M. E. and Powers, W. J., eds. (Raven Press, New York), pp. 317–330.
18. Laursen, H., Hansen, A. J., and Sheardown, M. (1993) Cerebrovascular permeability and brain edema after cortical photochemical infarcts in the rat. *Acta Neuropath.* 86, 378–385.
19. Grome, J. J., Gojowczyk, G., Hofmann, W., and Graham, D. I. (1998) Quantitation of photochemically induced focal cerebral infarction in the rat. *J. Cereb. Blood Flow Metab.* 8, 89–95.
20. Dietrich, W. D., Watson, B. D., Busto, R., Bethea, J. R., Scheinberg, P., and Ginsberg, M. D. (1987) Photochemically induced cerebral infarction: 2. Edema and blood-brain barrier disruption. *Acta Neuropath. (Berl.)* 72, 326–334.
21. Hakim, A. M. (1987) The cerebral ischemic penumbra. *Can. J. Neurol. Sci.* 14, 557–559.
22. Watson, B. D., Holets, V. R., Prado, R., and Bunge, M. B. (1993) Laser-driven photochemical induction of spinal cord injury in the rat: methodology, histopathology, and applications. *NeuroProtocols* 3, 3–15.
23. Watson, B. D., Dietrich, W. D., Prado, R., and Ginsberg, M. D. (1987) Argon laser-induced arterial photothrombosis: characterization and possible application to therapy of arteriovenous malformations. *J. Neurosurg.* 66, 748–754.
24. Nakayama, H., Dietrich, W. D., Watson, B. D., Busto, R., and Ginsberg, M. D. (1988) Photothrombotic occlusion of rat middle cerebral artery: histopathological and hemodynamic sequelae of acute recanalization. *J. Cereb. Blood Flow Metab.* 8, 357–366.
25. Tseng, S. C.-G., Feenstra, R. P. G., and Watson, B. D. Characterization of photodynamic actions of rose bengal on cultured cells. *Invest. Ophthalmol. Vis. Sci.*, 35, 3295–3307, 1994.
26. Brint, S., Jacewicz, M., Kiessling, M., Tanabe, J., and Pulsinelli, W. (1988) Focal brain ischemia in the rat: methods for reproducible neocortical infarction using tandem occlusion of the distal middle cerebral artery and ipsilateral common carotid arteries. *J. Cereb. Blood Flow Metab.* 8, 474–485.
27. Markgraf, C. G., Kraydieh, S., Prado, R., Watson, B. D., Dietrich, W. D., and Ginsberg, M. D. (1993) Comparative histopathologic consequences of photothrombotic occlusion of the distal middle cerebral artery in Sprague-Dawley and Wistar rats. *Stroke* 24, 286–293.
28. Yao, H., Ginsberg, M. D., Watson, B. D., Prado, R., Dietrich, W. D., Kraydieh, S., and Busto, R. (1993) Failure of MK-801 to reduce infarct volume in thrombotic middle cerebral artery occlusion in rats. *Stroke* 24, 864–870.
29. Morikawa, E., Huang, Z., and Moskowitz, M. (1992) L-arginine decreases infarct size caused by middle cerebral arterial occlusion in SHR. *Am. J. Physiol.* 263, H1632–1635.
30. Futrell, N., Watson, B. D., Dietrich, W. D., Prado, R., Millikan, C., and Ginsberg, M. D. (1988) A new model of embolic stroke produced by photochemical injury to the carotid artery in the rat. *Ann. Neurol.* 23, 251–257.
31. Futrell, N., Millikan, C., Watson, B. D., Dietrich, W. D., and Ginsberg, M. D. (1989) Embolic stroke from a carotid arterial source in the rat: pathology and clinical implications. *Neurology* 39, 1050–1056.
32. Futrell, N. (1991) An improved photochemical model of embolic cerebral infarction in rats. *Stroke* 22, 225–232.
33. Futrell, N. and Riddle, J. M. (1993) The ultrastructure of photochemically induced thrombi with embolization in a rat model. *Stroke* 24, 1983–1992.
34. Chopp, M., Glasberg, M. R., Riddle, J. M., Heitzel, F. W., and Welch, K. M. A. (1987) Photodynamic therapy of normal cerebral tissue in the cat: a noninvasive model for cerebrovascular thrombosis. *Photochem. Photobiol.* 46, 103–108.
35. Wester, P., Dietrich, W. D., Prado, R., Watson, B. D., Globus, M. Y.-T., and Leistra, H. (1992) Measurement of serotonin in plasma by *in vivo* microdialysis during photochemically induced thrombosis—methodological

aspects. In *Role of Neurotransmitters in Brain Injury*, Globus, M. Y.-T. and Dietrich, W. D., eds. (Plenum Press, New York), pp. 153–159.
36. Matsuno, H., Uematsu, T., Nagashima, S., and Nakashima, M. (1991) Photochemically induced thrombus model in rat femoral artery and evaluation of effects of heparin and tissue-type plasminogen activator with this model. *J. Pharmacol. Methods* 25, 303–318.
37. Matsuno, H., Uematsu, T., Umemura, K., Takiguchi, Y., Wada, K., and Nakashima, M. (1992) Effects of vapiprost, a novel thromboxane receptor antagonist, on thrombus formation and vascular patency after thrombolysis by tissue-type plasminogen activator. *Br. J. Pharmacol.* 106, 533–538.
38. Umemura, K., Wada, K., Uematsu, T., and Nakashima, M. (1993) Evaluation of the combination of a tissue-type plasminogen activator, SUN9216, and a thromboxane receptor antagonist, vapiprost, in a rat middle cerebral artery thrombosis model. *Stroke* 24, 1077–1082.
39. Boergen, K.-P., Birngruber, R., and Hillenkamp, E. (1981) Laser-induced endovascular closure as a possibility of selective vessel closure. *Opthalmic Res.* 13, 139–150.
40. Weichert, W., Breddin, H. K., and Staubesand, J. (1988) Application of a laser-induced endothelial injury model in the screening of antithrombotic drugs. *Sem. Thromb. Hemostasis (Suppl.)* 14, 106–114.
41. Wester, P., Watson, B. D., Prado, R., and Dietrich, W. D. (1993) A photothrombotic "ring" model of rat stroke-in-evolution displaying putative penumbral inversion *Stroke*. 26, 444–450, 1995.
42. Cameron, T., Prado, R., Watson, B. D., Gonzalez-Carvajal, M., and Holets, V. (1990) Photochemically induced cystic lesion in the rat spinal cord. I. Behavioral and morphological analysis. *Exp. Neurol.* 109, 214–223.
43. Hao, J.-X., Watson, B. D., Xu, X.-J., Wiesenfeld-Hallin, Z., Seiger, A., and Sundstrom, E. (1992) Protective effect of the NMDA antagonist MK-801 on photochemically induced spinal lesions in the rat. *Exp. Neurol.* 118, 143–152.
44. Bederson, J. B., Pitts, L. H., Tsuji, M., Nishimura, M. C., Davis, R. L., and Bartkowski, H. (1986) Rat middle cerebral artery occlusion: evaluation of the model and development of a neurologic examination. *Stroke* 17, 472–476.
45. Hicks, C. L. and Draper, J. (1981) Photo-oxidation by riboflavin in model systems for bovine milk. In *Oxygen and Oxy-Radicals in Chemistry and Biology* (Academic Press, New York), pp. 663–665.
46. Watson, B. D., Kanemitsu, H., Prado, R., and Dietrich, W. D. (1991) Properties of a reperfusible model of thrombotic stroke in the rat middle cerebral artery (MCA) territory. *N.A.T.O. ASI Series* 208, 294–295.
47. Herrmann, K. S. and Voigt, W.-H. (1984) Ultrastructural observations of an electron dense amorphous layer on selectively damaged endothelial cells, a possible trigger of thrombogenesis *in vivo*, and its inhibition by nafazatrom. *Thromb. Res.* 36, 205–215.
48. Leung, L. and Nachman, R. (1986) Molecular mechanisms of platelet aggregation. *Annu. Rev. Med.* 37, 179–186.
49. Huang, A. J.-W., Watson, B. D., Hernandez, E., and Tseng, S. C.-G. (1988) Induction of conjunctival transdifferentiation on vascularized corneas by photothrombotic occlusion of corneal neovascularization. *Ophthalmology* 95, 228–235.
50. Watson, B. D. (1991) Method for the Permanent Occlusion of Arteries. U.S. Patent No. 5,056,006 issued October 1, 1991.
51. Budzinski, A. Z., Olexa, S. A., Brizuela, B. S., Sawyer, R. T., and Stent, G. S. (1981) Anticoagulant and fibrinolytic properties of salivary proteins from the leech *Haementeria ghilianii. Proc. Soc. Exp. Biol. Med.* 168, 266–275.
52. Yasuda, T., Gold, H. K., Fallon, J. T., Leinbach, R. C., Guerrero, J. L., Scudder, L. E., Kanke, M., Shealy, D., Ross, M. J., Collen, D., and Coller, B. S. (1988) Monoclonal antibody against the platelet glycoprotein (GP) IIb/IIIa receptor prevents coronary artery reocclusion after reperfusion with recombinant tissue-type plasminogen activator in dogs. *J. Clin. Invest.* 81, 1284–1291.
53. Henderson, B. W., Owczarczak, B., Sweeney, J., and Gessner, T. (1992) Effects of photodynamic treatment of platelets or endothelial cells *in vitro* on platelet aggregation. *Photochem. Photobiol.* 56, 513–521.
54. Dietrich, W. D., Nakayama, H., Watson, B. D., and Kanemitsu, H. (1989) Morphological consequences of early reperfusion following thrombotic or mechanical occlusion of the rat middle cerebral artery. *Acta Neuropathol.* 78, 605–614.
55. Dietrich, W. D., Prado, R., Watson, B. D., and Nakayama, H. (1988) Middle cerebral artery thrombosis: acute blood-brain barrier consequences. *J. Exp. Neuropathol. Neurol.* 47, 443–451.
56. Dietrich, W. D., Prado, R., and Watson, B. D. (1988) Photochemically stimulated blood-borne factors induce blood-brain barrier alterations. *Stroke* 19, 857–862.
57. Dietrich, W. D., Prado, R., Watson, B. D., Busto, R., and Ginsberg, M. D. (1991) Hemodynamic consequences of common carotid artery thrombosis and thrombogenically activated blood in rats. *J. Cereb. Blood Flow Metab.* 11, 957–965.
58. Stieg, P. E., Sathi, S., Alvarado, S. P., Jackson, P. S., Pellegrini, J. W., Chen, H.-S., Lipton, S. A., and Jensen, F. E. (1993) Post-stroke protection by memantine minimally affects behavior and does not block LTP. *Soc.*

Neurosci. Abst. 19(Part 2), 1503.

59. Sathi, S., Edgecomb, P., Warach, S., Manchester, K., Donaghey, T., Stieg, P. E., Jensen, F. E., and Lipton, S. A. (1993) Chronic transdermal nitroglycerin (NTG) is neuroprotective in experimental rodent stroke models. *Soc. Neurosci. Abst.* 19(Part 1), 849.
60. de Ryck, M. (1990) Animal models of cerebral stroke: pharmacological protection of function. *Eur. J. Neurol.* (Suppl. 2) 30, 21–27.
61. Dietrich, W. D., Ginsberg, M. D., Busto, R., and Watson, B. D. (1986) Photochemically induced cortical infarction in the rat: 1. Time course of hemodynamic consequences. *J. Cereb. Blood Flow Metab.* 6, 184–194.
62. Dietrich, W. D., Feng, Z.-C., Leistra, H., Watson, B. D., and Rosenthal, M. (1994) Thrombotic infarction triggers multiple episodes of cortical spreading depression in distant brain regions. *J. Cereb. Blood Flow Metab.* 14, 20–28.
63. Domann, R., Hagemann, G., Kraemer, M., Freund, H.-J., and Witte, O. W. (1993) Electrophysiological changes in the surrounding brain tissue of photochemically induced cortical infarcts in the rat. *Neurosci. Lett.* 155, 69–72.
64. Dietrich, W. D., Watson, B. D., Busto, R., and Ginsberg, M. D. (1987) Metabolic plasticity following cortical infarction. A 2-deoxyglucose study in adult rats. In *Cerebrovascular Diseases—Fifteenth Research (Princeton) Conference,* Raichle, M. E. and Powers, W. J., eds. (Raven Press, New York), pp. 285–295.
65. Hurwitz, B. E., Dietrich, W. D., McCabe, P. M., Watson, B. D., Ginsberg, M. D., and Schneiderman, N. (1990) Sensory-motor deficit and recovery from thrombotic infarction of the vibrissal barrel-field cortex. *Brain Res.* 512, 210–220.
66. Wester, P., Dietrich, W. D., Prado, R., Watson, B. D., Globus, M. Y.-T., and Leistra, H. (1992) Serotonin release into plasma following common carotid artery thrombosis. *Stroke* 23, 870–875.
67. Dietrich, W. D., Busto, R., and Ginsberg, M. D. (1989) Effect of the serotonin antagonist ketanserin on the early remote hemodynamic consequences of brain infarction. In *Pharmacology of Cerebral Ischemia—Proceedings of the Second International Symposium on Pharmacology of Cerebral Ischemia,* Krieglstein, J., ed. (Elsevier, Amsterdam), pp. 311–315.
68. Dietrich, W. D., Dewanjee, S., Prado, R., Watson, B. D., and Dewanjee, M. (1993) Transient platelet accumulation in the rat brain after common carotid artery thrombosis. An ^{111}In-labeled platelet study. *Stroke* **24,** 1534–1540.
69. Dietrich, W. D., Prado, R., Halley, M., and Watson, B. D. (1993) Microvascular and neuronal consequences of common carotid artery thrombosis and platelet embolization in rats. *J. Neuropathol. Exp. Neurol.* 52, 351–360.
70. Cremer, J. E. and Seville, M. P. (1983) Regional brain blood flow, blood volume and hematocrit values in the adult rat. *J. Cereb. Blood Flow Metab.* 3, 254–256.
71. Klaassen, C. D. (1976) Pharmacokinetics of rose bengal in the rat, rabbit, dog, and guinea pig. *Toxicol. Appl. Pharmacol.* 38, 85–100.

Chapter 13

SUBARACHNOID HEMORRHAGE MODEL

Toru Matsui and Takao Asano

TABLE OF CONTENTS

I. INTRODUCTION

Ecker and Riemenschneider first described intracranial arterial spasm using arteriography.[36] The radiographic changes are ordinarily delayed in onset and are most frequent and striking during the second week. More than 30 to 40% of the patients with a ruptured intracranial aneurysm exhibit such intracranial arterial narrowing.[59,148–150] The arterial narrowing as shown arteriographically is developed in parallel with the amount of blood in the basal subarachnoid cistern revealed by a CT scan. Once radiographic evidence of intracranial arterial spasm has developed, it ordinarily resolves gradually over one to a few weeks. Patients with a ruptured intracranial aneurysm may subsequently develop clinical and pathological evidence of one or more cerebral infarcts. Delayed ischemic neurological deficits (DINDs) are apparently due to the delayed development of intracranial arterial spasm. Basically, this phenomenon was thought to be led by the liberation of factor(s) from the initial deposition of subarachnoid clotted blood.[34,35,45,99] However, many confusing discrepancies have been reported. The mere deposition of blood in the subarachnoid space does not always provoke the radiographic arterial narrowing and/or clinical syndrome of intracranial arterial spasm. And there is no animal model that accurately mimics what happens in the human patient who develops the syndrome.[14,17] The introduction of blood into the basal cistern of animals, even with concomitant arterial injury, ordinarily causes immediate arterial constriction but only a minor delayed arterial spasm with no cerebral infarction or delayed neurological deficits. There is even more confusion when one further considers the nature, the etiology, and the pathogenesis of the intracranial arterial narrowing called vasospasm.[34,35] According to definition, a spasm is a sudden but transitory contraction of muscle(s) that occurs involuntarily. Such a reaction

0-8493-8094-4/95/$0.00+$.50

can be produced both by chemical[9,23,109,113] and by mechanical stimulation.[14,17] But there is little evidence to link the brief transitory constriction of the cerebral artery that can be produced in the laboratory to the complex series of clinical and radiographic events that occur in humans following the rupture of intracranial aneurysms. No convincing evidence has been presented that long-lasting or propagated spasms due to involuntary constriction of cerebral arterial smooth muscles actually occurs.

So far, through a number of experimental efforts in an attempt to resolve the pathogenesis of this phenomenon, many spasmogen candidates, such as PGE_2,[147] $PGF_{2\alpha}$,[1,152] histamine,[3] serotonin,[1,2,4] metabolites of uric acid,[6] bilirubin,[88] and neuropeptides,[135] have been examined in different animal models. Ecker and Riemenschneider ascribed the possible mechanism of arterial narrowing following subarachnoid hemorrhage (SAH) to traction on the wall.[36] Echlin showed that direct application of fresh blood to the adventitia of the exposed basilar artery induces marked vasospasm.[34,35] The constriction, produced by serotonin and a number of blood-related products, is usually of short duration, but there is no clinically important delayed phase. The denervation hypersensitivity concept was proposed by Fraser et al.,[45] and Peerless and Yasargil.[99] They showed that experimental SAH induced the depletion of catecolamine in peripheral vascular nerves and vescicles. Unfortunately, an alpha-blocking agent completely reversed the acute vasospasm but not the clinically important late phase.

It was also conceived that chronic vasospasm might be related to an inflammatory response induced by the presence of blood in subarachnoid space. In this regard, prophylactic treatment of steroid and nonsteroid antiinflammatory agents was tried to prevent chronic vasospasm induced in a two-hemorrhage canine model.[21,22,55,95] These drugs reduced the severity of cerebral vasospasm experimentally but the protective effect has not been established in clinical cases.

In the 1980s, it was found that endothelial cells have an important role in the myogenic tonus of vascular smooth muscle,[47,48,74,75,136–138] and novel potent vasoconstrictor peptides, endothelin[153] and endothelium-derived relaxing factor (EDRF), have been thought to be associated with pathogenesis of cerebral vasospasm.[57,63,67–69,93,134]

The free radical chain reaction that plays a major role in the development of ischemic brain damage was linked to the pathogenesis of the cerebral vasospasm after SAH. So, free radical scavenging agents[7,46,154] have been tried experimentally and clinically to prevent the occurrence of late spasm.

Recent developments in molecular biology and biochemistry have led to a new concept concerning the contractile mechanism of smooth muscle, representing an important role of protein kinase C (PKC) in sustained phase of agonist-induced contraction.[106] The author linked this theory to the pathogenesis of the cerebral vasospasm.[86,124] Details will be given later.

This manuscript introduces a methodology producing experimental SAH with reference to the extensively associated literature. The author will not only introduce the SAH model in dogs, monkeys, cats, rabbits, and rats, but will also describe the historical course of concepts regarding the pathogenesis of cerebral vasospasm.

In the final section, representative techniques used in the *in vitro* studies will be introduced.

II. PRODUCTION OF SUBARACHNOID HEMORRHAGE

In any kind of basic experiment, the method should be chosen corresponding to its purpose. This is true for the selection of animal species as models, since different animals have different specificity.

From the viewpoint of methodology, the SAH models are classified into the following three categories; *in vivo, in vitro,* and *in situ.* In the *in vivo* study, an animal is subjected to SAH and then given a certain pharmacological or physiological treatment, followed by pathological or pathogenic assessment. In the *in vitro* model, arterial segments are excised from the animals, which are normal or subjected to experimental SAH. The segments are investigated with the biochemical, biological, immunochemical, physiological, or pathological techniques to be tested. In the *in situ* models, control normal animals or those subjected to SAH, are anesthetized and the subject artery is examined with the test techniques.

A. DOG

In 1968, the artery around the circle of Willis was torn or punctured to induce vasospasm by Brawly et al., Landau et al., Simeone et al., and Weir et al.[20,66,94,121,144] as shown in Figure 1. But with this method high mortality and morbidity such as uncontrollable bleeding, seizure, and respiratory arrest occurred.[34] Also, with the puncture method, intracranial hypertension and brain swelling, followed by uncontrollable hemorrhage, influenced the spasm, in addition to blood itself. So different ways to induce SAH were used. Blood was injected into the surgically exposed cisterna magna and a 25-gauge spinal needle held by a stereotaxic holder was advanced into the cisterna; 2 ml of unheparinized autologous fresh arterial blood was injected slowly at a rate of 0.4 ml/min with an exchange of cerebrospinal fluid. Thus, increased intracranial pressure and mechanical destruction of the vessels, which was found to be absent under microscopic observation, do not appear to be major factors in the production of vasospasm. In order to facilitate the contact of blood with the basilar artery the dog was tilted tail-up at approximately 30° for 10 to 20 min following the injection. In this model, the average arterial constriction appeared to be maximal within 30 min after the cisternal blood injection, and it recovered slowly. On the second day, vasospasm was again present. This biphasic phenomenon was first demonstrated by Brawley et al. in 1968.[20] A similar occurrence was noted in monkeys by Kagstrom et al. and Echlin et al. in 1969.[34,66] They named this "recurrent" and "delayed" spasm, respectively. The vasoconstriction induced by blood injection into the subarachnoid space recurred in most animals within 24 h, mostly in 2 to 4 d, and it frequently persisted for at least 7 d.[66,80,121,144] This study led to the thought that the acute stage of vasospasm might be induced by serotonin or the other vasoactive substances. This model has been used to elucidate the etiological factor(s) in the chronic stage. In the early 1970s, unknown substances derived from incubated serum or from hematoma fluids of ruptured cerebral aneurysms were reported to constrict smooth muscle. Fibrinogen degradation products were suggested by Kowalski[78] and Osbahr et al.[98]

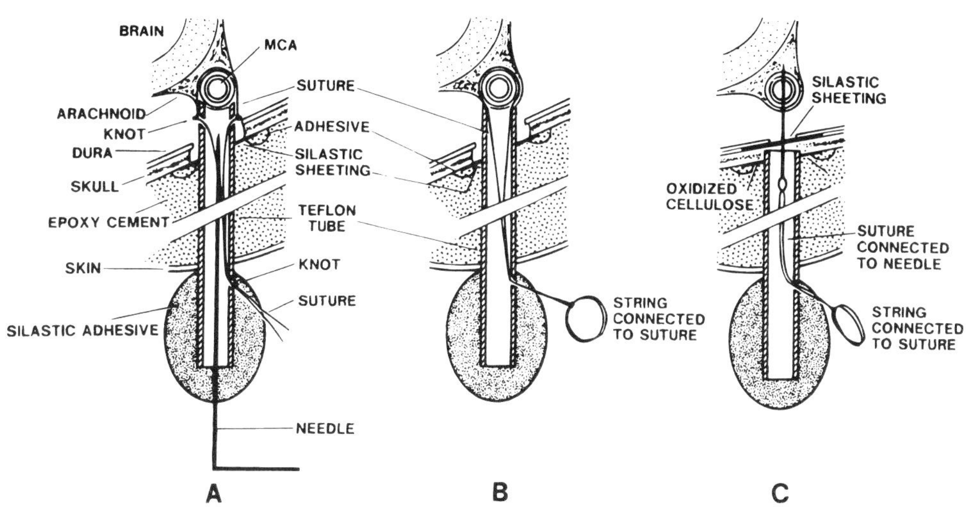

FIGURE 1. Implanted devices for experimental SAH. (A) puncture, (B) tearing, (C) withdrawal. (From Hayakawa, T. and Waltz, A. G., *Stroke*, 8, 421–426, 1977. With permission.)

Although the single-hemorrhage model produced biphasic vasospasms very similar to those in ruptured aneurysmal patients, the reduction in arterial diameter was not less than 40%. In this area, there were serious discrepancies between the two constrictions.

In 1983, Varsos et al.[139] reported two-hemorrhage models in dogs. The dog's head was placed in a stereotaxic frame with the animal in the prone position. A control angiogram was taken using 4 ml of Renographin 76 through the cannulated vertebral artery. And then a 22-gauge needle was placed into the cisterna magna, and 4 ml of cerebrospinal fluid was removed. Fresh autologous arterial unheparinized blood (4 ml) was injected at a rate of 2 ml/min. The dog was tilted tail-up for 20 to 30 min to facilitate settling of the blood around the basilar artery by gravity. One hour after the induction of SAH, a second angiogram was performed. On day 3, the dog was anesthetized and a third angiogram was performed. On days 7 and 14, fourth and fifth angiograms were carried out. All of the animals undergoing these procedures demonstrated vasospasm in the biphasic fashion observed in the single-hemorrhage model. The two-hemorrhage model offers an improvement on the single-hemorrhage model in the severity of spasm and its resistance to vasodilators such as aminophylline, nifedipine, or papaverine.[139] Since then, it has been generally accepted that the two-hemorrhage model resembles the situation found in patients with SAH. The method has provided a useful model for experimental research on chronic vasospasm.

Using this method, the discovery of a spasmogenic factor was attempted again. Monoamines[146] and prostaglandins[46,85,89,100,111,119] have been proposed as spasm-inducing factors. The breakdown products of erythrocytes,[96,97,128,145] the occurrence of lipid peroxidation or oxygen free radicals,[5,6,110] and alterations in the synthesis of cyclooxygenase or lypoxygenase products[83] have also been suggested. Occurrence of HETEs in the canine two-hemorrhage model indicated that the activity of 12-lipoxygenase of the platelets and 5-lipoxygenase in the arterial wall were enhanced and possibly correlated with the occurrence of the late vasospasm.[48] Correspondingly, antagonists or inhibitors of monoamine,[1–4,29] free radicals,[6] or lypoxygenase[154] have been used to prevent or treat chronic vasospasm. The effects of thromboxane synthetase inhibitor[112] and prostacyclin (PGI_2)[105] on vasospasm have been also investigated.

Based on the results obtained from the experimental findings mentioned, as well as various clinical trials, it was hypothesized that free radicals, hydroperoxides, and eicosanoids might be involved in the pathogenic mechanism underlying cerebral vasospasm.[8]

In place of the fresh arterial blood, 15-HPAA was injected into the cisterna magna in dogs, leading to the dose-dependent and biphasic sustained contraction of the basilar artery. Here, pronounced degenerative changes in the endothelium and the tunica media were observed by electron microscopy.[115] In accordance with the endothelial damage following SAH, the functional alteration of the endothelium was linked to pathogenesis of the late spasm. In the 1980s, two major vasoactive mediators derived from endothelium were discovered.[19] EDRF was first introduced by Furchgott and Zawadzki[47] and Furchgott[48] in 1980, who showed that the presence of the endothelial lining is obligatory for the vasorelaxant action of acethylcholine on aortic and arterial strips or rings from rabbits. Its biological and pharmacological properties are identical to those of nitric oxide, causing relaxation of the underlying smooth muscle by activation of the soluble guanilate cyclase. The production of EDRF depends on extracellular Ca^{2+}, though the activation of a voltage-dependent Ca^{2+} channel is not involved. Hence, Ca^{2+}-channel blocking agents do not inhibit the release of EDRF but rather act synergistically with it to relax vascular smooth muscle.[138] In this regard, a role of EDRF in the occurrence of cerebral arterial spasm has been suggested, using the canine two-hemorrhage model[67,68,134] or the rabbits[57,93] mentioned below.

Another mediator derived from endothelium is endothelin, a potent vasoconstrictor peptide, which was characterized and synthesized by Yanagisawa et al.[153] The strong and potent action of this substance has lent support to the suggestion of an important role in the pathogenesis of cerebral vasospasm.[62] The effect was thought to be highly dependent upon the extracellular calcium since it was blocked by calcium antagonists of the dihydropyridine groups.[153] Since prolonged vasospasm is resistant to treatment by calcium antagonists (nifedipine,[139] nimodipine,[38,50,92,94]) endothelin is not likely to be a relevant candidate for spasmogenic substance(s).

Nagasawa et al.[91,92] and Bevan et al.[15,16] ascribed the changes responsible for persistent narrowing in the large arteries to an increase in the wall stiffness, namely, decreased distensibility. The study was performed according to the method of Cox et al.,[25–28] who introduced the intriguing technique to investigate the mechanical property of vessels. Two days after the induction of SAH, the artery was more distensible, and then became progressively stiffer, approaching control by the seventh day. The progressive increase in stiffness, leading to the increased stiffness seen on the seventh day, is probably due to an increase in mural collagen. The artery in late spasm could be described as a persistent contracted artery with decreased distensibility.

In the mid-1980s, a new concept, that a "latch mechanism" occurs in the actin-myosin interaction, was hypothetized to explain the force persistence during the sustained phase of smooth muscle contraction.[104] Along this line, Rusmussen et al.[106] advocated that the initial phase of the contraction is mediated by a rise in cytosolic calcium concentration and the resulting calmodulin-dependent activation of both myosin light-chain kinase and the dissociation of caldesmon from the actin-caldesmon-tropomyosin-myosin fibrillar domain, leading to an interaction between actin and the phosphorylated light chains of myosin, and this is followed by the sustained phase due to phosphorylation of intermediate filament and actin-binding proteins other than myosin light chain by activation of PKC. This thesis seems especially pertinent to the pathogenesis of vasospasm as it primarily concerns the mechanism of sustained smooth muscle contraction.

First of all, we demonstrated that PKC-mediated smooth muscle contraction exists in the canine basilar artery, eliciting the partial dependency upon external Ca^{2+} and resistance to the Ca-channel blocking agent[124] as shown in Figure 2. Next, a topical application of H-7 or staurosporine, a PKC antagonist, to the transclivally exposed spastic basilar artery (day 7) subject to two-hemorrhage, almost completely reversed the late spasm, as observed by angiogram (Figure 3).[86] Furthermore, the 1,2-diacylglycerol (DAG), an intrinsic PKC activator, content of the basilar artery began to increase from day 3 (48 h after the first SAH) and

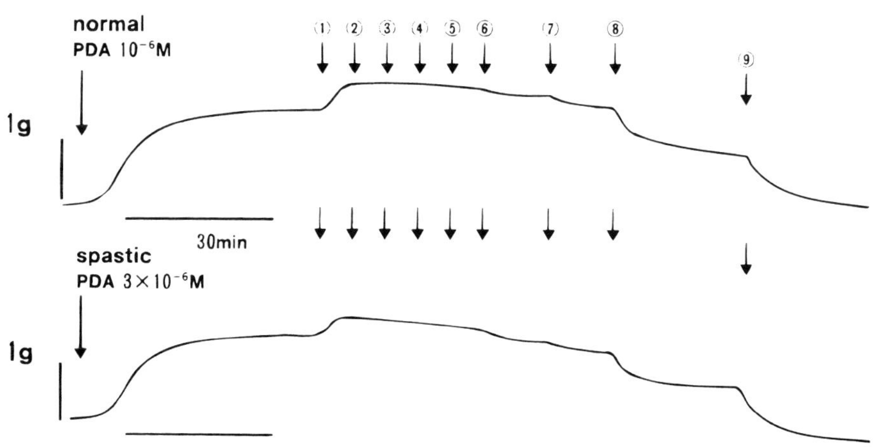

FIGURE 2. Effects of various antagonists on PDA-induced contraction of normal canine basilar arteries and those subjected to two-hemorrhage (day 7). 1, Methysergide; 2, atropine; 3, diphenhydramine; 4, phentolamine; 5, R24571; 6, W-7; 7; nicardipine; 8, H-7 (10^{-5} *M*); 9, H-7 (10^{-4} *M*).

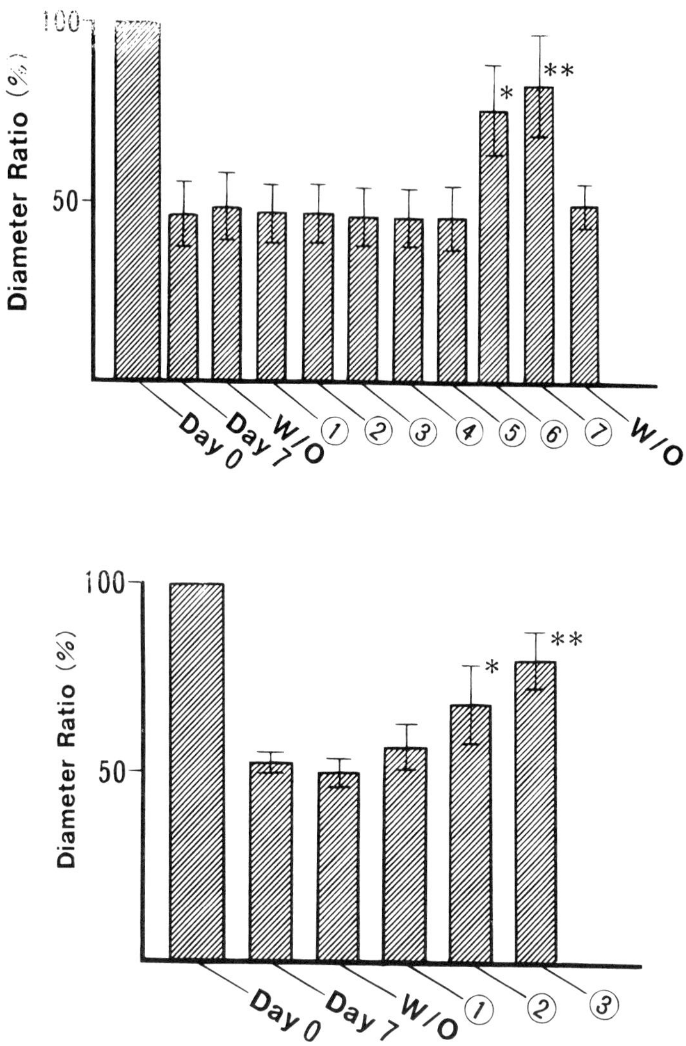

FIGURE 3. Effects of various antagonists on late vasospasm (day 7 of two-hemorrhage). Upper: 1, methysergide, atropine, diphenhydramine, and phentolamine; 2, R24571; 3, W-7; 4, nicardipine (10^{-7} *M*); 5, nicardipine (10^{-6} *M*); 6, H-7 (10^{-5} *M*); 7, H-7 (10^{-4} *M*). Lower: staurosporine was used at concentrations of 10^{-9}, 10^{-8}, and 10^{-7} *M*.

stayed in the range of 150 to 190% of the normal basilar artery until day 7, which is sufficient to activate PKC (Figure 4). Since it is documented that 12- and 15-HPETE inhibit DAG kinase, leading to an increased intracellular level of DAG,[117] the above results are in agreement with the description of Watanabe et al.[142] Our findings are also supported by the following two reports. Sakaki et al.[108] showed that the calmodulin contents in the canine basilar artery decreased following SAH. Suzuki et al.[126] demonstrated that no significant difference was found in canine basilar artery between myosin light-chain kinase, cAMP-, and cGMP-dependent protein kinase activities at any time after SAH.

B. MONKEY

Weir's group performed a right frontotemporal craniotomy under an operating microscope to open the Sylvian fissure.[52,53,70,144] The arachnoid membrane over the middle cerebral, internal carotid, anterior cerebral, and posterior communicating arteries was opened. The temporal lobe was then retracted and the remaining arachnoid over the posterior communicat-

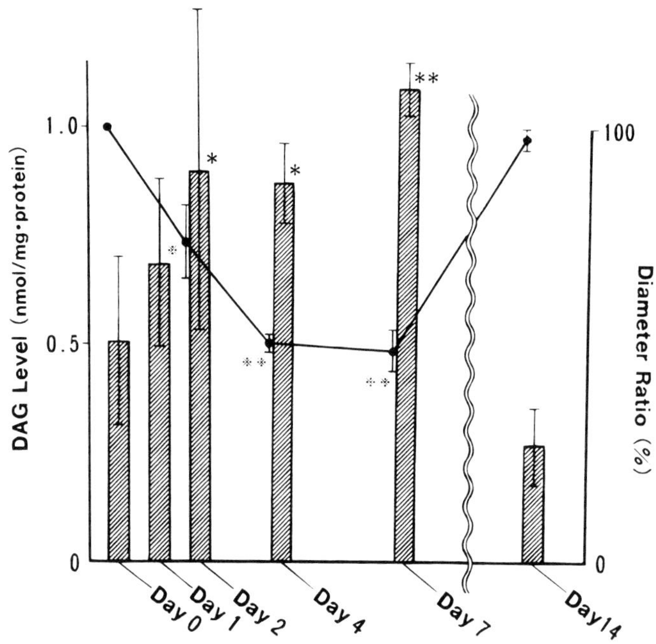

FIGURE 4. Time course of 1,2-diacylglycerol (shaded column) and angiographic diameters (closed circle) of canine basilar artery in the two-hemorrhage model. See Reference 86. (*,※,: $p < 0.05$ vs. Day 0. **,※※,: $p < 0.01$ vs. Day 0.)

ing artery was opened. Liliequst's membrane was incised and the arachnoid membrane over the tip of the basilar artery and the posterior cerebral artery was opened. An autologous blood clot was then carefully placed around the exposed arteries. The dura mater was closed with 6-0 silk sutures, and the scalp incision was closed in layers with 2-0 silk and 3-0 monofilament polyethylene sutures. The same procedure was then repeated immediately on the left side. The total amount of clot placed in the bilateral subarachnoid space is recorded. In the case of sham-operated animals, 5 ml of normal saline is instilled into the dissected subarachnoid space in place of the blood clot. In this model, sequential angiograms demonstrated early spasm 2 to 6 h after this procedure and a late spasm developed in the arteries attached to the blood clot on days 4 to 14 (mostly on days 6 to 8).

Bevan et al.[15] designed another method of producing the SAH model of monkeys. Under anesthesia, a 0.4-mm needle is placed through the intracranial portion of the internal carotid artery and percutaneously removed the following day, resulting in hemorrhage. Widespread vasospasm was usually present 5 d later, primarily in the ipsilateral vessels.

Using the former method, the effects of timing of clot removal,[52] U-74006F,[71] and antioxidants or Ca-channel blocking agents[79,82] on the changes in the cerebral arterial diameters were measured by angiogram. Effects on changes in the dilatory potency[70] and pathological findings[39,40] of the arterial wall were also measured. Bevan et al., using this model, demonstrated marked protection against vascular smooth muscle after SAH with diltiazem.[14,15]

C. RABBIT

According to Vorkapic et al.,[141] under general anesthesia a silastic tube is inserted with its tip in the prepontine cistern. Unheparinized autologous blood from central ear artery (3 ml/kg) is injected in selected portions around the basilar artery over 4 h. What ensues is that the basilar artery is persistently and closely encased by blood clot. In this model, diffuse cerebrovasospasm was seen on angiogram commencing on day 1 and continuing for 9 d.

Through intraarterial administration of papaverine, the papaverine-insensitive component of the angiographic narrowing correlated directly to the loss of contractility and with arterial wall stiffness.

Fortunately, the rabbit model is inexpensive as it is neccessary to work with a large number of animals. Others use rabbits in a different manner, in which rabbits exhibit a transient period of acute spasm following intracisternal injection of blood. This is followed in minutes to hours by a reversal of arterial narrowing to normal diameters. Beginning on day 2 arterial narrowing recurs lasting for approximately 48 h, after which the arterial diameters slowly return to normal.[140]

This model was used to study the pathological effect of SAH on the blood-brain barrier of the cerebral arteries[64,116,140] and the effects of certain antioxidants[140] on the chronic vasospasm.

D. CAT

In the cat SAH model, which is produced by a variety of methods, the reduction in the responsible arterial diameter is moderate compared with other animal models. The transcutaneous injection of the fresh autologous unheparinized blood (4 ml) into the cisterna magna, directly[33,87,95,143] or through an implanted microcatheter,[84,101] is followed by placing the animal in a head-down position, resulting in the distribution of the blood over the whole brain surface.

The mechanical rupture method, in which the artery concerned (middle cerebral artery) is microsurgically (transorbitally) exposed, is subgrouped into a needle puncture, tearing, and withdrawal.[56,151] These methods are convenient for producing chronic vasospasm.

Transclival or transcervical exposure of the basilar artery, which is irrigated with normal Krebs solution (KRS), is used for experimental acute cerebral vasospasm. Handa et al.,[51] and Kapp et al.[72] examined the effects of the vasodilating agent YC-93 on the acute cerebral vasospasm and spasmogenic qualities of platelet-derived substances, using the cat model.

Hubschmann[58] investigated the role of calcium in the parenchymal cell injury in SAH by injecting 0.5 ml of autologous blood via a 31-gauge needle into the cortical subarachnoid space.

E. RAT

A representative induction method of rat SAH was described by Delgado et al.[30] Cerebrospinal fluid (0.03 ml) was gently aspirated and then homologous blood (0.07 to 0.3 ml) was injected over 5 or 30 s, via a previously (3 to 5 d before) implanted cisternal catheter (opaque X-ray) or a directly punctured butterfly needle[121] through the exposed atlantooccipital membrane in SD rats anesthetized with barbiturates. During the injection of blood, the head was lowered by tilting the animal 20°. An angiography tube (Opti-100/12/15 HSG, Elema-Siemens, Sweden) with a 0.2 × 0.2 mm focus spot was used. Amipaque was chosen as the contrast medium because of its low toxicity. The vertebral and basilar arterial diameters were measured with a precision-calibrated measuring lens of fixed magnification and lens film distance. This showed that a reproducible biphasic vasospasm can be induced in the rat. Barry et al.[12] induced vasospasm in SD rats by puncturing the basilar artery via a transcervical and transclival approach. The basilar arterial diameter was measured photographically. In this model, the spasm peak was also seen on day 2, without any documentation of the presence of the acute phase. There was no mortality in the animals receiving blood or saline (sham). Ohta et al.,[96] using the same technique developed by Barry et al.,[12] induced the vasospasm in Wistar rats. In a comparative study, between Wistar and SD rats, vasospasm was found to be less pronounced and of shorter duration in Wistar rats.[30]

III. *IN VITRO* STUDY

The *in vitro* study of noncerebral vascular smooth muscle (NCVSM) has a long history. The *in vitro* investigation of cerebral vascular smooth muscle has been developed using the

NCVSM method. In research of vasospasm, an *in vitro* method using a small—volume chamber was first applied by Allen et al.[1–4] An isometric tension study using this method is useful in studies of the etiology and the receptor mechanism present in the cerebral arteries. It is introduced precisely later (Figure 5).

A. ISOMETRIC TENSION STUDY

After animals were sacrificed by exsanguination, the brain was removed and placed in an aerated nutrient Krebs solution at room temperature. The artery to be studied was then dissected free under an operating microscopy. A segment of the artery was excised by sectioning the vessel with razor blades fixed in parallel 3 mm apart (ring segment). This segment was then mounted on rigid parallel prongs, which is different from the instruments used by different authors. The time interval from sacrifice to mounting was never greater than 25 to 30 min. During mounting, the upper section of the chamber was several inches above the lower section to facilitate placement of the segment. After the segment was in place, the two sections of the chamber were clamped together. The chamber was then filled with 5 or 10 ml of nutrient solution, whose composition varies among authors. A mixture of 95% O_2 − 5% CO_2 was continuously bubbled through both the chamber and the buffer reservoir, with stirring by the vigorous bubbling of the gas mixture. The buffer was adjusted to pH 7.4 ± 0.05 and maintained throughout the experiments. The jacketed chamber and buffer reservoir were maintained at 37°C by a circulating, constant-temperature bath.

To make helical strips, a segment of the excised artery was helically cut at an angle of 45 to 60° into strips 20 to 25 mm long.[130,131,133] Each specimen was fixed vertically between hooks in a 20-ml muscle bath containing the nutrient solution. Hooks anchoring the upper end of the strips were connected to the lever of a force-displacement transducer. Isometric tension was measured using force-displacement transducers (Hewlett Packard FTA-10-1 and

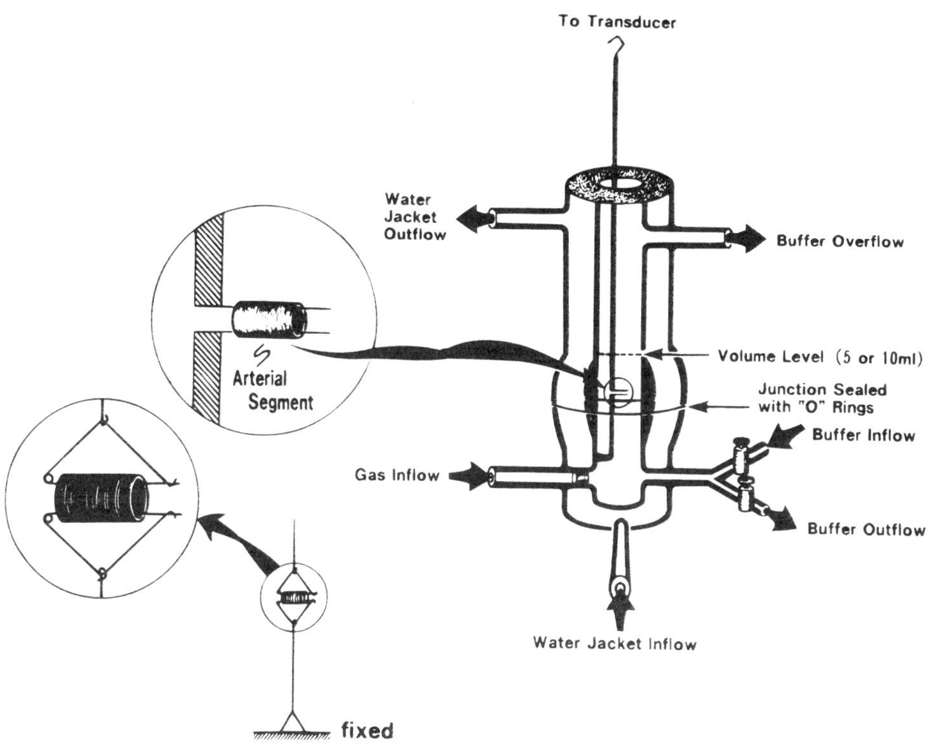

FIGURE 5. The chamber for isometric tension study. (Modified from Reference 1.)

FTA-100-1 or Nihon Kohden TB611 and MC 6000). This equipment was calibrated before each experiment with standard weights. The resting tension was adjusted to 1.5 to 3 g, which had proved optimal for inducing maximum contraction and was different among diferent animal species. The preparations were allowed to equilibrate for 90 to 120 min in the bathing media before the start of the experiments. All agents to be tested were added by means of a Bioquest or Gilson automatic pipette in increments of 10 to 500 μl. The total volume added for one test never exceeded 5% of the volume of the chamber contents. At the end of a test, the chamber and the artery were washed with 37°C buffer, which was pumped out through the chamber utilizing the buffer overflow outlet. From 20 to 30 min were allowed between tests for washing and stabilizing the artery.

The agents tested were added in a cumulative log-dose manner. When 5 ml of the chamber was filled, the initial 50 μl of the lowest-concentration test solution was added, each having a concentration one by one hundred as much as the original. The next 105 μl was added, each three times greater than the preceding solution. Subsequently 35 μl of the next lowest one was added, each having a concentration one over one-hundredth of the second solution. In this manner, the experiment was continued until a subsequent addition either did not increase the concentration or caused the artery to relax. The maximum contraction at each concentration was recorded and plotted as a cumulative dose-response curve. Then the ED_{50} value was computed for each test.

B. REMOVAL OF ENDOTHELIUM

Since Furchgott and Zwadzki[47] and Furchgott[48,49] showed that the vasodilator action of acetylcholine and A23187 is mediated by release of EDRF from the endothelial cells, the role of EDRF in the pathogenesis of chronic vasospasm has been investigated and envisaged[57,67–69,93,114,120] as mentioned in Section II.A. A representative method of endothelial cell removal has been described by many authors.[48,57,93] They have demostrated that endothelial cells of the arterial rings can be mechanically removed by a standard brief, gentle rubbing of the intimal surface with a 25- to 30-gauge stainless steel rod with a diameter equivalent to the lumen of the arteries. The endothelial removal enzymatically by collagenase also resulted in complete loss of the relaxing response to acetylcholine.[48,49]

C. ISOBARIC CONTRACTION STUDY

An isobaric contraction study was described in detail by Dobrin and Rovick[31] and Dobrin.[32] The vessels were cannulated, restored to *in situ* length, and immersed in a nutrient solution with a water jacket, which maintained the fluid at 37°C. The diameter was constantly monitored with a displacement transducer consisting of a differential transformer mounted to a nonrotating blade micrometer. The pressure was also measured and both diameter and pressure were recorded. Isobaric contraction induced by the agonists tested was examined by permitting the vessels to constrict while being distended by a constant intramural pressure. Isometric contraction was also performed by this method. So, if the pressure was maintained at each step level, the isobaric contraction could be measured. Similarly, if the diameter was maintained at a certain level, the isometric contraction could be measured. This method led to the investigation on the influence of initial length on the length-tension relationship of vascular smooth muscle, showing the highest pressure and strain whereby the reproducible stress-strain curve was given. This method might be useful in studying a change in the properties of the vascular muscle subjected to SAH. Along the same line of thinking, Cox designed a study of the three-dimensional mechanics of arterial segments and inflation-deflation techniques.[27]

D. PERFUSION TECHNIQUE

The perfusion technique was introduced by Hongo et al.[57] and used to study vascular responsiveness immunohistochemically. Under microscopy, approximately a 15- to 20-mm-long whole basilar artery was cannulated from both ends with 18-gauge stainless steel needles, stretched longitudinally to the approximate *in situ* length, and mounted in a vessel chamber filled with aerated KRS solution. All branches were tied off. The working volume of the chamber was 20 ml. The basilar arteries were intraluminally perfused with aerated KRS with a peristaltic pump at 1.5 to 2.5 ml/min to maintain the perfusion pressure at approximately 20 mmHg. Note that this method provides integrity of the separation of the intra- and extraluminal sides, which was confirmed by measuring the draining volume and monitoring the perfusion pressure. The *in vitro* experimental procedure is associated with the evaluation of the decreased distensibility in spasm arteries.[15] Basically, the arterial wall is composed of passive components (collagen and elastin) and an active component (smooth muscle). The force generated by the contraction of smooth muscle is transmitted to the passive components and detected as a developed tension.[10,13,18,31,32,40–43,65,103,104,107,123,129,130] Nagasawa et al.[90–92] demonstrated that the change in the passive elastic properties of arterial walls after blood injection (canine single-injection model) is one of the major factors affecting the development of cerebral vasospasm. This study uses the cylindrical segment of canine basilar artery, which is mounted horizontally in its *in vivo* axial length in a tissue bath containing KRS solution and oxygenated with 95% O_2 − 5% CO_2. Each segment was inflated with the solution from the reservoir under air pressure using a test apparatus.[90–92] While exchanging the KRS for a saline one containing no vasoactive substances, the pressure-diameter curve was obtained. It represents the passive elastic properties of the segments. From the parameters obtained from this experiment, the incremental elastic moduli (E_{inc}) was computed. It was closely related to the collagen/elastin ratio. Along this line of thinking, Kim et al.[76] measured the mechanical properties of spasm artery. These measurements were compared with those of normal controls, and passive compliances under total inhibition of spontaneous myogenic tone with diltiazem plus papaverine were calculated. The decrease in the value of the passive compliance, representing decreased distensibility, the increase in resting tension, and the shortening of length for maximal active contraction were obtained from this study. They favor a smaller diameter of the artery after SAH. The present authors also investigated the time course of passive compliance under treatment with H-7 plus the condition described by Kim et al.[76] Our study showed that the decrease in passive compliance became prominent

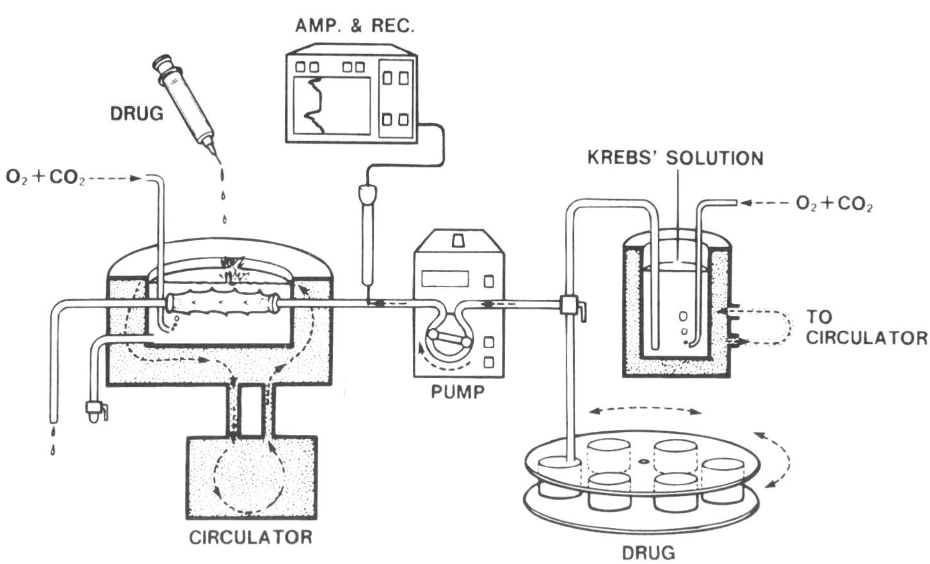

FIGURE 6. The perfusion model. (From Hongo, K. et al., *Stroke*, 19, 1550–1555, 1988. With permission.)

on day 4 through day 7 and was reversed on day 14 (unpublished data). Such a study may provide an ample explanation of the pathogenesis of chronic vasospasm (Figure 6).

CONCLUDING REMARK

As described above, a number of investigations have been attempted to elucidate the pathogenesis of cerebral vasospasm after subarachnoid hemorrhage. The authors believe that future research will be productive in this area.

REFERENCES

1. Allen, G. S., Henderson, L. M., Chou, S. N., French, L. A. (1974) Cerebral arterial spasm. Part 1: *in vitro* contractile activity of vasoactive agents on canine basilar and middle cerebral arteries. *J Neurosurg* 40: 433–441.
2. Allen, G. S., Gross, C. J., Henderson, L. M., Chou, S. N., French, L. A. (1974) Cerebral arterial spasm. Part 2: *in vitro* contractile activity of serotonin in human serum and CSF on the canine basilar artery, and its blockage by methysergide and phenoxybenzamine. *J Neurosurg* 40: 442–449.
3. Allen, G. S., Gross, C. J., Henderson, L. M., Chou, S. N., French, L. A. (1974) Cerebral arterial spasm. Part 3: *in vivo* intracisternal production of spasm by serotonin and blood and its reversal by phenoxybenzamine. *J Neurosurg* 44: 451–458.
4. Allen, G. S., Gross, C. J., Henderson, L. M., Chou, S. N. (1976) Cerebral arterial spasm. Part 4: *in vitro* effects of temperature, serotonin analogues, large nonphysiological concentrations of serotonin, and extracellular calcium and magnesium on serotonin-induced contractions of the canine basilar artery. *J Neurosurg* 44: 585–593.
5. Asano, T., Tanishima, T., Sasaki, T., Sano, K. (1979) Possible participation of free radical reactions initiated by clot lysis in the pathogenesis of vasospasm after subarachnoid hemorrhage. In *Cerebral Arterial Spasm*, Williams & Wilkins, Baltimore/London, pp. 190–201.
6. Asano, T., Sasaki, T., Koide, T., Takakura, K., Sano, K. (1984) Experimental evaluation of the beneficial effect of an antioxidant on cerebral vasospasm. *Neurol Res* 6: 49–53.
7. Asano, T., Sasaki, T., Koide, T., Takakura, K., Sano, K. (1984) Experimental evaluation of the beneficial effect of an antioxidant on cerebral vasospasm. *Neurol Res* 6: 49–53.
8. Asano, T. (1987) Delayed ischemic neurological deficits (DINDS), In *Acute Aneurysmal Surgery*, eds. Sano K, Asano T, Tamura A, Springer-Verlag, New York, pp. 27–52.
9. Awad, I. A., Carter, L. P., Spetzler, R. F., Medina, M., Williams, F. W., Jr. (1987) Clinical vasospasm after subarachnoid hemorrhage: response to hypervolemic hemodilution and arterial hypertension, *Stroke* 18: 365–372.
10. Burnstock, G., Prosser, C. L. (1960) Response of smooth muscles to quick stretch; relation of stretch to conduction. *Am J Physiol* 198: 921–925.
11. Baloon, B. J., Wei, E. P., Kontos, H. A. (1986) Superoxide anion radical does not mediate vasodilation of cerebral arterioles by VIP. *Stroke* 17: 1287–1290.
12. Barry, K. J., Gogjian, M. A., Stein, B. M. (1979) Small animal model for investigation of subarachnoid hemorrhage and cerebral vasospasm. *Stroke* 10: 538–541.
13. Bergel, D. H. (1961) The static elastic property of the arterial wall. *J Physiol* 156: 445–457.
14. Bevan, J. A., Bevan, R. D., Frazee, J. G. (1985) Experimental chronic cerebrovascula spasm in the monkey: an assessment of the functional change in the cerebral arteries and their protection by diltiazem. *Am J Cardiol* 56: 15H–20H.
15. Bevan, R. M., Bevan, J. A., Frazee, J. G. (1987) Functional arterial changes in chronic cerebrovasospasm in monkeys: an *in vitro* assessment of the contribution to arterial narrowing. *Stroke* 18: 472–481.
16. Bevan, J. A., Bevan, R. D., Frazee, J. G. (1988) Diltiazem protects against functional changes in chronic cerebrovasospasm in monkeys. *Stroke* 19: 73–79.
17. Bevan, J. A., Bevan, R. D. (1988) Arterial wall changes in chronic cerebrovasospasm: *in vitro* and *in vivo* pharmacological evidence. *Annu Rev Pharmacol Toxicol* 28: 311–329.
18. Bohr, D. F. (1973) Vascular smooth muscle updated. *Circ Res* 32: 665–672.
19. Botting, R., Vane, J. R. (1989) Vasoactive mediators derived from the endothelium. *Arch Mal Coeur* 82: 11–14.

20. Brawley, B. W., Strandness, D. E., Kelly, W. A. (1968) The biphasic response of cerebral vasospasm in experimental subarachnoid hemorrhage. *J Neurosurg* 28: 1–8.
21. Chyatte, D., Rusch, N., Sundt, T. (1983) Prevention of chronic cerebral vasospasm with ibuprofen and high dose methylpredonisolone. *J Neurosurg* 59: 925–932.
22. Chyatte, D., Sundt, T. (1984) Response of chronic experimental cerebral vasospasm to methylprednisolone and dexamethasone. *J Neurosurg* 60: 923–926.
23. Conway, L. W., McDonald, L. W. (1972) Structural changes of the intradural arteries following subarachnoid hemorrhage. *J Neurosurg* 37: 715–723.
24. Cox, R. H. (1974) Three-dimensional mechanics of arterial segments *in vitro:* methods. *J Appl Physiol* 36: 381–384.
25. Cox, R. H. (1975) Arterial wall mechanics and composition and the effects of smooth muscle activation. *Am J Physiol* 229: 807–812.
26. Cox, R. H., Jones, A. W., Swain, M. L. (1976) Mechanics and electrolyte composition of arterial smooth muscle in developing dogs. *Am J Physiol* 231: 77–83.
27. Cox, R. H. (1976) Mechanics of canine iliac artery smooth muscle *in vitro. Am J Physiol* 230: 462–470.
28. Cox, R. H. (1978) Comparison of carotid artery mechanics in the rat, rabbit and dog. *Am J Physiol* 243: H280–H288.
29. Cummins, B. H., Griffith, B. M. (1973) Intracarotid phenoxybenzamine for cerebral arterial spasm. *J Neurosurg* 38: 557–560.
30. Delgado, T. J., Brismar, J., Svendgaard, N. A. (1985) Subarachnoid hemorrhage in the rat: angiography and fluorescence microscopy of the major cerebral arteries. *Stroke* 16: 595–602.
31. Dobrin, P. B., Rovick, A. A. (1969) Influence of vascular smooth muscle on contractile mechanics and elesticity of arteries. *Am J Physiol* 217: 1644–1651.
32. Dobrin, P. B. (1973) Influence of initial length on length-tension relationship of vascular smooth muscle. *Am J Physiol* 225: 664–670.
33. Duff, T. A., Louine, J., Feilbach, J. A., Scott, G. (1988) Erythrocytes are essential for development of cerebral vasculopathy resulting from subarachnoid hemorrhage. *Stroke* 19: 68–72.
34. Echlin, F. A. (1965) Spasm of basilar and vertebral arteries caused by experimental subarachnoid hemorrhage. *J Neurosurg* 23: 1–11.
35. Echlin, F. A. (1968) Current concepts in the etiology and treatment of the vasospasm, *Clin Neurosurg* 15: 133–160.
36. Ecker, A., Riemenschneider, P. A. (1951) Arteriographic demonstration of spasm of the intracranial arteries with special reference to saccular arterial aneurysm. *J Neurosurg* 8: 660–667.
37. Edvinsson, L., Zygmunt, D. T., Jansen, I., Eckman, R., Svendgaard N-Aa, Uddman R. (1990) Involvement of perivascular sensory fibers in the pathophysiology of cerebral vasospasm following subarachnoid hemorrhage. *J Cereb Blood Flow Metab* 10: 602–607.
38. Espinosa, F., Weir, B., Overton, T., Castor, W., Grace, M., Boisvert, D. (1984) A randomized placebo-controlled double-blind trial of nimodipine after SAH in monkeys. *J Neurosurg* 60: 1176–1185.
39. Fein, J. M., Flor, W. J., Cohan, S. L., Parkhurst, J. (1974) Sequential changes of vascular ultrastructure in experimental cerebral vasospasm. *J Neurosurg* 41: 49–58.
40. Findley, J. M., Weir, B. K. A., Kanamaru, K., Espinosa, F. (1989) Arterial wall changes in cerebral vasospasm. *Neurosurgery* 25: 736–746.
41. Folkow, B., Lofving, B. (1956) The distensibility of the systemic resistance blood vessels. *Acta Physiol Scand* 38: 37–52.
42. Folkow, B. (1964) Description of the myogenic hypothesis. *Circ Res* (Supp I): 279–287.
43. Folkow, B. (1962) Transmural pressure and vascular tone—some aspects of a controversy. *Arch Int Pharmacodyn* 139: 455–469.
44. Fox, J. L. (1979) Intracranial vasospasm: a study with iron compounds. *Surg Neurol* 11: 363–368.
45. Fraser, R. A. R., Stein, B. M., Barret, R. E., Pool, J. L. (1970) Noradrenergic mediation of experimental cerebrovascular spasm. *Stroke* 1: 356-362.
46. Fukumori, T., Tani, E., Maeda, Y., Sukenaga, A. (1983) Effects of prostacyclin and indomethacin on experimental cerebral vasospasm. *J Neurosurg* 59: 829–834.
47. Furchgott, R. F., Zwadzki, J. (1980) The obligatory role of endothelin in the relaxation of arterial smooth muscle by acetylcholine. *Nature* 288: 373–376.
48. Furchgott, R. F. (1981) The requirement for endothelial cells in the relaxation of arteries by acetylcholine and some other vasodilators. *TIPS* 7: 173–177.
49. Furchgott, R. F. (1981) The requirement for endothelial cell in the relaxation of arteries by acetylcholine and some other vasodilators. *TIPS* 7: 173–176.
50. Grotenhuis, J. A., Bettag, W., Fiebach, B. J. O., Dabir, K. (1984) Intracarotid slow bolus injection of nimodipine during angiography for treatment of cerebral vasospasm after SAH. *J Neurosurg* 61, 231–240.

51. Handa, J., Yoneda, S., Koyama, T., Matsuida, M., Handa, H. (1975) Experimental cerebral vasospasm in cats: modification by a new synthetic vasodilator YC-93. *Surg Neurol* 3: 195–199.
52. Handa, Y., Weir, B. K., Nosko, M., Mosewich, R., Tsuji, T., Grace, M. (1987) The effects of timing of clot removal on chronic vasospasm in a primate model. *J Neurosurg* 67: 558–564.
53. Hara, H., Nosko, M., Weir, B. (1986) Cerebral perivascular nerves in subarachnoid hemorrhage. *J Neurosurg* 65: 531–539.
54. Harder, D. R. (1987) Pressure-induced myogenic activation of cat cerebral arteries is dependent on intact endothelium. *Circ Res* 60: 102–107.
55. Hashi, et al. (1980) Treatment of cerebral vasospasm with large doses of hydrocortisone. In *Cerebral Arterial Spasm,* Williams & Wilkins, Baltimore/London, pp. 611–618.
56. Hayakawa, T., Waltz, A. G. (1977) Experimental subarachnoid hemorrhgae from a middle cerebral artery. Neurological deficit, intracranial pressure, blood pressures, pulse rate. *Stroke* 8: 421–426.
57. Hongo, K., Ogawa, H., Kassell, N. F., Nakagomi, T., Sasaki, T., Tsukahara, T., Lehman, M. (1988) Comparison of intraluminal and extraluminal inhibitory effects of hemoglobin on endothelium-dependent relaxation of rabbit basilar artery. *Stroke* 19: 1550–1555.
58. Hubschmann, O. R. (1987) The role of calcium in parenchymal cell injury in subarachnoid hemorrhage. *Neurol Res* 9: 265–269.
59. Hughes, J. T., Schianchi, P. M. (1978) Cerebral artery spasm. A histological study at necropsy of the blood vessels in cases of subarachnoid hemorrhage. *J Neurosurg* 48: 515–525.
60. Hwa, J. J., Bevan, J. A. (1986) Stretch-dependent (myogenic) tone in rabbit ear resistance arteries. *Am J Physiol* 250: H87–H95.
61. Hwa, J. J., Bevan, J. A. (1986) Dihydropyridine calcium agonist selectively enhance resistance artery myogenic tone. *Eur J Pharmacol* 126: 231–238.
62. Ide, K., Yamakawa, K., Nakagomi, T., Sasaki, T., Saitoh, I., Kurihara, H., Yoshizumi, M., Yazaki, Y., Takakura. (1989) The role of endothelin in the pathogenesis of vasospasm following subarachnoid hemorrhage. *Neurol Res* 11: 101–104.
63. Jansen, I., Fallgren, B., Edvinsson, L. (1989) Mechanism of action of endothelin on isolated feline cerebral arteries: *in vitro* pharmacology and electrophysiology. *J Cereb Blood Flow Metab* 9: 743–747.
64. Jackowski, A., Crockard, A., Burnstock, G., Russell, R. R., Kristek, F. (1990) The time course of intracranial pathophysiological changes following experimental subarachnoid hemorrhage in the rat. *J Cereb Blood Flow Metab* 10: 835–849.
65. Johansson, B. (1974) Deteminants of vascular reactivity. *Fed Proc* 33: 121–126.
66. Kagstrom, E., Nelson, P. E., Svendgaard, N. A. (1969) Clinical and experimental spasm of the cerebral vessels. *Excerpta Med Int Congr Ser* 193: 60.
67. Kanamaru, K., Waga, S., Kojima, T., Fujimoto, K., Itoh, H. (1987) Endothelium-dependent relaxation of canine basilar arteries. Part I: Difference between acetylcholine- and A2187-induced relaxation and involvement of lypoxygenase metabolite(s). *Stroke* 18: 932–937.
68. Kanamaru, K., Waga, S., Kojima, T., Fujimoto, K., Niwa, S. (1987) Endothelium-dependent relaxation of canine basilar arteries. Part II: Inhibition by hemoglobin and cerebrospinal fluid from patients with aneurysmal subarachnoid hemorrhage. *Stroke* 18: 938–943.
69. Kanamaru, K., Waga, S., Kojima, T., Fujimoto, K., Itoh, H., Kubo, Y. (1989) Endothelium-dependent relaxation of human basilar arteries. *Stroke* 20: 1208–1211.
70. Kanamaru, K., Weir, B., Findley, J. M., Krueger, K. A., Cook, D. (1989) Pharmacological studies on relaxation of spastic primate cerebral arteries in subarachnoid hemorrhage. *J Neurosurg* 71: 909–915.
71. Kanamaru, K., Weir, B., Findley, J. M., Grace, M., Macdonald, R. L. (1990) A dosage study of the effects of 21-aminosteroid U74006F on chronic cerebral vasospasm in a primate model. *Neurosurgery* 27: 29–38.
72. Kapp, J. P., Robertson, J. T., White, R. P. (1976) Spasmogenic qualities of prostaglandin F2a in the cat. *J Neurosurg* 44: 173–175.
73. Katusic, Z. S., Shepherd, J. T., Vanhoutte, P. M. (1987) Endothelium-dependent contraction to stretch in canine basilar arteries. *Am J Physiol* 252: H671–H673.
74. Katusic, Z. S., Shepherd, J. T., Vanhoutte, P. M. (1984) Vasopressin causes endothelium-dependent relaxation of the canine basilar artery. *Circ Res* 55: 575–579.
75. Kim, P., Yaksh, T. L., Romero, S. D., Sundt, T. N. (1987) Production of uric acid in cerebrospinal fluid after subarachnoid hemorrhage in dogs: investigation of the possible role of xanthine oxidase in chronic vasospasm. *Neurosurgery* 21: 39–44.
76. Kim, P., Sundt, T. N., Vanhoutte, P. M. (1989) Alterations of mechanical properties in canine basilar arteries after subarachnoid hemorrhage. *J Neurosurg* 71: 430–436.
77. Kokubu, K., Tani, E., Nakano, M., Minami, N., Shindo, H. (1989) Effects of ML-9 on experimental delayed cerebral vasospasm. *J Neurosurg* 71: 916–922.

78. Kowalski, E. (1968) Fibrinogen derivatives and biological activities. *Semin Hemat* 5: 45–59.
79. Krueger, C., Weir, B., Nosko, M., Cook, D., Noris, S. (1985) Nimodipine and chronic vasospasm in monkeys: Part 1. Pharmacological studies of vessels in spasm. *Neurosurgery* 16: 137–140.
80. Kuwayama, A., Zervas, N. T., Belson, R., Shintani, A., Picekran, K. (1972) A model for experimental cerebral arterial spasm. *Stroke* 3: 49–56.
81. Landau, B., Ransohoff, J. (1968) Prolonged cerebral vasospasm in experimental subarachnoid hemorrhage. *Neurology* 18: 1056–1065.
82. Lewis, P. J., Weir, B., Nosko, M., Tanabe, T., Grace, M. (1988) Interathecal nimodipine therapy in a primate model of chronic cerebral vasospasm. *Neurosurg* 22: 492–500.
83. Liszczak, T. M., Varsos, V. G., Black, P. M., et al. (1983) Cerebral arterial constriction after experimental subarachnoid hemorrhage is associated with blood components within the arterial wall. *J Neurosurg* 58: 18–26.
84. Lobato, R., Marin, J., Salaices, M., Burgos, J., Ravilla, F., Garcia, A. (1980) Effects of experimental subarachnoid hemorrhage on the adrenergic innervation of cerebral arteries. *J Neurosurg* 53: 477–479.
85. Maeda, Y., Tani, E., Miyamoto, T. (1981) Prostaglandin metabolism in experimental cerebral vasospasm. *J Neurosurg* 55: 779–785.
86. Matsui, T., Takuwa, Y., Johsita, H., Yamashita, K., Asano, T. (1991) Possible role of protein kinase C-dependent smooth muscle contraction in the pathogenesis of chronic cerebral vasospasm. *J Cereb Blood Flow Metab* 11: 143–149.
87. Mayber, M. R., Hauser, W., Sundt, T. M. (1978) Ultrastructural changes in feline arterial endothelium following subarachnoid hemorrhage. *J Neurosurg* 48: 49–57.
88. Miao, F. J. P., Lee, T. J. F. (1989) Effects of bilirubin on cerebral arterial tone *in vitro*. *J Cereb Blood Flow Metab* 9: 666–674.
89. Morgan, H., White, R. P., Pennink, M., Robertson, J. A. (1972) Prostaglandins and experimental cerebral vasospasm. *Surg Forum* 23: 447–448.
90. Nagasawa, S., Handa, H., Okumura, A., Naruo, Y., Moritake, K., Hayashi, K. (1979) Mechanical properties of human cerebral arteies. Part I: Effects of age and vascular smooth muscle activation. *Surg Neurol* 12: 297–304.
91. Nagasawa, S., Handa, H., Okumura, A., Naruo, Y., Okamoto, S., Moritake, K., Hayashi, K. (1980) Mechanical properties of human cerebral arteries. Part II: Vasospasm. *Surg Neurol* 14: 285–290.
92. Nagasawa, S., Handa, H., Naruo, Y., Moritake, K., Hayashi, K. (1982) Experimetal cerebral vasospasm. Arterial wall mechanics and connective tissue composition. *Stroke* 13: 595–600.
93. Nakagomi, T., Kassell, N. F., Sasaki, T., Fujiwara, S., Lehman, M., Nazar, G., Torner, J. C. (1987) Effects of SAH on endothelial-dependent vasodilation. *J Neurosurg* 66: 915–923.
94. Nosko, M., Weir, B., Krueger, C., Cook, D., Noris, S., Overton, T., Boisvert, D. (1985) Nimodipine and chronic vasospasm in monkeys. Part 1. Clinical and radiological findings. *Neurosurgery* 16: 129–136.
95. Ohmoto, T., Yoshida, J., Shibata, K., et al. (1979) Cerebral vasospasm. *Neurol Med Chir* 19: 73–86.
96. Ohta, T., Kajikawa, H., Funatsu, N., Yoshikawa, Y., Someda, K. (1980) Cerebral vasospasm and its relaxation responses to vasodilators: pathological study of severe prolonged vasospasm. In *Cerebral Arterial Spasm*. Williams & Wilkins, Baltimore/London, pp. 132–138.
97. Osaka, K. (1977) Prolonged vasospasm produced by the breakdown products of erythrocytes. *J Neurosurg* 47: 403–411.
98. Osbahr, A. J., Gladner, J. A., Laki, K. (1964) Studies on the physiological activity of the peptide released during the fibrinogen-fibrin conversion. *Biochem Biophys Acta* 86: 535–542.
99. Peerless, S. J., Yasargil, M. G. (1971) Adrenergic innervation of the cerebral blood vessels in the rabbit. *J Neurosurg* 35: 148–154.
100. Peninnk, M., White, R. P., Crockerell, J. R. (1972) Role of PGF2a in genesis of experimental cerebral vasospasm. Angiographic study in dogs. *J Neurosurg* 37: 398–406.
101. Peterson, E. W., Cardoso, E. R. (1983) The blood brain barrier following experimental subarachnoid hemorrhage. *J Neurosurg* 58: 338–344.
102. Peterson, L. H., Jensen, R. E., Parnell, J. (1960) Mechanical properties of arteries *in vivo*. *Circ Res* 8: 622–639.
103. Peterson, L. H. (1962) Properties and behaviour of living vascular wall. *Physiol Rev* 42: 309–322.
104. Peterson, L. H. (1966) Physiological factors which influence vascular caliber and blood flow. *Circ Res* (Suppl): I-3–13.
105. Quintana, L., Konda, R., Ishibashi, Y., et al. (1982) The effect of prostacyclin on cerebral vasospasm. An experimental study. *Acta Neurochir* 62: 187–193.
106. Rasmussen, H., Takuwa, Y., Park, S. (1987) Protein kinase C in the regulation of smooth muscle contraction. *FASEB J* 1: 177–185.
107. Roach, M. R., Burton, A. C. (1957) The reason for the shape of the distensibility curves of arteries. *Can J Biochem Physiol* 35: 681–690.

108. Sakaki, S., Ohue, S., Kohno, K., Takeda, S. (1988) Impairment of vascular reactivity and changes in intracellular calcium and calmodulin levels of smooth muscle in canine basilar arteries after subarachnoid hemorrhage. *Neurosurgery* 25: 753–761.
109. Sasaki, T., Asano, T., Sano, K. (1980) Cerebral vasospasm and free radical reactions, *Neuro Med Chir* 20: 145–153.
110. Sasaki, T., Wakai, S., Asano, T., Watanabe, T., Kirino, T., Sano, K. (1981) The effects of lipid hydroperoxide of arachidonic acid on the canine basilar artery: an experimental study on cerebral vasospasm. *J Neurosurg* 54: 357–365.
111. Sasaki, T., Murota, S., Wakai, S., Asano, T., Sano, K. (1981) Evaluation of prostaglandin biosynthetic activity in canine basilar artery following subarachnoid injection of blood. *J Neurosurg* 55: 771–778.
112. Sasaki, T., Wakai, S., Asano, T., Takakura, K., Sano, K. (1982) Prevention of cerebral vasospasm after SAH with thromboxane synthetase inhibitor, OKY 1581. *J Neurosurg* 57: 74–82.
113. Sasaki, T., Asano, T., Takakura, K., Sano, K., Kassell, N. F. (1984) Nature of the vasoactive substance in CSF from patients with subarachnoid hemorrhage. *J Neurosurg* 60: 1186–1191.
114. Sasaki, T., Kassell, N. F., Torner, J. C., Maixner, W., Turner, D. M. (1985) Pharmacological comparison of isolated monkey and dog cerebral arteries. *Stroke* 16: 482–489.
115. Sasaki, T., Kassell, N. F., Yamashita, M., Fujiwara, S., Zuccarello, M. (1985) Barrier disruption in the major cerebral arteries following experimental subarachnoid hemorrhage. *J Neurosurg* 63: 433–440.
116. Sasaki, T., Kassell, N. F., Zuccarello, M., Nakagomi, T., Fujiwara, S., Colohan, A., Lehman, M. (1986) Barrier disruption in the major cerebral arteries during the acute stage after experimental subarachnoid hemorrhage. *Neurosurgery* 19: 177–184.
117. Seety, B. N. Y., Graeber, J. E., Stuart, M. J. (1987) The mitogenic effect of 12- and 15-hydroxyeicosatetraenoic acid on endothelial cells may be mediated via diacylglycerol kinase inhibition. *J Biol Chem* 262: 17613–17622.
118. Shibuya, M., Suzuki, Y., Takayasu, M., Asano, T., Harada, T., Ikegami, I., Satoh, S., Hidaka, H. (1988) The effects of intracellular calcium antagonist HA 1077 on delayed cerebral vasospasm in dogs. *Acta Neurochir* 90: 53–59.
119. Shimizu, T., Watanabe, T., Asano, T., Seyama, Y., Takakura K. (1988) Activation of the arachidonate 5-lipoxygenase pathway in the canine basilar artery after experimental subarachnoidal hemorrhage. *J Neurochem* 51: 1126–1131.
120. Shirahase, H., Usui, H., Shimaji, H., Kurahasi, K., Fujiwara, M. (19xx) Endothelium-dependent contraction induced by platelet-derived substances in canine basilar arteries. *J Pharmacol Exp Ther* 255: 182–186.
121. Simeone, F. A., Ryan, K. G., Cotler, J. R., et al. (1968) Prolonged experimental cerebral vasospasm. *J Neurosurg* 29: 357–366.
122. Simeone, F. A., Vinall, P. (1975) Mechanisms of contractile response of cerebral artery to externally applied fresh blood. *J Neurosurg* 43: 37–47.
123. Sparks, H. V., Bohr, D. F. (1962) Effects of stretch on passive tension and contractility of isolated vascular muscle. *Am J Physiol* 202: 835–840.
124. Sugawa, M., Koide, T., Naitoh, S., Takato, M., Matsui, T., Asano, T. (1991) Phorbol 12,13-diacetate-induced contraction of the canine basilar artery: role of protein kinase C. *J Cereb Blood Flow Metab* 11: 135–142.
125. Summerville, L. E., Hartshorne, D. J. (1986) Intracellular calcium and smooth muscle contraction. *Cell Calcium* 7: 353–364.
126. Suzuki, Y., Shibuya, M., Takayasu, Y., Asano, T., Ikegaki, I., Satoh, S., Saito, M., Hidaka, H. (1988) Protein kinase activity in canine basilar arteries after subarachnoid hemorrhage. *Neurosurgery* 22: 1028–1031.
127. Takayasu, M., Suzuki, Y., Shibuya, M., Asano, T., Kanamori, M., Okada, T., Kageyama, N., Hidaka, H. (1986) The effects of HA compound calcium antagonists on delayed cerebral vasospasm in dogs. *J Neurosurg* 65: 80–85.
128. Tanishima, T. (1980) Cerebral vasospasm: contractile activity of hemoglobin in isolated canine basilar arteries. *J Neurosurg* 53: 787–793.
129. Toda, N., Hayashi, S., Hatano, Y. (1976) Length-tension relationship in isolated canine cerebral arteries. *Jpn J Pharmacol* 26: 129–131.
130. Toda, N., Ozaki, T., Ohta, T. (1977) Subarachnoid hemorrhage and vasospasm in dogs. *J Neurosurg* 46: 296–303.
131. Toda, N., Miyazaki, M. (1978) Responses of isolated dog cerebral and peripheral arteries to prostaglandins after application of aspirin and polyphloretin phosphate (antagonize the effects PGE2). *Stroke* 9: 490–498.
132. Toda, N., Hatano, Y., Hayashi, S. (1978) Modifications by stretches of the mechanical responses of isolated cerebral and extracerebral arteries to vasoactive agents. *Pflugers Arch* 374: 73–77.
133. Toda, N., Shimizu, K., Ohta, T. (1980) Mechanism of cerebral arterial contraction induced by blood constituents. *J Neurosurg* 53: 312–322.
134. Tsukahara, T., Kassell, N. F., Hongo, K., Vollmer, D. G., Ogawa, H. (1989) Muscarinic cholinergic receptors on the endothelium of human cerebral arteries. *J Cereb Blood Flow Metab* 9: 748–753.

135. Uemura, Y., Sugimoto, T., Okamoto, S., Handa, H., Mizuno, N. (1987) Changes of neuropeptide immunoreactivity in cerebrovascular nerve fibers after experimentally produced SAH. *J Neurosurg* 66: 741–747.
136. Vanhoutte, P. M., Rubanyi, G. M., Miller, V. M., Houston, D. S. (1986) Modulation of vascular smooth muscle contraction by the endothelium. *Annu Rev Physiol* 48: 307–320.
137. Vanhoutte, P. M. (1987) Endothelium-dependent contractions in arteries and veins. *Blood Vessels* 24: 141–144.
138. Vanhoutte, P. M., Shimaoka, H. (1989) Endothelium-derived relaxing factor and coronary vasospasm. *Circulation* 80: 1–9.
139. Varsos, V. G., Liszczak, T. M., Han, D. H., Kistler, J. P., Vielma, J., Black, P., Hero, R. C., Zervas, N. T. (1983) Delayed cerebral vasospasm is not reversible by aminophylline, nifedipine, or papaverine in a 'two-hemorrhage' canine model. *J Neurosurg* 58: 11–17.
140. Vollmer, D. G., Kassell, N. F., Hongo, K., Ogawa, H., Tsukahara, T. (1989) Effect of the nonglucocorticoid 21-aminosteroid U74006F on experimental cerebral vasospasm. *Surg Neurol* 31: 190–194.
141. Vorkapic, P., Bevan, R. M., Bevan, J. A. (1990) Pharmacological irreversible narrowing in chronic cerebrovasospasm in rabbits is associated with functional damage. *Stroke* 21: 1478–1484.
142. Watanabe, T., Asano, T., Shimizu, T., Seyama, Y., Takakura, K. (1988) Participation of lipoxygenase products from arachidonic acid in the pathogenesis of cerebral vasospasm. *J Neurochem* 50: 1145–1150.
143. Waters, A., Harder, D. R. (1985) Altered membrane properties of cerebral vascular smooth muscle following subarachnoid hemorrhage: an electrophysiological study. I. Changes in resting membrane potential (E_m) and effects on the electrogenic pump potential contribution to E_m. *Stroke* 16: 990–997.
144. Weir, B., Erasm, R., Miller, J., et al. (1976) Vasospasm in response to repeated subarachnoid hemorrhage in the monkey. *J Neurosurg* 23: 395–406.
145. Wellum, G. R., Irvine, T. W., Jr., Zervas, N. T. (1980) Dose responses of cerebral arteries of the dog, rabbit, and man to human hemoglobin *in vitro*. *J Neurosurg* 53: 486–4xx.
146. White, R. P. (1979) Multiple origins of cerebral vasospasm. In Cerebrovascular Diseases, Eleventh Princeton Conference, eds. Price TK, Nelson E, Raven Press, New York, pp. 307–319.
147. White, R. P. (1986) Vasodilator proteins: role in delayed cerebral vasospasm. *Stroke* 17: 207–213.
148. White, R. P., Robertson, J. T. (1987) Pharmacodynamic evaluation of human cerebral arteries in the genesis of vasospasm. *J Neurosurg* 33: 260–269.
149. Wilkins, R. H. (1980) Attempted prevention or treatment of intracranial arterial spasm. A survey. *Neurosurgery* 6: 198–210.
150. Wilkins, R. H. (1986) Attempts at prevention or treatment of intracranial arterial spasm: an update. *Neurosurgery* 18: 808–825.
151. Yamaguchi, T., Waltz, A. G. (1971) Effects of subarachnoid hemorrhage from puncture of the middle cerebral artery on blood flow and vasculature of the cerebral cortex. *J Neurosurg* 35: 664–671.
152. Yamamoto, L., Feindel, W., Wolfe, L. S., Katoh, H., Hodge, C. P. (1972) Experimental vasoconstriction of cerebral arteries by prostaglandins, *J Neurosurg* 37: 385–396.
153. Yanagisawa, M., Hiroki, K., Kimura, S., Tamobe, Y., Kobayashi, M., Mitsui, Y., Yazaki, Y., Masaki, T. (1988) A novel potent vasoconstrictor peptide produced by vascular cells. *Nature* 332: 411–415.
154. Yokota, M., Tani, E., Maeda, Y., Kokubu, K. (1987) Effect of 5-lipoxygenase inhibitor on experimental delayed cerebral vasospasm. *Stroke* 18: 512–518.

B. Head Injury Models

Chapter 14

THE MOUSE HEAD INJURY MODEL: UTILITY IN THE DISCOVERY OF ACUTE CEREBROPROTECTIVE AGENTS

Edward D. Hall

TABLE OF CONTENTS

I. INTRODUCTION

Over the past several years there has been considerable drug discovery activity directed at the identification of agents for the early treatment of brain injury. Ideally, such an effort would include the availability of an *in vivo* model that could serve as a rapid screen for agents that would exert a cerebroprotective effect on the injured brain. Multiple models of acute head injury have been developed involving concussion,[1] contusion,[2] or fluid percussion[3] brain trauma in rats or cats. However, each of these, while having been employed to evaluate the therapeutic potential of several types of compounds, are too complicated and labor-intensive to be practical for routine application as a primary neuroprotection screen. Therefore, we developed a simple model of concussive head injury in mice that has proven quite useful for the rapid dose-response evaluation of significant numbers of compounds in terms of their ability to promote neurological recovery during the first posttraumatic hour. Moreover, it has been employed to evaluate longer term survival and the neurochemical and physiological mechanisms of brain injury together with the effects of selected neuroprotective drug candidates on those mechanisms including posttraumatic brain levels of hydroxyl radicals ($^{\bullet}OH$) and eicosanoids (e.g., $PGF_{2\alpha}$, PGE_2, 6-keto $PGF_{1\alpha}$, TXA_2) and alterations in blood-brain barrier (BBB) permeability.

0-8493-8094-4/95/$0.00+$.50

This chapter reviews the application of the model in the evaluation of the cerebroprotective action of pharmacological compounds using the novel 21-aminosteroid (lazaroid) lipid antioxidant tirilazad mesylate (U-74006F) as an example.[4]

II. METHODS

A. DESCRIPTION OF THE MODEL

Male CF-1 mice (Charles River, Portage, MI) weighing 17 to 21 g and fed (Ralston Purina Laboratory Chow #5001) *ad libitum* were employed. The head injury apparatus is depicted in Figure 1. Briefly, each mouse was held by the dorsal skin of the neck, and its head placed firmly on the metal baseplate of the injury apparatus. A Teflon impounder was lowered onto the center of the head. The striking surface of the impounder was flat and round and had a 6-mm diameter. A 50 g stainless steel weight was released by a pin. The weight fell a distance of 17, 18, or 19 cm depending on the desired magnitude of the injury. In the case of the 18-cm distance, this translated into a 900 g · cm injury force (18 cm × 50 g). Anesthesia was not required since this injury consistently caused an immediate loss of consciousness as judged by lack of righting reflex when the animal was placed on its back.

B. ASSESSMENT OF EARLY NEUROLOGICAL RECOVERY AND LONG TERM SURVIVAL

At 1 h postinjury, the neurological status of the mice was assessed by means of a grip test. The mice were picked up by the tail and their forepaws placed in contact with a taut

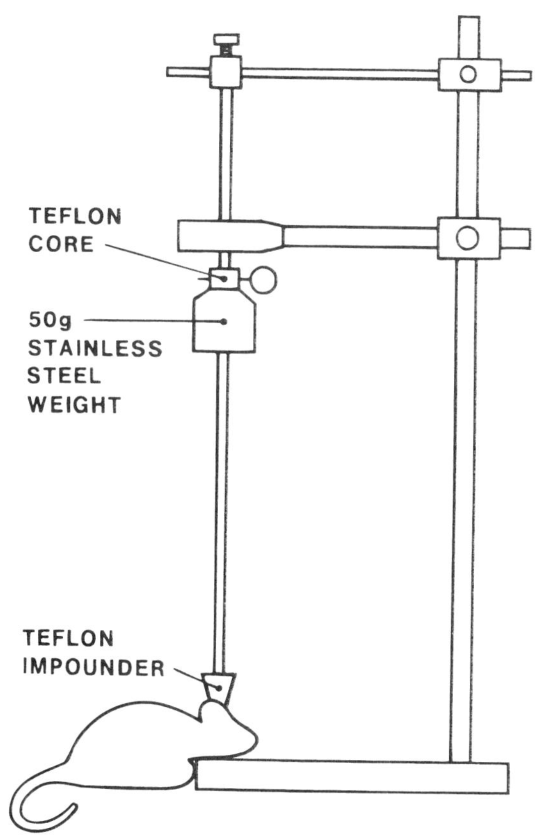

FIGURE 1. Diagram of the mouse concussive head injury apparatus.

string 60 cm in length that was tied to two upright metal bars 40 cm above a padded table. Care was taken so that both front paws came in contact with the string, allowing each mouse an equal chance to grasp the string. The tail was gently released, at which time the mouse either fell due to inability to hang on or remained on the string. The length of time that the mice could remain on the string in some manner (e.g., using one to four paws, tail, or paws plus tail) was measured with a 30-s maximum. The actual definition of injury severity was based on the 1-h postinjury grip score. A moderate injury was defined as a mean grip score in untreated (or vehicle treated) mice of >7 s but <12 s. A severe injury was defined by a mean grip score of ≤7 s. The severity of the injury could be manipulated by either changing the distance that the weight was dropped or the weight of the mice. Mice weighing 17 to 19 g manifest lower grip scores than mice weighing 19 to 21 g when subjected to the same weight-drop injury.

C. DRUG TESTING AND ADMINISTRATION

In testing potentially cerebroprotective compounds, a group of 20 to 25 mice was injured in rapid succession. In the case of severely injured mice, the immediate mortality was approximately 30%. At 3 to 5 min postinjury, the surviving mice were injected via one of the two tail veins with either vehicle or test drug. Dilation of the tail vein by gentle warming in a water bath facilitated IV injection using a 27-gauge needle. The volume of injection was kept constant at 0.1 ml. In the case of studies of long-term survival, subsequent injections were made proximal to the first injection site or in the opposite tail vein. Typically, a vehicle group was compared with two to five groups treated with varying doses of test compound. All drug administrations and neurological evaluations were carried out in a blinded fashion.

If water solubility of the test compound was not a problem, physiological saline (0.09% sodium chloride) is the most appropriate vehicle for use. However, U-74006F, which is a weak base that is minimally soluble at neutral pH, was dissolved in either 0.01 *N* hydrochloric acid or an acid buffer consisting of 0.02 *M* citric acid, 0.003 *M* sodium citrate, and 0.08 *M* sodium chloride.

D. SALICYLATE HYDROXYL RADICAL TRAPPING METHOD

Brain ˙OH levels were measured via the salicylate trapping method.[5] Salicylate (sodium salt, 300 mg/kg) was administered IP 15 min before sacrificing the animals. At set times after head injury (and 15 min after salicylate administration), the mice were decapitated and the brains quickly removed, frozen in liquid nitrogen and stored at −70°C until time of assay. Each frozen brain, with cerebellum and brainstem removed, was placed in a 1.5-ml microcentrifuge tube containing three 4-mm glass balls and 500 μl ethanol (1% ascorbate). The tissue was minced and homogenized in a vortex mixer for 2 min. The homogenate was centrifuged for 2 min at 14,000. The supernatant (100 μl) was injected onto a BAS ODS 5-μm column with a Perkin-Elmer ISS-100 autosampler. Each liter of mobile phase consisted of 15% acetonitrile, 10% tetrahydrofuran, and 14.15 g chloroacetic acid. All samples were measured against an external standard of 2,5-dihydroxybenzoic acid (DHBA) and salicylic acid. The column effluent flowed into a BAS electrochemical detector set at 650 mV, and then into a UV detector set at 295 nm. The DHBA was detected electrochemically, and salicylate was detected spectro-photometrically (UV). The data was expressed as the ratio of DHBA to salicylate to normalize the DHBA levels to differing brain concentrations of salicylate. Peaks were integrated with Waters Maxima 820 software.

E. MEASUREMENT OF EICOSANOIDS

The mouse brains (minus the cerebellum and brainstem) were weighed and homogenized in 2.4 ml methanol with an Ultra-Turrax Tissue-Mizzer (Tekmar Co., Cincinnati, OH). Added to the homogenate was 1.2 ml of 0.1 M potassium phosphate buffer (pH 7.4) and the suspension was rehomogenized, followed by centrifugation at 1500 *g* to remove the particulate matter. The supernatants were diluted with an appropriate volume of distilled water, adjusted to a final pH of 3.5 to 4.0 with 1 *M* phosphoric acid, and loaded onto C18 reverse-phase Sep-Pak cartridges (Waters, Bedford, MA), which had been prepared by washing with 1 ml methanol, followed by a slow percolation with 5 ml 0.5% EDTA (pH 7.4, in 15% aqueous methanol). The samples were washed onto the Sep-Pak with 5 ml of the 15% aqueous methanol, and extracted by passing 5 ml of 30% aqueous methanol, followed by 5 ml methanol through the cartridges. The 30% methanol and the methanol effluents were pooled and evaporated to dryness under vacuum in a centrifuge dryer (Savant Instruments, Hicksville, NY) and dissolved in 2 ml enzyme immunoassay (EIA) buffer (0.1 *M* potassium phosphate buffer, pH 7.4, with: 0.01% sodium azide, 0.4 *M* sodium chloride, 0.852 m*M* EDTA, and 0.1 *M* bovine serum albumin). The samples were stored frozen until analyzed by EIA for eicosanoids.

EIAs for PGF and TXB_2 were carried out using reagents and antibodies from Caymen Chemical Co. (Ann Arbor, MI). However, the radioimmunoassay kit for (^{125}I)-PGE_2 was purchased from Dupont NEN Research Products (Boston, MA). For all EIAs, 50 μl of the samples or appropriate amount of the standard eicosanoid solutions were pipetted into a 96-well immunoplate (MaxiSorbF96, Inter Med Nunc, Roskilde, Denmark), which had been previously coated with monoclonal mouse anti-rabbit IgG (2 μg per well). The coated plates were washed three times with a M96V Titertek microtiter plate washer (ICN Flow, McLean, VA). To each well, 50 μl of acetylcholine esterase (ACE)-linked eicosanoid conjugate and 50 μl of appropriate specific antibody were added, and the plates shaken and incubated overnight at room temperature. The following morning, the plates were washed as before. Added to each well were 200 ml of a freshly prepared mixture of 0.5 m*M* Ellman's reagent [5,5′-dithiobis(2-nitrobenzoate)] and 0.69 m*M* of ACE substrate (acetylthiocholine iodide) in 0.01 *M* potassium phosphate buffer, pH 7.4, and the plates were shaken in the dark for 1 to 18 h until adequate color was developed. The absorbance at 412 nm was determined using a Titertek 96-well plate reader, and the concentrations calculated by the Beckman software package EIA/RIA Immunofit. Concentrations of the eicosanoids were expressed as mean nanograms per gram wet weight.

F. MEASUREMENT OF BLOOD BRAIN BARRIER PERMEABILITY

Mice were injected via a tail vein with ^{14}C-methylated bovine serum albumin in a volume of 0.1 ml (1,000,000 DPM/ml) 5 min before sacrifice. Sacrifice was carried out via cervical dislocation and exsanguination 5 min after albumin injection. Separate groups of six mice were injured and sacrificed at 5, 15, 30, or 60 min postinjury. The brains were weighed, dried, and digested, and the amount of radiolabeled albumin measured in a scintillation counter. The amount of albumin was expressed as microgram-equivalents per gram wet weight.

III. RESULTS

A. EFFECTS OF U-74006F ON EARLY NEUROLOGICAL RECOVERY AND ONE-WEEK SURVIVAL

Initial studies of the efficacy of U-74006F in acute head injury were carried out to determine the ability of the compound to improve early neurological recovery and survival of severely head-injured mice.[6] Administration of a single IV dose of U-74006F produced a significant improvement in the 1-h postinjury neurological status (grip test score) over a broad range (0.003 to 30 mg/kg). A 1-mg/kg IV dose given within 5 min and again at 1.5 h after a severe injury, in addition to improving early recovery, also increased the one-week survival to 78.6% from 27.3% in vehicle-treated mice ($p < 0.02$). Figure 2 shows the dose-response curve for the acute effect of U-74006F on early neurological recovery of mice subjected to a severe concussive head injury.

B. EFFECTS OF U-74006F ON BRAIN HYDROXYL RADICAL AND EICOSANOID LEVELS

Recent data obtained from the mouse head injury model suggests that the compound may also reduce posttraumatic brain hydroxyl radical concentrations.[7–9] Administration of salicylate, a hydroxyl radical trap, leads to the posttraumatic formation of DHBA in brain tissue when it reacts with hydroxyl radical. Severe concussive head injury resulted in an increase in brain levels of salicylate-derived DHBA measured at 30 min postinjury. However, a 3-mg/kg IV dose of U-74006F produced a decrease in DHBA formation (Figure 3) indicative of either an attenuated formation or perhaps a chemical scavenging of hydroxyl radicals.

Much of the early posttraumatic increase in brain oxygen radical levels is a by-product of injury-induced arachidonic acid release and enzymatic oxidation by prostaglandin synthase.[10,11] Indeed, an increase in the brain levels of $PGF_{2\alpha}$ PGE_2, 6-keto $PGF_{1\alpha}$ and TXA_2 were observed at 1 h postinjury (Figure 4). However, U-74006F did not have any effect on

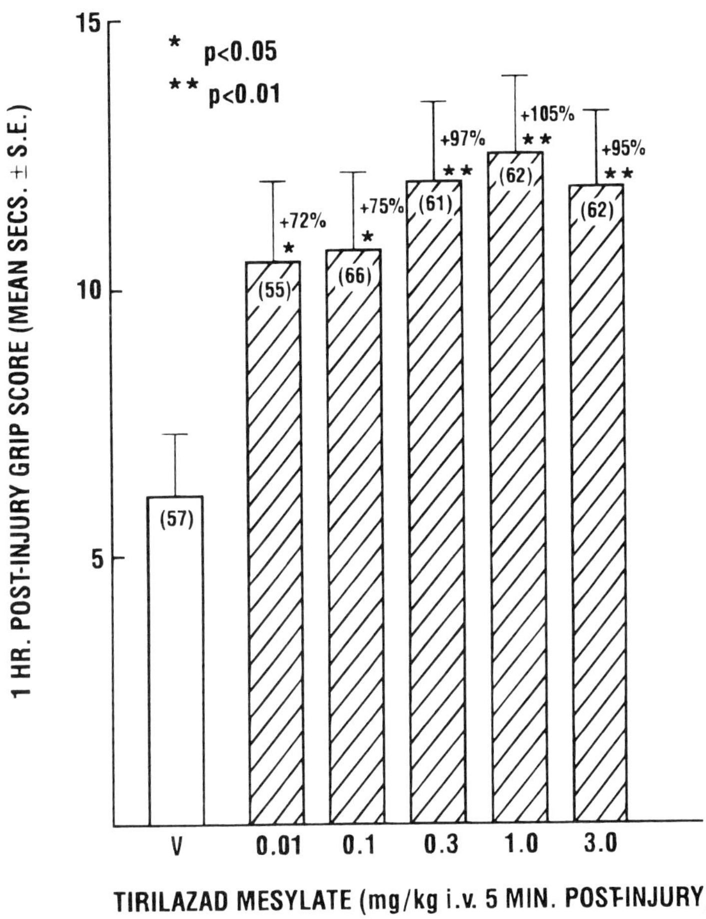

FIGURE 2. Dose-response relationship for the effect of U-74006F administered by IV bolus at 3 to 5 min postinjury on early neurological recovery (i.e., grip score) in mice subjected to a severe concussive head injury. All values are mean ± standard error. Numbers of animals are given in parentheses in each bar.

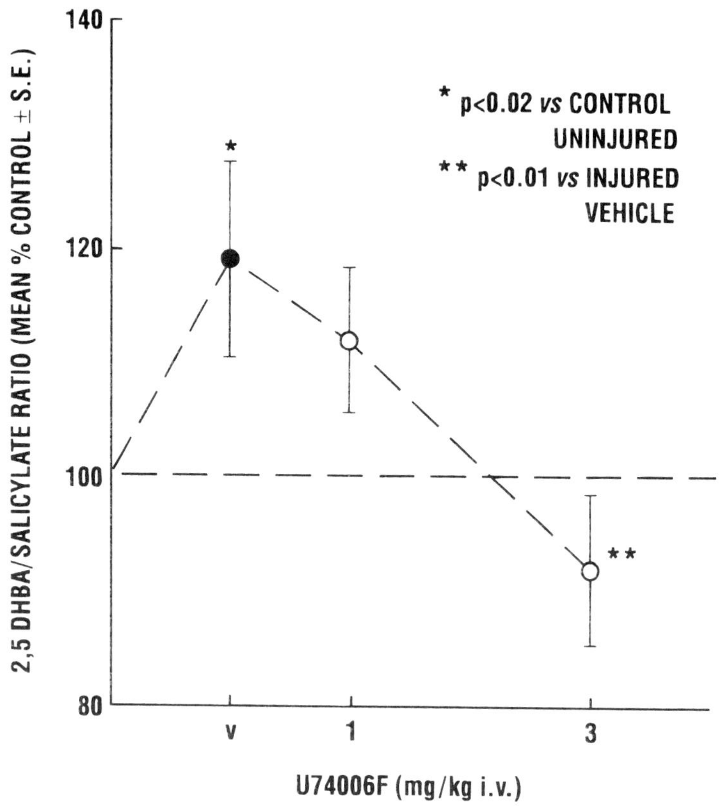

FIGURE 3. Dose-response analysis of the effect of U-74006F administered as an IV bolus at 3 to 5 min postinjury on forebrain levels of hydroxyl radical (i.e., 2,5-DHBA/salicylate) at 30 min after a severe concussive head injury. All values are mean ± standard error. N = 10 animals per group.

brain eicosanoid levels. Thus, the decrease in brain hydroxyl radical levels is independent of any effect on eicosanoid synthesis.

C. EFFECTS ON BLOOD-BRAIN BARRIER DISRUPTION

Coincident with the reduction in brain hydroxyl radical levels, U-74006F (3 mg/kg IV within 5 min postinjury) also acted to reduce posttraumatic opening of the BBB.[7] Figure 5 shows that severe concussive head injury in mice rapidly increased the uptake of ^{14}C-albumin into brain implying an increased permeability to protein. Acute U-74006F treatment, however, restored protein permeability of the barrier back to normal noninjured levels beginning with the 30 min postinjury measurement time, an effect which persists at least to 1 h postinjury.

The increased permeability of the BBB was also apparent in an increased brain uptake of the highly protein-bound U-74006F in head-injured mice compared to normal mice (Figure 6). The effect of U-74006F to reverse this leaky barrier may also be apparent in the fact that by 30 min after dosing, the brain levels of U-74006F are no longer different from those observed in normal uninjured brains. This effect of U-74006F to close the barrier may be related to the attenuation of hydroxyl radical levels or an antagonism of the damaging effects of free radicals on the barrier endothelium (e.g., lipid peroxidation). Indeed, free radicals are known to increase barrier permeability.[12] Consistent with the present data showing a reduction in posttraumatic BBB, U-74006F has been shown to attenuate post-traumatic brain edema in a rat model of fluid percussion head injury.[13]

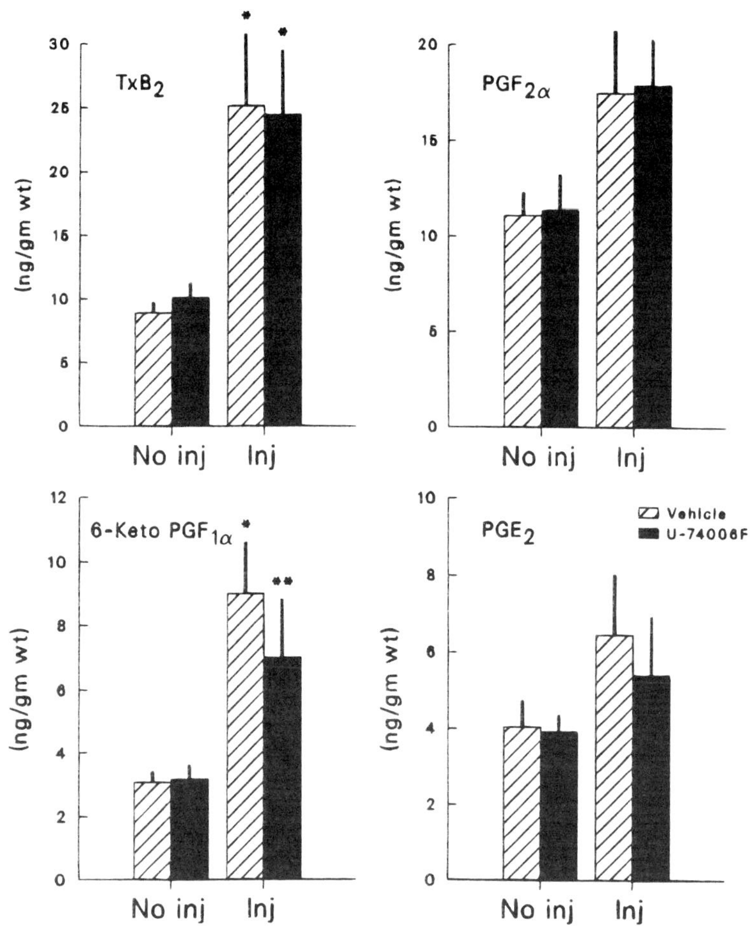

FIGURE 4. Effects of an acutely administered IV dose of U-74006F on the posttraumatic increase in the levels of $PGF_{2\alpha}$ PGE_2, 6-keto $PGF_{1\alpha}$, and TXA_2 in mouse forebrain at 60 min after a moderately severe concussive head injury. Values are mean (ng/g wet weight) ± standard error for ten animals. Asterisks indicate $p < 0.05$ vs. uninjured animals by one-way ANOVA.

IV. DISCUSSION

As the results show, the mouse concussive head injury model has been extremely useful in the determination of the cerebroprotective efficacy and mechanism of action of the 21-aminosteroid lipid antioxidant tirilazad mesylate (U-74006F). The compound improved early neurological recovery and one-week survival in the case of a severe level of injury. Mechanistic studies showed this effect to be paralleled by a reduction in brain microvascular hydroxyl radical levels and a preservation of normal BBB integrity.

In other drug discovery efforts in our laboratories, the cerebral antioxidant activity of the 21-aminosteroid U-74006F has been enhanced by replacing the steroid functionality, which possesses only weak antioxidant activity (without the complex amino substitution), with a more potent and effective antioxidant. A series of compounds were synthesized in which the steroid of U-74006F was replaced by the antioxidant ring structure (i.e., chromanol) of alpha tocopherol (vitamin E). One of these compounds, U-78517F, was demonstrated to have predictably more potent effects in regards to inhibition of lipid peroxidation *in vitro* and enhancement of early neurological recovery of head-injured mice.[14]

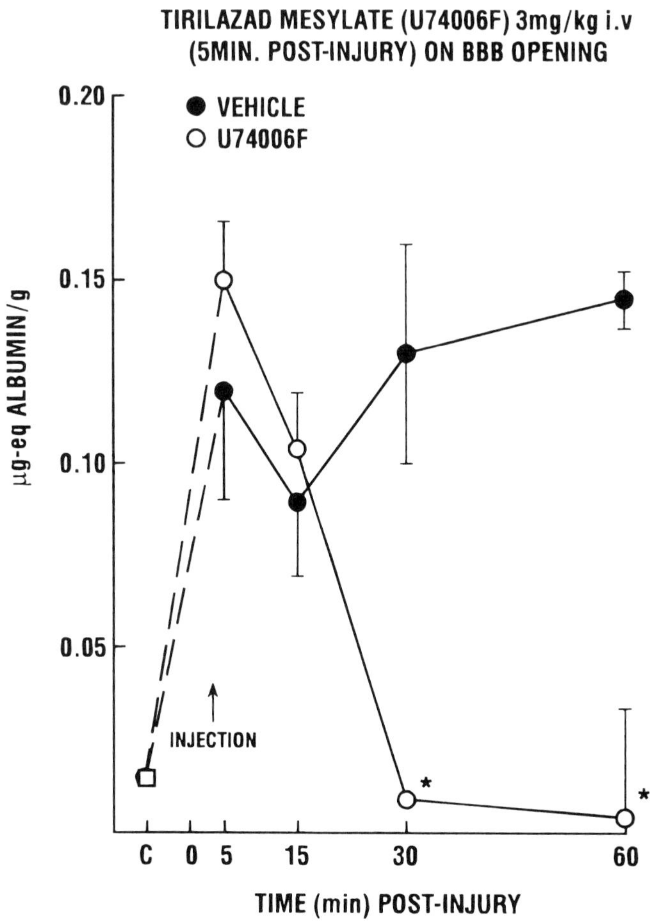

FIGURE 5. Effect of U-74006F (3 mg/kg IV at 3 min postinjury) to reverse the posttraumatic opening of the BBB (i.e., increased uptake of ^{14}C-albumin) in mice subjected to a severe concussive head injury. All values are mean ± standard error for six mice per group. Asterisks indicate $p < 0.05$ vs. vehicle-injured animals at same time point.

The mouse head injury model was first employed for evaluation of the cerebroprotective efficacy of glucocorticoid steroids such as methylprednisolone.[15,16] Methylprednisolone, when administered in 15 to 30-mg/kg IV doses, has been shown to attenuate early posttraumatic lipid peroxidation and preserve microvascular function in injured cat spinal cord.[17] Similarly, a 30-mg/kg IV dose was shown to improve early neurological recovery of mice subjected to a moderately severe concussive head injury.[15] In the same study, the glucocorticoid prednisolone was also effective, but the prototypical glucocorticoid hydrocortisone was ineffective over a wide range of doses, perhaps owing to its lack of ability to inhibit lipid peroxidation. In the case of a more severe level of injury, comparable to that employed during the evaluation of the 21-aminosteroid U-74006F[6] and the 2-methylamino chroman U-78517F,[14] a 60-mg/kg IV dose of methylprednisolone was required to produce a statistically significant increase in the 1-h postinjury grip score.[16] In that same study, the nonglucocorticoid lipid peroxidation-inhibiting steroid U-72099E was shown to be as effective as methylprednisolone.

The model has proven useful for differentiating the potency to two different prodrug forms of methylprednisolone. For instance, methylprednisolone 21-suleptanate, sodium salt

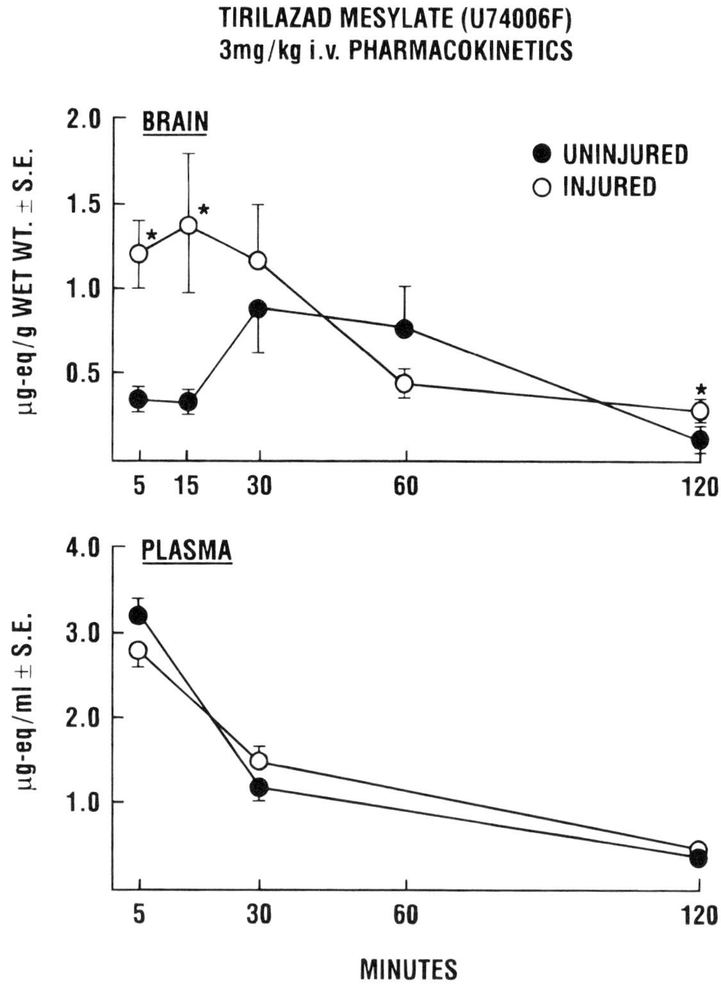

FIGURE 6. Comparative brain and plasma pharmacokinetics of ^{14}C-U-74006F in normal mice vs. mice subjected to a severe concussive head injury. In the injured mice, the compound was administered at 3 min postinjury. All values are mean ± standard error for six mice per group. Asterisks indicate $p < 0.05$ vs. uninjured animals at same time point.

(U-67590A) is four to eight times more potent than methylprednisolone 21-hemisuccinate, sodium salt in terms of improving early neurological recovery. This is probably due to the faster liberation of the free steroid from the suleptanate promoiety in comparison to the hemisuccinate.[18] An iron-chelating dextran-desferoxamine conjugate has also been observed to improve acute neurological recovery.[19]

In addition to utility in evaluating compounds with antioxidant activity, the mouse head injury model is also responsive to other types of compounds. In one study, we showed that the cyclooxygenase inhibitor ibuprofen improves early grip scores, which implies that blocking the posttraumatic increase in brain prostaglandin synthesis is associated with cerebroprotection. In contrast, the selective thromboxane synthetase inhibitor was ineffective, suggesting that blocking TXA_2 synthesis alone was insufficient to protect the injured brain.[20] Finally, the selective κ-opioid receptor agonist U-50488H was also shown to improve neurological recovery in head-injured mice.[21]

Thus, in summary, the mouse head injury model has proven useful for the initial screening of compounds that may beneficially affect the injured brain. The model is rapid, it allows for a large sample size, and is also appropriate for biochemical and physiological mechanism of action studies. In the case of the 21-aminosteroid antioxidant U-74006F, its activity in the mouse head injury as reviewed above has been confirmed in more sophisticated, labor-intensive head injury models in rats[13,22] and cats.[23] Thus, the model has good reliability for identification of compounds that may be useful for the acute treatment of head injury.

REFERENCES

1. Tornheim, P. A., Liwicz, B. H., Hirsch, C. S., Brown, D. L., McLaurin, R. L. (1983) Acute responses to blunt head trauma. Experimental model and gross pathology. *J. Neurosurg.* 59: 431–438.
2. Dixon, C. E., Clifton, G. L., Lighthall, J. W., Yaghmai, A. A., Hayes, R. L. (1991) A controlled cortical impact model of traumatic brain injury in the rat. *J. Neurosci. Methods* 39: 253–262.
3. Dixon, C. E., Lyeth, B. G., Povlishock, J. T. (1987) A fluid percussion model of experimental brain injury in the rat: neurological, physiological and histopathological characterizations. *J. Neurosurg.* 67: 110–119.
4. Hall, E. D., McCall, J. M., Means, E. D. (1994) Therapeutic potential of the lazaroids (21-aminosteroids) in CNS injury, ischemia and subarachnoid hemorrhage. *Adv. Pharmacol.* 28: 221–268.
5. Floyd, R. A., Henderson, R., Watson, J. J., Wong, P. K. (1986) Use of salicyclate with high pressure liquid chromatography and electrochemical detection (LCED) as a sensitive measure of hydroxyl free radicals in adriamycin treated rats. *Free Rad. Biol. Med.* 2: 13–18.
6. Hall, E. D., Yonkers, P. A., McCall, J. M., Braughler, J. M. (1988) Effect of the 21-aminosteroid U74006F on experimental head injury in mice. *J. Neurosurg.* 68: 456–461.
7. Hall, E. D., Yonkers, P. A., Andrus, P. K., Cox, J. W., Anderson, D. K. (1992) Biochemistry and pharmacology of lipid antioxidants in acute brain and spinal cord injury. *J. Neurotrauma* 9(Suppl. 2): 425–442.
8. Hall, E. D., Andrus, P. K., Yonkers, P. A. (1993) Brain hydroxyl radical generation in acute experimental head injury. *J. Neurochem.* 60: 588–594.
9. Althaus, J. S., Andrus, P. K., Williams, C. M., Von Voigtlander, P. F., Hall, E. D. (1993) The use of salicylate hydroxylation to detect hydroxyl radical generation in ischemic and traumatic brain injury: reversal by tirilazad mesylate (U-74006F). *Mol. Chem. Neuropathol.* 20: 147–162.
10. Kukreja, R. C., Kontos, H. A., Hess, M. L., and Ellis, E. F. (1986) PGH synthase and lipoxygenase generate superoxide in the presence of NADH and NADPH. *Circ. Res.* 59: 612–619.
11. Kontos, H. A., Povlishock, J. T. (1986) Oxygen radicals in brain injury. *CNS Trauma* 3: 257–263.
12. Greenwood, J. (1991) Mechanisms of blood-brain barrier breakdown. *Neuroradiology* 33: 95–100.
13. McIntosh, T. K., Thomas, M., Smith, D., Banbury, M. (1992) The novel 21-aminosteroid U74006F attenuates cerebral edema and improves survival after brain injury in rats. *J. Neurotrauma* 9: 33–46.
14. Hall, E. D., Braughler, J. M., Yonkers, P. A., Smith, S. L., Linseman, K. L., Means, E. D., Scherch, H. M., VonVoigtlander, P. F., Lahti, R. A., Jacobsen, E. J. (1991) U-78517F: a potent inhibitor of lipid peroxidation with activity in experimental brain injury and ischemia. *J. Pharmacol. Exp. Ther.* 258: 688–694.
15. Hall, E. D. (1985) High-dose glucocorticoid treatment improves neurological recovery in head-injured mice. *J. Neurosurg.* 62: 882–887.
16. Hall, E. D., McCall, J. M., Chase, R. L., Yonkers, P. A., Braughler, J. M. (1987) A non-glucocorticoid steroid analog of methylprednisolone duplicates its high-dose pharmacology in models of central nervous system trauma and neuronal membrane damage. *J. Pharmacol. Exp. Ther.* 242: 137–142.
17. Hall, E. D. (1992) The neuroprotective pharmacology of methylprednisolone: a review. *J. Neurosurg.* 76: 13–22.
18. Hall, E. D., Yonkers, P. A. (1989) Comparison of two ester prodrugs of methylprednisolone on early neurological recovery in a murine closed head injury model. *J. Neurotrauma* 6: 163–168.
19. Panter, S. S., Hall, E. D., Braughler, J. M. (1992) Dextran-coupled deferoxamine improves outcome in a murine model of head injury. *J. Neurotrauma* 9: 47–53.
20. Hall, E. D. (1986) Beneficial effects of acute intravenous ibuprofen on neurological recovery of head-injured mice: comparison of cyclooxygenase inhibition with inhibition of thromboxane A_2 synthetase or 5-lipoxygenase. *CNS Trauma* 2: 75–83.
21. Hall, E. D., Wolf, D. L., Althaus, J. S., VonVoigtlander, P. F. (1987) Beneficical effects of the kappa opioid receptor agonist U-50488H in experimental acute brain and spinal cord injury. *Brain Res.* 435:, 174–180.

22. Sanada, T., Nakamura, T., Nishimura, M. C., Isayama, K., Pitts, L. H. (1993) Effect of U74006F on neurologic function and brain edema after fluid percussion injury in rats. *J. Neurotrauma* 10: 65–71.
23. Dimlich, R. V. W., Tornheim, P. A., Kindel, R. M., Hall, E. D., Braughler, J. M., McCall, J. M. (1990) Effects of a 21-aminosteroid (U-74006F) on cerebral metabolites and edema after severe experimental head trauma. In: *Advances in Neurology, Vol 52,* D. Long et al., eds. Raven Press, New York, pp. 365–375.

Chapter 15

SIMPLE CONTROLLED CORTICAL IMPACT MODELS OF MOUSE HEAD INJURY

S. Tsuyoshi Ohnishi and John H. Mullen, Jr.

TABLE OF CONTENTS

I. INTRODUCTION

A closed head injury model in the mouse was developed by Hall, who employed a simple weight-drop technique to produce head injury and devised a motor function test (a grip test) to assess the degree of injury.[1] He used this model for the development of pharmacological treatments for head injury[1,2] (see also Chapter 14 of this book). Although the method is simple and effective, there were some problems in that anesthesia was not used, and that the investigator could not identify the exact location of injury because the impact was applied above the skin. In an attempt to improve the method, we developed two controlled cortical impact head injury models in the mouse—one that involves craniotomy and another that does not.

II. METHODS

A. DEVELOPMENT OF SIMPLE CONTROLLED CORTICAL IMPACT MODELS FOR THE STUDY OF MOUSE HEAD INJURY

The controlled impact method was successfully applied to the studies of rat spinal cord injury[3,4] and rat head injury.[5] We wanted to develop a controlled cortical impact method to study head injury in mice. While this method involves sophisticated mechanics and electronics,[3–5] we attempted to build a simpler instrument that could produce similar results. As

0-8493-8094-4/95/$0.00+$.50

shown in Figure 1, we used two different impounders (both made from steel nails), one with a diameter of 2.0 mm (sharp, pointed tip; 60°) and the other with a diameter of 2.5 mm (flat tip), which can freely move through holes made through the 3/8-in. thick top Plexiglas plate (acrylic sheet manufactured by Rohm and Haas, Philadelphia, PA) of the apparatus. The nail sits on the brain, either on the skull (Figure 1A) or through a hole made on the skull by craniotomy (Figure 1B). A scale was inscribed on the side of the impounder so that we could predetermine the stroke of the impounder. Then, the impounder was hit by a spring device with a velocity of 3 m/s. Since the movement of the nail was restricted by its head, the stroke could be accurately controlled. The spring device was taken from a toy called "Soft Shooters" (a toy to shoot a sponge arrow by compressing air using a spring-trigger mechanism; bought at a Kay-Bee toy store). The whole setup cost approximately $10.

B. SURGERY AND HEAD INJURY

A laboratory adjustable stand was used to support an animal holder that had an anesthesia mask (a polyethylene tube coupling), a Tygon tubing, a head holder, and a pillow (Figure 2). A mouse was anesthetized with 2% isoflurane (air as a carrier) and placed on the animal holder. A skin incision was made to expose the skull. The point of impact was located on the right parietal bone. From the midpoint of bregma and lambda (on the midline sagittal suture), we measured 2.5 mm toward the right. Muscle on the skull was scraped off and the position was marked by a marker pen. This is the position of injury (Figure 3). The head was resting on a pillow (Figure 1). A ring wire with a handle (made from a wire coat hanger) was used to push the skull onto the pillow to stabilize the head when the impact was made (Figure 4). We tested two methods of impact, i.e., impounding without and with craniotomy.

1. Noncraniotomy Method (Direct Method)

The animal holder was lifted by the laboratory stand until the sharp, pointed tip of the impounder (2 mm diameter; tip angle 60°) touched the marked position on the exposed skull.

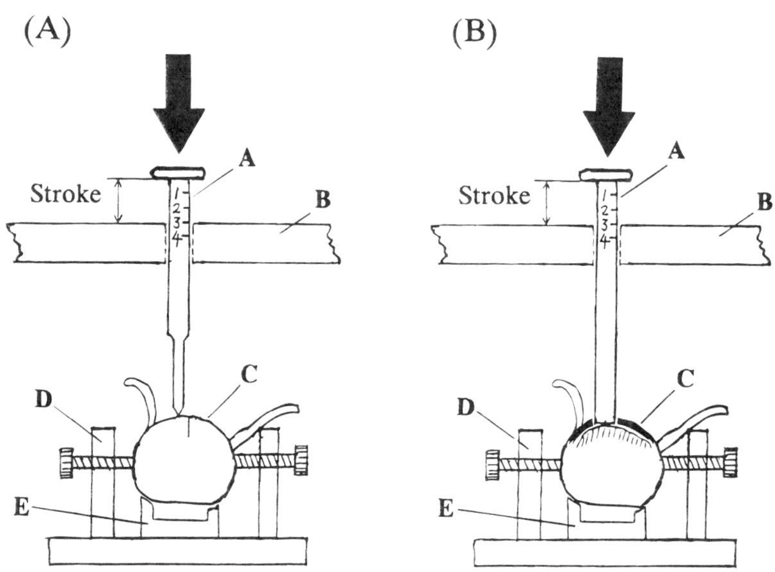

FIGURE 1. Schematic illustration of (A) a noncraniotomy method, and (B) a craniotomy method to produce controlled cortical impact injury. (Figures not to scale.) A: impounder; B: Plexiglas plate (3/8 in. thickness); C: exposed skull; D: head holder; E: pillow. Arrows on top of the impounder represent the impact derived from a spring-trigger device (not shown).

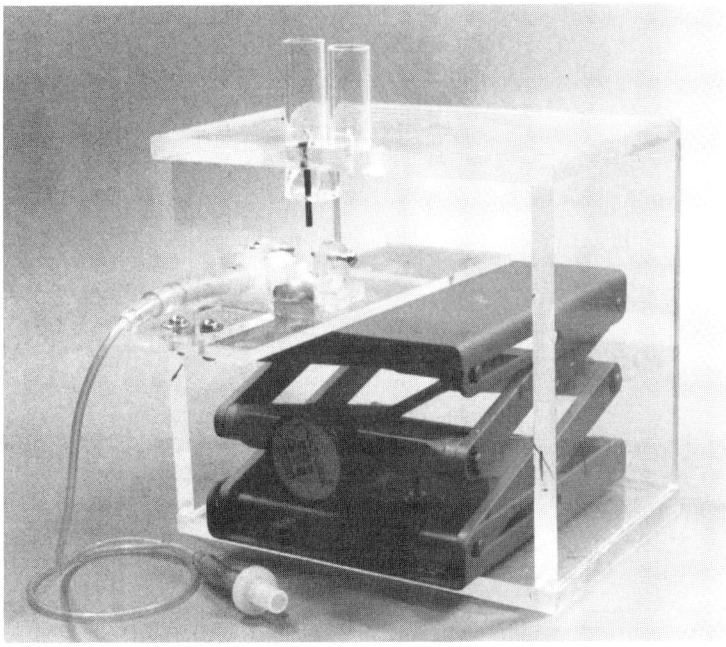

FIGURE 2. A simple controlled cortical impact apparatus for the mouse. The size of the plexiglas case was approximately 7 in. wide × 6 in. high × 5 in. deep. The animal holder with anesthesia mask tubing was supported by an adjustable laboratory stand. Two Plexiglas cylinders (for two different size impounders) on top of the case are supports for the spring device. The impounder for the craniotomy method is shown in the picture.

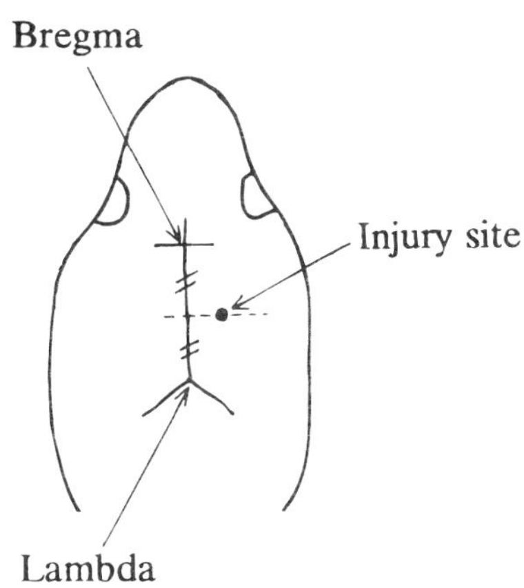

FIGURE 3. The position of injury was 2.5 mm to the right of the midpoint between bregma and lambda on the sagittal suture.

Then, the holder was further raised in order to give the impounder a desired stroke (Figure 1). The impounder was hit by the spring device. We then lowered the holder, the skin was sutured, and the animal was returned to the cage.

2. Craniotomy Method

A 3-mm-diameter hole was made through the skull using a Dremel high-speed drill. The dural membrane was left intact. The animal holder was raised until the tip of the flat-head

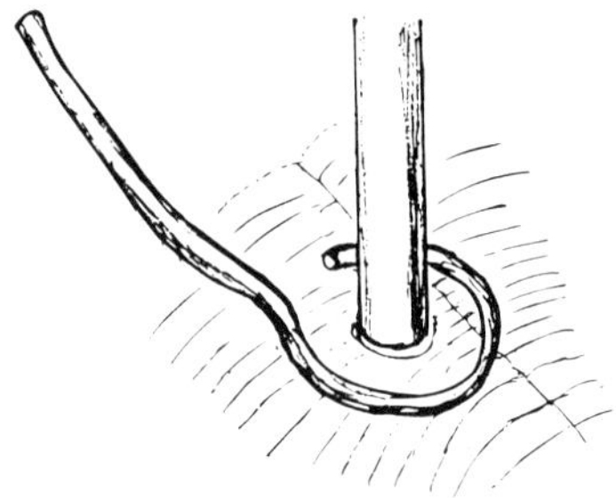

FIGURE 4. The head was pushed onto a pillow by pressing with a wire. The impounder for the craniotomy method is shown.

impounder (diameter 2.5 mm) touched the dural membrane. Then, the holder was gently lifted by the laboratory stand so that the brain pushed up the impounder to the desired stroke. The impounder was hit by the spring device to cause head injury. A small piece of gel-foam was placed to cover the craniotomy, and the skin was sutured.

C. CHRONIC MOTOR DEFICIT EVALUATIONS

The deficits in motor function were evaluated by two methods: a grip test and a beam test. The grip test (Figure 5 top) developed by Hall[1] was used with minor modifications. The modified portions of the criteria are shown in bold letters in Table 1. A beam test was developed that involves the measurements of balance and motor activities of a mouse on a slippery Plexiglas beam during the 60-s test period. The size of the testing portion of the beam was 4.5 mm wide × 150 mm high × 600 mm long (Figure 5 bottom), but the actual length of the beam was 1000 mm, because a mouse cage was attached on each end of the beam. The details of the test are described in Table 2 and Figure 6. In both tests, 5 is the normal motor function of the control mouse, and 0 of completely paralyzed mouse. The total motor score, which was the summation of both tests, gave the index of motor function, in which 10 was normal, and 0 was paralyzed.

III. RESULTS

A. APNEIC EPISODE

1. With the Noncraniotomy Method (Direct Method)

Mice did not normally suffer from an apneic episode after head injury without craniotomy. Occasionally, a few seconds of apnea occurred after impact, but the mouse recovered quickly to normal respiration. This suggests that the respiratory center was not located under our injury site. Ten minutes after injury, most animals walked around in their cages.

2. With the Craniotomy Method

Apnea was common in this model, especially when the stroke of the impounder was 3 mm. The duration of the apneic episode with a 3-mm stroke was 20 to 40 s. We applied CPR to the mouse while the subject was in apnea. After injury, most animals were quiet, and did not walk around at least for an hour.

B. CHRONIC MOTOR DEFICIT EVALUATIONS

The deficits in motor function of head-injured mice and their recovery are shown in Figure 7 (without craniotomy) and Figure 8 (with craniotomy). In both methods, the larger the impact, the lower the total motor score was. Data showed that the recovery after head injury was relatively quick in the noncraniotomy method, but it was slow in the craniotomy method.

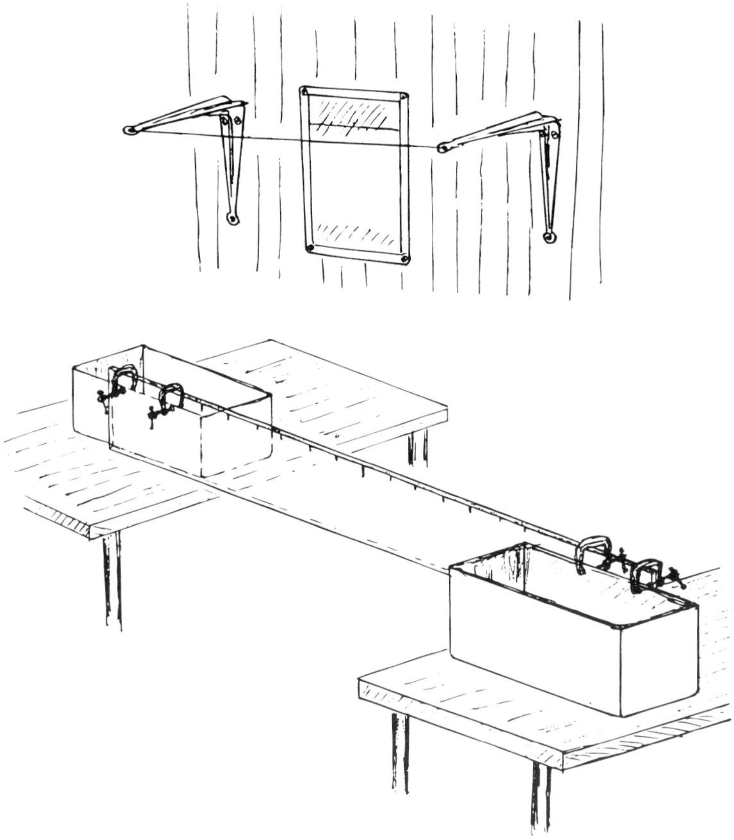

FIGURE 5. (Top panel) A grip test using a 600-mm wire spanning two shelf brackets. A mirror was attached on the wall to help observe the posture of the mouse during the test. (Bottom panel) A beam test using a Plexiglas plate (4.5 mm × 150 mm × 1000 mm) attached to two mouse cages with C-clamps. The length of the test portion of the beam was 600 mm. Marks on the beam show intervals of 50 mm. Both the wire and the beam were 90 cm above the floor. A box with sponge was placed underneath the testing devices.

IV. DISCUSSION

In the rat, the skull is thicker than that of the mouse, and therefore, it is difficult to apply the direct impounding method without craniotomy. On the contrary, the direct method can easily be applied for the mouse as long as we use a thin impounder (2.5 mm diameter) with a sharp tip (60°) to penetrate the skull. In this direct, noncraniotomy method in the mouse, apnea was not normally observed even when a stroke of 6 mm was applied. This suggests that the respiratory center was not located underneath the injury site. The injury produced by this method may be similar to the stabbing wound caused by a sharp object. Injury may be localized around the spot, and the impact may not propagate to other regions of the brain.

In contrast, similar to the case in the rat,[5] apnea was common in the craniotomy method. Since the dural membrane was left intact, and since a flat impounder was used, the impact may be propagated as pressure waves to other regions of the brain, including the respiratory center, to produce injury. Therefore, injury in this method may be similar to that in a liquid percussion model, which produces diffused axonal injury.

The recovery of motor function after injury in the craniotomy method (Figure 8) was similar to that in the rat controlled cortical impact model.[5] Our method in the mouse seems to produce similar results to those in the rat. At 24 h after injury, the motor function deficiency

TABLE 1
The Scoring System of the Grip Test Developed by Hall[1] with Slight Modifications (modified criteria are printed in bold)

Score	Criteria
0	Fell off string within the 30-s test period.
1	Held on in some way for 30 s. **(If traveled in either direction for at least 5 s, increase the score by 1.)**
2	Held on with four paws **or with three paws and tail** for at least 5 s. **(If traveled in either direction for at least 5 s, increase the score by 1.)**
3	Held on with four paws and able to place tail on the string for at least 5 s.
4	Held on with four paws, able to get tail on, and traveled along the string in either direction for at least 5 s.
5	Traveled to either end of the string within the 30-s test period.

A 60-cm string (vinyl-coated electric wire; 1.3 mm diameter) was spanned tightly between two shelf brackets (see Figure 5 top) 90 cm above the floor. A box laid with soft sponge was placed on the floor in order to protect the mouse if it fell. A mouse was picked up by the tail and placed at the midpoint of the string, making sure it held onto the string with both front paws. Then, the 30-s test started. Time to perform the string test—**The tests were performed 1 and 2 d following injury. Then, the test continued once in 2 or 3 d as needed for about a week.**

TABLE 2
The Scoring System of a Beam Balance Test in which Motor Function of a Mouse was Observed during the 60-s Test Period (see Figure 6 for illustrations)

Score	Criteria
0	Fell off within 30 s.
1	Held the beam some way (using two, three, or four paws) for at least 30 s, but the body is perpendicular to the direction of the beam (Figure 6A or E).
2	Held the beam with both front paws and one hind paw (by placing the body in an oblique position in relation to the span of the beam; Figure 6B), or with one front paw and one hind paw (by placing the body parallel to the beam, but not standing on the edge; Figure 6C) for at least a total of 10 s.
3	Maintained the same postures as those described for Score 2 for at least a total of 30 s; or stayed on top of the beam (parallel to the beam; Figure 6D) for at least 10 s, but could not walk.
4	Maintained the same postures as those described for Score 2, moved at least 15 cm in one direction; or stayed on top of the beam (Figure 6D) for at least 30 s, but could not walk.
5	Walked on the top edge (Figure 6D) at least a total of 15 cm in either direction.

A beam made of a Plexiglas plate (width 4.5 mm, height 150 mm, length 1000 mm) was attached to a mouse cage at each end, and held 90 cm above the floor (Figure 5 bottom). The actual beam part was 600 mm long. A box laid with a soft sponge was placed on the floor to protect the falling mouse. The mouse was picked up by the tail and placed at the midpoint of the beam, making sure it held the beam with front paws. Then, the 60-s test started. If the animal fell down within 30 s, the test was tried once more (after a few minutes rest) using a higher score. The tests were performed 1–2 d following injury. Then, the test continued once in 2 or 3 d as needed for about a week.

produced by a 3-mm stroke in the craniotomy method was similar to that produced by a 6-mm stroke in the noncraniotomy method (see Figures 7 and 8B). However, the recovery in the noncraniotomy method was much faster than that in the craniotomy method. The behavior of mice after injury was also different between these two methods; in the craniotomy method, animals were quiet for at least an hour after impact, while they were active in the direct method. It seems that the nature of injury in the direct, noncraniotomy method is different from that in the craniotomy method. In order to penetrate the skull in the noncraniotomy method, we used a thin impounder with a sharp tip. This might limit the injury to the site and to its immediate surroundings. This could be a reason for the quick recovery after

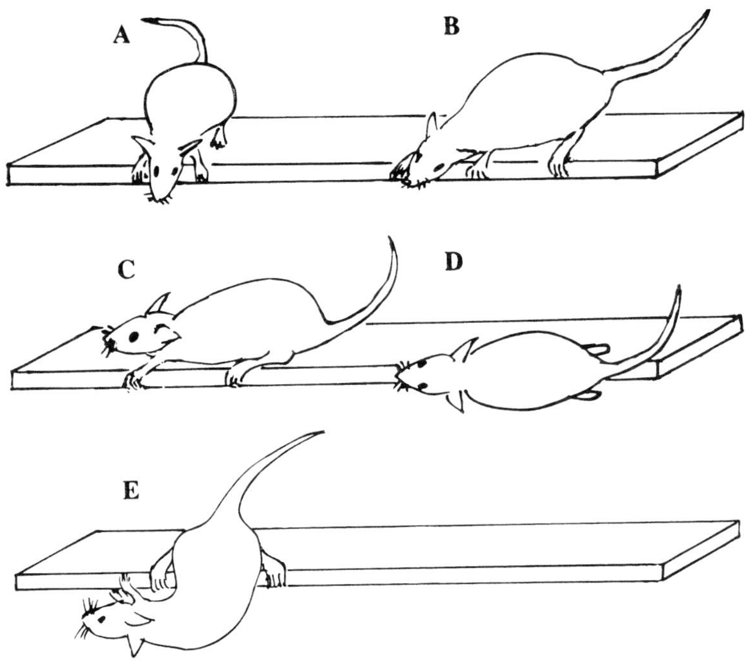

FIGURE 6. Illustration (top view) of the scoring method in the beam test. (A) Perpendicular, (B) oblique position, (C) parallel, but with one front paw and one hind paw, (D) on top of the beam. (E) If both hind paws were perpendicular to the beam and pointing toward the same direction, the animal was regarded as perpendicular to the beam, even if the body was in the oblique position or parallel to the beam.

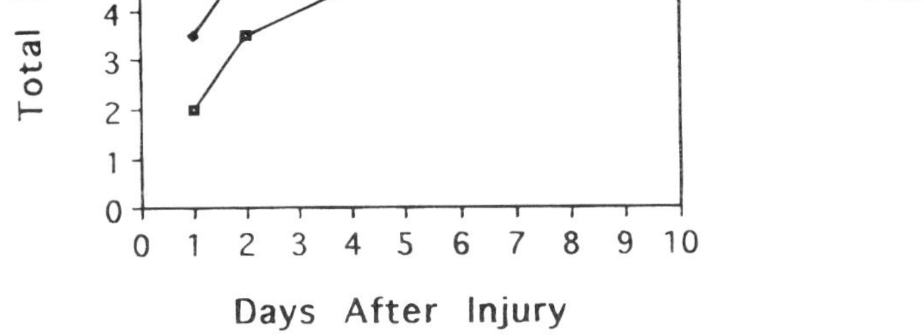

FIGURE 7. Effects of the impounding stroke on the recovery of the total motor score in the noncraniotomy method. Each data point is the average of measurements for six mice.

injury (Figure 7). Our direct method may be similar to a direct, weight-drop method in the rat[6] (see also Chapter 16 of this book). Further study is required to clarify these points.

We found that the grip test[1] is a good method to assess the deficiency in the motor function of injured mice. However, we thought that the introduction of a different test may further increase the reliability of the motor test. In searching for different tests, we found that a beam test using a slippery Plexiglas plate may be suitable for that purpose. While the grip

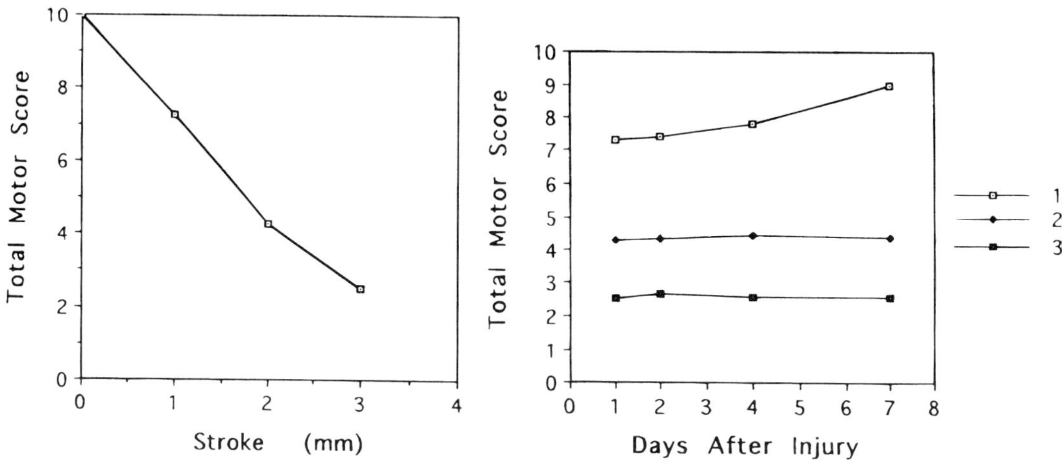

FIGURE 8. (left) Effects of the stroke on the total motor score on day 1 (the day following injury) in the craniotomy method. (right) Recovery of the total motor score after injury in the craniotomy method. Each data point is the average of measurements for four mice.

test indicates the ability of injured mice to hang on a wire and move, the beam test represents their ability to stand on a narrow beam and walk. Therefore, we can evaluate different motor functions of mice with these two tests. When the grip test score was improved in the days following injury, the beam test score was also found to be improved. Thus, we used the total motor score (the summation of both scores; 10 for normal and 0 for paralysis) to represent the motor function of mice.

A problem that is inherent to these tests is the difficulty in distinguishing whether an animal cannot physically move or does not want to move. This is the reason we attached empty mouse cages at the ends of the beam. Animals seemed to be more motivated to travel the beam when they saw the cage. The reproducibility of the test seems to be satisfactory. When the same investigator performed the test within an hour, the variability in the total motor score of the same animal was normally within ± 1.0. When two investigators tested the same animal, a difference between the scores obtained by two investigators was also within ± 1.0. In both the grip test and beam test, sometimes the motor score of an animal seemed to fall between two scores, for example, between 2 and 3. If this were the case, one could assign the score of 2.5. However, further subdivision between scores may not be appropriate.

In summary, we have developed simple controlled cortical impact models to study mouse head injury. We have also developed a method to assess the deficiency in the motor function of injured mice by combining a grip test and a beam test. We can produce injury either directly (without craniotomy) or after craniotomy. The nature of injury produced by these two methods is different. The direct method seems to produce a stabbing wound, while the craniotomy method may create diffused axonal injury.

ACKNOWLEDGMENT

The authors thank Dr. Shinya Sato, the Institute of Brain Diseases at the Tohoku University, Japan, for his valuable discussions. This work was supported by NIH grant NS30186.

REFERENCES

1. Hall, E. D. High-dose glucocorticoid treatment improves neurological recovery in head-injured mice. *J. Neurosurg.* 62: 882–887, (1985).
2. Hall, E. D., Yonkers, P. A., McCall, J. M., and Braughler, J. M. Effects of the 21-aminosteroid U74006F on experimental head injury in mice. *J. Neurosurg.* 68: 456–461, (1988).
3. Anderson, T. E. A controlled pneumatic technique for experimental spinal cord contusion. *J. Neurosci. Methods* 6: 327–333, (1982).
4. Somerson, S. K. and Stokes, B. T. Functional analysis of an electromechanical spinal cord injury device. *Exp. Neurol.* 96: 82–96, (1987).
5. Dixon, C. E., Clifton, G. L., Lighthall, J. W., Yaghmai, A. A., and Hayes, R. L. A controlled cortical impact model of traumatic brain injury in the rat. *J. Neurosci. Methods* 39: 253–262, (1991).
6. Shapira, Y., Shaufami, E., Sidi, A., Soffer, D., Freemam, S., and Cotev, S. Experimental closed head injury in rats: mechanical, pathophysiologic, and neurologic properties. *Crit. Care Med.* 16: 258–265, (1988).

Chapter 16

A CLOSED HEAD INJURY MODEL

Esther Shohami, Shamay Cotev, and Yoram Shapira

TABLE OF CONTENTS

I. INTRODUCTION

The incidence of severe head injury (SHI) in the Western world is estimated to be about 0.15% of the population annually. The majority of SHI cases occur in the age group of 15 to 24 years, mainly due to motor vehicle accidents.[1] In contrast to stroke the incidence of head injury is increasing.[2] Primary injury occurs when the initial traumatic force is applied to the head. The brain may suffer from either or all of the following: diffuse damage to the neurons and white matter, the tearing of arteries and veins, and the development of cerebral edema. Secondary injuries usually develop within minutes to hours following initial SHI and represent a cascade of pathophysiological changes promoted by the initial injury.

There are direct and indirect factors contributing to secondary injuries in SHI. Indirect (noncerebral) contributing factors include hypoxia and/or hypercapnia due to prolonged apnea after the impact, aspiration or concomitant chest trauma, and hypovulemia usually related to associated injury to other organs. Direct factors include raised intracranial pressure and endogenous mediators, which are massively released during the posttraumatic period, and are harmful to the brain. These mediators include free radicals, arachidonic acid and its metabolites (eicosanoids), excitatory amino acids (mainly glutamate), etc.

0-8493-8094-4/95/$0.00+$.50

Brain ischemia may result from secondary injury and is caused by a disturbance in the cerebral blood flow/cerebral metabolic rate ratio. It occurs when the blood flow decreases to an extent in which substrate supply for energy production (i.e., oxygen and glucose) cannot accommodate tissue energy demands.

Ischemia causes depolarization of cellular membranes and impairment of ionic homeostasis with a very significant increase of calcium ions in intracellular spaces. This increase is due to three factors:[3]

1. The decrease in the transmembrane potential during ischemia opens voltage-dependent calcium gates and the decrease in the transmembrane sodium gradient slows outward transport of Ca^{2+}.
2. Ca^{2+} is released from the endoplasmic reticulum due to a decrease in adenosine triphosphate (ATP) levels as a result of ischemia.
3. Excitatory amino acid levels increase during ischemia and cause an overstimulation of the glutamate receptors and the immediate intracellular effects, mainly, accumulation of intracellular calcium.

One major consequence of the rise in intracellular calcium is the activation of phospholipase A_2 (PLA_2) in the cell membrane. This enzyme hydrolyzes membrane phospholipids and releases free fatty acids, including arachidonic acid (AA). AA is a substrate for two distinct major metabolic pathways. Oxidation of AA by cyclooxygenase results in the production of prostaglandins (PGs) and oxidation by lipoxygenase yields leukotrienes (LTs). These compounds are vasoactive and might be involved in the development of delayed brain damage. The loss of membrane phospholipids disturbs the delicate balance between membrane proteins and lipids. This causes a further considerable increase in calcium penetration into the cell[4] and disturbs mitochondrial function.

II. BRAIN EDEMA

One of the immediate results of SHI is brain edema. This condition is defined as the abnormal accumulation of fluid inside the brain parenchyma causing an increase in brain tissue volume. These changes in water content and electrolytes in edema were classified by Klatzo[5] and Klatzo et al.[6] who described two types of brain edema: vasogenic and cytotoxic.

Vasogenic edema is caused by an injury to the capillaries' endothelial cells resulting primarily in extracellular water accumulation. The main pathogenic injury in this type of edema is an increase in permeability of the blood-brain barrier (BBB) associated with a leakage of serum proteins into the brain parenchyma.

Cytotoxic brain edema is characterized by the swelling of cellular elements of the brain parenchyma due to cytotoxic substances. These can selectively influence different cells and intracellular organelles. It may develop by exposure to various toxic compounds, such as excitatory amino acids, which cause accumulation of water and ions and lead to neuronal and glial swelling.

III. HEAD TRAUMA MODELS

The pathophysiological characteristics of the manifestations of head trauma (HT) are highly variable. Many animal models of HT emphasize only one or some combination of these manifestations. Animal models of HT are difficult to create. The data obtained should be reproducible and relevant as much as possible to human head injury. In the literature of head injuries a number of animal models have been described, represented by acceleration, compression, and penetrating injury.

Compression models include all those in which the head is maintained in a fixed position during impact. There are two types of compression models. In one type, the cranium is open and compression is delivered directly to the brain. With the other type, the cranium remains intact, and the energy is imparted over an intact skull.

In this chapter we describe a model of closed head injury in the rat, which was designed in an attempt to mimic, as closely as possible, a common clinical situation and to deliver a standard impact in order to elicit a controlled HT.

IV. CLOSED HEAD INJURY MODEL

A. INDUCTION OF TRAUMA

Male Sabra rats (strain of the Hebrew University) weighing 220 to 250 g are used for the study. HT is induced under ether anesthesia, which is confirmed by testing corneal and pupillary reflexes, according to the details described elsewhere by Shapira et al.[7] In brief, trauma is induced by a well-calibrated weight-drop device, which falls over the exposed skull covering the left cerebral hemisphere 1 to 2 mm lateral to the midline, in the mid-coronal plane (Figure 1). After recovery from anesthesia the rats are allowed to recover in their home cages with postoperative care and free access to food and water. A vertical view of the rat brain 10 d after the HT is shown in Figure 2.

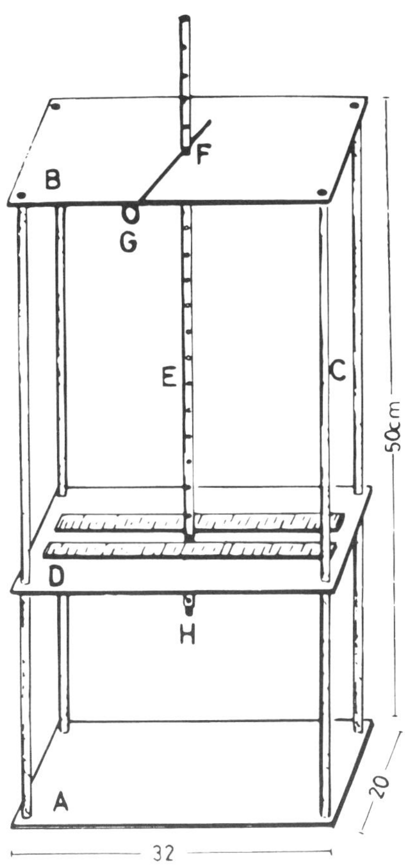

FIGURE 1. Weight-drop device used in this study to produce closed head trauma in rat. See text for details. (From Shapira, Y. et al., *Crit. Care Med.*, 16, 258–265, 1988. With permission.)

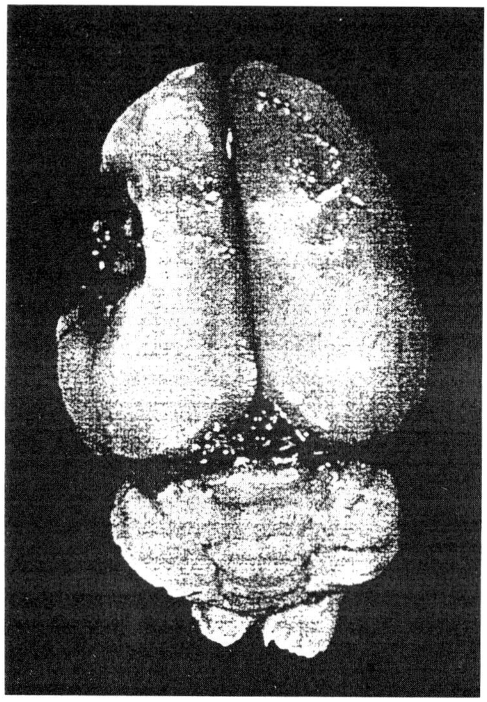

FIGURE 2. Vertical view of the brain of a rat, 10 d after the trauma, displaying a large, irregular crater in the left frontotemporal region. (From Shapira, Y. et al., *Crit. Care Med.*, 16, 258–265, 1988. With permission.)

B. EVALUATION OF CLINICAL STATUS

One hour after induction of trauma, the rats are tested by a set of criteria that evaluates their clinical status (Table 1). These criteria, referred to as the Neurological Severity Score (NSS), consist of reflexes and motor functions, and the maximal score at 1 h is 24 points. A rat with severe injury (a NSS of 15 to 20 points 1 h after trauma) will usually survive during the posttraumatic period, while with a score above 20 it is likely to die within 1 to 2 h. At 24 or 48 h the rats are reevaluated and their score is usually less than 20 points (Figure 3). For each rat, the difference between its score at 1 h and at the later time point is calculated, and referred to as DNSS. This difference reflects the recovery of the rats during the posttraumatic period, and the effect of a drug can be evaluated by comparing the spontaneous recovery to that achieved after drug treatment.

In our studies of this model we design the trauma (by adjusting the height of the fall of the weight) so that the initial NSS, determined at 1 h, will be in the range of 15 to 18. Figure 3 summarizes the NSS at different times and shows its spontaneous recovery. A residual deficit is still evident at 48 h after HT; during that period, however, the traumatized rats can eat and drink and they move freely in their cages.

C. EVALUATION OF EDEMA

1. Specific Gravity (SG) Measurements

The SG of brain tissue is determined according to the method of Marmarou et al.[8] Cortical tissue samples (~20 mg) from the contused hemisphere are placed on top of a linear gradient column of bromobenzene and kerosene. The columns are prepared as described in detail in

TABLE 1
Neurological Severity Score

	Points at 1h	Points other
Inability to exit from a circle (50 cm in diameter) when left in its center		
for 30 min	1	—
for 60 min	1	—
>60 min	1	1
Loss of righting reflex		
for 20 min	1	—
for 40 min	1	—
>60 min	1	1
Hemiplegia—inability of the rat to resist forced changes in position	1	1
Flexion of hindlimb when raised by the tail	1	1
Inability to walk straight when placed on the floor	1	1
Reflexes		
Pinna reflex	1	1
Corneal reflex	1	1
Startle reflex	1	1
Clinical grade		
Loss of seeking behavior	1	1
Prostation	1	1
Limb reflexes		
Loss of placing reflexes—		1
forelimbs left	1	1
right	1	1
hindlimbs left	1	1
right	1	1
Functional test		
Failure in beam-balancing task (1.5 cm wide)		
for 20 s	1	1
for 40 s	1	1
>60 s	1	1
Failure in beam-walking task		
8.5 cm wide	1	1
5 cm wide	1	1
2.5 cm wide	1	1
Maximum points	24	20

Shapira et al.[7] and calibrated with drops of K_2SO_4 at known SG. The tissue SG is calculated from the standard curve by measuring its equilibrium position in the column.

2. Water Content

Another approach to the measurement of edema is by determination of the tissue water content. A piece of tissue (~20 mg) is placed on a preweighed aluminum foil and weighed to yield wet weight (WW). After drying in a desiccated oven for 24 h at 95°C it is reweighed to yield dry weight (DW). Water percent in the tissue is calculated as $(WW - DW) \times 100/WW$.

The changes in the SG and percentage water are parallel, namely, the increase in water content and decrease of SG follow similar time courses. As early as 15 min after HT a significant edema is observed, as determined by both criteria. It gradually reaches a maximal value at 24 to 48 h after HT and by 7 to 10 d, it disappears.

It is interesting to note that while in the contused hemisphere there is accumulation of water as a result of the impact, the contralateral hemisphere retains its normal SG and water

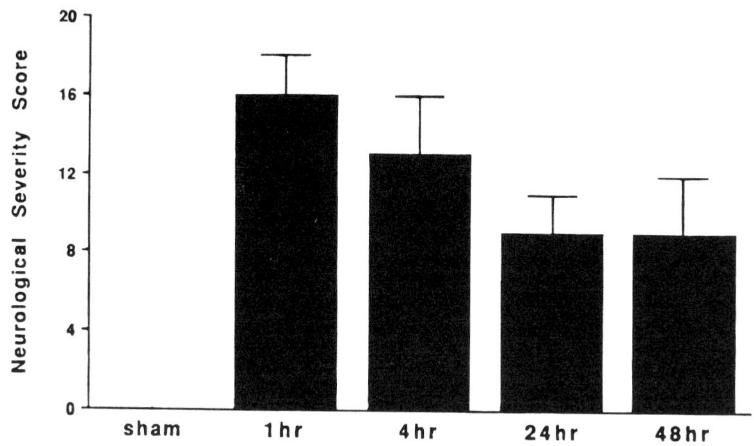

FIGURE 3. The NSS of rats, evaluated at different times after HT. At 1 h there is a maximal deficit, expressed by the high score, and it gradually declines as the rats spontaneously partially recover.

content during the whole post-HT period. Figure 4 shows the time course for edema formation in the contused hemisphere.

D. EVALUATION OF THE BLOOD-BRAIN BARRIER INTEGRITY

We showed that edema develops after HT, and the question of the role of a defect in permeability at the blood-brain barrier (BBB) was addressed. In order to test the integrity of the BBB the Evans blue dye extravasation method was used, according to Uyama et al.[9] Briefly, Evans blue (2% in saline) is injected IV to traumatized and control rats 1 h before sacrifice. Their chests are then opened under ether anesthesia and the rats are perfused with 40 to 50 ml of saline to clear the blood from the dye. The brains are removed, the hemispheres separated, weighed, and the extracted dye is quantified by fluorescence measurements.

The results of this study are depicted in Figure 5. They indicate a maximal disruption of the BBB integrity at 4 h after HT, followed by a decline toward normal at 24 to 48 h. However there is not a complete repair of the BBB even at 48 h. Our findings on the BBB and on edema show that the increase of BBB permeability is not directly temporally related to formation of cerebral edema. The peak of BBB breakdown precedes that of edema formation and one may suggest that this sequence of events points to the vasogenic type of edema in the pathophysiology of brain edema in our model of closed head injury in the rat.

V. CHEMICAL MEDIATORS IN CLOSED HEAD INJURY

The putative role of endogenous mediators in the development of delayed brain damage in the model of closed head injury was studied. Two classes of compounds that are normally found in the brain in low amounts and may be suspected to play a role in the pathogenesis of neuronal damage were studied.

A. EICOSANOIDS

The eicosanoids, the metabolic products of AA, have vasoactive properties that might lead to BBB impairment and to the development of edema. AA, which is normally found esterified in the phospholipid fraction of cell membranes, is liberated from the membranes upon activation of phospholipases (A_2, C) following various stimuli.[10,11] Free AA is the substrate for metabolic pathways that yield the PGs (the cyclooxygenase pathway), and the LTs, (the lipoxygenase pathway).

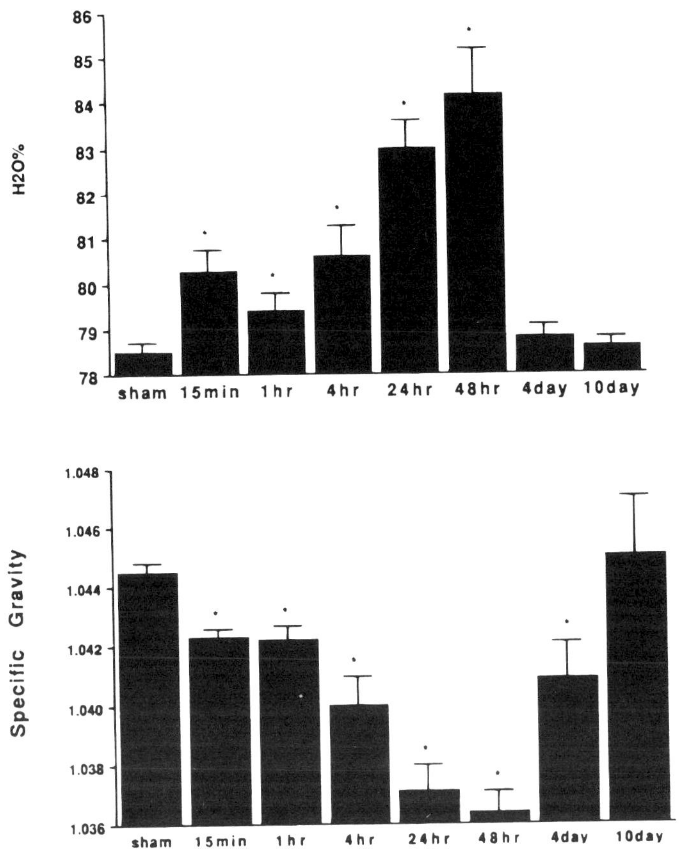

FIGURE 4. The time course of edema evolution and resolution. Edema develops between 15 min and 24 to 48 h, when it peaks. It is then gradually dissolved and by 7 to 10 d completely disappears. A similar pattern was obtained when edema was evaluated by SG or by water accumulation. Edema was evident only in the contused hemisphere, while the contralateral retained normal water content throughout the whole time course of the experiment. * $p < 0.05$ as compared to the same hemisphere of sham rats.

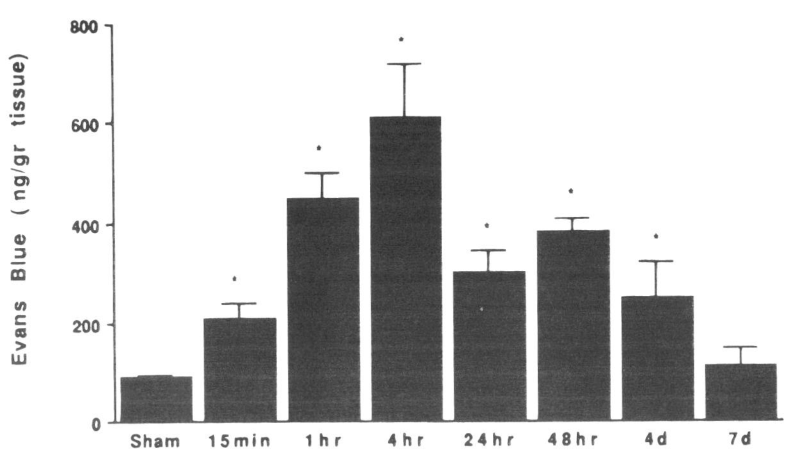

FIGURE 5. BBB dysfunction after HT, as measured by the extravasation of Evans blue dye into the brain parenchyma. A massive amount of dye was extracted from the contused hemisphere and to a lesser extent from the contralateral one as well. The peak of BBB defect was noticed at 4 h after HT with gradual decay, though no complete recovery followed.

TABLE 2
Summary of the Pharmacological Treatments in the Model of Closed Head Injury

Drug	Mode of action	Effect on Edema	Effect on NSS	Ref.
Dexamethasone	Glucocorticoid	–	–	19
Methylprednisolone	Glucocorticoid	–	–	20
Indomethacin	NSAID	–	–	19
SK&F 105809	Mixed CO & 5 LO inhibitor	+	–	22
MK801	NMDA-antagonist	+	+	27
HU-211	NMDA-antagonist	+	+	28
DFMO	ODC-inhibitor	–	+	18
Ketamine	NMDA-antagonist	–	+	26

+ Beneficial effect ($p < 0.05$ as compared to nontreated rats).
– No-effect.
Note: NSS was evaluated at 24 or 48 h, according to Table 1, and edema was evaluated at 4, 24, or 48 h. For details of timing of drug administration and doses, see references.

In our model of closed head injury we have shown that PLA_2, the rate-limiting enzyme in the liberation of AA from cell membranes, is activated 15 min after HT, and it remains activated even 24 h later.[12] In parallel, the accumulation of 5-HETE, the product of 5-lipoxygenase, was found at 15 min. At 24 h, PGE_2 levels were five- to sevenfold higher than in the controls.[13,14] Thus, the activation of the AA cascade after HT, which occurs with a similar time course to that of cerebral edema, may suggest that these compounds are indeed involved in the secondary mechanisms leading to delayed damage. In order to provide more direct evidence for the role of eicosanoids in the pathogenesis of brain injury, the effects of various specific drugs that inhibit their production were studied as described below.

B. POLYAMINES

The polyamines are the products of a biosynthetic pathway that is activated by a variety of stimuli, including ischemia and trauma.[15,16] The substrate for their synthesis is the amino acid ornithine, which is decarboxylated by the rate-limiting enzyme, ornithine-decarboxylase (ODC), to yield putrescine and subsequently spermine and spermidine. Their role as mediators of brain damage has been suggested to be related to their receptor site on the NMDA-receptor complex.[17] Activation of this glutamate receptor subtype has been shown to be involved in cytotoxicity, and since the polyamine site was recognized a number of studies have dealt with the possible involvement of this class of compounds in delayed brain injury. In our model of closed head injury we showed that the polyamine metabolic pathway is indeed activated in the posttraumatic period[18] and, therefore, we proceeded to study their role in the development of brain damage (see below).

VI. PHARMACOLOGICAL TREATMENTS

Measures to decrease mortality and morbidity after traumatic brain injury should consist of two major approaches. The first provides early, basic resuscitative management of the airway, adequate oxygenation and hyperventilation to reduce intracranial pressure, and thus external vascular compression and vasospasm, and blood volume replacement to maintain cerebral perfusion pressure. The aim of these measures is to maintain tissue oxygenation and prevent, as much as possible, secondary ischemic injury.

The second approach focuses on pharmacological intervention in the biochemical cascades that are triggered by the primary trauma (e.g., glutamate release, free radical formation, eicosanoids or polyamine biosynthesis, etc.). Such intervention can involve either a decrease in synthesis (i.e., by inhibiting specific enzymes) or impaired function (i.e., by blocking the effect via specific receptors). In our studies with the closed head injury model we have investigated the effects of several drugs, which represent a variety of mechanisms of action, on the outcome of trauma. The effects on clinical outcome (NSS) and on edema formation were studied.

A. GLUCOCORTICOID TREATMENT IN HEAD TRAUMA

Glucocorticoid (GC) treatment was initiated in head trauma and ischemia after encouraging results were obtained with such therapy in treating brain edema secondary to tumors and intracerebral abscesses.

A proposed mechanism of action of GC is their ability to inhibit the rate-limiting enzyme in eicosanoid production, PL-A_2. This enzyme releases arachidonic acid from membranous phospholipids, which, in turn, is oxidized to PGs and LTs.

In our model of closed head injury in rats dexamethasone[19] and methylprednisolone[20] did not affect the specific gravity of traumatized tissue, the neurological condition of the animal, or tissue prostaglandins. Thus the role of these mediators in the development of the delayed damage was questioned.

B. NONSTEROIDAL ANTIINFLAMMATORY DRUGS (NSAID)

The increased production of AA metabolites in the injured brain suggested the use of the more specific drugs, NSAID, in an attempt to improve the outcome. We studied the effect of indomethacin, which inhibits the cyclooxygenase in our model of closed head injury.[19] Despite a decrease in PGs, no reduction of edema, or improvement of NSS, were found.

One explanation for the ineffectiveness of indomethacin in reducing edema is that indomethacin specifically blocks the cyclooxygenase pathway. There is evidence that LTs, which are the products of the lipoxygenase pathway, are produced under pathological conditions.[21] Therefore, by inhibiting one metabolic pathway, more substrate may be diverted to the alternative pathways, namely, lipoxygenase.

Indeed, in a recent study[22] we demonstrated a reduction of edema in traumatized rats following treatment with a novel mixed inhibitor of cyclooxygenase and 5-lipoxygenase, SK&F-105809. Thus, the role of the eicosanoids in the production of posttraumatic edema is supported.

C. NMDA RECEPTOR ANTAGONISTS

Extracellular accumulation of glutamate and aspartate was found in experimental models of ischemia[23] and head trauma.[24] Thus, NMDA receptor antagonists that are able to cross the BBB have potential therapeutic use. One occasionally used intravenous anesthetic is ketamine, which is a noncompetitive NMDA antagonist. It was found to be protective in cultured astrocytes.[25]

In the present closed head injury model, Shapira et al.[26] observed that ketamine significantly improved the neurological outcome and reduced cerebral infarct volume. In another study the effect of MK801 was investigated in the same model.[27] This treatment was found to be effective in reducing edema formation 24 h after HT, and in improving the neurological status at 48 h. These results support the role of glutamate in the development of cerebral edema and neuronal damage after HT.

D. POLYAMINE SYNTHESIS INHIBITORS

Based on our findings that polyamines accumulate during the posttraumatic period, we treated rats with difluoromethylornithine (DFMO), which inhibits the enzyme ODC. This inhibition, though not effective in reducing brain edema, had a significant beneficial effect on the NSS measured 48 h post-HT.[8] We have also shown (unpublished data) that this treatment reduced the infarcted area in the brains of the traumatized rats. Thus, we proposed that at least in part, the delayed brain damage is mediated via the polyamines.

E. HU-211, A NOVEL NONPSYCHOTROPIC CANNABINOID

In a recent study,[28] the effect of HU-211 on the outcome of HT in our model was investigated. It was found to have a very pronounced beneficial effect on the NSS, on edema, and on the integrity of the BBB. Threefold reduction in Evans blue extravasation at 4 and 24 h after HT and about 50% reduction in the amount of water accumulated in the brain are among the best results obtained in this model with any therapy. Moreover, the effect was evident even when the drug was given up to 3 h after the impact, namely, it has a "therapeutic window" of at least 3 h. The mechanism of action is not yet elucidated; however, some noncompetitive NMDA antagonist properties were reported for that drug.[29]

These various investigations of the effects of different classes of drugs in this model were aimed at providing new clinically applicable therapeutic agents in order to block or diminish the extent of the injurious events that occur following head injury in patients. Such studies require an animal model that mimics, as closely as possible, the clinical situation. The closed head injury model described in this chapter is an appropriate model for this type of study and we hope that eventually, these and other studies will provide the clinicians caring for patients with severe head injuries new therapies that will lower the devastating effects that are inherent to these injuries.

REFERENCES

1. Grabow, J. D., Offord, K. P., Rieder, M. E. The cost of head trauma in Olmsted County, Minnesota, 1970–74. *Am J Public Health* 1984; 74:710–712.
2. Goldstein, M. Traumatic brain injury: a silent epidemic. *Ann Neurol* 1990; 27:327.
3. Hansen, A. J. Extracellular ion concentration in cerebral ischemia. In: Zeuther J, ed. *The Application of Ionselective Microelectrodes*. New York: Elsevier/North-Holland Biomedical Press, 1981:239–254.
4. Farber, J. L., Chien, K. R., Mittnacht, S. J. R. The pathogenesis of irreversible cell injury in ischemia. *Am J Pathol* 1981; 102:271–281.
5. Klatzo, I. Neuropathological aspects of brain oedema. *J Neuropath Exp Neurol* 1967; 26:1–14.
6. Klatzo, I., Wisniewski, H., Steinwall, O. Dynamics of cold injury edema. In: Klatzo I, Seitelberger F, ed. *Brain Edema*. New York: Springer Verlag, 1967:554–563.
7. Shapira, Y., Shohami, E., Sidi, A., Soffer, D., Freeman, S., Cotev, S. Experimental closed head injury in rats: Mechanical pathophysiologic, and neurologic properties. *Crit Care Med* 1988; 16:258–265.
8. Marmarou, A., Werner, P., Shulman, K., Bhagavan, H. A simple gravimetric technique for measurement of cerebral edema. *J Neurosurg* 1978; 49:530–537.
9. Uyama, O., Okamura, N., Yanase, M., Narita, M., Kawabata, K., Sugita, M. Quantitative evaluation of vascular permeability in the gerbil brain after transient ischemia using Evans blue fluorescence. *J Cereb Blood Flow Metab* 1988; 8:282–284.
10. Wei, E. P., Lamb, R. G., Kontos, H. A. Increased phospholipase C activity after experimental brain injury. *J Neurosurg* 1982; 56:695–698.
11. Abe, K., Kogure, K., Yamamoto, H., Imazawa, M., Miyamoto, K. Mechanism of arachidonic acid liberation during ischemia in gerbil cerebral cortex. *J Neurochem* 1987; 48:503–509.
12. Shohami, E., Shapira, Y., Yadid, G., Reisfeld, N., Yedgar, S. Brain phospholipase A_2 is activated after experimental closed head injury in the rat. *J Neurochem* 1989; 53:1541–1546.
13. Shohami, E., Shapira, Y., Yadid, G. Cotev, S., and Feuerstein, G. Increased 5-HETE production in the brain following head injury. *Ann NY Acad Sci* 1989; 559:485–487.
14. Shohami, E., Shapira, Y., Sidi, A., Cotev, S. Head injury induced increased prostaglandin synthesis in rat brain. *J Cereb Blood Flow Metab* 1987; 7:58–63.

15. Koenig, H., Goldstone, A. D., Lu, C. Y., Blood-brain-barrier breakdown in cold injured brain is linked to a biphasic stimulation of ornithine decarboxylase acitvity and polyamine synthesis; both are coordinately ihibited by verapamil, dexamethasone and aspirin. *J Neurochem* 1989; 52:101–109.
16. Paschen, W., Hallmayer, J., Mies, G. Regional profile of polyamines in reversible cerebral ischemia of Mongolian gerbils. *Neurochem Pathol* 1987; 7:143–156.
17. Williams, K., Romano, C., Molinoff, P. B. Effects of polyamines on the bindig of 3H-MK-801 to the N-methyl-D-aspartate receptor: pharmacological evidence for the existence of a polyamine recognition site. *Moll Pharmacol* 1989; 36:575–581.
18. Shohami, E., Nates, J. L., Glantz, L., Trembovler, V., Shapira, Y., Bachrach, U. Changes in polyamine levels following head injury. *Exp Neurol* 1992; 117:189–195.
19. Shapira, Y., Davidson, E., Weidenfeld, Y., Cotev, S., Shohami, E. Dexamethason and indomethacin do not affect brain edema following head injury in rats. *J Cereb Blood Flow Metab* 1988; 8:395–402.
20. Shapira, Y., Yadid, G., Cotev, S., Shohami, E. Methylprednisolone does not affect brain edema following head trauma in rats. *Anesth Analg* 1991; 68:S247.
21. Dempsey, R. J., Roy, M. W., Cowen, D. E., Maley, M. E. Lipoxygenase metabolites of arachidonic acid and the development of ischemic cerebral edema. *Neurol Res* 1986; 8:53–59.
22. Shohami, E., Glantz, L., Nates, J., and Feuerstein, G. The mixed lipoxigenase/cycloxygenase inhibitor SK&F 105809 reduces cerebral edema after closed head injury in rat. *J Basic Clin Physiol Pharmacol* 1992; 3:99–107.
23. Beneviste, H., Drejer, J., Schousboe, A., Diemer, N. H. Elevation of extracellular concentration of glutamate and aspartate in rat hippocampus during transient ischemia monitored by intracerebral microdialysis. *J Neurochem* 1984; 43:1369–13.
24. Faden, A. I., Demediuk, P., Panter, S. S., Vink, R. The role of excitatory amino acids and NMDA receptors in traumatic head injury. *Science* 1989; 244:798–800.
25. Chan, P. H., Chu, L. Ketamine protects cultured astrocytes from glutamate-induced swelling. *Brain Res* 1989; 487:380–383.
26. Shapira, Y., Artru, A. A., Lam, A. M. Ketamine decreases cerebral infract volume and improves neurological outcome following experimental head trauma in rats. *J Neurosurg Anesth* (submitted).
27. Shapira, Y., Yadid, G., Cotev, S., Niska, A., Shohami, E. Protective effect of MK-801 in experimental brain injury. *J Neurotrauma* 1990; 7:131–139.
28. Shohami, E., Novikov, M., Mechoulam, R. A non-psychotropic cannabinoid, HU-211, has cerebroprotective effects after closed head injury. *J. Neurotrauma* 1993; 10:109–119.

Chapter 17

WEIGHT-DROP MODEL OF TRAUMATIC BRAIN INJURY: ASSESSMENT OF THE ACUTE INFLAMMATORY RESPONSE TO CEREBRAL TRAUMA

Patrick M. Kochanek, Robert S.B. Clark, Joanne K. Schiding, and Edwin M. Nemoto.

TABLE OF CONTENTS

I. INTRODUCTION

Many pathophysiologic events are involved in the extension of tissue damage in traumatic brain injury in humans including contusion, diffuse and local axonal injury, cytotoxic and vasogenic cerebral edema, parenchymal and extraparenchymal hemorrhage, ischemia, hyperemia, intracranial hypertension, and cerebral infarction. Numerous laboratory models have been developed to simulate traumatic brain injury and investigate the various pathophysiologic mechanisms, and much of this work has been done in rats.[1–5] We developed a modification of the Feeney weight-drop model of traumatic brain injury[1] in rats to study the inflammatory response to trauma. This chapter outlines our studies characterizing the model and investigating the hypothesized role of local acute inflammation in exacerbating tissue injury after cerebral trauma.

II. WEIGHT-DROP MODEL OF TRAUMATIC BRAIN INJURY IN RATS

A. BACKGROUND

In 1911, Allen[6] first described an experimental model of weight-drop injury to the spinal cord. Feeney et al.[1] first modified this model in 1981 to produce a concussive brain injury in rats. Recently, we developed a simplified version of the weight-drop model and showed it is characterized by a very reproducible cerebral contusion and patterns of posttraumatic

0-8493-8094-4/95/$0.00+$.50

cerebral edema, intracranial hypertension, and cerebral blood flow (CBF) that mirror those often observed after human head injury.[7–11]

B. ANESTHESIA AND SURGICAL PROCEDURE

Male Wistar rats (375 to 475 g) are initially anesthetized with 5% isoflurane or halothane in O_2, and intubated with an angiocatheter (Baxter, Deerfield, IL). The anesthetic is then reduced to 1.5% halothane or 2% isoflurane in 66% N_2O/balance O_2, and the lungs are mechanically ventilated. Rats are paralyzed with pancuronium bromide (0.6 mg/kg/h). Bicillin (100,000 U) and gentamicin (5 mg) are administered IM. All surgical procedures are performed aseptically using betadine scrub, surgical drapes, sterile gloves, and mask. Femoral venous and arterial catheters (PE-50) are inserted. Arterial blood pressure is continuously monitored. Rectal temperature is maintained between 36.5 and 37.5°C by a heated water blanket.

The rat's head is fixed in a stereotaxic frame (David Kopf, Tujunga, CA). After retraction of the scalp, a craniotomy (1.0 × 0.5 cm in size) is made over the right parietal cortex using the coronal and interparietal sutures as margins. A high-speed dental drill is used under a binocular operating microscope. The bone flap is left in place for 30 min of equilibration on 0.4% halothane or 1.1% isoflurane in 66% N_2O/balance O_2. If brain temperature monitoring is desired, a burr hole is made over either the contralateral parietal cortex or over the ipsilateral frontal cortex (5.0 mm anterior and 2.0 mm lateral to bregma) and a 23-μm temperature probe (Physitemp) is inserted 1 to 2 mm into the cortex. PCO_2 is maintained between 35 and 45 torr and the PO_2 greater than 100 torr.

C. BRAIN INJURY

The bone flap is removed and brain injury is produced by dropping a 10-g brass rod (3.0 mm diameter and 15.2 cm length) through a glass guide tube (12.5 cm length, 3.5 mm inner diameter, 5.0 mm outer diameter) onto the exposed right parietal cortex (dura intact). After trauma, the bone flap is immediately replaced and sealed with dental cement (Koldmount, Vernon Benshoff Co., Albany, NY). The scalp incision is closed, the femoral catheters are removed, and the rats allowed to awaken before tracheal extubation. After 30 min in supplemental O_2, the rats are returned to their cages and allowed free access to food and water until subsequent studies.

D. POSTTRAUMA SURVIVAL

Using this protocol, over 95% survival to 7 d posttrauma is achieved, and although the rats exhibit mild lethargy and a transient 5% body-weight loss (over approximately 48 h), they eat, drink, and ambulate.[7–9] Between four and five rats can be traumatized in a single day by one technician using this protocol.

E. CHARACTERISTICS OF THE LESION

A clearly defined and severe traumatic contusion is produced. The contusion is hemorrhagic in nature and at its center extends from the cortical surface down to the level of the amygdala (Figure 1). The hippocampus and thalamus are consistently involved. Analysis of serial coronal sections through the injured hemisphere from rats at two weeks after trauma reveals that the total lesion volume represents 16.0 ± 2.7 mm^3 (mean ± SEM), which is approximately 4.7% of the hemispheric volume.[12] Cerebral edema, as quantified by percent brain water in the injured (right) hemisphere (wet-dry weight method) increases approximately 1% during the initial 24 h posttrauma, which is sutained at 48 h and resolves by 7 d.[7,8] Hemispheric edema is dose-dependent in the model; a more severe insult (10 g × 10 cm) results in a larger (1.5%) increase in brain water at 24 h posttrauma.[7] Brain water in the contralateral hemisphere is not affected. A more regional assessment of brain water made by sampling a

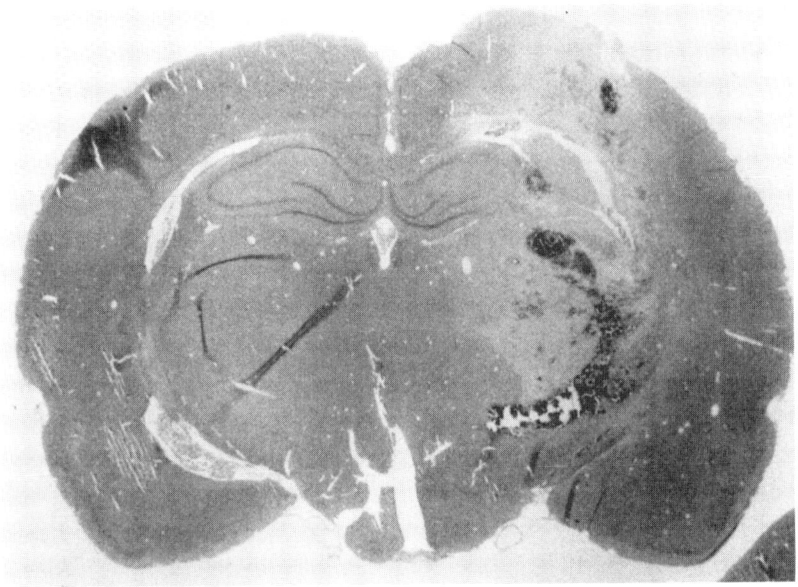

FIGURE 1. Representative hematoxylin and eosin-stained coronal section through the center of the lesion from a rat brain perfusion-fixed at 24 h after weight-drop trauma (10 g × 5 cm). A hemorrhagic contusion is observed that extends though the parietal cortex, hippocampus, and thalamus. Intraparenchymal and intraventricular hemorrhage are noted.

6-mm coronal section taken through the lesion at 24 h posttrauma reveals that the increase in brain water is approximately 3%.[13] Intracranial pressure, measured with a cisternal catheter via the method of Bouman and Van Wimersma[14] is unaltered at 2 h posttrauma, but is increased from control values of about 4 torr to between 11 and 15 torr at 24 h posttrauma.[8]

Regional CBF (Measured using ^{14}C-iodoantipyrine autoradiography) is reduced at 2 h posttrauma,[8] similar to that observed in other models of traumatic brain injury in rodents, such as fluid percussion[15] and controlled cortical impact.[16] The reduction is most marked in the impact zone, but is also significant throughout both the injured and contralateral hemispheres.[8]

At 24 h posttrauma, a heterogeneous regional CBF pattern is observed, with a low-flow core in the contused area surrounded by regions of hyperemia, and accompanying hyperemia in more distant regions such as the ipsilateral frontal cortex, and contralateral parietal cortex and amygdala.[9,10] These hyperemic regions exhibit increases in regional CBF in the range of 60 to 160% above control values.[9,10] Delayed postinsult hyperemia has been observed in other injury models such as cortical freeze injury and cortical compression.[17,18] In addition, the time course of posttraumatic alterations in CBF in this model are remarkably similar to those observed by Obrist et al.[11,19] in humans after closed head injury. We have concentrated our description of the weight-drop model on quantitation of the cerebrovascular disturbances produced (i.e., edema, CBF, intracranial hypertension), in contrast to the work of Feeney et al.,[1,20,21] who have already investigated some of the functional and metabolic disturbances observed in their weight-drop model.

F. DISCUSSION OF THE MODEL

Although this model lacks some of the sophistication of contemporary models such as the controlled cortical impact and fluid percussion,[2,3] it is a simple, reproducible, and very inexpensive model that produces a consistent lesion and a spectrum of pathophysiologic changes in brain, mirroring many (but certainly not all) of those observed in human head injury with contusion. It is particularly useful as a screening model since relatively large numbers of studies can be performed in a rather short period of time, and it can be established by almost any laboratory familiar with basic surgical skills in rodents. In addition, very similar results are being obtained in both the weight-drop model and the controlled cortical impact model for changes in both routine pathophysiologic parameters (brain water, CBF) and novel cellular and molecular markers of traumatic brain injury such as microglial activation, tissue myeloperoxidase activity, and nerve growth factor.[12,22,23] Thus, this model is quite useful as a simple tool to investigate the biochemical events accompanying cerebral contusion.

Several differences between this method and the Feeney weight-drop model are worthy of discussion, even though both models use the same free weight-drop mechanism. We attempted to simplify the injury device by eliminating the air holes in the guide tube, the footplate on the dural surface, and the string suspension of the weight. The weight is hand-held to a predefined level on the glass guide tube, and the impact is made by the brass rod directly onto the dural surface. This produces an injury with more of a penetrating component than the Feeney weight-drop model. We use a force of 50 gm · cm (a 10 g weight dropped 5 cm), in contrast to the 200, 600, or 1000 gm · cm impacts used by Feeney. Despite this, the lesion produced with our model is larger and is not limited to the cortex, most likely because of our elimination of the footplate and resultant direct impact and parenchymal penetration by the weight. One potential criticism of the weight-drop method has been the effect of bouncing of the rod after impacting onto the footplate. Penetration of the rod into the parenchyma in our modification is not accompanied by any bounce as assessed visually, and almost certainly precludes this problem. The actual contribution of bounce to the observed lesion in the Feeney model is unclear, and may be minimal. Our modification does, however, produce injury to the dura. To summarize, although both models use the free weight-drop mechanism, there are differences in the mechanism of insult and the resultant pathobiology.

G. ACUTE INFLAMMATORY RESPONSE TO TRAUMATIC BRAIN INJURY

One of the two major areas that we have used this weight-drop model to investigate is the nature of the acute inflammatory response to traumatic brain injury. A hypothetical scheme for the role of acute inflammation in the evolution of brain injury after trauma is shown in Figure 2, and specific aspects of this scheme are beginning to be studied. In the following section, some of the current findings in the area of neutrophils and the acute inflammatory response to traumatic brain injury are presented.

1. Posttraumatic Neutrophil Accumulation

Neutrophil infiltration was reported by Persson[24] at 12 h after stab wounds in rat brain, and by Means and Anderson[25] in venules at 4 h after spinal cord injury in cats. However, neither study addressed the possibility of a role for neutrophils in cerebrovascular disturbances or secondary injury. Using autologous ^{111}In-labeled neutrophils and ^{51}Cr-labeled erythrocytes, neutrophils accumulate in the injured hemisphere after weight-drop trauma.[7] Accumulation was explained by an increase in cerebral blood volume (presumably largely due to hemorrhage) between 0 and 2 h posttrauma, but between 4 and 8 h, it was not due to hemorrhage and represented acute inflammation (Figure 3). Horner et al.[22] also studied neutrophil accumulation in an identical weight-drop trauma model (10 g × 10 cm impact) using the method of Barone et al.[26] (quantifying neutrophil accumulation with a tissue myeloperoxidase [MPO] assay). Horner et al.[22] also used immunohistochemical methods to define further the type and location of the accumulated leukocytes. Brain homogenates were evaluated for MPO activity and sections were examined for leukosialin staining at 4 h, 16 h, 3 d, and 7 d posttrauma. Neutrophil accumulation at 4 h was limited to areas of hemorrhage, confirming our prior study.[7] Neutrophils at 16 h, however, were in a striking perivascular location—representing acute inflammation.[22] By 3 d, neutrophils were not seen, and monocyte accumulation was noted, which peaked at 7 d.

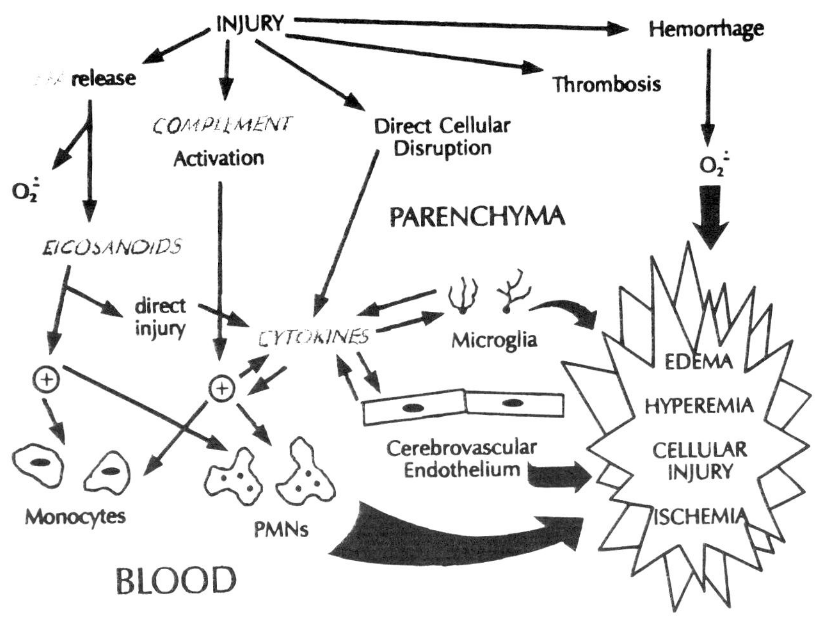

FIGURE 2. Schematic illustration of the hypothesized acute inflammatory pathways in the pathogenesis of traumatic brain injury and how they may influence important pathophysiologic processes such as posttraumatic edema, hyperemia, cellular injury, and ischemia.

Clark et al.[12] applied the MPO assay to determine the time course of neutrophil accumulation in the weight-drop model and confirmed the results of Horner et al.[22] Brain MPO activity was not increased at 2 h posttrauma compared to normal control, but marked increases in brain MPO activity were noted at both 24 and 48 h posttrauma. These increases corresponded to approximately 4×10^6 neutrophils per gram brain tissue at 24 h posttrauma.[12] Neutrophil accumulation resolved by 7 d posttrauma. Clark et al.[12] also studied neutrophil accumulation in the controlled cortical impact model, a standardized and contemporary model of traumatic brain injury developed by Lighthall[27] and modified by Palmer et al.,[28] which produces contusion and axonal injury. Using a moderate insult (4 m/s velocity and a 2.5 mm depth), the time course of accumulation was identical to that observed in weight drop, but the amount of accumulation was approximately 30 to 40% less than that observed in our weight-drop model at both 24 and 48 h posttrauma.[12]

2. Inhibition of Posttraumatic Neutrophil Accumulation

Biagas et al.[29] reported that induction of circulating neutropenia using vinblastine sulfate (0.5 mg/kg IV, given 5 d before trauma) results in profound and relatively selective neutropenia in rats (24 ± 24 neutrophils per microliter in peripheral blood) with no appreciable effects on platelets, monocytes, lymphocytes, or hematocrit. Vinblastine treatment reduced posttraumatic neutrophil accumulation (MPO activity) to control values at 24 h posttrauma.[29] Recently, Kaczorowski et al.[30] reported that posttraumatic neutrophil accumulation (MPO activity at 24 h posttrauma) is attenuated by over 40% in our weight-drop model by pre- and posttrauma treatment with recombinant soluble complement receptor-1 (sCR-1)—a clinically relevant and reversible inhibitor of both the intrinsic and extrinsic pathways of complement activation. Products of complement activation such as $C3_a$ and $C5_a$ are potent neutrophil chemotaxins and are involved in numerous models of inflammation.[31] Studies with sCR-1 treatment in trauma models outside of the CNS (such as cutaneous thermal injury in rats) demonstrate a very similar level of inhibition of neutrophil accumulation (about 40%).[31] Similarly, Hines

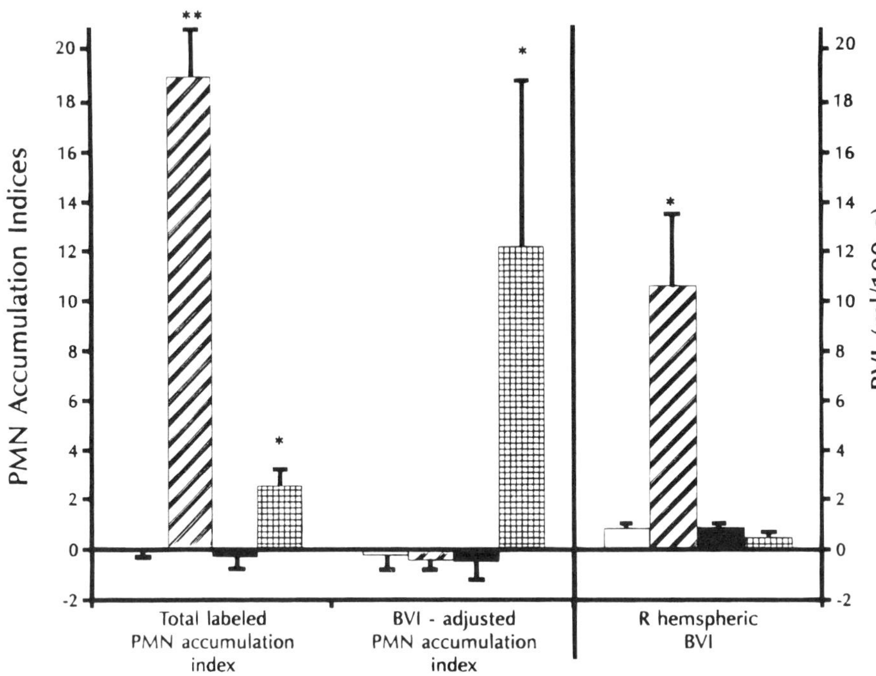

FIGURE 3. Total labeled neutrophil (PMN) accumulation and blood-volume-adjusted (BVI) neutrophil accumulation in sham and trauma groups between 0 and 2 h posttrauma (sham □, trauma ▨) and between 4 and 8 h posttrauma (sham ■, trauma ▦). $*p < 0.05$, or $**p < 0.01$ vs. respective sham group. Neutrophils were labeled with ^{111}In and erythrocytes were labeled with ^{51}Cr. Neutrophil accumulation was related to an expanded blood volume (presumably due to hemorrhage) between 0 and 2 h posttrauma: however, between 4 and 8 h posttrauma, neutrophil accumulation was not due to hemorrhage and presumably represented the initial phases of the acute inflammatory response. (From Schoettle, R. J. et al., *J. Neurotrauma,* 7, 207–217, 1990. With permission.)

et al.[32] recently demonstrated a 37% reduction in posttraumatic (weight-drop) neutrophil accumulation in rats treated with an inhibitor of the endothelial intercellular adhesion molecule ICAM-1.

Based on all of these studies, we conclude that: (a) neutrophil accumulation occurs after cerebral trauma (weight-drop) in rats, and is quantifiable using the MPO assay: (b) immediately posttrauma, neutrophils are associated with hemorrhage; however, based on the MPO data, the magnitude of this accumulation is small compared to that from inflammation during the initial 48 h posttrauma; (c) accumulation associated with inflammation begins between 4 and 8 h posttrauma and peaks between 24 and 48 h; (d) accumulation in the injured hemisphere is perivascular; (e) neutrophil accumulation also occurs in the controlled cortical impact model, and has a similar time course to that observed in the weight-drop model; and (f) the complement cascade and leukocyte adhesion molecules (ICAM-1) are involved in this process, accounting for at least 40% of the posttraumatic neutrophil accumulation in brain.

3. Neutrophil Accumulation and Alterations in Posttraumatic CBF

The possibility of a causal relationship between neutrophil accumulation and posttraumatic cerebrovascular disturbances was again investigated, using vinblastine-induced neutropenia. Based on the role of neutrophils in inflammation outside of the CNS, we tested the hypothesis that neutrophils contribute to posttraumatic hyperemia (24 h) in our weight-drop model. Neutrophil depletion with vinblastine reduced posttraumatic CBF (autoradiography), but only in the injured hemisphere—where neutrophils accumulate.[10] Surprisingly, CBF was reduced not only in hyperemic areas, but also in areas with normal CBF. In that cerebral metabolic rate is reduced at 24 h posttrauma in animal models and in humans, it is anticipated that even normal CBF represents luxury flow (relative hyperemia).[10,11] These are the first data to suggest that the acute inflammatory process contributes to the genesis of posttraumatic cerebral hyperemia.

Neutrophils are also implicated in producing hypoperfusion in brain. In certain models of cerebral ischemia, they contribute to flow reductions by plugging vessels in areas of low flow (rheologic effects).[33] Thus, the hypothesis that neutrophils contribute to early posttraumatic hypoperfusion was also tested by comparing CBF (autoradiography) at 2 h posttrauma in vinblastine- and saline-treated rats. CBF was similarly reduced at 2 h posttrauma in both groups.[10] Thus, in contrast to the amelioration of hyperemia at 24 h posttrauma by vinblastine treatment in our model, no effect was observed on the hypoperfusion at 2 h posttrauma. Neutrophils do not contribute to early posttraumatic hypoperfusion.

III. CONCLUSION

Possibly more than just coincidence, in the original description by Feeney et al.[1] of the weight-drop model, macrophages were noted frequently along the lining of the contusion at 4 d posttrauma. These studies using this modified weight-drop model of traumatic brain injury have begun to build on this work, and on the work of Persson,[24] Means and Anderson,[25] and Hallenbeck et al.[34] to define the acute inflammatory response to traumatic brain injury.

REFERENCES

1. Feeney, D. M., Boyeson, M. G., Linn, R. T., Murray, H. M., and Dail, W. G. (1981). Responses to cortical injury: I. Methodology and local effects of contusions in the rat. *Brain Res.* 211, 67–77.
2. McIntosh, T. K., Vink, R., Noble, L., Yamakami, I., Fernayak, S., Soares, H., and Faden, A. L. (1989). Traumatic brain injury in the rat: characterization of a lateral fluid-percussion model. *Neuroscience* 28, 233–244.
3. Dixon, C. E., Clifton, G. L., Lighthall, J. W., and Yaghmai, A. A. (1991). A controlled cortical impact model of traumatic brain injury in the rat. *J. Neurosci. Methods* 39, 253–262.
4. Abd-Elfattah Foda, M. A. and Marmarou, A. (1993). Early and late morphological changes following a new model of brain injury in rats: similarities to diffuse axonal injury in man. *Proceedings of the Eleventh Annual Neurotrauma Symposium,* November, Washington, D.C.
5. Povlishock, J. T., Erb. D. E., and Astruc, J. (1992). Axonal response to traumatic brain injury: reactive axonal change, deafferentation, and neuroplasticity. *J. Neurotrauma* 9, S189–200.
6. Allen, A. R. (1911). Surgery of experimental lesions of spinal cord equivalent to crush injury of fracture dislocation of spinal column. *J. Am. Med. Assoc.* 57, 878–880.
7. Schoettle, R. J., Kochanek, P. M., Magargee, M. J., Uhl, M. W., and Nemoto, E. M. (1990). Early polymorphonuclear leukocyte accumulation correlates with the development of posttraumatic cerebral edema in rats. *J. Neurotrauma* 7, 207–217.
8. Grundl, P. D., Biagas, K. V., Kochanek, P. M., Schiding, J. K., Barmada, M. A., and Nemoto, E. M. (1994). Early cerebrovascular response to head injury in immature and mature rats. *J. Neurotrauma* (in press).
9. Biagas, K. V., Grundl, P. D., Schiding, J. K., and Kochanek, P. M. (1992). Hyperemic response to percussive trauma in immature, mature, and aged rats. *Soc. Neurosci. Abstr.* 18, 1088.
10. Kochanek, P. M., Uhl, M. W., Biagas, K. V., Schiding, J., Grundl, P. D., and Nemoto, E. M. (1993). Effect of neutrophil depletion on regional cerebral blood flow (rCBF) after trauma in rats. *J. Cereb. Blood Flow Metab.* 13, S121.
11. Obrist, W. D., Marion, D. W., Aggarwal, S., and Darby, J. M. (1993). Time course of cerebral blood flow and metabolism in comatose patients with acute head injury. *J. Cereb. Blood Flow Metab.* 13, S571.
12. Clark, R., Schiding, J., Kaczorowski, S., Marion, D., and Kochanek, P. (1994). Neutrophil accumulation after traumatic brain injury in rats: comparison of weight drop and controlled cortical impact models. *J. Neurotrauma,* 11, 499–506.
13. Uhl, M. W., Schoettle, R. J., Kochanek, P. M., Margargee, M. J., and Nemoto, E. M. (1990). Neutropenia does not prevent post-traumatic cerebral edema. *Crit. Care Med.* 18, S274.

14. Bouman, H. J. and Van Wimersma Greidanus, T. B. (1979). A rapid and simple cannulation technique for repeated sampling of cerebrospinal fluid in freely moving rats. *Brain Res. Bull.* 4, 575–577.
15. Yamakami, I. and McIntosh, T. K. (1989). Effects of traumatic brain injury on regional cerebral blood flow in rats as measured with radiolabeled microspheres. *J. Cereb. Blood Flow Metab.* 9, 117–124.
16. Bryan, Jr., R. M., Cherian, L., and Robertson, C. (1993). Regional cerebral blood flow after controlled cortical impact injury in rats. *Proceedings of the Eleventh Annual Neurotrauma Symposium,* November, Washington, D.C.
17. Pappius, H. (1981). Local cerebral glucose utilization in thermally traumatized rat brain. *Ann. Neurol.* 9, 484–491.
18. Busto, R., Yoshida, S. Ginsberg, M. D., Alonso, O., Smith, D. W., and Goldberg, W. J. (1984). Regional blood flow in compression-induced brain edema in rats: effect of dietary vitamin E. *Ann. Neurol.* 15, 441–448.
19. Obrist, W. D., Langfitt, T. W., Jaggi, J. L., Cruz, J., and Gennarelli, T. A. (1984). Cerebral blood flow and metabolism in comatose patients with acute head injury. *J. Neurosurg.* 61, 241–253.
20. Feeney, D. M. and Westerberg, V. S. (1990). Norepinephrine and brain damage: alpha noradrenergic pharmacology alters functional recovery after cortical trauma. *Can. J. Psychol.* 44, 233–252.
21. Dail, W. G., Feeney, D. M., Murray, H. M., Linn, R. T., and Boyeson, M. G. (1980). Responses to cortical injury: II. Widespread depression of the activity of an enzyme in cortex remote from a focal injury. *Brain Res.* 211, 79–89.
22. Horner, H. C., Setler, P. E., Fritz, L. C., and Hines, D. (1992). Characterization of leukocyte infiltration in traumatic brain injury in the rat. *Soc. Neurosci. Abstr.* 18, 173.
23. DeKosky, S., Goss, J., Miller, P., Styren, S., Kochanek, P., and Marion, D. (1994). Upregulation of nerve growth factor following cortical trauma, *Exp. Neurol.* 130, 173–177.
24. Persson, L. (1976) Cellular reactions to small cerebral stab wounds in the rat frontal lobe. *Virchows Arch. B. Cell. Path.* 22, 21–37.
25. Means, E. D. and Anderson, D. K., (1983). Neuronophagia by leukocytes in experimental spinal cord injury. *Neuropathol. Exp. Neurol.* 42, 707–719.
26. Barone, F. C., Hillegass, L. M., Price, W. J., White, R. F., Lee, E. V., Feuerstein, G. Z., Sarau, H. M., Clark, R. K., and Griswold, D. E. (1991). Polymorphonuclear leukocyte infiltration into cerbral focal ischemic tissue: myeloperoxidase activity assay and histologic verification. *J. Neurosci. Res.* 29, 336–345.
27. Lighthall, J. (1988). Controlled cortical impact: a new experimental brain injury model. *J. Neurotrauma* 5, 1–15.
28. Palmer, A. M., Marion, D. W., Botscheller, M. L., and Redd, E. E. (1993). Therapeutic hypothermia is cytoprotective without attenuating the traumatic brain injury-induced elevations in interstitial concentrations of aspartate and glutamate. *J. Neurotrauma* 10, 363–372.
29. Biagas, K. V., Uhl, M. W., Schiding, J. K., Nemoto, E. M., and Kochanek, P. M. (1992). Assessment of posttraumatic polymorphonuclear leukocyte accumulation in rat brain using tissue myeloperoxidase assay and vinblastine treatment. *J. Neurotrauma* 9, 363–371.
30. Kaczorowski, S., Schiding, J., Toth, C., and Kochanek, P., Effect of soluble complement receptor-1 on neutrophil accumulation after traumatic brain injury in rats. *J. Cere. Blood Flow Metab.* (in press)
31. Mulligan, M. S., Yeh, C. G., Rudolph, A. R., and Ward, P. A. (1992). Protective effects of soluble CR1 in complement and neutrophil mediated tissue injury. *J. Immunol.* 148, 1479–1485.
32. Hines, D., Fritz, L., Yednock, T., Setler, P., and Horner, H. (1993). Anti-intercellular cell adhesion molecule (anti-ICAM-1) antibody reduces myeloperoxidase activity in rat head injury. *Soc. Neurosci. Abstr.* 19, 1875.
33. Kochanek, P. M. and Hallenbeck, J. M. (1992). Polymorphonuclear leukocytes and monocytes/macrophages in the pathogenesis of cerebral ischemia and stroke. *Stroke* 23, 1367.
34. Hallenbeck, J. M., Dutka, A. J., Tanishima, T., Kochanek P. M., Kumaroo, K. K., and Thompson, C. B. (1986). Polymorphonuclear leukocyte accumulation in brain regions with low blood flow during the early postischemic period. *Stroke* 17, 246–253.

Chapter 18

RODENT MODELS OF TRAUMATIC BRAIN INJURY AND MOTOR FUNCTION ASSESSMENT: CONTROLLED CORTICAL IMPACT AND FLUID PERCUSSION MODELS OF TRAUMATIC BRAIN INJURY

C. Edward Dixon and Ronald L. Hayes

TABLE OF CONTENTS

I. INTRODUCTION

The consequences of mechanical brain injury range from a concussive syndrome characterized by brief loss of consciousness in the absence of significant histopathology to prolonged coma or death usually associated with extensive brain pathology.[1,2] Several investigators have developed animal models of traumatic brain injury (TBI) in an attempt to reproduce

0-8493-8094-4/95/$0.00+$.50

various aspects of the biomechanical responses, neurological syndromes, and pathology observed in human closed head injury. These models use a variety of experimental techniques that have been developed to study these syndromes, including devices that accelerate or rotate the skull,[3] impact to the intact and freely moving cranium,[4,5] rapid injection of fluid into a closed cranium (fluid percussion),[6,7] and direct brain impact by either weight-drop[8] or controlled cortical impact.[9,10] In the first section of this chapter, we will describe commonly used methods to produce fluid percussion and controlled cortical impact injury in rats. In the second section, we will describe procedures for the evaluation of motor deficits following experimental TBI in rats.

II. FLUID PERCUSSION

A. RATIONALE

Fluid percussion models produce graded levels of brain injury associated with rapid injection of small volumes of saline into the closed cranial cavity. The classic technique uses a pendulum hammer to hit the end of a plunger within a fluid-filled tube.[6,7,11] The pressure pulse is delivered through a craniotomy inducing a brief intracranial pressure increase and associated deformation of neural tissue. The technique does not attempt to reproduce precisely the array of events following closed head injury in humans, in which injury results from rapid acceleration-deceleration of the head, usually with impact to intact skull. Rather, fluid percussion brain injury successfully produces graded levels of injury associated with predictable neurologic, physiologic, and histologic changes that are comparable to those observed in human brain injury.[7,12–17] As others have pointed out, efforts to reproduce directly the mechanisms of human brain injury usually resulted in models that do not reliably reproduce the neurologic deficits or pathologies seen clinically.[18] Furthermore, when other techniques are used, such as extracranial impact to the intact skull or nonimpact acceleration of the head, it is more difficult to quantify directly and control the forces producing brain deformation, and measurements of brain displacement have not been precise.[18,19] Our laboratories have developed and thoroughly characterized a rat model of fluid percussion brain injury.[7] Fluid percussion injury in the rat reproduces many of the features of head injury observed in other models and species. Thus, this animal model represents a useful experimental approach to studies of pathologic changes similar to those seen in human head injury.

B. INJURY DEVICE

The fluid percussion device consists of a Plexiglas cylindrical reservoir, 60 cm long and 4.5 cm in diameter, bounded at one end by a Plexiglas piston mounted on O-rings (Figure 1). The O-rings are lubricated with white petroleum jelly. The opposite end of the reservoir is fitted with a transducer housing ending with a 2.6-mm-inside-diameter male Leur-loc fitting. The entire system is filled with isotonic saline. Care is taken to ensure that no air bubbles are present within the system. Injury is induced by the descent of a metal pendulum into the piston. This injects a small volume of saline epidurally into the closed cranial cavity, producing a brief displacement and deformation of neural tissue. The resulting pressure pulse is measured in atmospheres (atm) by an extracranial transducer (Statham PA 85-100) and recorded on a storage oscilloscope (Textronix 5111). The magnitude of the pressure pulse can be regulated by varying the height from which the pendulum is released.

C. SURGICAL PROCEDURES

Animals are surgically prepared for fluid percussion injury under sodium pentobarbital anesthesia (54 mg/kg, IP) 48 h prior to trauma. Surgical procedures are performed under sterile conditions in a laboratory site dedicated to rat surgery. A 4.8-mm diameter craniotomy is performed over the sagittal suture midway between lambda and bregma. Two stainless

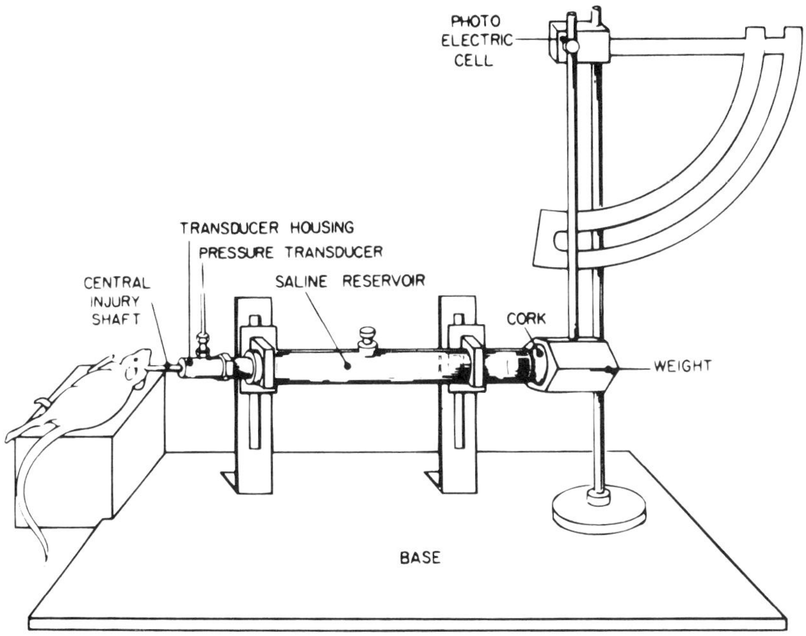

FIGURE 1. Diagram of the fluid percussion model of brain injury in the rat. A 4-mm central craniotomy is connected to one end of a Plexiglas cylinder filled with physiological saline. At the other end in the cylinder is a Plexiglas cork mounted on O-rings. Injury is produced by striking the cork with a 4.8-kg pendulum dropped from a specific height. The pressure transient is recorded on a storage oscilloscope with an extracranial transducer.

steel skull screws (2–56 × 9.5 mm) are placed into burr holes 1 mm rostral to bregma and 1 mm caudal to lambda. A rigid plastic injury tube (modified Leur-loc syringe hub, 2.6 mm inside diameter) is placed over the exposed dura and bonded to the skull with cyanoacrylate adhesive. Dental acrylic is then poured around the injury tube and skull screws and the tube is then sealed with Gelfoam. The outer surface of the tube is abraded to improve the adherence of the acrylic. The scalp is sutured closed and Bacitracin ointment is applied over the incision. Animals are returned to their home cage (one animal per cage) after they have recovered from anesthesia.

Animals are injured 48 h after surgery at an injury magnitude determined by the specific needs of an investigator. All rats are anesthetized with an inhalant anesthesia such as isoflurane (4%) prior to injury. Prior to injury, the injury level is preset by blocking the end of the connecting tube and adjusting the height of the pendulum drop to produce the desired pressure level. After injury, animals are then observed for the presence of convulsions and apnea. If desired, ventilatory support (room air) can be given by connecting the animal to a rodent respirator with a 60-cc syringe securely placed over their face.

D. METHODOLOGICAL ALTERNATIVES

In recent years, a number of variations have been made to the basic rat fluid percussion model. These variations include: (a) site of injury (e.g., central vs. lateral), (b) anesthetic regimen, including ventilatory support, (c) method of affixing the injury tube over the craniotomy, (d) interval from injury tube implantment and injury, (e) length and material of the connecting tube, and (f) the aperture of the injury tube. Different combinations of these variations can result in profoundly different injury profiles. For example, relocating the site of injury from the vertex to a more lateral orientation can change the model from one that produces a diffuse moderate brain injury with little detectable histopathology,[20] to one that produces a more focal contusion and cell loss.[11]

III. CONTROLLED CORTICAL IMPACT

A. RATIONALE

Recently, the controlled cortical impact technique has been developed in which the biomechanical events contributing to injury can be analyzed by establishing a quantifiable relationship between measurable engineering parameters, such as force, velocity, or deformation, and the magnitude of tissue damage and/or functional impairment. The application of a stroke-constrained pneumatic impactor allows accurate, reliable, and independent control of the deformation parameters over a wide range of contact velocities, and has been demonstrated in spinal cord injury models to produce contusion and allow a reproducible gradation of injury outcome severity.[21,22]

Experimental TBI, employing a pneumatic impactor, has been well characterized in the laboratory ferret[9,23] and rat.[10] In general, the injury responses resembled aspects of severe closed head injury in humans. In contrast to fluid percussion models of traumatic brain injury,[7,24–26] the cortical impact model of experimental brain injury uses a known impact interface, and a measurable, controllable impact velocity and cortical deformation. These controlled mechanical variables enable the amount of deformation and the change in deformation over time to be accurately determined.

B. INJURY DEVICE

Our injury device is modified from similar devices developed at the Biomedical Science Department of the General Motors Research Laboratories. Built in the Bioengineering Department at the Medical College of Virginia, the pneumatic impactor consists of a small (1.975 cm) bore, double-acting, stroke-constrained pneumatic cylinder with a 5.0-cm stroke (see Figure 2). The cylinder is rigidly mounted in a vertical position on a horizontal crossbar. An impactor tip (i.e., that part of the shaft that contacts the exposed dura mater) is attached to the lower rod end. The upper rod end is attached to the transducer core of a linear variable

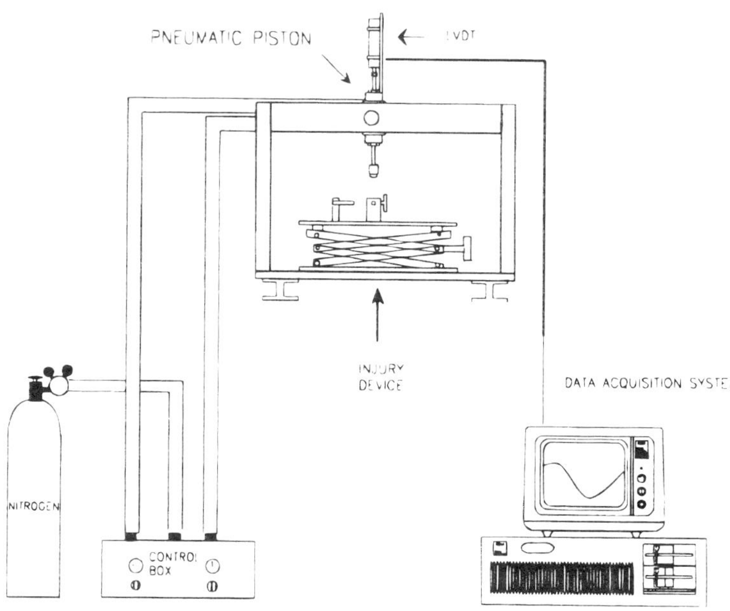

FIGURE 2. Diagram of the controlled cortical impact model of traumatic brain injury in the rat. A control box controls the flow of high-pressure gas to the pneumatic piston. An impactor tip is attached to the lower piston rod. A linear variable differential transducer (LVDT) is attached to the upper piston rod to precisely measure the velocity of the stroke. Output from the LVDT is stored and analyzed on a PC-based data-acquisition system.

displacement transducer (LVDT). The impact velocity can be adjusted between 1.0 and 7.0 m/s by controlled gas pressure. Impact velocity is directly measured by the LVDT (Shaevitz Model 500 HR), which produces an analog signal recorded by a PC-based data-acquisition system (R.C. Electronics) for analysis of time/displacement parameters of the impact. The animals are rigidly mounted in a stereotaxic head holder that is attached to a laboratory jack that can be adjusted in the vertical axis beneath the impactor. Thus, the amount of tissue deformation can be precisely controlled by adjusting the height of the animal's brain surface relative to the impactor tip at full extension. Zeroing the device is the most subjective part of the procedure. Our group has attempted to resolve this problem by employing an audible baseline monitor (World Precision Instruments, Model ABM-C), which generates a change in pitch when the impactor tip touches the dura during the zeroing process.

C. SURGICAL PROCEDURES

All animals are initially anesthetized with 4% isoflurane with a 2:1 N_2O/O_2 mixture in a vented anesthesia chamber. Following endotracheal intubation, rats are mechanically ventilated with a 2% isoflurane mixture. The rats are then mounted in the injury device stereotaxic frame in a prone position secured by ear bars and incisor bar. The head is held in a horizontal plane with respect to the interaural line. A midline incision is made, the soft tissues reflected, and a 10-mm craniotomy is made centrally between bregma and lambda. Core body temperature can be monitored continuously by a rectal thermistor probe and maintained at 37 to 38°C.

D. METHODOLOGICAL ALTERNATIVES

Like the fluid percussion model, a number of variations have been made to the basic rat controlled cortical impact model. These variations include: (a) site of injury (central vs. lateral), (b) size and shape of the impact tip, (c) amount of deformation, (d) impact velocity, (e) use of secondary craniotomies to direct the movement of tissue, (f) angle of impact, and (g) anesthetic regimen, including ventilatory support. Figure 3 demonstrates that adding a second bilateral craniotomy can increase the beam-walking deficits over those produced by a lateral impact with a unilateral craniotomy. Figure 4 compares the logistic regression plots between midline and lateral impacts on the probability of occurrence of contusion as a function of impact velocity. Lateral impacts appear more likely to produce cortical contusions that midline impacts as velocity increases.

IV. ASSESSMENTS OF MOTOR FUNCTION FOLLOWING TBI

Motor deficits are frequently used end points for assessing functional deficits in rats following TBI. Initially, we will describe a battery of tests to measure loss of responsiveness following TBI. Acute loss of responsiveness correlates highly with the loading parameters of both fluid percussion and cortical impact models. Therefore, if the experimental manipulations are performed after recovery of the acute responses, the acute responses can be used to ensure the consistency of the magnitude of injury. We next describe protocols with detailed instructions for conducting the beam-balance and beam-walking tests that were employed in the initial characterization of both the fluid percussion[7] and cortical impact models.[10] These tests require that the animals be trained and tested prior to injury. The beam tests have been used extensively to study mechanisms of TBI and for preclinical testing. Last, we describe a modified rotarod test, developed by Hamm et al.[27] The modified rotarod test is a promising new technique that may assess domains not measured by beam-balance or beam-walking tests.

A. ACUTE NEUROLOGICAL ASSESSMENTS

If a specific protocol employs short-acting anesthetics, then acute neurological suppression can be assessed by using a battery of tests that have been shown to be sensitive to varying

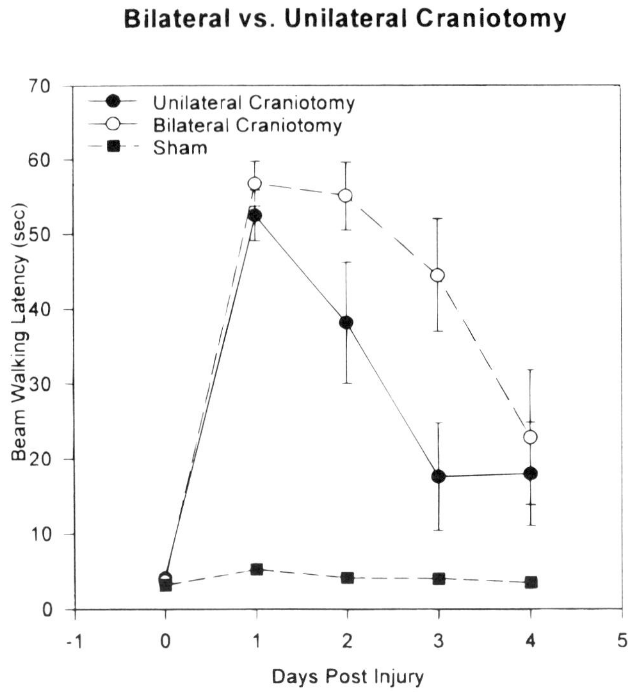

FIGURE 3. Differences in beam-walking performance following a lateral cortical impact with either a unilateral or a bilateral craniotomy. All rats were trained prior to injury on the beam-walking task. The unilateral group (n = 8) was impacted through a lateral craniotomy at a velocity of 6 m/s and a depth of 2 mm. The bilateral group (n = 8) was impacted identically, but rats were given a second contralateral craniotomy. A sham group (n = 8) underwent the unilateral or bilateral craniotomy but were not impacted. Results indicate that the bilateral group exhibited longer beam-walking deficits than the unilateral group. $p < 0.05$.

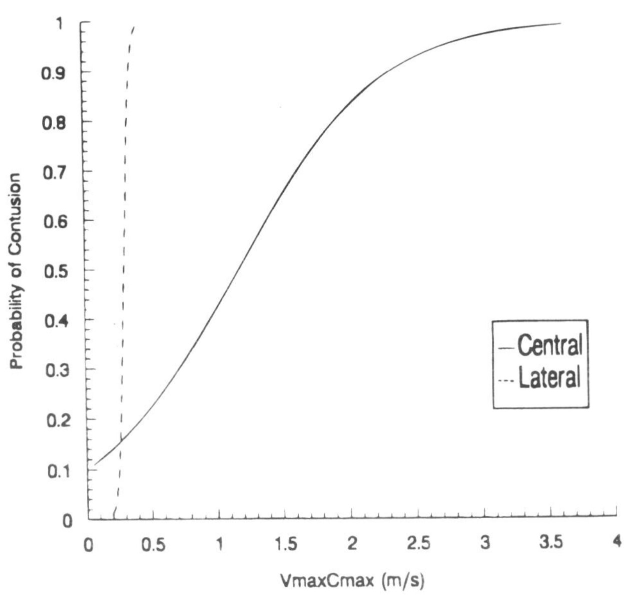

FIGURE 4. A comparison between midline and lateral impacts. LOGISTS plots, generated by SAS, showing the functions by which the probability of a cortical contusion increases as velocity increases.

magnitudes of both fluid percussion and controlled cortical impact models of TBI. This battery, which includes tests analogous to motor components of the Glasgow Coma Scale,[28] allow quantification of various features of the suppression of the animal's responsiveness to external stimuli and relationships between durations of behavioral suppression and magnitudes of injury.

Assessments of simple nonpostural somatosensory functions are conducted by recording the duration of suppression of two responses to stimulation. The pinna reflex is assessed by touching the auditory meatus of the animal's ear to elicit a vigorous head shake. The corneal reflex is evaluated by lightly touching the cornea with a cotton swab to elicit an eyeblink. Assessments of simple postural somatomotor functions include measurements of the duration of suppression of flexion reflexes to stimulation. The assessment of the paw and tail flexion reflexes, consist of the gradual application of pressure on the hindpaw or tail until paw or tail withdrawal is noted. Another simple postural somatomotor function, the head support, is assessed by noting the duration of suppression of the animal's ability to support the weight of its head. Inability of the animal to support its head is associated with a generalized loss of muscle tone.

More complex postural somatosensory function is assessed by recording the duration of suppression of the righting and escape responses. The righting response will be defined as the animal's ability to right itself three times consecutively after being placed on its back. An escape response will be assessed by briefly pinching the tail of the animal to elicit locomotive activity away from the noxious stimulus.

B. BEAM-BALANCE PROCEDURE

The beam-balance task is used to assess motor and vestibular functioning by observing the animal's ability to balance on a stationary beam. The beam is approximately 2 cm wide, 2 cm high, and 30 cm long. The balance beam is attached to the top edge of a bench so that it extends over a cushion placed on the floor.

1. Training

The animal is set on the beam balance with its body placed lengthwise along the beam. Time the animal for a 60-s interval after the animal has initially secured its balance on the beam. Should the animal fall, place him on the beam again, and restart the timer. Should the animal fall more than four times, remove the animal from the beam for at least a 10-min rest before starting again. Place a foam cushion beneath the balance beam to protect the animal should it fall off. After a 60-s interval, remove the animal for a moment to disorient him, and replace him on the beam for another 60-s trial. Repeat this procedure until the animal has spent three consecutive trials on the beam at what is a score of two or better (see Table 1).

TABLE 1
Scale for Beam Balancing

Scores	Behaviors
1	Balances with steady posture
2	Grasps sides of beam and/or has shaky movement
3	Hugs the beam, or slips or spins on beam
4	Attempts to balance on beam but falls off
5	Drapes over beam or hangs on beam, and falls off
6	Falls off: no attempt to balance or hang on beam

2. Preinjury Assessment

The animal is set on the beam balance with its body placed lengthwise along the beam and timed for a 60-s interval when it has secured its balance on the beam. Two more 60-s trials are run, momentarily removing him from the beam between trials to disorient him. The score and the number of seconds the animal stayed on the beam is recorded. For pre-assessment, the animal must stay on the bar 60 s each trial at a score of two or better in the first three consecutive trials. The score and the latency for each of the three trials in the beam-balance section is recorded as the animal's preinjury performance.

3. Postinjury Assessment

Set the animal on the beam balance. Begin timing the animal for a 60-s interval when the animal has secured its balance on the beam. For each trial, record the score and number of seconds that it stayed on the beam. If the animal falls, record the score and latency, and begin the next trial. If the animal remains on the beam for the whole 60 s, remove him to disorient him as before, and place him on the beam to begin the next trial. Run three trials only, regardless of whether the animal achieves three full 60-s trials. Remember to be quiet during all testing and training procedures, and to not rush.

C. BEAM-WALKING PROCEDURE

The beam-walking task allows the assessment of more refined locomotor activity in the context of a learned avoidance task. Prior to injury, rats are trained to escape a bright light and loud white noise (Heath Model AD-1309) by traversing a narrow wooden beam (2.5 × 100.0 cm) to enter a darkened goal box at the opposite end of the beam.

1. Training

Place the rat in the goal box for 60 s. Turn on white noise at the end of 60 s. Remove the rat from the box and place it on the beam with its nose close to the goal box. Turn the noise off when the animal's two front paws have entered the goal box. Leave the rat in the goal box for 30 s after each trial. On successive trials, place the animal further away from the goal box each time, moving the distance to the next peg hole. When the animal can run the full length of the beam well, place the four pegs in the peg holes of the beam.

For the remainder of training, run several trials timing the rats using the following methods. Turn on noise just before removing the animal from the box. Start the timer when the animal has been placed on the beam and achieved good footing. Stop the timer and noise when both front paws have entered the goal box. Each time the rat is removed from the goal box and placed on the beam, help it to get a good footing on the beam before moving your hand away. Run trials until the animal can run the beam well at an average latency of 10 s or less. Most rats only require a maximum of 20 trials to train appropriately, including three to five timed trials. If the rat is difficult to train, give it a 10- to 20-min break before continuing. Do not exhaust the rat.

2. Preinjury Assessment

Begin running the animal trials on the full length of the beam with the pegs in place. Prior to testing place the rat in the goal box for 60 s. Time each trial, using this procedure until the animal achieves three consecutive trials at or below the 5-s mark. After the final trial, leave the rat in the goal box for 30 s. Rats that do not reach the criteria in 15 or fewer trials should not be used for an experiment.

3. Postinjury Assessment

Begin running the animal trials on the full length of the beam with the pegs in place. Run and record latencies for the first three trials only. Start by placing the rat in the goal box for 60 s. If a rat takes longer than 60 s to run the beam, place it in the goal box from the front of box. Record any slips, pauses, or unusual behavior. After the final trial, leave the rat in the goal box for 30 s.

It is important to form a strong association between the animal reaching the goal box and the aversive noise being turned off, so always turn the noise off as soon as the front paws have entered the box, but not any sooncr. Do not rush or fatigue an injured rat. Handle the animal with care at all times.

D. ROTAROD

1. Rationale

The rotarod task first developed by Dunham and Miya[29] is a frequently used screening test for evaluating neurological effects of various drugs. Recently, the rotarod test has been modified to assess chronic neurological deficits following TBI in rats (Hamm et al., 1994). Initial findings suggest that, compared to the beam-balance and beam-walking tasks, the rotarod task is a more sensitive and efficient index for assessing motor impairment produced by varying magnitudes of brain injury. Whether it is a more sensitive measure for theraputic efficacy is under investigation.

2. Rotarod Device

The rotarod device consists of a metal frame with a motorized (Gearmotor, Bodine Electric Company), rotating assembly of rods. Rods were chosen for the walking surface rather than a continuous surface used on commercially available rotarod units. The rationale for using rods are that grip strength has been found to be decreased following TBI,[30] and it was felt that requiring the rats to grip the rods in order to walk on the rotarod may increase the sensitivity of the apparatus. Two large, clear Plexiglas disks are positioned on each side of the frame. The walking area of the apparatus is a cylinder that consists of 18 rungs (1-mm stainless steel rods). The animal's task is to walk on the rungs while they rotate. A control box (Bodine Electric Company) located adjacent to the device contains controls for speed, direction of rotation (forward or reverse), and a brake. Rotational speeds ranged from 0 to a maximum speed of 30 rpm.

3. Rotarod Procedure

The rotarod procedure used to assess injury-induced motor deficits involves placing each animal on the device either facing the wall (forward direction) or facing the experimenter (reverse) and the device rotated in the opposite direction of the animal. Directions should be alternated between animals. The animal is allowed to remain stationary for 10 s at 0 rpm. The speed is slowly increased to 3 rpm for 10 s and then steadily increased by 3 rpm in 10-s intervals until the maximum speed of 30 rpm is reached. The animal remains on the device at this speed for 20 s until the 2-min test period elapses. A trial also ends if the animal falls completely off the rungs or grips the device and is spun around for two consecutive revolutions without attempting to walk on the rungs. Animals are tested for two trials per day and the mean duration (in seconds) on the device is recorded.

ACKNOWLEDGMENTS

This work was supported by grants CDC-R49-CCR303547, NIDRR H133B80029, NIH-NS21458 and NS31998. All animal studies carefully conformed to the *Guide for the Care and Use of Laboratory Animals* from the U.S. Department of Health and Human Services and were approved by the University of Texas Health Science Center at Houston Animal Welfare Committee.

REFERENCES

1. Denny-Brown, D. and Russel, W. R. (1941) Experimental cerebral concussion, *Brain,* 64: 93–164.
2. Ommaya, A. K. and Gennarelli, T. A. (1974) Cerebral concussion and traumatic unconsciousness. Correlation of experimental and clinical observations on blunt head injuries, *Brain,* 97: 633–654.
3. Gennarelli, T. A. and Thibault, L. E. (1985) Biological models of head injury, in Becker, D. P. and Povlishock, J. E. (eds): *Central Nervous System Trauma Status Report, 1985,* Bethesda, MD: National Institute of Neurological and Communicative Disorders and Stroke, pp. 391–404.
4. Nilsson, B., Ponten, U., and Voigt, G. (1977) Experimental head injury in the rat. Part 1: Mechanics, pathophysiology, and morphology in an impact acceleration trauma model, *J. Neurosurg.,* 47: 241–251.
5. Marmarou, A., Montasser, A., Abd-Elfattah, F. Van Den Brink, W., Campbell, J., and Demetridou, K. (1994) A new model of diffuse brain injury in rats, *J. Neurosurg.,* 80:, 291–300.
6. Sullivan, H. G., Martinez, J., Becker, D. P., Miller, J. D., Griffith, R., and Wist, A. O. (1976) Fluid percussion model of mechanical brain injury in the cat, *J. Neurosurg.,* 45: 520–534.
7. Dixon, C. E., Lyeth, B. G., Povlishock, J. T., Findling, R. L., Hamm, R. J., Marmarou, A., Young, H. F., and Hayes, R. L. (1987) A fluid percussion model of experimental brain injury in the rat, *J. Neurosurg.,* 67: 110–119.
8. Feeney, D. M., Boyeson, M. G., Linn, R. T., Murray, H. M., and Dail, W. G. (1988) Responses to cortical injury. I: Methodology and local effects of contusions in the rat. *Brain Res.,* 211: 67–77.
9. Lighthall, J. W. (1988) Controlled cortical impact: a new experimental brain injury model, *J. Neurotrauma,* 5(1): 1–15.
10. Dixon, C. E., Clifton, G. L., Lighthall, L. W., Yaghmai, A. A., and Hayes, R. L. (1991) A controlled cortical impact model of traumatic brain injury in the rat, *J. Neurosci. Methods,* 39: 253–262.
11. McIntosh, T. K., Nobel, L., Andrews, B., and Faden, A. I. (1987) Traumatic brain injury in the rat: characterization of a midline fluid-percussion model, *Cent. Nerv. Syst. Trauma,* 4(2): 119–134.
12. Becker, D. P. (1984) The temporal genesis of primary and secondary brain damage in experimental and clinical head injury. NATO Advanced Workshop on Mechanisms of Secondary Brain Damage, Mauls:Sterzing, Italy.
13. Bruce, D. A., Langfitt, T. W., Miller, J. D., et al. (1973) Regional cerebral blood flow, intracranial pressure and brain metabolism in comatose patients, *J. Neurosurg.,* 38: 131–144.
14. Fieschi, C., Beduschi, A., Agnoli, A., et al. (1972) Regional cerebral blood flow and intraventricular pressure in acute brain injuries. *Eur. Neurol.,* 8: 192–199.
15. Hayes, R. L., Pechura, C. M., Katayama, Y., et al. (1984) Activation of pontine cholinergic sites implicated in unconsciousness following cerebral concussion in the cat, *Science,* 233: 301–303.
16. Lewelt, W., Jenkins, L. W., and Miller, J. D. (1980) Autoregulation of cerebral blood flow after experimental fluid-percussion injury of the brain. *J. Neurosurg.,* 53: 500–511.
17. Wei, E. P., Dietrich, W. D., Povlishock, J. T., et al. (1980) Functional, morphological, and metabolic abnormalities of the cerebral microcirculation after concussive brain injury in cats, *Circ. Res.,* 46: 37–47.
18. Langfitt, T. W. and Obrist, W. D. (1981) Cerebral blood flow and metabolism after intracranial trauma, in: *Progress in Neurosurgical Surgery.* H. Krayenbuhl (ed). S. Karger: Basel, Switzerland, Vol. 10, pp. 14–48.
19. Gurdjian, E. S., Lissner, H. R., Webster, J. E., et al. (1954) Studies on experimental concussion. Relation of physiologic effect to time duration of intracranial pressure increase at impact, *Neurology,* 4: 674–681.
20. Lyeth, B. G., Jenkins, L. W., Hamm, R. J., Dixon, C. E., Phillips, L. L., Clifton, G. L., Young, H. F., and Hayes, R. L. (1990) Prolonged memory impairment in the absence of hippocampal cell death following traumatic brain injury in the rat, *Brain Res.,* 526: 249–258.
21. Anderson, T. E. (1982) A controlled pneumatic technique for experimental spinal cord contusion, *J. Neurosci. Methods,* 6: 327–333.
22. Kearney, P. A., Ridella, S. A., Viano, D. C., and Anderson, T. W. (1988) Interaction of contact velocity and cortical compression in determining the severity of spinal cord injury, *J. Neurotrauma,* 5(3): 187–208.
23. Lighthall, J. W., Goshgarian, H. G., and Pinderski, C. R. (1990) Characterization of axonal injury produced by controlled cortical impact, *J. Neurotrauma,* 7(2): 65–76.
24. Lindren, S. and Rinder, L. (1966) Experimental studies in head injury. II. Pressure propagation in "percussive concussion", *Biophysik,* 3: 174–180.
25. Hayes, R. L., Stalhammar, D., Povlishock, J. T., Allen, A. M., Galinat, B. J., Becker, D. P., and Stonnington, H. H. (1987) A new model of concussive brain injury in the cat produced by extradural fluid volume loading: I. Physiological and neuropathological observations, *Brain Injury,* 1(1): 93–112.
26. Dixon, C. E., Lighthall, J. W., and Anderson, T. E. (1988) A physiologic, histopathologic, and cineradiographic characterization of a new fluid percussion model of experimental brain injury in the rat, *J. Neurotrauma,* 5(2): 91–104.
27. Hamm, R. J., Pike, B. R., O'Dell, D. M., Lyeth, B. G., and Jenkins, L. W. (In press) The rotarod test: an evaluation of its effectiveness in assessing motor deficits following traumatic brain injury, *J. Neurotrauma.*

28. Teasdale, G. and Jennett, B. (1974) Assessment of coma and impaired consciousness. A practical scale. *Lancet,* 81: 84.
29. Dunham, N. W. and Miya, T. S. (1957) A note on a simple apparatus for detecting neurological deficit in rats and mice. *J. Am. Pharm. Assoc.,* 46: 208.
30. Hall, E. D., Yonkers, P. A., McCall, J. M., and Braughler, J. M. (1988) Effects of 21-aminosteriod U74006F on experimental head injury in mice. *J. Neurosurg.,* 68: 456–461.

C. Spinal Cord Study

Chapter 19

A MODEL OF SPINAL CORD ISCHEMIA

Paul C. Francel, John A. Jane, and Irvin Kron

TABLE OF CONTENTS

ABSTRACT

A new model of spinal cord ischemia is introduced that enables further analysis of potential interventions that may improve functional outcome after acute spinal cord injury. This model is helpful not only when direct spinal cord ischemia (i.e., vascular occlusion from thrombus, embolus, or iatrogenic occlusion) occurs during thoracic or abdominal aortic surgery, but also when ischemia occurs during traumatic spinal cord injury. The model also allows the study of other surgical, pharmacological, and pathophysiological interventions in the future. In this report we discuss its use to analyze the pathophysiology of ischemic spinal cord injury. Ischemia of the spinal cord increases the intraspinal pressure and, in particular, the cerebrospinal fluid pressure, resulting in a decrease in perfusion pressure to the cord, which culminates in pathological destruction. These events can be interrupted to a significant degree using either cerebrospinal fluid drainage or free radical scavengers such as tirilazad, a 21-aminosteroid (Upjohn U-74006F).

I. INTRODUCTION

Mechanical injury to the spinal cord does not usually result in complete anatomical dissection, but a complete functional loss can occur. Since spinal cord function worsens after

0-8493-8094-4/95/$0.00+$.50

injury, a concept of secondary injury has evolved to explain this phenomenon.[21,22,81,86,90] One major aspect of the secondary injury is spinal cord ischemia, which is probably due to multiple factors, including local loss of autoregulation in the injured segment of the spinal cord, major systemic reduction of blood flow,[1,2,24,35,39] and marked reduction in microcirculatory flow in gray and white matter.[5,8,19,25,34,68,74,76,77,79,82] There is often injury resulting from reperfusion after a period of delayed ischemia. In the blunt trauma model of spinal cord injury, secondary mechanical injury cannot be easily separated from damage that is secondary to delayed ischemic injury and probable reperfusion injury.[29,66] However, in this model, any type of spinal cord ischemia can be studied without the superimposition of mechanical trauma.

In this chapter a model of intraaortic balloon occlusion is presented that allows the study of the ischemic component of spinal cord injury in isolation. Such a model provides a testing ground for numerous surgical and pharmacological (medical) interventions that may have clinical utility in the future.

II. EXPERIMENT DESIGN

The details of experimental design depend somewhat on the drugs tested, their biological half-life, and their means of introduction. Any medical interventions must also be timed to be consistent with the use of the test drugs in clinical practice. To illustrate this technique, the experimental design, using the surgical intervention of cerebrospinal fluid (CSF) drainage and the medical (pharmacological) intervention of tirilazad, evaluates their independent and combined effect on limiting ischemic spinal cord injury. Experimenters can change interventions, based on their particular interest, without negating the following basic experiment.

Eight hours before occlusion, New Zealand albino rabbits each received 1 mg/kg of tirilazad mesylate or an equivalent volume of the vehicle intraperitoneally and were anesthetized with 40 mg/kg of ketamine and 10 mg/kg of xylozine intramuscularly. To rule out a possible protective effect of ketamine or xylozine, researchers used control animals undergoing just operation and animals undergoing operation and balloon occlusion but no drug agents. These animals were then compared to those that underwent CSF drainage, tirilazad treatment, or CSF drainage and tirilazad combined. All animals were prepped and draped in a sterile fashion, placed under a halothane hood at 1.5% volume flow rates for maintenance of anesthesia, and hypothermia was controlled with a heating pad. The animals were allowed to ventilate spontaneously. Antibiotics were given preoperatively and an intravenous access was established via an ear vein.

III. OPERATIVE TECHNIQUE

Bilateral inguinal incisions were made to isolate both femoral arteries (Figure 1). A 20-gauge angiocatheter was placed in one artery to monitor the distal aortic pressure. In the contralateral artery, a 3-French vascular catheter was introduced and directed to approximately the level of the inguinal ligament. Both catheters were subsequently transduced. After performing a celiotomy, the surgeons isolated the aorta at the level of the left renal artery. Five minutes prior to balloon occlusion, each animal received either 1 mg/kg of tirilazad mesylate or an equivalent volume and also received heparin intravenously. Two additional doses of 1 mg/kg of test drug or vehicle were then given intravenously at 10 and 20 min after the balloon occlusion. The aorta was occluded using a 3-French catheter advanced proximally within the infrarenal aorta 2 cm above the left renal artery. A vascular loop was then placed via the celiotomy approximately 1 cm inferior to the left renal artery. The loop was then secured tightly with a hemostat and the celiotomy was closed. Since the spinal cord and lumbar spine are fed by radicular arteries branching from the aorta, any aortic occlusion

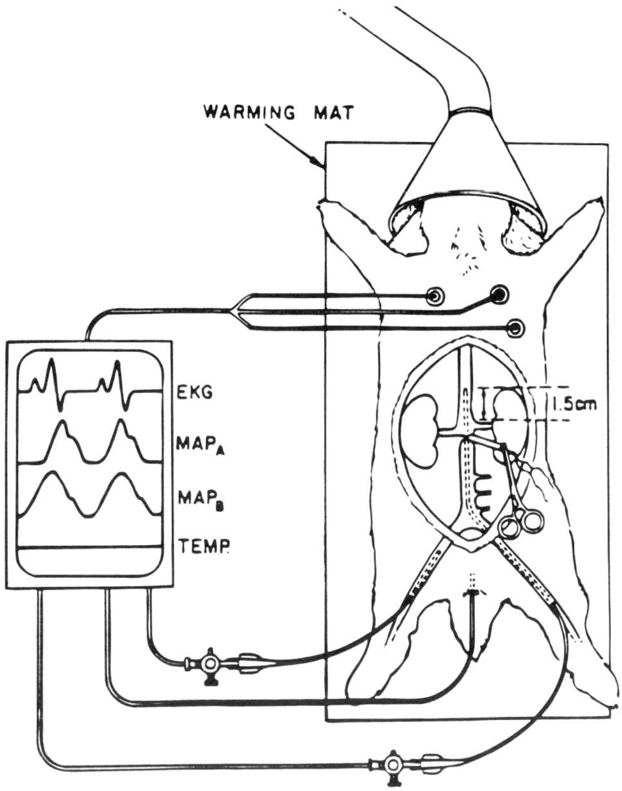

FIGURE 1. Schematic diagram of the experimental design. This system for infrarenal aortic balloon occlusion produces ischemia to the lower portion of the spinal cord in a reversible and controlled manner. Pressure can be monitored within a catheter passed suprarenal, and compared to a catheter placed in the contralateral femoral artery. This gives a measure of the mean arterial pressure, which then can be used in conjunction with a measured cerebrospinal fluid pressure to calculate a perfusion pressure. (From Francel PC, Long BA, Malik JM, Tribble C, Jane JA, Kron IL. *Journal of Neurosurgery,* Nov 1993. With permission.)

would induce ischemia below this level. The heart rate, temperature, and proximal and distal aortic pressures were recorded throughout the 40-min occlusion period.

The evaluation was divided into two phases. Phase one determined the animals' functional neurological outcome. Ten animals given tirilazad mesylate were compared to ten vehicle control animals. Forty minutes of infrarenal aortic occlusion was performed, followed by a strict evaluation of the neurological outcome at 12, 24, 48, 72, and 96 h. The neurological outcome was then graded according to the following scale: 0 = no movement, 1 = trace movement, 2 = sits with assistance, 3 = sits unassisted but cannot hop, 4 = weak hop, and 5 = normal hop. In some studies, neurological outcome was also combined as either paraplegic or nonparaplegic, depending on the purpose of the evaluation.

In phase two, the animals were divided into five groups: (a) sham surgery (controls), (b) CSF drainage, (c) tirilazad mesylate treatment, (d) tirilazad mesylate treatment plus CSF drainage, and (e) surgery with no aortic occlusion but received tirilazad mesylate. Functional outcome, CSF pressure, and calculated spinal cord perfusion pressure were analyzed and the relationship between these parameters and the animals' functional neurological outcome was examined.

The rabbits were turned with their shoulders upright, exposing the posterior neck and the base of the skull, to alter and measure CSF pressure. Through a posterior midline incision, a 24-gauge angiocatheter was placed into the cisterna magna; correct placement was checked by observing the CSF pressure and, in particular, respiratory and cardiovascular variations. Any animal that had a bloody CSF tap was excluded from the study because the intrathecal blood would affect measured CSF pressure. Animals in which CSF was drained had angiocatheters left open to atmospheric pressure. The pressure was also transduced on a continuous basis.

Perfusion pressure was calculated as the difference between CSF pressure and mean arterial pressure measured distal to the balloon. This mean arterial pressure gave a rough approximation of the arterial pressure below that level and was determined from a separate catheter placed in the artery contralateral to the balloon-containing catheter. Cerebrospinal fluid and mean arterial blood pressure was measured at the same level. The perfusion pressure was calculated and used as a relative measurement to approximate perfusion pressure at the spinal cord because the exact pressure could not be directly measured. No attempt was made to keep the amount of CSF drainage constant, but rather CSF drainage was allowed to occur to equilibrate with the atmospheric pressure. The total amount of CSF drainage was never more than a few cubic centimeters. In a sense, the pressure was thereby zeroed at the start of the experiment in the CSF drainage group. The hemostatic vessel loop was then removed after 40 min and the 3-French catheter was withdrawn. In this testing protocol, each animal received 1 mg/kg of tirilazad mesylate or vehicle intravenously 2 h after the occlusion and then every 6 h for a total of 38 h.

IV. PATHOLOGICAL ANALYSIS

The spinal cord preparation was then evaluated and preliminary pathological analysis was performed. Entire spinal columns were removed immediately after the animals had been sacrificed and total laminectomies allowed optimal penetration of the fixative to the spinal cord. Ten-percent phosphate-buffered formula and solution were used as a fixative for at least 10 d. The spinal cords with their nerve roots and dorsal ganglia were then removed and specimens were obtained from the thoracic, lumbar, and sacral spinal cords, including the cauda equina. The specimens underwent paraffin embedding and were stained with hematoxylin-eosin. Specimens that were obtained from the upper thoracic cord were used as controls to differentiate any ischemic changes from nonspecific postmortem changes. Several criteria were evaluated in the pathological analysis, including the extent of neural damage, intensity and character of the inflammatory response, changes in the blood vessels, and changes in the white matter. For this purpose, a simplified scale of neural damage was developed that included five grades: 0 = extensive generalized neural loss, 1 = extensive neural loss with focal changes at different levels, 2 = focal neuronal loss, 3 = eosinophilia and/or pyknosis, and 4 = normal.

V. RESULTS

Significant spinal cord ischemia protection was obtained by using both the medical intervention, tirilazad, and the surgical intervention, CSF drainage. Animals receiving tirilazad mesylate in the phase 1 study showed far better neurological function than their vehicle-treated controls (Figure 2). Neurological function at 48 or 96 h was chosen to evaluate function on a more subacute scale. Results in the phase 2 animals showed that tirilazad mesylate produced a marked decrease in the percentage rate of paraplegia (Figure 3). An evaluation of whether the drug alone had any negative neurological effects on animals, whether or not they underwent occlusion, showed no deleterious effect.

To investigate the effect of a surgical intervention, we subsequently evaluated the effect of CSF drainage on CSF pressure and perfusion pressure (Figure 4). As expected, a decreased CSF pressure translated into increased perfusion pressure to the spinal cord. If both surgical

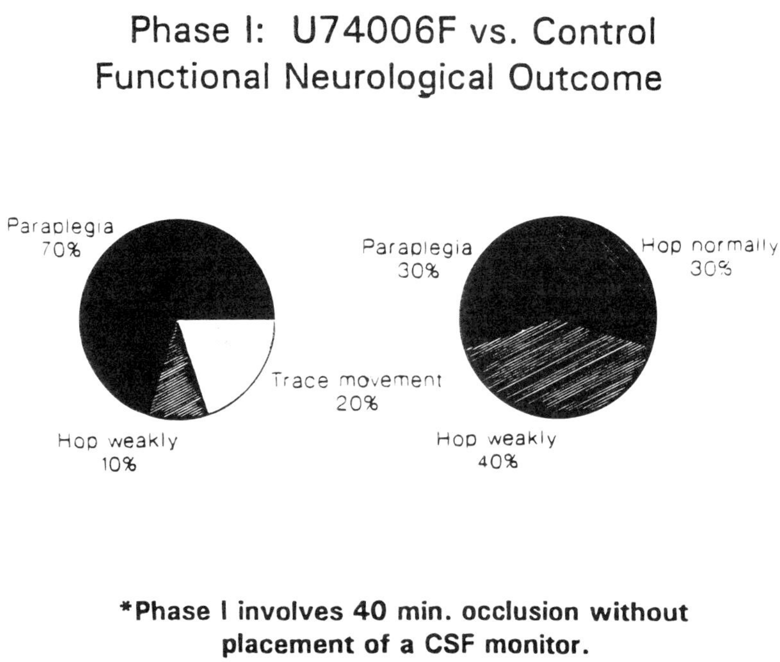

FIGURE 2. Tirilazad mesylate vs. control and neurological outcome. This phase 1 study presents neurological outcome on the 0 to 5 scale mentioned in the text. Forty minutes of infrarenal aortic occlusion was performed without placement of a cerebrospinal fluid (CSF) monitor. Data presented are the functional neurological outcome obtained at 96 h after occlusion. (From Francel PC, Long BA, Malik JM, Tribble C, Jane JA, Kron IL. *Journal of Neurosurgery*, Nov 1993. With permission.)

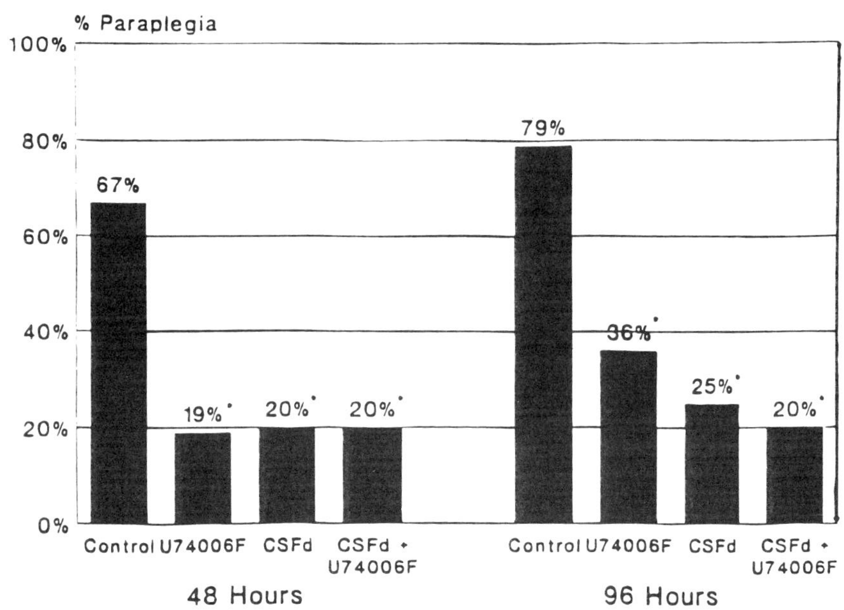

FIGURE 3. Neurological outcome in the phase 2 trials. Percent paraplegia rate is shown. Statistical significance was noted in all of the experimental groups. (From Francel PC, Long BA, Malik JM, Tribble C, Jane JA, Kron IL. *Journal of Neurosurgery*, Nov 1993. With permission.)

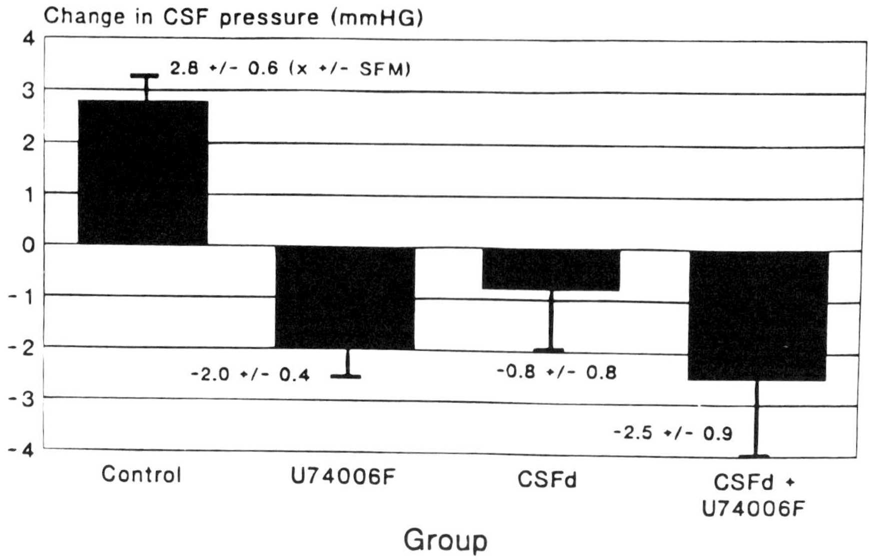

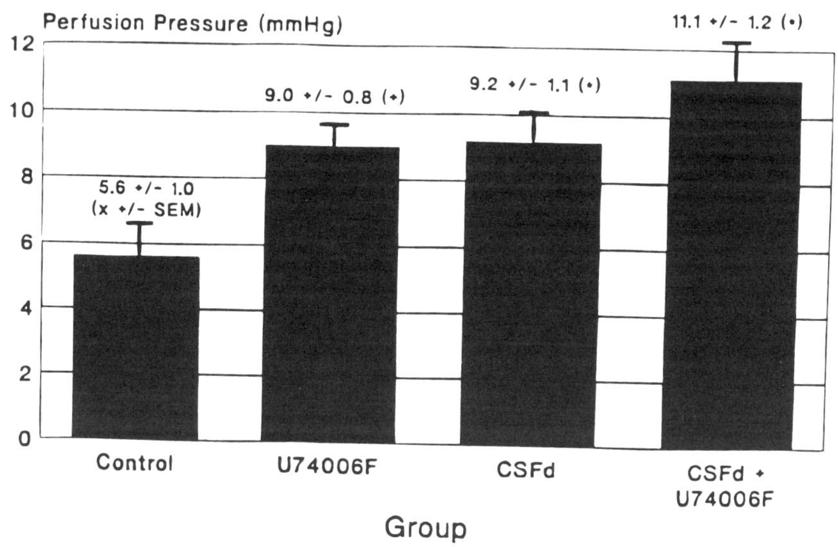

FIGURE 4. Changes in CSF pressure (A) and calculated perfusion pressure (B) in tirilazad mesylate-treated animals, CSF drainage animals, and animals treated under combination treatment in the phase 2 studies. Standard error is indicated by the error bars. (From Francel PC, Long BA, Malik JM, Tribble C, Jane JA, Kron IL. *Journal of Neurosurgery,* Nov 1993. With permission.)

and medical treatments were performed together, an even larger drop of CSF pressure was obtained, translating into an even larger increase in perfusion pressure. In evaluating functional neurological outcome in animals treated both with tirilazad and CSF drainage, the animals that received both treatments clearly had the best functional neurological outcome. However, the tirilazad group and the CSF drainage group each showed significant improvement in neurological outcome.

The data obtained can also be analyzed by comparing paraplegia with nonparaplegia. Figure 5 illustrates the change in CSF pressure and the calculated perfusion pressure in paraplegia and nonparaplegia. Comparisons were taken from the entire group of experimental animals, not just treatment groups or control groups. The paraplegia group showed an increase

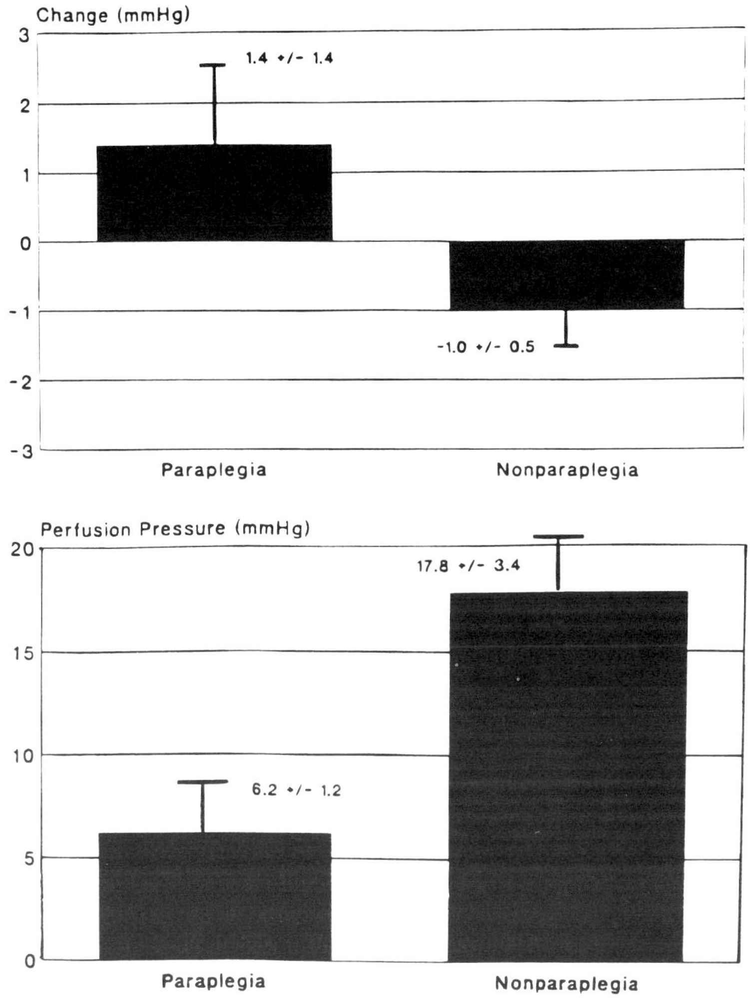

FIGURE 5. Functional neurological outcome relative to CSF pressure (A) and calculated perfusion pressure (B). Paraplegia and nonparaplegia were compared as indices of functional neurological outcome related to CSF pressure and calculated perfusion pressure. A statistically significant difference was noted for all comparisons. (From Francel PC, Long BA, Malik JM, Tribble C, Jane JA, Kron IL. *Journal of Neurosurgery*, Nov 1993. With permission.)

in CSF pressure; the nonparaplegia group showed a drop in CSF pressure and a resulting increase in the calculated perfusion pressure. The paraplegic animals did not show the same magnitude of change. This closely correlated the functional result with the physiological result of CSF pressure and perfusion pressure. The pathological result closely correlated with both the functional result and the physiological measurements. The animals that underwent sham operation remained neurologically intact. The animals in the occlusion control group receiving no further treatment showed the worst result. The animals in the tirilazad group and in the CSF drainage group both had fairly good preservation of function; however, the best result, a neurological outcome that approximated normal or nearly normal, was achieved in animals that received both treatments. In comparing these clinical results with pathology, the pathological specimens correlated very well with functional outcome.

The control group animals showed the greatest microscopic changes, characterized by extensive and diffuse neuronal loss, white matter edema, and capillary endothelial swelling and proliferation. The sham-operated group had normal spinal cords. The tirilazad mesylate group without CSF drainage showed only occasional focal damage to the gray matter and more limited mononuclear inflammatory cell reaction. Only eosinophilia and/or pyknosis of some motor neurons was found in animals with CSF drainage alone. The tirilazad mesylate plus CSF drainage group had the best pathological results, showing normal or near normal spinal cords. When correlating the CSF pressures of animals that underwent pathological analysis with those of control group animals, a significant difference was noted in CSF pressures, with 4.7 torr noted in the tirilazad group, 1.7 torr in the CSF drainage group, and 11 torr in the control group.

VI. A MOLECULAR AND CELLULAR MODEL OF SPINAL CORD ISCHEMIA

Using the above experimental design, lower spinal cord ischemia was selectively induced via an infrarenal aortic balloon occlusion system. Surgical intervention (CSF drainage) and medical intervention (tirilazad) were tested using this system. Both methods appeared to lower CSF pressure, increasing perfusion pressure to the spinal cord. The increased perfusion pressure correlated with an improvement in the functional neurological outcome. These functional and physiological measures correlated with ongoing biochemical and metabolic changes that occurred in the spinal cord secondary to spinal cord ischemia. It is known that both an ischemic and a reperfusion phase occur in spinal cord ischemia and, in particular, spinal cord injury, and these phases may play the dominant role in the secondary injury component of spinal cord injury. It is particularly important to study ischemia and secondary injury because it may be the only phase of the injury in which medical or surgical intervention might prevent further spinal cord damage.[3,14,17,65,78]

Figure 6 indicates that the ischemia and reperfusion injury to the central nervous system are probably caused by the production of free radicals that alter lipid membranes, inducing lipid peroxidation that destroys cells.[4,6,13,18,20,32,53,91] All cell types are sensitive to this event. Initially, there is a short phase of hyperemia because of the microvasculature's loss of autoregulation. However, this normalizes approximately 30 min after injury with a progressive decrease in spinal cord blood flow culminating in irreversible damage.[44] This decrease could be reversed using tirilazad. Figure 6 explains what happens during spinal cord injury and ischemia.

Hall suggested that blunt injury to the spinal cord induces hemorrhage in the traumatized zone and also causes significant influx of calcium into the traumatized cells.[26,31,40–55,57–61,71,80,87,92] The calcium influx activates other mechanisms including various phospholipases, which results in the release of arachidonic acid and increased prostaglandin synthetase activity.[12,27,28,30,37,38,64,84,93] This initial cascade also induces the production of free radicals and lipid peroxidation products that activate phospholipases, resulting in the further release of arachidonic acid. Arachidonic acid is then transformed into various metabolites, many of which, including thromboxane E_2 and prostaglandin $F_{2\alpha}$, are potent mediators of ischemia.

The iron released from hemoglobin present in damaged red blood cells at the site of trauma is also a potent catalyst for free radical production and lipid peroxidation. The arachidonic acid metabolites promote further ischemia in the gray matter, and this gray matter ischemia produces fairly significant intracellular lactic acidosis and free radical generation that then generates further lipid peroxidation. The central mediators of the ischemic damage to the traumatized tissues, therefore, appear to be the oxygen-containing free radicals and lipid peroxidation products. Unfortunately, the free radicals and lipid peroxidation products promote further release of arachidonic acid and further production of ischemia-producing arachidonic acid metabolites. Eventually, the ischemia spreads from the gray matter into the

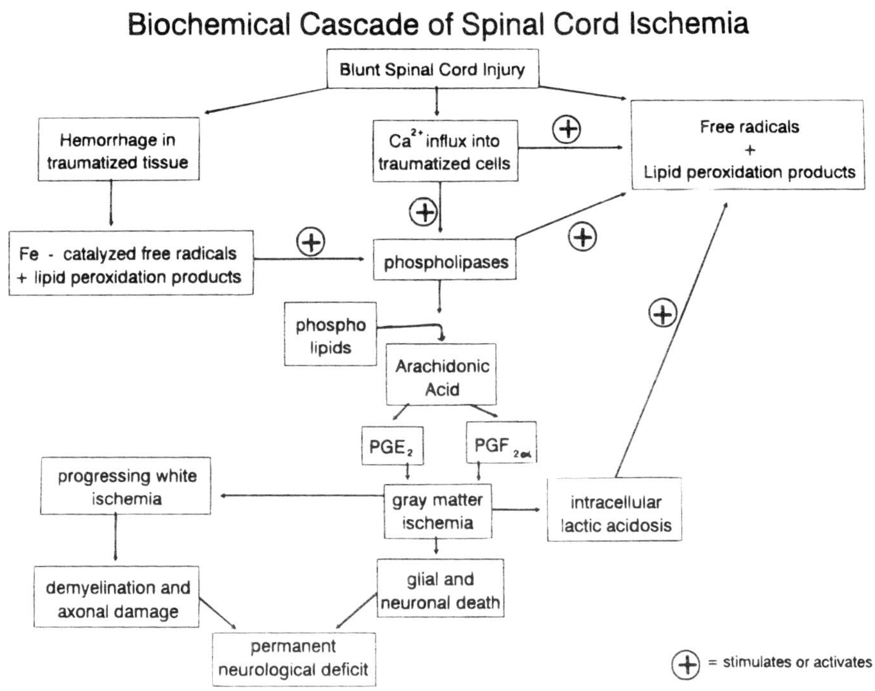

FIGURE 6. Schematic diagram of the biochemical cascade induced by spinal cord ischemia, culminating in spinal cord injury. Multiple mediators of ischemia are produced during spinal cord ischemia and, in particular, during spinal cord mechanical injury. The generation of free radical scavengers, lipid peroxidation products, and arachidonic acid metabolites appears to play a dominant role in perpetuating ischemia and producing cell death. Ultimately, cell death is mediated via an influx of calcium caused by raising intracellular calcium, which stimulates autodestructive processes. See text for further details.

white matter, producing a combination of both demyelination and axonal damage in the white matter. The sum total of these ischemic events produces the permanent neurological deficits.

VII. MECHANISMS OF ACTION OF FREE RADICAL SCAVENGERS

Tirilazad mesylate has been shown to be a very potent free radical scavenger and inhibitor of lipid peroxidation.[3,12,13,18,41–43,45,46,48,51,53,55,56,60,61,65] It effectively blocks reactions catalyzed by iron and appears to decrease the amount of arachidonic acid released at the trauma site. These actions are not related to any glucocorticoid activity of the compound, even though it was originally designed as a modification of the cortisol molecule. The drug does not have any effect on the hypothalamic-pituitary-adrenal axis, and does not appear to act pharmacologically on the steroid receptors. The molecular and cellular action seems to culminate in changes observed at the physiological level. For example, tirilazad blunts the intracranial pressure (ICP) rise noted in subarachnoid hemorrhage in cats.[56] It appears that this drug could exert an extremely important beneficial effect on ICP or intraspinal pressure by lowering CSF pressure, which improves perfusion pressure.[36,62,63,88] As noted, an improved perfusion pressure appears to improve functional outcome. This lowering of CSF pressure is probably significant because the effect can be mimicked by lowering the CSF pressure via a surgical intervention, namely CSF drainage. If both pharmacological and surgical interventions are performed, an even larger increase in perfusion pressure can be brought about, culminating in an improved neurological outcome.

The fact that both surgical and medical intervention produced improvement in spinal cord perfusion in a model of spinal cord ischemia suggests that several avenues of intervention may improve outcome and that such treatment may be used in a variety of situations.[7,9,10,67,69,70,72,73,75,83,85,89] These interventions are not limited to traumatic spinal cord injury, but may also be used in chronic spinal cord ischemia, which occurs during bone- or tumor-induced spinal cord compression. They may also be used when performing intradural surgery (e.g., during intramedullary spinal cord explorations), during repair of abdominal aortic aneurysms, or during retroperitoneal explorations in which the blood supply from the aorta to the spinal cord could be diminished. In these instances, the interventions could be applied prior to the potential insult to the spinal cord.

VIII. SUMMARY

In this chapter we have presented an experimental model for studying ischemia to the spinal cord. This model utilized infrarenal aortic occlusion so that the induced ischemia could be studied in isolation. Other models of spinal cord injury have had both an ischemic component and biochemical disruption; the latter usually cannot be affected by pharmacological or surgical means. The experimental model isolates the study of ischemia and reperfusion from biochemical injury and enables a study of spinal cord ischemia extending from the biochemical and cellular level through the organ and system physiological level, finally to the functional organism level.

This spinal cord injury model was used to study the protective effect of medical intervention (tirilazad) and surgical intervention (CSF drainage). This model made it possible to study the protective effect of these two interventions at several levels. At the organism level, the tested animals showed a marked improvement in functional outcome. When analyzing the results at the physiological level, both interventions produced a decrease in CSF pressure, which translated into an increase in perfusion pressure. This increased perfusion pressure improved functional neurological outcome.

These improvements can be understood when the effect of these agents at the molecular and cellular level is examined. Tirilazad protects against ischemia by blocking the deleterious effects of free radicals and lipid peroxidation metabolites, and, in some unknown way, this pharmacological blockade protects the organism via a mechanism that decreases CSF pressure and increases perfusion pressure. That decrease in CSF pressure is an important effect of this drug is shown when CSF drainage is performed. By "jumping" to this stage of protection, CSF drainage likewise produces an improvement in the perfusion pressure to the spinal cord, which translates into improved functional outcome without intervening at the free radical/lipid peroxidation stage. The model does not address one major unanswered question: how do free radicals and lipid peroxidation metabolites increase CSF pressure and decrease perfusion pressure?

This model should encourage further experimental studies in spinal cord ischemia and the testing of further pharmacological and surgical interventions that will generate new and more effective ways to treat spinal cord ischemia.

ACKNOWLEDGMENTS

The authors thank Dr. Edward Hall with the Upjohn Company for providing the tirilazad used in these studies, and Kim Mann and Elizabeth Fisher for editorial assistance.

REFERENCES

1. Alderman, J. L., Osterholm, J. L., D'Amore, B. R., et al. Influence of arterial blood pressure upon central hemorrhagic necrosis after severe spinal cord injury. *Neurosurgery* 4, 53–55, 1979.
2. Alexander, S., Kerr, F. W. L. Blood pressure responses in acute compression of the spinal cord. *J Neurosurg* 21, 263–270, 1964.
3. Anderson, D. K., Braughler, J. M., Hall, E. D., et al. Effects of treatment with U74006F on neurological outcome following experimental spinal cord injury. *J Neurosurg* 69, 562–567, 1988.
4. Anderson, D. K., Means, E. D. Iron-induced lipid peroxidation in spinal cord: protection with mannitol and methylprednisolone. *J Free Rad Biol Med* 1, 59–64, 1985.
5. Anderson, D. K., Means, E. D., Waters, T. R., et al. Microvascular perfusion and metabolism in injured spinal cord after methylprednisolone treatment. *J Neurosurg* 56, 106–113, 1982.
6. Anderson, D. K., Saunders, R. D., Demediuk, P., et al. Lipid hydrolysis and peroxidation in injured spinal cord: partial protection with methylprednisolone or vitamin E and selenium. *Cent Nerv Syst Trauma* 2, 257–267, 1985.
7. Barone, G. W., Joob, A. W., Flanagan, T. L., et al. The effect of hyperemia on spinal cord function after temporary thoracic occlusion. *J Vasc Surg* 8, 535–540, 1988.
8. Bingham, W. G., Goldman, H., Friedman, S. J., et al. Blood flow in normal and injured monkey spinal cord. *J Neurosurg* 43, 162–171, 1975.
9. Blaisdell, F. W. The mechanism of paraplegia after temporary thoracic aortic occlusion in its relationship to spinal fluid pressure. *Surgery* 51, 351–355, 1990.
10. Bower, T. C., Murray, M. J., Gloviczki, P., et al. Effects of thoracic aortic occlusion and cerebrospinal fluid drainage on regional spinal cord blood flow in dogs: correlation with neurologic outcome. *J Vasc Surg* 9, 135–144, 1988.
11. Bracken, M. D., Shepard, M. J., Collins, W. F., et al. A randomized, controlled trial of methylprednisolone or naloxone in the treatment of acute spinal-cord injury. Results of the second National Acute Spinal Cord Injury Study. *N Engl J Med* 322, 1405–1411, 1990.
12. Braughler, J. M., Duncan, L. A., Chase, R. L. Interaction of lipid peroxidation and calcium in the pathogenesis of neuronal injury. *Cent Nerv Syst Trauma* 2, 269–283, 1985.
13. Braughler, J. M., Hall, E. D. Central nervous system trauma and stroke. I. Biochemical considerations for oxygen radical formation and lipid peroxidation. *Free Rad Biol Med* 6(3), 289–301, 1989.
14. Braughler, J. M., Hall, E. D. Current application of "high-dose" steroid therapy for CNS injury. A pharmacological perspective. *J Neurosurg* 62, 806–810, 1985.
15. Braughler, J. M., Hall, E. D. Effects of multi-dose methylprednisolone sodium succinate administration on injured cat spinal cord neurofilament degradation and energy metabolism. *J Neurosurg* 61, 290–295, 1984.
16. Braughler, J. M., Hall, E. D. Lactate and pyruvate metabolism in injured cat spinal cord before and after a single large intravenous dose of methylprednisolone. *J Neurosurg* 59, 256–261, 1983.
17. Braughler, J. M., Hall, E. D., Means, E. D., et al. Evaluation of an intensive methylprednisolone sodium succinate dosing regimen in experimental spinal cord injury. *J Neurosurg* 67, 102–105, 1987.
18. Braughler, J. M., Pregenzer, J. F., Chase, R. L., et al. Novel 21-amino steroids as potent inhibitors of iron-dependent lipid peroxidation. *J Biol Chem* 262, 10438–10440, 1987.
19. Chehrazi, B. B., Scremin, O., Decima, E. E. Effect of regional spinal cord blood flow and central control in recovery from spinal cord injury. *J Neurosurg* 71, 747–753, 1989.
20. Coles, J. C., Ahmed, N., Mehta, H. V., et al. Role of free radical scavenger in protection of the spinal cord during ischemia. *Ann Thorac Surg* 42, 551–556, 1986.
21. Collins, W. F. A review and update of experimental and clinical studies of spinal cord injury. *Paraplegia* 21, 204–219, 1983.
22. De la Torre, J. C. Spinal cord injury. Review of basic and applied research. *Spine* 6, 315–335, 1981.
23. Demopoulos, H. B., Flamm, E. S., Seligman, M. L., et al. Further studies on free radical pathology in the major central nervous system disorders: effects of very high doses of methylprednisolone on the functional outcome, morphology, and chemistry of experimental spinal cord impact injury. *Can J Physiol Pharmacol* 60, 1415–1424, 1982.
24. Dolan, E. J., Tator, C. H. The treatment of hypotension due to acute experimental spinal cord compression injury. *Surg Neurol* 13, 380–384, 1980.
25. Ducker, T. B., Perot, P. L., Jr. Local tissue oxygen and blood flow in the acutely injured spinal cord. *Proc Veterans Admin Spinal Cord Inj Conf* 18, 29–32, 1971.
26. Faden, A. I., Jacobs, T. P. Effect of TRH analogs on neurologic recovery after experimental spinal trauma. *Neurology* 35, 1331–1334, 1985.
27. Faden, A. I., Jacobs, T. P., Smith, M. T. Evaluation of the calcium channel antagonist nimodipine in experimental spinal cord ischemia. *J Neurosurg* 60, 796–799, 1984.

28. Faden, A. I., Simon, R. P. A potential role for excitotoxins in the pathophysiology of spinal cord injury. *Ann Neurol* 23, 623–626, 1988.
29. Fehlings, M. G., Tator, C. H. A review of experimental models of acute spinal cord injury, in Illis L (ed): *Spinal Cord Dysfunction: Assessment.* New York: Oxford University Press, 1988, pp. 3–33.
30. Fehlings, M. G., Tator, C. H., Linden, R. D. The effect of nimodipine and dextran on axonal function and blood flow following experimental spinal cord injury. *J Neurosurg* 71, 403–416, 1989.
31. Flamm, E. S., Young, W., Demopoulos, H. B. Experimental spinal cord injury. Treatment with naloxone. *Neurosurgery* 10, 227–231, 1982.
32. Fleckenstein, A. E., Smith, S. L., Linseman, K. L., et al. Comparison of the efficacy of mechanistically different antioxidants in the rat hemorrhagic shock model. *Circ Shock* 35, 223–230, 1991.
33. Francel, P. C., Long, B. A., Malik, J. M., Tribble, C., Jane, J. A., Kron, I. L. Limiting ischemic spinal cord injury using a free radical scavenger 21-aminosteroid (Tirilazad): the possible protective role of decreased cerebrospinal fluid pressure. *J Neurosurg* 79, 742–751, 1993.
34. Griffiths, I. R. Spinal cord blood flow after acute experimental cord injury in dogs. *J Neurol Sci* 27, 247–259, 1976.
35. Griffiths, I. R. Spinal cord blood flow in dogs: 1. The effect of blood pressure. *J Neurol Neurosurg Psychiatry* 36, 914–920, 1973.
36. Grubbs, P. E., Marini, C., Toporoff, B., et al. Somatosensory evoked potentials and spinal cord perfusion pressure are significant predictors of postoperative neurologic dysfunction. *Surgery* 104, 216–223, 1988.
37. Guha, A., Tator, C. H., Piper, I. Effect of a calcium channel blocker on posttraumatic spinal cord blood flow. *J Neurosurg* 66, 423–430, 1987.
38. Guha, A., Tator, C. H., Piper, I. Increase in rat spinal cord blood flow with the calcium channel blocker nimodipine. *J Neurosurg* 63, 250–259, 1985.
39. Guha, A., Tator, C. H., Rochon, J. Spinal cord blood flow and systemic blood pressure after experimental spinal cord injury in rats. *Stroke* 20, 372–377, 1989.
40. Guha, A., Tator, C. H., Smith, C. R., et al. Improvement in post-traumatic spinal cord blood flow with a combination of a calcium channel blocker and a vasopressor. *J Trauma* 29, 1440–1447, 1989.
41. Hall, E. D. Effects of the 21-aminosteroid U74006F on posttraumatic spinal cord ischemia in cats. *J Neurosurg* 68(3), 462–465, 1988.
42. Hall, E. D. Free radicals and CNS injury. *Crit Care Clin* 5(4), 793–805, 1989.
43. Hall, E. D. Pathophysiology of spinal cord injury. Current and future therapies. *Minerva Anestesiol* 55(3), 63–66, 1989.
44. Hall, E. D. Steroids and neuronal destruction or stabilization. *Ciba Found Symp* 153, 206–214, 1990.
45. Hall, E. D. Inhibition of lipid peroxidation in CNS trauma. *J Neurotrauma* 8 (Suppl 1), S31–S40, 1991.
46. Hall, E. D. The neuroprotective pharmacology of methylprednisolone. *J Neurosurg* 76, 13–22, 1992.
47. Hall, E. D., Braughler, J. M. Central nervous system trauma and stroke. II. Physiological and pharmacological evidence for involvement of oxygen radicals and lipid peroxidation. *Free Rad Biol Med* 6(3), 303–313, 1989.
48. Hall, E. D., Braughler, J. M. Role of lipid peroxidation in post-traumatic spinal cord degeneration: a review. *Cont Nerv Syst Trauma* 3, 281–293, 1986.
49. Hall, E. D., Braughler, J. M. Effects of intravenous methylprednisolone on spinal cord lipid peroxidation and (Na^+ + K^+)-ATPase activity. Dose-response analysis during 1st hour after contusion injury in the cat. *J Neurosurg* 57, 247–253, 1982.
50. Hall, E. D, Braughler, J. M. Glucocorticoid mechanisms in acute spinal cord injury: a review and therapeutic rationale. *Surg Neurol* 18, 320–327, 1982.
51. Hall, E. D., Braughler, J. M. Acute effects of intravenous glucocorticoid pretreatment on the *in vitro* peroxidation of cat spinal cord tissue. *Exp Neurol* 73, 321–324, 1981.
52. Hall, E. D., Braughler, J. M., McCall, J. M. New pharmacological treatment of acute spinal cord trauma. *J Neurotrauma* 5(1), 81–89, 1988.
53. Hall, E. D., Braughler, J. M., McCall, J. M. Role of oxygen radicals in stroke: effects of the 21-aminosteroids (lazaroids). A novel class of antioxidants. *Prog Clin Biol Res* 361, 351–362, 1990.
54. Hall, E. D., McCall, J. M., Chase, R. L., et al. A nonglucocorticoid steroid analog of methylprednisolone duplicates its high-dose pharmacology in models of central nervous system trauma and neuronal membrane damage. *J Pharmacol Exp* 242, 137–142, 1987.
55. Hall, E. D., Pazara, K. E., Braughler, J. M., et al. Nonsteroidal lazaroid U78517F in models of focal and global ischemia. *Stroke* 21(11 Suppl), III83–87, 1990.
56. Hall, E. D., Wolf, D. L. A pharmacological analysis of the pathophysiological mechanisms of posttraumatic spinal cord ischemia. *J Neurosurg* 64, 951–961, 1986.
57. Hall, E. D., Wolf, D. L., Braughler, J. M. Effects of a single large dose of methylprednisolone sodium succinate on experimental posttraumatic spinal cord ischemia. Dose-response and time-action analysis. *J Neurosurg* 61, 124–130, 1984.

58. Hall, E. D., Yonkers, P. A., McCall, J. M. Attenuation of hemorrhagic shock by the non-glucocorticoid 21-aminosteroid U74006F. *Eur J Pharmacol* 147(2), 299–303, 1988.
59. Hall, E. D., Yonkers, P. A. Mechanisms of neuronal degeneration secondary to central nervous system trauma or ischemia [letter]. *J Neurotrauma* 6, 227–228, 1989.
60. Hall, E. D., Yonkers, P. A., Horan, K. L., et al. Correlation between attenuation of posttraumatic spinal cord ischemia and preservation of tissue vitamin E by the 21-aminosteroid U74006F: evidence for an *in vivo* antioxidant mechanism. *J Neurotrauma* 6, 169–176, 1989.
61. Hall, E. D., Travis, M. A. Effects of the non-glucocorticoid 21-aminosteroid, U74006F, on acute cerebral hypoperfusion following experimental SAH. *Exp Neurol* 102, 244–248, 1988.
62. Holtz, A., Nyström, B., Gerdin, B. Effect of methylprednisolone on motor function and spinal cord blood flow after spinal cord compression in rats. *Acta Neurol Scand* 82, 68–73, 1990.
63. Hsu, C. Y., Dimitrijevic, M. R. Methylprednisolone in spinal cord injury: the possible mechanism of action. *J Neurotrauma* 7, 115–119, 1990.
64. Jacobs, T. P., Shohami, E., Baze, W., et al. Deteriorating stroke model: histopathology, edema, and eicosanoid changes following spinal cord ischemia in rabbits. *Stroke* 18, 737–742, 1987.
65. Jacobsen, E. J., McCall, J. M., Ayer, D. E., et al. Novel 21-aminosteroids that inhibit iron-dependent lipid peroxidation and protect against central nervous system trauma. *J Med Chem* 33, 1145–1151, 1990.
66. Khan, M., Griebel, R. Acute spinal cord injury in the rat: comparison of three experimental techniques. *Can J Neurol Sci* 10, 161–165, 1983.
67. Kirshner, D. L., Kirshner, R. L., Heggeness, L. M., et al. Spinal cord ischemia: an evaluation of pharmacologic agents in minimizing paraplegia after aortic occlusion. *J Vasc Surg* 9, 305–308, 1989.
68. Kobrine, A. I., Doyle, T. F., Martins, A. N. Local spinal cord blood flow in experimental traumatic myelopathy. *J Neurosurg* 42, 144–149, 1975.
69. Laschinger, J. C., Izumoto, H., Kouchoukos, N. T. Evolving concepts in prevention of spinal cord injury during operations on the descending thoracic and thoracoabdominal aorta. *Ann Thorac Surg* 44, 667–674, 1987.
70. Laschinger, J. C., Owen, J., Rosenbloom, M., et al. Direct noninvasive monitoring of spinal cord motor dysfunction during thoracic aortic occlusion. *J Vasc Surg* 7, 161–171, 1988.
71. Lim, K. H., Connally, M., Rose, D., et al. Prevention of reperfusion injury of the ischemic spinal cord: use of recombinant superoxide dismutase. *Ann Thorac Surg* 42, 282–286, 1986.
72. Massad, M., Joob, A. W. Ischemic spinal cord injury during surgery on the descending aorta. *Cardiac Surg* 4, 65–75, 1990.
73. McCullough, J. L., Hollier, L. H., Nugent, M. Paraplegia after thoracic aortic occlusion: influence of cerebrospinal fluid drainage. *J Vasc Surg* 7, 153–160, 1988.
74. Nemecek, S. Morphological evidence of microcirculatory disturbances in experimental spinal cord trauma. *Adv Neurol* 20, 395–405, 1978.
75. Piano, G., Gerwertz, B. L. Mechanisms of increased cerebrospinal fluid pressure with aortic occlusion. *J Vasc Surg* 11, 695–701, 1990.
76. Rivlin, A. S., Tator, C. H. Regional spinal cord blood flow in rats after severe cord trauma. *J Neurosurg* 49, 844–853, 1978.
77. Robertson, C. S., Foltz, R., Grossman, R. G., et al. Protection against experimental ischemic spinal cord injury. *J Neurosurg* 64, 633–642, 1986.
78. Sandler, A. N., Tator, C. H. Review of the effect of spinal cord trauma on the vessels and blood flow in the spinal cord. *J Neurosurg* 45, 638–646, 1976.
79. Sandler, A. N., Tator, C. H. Effect of acute spinal cord compression injury on regional spinal cord blood flow in primates. *J Neurosurg* 45, 660–676, 1976.
80. Schanne, F. A. X., Kane, A. B., Young, E. E., et al. Calcium dependence of toxic cell death: a final common pathway. *Science* 206, 700–702, 1979.
81. Senter, H. J., Venes, J. L. Altered blood flow and secondary injury in experimental spinal cord trauma. *J Neurosurg* 49, 569–578, 1978.
82. Senter, H. J., Venes, J. L. Loss of autoregulation and posttraumatic ischemia following experimental spinal cord trauma. *J Neurosurg* 50, 198–206, 1979.
83. Spence, P. A., Lust, R. M., Sun, Y. S., et al. Prevention of reflex-mediated increases in cerebrospinal fluid pressure during aortic surgery. *Surg Forum* 41, 297–300, 1990.
84. Stokes, B. T., Fox, P., Hollinden, G. Extracellular calcium activity in the injured spinal cord. *Exp Neurol* 80, 561–572, 1983.
85. Svensson, L. G., Von Ritter, C. M., Groeneveld, H. T., et al. Cross-clamping of the thoracic aorta: influence of aortic shunts, laminectomy, papaverine, calcium channel blocker, allopurinol, and superoxide dismutase on spinal cord blood flow and paraplegia in baboons. *Ann Surg* 204, 38–47, 1986.
86. Tator, C. H., Fehlings, M. G. Review of the secondary injury theory of acute spinal cord trauma with emphasis on vascular mechanisms. *J Neurosurg* 75, 15–26, 1991.

87. Wallace, M. C., Tator, C. H. Failure of blood transfusion or naloxone to improve clinical recovery after experimental spinal cord injury. *Neurosurgery* 19, 489–494, 1986.
88. Wallace, M. C., Tator, C. H. Spinal cord blood flow measured with microspheres following spinal cord injury in the rat. *Can J Neurol Sci* 13, 91–96, 1986.
89. Woloszyn, T. T., Marini, C. P., Coons, M. S., et al. Cerebrospinal fluid drainage and steroids provide better spinal cord protection during aortic cross-clamping than does either alone. *Ann Thorac Surg* 49, 78–83. 1990.
90. Woloszyn, T. T., Marini, C. P., Coons, M. S., et al. A multimodality approach lengthens warm ischemic time during aortic cross-clamping. *J Surg Res* 52, 15–21, 1992.
91. Young, W. The post-injury response in trauma and ischemia. Secondary injury or protective mechanisms? *Cent Nerv Syst Trauma* 4, 27–51, 1987.
92. Young, W., DeCrescito, V., Flamm, E. S., et al. Pharmacological therapy of acute spinal cord injury: studies with high-dose methylprednisolone and naloxone. *Clin Neurosurg* 34, 675–697, 1988.
93. Young, W., Flamm, E. S. Effect of high-dose corticosteroid therapy on blood flow, evoked potentials, and extracellular calcium in experimental spinal injury. *J Neurosurg* 57, 667–673, 1982.

Chapter 20

SPINAL CORD INJURY MODELING AND FUNCTIONAL ASSESSMENT

Bradford T. Stokes, Philip J. Horner, and Minoru Akino

TABLE OF CONTENTS

I. INTRODUCTION

Experimental spinal injury models have improved vastly over the last decade. It is generally agreed that small groups of animals can be used in experimental procedures because of improvements in the design of devices that produce the injuries. The reliability and predictability of such devices is clearly the major factor in the enhanced statistical power achieved in the variety of such designs now employed in experimental neural trauma. It is also clear that choices must be made in the various models between modes of injury that allow certain biomechanical measures (e.g., force, displacement, energy, impulse, etc.) to be predetermined or others in which injury descriptors are passively monitored as the injury sequence is taking place. Furthermore, it is known that the precision of the monitoring of these various measures of the injury plays an important role in the determination of the variability of outcome measures that characterize the specific injury paradigm.

In the following material, an attempt is made to summarize the various choices one must make on the way to creating a valid *in vivo* spinal cord injury model. While a comprehensive discussion of several models takes place (e.g., weight drop, controlled contusion), certain others are mentioned only briefly (e.g., photochemical injury,[50] or spinal ischemia[54]). This is mostly predicated by their paucity of use by a wide variety of investigators, lack of a sufficient variety of outcome measures needed to judge the efficacy of the model, or the

0-8493-8094-4/95/$0.00+$.50

lack of demonstrated predictability in histopathological outcomes in chronic preparations. After describing the devices used to produce the injuries, a variety of biological characteristics of the tissue to be injured that directly impact on the outcomes of various injury protocols is also examined. Finally, a consideration of the selection of behavioral outcome measures that characterize the injuries is attempted. Newer alternative approaches that might allow more discriminative indices of behavioral outcomes (electromyography, etc.) are also discussed.

II. MODELS

Here the emphasis is placed on a number of spinal injury models that allow variable degrees of control and monitoring of the relevant biomechanical events that characterize the injury sequence. Since one assumes that such paradigms mimic events occurring during human injury, modeling issues discussed here have relevance to issues of therapeutic intervention, biomechanics, and pharmacological treatments discussed elsewhere in this volume. Readers are also referred to more a more detailed discussion of these modeling issues in some reviews and device papers.[10,17,26,37,46,47]

A. COMPRESSION INJURIES

The spinal cord can be compressed at relatively slow velocities with predictable results that in some ways mimic outcomes achieved with rapid contusion. This may be done using small vascular clips, static application of weights, or inflation of catheter balloons in the spinal canal.[9,18,25,27,41,49] It does not seem to matter whether one uses a dorsal or ventral approach to the spinal compartment during injury and in many cases the resulting slower compressions produce a profound spinal ischemia and the loss of axonal conductivity at about 80% compression. A more rapid compression (80% of cord diameter in <2 s) yields immediate loss of axonal conduction, a large diminution in blood flow, prompt alteration of the adenine nucleotide pool, and consistent central anatomical changes with permanent paralysis in most animals.[1,11,26] It has been proposed that the outcomes produced by the quicker mode of injury are only partially accounted for by spinal ischemia since the more rapid compression results in permanent deficits not characteristic of the slow compression approach. The role of blood flow in these injuries, albeit less at the more severe magnitudes, makes this model an ideal one in which to evaluate therapeutic agents known to provoke changes in spinal blood flow. However, it is yet to be shown that the entire cascade of degradative events takes place in these models in the same time frame or with similar pathological outcomes as in contusion injuries discussed below.

B. WEIGHT DROP

Recent developments in weight-drop technology have considerably advanced the utility of this approach since it was proposed by Allen in 1911. In particular, the work of Noble and Wrathall,[32] Wrathall,[52] and Wrathall et al.[53] considerably advanced the field using a relatively inexpensive version of this device. This approach has shown that simple control of vertebral stability produces reasonably reproducible trauma biomechanics,[37] chronic behavioral outcomes,[20] functional connectivity as assessed by evoked potentials,[40] and histopathological changes.[31,33] Others interested in pharmacological therapeutics after spinal injury have used similar devices to characterize outcome measures and recovery profiles.[26,36,42] Significant variability in outcome measures, particularly in mildly injured animals and/or the inability to monitor relevant biomechanical events during the injury sequence have, however, hampered further development of this particular approach.

One recent exception to this caveat, which partially solves this dilemma, is the weight-drop device pioneered by the New York University (NYU) group.[23] It is currently being used in the MASCIS trials with a rodent model to develop pharmacological strategies in spinal cord trauma. Briefly, the device, while still dependent on standard weight-drop technology, uses a series of optical potentiometers to measure both the movement of the spinal column and the weight itself. By subtracting these two measures, an accurate estimate of the degree and time course of compression of the spinal compartment may be obtained. The preimpact downward velocity of the impact probe may also be measured by these monitoring devices. Using such approaches, this group has established that the empirically derived ratio of the depth and time course of compression is an accurate predictor of ionic shifts, which are presumably associated with 24-h lesion volumes in these injured animals.[16] The use of this device in acute phase pharmacological trials in experimental spinal cord injury (MASCIS) is currently in progress. Such approaches may result in much quicker ways to bring promising drugs to trial in injured human populations. The NYU device, however, does not provide direct measures of the biomechanics (e.g., stiffness, etc.) of the spinal compartment before and during the injury sequence. Such biomechanical information may be of importance in assessing the starting conditions of the pending insult, the contribution of various meningeal compartments to injury variability, and in assessing the possibility that the impactor has struck adjacent tissue (e.g., bone, muscle) on its downward trajectory. Several other devices, here referred to as controlled contusion devices, either predetermine what these biomechanical variables will be or assess them during the injury sequence itself.

C. CONTROLLED CONTUSION DEVICES

As discussed earlier, extensive progress has been made in the area of spinal cord modeling with devices using the weight-drop principle. A careful control of vertebral column stability, monitoring of the relevant variables such as spinal cord displacement during the injury sequence, and avoidance of impounder devices that complicate assessment of injury biomechanics all are characteristic of the advances in this technology. Newer generation devices that provide control during the injury itself and therefore allow greater precision or actual determination of injury parameters are recent innovations in this technology. With such control, variables of interest such as probe contact velocity, probe displacement, spinal preloading, etc. can be preset. Other monitored variables (e.g., force) can therefore be used to assess preinjury compartmental biomechanics, meningeal stiffness, and a variety of other important injury determinants.

The first of these devices (the stroke-constrained pneumatic impactor) was pioneered by the research group in the General Motors Research Laboratories.[1,3,26] It has generally been used to injure the cervical spinal region (generally C6 or C7) in the closed, spinal configuration (no laminectomy). By maintaining the normal structure of the spinal column, certain artifacts associated with the laminectomy procedure (e.g., size of laminectomy site, etc.) are avoided and close approximation of the human injury process is achieved. The technique allows independent control of either rate or magnitude of vertebral (and thus spinal cord) displacement.

In this model, the pneumatically controlled shaft of an impounder device is adjusted by a crosshead frame to control the amount of vertebral dislocation. Solenoid devices independently control the rate (0.5 to 7 m/s) and degree (10 to 75% compression) of displacement of the impactor with a reasonable degree of reproducibility ($\pm$ 5%). Duration of compression can also be preselected from 10 ms to several seconds. The device could therefore be used for slow compression injuries as well. Studies have also shown that the rate of compression is relatively constant even after tissue impact, and, thus, a stable velocity of impactor trajectory can be achieved after vertebral contact. Outcome measures have also been assessed to insure that adjustments in injury severity result in a predictable variety of injury deficits.

One of the critical measures evaluated with the pneumatic device has been the somatosensory evoked response. It has been possible on the basis of such evoked potential analysis to separate axonal conduction block from the degree of hemorrhagic necrosis by selecting appropriate injury parameters.[2] The wide range of possible compression rates and the possibility of combining sustained compression with more dynamic injuries make this an instrument of choice for those who wish to model combinations of these factors in the acute and chronic injury process. Such an approach may be appropriate for those whose models require that vertebral displacement be followed by static compression, a process directly relevant to the clinical situation.

Alternative methodologies to this pneumatic device are limited but include a system developed at the Ohio State University that allows a range of independent variables to be predetermined (i.e., velocity, displacement, preload duration and magnitude, starting point biomechanics, etc.) and directly monitored during injury (See Figure 1). The device has been used for a wide variety of studies over the last decade that collectively support the hypothesis that quantifiable and reproducible injuries are possible in rodents.[5,8,12,26] Furthermore, by using sensitive monitoring devices in the current device configuration (open-loop), it has been possible to monitor essential elements of the biomechanical tissue response during the critical period of injury.[4,46,47] The inadequate biomechanical information provided by weight-drop devices during this period of time is thus largely avoided.[5,26,30,34,37] The definition of the inherent biomechanical characteristics of the tissues being compressed or injured by the advancing probe under these circumstances is therefore possible and should provide an understanding of how various tissue compartments (e.g., white and gray matter) are differentially affected by the injury process.

The first experiences with the Ohio State device involved the use of a closed-loop analog computer to preprogram either displacement or force profiles. Based on these trials, displacement was chosen as the independent variable since there was little difference in predictability of outcome measures when using either force or displacement. A mechanical stop was added to a later version of the device to insure the accuracy of probe movements.[47] Force feedback from transducer elements, under these displacement-controlled conditions, thus allows an accurate measurement of tissue characteristics (see References 47 for an example) before, during, and after the injury sequence (Figure 2).

In this configuration, the investigator is able to first select from three separate modes of operation of the pattern generator. The first allows repetitive movement of the impactor probe to check the accuracy of the physical movement of the impactor probe during a desired hit sequence (1—calibration); the second involves a minimal displacement of the probe ($\pm$ 15 μm) at 60 Hz to set the start point for the injury sequence (2—touch signal). Finally, the actual injury itself is produced by a rapid downward displacement of the impactor (3—hit sequence) followed by rapid removal of the injury probe. Impactor movement is therefore produced by the pattern generator through the power amplifier and its attached impact probe. Displacement and force are followed by sensitive tranducers mounted on the impact probe itself and all relevant data is stored for later analysis by available computer software.

This device therefore incorporates a number of important design innovations. Reproducible biomechanical starting points were determined since initial experiments revealed up to 50% variability in the stiffness of the meningeal envelope in animals selected for a single experimental group. Because it was also known that small variations (<0.1 mm) in displacement of the spinal compartment could result in large errors in outcome assessments (particularly in chronic open-field behaviors) and the small size of the laminectomy site made surface determination difficult, visual determination of the dural surface was also avoided. Instead, the dural surface is actively sensed during the minimal probe vibration mentioned above (touch signal) to a start point (3000 dyn peak-to-peak) achieved just before the injury sequence is begun. The small dimpling that this produces on the dural surface does not produce detectable injury to the underlying spinal cord. Because the surface of the spinal compartment can be monitored in this way, one can also avoid the necessity for making a large laminectomy site, a major contributor to spinal instability during the injury process itself. With a small

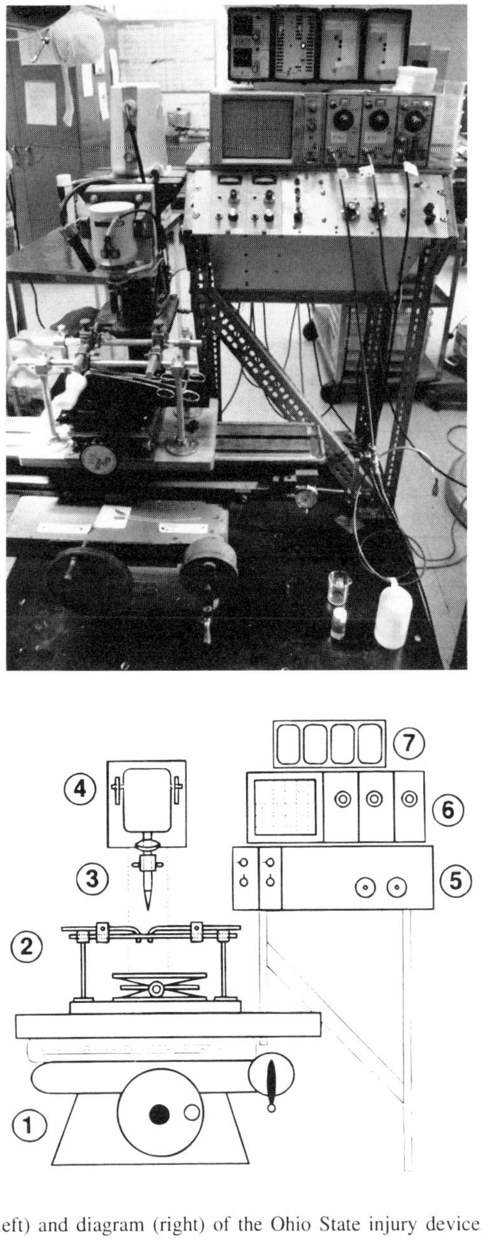

FIGURE 1. Photograph (left) and diagram (right) of the Ohio State injury device. (1) Lateral/A-P adjustment provided by a tool grinder platform, (2) spinal frame with modified Allis clamps to rigidly fix vertebral processes and adjustable platform for vertical positioning of the animal, (3) impactor transducer assembly with force transducer and noncontacting displacement transducer, (4) power amplifier (Ling Shaker), (5) pattern generator, (6) oscilloscope for touch adjustment and instrument calibration, (7) transducer amplifiers.

opening at a single laminectomy site and rigid fixation of the rostral and caudal vertebral processes with Allis clamps, this procedure also eliminates large movements of the vertebral column during the injury sequence.[44] The Ohio State device has therefore allowed direct estimates of the contribution of some of these biomechanical factors to the injury process. In addition, the rapid (<4.0 ms) removal of the probe from the spinal surface is effected by fast reversal of the impactor probe. This eliminates the possibility of any secondary bounces or probe contact with the recoiling spinal compartment, a known source of error in spinal

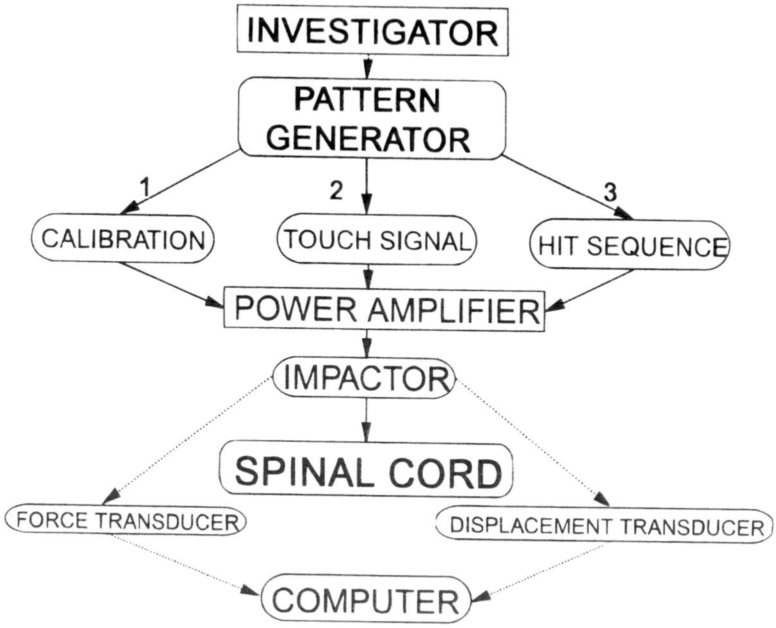

FIGURE 2. Schematic diagram of the Ohio State injury device. Investigator control of all procedures (calibration, touch sensing, or hit sequence) is possible through the power amplifier (Ling Shaker). This in turn controls the impactor probe itself, which compresses the spinal compartment to a peak displacement of approximately 1.0 mm over 20 ms. All events, including probe removal, are monitored by a series of sensitive transducers mounted on the probe itself. Data is simultaneously recorded on a digital storage oscilloscope and a computer system through an A-D converter.

injury.[19,28] All these factors combine to allow a control of injury parameters not previously achieved. Furthermore, this device has substantially decreased the coefficients of variation in behavioral or neuropathological outcome measures. In a practical sense, this results in the ability to use small groups of animals to achieve the statistical power necessary to test various therapeutic stratagems after spinal cord injury.[8]

III. INJURY PROTOCOLS

A. SELECTION OF DEVICE

Numerous considerations determine the type of injury device to be used in a given protocol. While cost may be a factor, one must also consider the variability of outcome measures, the necessity of titering the degree of injury, the ability to cross-correlate outcome measures with biomechanical estimates of tissue injury, and the acute/chronic nature of the injury model. Weight-drop devices certainly meet the criteria of being the least expensive, but often produce unacceptable variability in outcome measures and thus reduce the inherent advantage of using small groups of animals to determine the success of a given therapeutic intervention. Furthermore, weight-drop injury often induces much greater variability of outcome parameters with injury severities in the range of most interest, i.e., mild injury with partial survival of peripheral white matter and significant but incomplete recovery of motor function. Conversely, the study of chronic outcome measures may have an inherent variability, precipitated by secondary injury factors, that is not related to variability in the biomechanics of the initial insult. The ability to control such factors initially may therefore not play an important role in later chronic changes in outcome variables. However, all evidence suggests that the physical nature of the injury must be monitored with any of the devices above and an attempt must be made to insure that significant predictability is provided by cross-correlation of behavioral outcome measures with histopathology and/or biomechanics. Under such circumstances, limitations on the biological variability will be achieved as a precursor to development of injury protocols that are standardized between laboratories.

B. BIOLOGICAL VARIABILITY IN MODELING

Given the need for standardization of spinal injury models, selection of species and injury severity therefore become important parameters. As an important addendum to such issues, however, one must remember that other aspects of modeling (choice of anesthetic, surgical approaches, monitoring of physiological parameters, etc.) are of significance when attempting to establish a model, particularly in novel surroundings. Anesthetic considerations are of importance since all experimental spinal injuries are produced under such conditions. One must consider the type of anesthetic, its route of administration, how long it is necessary to keep the animal at a deep surgical plane, what the immediate and prolonged physiological effects are of the anesthetic agent, and how a particular anesthetic might interact with an acute or chronic therapeutic regimen. As surgery is begun, it is clearly necessary to monitor and stabilize body temperature to avoid the neuroprotective effects of low body temperature (e.g., lowering brain temperature by only a few degrees markedly attenuates NMDA toxicity). Prior to the actual injury itself, obvious precautions should be observed. These include insuring that minimal bleeding has occurred and that care is taken to avoid touching the spinal cord surface with surgical instruments. Monitoring of blood gas levels may also assist in the determination of physiological stability of the preparation. One further caveat is to consider the size of the laminectomy site. Certain procedures (e.g., weight-drop injuries) require a relatively large laminectomy site to allow accurate placement of the drop tube on the dural surface. Such two- to three-segment laminectomies will provide a relatively unstable vertebral column, the movement of which must be monitored in order to estimate the compression of the spinal compartment. Other devices allow single-segment laminectomies (and thus significant spinal stability) because they directly measure spinal displacement and the resultant tissue response to avoid injury artifacts.[4,47]

C. ANIMAL CARE

Animal care is perhaps the most important aspect of successful modeling of the injured spinal cord. Centers that study these problems have developed standardized procedures that largely avoid the problems of limb autotomy, bladder infections, and excessive pain syndromes. A number of centers also use self-contained animal care carts (Bioclean) that constantly filter the air and thus avoid the respiratory infections these injured animals often suffer if caged with general rodent populations. While these advances in modeling are at least somewhat due to the ability to mimic clinically relevant lesions, the emphasis on individual care of injured subjects has undoubtedly contributed to success in modeling and limitations of variability in outcome measures. Such plans of course begin with appropriate handling of animals before operative procedures are begun.

In our laboratory, daily handling of subjects for up to three weeks prior to injury is necessary to achieve gentling and allow time for training in a variety of behavioral paradigms. One can begin formal testing (pretesting) as early as two weeks prior to surgery so that appropriate baseline data may be collected and evaluated. The willingness of animals to perform generalized behavioral evaluations (e.g., open-field walking, inclined plane analysis, or grid walking—see below) after injury procedures is obviously affected by prior training.

Following injury the laminectomy site may be covered with protective materials such as Durafilm (particularly when the site is to be reopened) and the overlying muscle sutured with 4-0 chromic gut to protect the exposed spinal column. Wound clips are normally used to close the skin incision. The incision can then be sprayed with an appropriate antiseptic (e.g., Furazolidone) aerosol powder and lactate Ringer's solution (5 to 10 cc) administered subcutaneously to prevent dehydration. The animal is then placed in a cage with adequate bedding covered with paper towels to avoid aspiration in the early postoperative period. When sufficiently awake from anesthesia, the animal is provided with a water bottle hung low from the side of the cage and soft food is provided. About one third teaspoon of sugar is usually added to the water to encourage postoperative drinking. The following morning the animal is returned to a regular cage with another injured animal since it appears to reduce autotomy. Bladders are expressed two to three times daily postoperatively until the animal regains bladder control; this may be somewhat variable depending on severity of injury and sex of animal. Antibiotics (Gentocin, 4.0 mg, IP) are usually administered prophylactically before surgery begins and to prevent urinary tract infections until bladder control returns. All animals may also be maintained on chronic doses of vitamin C (10 mg/d) while in the laboratory to help with general health and accelerate wound healing. Animals should also be observed closely for any sign of infection and appropriate antibiotic administered when necessary. A recording system that closely monitors changes in animal weight or inability to feed or drink is also mandatory. Postoperative alterations in these measures can be easily handled with either nutritional supplements and/or lactated Ringer's solution as necessary.

IV. OUTCOME VARIABLES

A. SELECTION OF BEHAVIORAL TRIALS

The need for an increasing variety of methods to assess recovery of function has been previously described.[21,22] In this context, it is important to remember that maintaining a separation of scoring in different trials may be necessary (as opposed to an overall combination neuroscoring system) since any particular behavioral trial may correlate best with other pathophysiological changes only at a specific postinjury time.[5] The concept of following and explaining inherent patterns of behavioral change in the face of ongoing plasticity in the injured CNS, however, still proves mechanistically difficult but is valuable as long as this caveat is recognized. Attempts to follow behavioral alterations in both the acute and chronic stages is to be encouraged as are concomitant assessments of spinal cytopathology. Cross-correlation of behavioral outcome measures to other injury descriptors is therefore of great importance even though it is often difficult to associate a given behavior with loss of particular pathways at this stage.[5,8] Given these limitations, however, a variety of approaches have been taken to assess motor capacities (or sensori-motor activities). These most often are described as generalized tests of motor function or motivated testing paradigms; alternatives to assessing somatosensory function are limited and discussed elsewhere.[52]

B. GENERALIZED MOTOR ASSESSMENT

Testing after spinal injury usually begins on the first postoperative day with several behavioral measures being assessed: (a) a modified Tarlov scale to indicate general locomotor abilities, (b) an inclined plane test, which is sensitive to postural abilities, and (c) a grid-walking test designed to assess tactile and proprioceptive abilities. Variations on these general schemes include the videotaping of animals crossing a static beam (gait analysis) or reaching with forelimbs for food rewards after cervical cord lesions.[29,43] The first three of these are emphasized here since they have become somewhat standardized in most laboratories and are relatively simple to perform.

The open-field scoring assessment is quickly becoming the most reliable, albeit somewhat subjective, way to characterize general motor behavior after spinal injury. It is currently being made even more quantifiable for use in the MASCIS pharmacological trials in experimental spinal cord injury. In a more general form, animals are adapted to the test equipment and on two separate days prior to surgery and behavioral pretests are conducted and scored.

As before,[6,13,45] open-field walking is used as a general assessment of mobility and is graded on a modified Tarlov scale from 0 to 5; 0 representing flaccid paralysis, and 5 a normal gait and base of support without toe drags. As indicated below (Table 1), an updated variation of this general theme has allowed more discriminative assessment of hindlimb motor capacities.

Two examiners, blind to the various groups to be tested score all animals every 2 to 4 d postinjury for the first week and weekly thereafter. The second test (inclined plane) is easily conducted to test the animals ability to posturally compensate for displacement in space. The animal is placed in three positions (head up, right side up, and left side up) and the score (maximal angle achieved) reported is the average for the three positions. The third examination, designated as grid-walking (GW) analysis, tests the animals ability to properly place the hindlimbs on the grid bars during locomotion. This test requires fore- and hindlimb coordination and, therefore, hindlimb support is necessary before this behavioral test can be used. The animal is placed on a wire mesh (4 × 4 cm grid spaces) for 3-min periods and the number of times the hindfoot falls through the mesh is recorded. The quotient of footsteps per second of ambulation is used as the final measure of this task. If the animals do not accumulate at least 30 s of locomotion time, the test is repeated. Animals are evaluated on postoperative days 3, 8, 14, and 21 for this and the inclined plane test. If injury parameters are carefully selected (see earlier discussion of the Ohio State device), it is possible to create distinct populations of experimental animals that show a titering of behavioral responses according to injury severity (Figure 3).

Here animals were injured in three groups (n = 8 for each) with varying degrees of peak displacement (0.7 to 1.1 mm) with the Ohio State injury device described earlier. Differences in degree of initial injury are obvious when compared to preinjury values (PRE-INJ) for all three injury severities as are the rate and final degree of outcome estimated by these trials. Note that grid walking was not possible for many animals in the most severely injured group (1.1 mm displacement) and therefore the small group whose data are represented (n = 3) appear similar to the 0.9-mm displacement animals. All groups were otherwise significantly different at high levels of significance ($p < 0.001$) based on a repeated-measures ANOVA over the time period illustrated.

In all such trials, nonparametric statistics (e.g., Kruskall-Wallis ANOVA) should be used to analyze open-field scores. Standard repeated-measures ANOVA and post hoc comparisons

TABLE 1
Open-Field Walking Scale

0	No spontaneous movement
	0.7 Slight movement
1	Movement in the hip and/or knee, none at ankle
	1.3 Active movement at hip or knee, none at ankle
	1.7 Questionable synergy at ankle
2	Movement of limb in all three major joints
	2.3 Attempts at support
	2.7 Support in stance only
3	Active support, uncoordinated gait
	3.3 Intermittent bouts of coordinated gait
	3.7 Lack of control of ankle or foot, walks on knuckles or on medial surface of foot
4	Coordination of forelimbs and hindlimbs in gait
	4.3 Improved hindlimb postural support, abdomen not low to ground few toe drags
	4.7 One or two toe drags, slight unsteadiness turning at full speed
5	Normal gait and base of support, no loss of balance on fast turns, no toe drags

From Behrmann, D.L. et al., *Exp. Neurol.*, 119, 258–267, 1993. With permission.

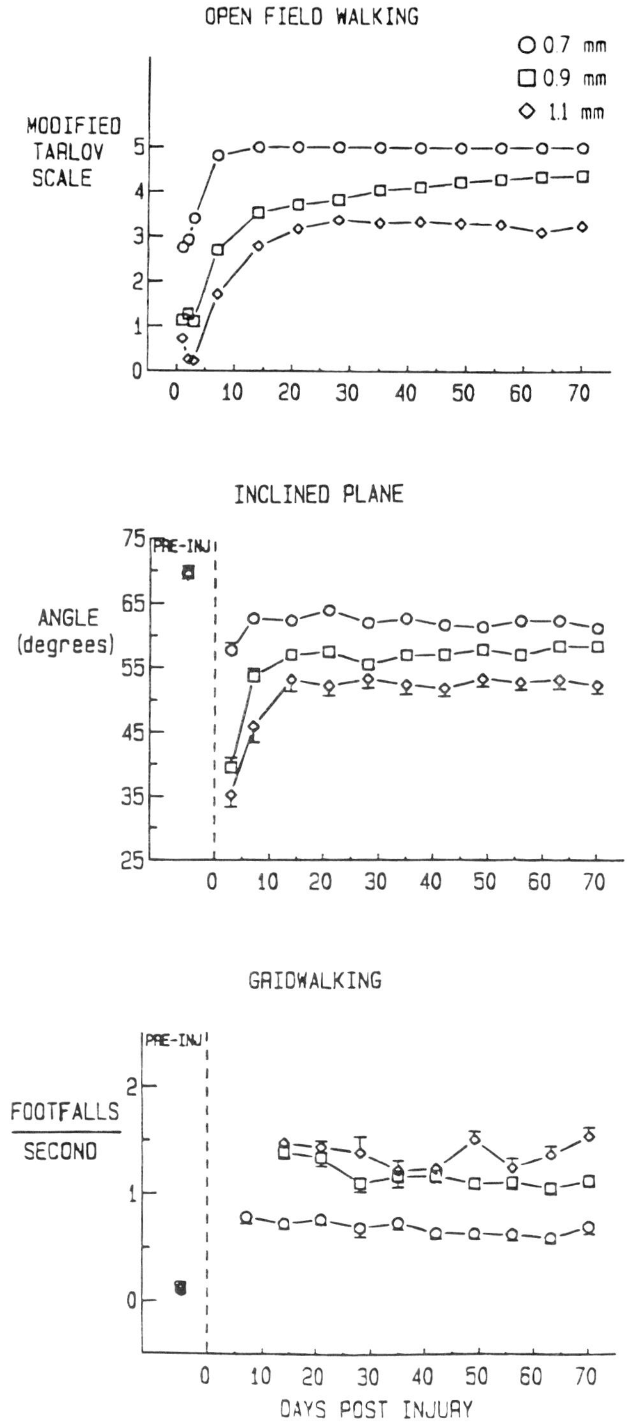

FIGURE 3. Behavioral outcomes of three groups of different injury severity. Peak displacement of the impact probe was varied (0.7, 0.9, and 1.1 mm) in an attempt to create model experimental populations that showed variable outcomes dependent on injury intensity. Initial outcomes were markedly different as were chronic recovery profiles for each behavioral assessment (open-field analysis, inclined plane stability, and grid-walking footsteps) over the time period considered. Control over small variations in peak displacement thus creates large, but predictable, changes in biological outcome measures.

are used for other measures as necessary. In this way, a highly reproducible analysis of motor capacities may be conducted. Limitations of these behavioral approaches and their analysis, however, are obvious in that they fail to maximize motivational aspects of behavioral trials and therefore run the risk, particularly in injured animals, of allowing masking of possible behavioral capacities. We have found, for instance, that such trials quite often are not sensitive enough to reveal the underlying pathology.[24]

C. ADVANCED EVALUATION OF BEHAVIORAL POTENTIAL

Numerous attempts have been made to enhance the discrimination of behavioral trials involving spinal injured animals. Recent attempts using video analysis of beam walking[29] or attempts to obtain food rewards[43] are examples of such approaches. Others have attempted a combination of food rewards and gait analysis[48] in situations that evoke a more goal-directed behavior. Such attempts, while valuable, may also be specific for only certain ranges of injury severity and require larger numbers of animals for appropriate power analysis.[8]

The use of electromyography to assess the specific roles of certain muscles in hindlimb responses after spinal injury has recently allowed the description of more definitive deficits.[24] The transition from such individual assessment of evoked muscle responses to behavioral tasks of longer duration has also begun with the assessment and/or conditioning of H-reflexes in the chronically injured animal.[51] The potential of monitoring such changes 24-h per day is exciting in that it holds enormous promise for specific evaluation of a variety of pharmacological interventions over acute and chronic time periods. Additional stimulus response paradigms in awake animals that involve known descending pathways (e.g., acoustic startle) are also being currently developed.

Although the startle test is itself noninvasive, it involves an additional operative procedure to prepare the animals for simultaneous hindlimb electromyography. To test such behavioral startle reflex responses, the animal is placed in a small cage mounted on a force transducer. Amplitude of the movement and the time needed for a response to the tone are measured on an oscilloscope. The cage is fixed in a ventilated soundproof chamber, which minimizes external interference. The small cage must be of adequate size for the animal to stand comfortably and preconditioning to this environment over several weeks is absolutely necessary to avoid undue stressors. Tests are performed weekly when hindlimb support is adequate.

Electromyography of this response is conducted as in Reference 24. It provides a direct test of the hypothesis that brainstem pathways are intact in injured animals and whether or not their improvement after therapeutic protocols involves the preservation or regrowth of selected, injured descending pathways. In this group of animals, hindlimb EMG responses are monitored using small surgical staples (Ethicon) placed transcutaneously, bilaterally over the tibialis anterior and vastus lateralis muscles. Maximum responses are elicited during a variety of testing paradigms in awake animals and the responses rectified and averaged. An example of the utility of such a response is illustrated in Figure 4.

Here both tibialis anterior muscles are monitored for acoustic startle responses prior to and after mild contusion injury with the Ohio State injury device. The large and reproducible responses occurring before injury with a latency of approximately 6 to 8 ms are eliminated at day 1 postinjury. Longer latency responses ($>$20 ms) appear after injury at day 15 and increase in amplitude up to two months postinjury. The mechanisms of this release phenomena are largely unknown but may include rearrangement of existing lumbar spinal circuitry or delay-induced conduction in thoracic spinal pathways. Since these muscles play an important role in limb position during stance and swing phases of hindlimb locomotion, finding effective ways to reverse such responses may have important consequences for injured animals. The ability to monitor these responses in awake animals thus has considerable advantages as we seek to study and modify such behaviors. Our ability to find new ways to elicit reflex responses in muscles affected by these injuries (see Reference 51 for an example) and to

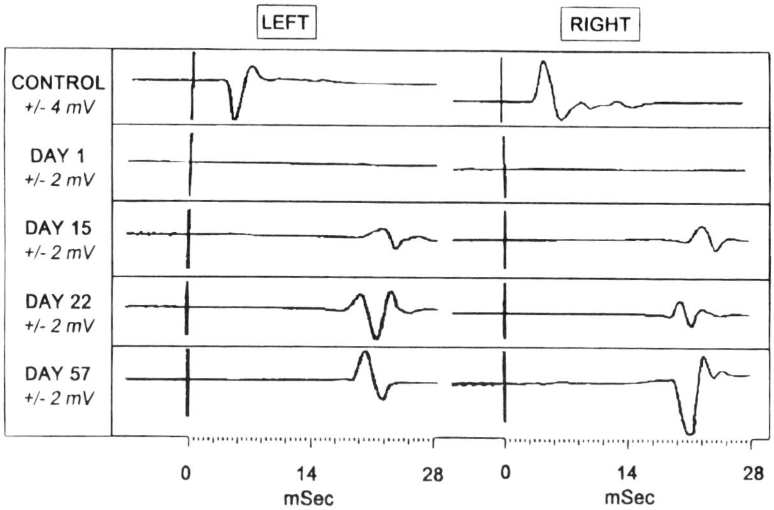

FIGURE 4. Acoustic startle responses in awake normal or injured (0.9 mm peak displacement) animals enclosed in a soundproof chamber. Electromyographic (EMG) recordings were made bilaterally from tibialis anterior musculature at time periods after the termination of the 4 ms startle signal (vertical bar in each recording). Note the change in scale between control responses and all subsequent injury EMG recordings. A unique response appears after injury with a latency of over 20 ms from tone termination at two weeks postinjury. Since this response does not occur in the normal animal, its characteristics in this important muscle controlling hindlimb movements may tell us something about reorganization of the spinal cord after impact injury.

adopt new methods of noninvasively stimulating motor responses (e.g., magnetic stimulators) will undoubtedly contribute to mechanistic explanations of a variety of present interventions (e.g., transplants, pharmaceuticals, etc.) into the injury process.

D. PREDICTABILITY OF OUTCOME MEASURES FROM BIOMECHANICAL DATA

The issues of relative determinations of the biomechanics of spinal injury, specific modeling approaches, and selection of outcome measures have been addressed. In an attempt to control the variance of such predictors, one often makes the assumption that careful selection of a model and injury methodology will result in a significant decrease in the variability of biological outcomes (i.e., pathophysiology, neuropathology, and behavior). This would of course translate into experimental groups with predictable and reproducible outcomes that could be cross-correlated and examined to see which biomechanical indices independently predicted each of the outcome measures. Such reasoning, however, might be flawed if the initial impact parameters were only predictive of acute injury outcomes and not those associated with delayed or secondary injury that might be produced by a separate set of mechanisms (e.g., immune-mediated autolytic destruction).[38,39] In this regard, biomechanical analysis has revealed that the lesion shape predicted from modeling the approximate distribution of tissue forces should be roughly biconical and distributed around the peak of impact epicenter. Mapping of the three-dimensional destruction at the injury site has revealed, however, that late lesion damage spreads into the base of the dorsal funiculi rostral and caudal to the lesion site. Such unpredicted lesion shapes might imply that alternative mechanisms determine the late postlesion autolysis of spinal tissue.

It is known, however, that the overall lesion shape is nearly biconical and that impact parameters, as measured above are very predictive of the percent cross-sectional area spared at the lesion epicenter, lesion volume, and lesion length.[8,12] Experience with the Ohio State device has also allowed the creation of animal groups that show both acute and chronic behavioral outcomes that are consistently predicted by the measured impact parameters.[8] Experimental approaches that attempt to arrest or reverse this pathological process should be aided by the small groups used in these experiments since each trial of a potential therapy requires dose responses of the injury and the intervention as suggested above. In this regard, variability of certain individual measures of behavioral outcome (e.g., inclined plane analysis) in these experiments has been reduced from 21% (weight-drop experiments)[1] to 5.4%.[7] Consistency of outcome has thus been achieved using animal models and devices of the type described earlier.

V. SUMMARY

It is now clear that, by taking into account the variables discussed earlier, models can be created that closely mimic spinal cord injuries occurring in human populations. With such approaches and devices a wide and predictable range of spinal cord injuries and behavioral outcomes can be produced.[5,8,13,34,35,45,47] It is therefore possible to reduce, if not eliminate, many of the major design disadvantages found in older methods of spinal injury. For example, one can avoid the use of weight-drop height as a predictor of subsequent outcomes and eliminate the adverse effects documented by others due to postimpact bounce and tube friction.[19,28] The addition of touch circuitry in newer models has also allowed a determination of a known starting point for our subsequent impact paradigms, a factor that contributes greatly to injury consistency and reliability. Furthermore, the necessity of monitoring two mechanical parameters (displacement and force) has been supported, as before,[14,15] to insure a reliable hit paradigm and give information about the variability in stiffness of the individual spinal compartments.

In this review, a number of relatively simple impact devices and behavioral approaches were described. Such devices provide the needed degree of precision and impact strength to apply such techniques to a range of species and impact parameters to mimic open and closed injuries that might occur in the human population. Furthermore, they provide the flexibility to also predetermine temporal characteristics of injury paradigms in a more precise way than previously possible. Each has its own advantages for evaluating specific aspects of the biomechanical events that occur during contusion injury. As newer behavioral paradigms arise, they will undoubtedly also contribute to modeling the specific outcome measures that characterize and distinguish spontaneous recovery of function from modes of recovery precipitated by therapeutic interventions.

REFERENCES

1. Anderson, T. E. 1982. A controlled pneumatic technique for experimental spinal cord contusion. *J. Neurosci. Methods* 6, 327–333.
2. Anderson, T. E. 1985. Spinal cord injury: experimental dissociation of hemorrhagic necrosis and subacute loss of axonal conduction. *J. Neurosurg.* 62, 115–119.
3. Anderson, T. E. 1986. Effects of acute alcohol intoxication on spinal cord vascular injury. *CNS Trauma* 3, 183–192.
4. Anderson, T. E. and B. T. Stokes. 1992. Experimental models for spinal cord injury research: physical and physiological considerations. *J. Neurotrauma* 9 Suppl 1, S135–S142.
5. Beattie, M. S., B. T. Stokes, and J. C. Bresnahan. 1988. Experimental spinal cord injury: strategies for acute and chronic intervention based on anatomic, physiologic and behavioral studies. Pages 43–74 in D. G. Stein and B. A. Sabel, Eds., *Pharmacologic Approaches to the Treatment of Brain and Spinal; Cord Injury.* Plenum Publishing Corp., New York.

6. Behrmann, D. L., J. C. Bresnahan, and M. S. Beattie. 1993. A comparison of YM-14673, U-50488H, and Nalmefene after spinal cord injury in the rat. *Exp. Neurol.* 119, 258–267.
7. Behrmann, D. L., J. C. Bresnahan, M. S. Beattie, and B. R. Shah. 1991. Spinal cord injury in rats, using a displacement controlled impact device: effects of nalmefene, U-50488H, and YM-14673. *Soc. Neurosci. Abst.* (Abstract).
8. Behrmann, D. L., J. C. Bresnahan, M. S. Beattie, and B. R. Shah. 1992. Spinal cord injury produced by consistent displacement of the cord in rats behavioral and histologic analysis. *J. Neurotrauma* 9(3), 197–217.
9. Black, P., R. S. Markowitz, V. Cooper, A. Mechanic, H. Kushner, I. Damjanov, and S. D. A. U. W. Finkelstein. 1986. Models of spinal cord injury: Part 1. Static load technique. *Neurosurgery* 19, 752–762.
10. Blight, A. 1988. Mechanical factors in experimental spinal cord injury. *J. Am. Paraplegia. Soc.* 11, 26–34.
11. Bohlman, H. H., E. Bahniuk, G. Raskulinecz, and G. Field. 1979. Mechanical factors affecting recovery from incomplete cervical. *Johns Hopkins Med. J.* 145, 115–125.
12. Bresnahan, J. C., M. S. Beattie, B. T. Stokes, and K. M. Conway. 1991. Three-dimensional computer-assisted analysis of graded contusion lesions in the spinal cord of the rat. *J. Neurotrauma* 8(2), 91–101.
13. Bresnahan, J. C., M. S. Beattie, F. D. Todd, III, and D. H. Noyes. 1987. A behavioral and anatomical analysis of spinal cord injury produced by a feedback-controlled impaction device. *Exp. Neurol.* 95, 548–570.
14. Choulli, K., J. P. Herman, J. M. Rivet, H. Simon, and M. Le Moal. 1987. Spontaneous and graft-induced behavioral recovery after 6-hydroxydopamine lesion of the nucleus accumbens in the rat. *Brain Res.* 407, 376–380.
15. Collier, T. J., D. M. Gash, and J. R. Sladek, Jr. 1987. Norepinephrine deficiency and behavioral senescence in aged rats. Transplanted locus ceruleus neurons as an experimental replacement therapy. *Ann. N.Y. Acad. Sci.* 495, 396–403.
16. Constantini, S. and W. Young. 1994. The effects of methylprednisolone and the ganglioside GM1 on acute spinal cord injury in rats. *J. Neurosurg.* 80, 97–111.
17. NIH Workshop on Experimental Spinal Cord Injury: 1992 Eds. D. A. Anderson, C. Y. Hsu, M. E. Michel, and B. T. Stokes. Modeling and Criteria. *J. Neurotrauma* 9:2, 113–186.
18. Fehlings, M. G., C. H. Tator, and R. D. Linden. 1989. The relationships among the severity of spinal cord injury, motor and somatosensory evoked potentials and spinal cord blood flow. *Electroencephalogr. Clin. Neurophysiol.* 74, 241–259.
19. Ford, R. W. 1983. A reproducible spinal cord injury model in the cat. *J. Neurosurg.* 59, 268–275.
20. Gale, K., H. Kerasidis, and J. R. Wrathall. 1985. Spinal cord contusion in the rat: behavioral analysis of functional neurologic impairment. *Exp. Neurol.* 88, 123–134.
21. Goldberger, M. E. 1991. The use of behavioral methods to predict spinal cord plasticity. *Res. Neurol. Neurosci.* 2, 339–350.
22. Goldberger, M. E., B. S. Bregman, C. J. Vierck, Jr., and M. Brown. 1990. Criteria for assessing recovery of function after spinal cord injury: behavioral methods. *Exp. Neurol.* 107, 113–117.
23. Gruner, J. A. 1992. A monitored contusion model of spinal cord injury in the rat. *J. Neurotrauma* 9, 123–6; discussion 126–8.
24. Gruner, J. A., C. K. Wade, G. Menna, and B. T. Stokes. 1993. Myoelectric evoked responses versus locomotor recovery in chronic spinal cord injured rats. *J. Neurotrauma* 10, 327–347.
25. Holtz, A. and B. Gerdin. 1991. Blocking weight-induced spinal cord injury in rats: therapeutic effect of the 21-aminosteroid U74006F. *J. Neurotrauma* 8, 239–245.
26. Kearney, P. A., S. A. Ridella, D. C. Viano, and T. E. Anderson. 1988. Interaction of contact velocity and cord compression in determining the severity of spinal cord injury. *J. Neurotrauma* 5, 187–208.
27. Kobrine, A. I., D. E. Evans, and H. V. Rizzoli. 1979. Experimental acute balloon compression of the spinal cord. *J. Neurosurg.* 51, 841–845.
28. Koozekanani, S. H., W. M. Vise, R. M. Hashemi, and R. B. McGhee. 1976. Possible mechanisms for observed pathophysiological variability in experimental spinal cord injury by the method of Allen. *J. Neurosurg.* 44, 429–434.
29. Kunkel-Bagden, E. and B. S. Bregman. 1990. Spinal cord transplants enhance the recovery of locomotor function after spinal cord injury at birth. *Exp. Brain Res.* 81, 25–34.
30. Molt, J. T., L. R. Nelson, D. A. Poulos, and R. S. Bourke. 1979. Analysis and measurement of some sources of variability in experimental spinal cord trauma. *J. Neurosurg.* 50, 784–791.
31. Noble, L. J. and J. R. Wrathall. 1985. Spinal cord contusion in the rat: morphometric analyses of alterations in the spinal cord. *Exp. Neurol.* 88, 135–149.
32. Noble, L. J. and J. R. Wrathall. 1987. An inexpensive apparatus for producing graded spinal cord contusive injury in the rat. *Exp. Neurol.* 95, 530–533.
33. Noble, L. J. and J. R. Wrathall. 1989. Correlative analyses of lesion development and functional status after graded spinal cord contusive injuries in the rat. *Exp. Neurol.* 103, 34–40.
34. Noyes, D. H. 1987. An electromechanical impactor for producing experimental spinal cord injury in animals. *Med. Biol. Eng. Comp.* 25, 335–340.

35. Noyes, D. H. 1987. Correlation between parameters of spinal cord impact and resultant injury. *Exp. Neurol.* 95, 535–547.
36. Ohnishi, S. T., J. K. Barr, K. K. Sadanaga, and C. Katagi. 1988. Charybdotoxin improves motor recovery of the rat after spinal cord injury. *Pharmacol. Biochem. Behav.* 31, 187–191.
37. Panjabi, M. M. and J. R. Wrathall. 1988. Biomechanical analysis of experimental spinal cord injury and functional loss. *Spine* 13, 1365–1370.
38. Popovich, P. G., J. F. Reinhard, Jr., E. M. Flanagan, and B. T. Stokes. 1994. Elevation of the neurotoxin quinolinic acid occurs follwing spinal cord trauma. *Brain Res.* 633, 348–352.
39. Popovich, P. G., W. J. Streit, and B. T. Stokes. 1993. Differential expression of MHC Class II antigen in the contused rat spinal cord. *J. Neurotrauma* 10, 37–46.
40. Raines, A., K. L. Dretchen, K. Marx, and J. R. Wrathall. 1988. Spinal cord contusion in the rat: somatosensory evoked potentials as a function of graded injury. *J. Neurotrauma* 5, 151–160.
41. Rivlin, A. S. and C. H. Tator. 1978. Effect of duration of acute spinal cord compression in a new acute cord injury model in the rat. *Surg. Neurol.* 10, 38–43.
42. Sadanaga, K. K. and S. T. Ohnishi. 1989. Chlorpromazine protects rat spinal cord against contusion injury. *J. Neurotrauma* 6, 153–161.
43. Schrimsher, G. W. and P. J. Reier. 1992. Forelimb motor performance following cervical spinal cord contusion injury in the rat. *Exp. Neurol.* 117, 287–298.
44. Somerson, S. K. 1986. *Analysis of an Electro-mechanical Spinal Cord Injury Device,* Thesis, The Ohio State University.
45. Somerson, S. K. and B. T. Stokes. 1987. Functional analysis of an electromechanical spinal cord injury device. *Exp. Neurol.* 96, 82–96.
46. Stokes, B. T. 1992. Experimental spinal cord injury: a dynamic and verifiable injury device. *J. Neurotrauma* 9(2), 129–134.
47. Stokes, B. T., D. H. Noyes, and D. L. Behrmann. 1992. An electromechanical spinal injury device with dynamic sensitivity. *J. Neurotrauma* 9, 187–195.
48. Stokes, B. T. and P. J. Reier. 1992. Fetal grafts alter chronic behavioral outcome after contusion damage to the adult rat spinal cord. *Exp. Neurol.* 116, 1–12.
49. Tarlov, I. M. and H. Klinger. 1954. Spinal cord compression studies. II. Time limits for recovery after acute compression in dogs. *AMA Arch. Neurol. Psychiatry* 71, 271.
50. Watson, B. D., R. Prado, W. D. Dietrich, M. D. Ginsberg, and B. A. Green. 1986. Photochemically induced spinal cord injury in the rat. *Brain Res.* 367, 296–300.
51. Wolpaw, J. R. and J. S. Carp. 1993. Adaptive plasticity in the spinal cord. *Adv. Neurol.* 59, 163–174.
52. Wrathall, J. R. 1989. Behavioral methods for evaluating rats with contusive spinal cord injury: the combined behavioral score. Pages 26–33 in M. Brown and M. E. Goldberger, Eds., *Conference Proceedings: Criteria for Assessing Recovery of Function: Behavioral Methods.* American Paralysis Association, Springfield, NJ.
53. Wrathall, J. R., R. K. Pettegrew, and F. Harvey. 1985. Spinal cord contusion in the rat: production of graded, reproducible, injury groups. *Exp. Neurol.* 88, 108–122.
54. Zivin, J. A. and U. Degirolami. 1980. Spinal cord infarction: a highly reproducible stroke model. *Stroke* 11, 200–202.

Chapter 21

A WEIGHT FREE-DROP TECHNIQUE TO PRODUCE SPINAL CORD INJURY AND A SIMPLE MOTOR FUNCTION TEST

S. Tsuyoshi Ohnishi, Kenneth K. Sadanaga, and John H. Mullen, Jr.

TABLE OF CONTENTS

I. INTRODUCTION

A. EXAMINATION OF A WEIGHT-DROP METHOD

A weight-drop technique, which was originally designed by Allen[1] and modified by Wrathall et al.,[2,3] has been used by many investigators including our group[4–6] for the study of spinal cord injury in the rat. This method has the advantage that the apparatus for the study is simple and inexpensive. The instrumentation consists of a weight, a vertical rod, and an impounder. The spinal cord is exposed at the thoracolumbar region (T9) by a dorsal laminectomy. The impounder, which is attached to the vertical rod, rests on top of an exposed spinal cord. The weight is designed to slide along the rod from a specified height and hit the impounder (Figure 1). There are several problems with this method:

1. Since the weight slides down a rod, friction between the weight and rod could be a source of error.
2. The impounder weighs about 5 g. This weight of the impounder can apply pressure and may contribute to additional damage to the spinal cord.
3. The dropping weight is about 5–10 g, while the impounder is about 5 g. Upon contact, the impounder may absorb a part of the energy of the dropping weight.
4. Because of the friction, and because of the inertia of the impounder, the exact impounding force cannot be calculated.

0-8493-8094-4/95/$0.00+$.50

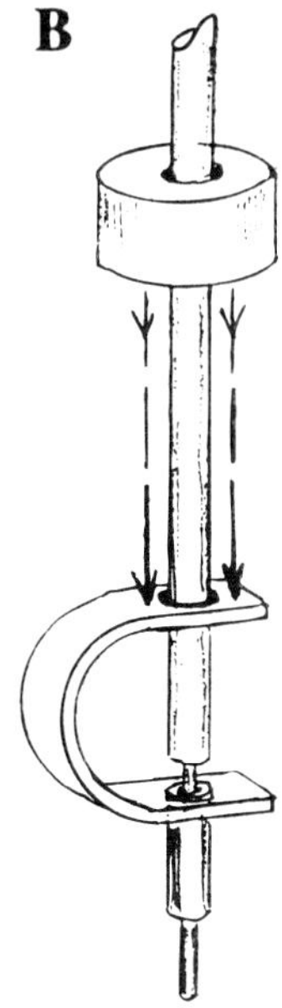

FIGURE 1. The principle of a conventional weight-drop method to produce spinal cord injury. A: sliding weight; B: metal rod; C: C-shaped metal plate to drive an impounder; D: plastic impounder. The combined weight of C and D is about 5 g.

5. Even under anesthesia, when the impounder hits the spinal cord, the animal may move or jump up 0.5 cm because of a natural reflexive response to impounding. Since the impounder and the weight are still sitting on top of the exposed spinal cord, and since there is friction between the weight and the rod, if the animal jumps up, the impounder tip can compresses the spinal cord further, creating additional damage to the spinal cord.

Although each of these problems may be small, when added together they may produce a substantial error and could be a cause for variability in the results. Anderson[7] and Somerson and Stokes[8] have developed sophisticated apparatuses that could eliminate these problems. We wanted to develop a new method that still preserves the simplicity of the weight-drop method, but is free of these problems.

B. MOTOR FUNCTION TEST

Gale et al. developed a method to measure the degree of damage caused by spinal injury and also the degree of recovery after the injury.[3] Although their method can produce valuable information regarding the nature and degree of damage on the spinal cord, to conduct all of the motor function tests may be time-consuming. For that reason, we wanted to develop a simple test in which we can estimate motor function of an injured rat more quickly and efficiently.

II. METHODS

A. DEVELOPMENT OF A NEW FRICTION-FREE WEIGHT-DROP METHOD WITHOUT INITIAL PRESSURE ON THE SPINAL CORD

In order to solve all of the problems described earlier, we developed a method that has the following features:

1. Friction between the dropping weight and the rod is eliminated.
2. The pressure on the exposed spinal cord before applying the weight drop is near zero.

3. The inertia of the impounder is very small (0.7 g).
4. No additional damage is imposed on the spinal cord even if an animal moves upward.

In order to remove the friction between the weight and the rod, we developed a weight free-drop method (Figure 2). We used a steel ball bearing (5/16 in. diameter), **A,** which is held by an electromagnet, **B.** When the electric current applied to the electromagnet **B** from a power supply, **C,** is shut off, the ball makes a free-fall hitting an impounder (2.5 mm tip diameter), **G.** In order to make the initial pressure of the impounder on the spinal cord zero, we developed a seesaw type of mechanism, **H,** in which the weight of the impounder is counterbalanced by a weight, I. Since the ball bounces away immediately after hitting the

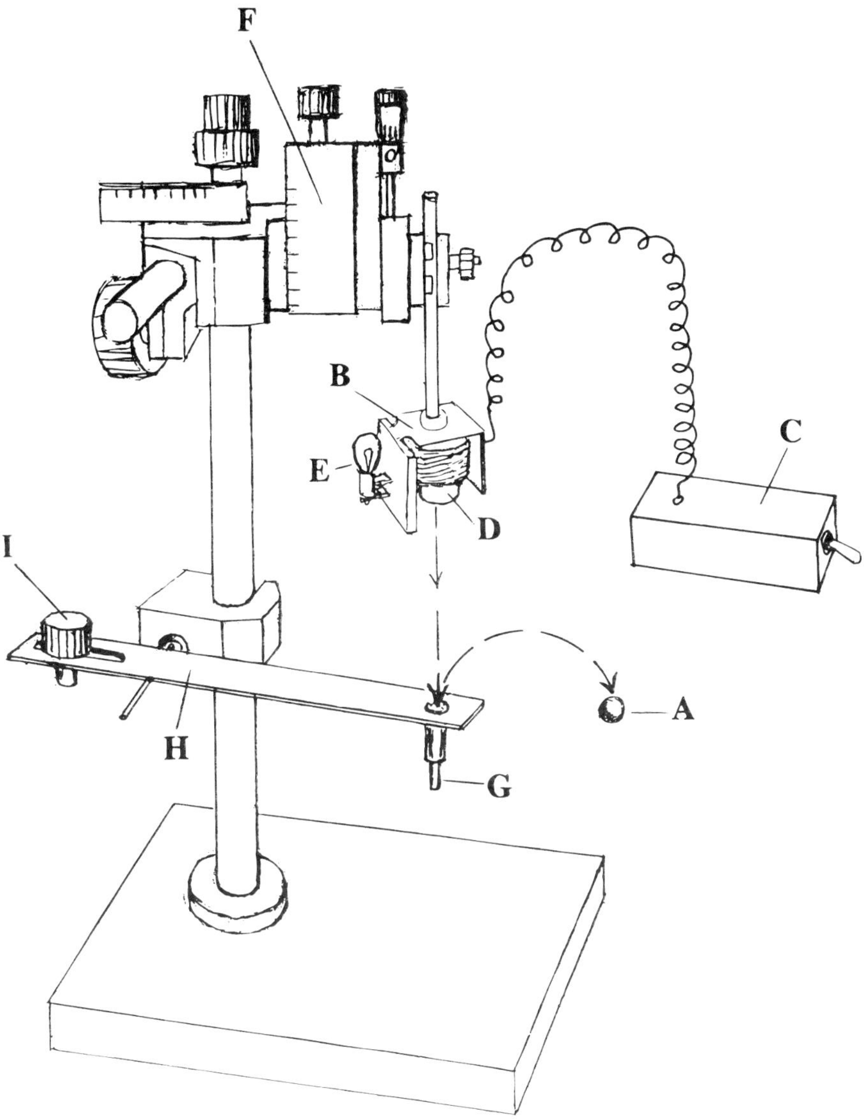

FIGURE 2. A schematic illustration of a friction-free weight-drop apparatus. A: ball bearing ($^5/_{16}$ in. diameter); B: electromagnet taken out from a relay; C: power supply with a shut-off switch; D: Plexiglas tubing to fix the position of the ball; E: lamp to indicate that the magnet is activated; F: micromanipulator, G: plastic impounder with a tip diameter of 2.5 mm; H: seesaw mechanism; I: counterbalance.

impounder (as shown by a dotted curve in the figure), the impounder will not damage the spinal cord even if the animal jumps up.

B. ANIMALS

Sprague-Dawley female rats (body weight 250 to 350 g) were purchased from Ace Animals (Boyertown, PA). IP injection of Nembutal (sodium pentobarbital; 50 mg/kg) was used for anesthesia. The anesthetized animal was placed in ventral recumbency on a small board. The dorsal laminectomy (at T9 site) was performed and the contusion injury was applied while the animal was staying on the board. The use of female animals reduced the need of compressing the bladder after spinal cord injury (see Section III).

C. MOTOR SCORE MEASUREMENT

We used three different tests; walking ability (which is similar to the Tarlov test),[9] balancing on a rod, and the ability to stay on or climb an inclined plate covered with rubber. Each test has four motor scores; 0 being complete paralysis, and 3 normal (Table 1).

III. RESULTS

A. SELECTION OF A STEEL BALL

As shown in Table 2, we used three different ball bearings with diameters 9.53, 7.94, and 7.0 mm (weights 3.5, 2.04, and 1.4 g, respectively). Since we had been using a 5-g weight from a 5-cm height in the conventional weight-drop method, we first tried a 3.5-g ball from a 5-cm height to see how this method produces injury. To our surprise, two of four animals died, suggesting that this impact was too harsh. In other words, this free-fall method is more efficient in creating injury than a slide-down type weight-drop method. Then, we tested a 1.4-g ball. With this weight, we could not produce a sufficient degree of injury. Therefore, we decided to use a 2.04 or 3.5 g ball for the rest of the experiment.

TABLE 1
A Simple Motor Function Test for Spinal Cord Injury Study

Score	On a flat surface	Horizontal wooden rod (1.25 in. diameter)	Inclined rubber surface
0	Hindlegs cannot support body weight	Hindlegs cannot grab the rod	Drops on the 55° slope
1	Can support, but cannot walk	Hindlegs try to grab the rod	Stays for >10 s on 55° slope
2	Walks with much deficiency of hind legs	Using hindlegs, walks with difficulty	Climbs on 55° slope
3	Walks quickly	Can walk	Climbs on 60° slope

Note: The summation of three test scores gives the total motor score.

TABLE 2
Ball Bearings Used for the Test

Diameter (mm)	Diameter (in.)	Weight (gr)
9.53	3/8	3.5
7.94	5/16	2.04
7.00	—	1.4

B. EFFECTS ON URINATION

In our previous experiments in which we used male rats,[4–6] we had to compress their bladders twice a day for several days after injury. However, in this experiment, of 40 animals, in only three animals did we have to compress the bladder a day or two after injury. By 3 d postinjury, we did not have to compress the bladder; the rats urinated spontaneously.

C. THE DEGREE OF INJURY AND RECOVERY AFTER INJURY

Our dropping heights were 2, 3, 5, and 7 cm, respectively. We used eight animals for each height, and measured their motor function 1, 7, 14, 21, and 28 d after injury. As shown in Figure 3, both the degree of injury and the recovery are dependent on the height of the weight drop.

IV. DISCUSSION

We developed a new weight-drop method to produce spinal cord injury in the rat. The method seems to be more effective than the conventional weight-drop method for two reasons. First, our method did not involve any friction in the weight movement. Second, we used a very light impounder (0.7 g). Therefore, it absorbed very little energy, thereby making the method more efficient than the previous method in which the impounder itself weighed about 5 g.

We devised a method in which the pressure of the impounder on the spinal cord is zeroed by using a counterbalance. Therefore, before the impounding, no pressure was applied to the spinal cord, eliminating a possibility of injuring the spinal cord before the impounding. In this method, a free-falling ball hits the impounder and immediately falls on the lab bench (we placed the whole setup in a plastic tray so that we do not lose the ball). Since the weight did not rest on top of the impounder as it did in the old method, the impounder did not produce any extra damage after injury, even though the animal may occasionally move upon impact.

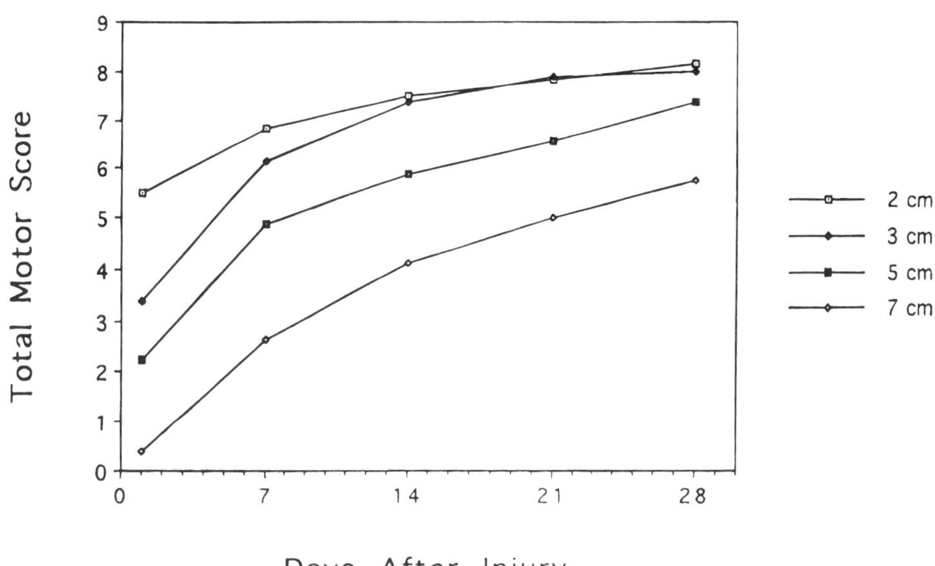

FIGURE 3. Recovery of the total motor score of injured rats. The level of the motor score of sham control rats is 9, and 0 is the level of complete paralysis. Each point indicates the average value calculated from the measurements of eight rats. Numbers on the right side of the figure indicate the heights of the falling ball. The weight of the ball was 2.04 g.

We have devised a simple method to measure the total motor function by measuring (a) the ability of the hind legs to support the body weight or the ability to walk on a flat surface, (b) the ability to hold a rod or to walk on a rod, and (c) the ability to stay on an inclined rubber surface or to climb the slope. This scoring system for motor function was found to be sensitive to motor deficiency when the total motor score was low (from 0 to 6). However, when the score approached 9 (in other words, when an animal almost recovered), it was difficult to distinguish whether an animal was almost normal or completely normal. Another problem, which is inherent to motor score tests, is the difficulty in judging whether animals cannot physically move or do not want to move. Nevertheless, we prefer this method because it is simple and quick. The total time to perform the three tests on one animal was less than 2 min. The reproducibility of the test seems to be satisfactory. When the same investigator performed the test within 1 h, the variability in the total motor score of the same animal was less than ± 1.0. When two investigators tested the same animal, the difference of the score between the investigators was also within ± 1.0. In each individual motor function test, sometimes the evaluation seems to fall between two scores, for example, between 0 and 1. If this were the case, one could assign the score of 0.5. However, further subdivision between scores may not be appropriate.

In this experiment, we did not have to compress the bladder of injured animals as much as we did in our previous experiments.[4–6] There may be at least two reasons for this. First, we used female rats in contrast to the previous experiment where we used the male. Physiologically, female rats had more of a tendency, when compared to males, to spontaneously urinate after injury. Second, since we chose a relatively light injury, the damage to the spinal cord, and thus to the bladder, may not be as severe as that in our previous experiments. As shown in Figure 3, our new apparatus and new motor function test yielded satisfactory results. We were able to produce graded levels of spinal cord injury in the rat. Their recovery after injury was also related to the severity of the injury.

In conclusion, the combination of a newly designed free-falling ball apparatus, a simple motor function test, and the use of female rats made the spinal cord injury study simple and easy. The method may be used for the study of the effects of pharmacological treatments on the recovery after spinal cord injury.

ACKNOWLEDGMENTS

The authors thank Dr. Shinya Sato from the Institute of Brain Diseases at the Tohoku University for his valuable suggestions. The study was supported by NIH grant NS30186.

REFERENCES

1. Allen, A. R. Surgery of experimental lesion of spinal cord equivalent to crush injury of fracture dislocation of spinal column. A preliminary report. *JAMA* 57:878–880 (1911).
2. Wrathall, J. R., Pettegrew, R. K., and Harvey, F. Spinal cord contusion in the rat: production of graded reproducible injury groups. *Exp. Neurol.* 88:108–122 (1985).
3. Gale, K., Kerasidis, H., and Wrathall, J. R. Spinal cord contusion in the rat: behavioral analysis of functional neurologic impairment. *Exp. Neurol.* 88:123–134 (1985).
4. Ohnishi, S. T., Barr, J. K., Katagi, C., and Katsuoka, M. New prostaglandin derivatives protect rat spinal cord against contusion injury. *Arzneim. Forsch.* 39:236–239 (1989).
5. Ohnishi, S. T., Barr, J. K., Sadanaga, K. K., and Katagi, C. Charybdotoxin improves motor recovery of the rat after spinal cord injury. *Pharmacol. Biochem. Behavior.* 31:187–191 (1988).
6. Sadanaga, K. K. and Ohnishi S. T. Chlorpromazine protects rat spinal cord against contusion injury. *J. Neurotrauma* 6:153–161 (1989).

7. Anderson, T. E. A controlled pneumatic technique for experimental spinal cord contusion. *J. Neurosci. Methods* 6:327–333 (1982).
8. Somerson, S. K. and Stokes, B. T. Functional analysis of an electromechanical spinal cord injury device. *Exp. Neurol.* 96:82–96 (1987).
9. Tarlov, I. M. Spinal cord compression studies. III. Time limits for recovery after gradual compression in dogs. *Arch. Neurol. Psychiatry* 71:588–597 (1954).

IV. Behavioral Study

Chapter 22

THE ASSESSMENT OF COGNITIVE FUNCTION FOLLOWING TRAUMATIC BRAIN INJURY

Robert J. Hamm

TABLE OF CONTENTS

I. INTRODUCTION

The sequelae of traumatic brain injury (TBI) in humans encompass a wide range of disabilities (e g., Binder, 1986; Ewen et al., 1989; Gronwall and Wrightson, 1974; Hall and Bornstein, 1991; Levin et al., 1986; 1988; 1992; Stuss et al., 1985). One of the most important of these is cognitive dysfunction. The cognitive disturbances that follow TBI include almost every aspect of information processing. Information processing speed, reaction time, vigilance, and attention are altered following injury. The effect of brain injury on memory is most reliable and impressive on tests of declarative memory, especially on tasks that require high levels of mental effort or retention over an extended time. The amnesias observed after TBI can be either anterograde or retrograde or both. Retrograde amnesia is defined as a memory loss for events (facts) that were acquired before the injury. It is temporally graded with recent time periods more affected than remote time periods. Anterograde amnesia is marked by an impairment in the ability to acquire new information after the injury.

Although cognitive impairment is one of the most significant and enduring features of TBI in humans, the assessment of cognitive function in animal models of TBI has been ignored until recently. Consequently, the description of TBI-induced cognitive deficits in animals has just begun, and the investigation of therapeutic interventions designed to ameliorate the impairment of cognitive function is in its infancy.

0-8493-8094-4/95/$0.00+$.50

II. UTILITY OF COGNITIVE ASSESSMENT

Given that the assessment of cognitive function is typically a labor-intensive task, why should it be evaluated? What are the benefits? There are a number of answers to these questions:

1. **Validity of the animal model of TBI**—Since cognitive impairment is a hallmark of human TBI, any animal model of TBI should reproduce this major sequelae of human TBI. If an animal model does not produce an enduring deficit in learning or memory, one could question the relevance of the animal model to the human condition.
2. **Neuroanatomical specificity**—The field of animal cognition has advanced significantly in the last few years. There is growing evidence that suggests that the mammalian nervous system contains multiple memory systems that differ in the "kind" of memory they mediate. By using animal learning and memory tasks that are able to dissociate the contribution of various brain regions to the performance of a task, inferences can be made regarding the functional status of restricted brain regions following TBI. For example, the hippocampus is known to be a critical brain region for the performance of certain types of cognitive tasks (e.g., the Morris water maze). Observing a behavioral deficit on a task with an established neuroanatomical specificity allows conclusions to be drawn about the vulnerability of certain brain regions to damage following TBI. This information can be used to corroborate physiological measurements of regional dysfunction (e.g., metabolic, receptor binding, etc.) or may suggest new brain regions for further analysis.
3. **Neurotransmitter specificity**—Just as certain tasks have been shown to be mediated by discrete brain regions, animal tasks have also been found to involve specific neurotransmitters. For example, the cholinergic system is known to mediate performance on several cognitive tasks. Thus, if deficits are observed in a task that is known to be mediated by a specific neurotransmitter system, inferences can be made about the functional status of that neurotransmitter system following TBI.
4. **Time course of dysfunction and recovery**—Although cognitive assessment is not unique in providing information on the duration of dysfunction and recovery of function, cognitive evaluation does offer some distinct advantages. For example, the assessment of motor impairment is the most widely used outcome measure following experimental TBI. However, motor deficits typically recover within a few weeks following injury. On the other hand, as described later, cognitive deficits can persist for several months following TBI. This extended period of cognitive impairment provides an opportunity to test the efficacy of treatments that are designed to enhance the long-term recovery of function following TBI. Results from these studies would be especially relevant to the rehabilitation of human head injured patients.

III. INTERPRETATIONAL CAUTIONS

Although the assessment of cognitive function offers a number of advantages, great care must be exercised in the selection and design of cognitive assessment tasks and in the interpretation of the resulting behavioral data. Investigators must realize that the animal behavior that is recorded on a cognitive task is the end product of a number of processes, some of which are **not** cognitive in nature. In a simplified way, Figure 1 illustrates the number of processing stages that are involved in most cognitive tasks. Although the sensory register and motivational components of the flow chart may be considered to be cognitive in nature, for the purposes of this chapter, cognitive is meant to refer to the process directly

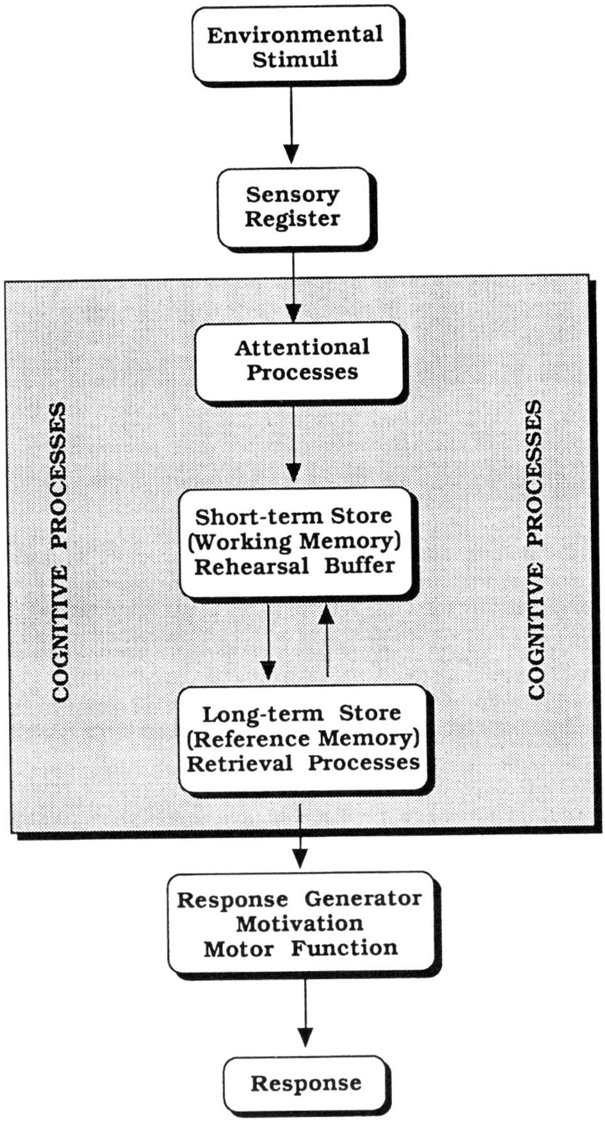

FIGURE 1. Simplified flowchart of the stages of information processing. Both cognitive and noncognitive processes intervene between the experimental setting and the behavioral output.

involved in learning and memory. In the first stage of Figure 1, environmental stimuli enter the sensory register of the animal. As certain aspects of the environmental stimuli are associated with important outcomes (e.g., reward), the animal will selectively attend to those stimuli. The association between a stimulus and an important outcome will then enter a short-term storage stage (sometimes referred to as working memory) that is capable of storing information only for a limited duration (generally less than 1 min). If the information is rehearsed sufficiently, it will be transferred to permanent long-term storage (or reference memory). Once the information is stored, if the animal is presented with the proper cues for the retrieval of the information, it will recover the information. Then depending on the animal's ability to respond, and its motivation to respond, it will generate an appropriate response. Applied to the Morris water maze task, the sensory register would record the intra- and extramaze stimuli. In the next stage, the animal attends to selected aspects of the test environment (e.g., extramaze cues), and it experiences an important event (e.g., it finds the escape platform). The location of the platform will then be processed (working memory). After sufficient processing or rehearsal, the location of the platform will be transferred to more permanent storage (long-term or reference memory). This information can then be retrieved for use in the future. However, as is indicated in the figure, motor and motivational factors intervene between the cognitive processes and the observed behavior. In assessing cognitive performance, it is necessary to ensure that these noncognitive processes do not confound cognitive task performance.

The importance of motivational and motor concerns are especially relevant following TBI. If food is used as a reward to motivate an animal to perform a cognitive task, food must be an equally effective reward for injured and uninjured animals. Otherwise the difference in performance on the task may be the result of motivational and not cognitive factors. It is common for injured animals to have a reduction in food appetite and loss of weight for several days following TBI. If injured animals are tested on a food-motivated task during the time of weight loss following TBI, cognitive task performance will likely be confounded by motivational differences. Similarly, motor impairment may also confound cognitive assessment. Motor deficits are also a common sequelae of TBI, and since most cognitive tasks require running or swimming, motor impairment might also confound the valid assessment of cognitive function following TBI.

In reviewing the experimental data on cognitive function below, the possibility of confounding by noncognitive factors will be discussed.

IV. REVIEW OF CURRENT DATA

A. LONG-TERM POTENTIATION

Long-term potentiation (LTP) in the hippocampus is a form of synaptic plasticity and is generally considered to be a cellular basis of learning and memory. If LTP represents a memory storage mechanism, its induction and expression characteristics following TBI provide a window into the operation of neural circuitry that encodes memory. Miyazaki et al. (1992) examined changes in synaptic responses (population spike and population EPSP) of CA_1 pyramidal cells to stimulation of the Schaffer/commissural pathways 2 to 3 h after central fluid percussion injury. Prior to tetanic stimulation, TBI was found to decrease population spike threshold but not EPSP thresholds. Also, following TBI there was a decrease in maximal amplitude of population spikes as well as EPSPs. TBI was found to suppress LTP as assessed by reductions in post tetanic increases in population spikes as well as EPSPs.

Although the mechanics by which TBI suppresses the induction of LTP are unknown, several features of the data are of interest. The suppression of LTP following TBI was not the result of a nonspecific suppression of neural function. In fact, graphs of the population spike responses as a function of EPSP slope suggest increased excitability of pyramidal neurons. Since prior induction of LTP-like synaptic changes can suppress subsequent induction of LTP (Malenka et al., 1986), the authors speculate that the activation of extrinsic inputs to the hippocampus by TBI may contribute to the suppression of LTP observed following injury. In other words, LTP-like synaptic activity produced by TBI may underlie the suppression of LTP observed after TBI.

It is generally accepted that the *N*-methyl-D-aspartate (NMDA) receptor complex plays an important role in LTP (Collingridge and Bliss, 1987; Cotman and Iverson, 1987; Morris, 1989), and it is also well established that the NMDA receptor system contributes to the pathophysiology associated with TBI (for reviews, see Hayes et al., 1992 and McIntosh, 1993). How TBI-induced alterations in the NMDA receptor system affect LTP are uncertain. However, NMDA antagonists have been shown to reduce the motor and learning deficits produced by TBI (Faden et al., 1989; Hamm et al., in press b; McIntosh et al., 1989). Would

NMDA antagonists also be beneficial in reducing the suppression of LTP following TBI? Many important questions remain to be answered about the mechanisms responsible for the TBI-induced suppression of LTP.

B. PASSIVE AVOIDANCE

The passive avoidance (PA) task, also referred to as inhibitory avoidance, is a widely used task to assess memory dysfunction. The popularity of the task is partly due to the ease of the training and testing procedure. The PA task is usually performed in an apparatus that consists of two compartments (illuminated and darkened) divided by a guillotine door (see Figure 2 for an illustration of a typical passive avoidance apparatus). Animals are placed in the lighted compartment. As soon as the animal crosses into the dark compartment, the door is closed, and the animals receive an electric foot-shock. One training trial is sufficient to produce learning (i.e., learning to avoid the dark compartment). Animals are tested for the retention of the passive avoidance response by placing them in the illuminated compartment and recording the latency for the animal to cross into the previously shocked, dark compartment. Long latencies indicate good learning and retention of the task.

Performance on the PA task is unlikely to be confounded by noncognitive processes. The motivation for the PA task is the avoidance of an electric foot-shock. As assessed by the tail-flick procedure, TBI does not alter nociception (Lyeth, personal communication). Motor deficits that are frequently associated with TBI will not confound cognitive assessment of the PA task. If motor deficits are present at the time of testing, they most likely increase the latency to move into the dark compartment. This increase in latency would be interpreted as evidence of the retention of the association. Thus, if motor performance were impaired, it would negatively bias the observation of a cognitive deficit after TBI. And finally, motor/activity levels of injured vs. sham-injured animals can be examined by comparing the initial crossing latency on the training trial.

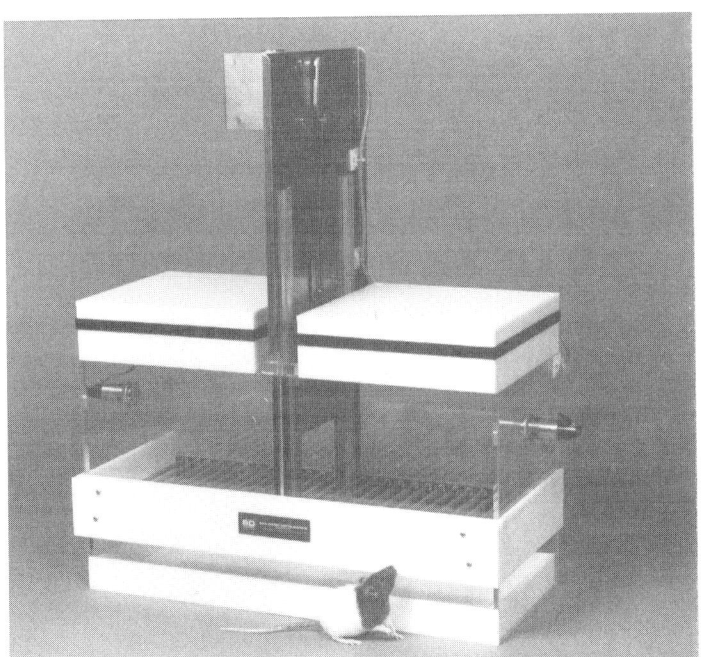

FIGURE 2. Illustration of a typical passive avoidance apparatus. (Courtesy of San Deigo Instruments.)

Lyeth (personal communication) examined the effect of central fluid percussion TBI on the acquisition and retention of the PA task. Rats were given one training trial 24 h after TBI and were tested for retention 24 h later. The injured animals crossed into the previously shocked compartment with a shorter latency than sham-injured animals. This finding demonstrates that the injured animals had not learned or remembered the association between the compartment and electric foot-shock. Unfortunately, the PA task does not differentiate between deficits due to acquisition and storage (i.e., the ability to learn and store the association) or retrieval (i.e., the ability to recall the association) of information.

Another recent experiment (Hamm et al., 1993) examined PA task performance after a longer recovery interval following TBI. Rats were either injured at a moderate level of central fluid percussion brain injury or surgically prepared but not injured (sham-injured control group). Nine days after fluid percussion injury, injured and sham-injured rats were trained on the one-trial PA task with retention assessed 24 h later. The results of this experiment revealed that performance of the PA task was not significantly impaired by fluid percussion TBI. Integrating the findings obtained with the PA task, it appears that TBI produces a transitory impairment of PA performance (i.e., anterograde amnesia) that is no longer present 9 d after fluid percussion TBI.

Lesions of a number of brain regions have been found to disrupt PA performance (e.g., nucleus basalis, frontal cortex, amygdala, nucleus accumbens). Interestingly, however, hippocampal lesions do not affect PA performance at short retention intervals (Winocur, 1985; Mundy and Tilson, 1988). The results from the research that has employed the PA task suggest that the neural circuitry that mediates PA performance is impaired for a short period of time after TBI but recovers rather quickly.

C. VISUAL DISCRIMINATION

Using an operant conditioning procedure, Gorman et al. (1993) trained rats prior to TBI to learn the association between a stimulus (light position) and food reinforcement (a representative operant chamber is presented in Figure 3). In an operant chamber equipped with two response bars and a light above each bar, rats were trained to press the bar directly under a light. Thus, a correct response and its reinforcement was contingent on the position of the light. Rats were trained to a criterion of at least 80% correct responding prior to TBI.

FIGURE 3. Illustration of a typical operant conditioning apparatus. (Courtesy of San Deigo Instruments.)

Postinjury testing indicated that TBI produced a deficit in this visual discrimination task only on the first post-TBI day. The decline in the percentage of correct responses occurred without changes in bar-pressing rates, suggesting that impairment of performance was not due to motivational or motor deficits. The duration of anterograde amnesia observed on this task is similar to that seen with the PA task described earlier.

D. RADIAL-ARM MAZE

The radial-arm maze (RAM, see Figure 4 for an illustration) has been used to study a number of different aspects of learning and memory. The maze can be configured to study spatial learning, nonspatial learning, working memory (storage of trial-specific information), or reference memory (storage of information relevant to all trials). The typical RAM is elevated and has eight equally spaced arms that project from a central choice area. Each arm of the maze is separated from the central choice area by guillotine doors that can be raised to allow the rat access to each maze arm. A food reward can be placed at the end of each arm. On any particular trial, the rat is allowed to visit several arms of the maze. The optimal strategy is to visit each baited arm only once (i.e., there is no reward for returning to an arm in which the rat has already eaten the food). By not baiting all the arms of the maze (usually two to four of the arms are not baited; the unbaited arms remain constant for each rat), it is possible to assess both spatial working and reference memory. Reference memory errors occur when the rat enters one of the arms that is never baited. Working memory errors occur when the rat reenters an arm that it had previously visited.

Several brain regions are critical for spatial RAM performance. The most important of these is the hippocampus. Numerous experiments have shown that destruction of the hippocampus impairs the working memory component of the RAM task but does not impair reference memory (Becker et al., 1980; Jarrard, 1978; 1986; Olton et al., 1978; 1982).

In one of the first investigations of cognitive function after TBI, Lyeth et al. (1990) examined the performance of rats in the RAM after two levels of severity of central fluid percussion injury. By baiting six of the eight arms of the RAM, they were able to evaluate both working and reference memory following TBI. Animals were trained to a learning criterion prior to TBI and were tested for 25 d after injury. Results revealed that reference

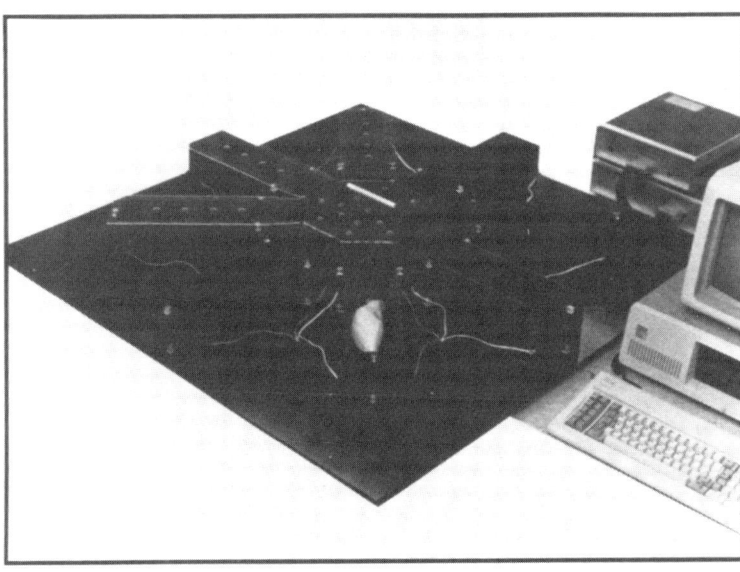

FIGURE 4. Illustration of a radial-arm maze apparatus. (Courtesy of San Deigo Instruments.)

memory errors were not increased by either injury level. The impairment of working memory was related to the severity of the injury. In the mild injury condition, working memory errors returned to preinjury levels by 10 d after TBI. In the moderate injury condition, working memory deficits were present until 20 d after TBI. This pattern of results (i.e., working memory impairment without reference memory errors) corresponds exactly to the memory deficits observed after hippocampal damage. Interestingly, however, the working memory deficits produced by central fluid percussion injury occurred in the absence of neural cell death in the hippocampus or overt axonal injury in hippocampal pathways. The authors suggest that fluid percussion injury, as it was employed in this study, produces a sublethal insult that disrupts normal neuronal function in sensitive brain regions. The results of this experiment further document that the hippocampus is especially vulnerable to TBI. This conclusion is consistent with other research that has documented the vulnerability of the hippocampus to other brain insults (Ben-Air, 1985; Mundy and Tilson, 1988; Pulsinelli et al., 1982).

Because the RAM procedure places a high demand on the motor system and uses food to motivate the animal's behavior, there is a possibility that RAM performance may be influenced by these noncognitive processes. In the case of the Lyeth et al. (1990) study, these confounding effects appear unlikely. First, working memory errors were still present at a time when motor deficits (as measured by beam balance and beam walking) had returned to preinjury levels. Similarly, working memory deficits were observed when animals had resumed eating and were gaining weight. And finally, TBI did not increase reference memory errors. If motor or motivational factors disturbed the performance of the injured animals, it would be expected that both working and reference memory would be equally affected. The Lyeth et al. (1990) data demonstrate that only working memory was impaired. For these reasons, the observed deficits in working memory are in all probability the result of a memorial process.

E. SPATIAL DELAYED DISCRIMINATION

Using an operant conditioning apparatus, Gorman et al. (1993) trained animals to bar-press for food reinforcement. On a given trial, one of two response bars was presented. After a bar-press, the bar was retracted and a light in the center of the chamber was illuminated for a variable period of time (2, 4, 8, or 12 s). The light was then extinguished, a food pellet delivered, and both response bars were presented. To receive the second food pellet, rats were required to press the same bar that was pressed earlier. A response on the incorrect bar terminated the trial without a food pellet. There was a 30 s intertrial interval. Rats were trained to a criterion of at least 90% correct responding at each delay interval prior to TBI or sham injury.

An analysis of the percentage of correct responses on this spatial delayed discrimination task indicated that TBI did not impair the performance with a 2-s delay. However, as the delay interval increased injured animals evidenced a deficit in performance. As was the case in the visual discrimination task described above, this disruption on memory function occurred in the absence of a change in response rate.

F. MORRIS WATER MAZE

In the Morris water maze (MWM, see Figure 5) procedure, rats are placed in a large pool of water (typically 2 m in diameter) and are required to find a hidden goal platform in order to escape the water. The goal platform is hidden by being slightly below the water's surface and by making the water opaque. In the spatial version of the MWM procedure, it is important that there are no local (i.e., intramaze) cues to the location of the goal platform. Normal rats learn quickly to find the goal platform from any location in the maze. Several measures of MWM performance can be used: latency to find the goal platform; length of the swimming

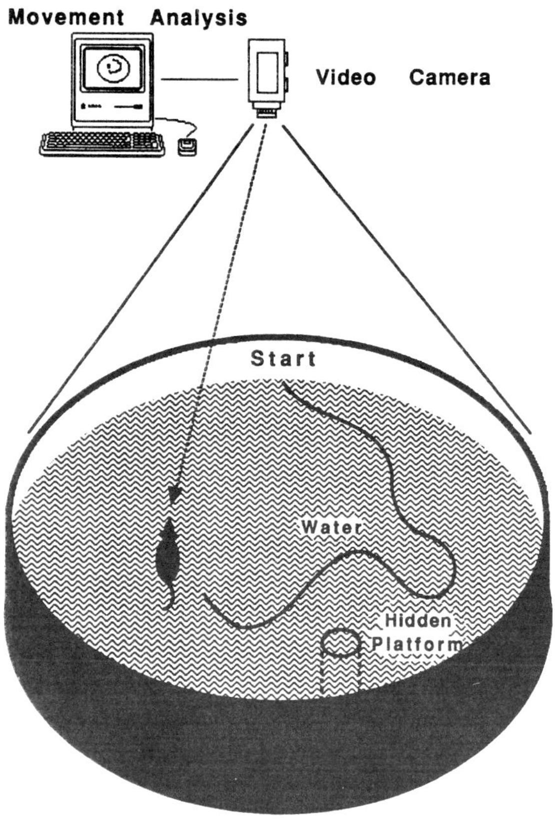

FIGURE 5. Illustration of a Morris water maze apparatus.

path used to find the goal platform; the proportion of time spent by the rat in the quadrant of the maze that contains the goal platform. The traditional MWM procedure can also be modified to evaluate working memory and nonspatial visual discrimination learning (Morris et al., 1986). By using various versions of the MWM, it is possible to dissociate the contribution of the hippocampus vs. other brain regions to the performance of the task.

The MWM procedure avoids or minimizes the confounding of motor and motivational processes. The MWM procedure does not use food to motivate task performance. Research has also demonstrated that MWM performance is relatively insensitive to the motor capacity of the animal (Rapp et al., 1987). In addition, the MWM provides a direct index of the animal's motor ability by measuring the swimming speed of the animal. In most of the studies reviewed below, explicit measurements of swimming speed indicated that the injured and uninjured animals swam at the same speeds. This finding indicates that both injured and uninjured animals were equally capable and equally motivated to escape the water.

Performance on the MWM task has been examined following several forms of TBI. Using a lateral fluid percussion injury model, Smith et al. (1991) investigated the effect of TBI on the retention of MWM performance (i.e., retrograde amnesia). Rats were trained prior to TBI and were tested for retention 42 h after moderate and severe injury magnitudes. They found that TBI produced a significant impairment of retention of the MWM performance, and the degree of impairment was correlated to the injury severity. These findings confirm that TBI produces a retrograde amnesia following injury.

MWM performance has also been investigated after controlled cortical impact TBI (Hamm et al., 1992). In this experiment rats were tested on their ability to learn the MWM at various times following TBI (i.e., anterograde amnesia). Results revealed that injured rats displayed a significant deficit in MWM performance for more than one month after TBI. More recent findings with the cortical impact model indicate that deficits in MWM performance are present three months after a more severe level of injury (Dixon, personal communication). This duration of cognitive impairment is quite impressive and will be commented on below.

Two additional studies have examined MWM performance following central fluid percussion TBI. Hamm et al. (1992) employed a mild level of injury that did not produce significant motor impairment following injury. Although there were no significant motor deficits observed, significant MWM deficits were present 11 to 15 d after injury. This result demonstrates that even a mild severity of TBI can produce cognitive impairment, and that the MWM procedure is a very sensitive outcome measure to employ in the study of TBI.

The final experiment (Hamm et al., 1993) investigated two different versions of the MWM in order to examine the selectivity of the cognitive impairment observed after TBI. The first version of the MWM procedure assessed was the standard MWM task in which the animal starts each trial from a different starting location. Because the animal starts each trial from various locations, the extramaze cues associated with the goal platform on any trial depend on the start location. Consequently, the animal must learn the relationship among multiple spatial cues in order to quickly locate the platform. Numerous experiments have shown that performance of this task is severely disrupted by hippocampal lesions (Morris et al., 1992; Schenk and Morris, 1985; Sutherland et al., 1982). The second version of the MWM used was one in which the animal started each trial from the same starting location. In the constant starting position version of the Morris water maze, the animal must learn to swim from a fixed start location to a fixed goal location. Since this task can be solved by learning a specific path to the goal (e.g., swimming toward a particular extramaze cue), it does not require spatial configural learning that would require the involvement of the hippocampus. Research has documented that disruption of hippocampal circuitry by lesions does not impair the acquisition of the constant-start version of the MWM task (Eichenbaum et al., 1990). By examining the effect of TBI on the performance of these two tasks, this investigation provides direct information on the regional generality or specificity of cognitive dysfunction following TBI. That is, if the TBI, as employed in this study, produces a widespread disruption of neural function, performance on both versions of the MWM would be impaired. If, however, TBI is associated with a more restricted dysfunction in especially sensitive brain regions, in this case the hippocampus, then only the variable-start version of the MWM should be impaired after TBI. Results revealed that only the variable-start version of the MWM was impaired 11 to 15 d after TBI. This behavioral evidence implies that the source of the cognitive deficits observed after a moderate level of TBI is primarily the result of hippocampal dysfunction.

V. SUMMARY OF FINDINGS AND FUTURE DIRECTIONS

TBI has been found to produce impairment of cognitive function on a number of different tasks. The PA and the visual discrimination tasks, tasks that do not depend on hippocampal processing, are only impaired for a short time immediately after injury. Another task that also does not depend on hippocampal processing, the constant-start version of the MWM, is also not impaired after TBI. Tasks that are sensitive to hippocampal damage, hippocampal LTP, the RAM, and the standard variable-start MWM, reveal substantial and enduring deficits in cognitive function following TBI. Thus, it appears that tasks that are known to be impaired after hippocampal damage will also reveal significant and long-lasting deficits following TBI.

As is the case in the assessment of human cognitive function, a number of tasks can be used to evaluate the cognitive ability of rats after TBI. New tasks should be evaluated for their sensitivity to TBI-induced deficits. In this regard, a modified probe test for spatial memory (Olton, 1993) may be useful. In addition, the learning set version of the MWM (Whishaw, 1985) may also be useful. This procedure does not confound acquisition with recovery as the typical MWM procedure does. Lastly, by employing multiple measures of performance (Auer et al., 1989; Packard and McGaugh, 1992), it may be possible to dissociate and specify what neuroanatomical regions are damaged following TBI.

ACKNOWLEDGMENTS

This work was partially supported by NIH grant NS12587. The author would like to thank San Deigo Instruments for preparing the illustrations of the various pieces of equipment.

REFERENCES

Auer, R. N., Jensen, M. L., and Whishaw, I. Q. (1989). Neurobehavioral deficits due to ischemic brain damage limited to half of the CA1 sector of the hippocampus. *The Journal of Neuroscience,* 9, 1641–1647.

Becker, J. T., Walker, J. A., and Olton, D. S. (1980). Neuroanatomical basis of spatial memory. *Brain Research,* 200, 307–320.

Ben-Air, Y. (1985). Limbic seizures and brain damage produced by kainic acid: mechanisms and relevance to human temporal lobe epilepsy. *Neuroscience,* 14, 375–403.

Binder, L. M. (1986). Persisting symptoms after mild head injury: a review of the postconcussive syndrome. *Journal of Clinical and Experimental Neuropsychology,* 8, 323–346.

Collingridge, G. L. and Bliss, T. V. P. (1987). NMDA receptors—their role in long-term potentiation. *Trends in Neuroscience,* 10, 288–294.

Cotman, C. W. and Iverson, L. L. (1987). Excitatory amino acids in the brain—focus on NMDA receptors. *Trends in Neuroscience,* 10, 263–265.

Eichenbaum, H., Stewart, S., and Morris, R. G. M. (1990). Hippocampal representation in place learning. *Journal of Neuroscience,* 10, 3531–3542.

Ewert, J., Levin, H. S., Watson, M. G., and Kalisky, K. (1989). Procedural memory during posttraumatic amnesia in survivors of severe head injury. *Archives of Neurology,* 46, 911–916.

Faden, A. I., Demediuk, P., Panter, S. S., and Vink, R. (1989). The role of excitatory amino acids and NMDA receptors in traumatic brain injury. *Science,* 244, 798–800.

Gorman, L. K., Shook, B. L., and Becker, D. P. (1993). Traumatic brain injury produces impairments in long-term and recent memory. *Brain Research,* 614, 29–36.

Gronwall, D. and Wrightson, P. (1974). Delayed recovery of intellectual function after minor head injury. *Lancet,* 2, 605–609.

Hall, S. and Bornstein, R. A. (1991). The relationship between intelligence and memory following minor or mild closed head injury: greater impairment of memory than intelligence. *Journal of Neurosurgery,* 75, 378–381.

Hamm, R. J., Dixon, C. E., Gbadebo, D. M., Singha, A. K., Jenkins, L. W., Lyeth, B. G., and Hayes, R. L. (1992). Cognitive deficits traumatic brain injury produced by controlled cortical impact. *Journal of Neurotrauma,* 9, 11–20.

Hamm, R. J., Lyeth, B. G., Jenkins, L. W., O'Dell, D. M., and Pike, B. R. (1993). Selective cognitive impairment following traumatic brain injury in rats. *Behavioural Brain Research,* 59, 169–173.

Hamm, R. J., O'Dell, D. M., Pike, B. R., and Lyeth, B. G. (1993). Cognitive impairment following traumatic brain injury: The effect of pre- and post-injury administration of scopolamine and MK-801. *Cognitive Brain Research,* 1, 223–226.

Hayes, R. L., Jenkins, L. W., and Lyeth, B. G. (1992). Neurotransmitter mediated mechanisms of traumatic brain injury: acetylcholine and excitatory amino acids. *Journal of Neurotrauma,* 9(Suppl.), S173–S187.

Hayes, R., Jenkins, L., Lyeth, B., Balster, R., Robinson, S., Clifton, G., Stubbins, J., and Young, H. (1988). Pretreatment with phencyclidine, an N-methyl-D-aspartate receptor antagonist, attenuates long-term behavioral deficits in the rat produced by traumatic brain injury. *Journal of Neurotrauma,* 5, 287–302.

Jarrard, L. E. (1978). Selective hippocampal lesions: differential effects on performance by rats of a spatial task with preoperative versus postoperative training. *Journal of Comparative and Physiological Psychology,* 92, 1119–1127.

Jarrard, L. E. (1986). Selective hippocampal lesions and behavior. In R. L. Isaacson and K. H. Pribram (Eds.), *The Hippocampus* (Vol. 4, pp. 93–126). New York: Plenum Press.

Levin, H. S. (1992). Neurobehavioral recovery. *Journal of Neurotrauma,* 9(1), S359–S373.

Levin, H. S. and Goldstein, F. C. (1986). Organization of verbal memory after severe head injury. *Journal of Clinical and Experimental Neuropsychology,* 8, 643–656.
Levin, H. S., High Jr., W. M., and Eisenberg, H. M. (1988). Learning and forgetting during posttraumatic amnesia in head injured patients. *Journal of Neurology, Neurosurgery, and Psychiatry,* 51, 14–20.
Lyeth, B. G., Jenkins, L. W., and Hamm, R. J. (1990). Prolonged memory impairment in the absence of hippocampal cell death following traumatic brain injury in the rat. *Brain Research,* 526, 249–258.
Malenka, R. C., Madison, D. V., and Nicoll, R. A. (1986). Potentiation of synaptic transmission in the hippocampus by phorbol esters. *Nature,* 321, 175–177.
McIntosh, T. K. (1993). Novel pharmacologic therapies in the treatment of experimental traumatic brain injury: a review. *Journal of Neurotrauma.* 10(3), 215–261.
McIntosh, T. K., Vink, R., Soares, H., Hayes, R., and Simon, R. (1989). Effects of the N-methyl-D-aspartate receptor blocker MK-801 on neurologic function after experimental brain injury. *Journal of Neurotrauma,* 6, 247–259.
Miyazaki, S., Katayama, Y., Lyeth, B. G., Jenkins, L. W., DeWitt, D. S., Goldberg, S. J., Newlon, P. G., and Hayes, R. L. (1992). Enduring suppression of hippocampal long-term potentiation following traumatic brain injury in rats. *Brain Research,* 585(1–2), 335–339.
Morris, R. G. (1989). Synaptic plasticity and learning: Selective impairment of learning rats and blockade of long-term potentiation *in vivo* by the N-methyl-D-aspartate receptor antagonist AP5. *Journal of Neuroscience,* 9(9), 3040–3057.
Morris, R. G. M., Garrud, J., Rawlins, N. P., et al. (1982). Place navigation in rats with hippocampal lesions. *Nature,* 297, 681–683.
Morris, R. G. M., Hagan, J. J., and Rawlins, J. N. P. (1986). Allocentric spatial learning by hippocampectomized rats: a further test of the "spatial mapping" and "working memory" theories of hippocampal function. *Quarterly Journal of Experimental Psychology,* 388, 365–369.
Mundy, W. R. and Tilson, H. A. (1988). Behavioral impairment in the rat after colchicine lesions of the hippocampus and nucleus basalis. *Neurotoxicology,* 9, 511–520.
Olton, D. S. (1993) Age-related behavioral impairments: benefits of multiple measures of performance. *Neurobiology of Aging,* 14, 637–638.
Olton, D. S., Walker, J. A., and Gage, F. H. (1978). Hippocampal connections and spatial discrimination. *Brain Research,* 139, 295–308.
Olton, D. S., Walker, J. A., and Wolf, W. A. (1982). A disconnection analysis of hippocampal function. *Brain Research,* 233, 241–253.
Parkard, M. G. and McGaugh, J. L. (1992). Double dissociation of fornix and caudate nucleus lesions on acquisition of two water maze tasks: further evidence for multiple memory system. *Behavioral Neuroscience,* 106, 439–446.
Pulsinelli, W., Breirley, J., and Plum, F. (1982). Temporal profile of neuronal damage in a model of transient forebrain ischemia. *Annals of Neurology,* 491–498.
Rapp, P. R., Rosenberg, R. A., and Gallagher M. (1987). An evaluation of spatial information processing in aged rats. *Behavioral Neuroscience,* 101, 3–12.
Schenk, F. and Morris, R. G. M. (1985). Dissociation between components of spatial memory in rats after recovery from the effects of retrohippocampal lesions. *Experimental Brain Research,* 58, 11–28.
Smith, D. S., Okiyama, K., Thomas, M. J., Claussen, B., and McIntosh, T. K. (1991). Evaluation of memory dysfunction following experimental brain injury using the Morris water maze. *Journal of Neurotrauma,* 8, 259–269.
Stuss, D. T., Ely, P., Hugenholtz, H., Richard, M. T., LaRochelle, S., Poirier, C. A., and Bell, I. (1985). Subtle neuropsychological deficits in patients with good recovery after closed head injury. *Neurosurgery,* 17, 41–47.
Sutherland, R. J., Kolb, B., and Whishaw, I. Q. (1982). Spatial mapping: definitive disruption by hippocampal of medial frontal cortical damage in the rat. *Neuroscience Letters,* 31, 271–276.
Whishaw, I. Q. (1985) Formation of a place learning-set in the rat: a new procedure for neurobehavioral studies. *Physiology and Behavior,* 35, 845–851.
Winocur, G. (1985). The hippocampus and the thalamus: their roles in short- and long-term memory and the effects of interference. *Behavioural Brain Research,* 16, 135–154.

V. *Tissue Culture and Tissue Slice Models*

Chapter 23

IN VITRO MODELS OF ISCHEMIC INJURY

Cheolsu Shin and Xiaofeng Sun

TABLE OF CONTENTS

I. INTRODUCTION

Cerebral ischemia with subsequent neuronal damage is a major problem that affects people of all ages and is one of the leading causes of morbidity and mortality at the present time. Animal models have been used to examine the mechanisms underlying the damage following acute ischemic insult and reperfusion, and have yielded many valuable insights and understanding. However, because of the inherent complexity *in vivo* of a whole-animal system, more detailed study of the basic mechanisms is only possible with *in vitro* preparation. Since the hippocampus is a structure especially sensitive to ischemic injury in humans and animal models, acute hippocampal slice preparation has been a useful tool in studying the changes either surrounding an acute ischemic insult *in vitro* or at several time points after an ischemic insult *in vivo*.

This chapter is aimed at explaining the chronic *in vitro* hippocampal preparation to study the long term consequences of ischemic insult. Organotypic explant cultures preserve the basic structural and intrinsic connective organization of the tissue of origin, while maintaining the advantage of long-term survival and easy accessibility.

Our laboratory has developed the hippocampal organotypic culture system over the last few years in the study of mechanisms of neuronal hyperexcitability. We have preliminary data showing that, when exposed to ischemic insult, Ca1 pyramidal neurons do indeed develop the characteristic delayed neuronal damage. This system allows testing of hypotheses

0-8493-8094-4/95/$0.00+$.50

that are not easily testable either in whole animals or in acute *in vitro* preparation and should complement the studies using dissociated cultures.

II. EXPERIMENTAL MODELS OF TRANSIENT ISCHEMIA

Animal models are needed to study the pathophysiologic mechanisms and to attempt various interventional measures. The use of intact animal in the study of mechanisms is somewhat limited because of the complexity of the whole organism and the number of known and unknown variables that have to be considered in the interpretation of results. One good example of the difficulties encountered in whole-animal studies is the neuroprotective effect of the systemically administered NMDA antagonist MK801.[1] However, MK801 had a hypothermic effect on the animal's body temperature. Indeed, when the body temperature was artificially lowered in the experimental animal, similar degrees of neuroprotection were found, suggesting that the neuroprotective effect of MK801 may be due to hypothermia.[2] In addition, the ischemic insult itself is usually delivered with the experimental animal under anesthesia, adding more complicating factors. Furthermore, all the systemically administered drugs are metabolized and have to cross the blood-brain barrier. Intraventricular administration of drugs may circumvent some of these problems but the experimental protocol becomes much more cumbersome.

Acute *in vitro* slice preparation circumvents some of these difficulties, and has been used to study changes that occur during the actual ischemic/hypoxic period and a few hours afterwards.[3] It has also been used to study various electrophysiologic parameters at certain time points after an ischemic insult *in vivo.*[4] However, the process of slice preparation causes acute trauma to the tissue and its short survival imposes a substantial restriction.

Dissociated cultures of either neocortical or hippocampal neurons have been used extensively in the study of neuronal damage, especially due to excitotoxicity.[5,6] Much of the knowledge on the ionic and excitotoxic mechanisms has been gained through this approach.[7] However, since dissociated cultures have drastically altered glial environment, synaptic connections, and cytoarchitecture, interpretation and application of these results to the *in vivo* situation require much caution. For instance, NMDA receptor antagonists are highly effective *in vitro* in dissociated culture preparation but not in intact normothermic animals.[2,8] AMPA receptor antagonists are highly protective in intact animals but not nearly so in dissociated cultures.[9,10]

Organotypic explant cultures preserve the basic structural and connective organization of the tissue of origin, while maintaining the advantage of long-term survival and easy accessibility. The roller-tube method[11] has been used extensively in the neural tissue culture field, since the explanted tissue flattens to a very thin layer of cells, almost to monolayer, allowing easy access to neurons using differential interference optics or phase contrast microscopic guidance. More recently, biologically permeable membrane has been used to support explant culture in an interface between the media and atmosphere with similar results.[12]

In the organotypic explant culture of hippocampus, part of the traditional trisynaptic connections are preserved from dentate to CA3, then CA3 to CA1. Since the dentate granule cells are still undergoing mitosis at the time of explant (5 to 7 d postnatal), and since the entorhinal cortical afferents are not always preserved, dentate gyrus remains relatively underdeveloped. In addition, some degree of reorganization occurs in recovery from dissection and mechanical trauma of slicing. Anatomically, however, both pyramidal and granule neurons appear to develop and mature as though *in situ* as demonstrated by intracellular injection of Lucifer yellow and HRP as well as by electron microscopy.[13,14] The dendritic arborizations closely resemble those of adult hippocampal neurons *in situ.* This is in sharp contrast to cerebellar explant culture, which remain quite immature without the full arborizations seen in mature Purkinje neurons.[15] Intrinsic connections are maintained as evidenced by Timm staining patterns of mossy fibers and by transport of HRP, although evidence of reorganization is seen with supragranular Timm staining in dentate molecular layer.[14,16]

The thickness of the explant becomes close to monolayer allowing easy access of the recording electrode under direct visual guidance, enabling even patch-clamp recording. The thin layer of cells in the explant also allows rapid diffusion of drugs and ions as well as oxygen. Thus electrophysiologically, many intracellular studies have been done to characterize the behavior of CA3 and CA1 pyramidal cells in culture. Intrinsic properties of the cells and the synaptic responses appear to be very similar to those in acute hippocampal slice preparation.[11] Responses to bath-applied neurotransmitter receptor agonists and antagonists are also observed to be quite similar.

To date, the explant culture system has not been fully exploited for the study of ischemic neuronal injury.[17,18] One factor is the time and effort that go into the successful development and maintenance of high-quality explant cultures. Once developed, however, it provides a powerful tool for studying neural network excitability and its responses to exogenous stimuli.

III. CULTURE METHODS

For actual details on the tissue culture methods, the readers are referred to books dealing with culturing techniques such as *Culturing Nerve Cells,* edited by Gary Banker and Kimberly Goslin.[19] The description below is meant to give an overview of the preparation of the primary neuronal cultures.

A. DISSOCIATED CULTURE

Dissociated cultures can be prepared from rat brains from embryos and pups of approximately 18 d of gestation up to 6 to 10 d postnatal age. Embryos are delivered from the anesthetized pregnant dam through an abdominal incision after the skin has been sterilized with copious amounts of ethanol. The uterine sac with a series of rat embryos is removed from the incision, being careful not to allow it to touch the unsterile surface. Each embryo is then removed from the uterine sac and its thin skull opened with fine dissecting scissors. The brain is removed with a spatula and placed in an ice-cold buffer. For rat pups, they are first sacrificed by decapitation and the scalp is sterilized with alcohol. The skin is removed and the skull is opened using fine dissecting scissors for the removal of the brain.

Under a dissecting microscope in a horizontal laminar flow hood, meninges and blood vessels are removed. Further dissection can then be performed to isolate the forebrain or hippocampus. In larger pups, subregion microdissection can be performed in the hippocampus by first slicing the hippocampus using McIlwain tissue chopper at 400 to 500 μm thickness, and then using fine-point knife to separate the dentate from the hippocampus proper. Dissected tissue is then chopped into smaller pieces of 1 mm or less and subjected to trypsin or papain digestion to dissociate the neurons. Since cell lysis releasing DNA into solution could make the mixture very viscous, one may also add DNase (1 mg/ml). After enzymatic digestion, trypsin is removed and the tissue is triturated with a Pasteur pipette with a fire-polished tip for mechanical dissociation. Nylon mesh could be used to remove the remaining clumps of tissue. Cell density is determined by using a hemacytometer and the percentage of viable cells, by trypan blue dye exclusion. Depending upon the various needs, plating density may vary from 1000 to several hundred thousand viable cells per square centimeter. Polylysine-coated culture dishes or coverslips are usually used for plating. Glia-conditioned media made from essential medium containing 5 to 20% heat-inactivated serum is usually used with various concentrations of supplemented glutamine, glucose, and bicarbonate. Initially, an antimitotic agent such as cytosine arabinoside is added in micromolar concentration to reduce glial proliferation. Humidified 5 to 9% CO_2 environment at 37°C is used to maintain the cultures up to several weeks with weekly or biweekly feedings. As neurons mature in culture and develop processes and synaptic connections, it is important that same-age cultures be used for experiments with appropriate controls.

B. EXPLANT CULTURE

Organotypic cultures are prepared from 6-d-old rat pups largely according to the methods described by Gähwiler.[11] Pups are sacrificed by decapitation. In a laminar flow hood, the scalp is sterilized with alcohol and the brain is removed rapidly and, using a sterile technique, the hippocampus is microdissected under a dissecting microscope. McIlwain tissue chopper is used to make 400- to 500-μm-thick transverse slices.

For the coverslip method, slices are placed on collagen-coated plastic coverslips (Aclar) and incubated in a humidified 5% CO_2 incubator for 24 h to allow attachment to collagen substrate. An alternative is to use the chick plasma clot method to attach the slice to the coverslip, which can then be rotated immediately. Coverslips are placed in tightly capped 15-ml test tubes (Falcon) with 1.5 ml of Gähwiler's media consisting of 50% Eagle's basal medium (BME), 25% heat-inactivated horse serum, and 25% Hank's buffered salt solution (HBSS) with supplemented glucose, at 6.5 mg/ml (36 m*M*) final concentration. A roller-tube culture system (New Brunswick) is then used to rotate the coverslips at 12 rph in a 37°C incubator. Media in the culture tubes are exchanged at weekly intervals.

For the membrane insert method, several slices can be placed on Millicell-CM tissue culture membranes (Millipore) with 1 ml of Gähwiler's medium in each well of six-well tissue culture plates and placed in a humidified 5% CO_2 incubator at 37°C.[12] Media are changed every 3 to 4 d. Most experiments are done with cultures of 10 to 20 d *in vitro*.

IV. METHODS OF ISCHEMIA

Ischemia can be simulated in different ways. In actual ischemic situation, there is cessation of blood flow to the tissue resulting in: (a) hypoxia; (b) lack of nutrients or, in practical terms, hypoglycemia; (c) acidosis; and (d) lack of removal of waste products. Of these, the last factor is difficult to simulate reliably. The first three can easily be manipulated.

Hypoxia can be achieved by changing the ambient gas composition of the culture environment. Many investigators use custom-made apparatus that allow a humidified anoxic or hypoxic gas mixture to flow freely through a sealed incubator to flush out ambient oxygen (Figure 1). It is crucial that the temperature of the cultures be maintained at a desired level as hypothermia itself can be protective. Thus, fluctuating temperature will result in varying severity of ischemic neuronal damage. It is also advised that the media be previously deoxygenated by bubbling with an anoxic gas mixture, although the oxygen content of the media is not high due to the absence of hemoglobin, which carry the bulk of oxygen in the blood. There are also specially sealed anoxic chambers that remove any residual oxygen by chemical catalytic means, but such strict anoxia may not be necessary from a practical point.

Chemical hypoxia is also a simple method to achieve a similar result. By using NaCN (1 to 3 m*M*), which blocks oxidative phosphorylation pathway, oxidative energy production is halted.

Hypoglycemia is achieved by omitting D-glucose from the media. Osmotic content should be maintained by substituting equimolar 2-deoxyglucose for D-glucose. 2-Deoxyglucose is taken up by the cell through a glucose transport mechanism, but is not further metabolized in the glycolytic pathway.

Acidosis is achieved usually by altering the buffering condition of the media. Cultures normally face 5% CO_2 atmosphere during maintenance. Despite the removal of the oxygen, as CO_2 content is not changed, pH will remain near baseline unless the media buffers are changed appropriately.

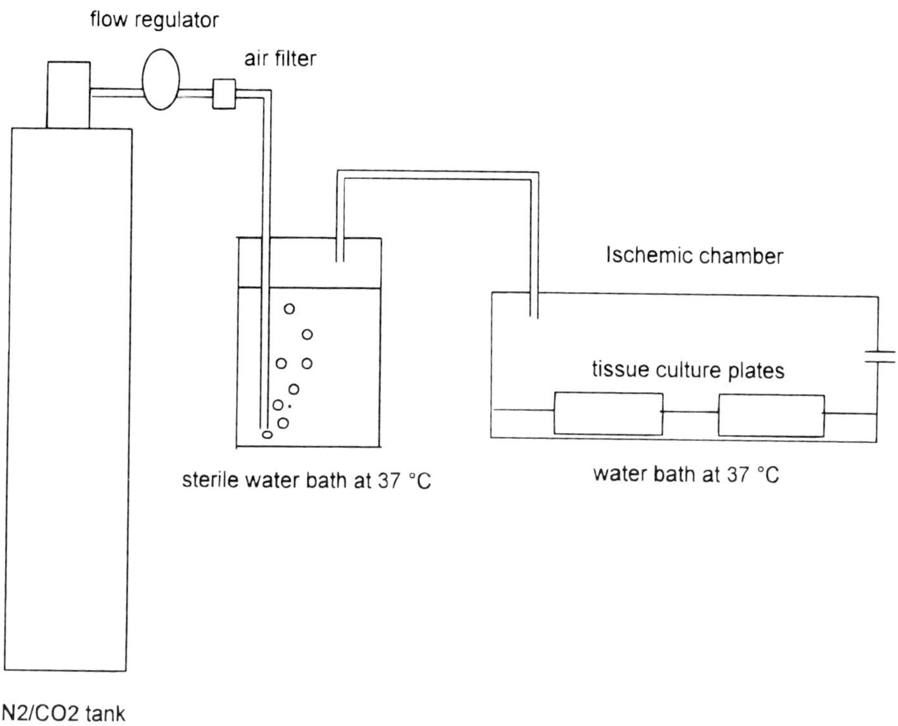

FIGURE 1. Ischemic chamber diagram. Schematic representation of the ischemic apparatus. Anoxic gas mixture is bubbled through the sterile water bath to be warmed and humidified before entering the sealed ischemic chamber. Cultures need to be maintained at a constant temperature in a water bath.

V. ASSESSMENT OF ISCHEMIC INJURY

One obvious way to assess ischemic injury is by morphologic-anatomic observation. Qualitative assessment can be made by observing the morphology of cultures using phase-contrast or interference optics. An exclusion dye such as trypan blue can be used to assess damaged neurons and quantitative measurements of proportion of damaged neurons can be obtained. Fluorescent dyes such as propidium iodide (PI) can also indicate damage to the neurons. PI enters the cell through damaged cell membranes and bind to nucleic acid. Bound PI will fluoresce and qualitative damage assessment can be made reliably through epifluorescent microscopy (Plate 1).*

The most frequently used biochemical index of cell damage is the measurement of lactate dehydrogenase (LDH) efflux from the cells.[20] This necessitates the use of serum-free media for the duration of experiments, which is usually well tolerated by the primary neuronal cultures. Released LDH is quite stable and can be reliably measured from the media by calorimetric assay in a spectrophotometer using kits from any commercial source (e.g., Sigma).

VI. PRACTICAL PROTOCOL

A. ISCHEMIA PROTOCOL USING EXPLANT CULTURES

Ischemia is simulated by using 95% N_2 and 5% CO_2 gas with 2-deoxyglucose (2-DG)-substituted media. Chemical ischemia using NaCN and 2-DG can also be done similarly.

*Plate 1 is part of the frontispiece.

All the solutions are made up to 36 m*M* glucose or 36 m*M* 2-DG to maintain osmolality equal to that in which cultures are grown.

1. Solutions

A. HBSS (Gibco with 4 meq HCO_3) with 36 m*M* 2-DG and 4.6 μg/ml PI, bubbled previously for 2 h with 95% N_2 and 5% CO_2 to reduce the oxygen tension of the media itself (pH of this solution after 2 h is measured at 6.6 to 6.7).
B. HBSS with 36 m*M* glucose and 4.6 μg/ml PI.
C. Serum-free media: 75% BME, 25% HBSS, 36 m*M* glucose, and 4.6 μg/ml PI.
D. Gähwiler's media: 50% BME, 25% heat-inactivated horse serum, 25% HBSS, and 6.5 mg/ml (36 m*M*) glucose (this media composition contains approximately 21 to 22 meq of HCO_3, pH of 7.2 at equilibrium with 5% CO_2).

2. Protocol

1. Select morphologically intact explant by phase contrast microscopy.
2. Incubate 15 min in solution C to screen out cultures with preexisting damage, if any.
3. Experimental: In gas chamber (see Figure 1) (airtight container with an inlet and an outlet, with 5 to 6 l/min flow of humidified 95% N_2 and 5% CO_2) in glucose-free solution A for varying times.
   Control: In regular CO_2 incubator in solution B.
4. Change all media to solution C and return to regular CO_2 incubator.
5. Make 2-, 4-, and 6-h observation.
6. Return to Gähwiler's media, solution D.
7. Make 24-h observation with 30 min PI preincubation in solution C.
8. Make 48-, 72-, 96-, and 120-h histologic observation without PI.

B. ISCHEMIA PROTOCOL USING DISSOCIATED CULTURES

A similar airtight container with anoxic condition is used. Strategies may differ in how to eliminate oxygen from the environment and the media. Continuous flow of humidified anoxic gas mixture as above can be used or the chamber can be sealed after flushing out ambient oxygen-containing atmosphere. A completely sealed anaerobic incubator (Forma Scientific) with 5% CO_2, 10% H_2, and 85% N_2 and a chemical catalytic reaction with palladium to eliminate any residual oxygen can also be used, but requires special equipment.[21]

VII. APPLICATION OF THE PROTOCOL

Following are the descriptions of some of the ischemic experiments done using the above setup and experimental protocol to study the mechanisms of ischemic neuronal injury.

Our objective was to examine the role of the Ca^{2+}-calmodulin system in ischemic neuronal damage according to the scheme presented (Figure 2). NMDA receptor has been implicated in ischemia-induced neuronal injury.[7] It gates cation channels permeable to Ca^{2+}, which may mediate subsequent injury by activating the Ca^{2+}-calmodulin system intracellularly.

Hippocampal explant cultures from 6-d-old pups were exposed at 10 d *in vitro* to experimental ischemia as detailed in the above protocol. Calmidazolium, an inhibitor of Ca^{2+}-calmodulin, was applied in the media for 30 min before, during, and for 24 h after the exposure. At 24 h, severity of injury was assessed by PI fluorescence (0 = none, 1 = minimal, 2 = mild, 3 = moderate, 4 = severe). With 30-min exposure to ischemia, CA_1-specific pattern of injury to the pyramidal neurons in the hippocampal formation was observed (Plate 2).* Calmidazolium reduced the injury index to 0.4 ± 0.13 (1 μ*M*; n = 13) from

*Plate 2 is part of the frontispiece.

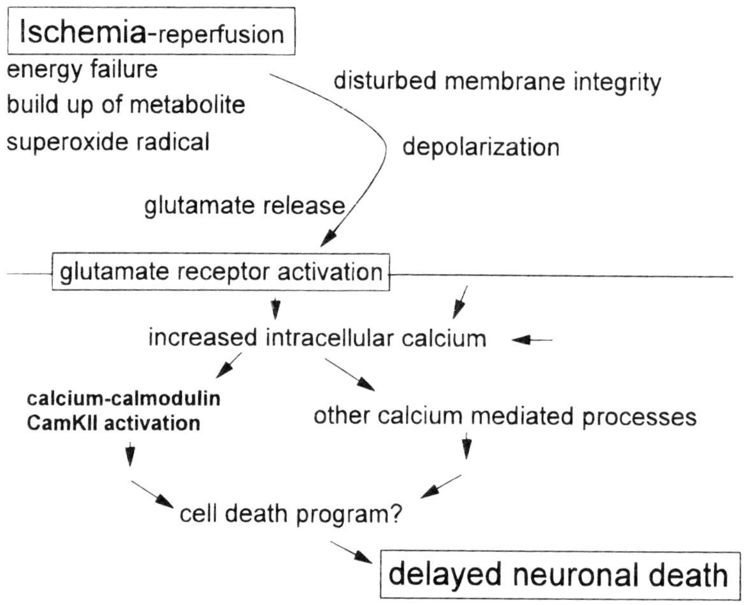

FIGURE 2. Possible mechanisms of ischemic neuronal injury. Speculative simplified schematic diagram of possible mechanisms of ischemic injury. Many other factors could be included such as nitric oxide, carbon monoxide, and arachidonate pathways.

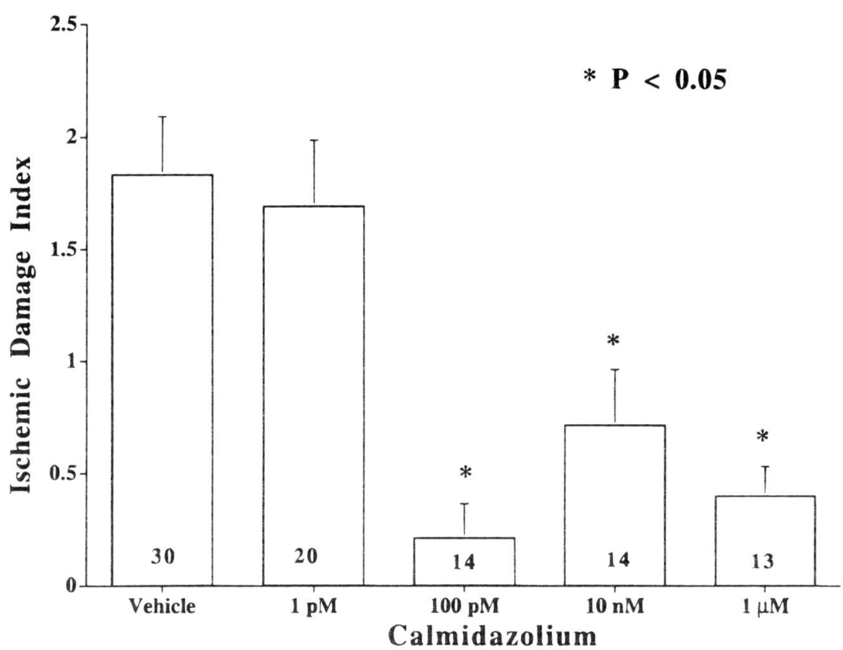

FIGURE 3. Calmidazolium prevents ischemic neuronal injury. Bath-applied calmidazolium, a calcium-calmodulin inhibitor, suppresses ischemia-induced neuronal injury in CA1 of the hippocampal explant culture. Asterisks denote statistical significance ($p < 0.05$, Mann-Whitney U). The number of cultures examined is indicated within each bar.

1.83 ± 0.26 (vehicle; n = 30) ($p < 0.05$; Mann-Whitney U). There was dose-dependent protection at even lower doses (Figure 3). W7, another inhibitor of Ca^{2+}-calmodulin, also protected in a dose-dependent fashion.

The results support the hypothesis that activation of the Ca^{2+}-calmodulin pathway plays a key role in the development of brain injury induced by ischemic insults. Inhibitors of the Ca^{2+}-calmodulin pathway could be useful as neuroprotective agents. This example also demonstrates the usefulness of the explant culture of hippocampus in the mechanistic study of ischemic neuronal injury and related phenomena.

REFERENCES

1. Gill, R., Foster, A. C., and Woodruff, G. N. (1988) MK-801 is neuroprotective in gerbils when administered during the postischaemic period. *Neuroscience* 25, 847–855.
2. Buchan, A. and Pulsinelli, W. L. (1990) Hypothermia but not the N-methyl-D-aspartate antagonist, MK-801, attenuates neuronal damage in gerbils subjected to transient global ischemia. *J. Neurosci.* 10, 311–316.
3. Morris, M. E., Leblond, J., Agopyan, N., and Kmjevic, K. (1991) Temperature dependence of extracellular ionic changes evoked by anoxia in hippocampal slices. *J. Neurophysiol.* 65, 157–167.
4. Urban, L., Neill, K. H., Crain, B. J., Nadler, J. V., and Somjen, G. G. (1989) Postischemic synaptic physiology in area CA1 of the gerbil hippocampus studied *in vitro*. *J. Neurosci.* 9, 3966–3975.
5. Choi, D. W. and Rothman, S. M. (1990) The role of glutamate neurotoxicity in hypoxic-ischemic neuronal death. *Annu. Rev. Neurosci.* 13, 171–182.
6. Dubinsky, J. M. and Rothman, S. M. (1991) Intracellular calcium concentrations during "Chemical Hypoxia" and excitotoxic neuronal injury. *J. Neurosci.* 11, 2545–2551.
7. Choi, D. W. (1990) Cerebral hypoxia: some new approaches and unanswered questions. *J. Neurosci.* 10, 2493–2501.
8. Choi, D. W., Peters, S., and Viseskul, V. (1987) Dextrophan and levorphanol selectively block *N*-methyl-D-aspartate receptor-mediated neurotoxicity on cortical neurons. *J. Pharmacol. Exp. Ther.* 242, 713–720.
9. Koh, J. and Choi, D. W. (1991) Selective blockade of non-NMDA receptors does not block rapidly triggered glutamate-induced neuronal death. *Brain Res.* 548, 318–321.
10. Sheardown, M. J., Nielsen, E. O., Hansen, A. J., Jacobsen, P., and Honore, T. (1990) 2,3-Dihydroxy-6-nitro-7-sulfamoyl-benzo (F) quinoxaline: a neuroprotectant for cerebral ischemia. *Science* 247, 571–574.
11. Gähwiler, B. H. (1988) Organotypic cultures of neural tissue. *TINS* 11, 484–489.
12. Stoppini, L., Buch, P.-A., and Muller, D. (1991) A simple method for organotypic cultures of nervous tissue. *J. Neurosci. Methods* 37, 173–182.
13. Frotscher, M. and Gähwiler, B. H. (1988) Synaptic organization of intracellularly stained CA3 pyramidal neurons in slice cultures of rat hippocampus. *Neuroscience* 24, 541–551.
14. Zimmer, J. and Gähwiler, B. H. (1984) Cellular and connective organization of slice cultures of the rat hippocampus and fascia dentata. *J. Comp. Neurol.* 228, 432–446.
15. Gähwiler, B. H. (1984) Slice cultures of cerebellar, hippocampal and hypothalamic tissue. *Experientia* 40, 235–243.
16. Keller, F, Lipp, H. P., and Waser, P. G. (1986) The organization of intrinsic hippocampal connections in explants of rat hippocampus studied by topical application of HRP crystals. *Brain Res.* 380, 191–195.
17. Newell, D. W., Malouf, A. T., and Franck, J. E. (1990) Glutamate-mediated selective vulnerability to ischemia is present in organotypic cultures of hippocampus. *Neurosci. Lett.* 116, 325–330.
18. Vornov, J. J., Tasker, R. C., and Coyle, J. T. (1991) Early neuronal membrane injury precedes very delayed histologic degenration after "ischemia" in organotypic hippocampal culture. *Soc. Neurosci. Abstr.* 17, 1093.
19. *Culturing Nerve Cells*, Eds. Banker, G. and Goslin, K. (1991), MIT Press, Cambridge, MA.
20. Koh, J. Y., and Choi, D. W. (1987) Quantitative determination of glutamate-mediated cortical neuronal injury in cell culture by lactate dehydrogenase efflux assay. *J. Neurosci. Methods* 20, 83–90.
21. Goldberg, M. P. and Choi, D. W. (1993) Combined oxygen and glucose deprivation in cortical cell culture: calcium-dependent and calcium-independent mechanisms of neuronal injury. *J. Neurosci.* 13, 3510–3524.

Chapter 24

BRAIN SLICES IN THE STUDY OF CNS TRAUMA

Peter G. Aitken

TABLE OF CONTENTS

I. INTRODUCTION

Brain slices have been a part of the neurobiologist's arsenal of tools for at least 70 years. Perhaps the first recorded use of brain slices is the report by Warburg et al.[17] of metabolic comparisons between retina and thin slices of rat brain. Early studies used brain slices almost exclusively for the study of metabolic and biochemical processes. Although Libet and Gerard[9] recorded spontaneous potentials from isolated frog brain as early as 1939, recording of electrical signals from mammalian brain slices did not get its start until considerably later, with the recording of resting membrane potentials by Li and McIlwain[8] and of neural activity by Yamamoto and McIlwain.[8] Since these pioneering experiments, and particularly in the last two decades, the use of brain slice preparations in almost every subfield of neurobiology has grown explosively. While the hippocampus is certainly the most popular brain region for slice studies, many other brain regions have been successfully used, including neocortex,[3] olfactory bulb,[12] spinal cord,[16] hypothalamus,[6] thalamus,[7] and cerebellum.[10] As regards species, the emphasis has been on rats and guinea pigs, but tissue from various other species have been used as well, including humans (tissue removed during surgery).[15]

Compared with intact animals, brain slices offer significant advantages in experimental control and accessibility. Experiments that are difficult or impossible with whole animals become trivial or at least possible with a slice preparation. There are disadvantages as well, of course. Being isolated, slices cannot provide information about contributions of systemic factors, such as blood flow and intracranial pressure, to the phenomena under study. The

0-8493-8094-4/95/$0.00+$.50

design and interpretation of experiments must take these factors into account, and at times brain slices must be used in conjunction with whole-animal experiments to obtain the complete picture. This is particularly true in the field of CNS trauma, where such systemic factors are more likely to play a role in the response of the intact organism to injury.

While the experimental techniques applied to brain slices can be extremely complex, the techniques for maintaining brain slices are relatively simple. In this chapter an overview of the materials and procedures for maintaining brain slices is presented, and some work done in this laboratory on the role of membranes in CNS trauma is described.

II. MATERIALS

A. SOLUTIONS

During all stages of preparation, maintenance, and experimentation brain tissue must be kept submerged or moistened with an appropriate bathing medium. This solution is called artificial cerebrospinal fluid (ACSF) as it is intended to mimic the most important characteristics of real cerebrospinal fluid: osmotic pressure, ionic constitution, pH, and metabolic substrates. There are minor variations in the composition of control ACSF between laboratories. These variations are most typically in the concentrations of K^+, Ca^{2+}, and Mg^{2+}. The values for these ions in the recipes below are toward the low end of the scale. In the literature, ACSF K^+ concentrations as high as 5 or 5.5 m*M*, and Ca^{2+} and Mg^{2+} concentrations as high as 2.4 m*M*, are not uncommon. These variations are justified by the fact that reported concentrations of these ions in real CSF vary, and that for most types of experiments the minor differences are unlikely to have any important effects on the phenomena under study.

Obviously, dissolving compounds in the ACSF is an important method of delivering drugs to the tissue. Beyond this, variations in the basic ACSF composition can be used to elicit specific effects in the tissue. The most common alterations are removal of Ca^{2+} (replacing it with an equal amount of Mg^{2+}) to block synaptic transmission, and increasing K^+ to 50 m*M* or higher (deleting an equal amount of Na^+) to depolarize neurons for the study of neurotransmitter release. When making ionic substitutions it is important to keep osmotic pressure changes in mind. With normal osmolarity in the 300 to 310 mOsm range, changes of a few millimoles in the concentration of an ion are unlikely to have a measurable effect in most situations. Since larger osmotic changes can have significant effects on the excitability of neurons, it is important to compensate for ionic additions or deletions. This is most commonly done by adjusting the Na^+ concentration (or, for anion modifications, the Cl^- concentration).

Our practice is to make ACSF fresh each day. If ACSF is kept cold it can be refrigerated for a day or two. Once warmed and oxygenated it should not be kept for more than a day because the sugar in it provides good conditions for the growth of microorganisms.

1. Bicarbonate-Buffered ACSF

This is the most commonly used type of ACSF. It has the advantage of using bicarbonate and, to a lesser degree, phosphate to buffer the pH, avoiding the use of any artificial substances. The unavoidable disadvantage is that the actual pH, being dependent on the balance between HCO_3^+ and dissolved CO_2, will change with temperature and gas saturation. When saturated with 95% O_2 − 5% CO_2 at room temperature the pH is 7.35 to 7.40. If the solution is bubbled while cold the pH will be lower; if allowed to stand uncovered after bubbling the pH will rise as CO_2 leaves the solution. The pH is known to affect neural responses, with lower pH decreasing and higher pH increasing excitability.[2,14] It is likely that ACSF pH can affect intracellular processes as well, although the relationship between extracellular and intracellular pH is poorly understood for neurons. It is clearly important, particularly in studies of electrical activity, to control the ACSF pH.

Final composition (in mM)

NaCl:	130
KCl:	3.5
NaH_2PO_4:	1.25
$NaHCO_3$:	24
$CaCl_2$:	1.2
$MgSO_4$:	1.2
Dextrose:	10

Stock solutions

The following stock solutions can be prepared ahead of time and kept refrigerated for several weeks:

Solution 1:	1 *M* dextrose
Solution 2:	240 m*M* $CaCl_2$
Solution 3:	240 m*M* $MgSO_4$
Solution 4:	1.3 *M* NaCl
	35 m*M* KCl
	12.5 m*M* NaH_2PO_4
	240 m*M* $NaHCO_3$

Preparation

For each liter of ACSF start with approximately 850 cc H_2O.

Add: 10 cc solution 1
5 cc solution 2
5 cc solution 3

Stir briefly.

Add: 100 cc solution 4.

Bring volume to 1 l with H_2O and stir thoroughly. Bubble at room temperature with 95% O_2 − 5% CO_2 for 15 to 20 min. For bubbling the solutions we use air stones (available at aquarium shops) on Tygon tubing, with a gas flow of approximately 250 cc/min (for a 1 l container).

2. Hepes-Buffered ACSF

An alternate formulation of ACSF uses *N*-[2-hydroxyethyl]piperazine-n′-[2-ethanesulfonic acid] (Hepes) as the pH buffering agent. This ACSF has the advantage that its pH is not dependent on dissolved CO_2. We sometimes use it for dissection, then switch to bicarbonate-buffered ACSF for the actual experiment. Hepes-buffered ACSF can be used throughout an experiment, however.[11] Note that the recipe below uses dry chemicals almost exclusively in preference to stock solutions. This is due primarily to habit and to the fact that we use this solution relatively infrequently. Stock solutions should work perfectly well, with one solution containing the first four ingredients at 10× concentration, another containing 240 m*M* $CaCl_2$, and the third 1 *M* dextrose. The dry Hepes would then be dissolved last in the final solution.

Final composition (in mM)

NaCl:	122
KCl:	3.1

$MgSO_4$:	1.2
KH_2PO_4:	0.4
$CaCl_2$:	1.2
Hepes:	25
Dextrose:	10

Preparation

For each liter, dissolve in approximately 900 cc H_2O:

7.13 g	NaCl
1.8 g	dextrose
0.231 g	KCl
0.145 g	$MgSO_4$
0.054 g	KH_2PO_4
5.96 g	Hepes

Add: 1 cc of 1.2 *M* $CaCl_2$ stock solution.

Bring volume to 1 l. Titrate to pH 7.4 with 25% NaOH at room temperature. While this solution does not need to be gassed with 95% O_2 − 5% CO_2 to obtain the proper pH, it will require oxygenation if the ACSF is to be a source of oxygen for the tissue.

B. TISSUE SLICER

Early brain slice experiments utilized a hand-held razor blade to take "thin" slices of tissue from a whole or blocked brain. While some investigators with steady hands may still use this technique, the current practice is to use a mechanical device that provides precise slice thickness and greater convenience. Tissue slicers fall into two categories. Choppers, or guillotine slicers, use a blade that travels rapidly in the vertical plane, striking the tissue that is placed on a horizontal surface. Between strokes either the blade or the tissue is advanced horizontally by a distance equal to the desired slice thickness. Vibrating-blade cutters employ a blade that vibrates parallel to its sharp edge while advancing horizontally, at a relatively slow rate, through the tissue which has been fastened to the cutting stage, usually with cyanoacrylic cement. Between strokes the tissue is raised, or the blade lowered, by the slice thickness.

Choppers are suitable for relatively small pieces of tissue, such as a rodent hippocampus. They have the advantage of speed, and can produce a dozen slices in less than a minute. On the negative side, the chopping action appears to cause more tissue damage, resulting in a more a substantial layer of dead or damaged cells on the cut surfaces of the slices. For many types of experiments this presents no problem, but for some procedures, such as patch clamping, it is generally believed to be deleterious.

Vibrating-blade cutters are most suitable for larger tissue samples, such as a whole or half rodent brain. Because the tissue must be glued to the cutting stage, the tissue of interest must be left embedded in a larger block of brain for slicing. After cutting, the desired tissue can be dissected free of the larger slice if needed (while submerged in cold ACSF). The slower, vibrating blade results in less tissue damage, but the cutting process is slow—20 to 40 s per slice is typical.

C. EXPERIMENTAL CHAMBER

The experimental chamber is the central component of experiments using a brain slice preparation. This chamber must provide (a) physical support of the tissue, (b) exposure to ACSF, (c) temperature control, (d) illumination and visual access, and (e) proper O_2 and

CO_2 levels. In addition, other characteristics of the chamber will depend on the type of experiment being performed. Electrophysiological studies, for example, require a chamber that provides access for stimulating and recording electrodes as well as a chlorided silver ground wire in the bath. Pharmacological studies of transmitter release, on the other hand, require a chamber with a small fluid volume.

Slice chambers are of four types, as illustrated in Figure 1. Three of these types are referred to as interface chambers because the tissue is maintained at the interface between the ACSF and an atmosphere of warmed, humidified 95% O_2 − 5% CO_2. This type provides superior oxygenation, permitting the temperature to be higher (35 to 37°C compared with 30 to 34°C for the submersion chamber), provides better access for precise electrode placement, and permits drug delivery by application of small "nanodrops" directly to the slice surface. In the fourth type, the submersion chamber, the slice is completely immersed in flowing ACSF. Submersion chambers provide better delivery of drugs administered in the ACSF, and avoid the problems of the tissue drying out that can occur with interface chambers that are not properly set up. To avoid drying of the tissue in an interface chamber it is essential to maintain a proper fluid level so that there is a thin capillary layer of fluid over the slices, and a fully humidified atmosphere over the slices.

D. HOLDING CHAMBER

A single piece of brain tissue will usually provide several individual slices. In some experimental protocols it is necessary to keep all but one slice isolated from experimental manipulations. This is done by means of a holding chamber from which slices are moved as needed, one at a time, to the experimental chamber. Because a holding chamber does not have the requirements of experimental access, drug application, etc. it can be relatively simple in design. We place slices on a net support in a beaker of ACSF in a 30 to 32°C water bath. The ACSF is bubbled with 95% O_2 − 5% CO_2 at a rate sufficient to keep the fluid circulating without disturbing the slices.

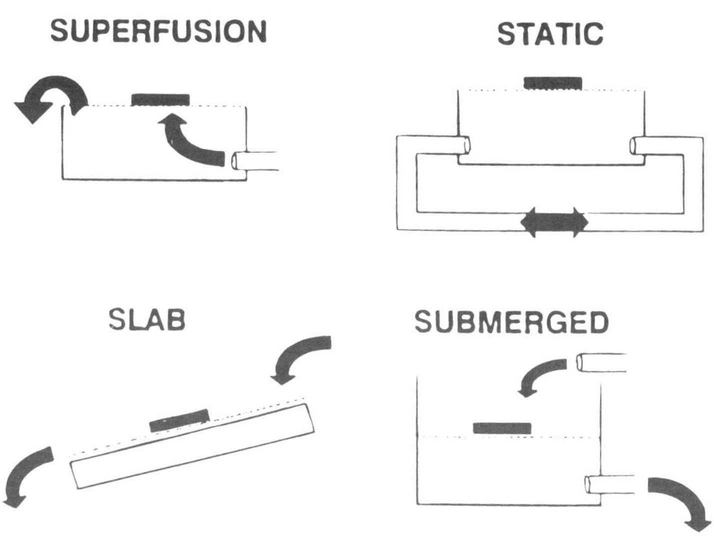

FIGURE 1. Diagrammatic representations of four types of experimental chambers used for brain slices. Arrows indicate flow of ACSF. In all chamber types except submerged the tissue slice is supported, on a mesh of stretched nylon panty-hose material, at the interface between the ACSF and the warmed, humidified 95% O_2 − 5% CO_2. The static chamber can use a push-pull pump to circulate the ACSF (as illustrated) or can be totally static with no forced ACSF circulation. The slice chamber is mounted on a thermostatically controlled water bath; the gas mixture (when used) is bubbled through this bath. (From Aitken, P. G., in *Methods in Toxicology: In Vitro Biological Systems*, Academic Press, New York, 1993, pp. 1–6. With permission.)

III. PROCEDURES

A. ANESTHESIA

When using rats or guinea pigs, our preference is for anesthesia with diethyl ether. Our rationale is that because of its volatility, any residual ether in the tissue following dissection will quickly dissipate and is therefore unlikely to interfere with the experiment. Other anesthetics, such as barbiturates or halothane, can be used if possible residual anesthetic in the tissue will not cause problems. If consistent with animal care regulations, unanesthetized animals can be killed by decapitation or cervical dislocation.

B. DISSECTION

Once the animal is anesthetized it can be decapitated or not. This is largely a matter of personal preference, although we believe that decapitation simplifies the next step of removing the brain from the skull. Brain removal should be done quickly, because during this period the brain is ischemic and subject to damage. Once the brain is removed it is placed in ice-cold oxygenated ACSF and allowed to remain there for a minute or two to permit the deeper portions of tissue to become chilled.

The dissection procedures depend on the brain region of interest and the cutting method to be used. The goal is to isolate, in a form that will permit slicing, the brain structure to be studied. Some structures, notably the hippocampus, can be isolated from the remainder of the brain. Others must be left embedded in a block of surrounding tissue, which should have at least one flat surface that will rest on the cutting stage. In either case, surface vasculature and membranes should be removed because they can interfere with the cutting process. During the dissection process the tissue should be kept moistened with chilled ACSF.

Once the tissue is ready it is placed on the stage of the cutter. When using a chopper it is our practice to glue a piece of filter paper to the Teflon stage using rubber cement. This is done at least 30 min prior to dissection so that the cement solvents have completely dissipated; the paper is moistened with ACSF before the tissue is placed on the stage. The blade travel is adjusted so that the cut of each stroke extends just barely into the filter paper. For a vibrating-blade cutter a drop of cyanoacrylic cement (Superglue) is placed on the stage (also Teflon). The flat surface of the tissue block is dried by resting it on dry filter paper momentarily, then placed on the glue. The stage is placed in the cutting boat and the tissue covered with cold ACSF.

When using a vibrating-blade cutter it is sometimes necessary to provide some mechanical support for the tissue block. One method is to glue small blocks of solid agar behind and/or to the sides of the tissue. Another is to glue the tissue to the cutting stage and surround it with a small dam of filter paper (e.g., a 2-cm-diameter cylinder, 1 cm high), then fill the dam with semisolid agar, which sets when the tissue is covered with ACSF. For solid blocks of agar we dissolve 2 to 3% agar in hot 150 m*M* NaCl, pour into Petri dishes and chill until set. For softened agar it is necessary to find a concentration that is liquid at a temperature that will not damage the tissue (perhaps 30°C) but is relatively solid when chilled. We found 0.7% agar in 150 m*M* NaCl to work well.

As the slices are cut they are removed with a sable-hair artist's brush to a holding container containing cold ACSF. From there they can be dissected further if desired, then transferred directly to the experimental chamber or to a holding chamber. For transferring larger slices, artist's brushes work well. Smaller slices can be moved using a blunt pipette, such as a reversed Pasteur pipette.

C. MAINTENANCE OF SLICES

After dissection, slices need some time to recover before experimentation can begin. For electrophysiological studies at 35.5°C we have found 90 min to provide essentially full recovery of neural responses. Other types of experiments may require more or less time. It is advisable to conduct pilot experiments to determine the recovery time required for a particular situation. When a slice has been in a holding chamber for 90 min or more, transferring it to the experimental chamber necessitates a much shorter rest. Maintaining viable slices for 6 to 8 h after dissection is routine, and when required by the experimental protocol we have found 16 to 18 h to be possible. A schematic illustration of the setup for brain slice experiments is shown in Figure 2 and a diagram of a hippocampal slice is shown in Figure 3.

IV. STUDIES OF MEMBRANE CURRENTS IN GLIAL CELLS DURING HYPOXIA

A common type of CNS trauma, arising from a variety of causes, is ischemia. Whether global as in the case of cardiac arrest, or focal as in a stroke, cerebral ischemia beyond a relatively short duration kills neurons and can result in severe neurological deficits and death. Various membrane-related phenomena have been implicated as participants in this process. For several years our interest has been focused on the severe neural depolarization that occurs during ischemia. A similar depolarization can be provoked in hippocampal tissue slices by withdrawal of oxygen; flow of ions across the membrane during this depolarization, and in particular entry of Ca^{2+} into cells, has been identified as an important factor in determining whether or not neurons will recover from the insult. (See Reference 13 for a review.)

Our work has used the whole-cell patch clamp technique in an attempt to clarify the membrane mechanisms that underlie the current that flows during hypoxic depolarization. Experiments on hippocampal CA_1 pyramidal neurons have been reported elsewhere.[4,5] Reported here are some results obtained from presumed glial cells that were encountered occasionally in the search for neurons.

We prepared 400-μm-thick transverse slices from hippocampi of young rats and maintained them at 35.5°C in an interface chamber using the bicarbonate-buffered ACSF described

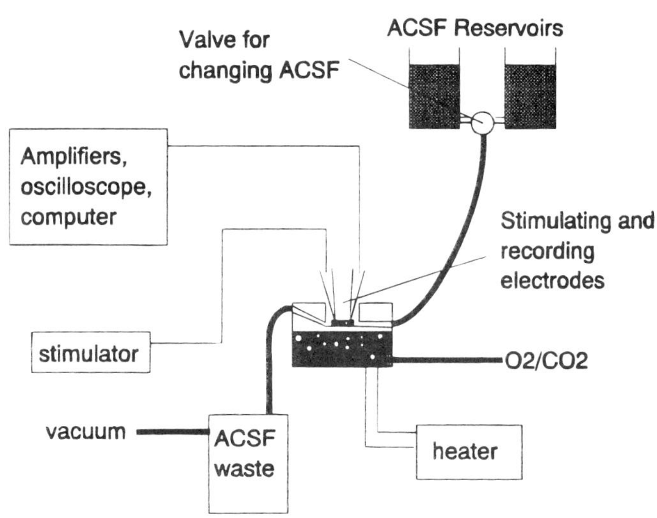

FIGURE 2. A brain slice chamber with its supporting peripheral equipment. ACSF is usually gravity-fed via intravenous drips; if precise flow is required syringe pumps can be used instead. Waste ACSF is drawn into a collection flask by vacuum, as illustrated, or in some chamber designs, by gravity alone. A heater controlled by a thermistor controls the chamber temperature. O_2-CO_2 is required only for interface-type chambers. Stimulating and recording equipment for electrophysiological procedures are shown; other types of experiments will require their own specialized apparatus.

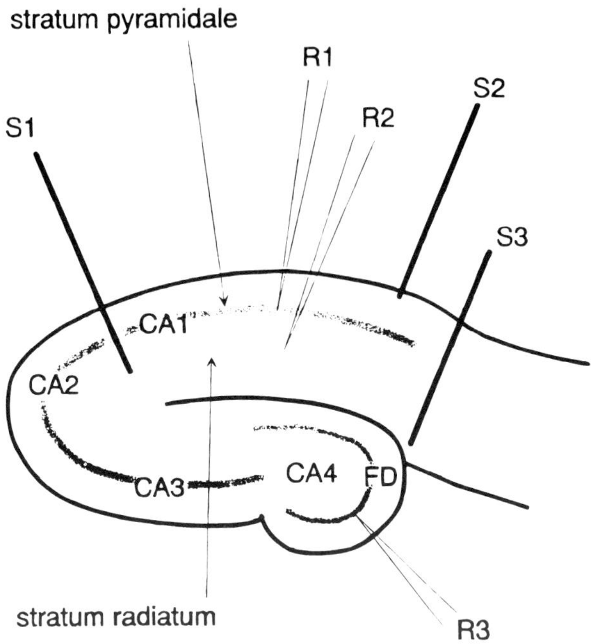

FIGURE 3. Schematic diagram of a transverse hippocampal slice showing the major cytoarchitectonic cell fields and positions of stimulating and recording electrodes for commonly used electrophysiological paradigms. CA1 through CA4: cell fields of the hippocampus proper. FD: fascia dentata (dentate gyrus). S1 and S2: stimulating electrodes for delivering orthodromic and antidromic stimulation, respectively, to the CA1 field. R1 and R2: extracellular electrodes for recording the somatic (R1) and dendritic (R2) field responses in CA1. S3 and R3: Stimulating and recording electrodes for evoking and recording orthodromic responses in FD.

earlier. The extracellular DC potential was recorded in stratum pyramidale of the CA_1 region with a glass micropipette electrode. A seal with cells in the stratum pyramidale was established within 0.5 mm of the extracellular electrode using a "blind" search guided by electrical signals. The patch pipette solution contained either cesium gluconate or, in one experiment, potassium gluconate as the major electrolyte, with Hepes buffer and 2 m*M* $MgCl_2$. Seals of at least 1 GΩ were required, and the patch was broken by gentle suction. A cell was presumed to be a glia if it did not fire action potentials either spontaneously or in response to depolarizing current pulses, and did not exhibit synaptic currents in response to orthodromic stimulation.

The holding potential was slightly negative to resting potential, and voltage clamp was maintained relative to the nearby extracellular potential rather than to a distant bath ground, ensuring clamp to actual transmembrane potential. The holding current (I_m) was monitored continuously. Hypoxia was caused by replacing the gas above the slices with 95% N_2 − 5% CO_2 for periods of several minutes. Current-voltage (I-V) relationships were determined by delivering ramp commands to the cell. Each ramp consisted of a step to −40 mV from the static holding potential, a linear ramp over 300 ms to 100 mV positive to that potential, then a step back to the static holding potential. From the I-V plots we calculated the cell's input membrane resistance (slope resistance, R_m) and the zero-current potential (ZCP), the potential at which the I-V curve crosses the zero-current line. We interpret the ZCP to be a reflection of the membrane potential the cell would exhibit at a given moment were it not under voltage clamp.

Results were obtained from 14 presumed glial cells, and responses during a hypoxic episode in one experiment are shown in Figure 4. I-V relationships were purely ohmic with

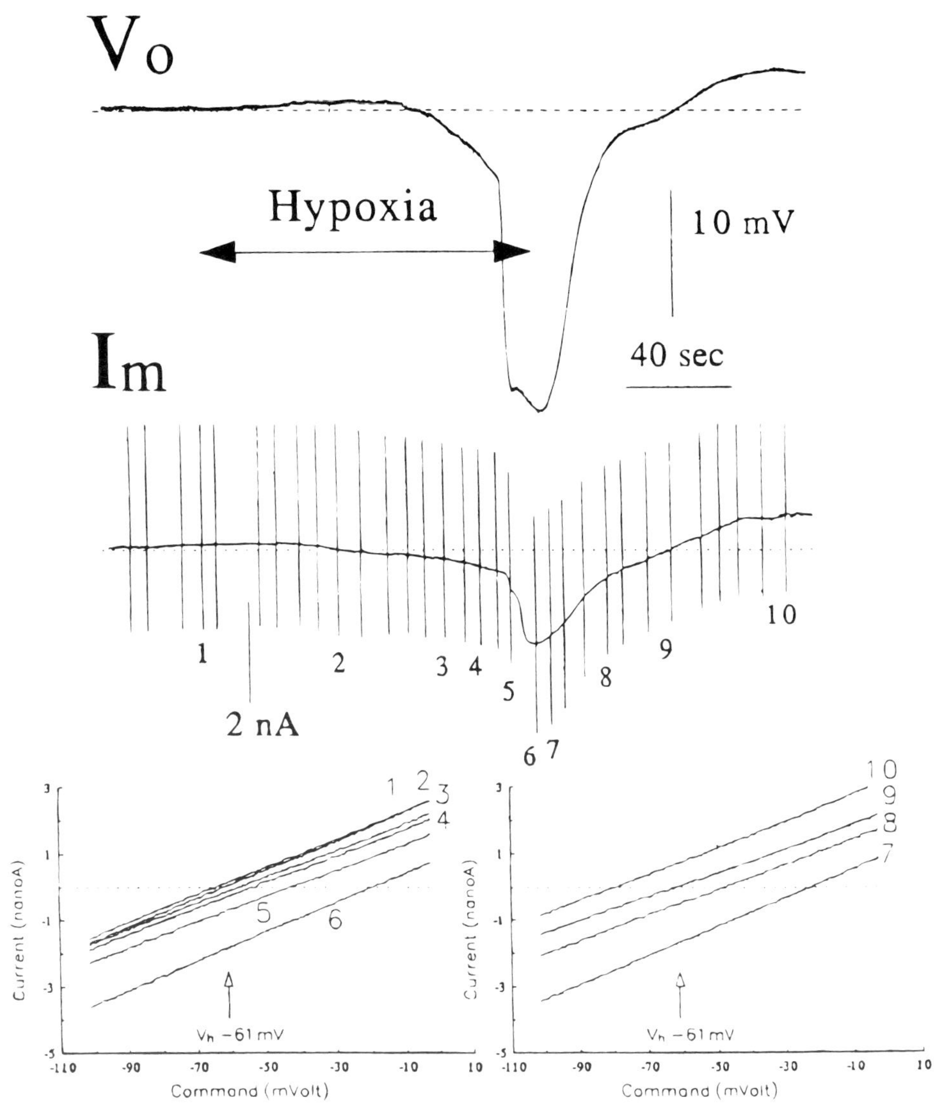

FIGURE 4. Responses of a voltage clamped glial cell during hypoxia. The tracing labeled V_0 shows the extracellular DC potential and the severe negative V_0 shift that is characteristic of hypoxic depolarization. The I_m tracing shows the steady-state current required to hold the cell at −61 mV. Vertical deflections on the I_m tracing are ramp commands used to test membrane resistance. The bottom two graphs show the I-V responses to the ramp portions of the corresponding numbered ramp commands. (Unpublished data of G. Czéh.)

no sign of rectification or voltage-dependent currents. During hypoxic depolarization the responses remained ohmic; there was a dramatic increase in I_m a positive shift of ZCP, and (in most cells) decrease of R_m. In several cells there was a transient increase in R_m preceding its decrease. These changes are similar to, but much less severe, than changes we found in CA_1 pyramidal neurons under the same conditions.[4,5] Note that the finding of no active currents is weakened by our use of Cs^+, known to block various K^+ currents, in the patch pipette. The literature does contain reports of voltage-gated currents in glia. Of more interest is our finding that glia do not appear to play an active role in the initiation of hypoxic depolarization; rather they are passive followers in the chain of events, a chain that is probably initiated and maintained by membrane mechanisms of neurons.

SOURCES OF EQUIPMENT

The only specialized equipment needed for basic brain slice preparations are the tissue slicer and the slice chamber. The following list of suppliers is not intended to be exhaustive, but we know their products to have been found suitable by numerous investigators.

Tissue chopper, vibrating-blade slicer, tissue chambers:
Stoelting Corp.
620 Wheat Lane
Wood Dale, IL 60191
(708) 860-9700

Vibrating-blade slicer:
World Precision Instruments
International Trade Center
175 Sarasota Center Blvd.
Sarasota, FL 34240
(813) 371-1003

Slice chambers:
Fine Science Tools
373-G Vintage Park Drive
Foster City, CA 94404
(800) 521-2109

Slice chambers:
Medical Systems Corp.
One Plaza Rd.
Greenvale, NY 11548
(516) 621-9190

REFERENCES

1. Aitken, P. G. Hippocampal slices. In: *Methods in Toxicology:* In Vitro *Biological Systems,* edited by Tyson, C. A. and Frazier, J. M. New York: Academic Press, 1993, pp. 1–6.
2. Balestrino, M. and Somjen, G. G. Concentration of carbon dioxide, interstitial pH and synaptic transmission in hippocampal formation of the rat. *J. Physiol.* 396, 1988, 247–266.
3. Blanton, M. G., Turco, J. L. L., and Kriegstein, A. R. Whole cell recording from neurons in slices of reptilian and mammalian cerebral cortex. *J. Neurosci. Methods* 30, 1989, 203–210.
4. Czéh, G., Aitken, P. G., and Somjen, G. G. Whole-cell membrane current and membrane resistance during hypoxic spreading depression. *NeuroReport* 3, 1992, 197–200.
5. Czéh, G., Aitken, P. G., and Somjen, G. G. Membrane currents in CA1 pyramidal cells during spreading depression (SD) and SD-like hypoxic depolarization. *Brain Res.* 632, 1993, 195–208.
6. Hatton, G. I., Doran, A. D., and Salm, A. K. Brain slice preparation: hypothalamus. *Brain Res. Bull.* 5, 1980, 405–414.
7. Kato, H. Types of neurons in the sliced LGB of cat. *Brain Res.* 66, 1974, 332–336.
8. Li, C.-L. and McIlwain, H. Maintenance of resting membrane potentials in slices of mammalian cerebral cortex and other tissues *in vitro. J. Physiol.* 139, 1957, 178–190.
9. Libet, B. and Gerard, R. W. Control of potential rhythm of isolated frog brain. *J. Neurophysiol.* 2, 1939, 153–169.

10. Linas, R. and Sugimori, M. Electrophysiological properties of *in vitro* Purkinje cell somata in mammalian cerebellar slices. *J. Physiol. (London).* 305, 1980, 171–195.
11. Pittman, Q. J., Hatton, J. D., and Bloom, F. E. Spontaneous activity in perfused hypothalamic slices: dependence on calcium content of perfusate. *Exp. Brain Res.* 42, 1981, 49–52.
12. Richards, C. D. and Sercombe, R. Electrical activity observed in guinea-pig olfactory cortex maintained *in vitro.* J. *Physiol.* 211, 1970, 571–587.
13. Somjen, G. G. Basic mechanisms is cerebral hypoxia and stroke: background, review, and conclusions. In: *Mechanisms of Cerebral Hypoxia and Stroke,* edited by Somjen, G. G. New York: Plenum, 1988, pp. 447–466.
14. Tombaugh, G. C. Mild acidosis delays hypoxic spreading depression and improved neuronal recovery in hippocampal slices. *J. Neurosci.* 14(9), 1994, 5635–5643.
15. Urban, L., Aitken, P. G., Friedman, A., and Somjen, G. G. An NMDA-mediated component of excitatory synaptic input to dentate granule cells in 'epileptic' human hippocampus studied *in vitro. Brain Res.* 515, 1990, 319–322.
16. Urban, L., Aitken, P. G., and Somjen, G. G. Interstitial potassium concentration, slow depolarization, and focal potential responses in the dorsal horn of the rat spinal slice. *Brain Res.* 331, 1985, 168–171.
17. Warburg, D., Posener, K., and Negelein, E. Über den Stoffwechsel der Carcinomzelle. *Biochem. Z.* 152, 1924, 309.
18. Yamamoto, C. and McIlwain, H. Electrical activities in thin sections from the mammalian brain maintained in chemically defined media *in vitro. J. Neurochem.* 13, 1966, 1333–1343.

Chapter 25

ORGANOTYPIC SPINAL CORD CULTURES AND A MODEL OF CHRONIC GLUTAMATE-MEDIATED MOTOR NEURON DEGENERATION

Andrea M. Corse and Jeffrey D. Rothstein

TABLE OF CONTENTS

I. INTRODUCTION

In our laboratory, we have developed an *in vitro* model of the *mature* spinal cord using organotypic spinal cord cultures prepared from postnatal rats.[1] Much of the previous work on cultured motor neurons involves embryonic neurons. Spinal cord neurons from 2- to 16-d-old postnatal rats have successfully been cultured using a roller-tube methodology with long-term (eight weeks) survival but suboptimal organotypic preservation. The organotypic spinal cord cultures developed in our laboratory are based on modifications of other organotypic tissue culture methodologies[2,3] and combine the following advantages:

- *postnatal* motor neurons
- *long-term* survival
- well-preserved *organotypic morphology*

Subsequently, we have combined our interests in chronic glutamate neurotoxicity with our organotypic culture methodology and developed an *in vitro* model of chronic glutamate toxicity in *postnatal* rat spinal cord.

Chronic glutamate toxicity is implicated in the pathogenesis of a number of neurodegenerative disorders. In amyotrophic lateral sclerosis (ALS), a devastating progressive neurological disorder characterized by selective degeneration of motor neurons, glutamate metabolism is abnormal. Glutamate levels in the cerebrospinal fluid of patients with ALS are elevated compared to disease-matched controls.[4] More recently, a selective loss of high-affinity glutamate transport has been identified in the spinal cord and selected brain regions of patients with ALS.[5] The neurotransmitter glutamate is normally rapidly cleared from the synaptic cleft by high-affinity, high-capacity transporters on astrocytes and neurons. We hypothesized

0-8493-8094-4/95/\$0.00+\$.50

that a defect in the glutamate transporter, which would produce sustained levels of extracellular glutamate, could result in motor neuron death (Figure 1).

To test this hypothesis, we developed a model of chronic glutamate toxicity using organotypic spinal cord cultures and selective inhibitors of glutamate transport, such as threohydroxyaspartate (THA). Inhibition of glutamate transport in this model continuously raised the concentration of glutamate in the culture medium. Using this paradigm we have discovered two significant characteristics of glutamate toxicity in spinal cord neurons: (a) chronic exposure to excessive extracellular glutamate (3 to 25 μm) can lead to slow loss (over weeks to months) of motor neurons (Figure 2), and (b) motor neurons are uniquely sensitive to non-NMDA but not NMDA receptor-mediated toxicity (Figure 3).

Additionally, this culture system provides a model in which to study the effects of neurotrophic factors on postnatal motor neuron survival and the potential neuroprotective effects of neurotrophic factors on chronic, glutamate-mediated motor neuron degeneration.[6]

We routinely prepare organotypic spinal cord cultures from 8- to 10-d-old rat pups, which have the advantage of providing relatively *mature* spinal cord tissue, more analogous to the *mature* nervous system afflicted in human neurodegenerative motor neuron diseases. Recently, organotypic spinal cord cultures from postnatal mice have also successfully been prepared in our laboratory. Unlike the standard roller-tube technique for organotypic or explant cultures, the method described below utilizes semipermeable membrane inserts, e.g., Millipore, Millicell-CM. With this technique, organotypic spinal cord cultures are reliably maintained for more than three months with >95% viability and remarkably preserved organotypic morphology, including partially preserved synaptic circuitry and critical cell-cell (i.e., neuronal-glial) interactions. Additionally, these cultures are maintained without the use of antibiotics or antifungal agents. The technique described below is a modification of the method described by Stoppini et al.[3]

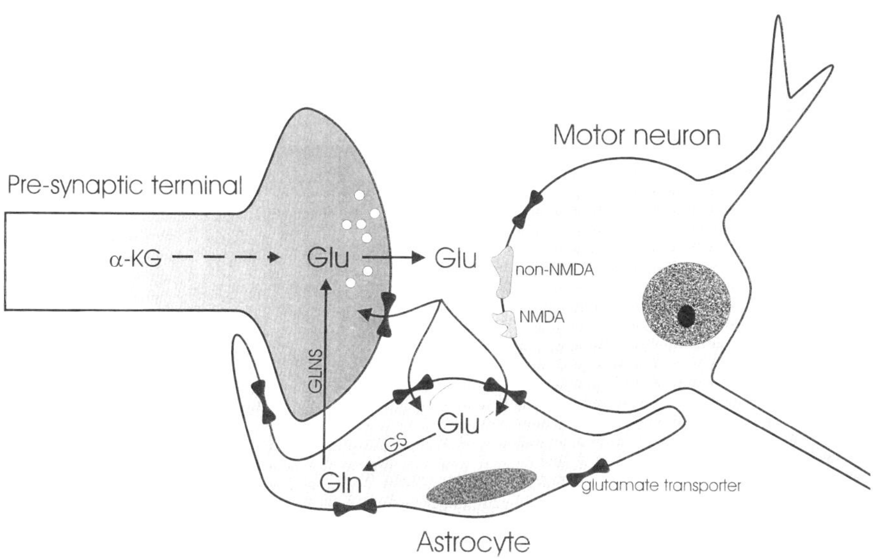

FIGURE 1. The neurotransmitter glutamate is released from the presynaptic nerve terminal into the synaptic space. Postsynaptic ionotropic glutamate receptors on motor neurons include NMDA and non-NMDA receptors. Glutamate is cleared from the synaptic space by glutamate transporters located on pre- and postsynaptic nerve terminals and surrounding glial cells. A defect in the glutamate transporter could result in increased synaptic glutamate levels and motor neuron degeneration.

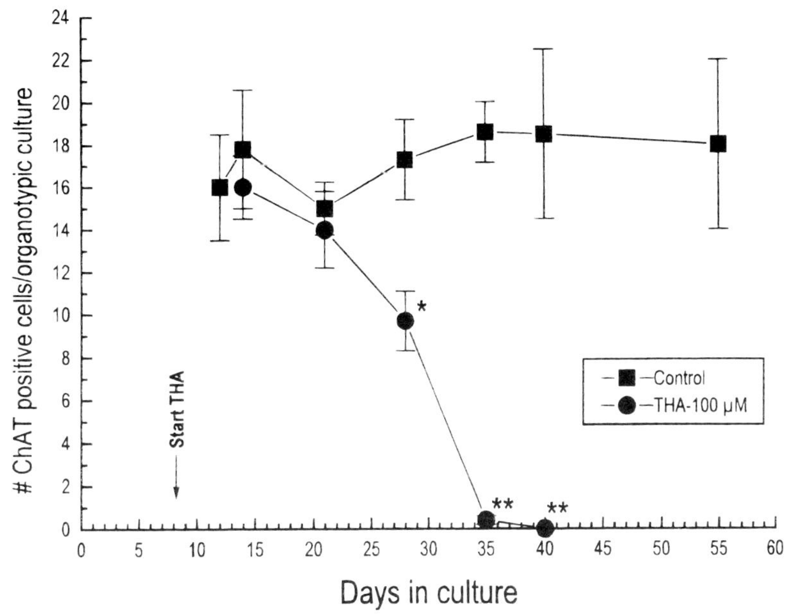

FIGURE 2. Exposure of organotypic spinal cord cultures to THA (100 μM), a glutamate transport inhibitor, results in elevated levels of synaptic glutamate and slow motor neuron loss demonstrated by a gradual reduction in counts of ChAT-positive motor neurons (shown here), which is paralleled by counts of motor neurons in plastic-embedded semithin sections. (From Rothstein, J. D. et al., *Proc. Natl. Acad. Sci. U.S.A.*, 90, 6593, 1993. With permission.)

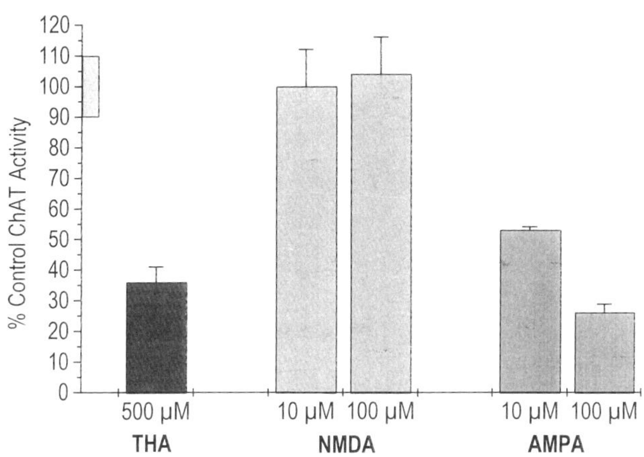

FIGURE 3. The toxic effects of glutamate receptor agonists are compared to the effects of THA. At 25 d, cultures incubated in the presence of the glutamate receptor agonist AMPA (10 or 100 μM) show a reduction in the ChAT activity to levels comparable to THA (500 μM). Cultures exposed to the glutamate receptor agonist NMDA (10 or 100 μM) show no significant change in ChAT activity compared to control cultures. These findings suggest slow, glutamate-mediated motor neuron degeneration in this model results from non-NMDA receptor-mediated toxicity.

II. CULTURE PREPARATION

Culture preparation requires approximately 20 min per animal. One rat pup provides a maximum of 10 to 15 cross sections of lumbar cord. Five cross sections are placed on one Millipore membrane in a culture plate well; therefore, one animal usually provides enough tissue for two to three culture wells.

The following preparatory steps can be performed the day prior to culturing: autoclave materials and prepare incubation medium.

1. Autoclave:

   Operating scissors 5 $^1/_2$″, straight sharp-blunt (Roboz RS-6780)
   Micro dissecting forceps 7″, serrated (Roboz RS-5280)
   Micro dissecting tweezers 4 $^3/_8$″, straight (Roboz RS-5045)
   Micro spatula 6 $^1/_4$″, spoon end $^1/_2$″ × $^1/_8$″
   Micro dissecting scissors 4 $^1/_2$″, straight, sharp (Roboz RS-5676)
   Two micro dissecting tweezers 4 $^1/_8$″, tips on 45° angle (Roboz RS-5005)
   Modified glass "transfer" pipettes (see below)
   Aclar plastic film (1–5 $^1/_2$ × 3 $^1/_2$ cm piece per animal plus one to two for setup)

   Transfer pipettes are used to transfer the spinal cord sections to the Millipore membrane as described below. We make transfer pipettes from 10-ml glass pipettes cut at the distal and proximal ends to create a modified pipette approximately 5 cm long and slightly tapered at the distal tip. The ends are flamed for smoothness.
2. Incubation medium:

   25% Horse serum, heat-inactivated (Gibco 26050-013)
   25% Hank's balanced salt solution (HBSS) (Gibco 24020-059) supplemented with D-glucose (25.6 mg/ml)
   50% Modified essential medium (MEM) with 25 m*M* HEPES (Gibco 12370-029)
   2 m*M* L-glutamine (Sigma G-5763)

   We routinely prepare 0.2 *M* L-glutamine in bulk by dissolving 1.46 g L-glutamine in 50 ml double-distilled H_2O and preparing 4-ml aliquots for storage at −20°C. We prepare 400 ml incubation medium weekly as follows:

   100 ml horse serum
   100 ml HBSS with glucose
   200 ml MEM
   4 ml 0.2 *M* L-glutamine

   The final pH of the medium is 7.2, though initial studies using pH 7.4 or 7.8 revealed similar results on motor neuron survival. Under the laminar flow hood, sterile-filter the incubation medium (Nalgene 500 ml disposable filter). The medium is stored at 4°C for up to one week.

On the culture day, prepare the following in the laminar flow hood: culture plates, petri dishes, tissue chopper, and selected dissection tools.

1. Culture plates

   Five 35-mm six-well plates (Falcon #3046)
   Thirty millicell-CM 30-mm inserts (Millipore, PICM 03050)

   Using sterile technique, in the laminar flow hood, pipette 1 ml medium at 37°C into each Falcon plate well. Using sterile forceps, place a Millipore insert into each medium-filled well. Cover the prepared plates and set aside in the hood.

   We use only the Millipore brand membranes. Our attempts to culture on other brand inserts, i.e., Costar, Falcon, and Nunc, resulted in inconsistent tissue adherence and loss of spinal cord organotypic features after one to two weeks in culture. A disadvantage of this technique is the expense, specifically of the single-use Millipore membranes.

We have attempted to reuse Millipore inserts for chronic cultures without success.

2. Petri dishes

   35 × 10 mm Nunclon dishes (Nunc), two per animal
   Gey's balanced salt solution (GBSS) (Gibco 14260-020) with glucose (6.4 mg/ml)

   Sterile-filter the GBSS with glucose solution (Falcon 100-ml filter). Fill each petri dish, cover, and set aside in the hood.

3. Tissue chopper

   McIlwain tissue chopper
   Double-edged razor blade
   Aclar plastic film (autoclaved)

   Wipe tissue chopper with 70% ethyl alcohol (ETOH). Set slice thickness at 350 μm. Attach double-edged razor blade to chopper and position the blade for even cuts with sufficient force to partially, but not completely, cut through the Aclar.

4. Dissection tools

   Two 100-ml beakers with 25 to 50 ml 70% ETOH

   Place straight dissecting tweezers, angled dissecting tweezers (two), and spatula in one beaker and modified pipettes for tissue transfer in a second beaker.

III. ANIMAL DISSECTION

Initial animal dissection is performed outside the hood and completed inside the hood. Wear gown, gloves, hood, and mask.

Dissecting bench setup:

1-l beaker with 5% betadine
250-ml beaker with 70% ETOH
Spray bottle with 70% ETOH
Spray bottle with sterile saline
Single-edged razor blade
Styrofoam dissecting surface and pins
Numerous 4 × 4 cotton gauze soaked in 70% ETOH
Magnifying dissecting glasses

Place forceps and scissors in betadine solution and straight tweezers and single-edged razor blade in 70% ETOH solution. Wipe the styrofoam dissection surface with 70% ETOH.

We routinely dissect two to three animals at a time. Quickly dip each rat pup in betadine then ETOH and rapidly decapitate with scissors. Pin pups supine on ETOH-soaked gauze on the dissecting board. Make a single longitudinal midline cut with the razor blade extending from the suprasternal notch to the pubic symphysis in each animal (Figure 4A). Expose the thoracic cavities by removing ventral ribs and overlying skin with forceps and scissors (Figure 4B). Remove the contents of thoracic and abdominal cavities using forceps, exposing the full length of the vertebral column (Figure 4C). Rinse the cavities with saline. Remove the carcasses from the board, dip into betadine solution, and rinse the cavities with 70% ETOH spray. Pin pups supine onto clean 70% ETOH-soaked gauze on the dissecting board and place the board in the laminar flow hood.

In the hood, make a transverse cut with the micro dissecting scissors between the two most caudal full-length ribs, through the vertebral column and spinal cord. Carefully unroof the ventral vertebral column from the lumbosacral spinal cord by making small cuts with

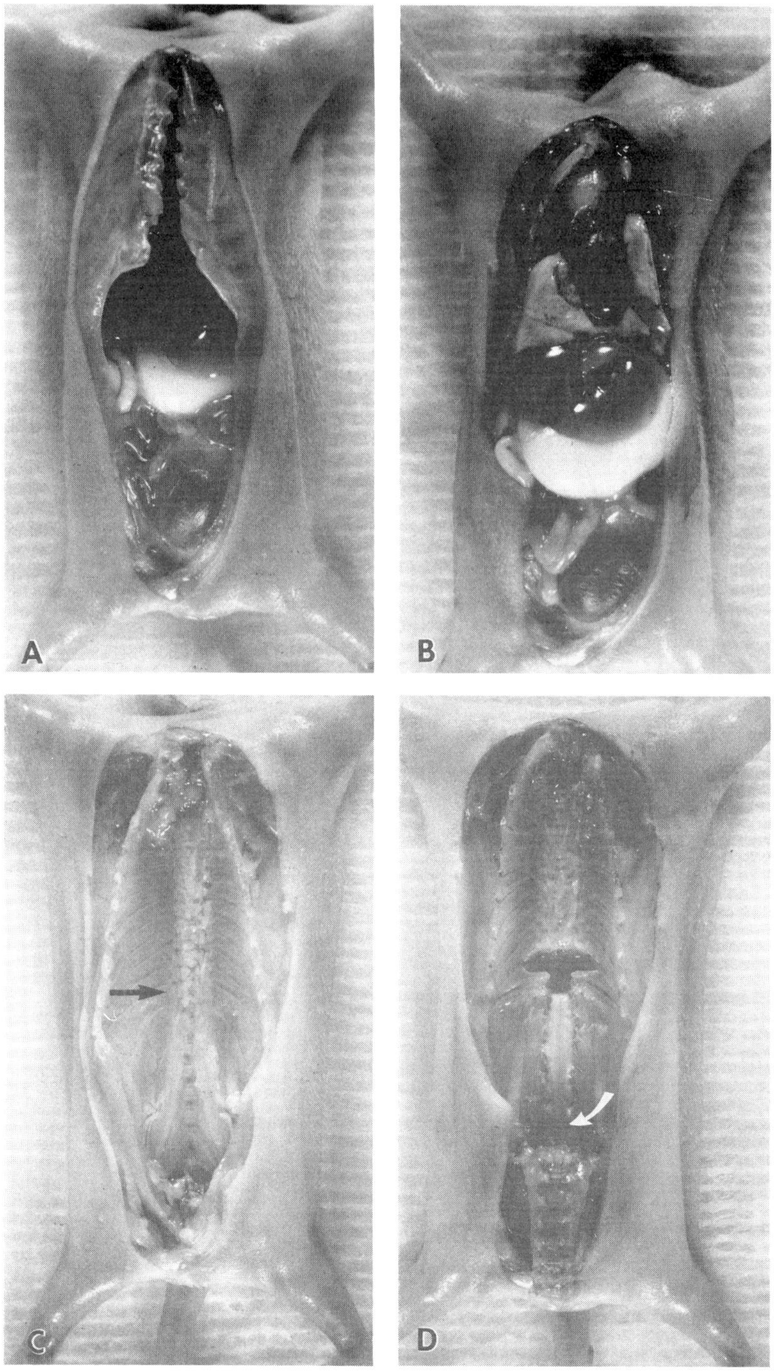

FIGURE 4. The dissection and removal of the lumbar spinal cord in an 8-d postnatal rat is shown. The thoracic and peritoneal cavities are exposed and the contents removed (A,B,C). A transverse cut is made through the vertebral column at the level between the two most caudal full-length ribs (black arrow). The spinal cord retracts caudally. The ventral vertebral column is cut away from the lumbosacral cord and lifted caudally (D). A transverse cut is made at the distal spinal cord near the conus (white arrow), and the spinal cord is gently removed from the canal.

the micro dissecting scissors held parallel to the spinal cord and directed distally (Figure 4D). Make a second transverse cut through the distal end of the spinal cord near the conus and carefully lift the cord from the vertebral column using curved tweezers. Meninges attached to the cord can be used to lift the cord. Place the cord into a GBSS-filled petri dish (Figure 5A). Remove the cords from the other partially dissected pups and place in separate petri dishes.

With the magnifying glasses, enter the hood with your head and shoulders, carefully dissect the cord free of surrounding meninges, and transect the remaining nerve roots adjacent to the cord (Figure 5B). Remove the magnifying glasses. Lift the cord onto Aclar using the spatula. Align the cord longitudinally along the length of the Aclar with the dorsal spinal cord vessel on top, removing any excess GBSS away from the cord with the spatula.

Place Aclar on the tissue chopper and cut the cord transversely in 350-μm sections. Stop the chopper at the point the cord narrows distally. Cross sections of the cord at these lower levels have very few motor neurons; therefore, we do not culture distal cord. However, by

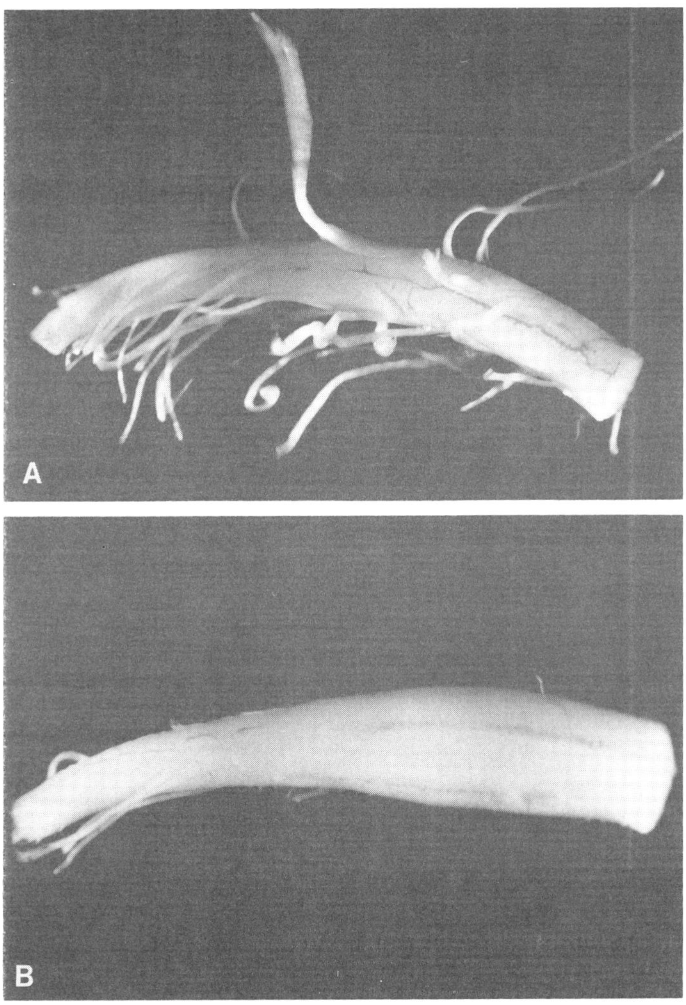

FIGURE 5. The dissected spinal cord is placed in GBSS (A). Under magnification the meninges are dissected away from the cord, and the spinal roots are transected from the lumbar (rostral) portion of the cord (B).

including the distal portion of the cord in the dissection, the cord has increased weight and surface area on the Aclar, increasing stability during the chopping process. If the dissected cord is too short, the cord frequently sticks to the razor blade and lifts off the Aclar during the chopping process. Because Aclar often wrinkles during autoclaving, hold the corners of the Aclar down during the chopping process.

Lift the Aclar from the chopper and place the cord sections into a fresh GBSS-filled petri dish by turning the Aclar upside down against the petri dish and "washing" the sectioned cord into the dish.

Again, using magnification and positioned in the hood, carefully separate each transverse section of cord with curved tweezers (Figure 6). Transfer individual sections of cord onto Millipore membranes using the modified pipette as follows: creating suction on the top of the pipette with the thumb, lift a single cord section into the pipette; let the cord section settle to the bottom of the GBSS in the pipette; and lightly touch the surface of the membrane with the section, releasing a **minimal** amount of GBSS with the section. If too much GBSS is released with the cord section, the tissue will not adhere adequately to the membrane, and the tissue will not survive. Excess GBSS on the membrane can be partially removed with bulb suction and a sterile Pasteur pipette. Only transfer spinal cord sections that appear grossly intact and undamaged. Occasionally, a cord will twist during chopping, and no sections will be appropriate for culturing. The transfer of the tissue is the most difficult technique in this culturing process but quickly becomes easier with practice.

Place five cord sections circumferentially on the membranes in each well (Figure 7A). Each cord provides approximately 10 to 15 sections. Complete the process on the remaining dissected spinal cords. Incubate the culture plates at 37°C in a 5% CO_2–95% O_2 humidified incubator (Forma Scientific, Marietta, OH).

In all experiments, cultures are allowed an 8-d recovery period after initial preparation, prior to the addition of drugs (see below).

The culture medium, along with any added pharmacological agents, is changed every 3 d. To change the culture medium, autoclave Pasteur pipettes and warm the incubation medium to 37°C. Remove one to two plates at a time from the incubator and place inside

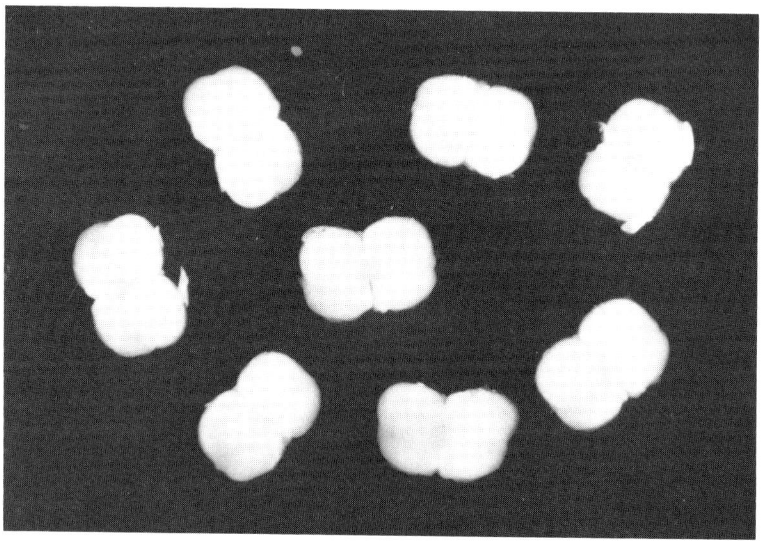

FIGURE 6. The 350-μm cross sections of lumbar cord are carefully separated under magnification in GBSS, and any damaged sections are removed. Each section is transferred individually with the transfer pipette to the Millipore membrane.

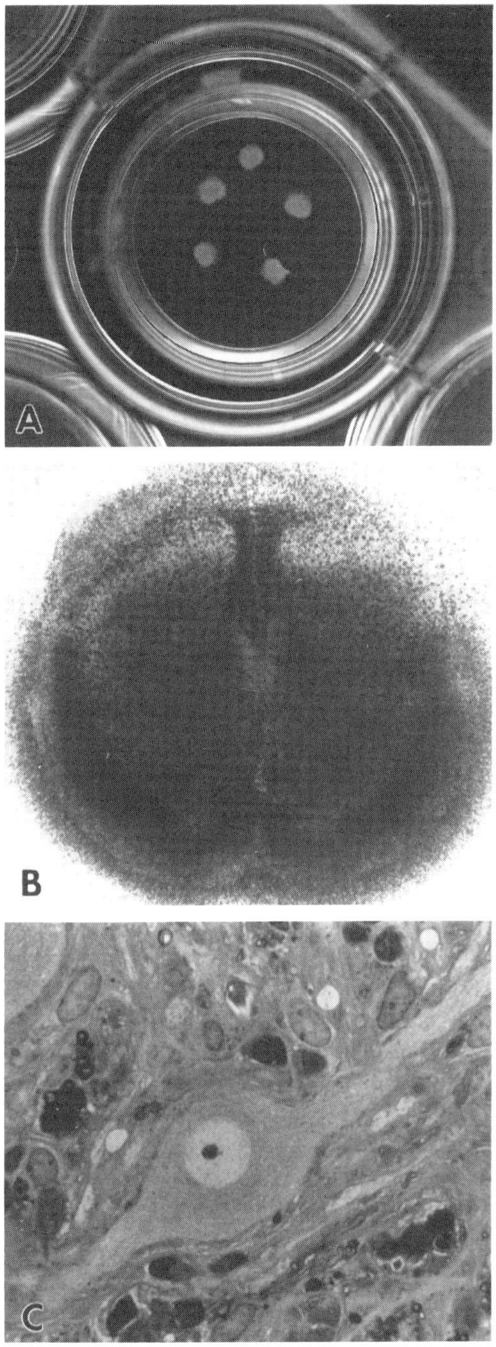

FIGURE 7. Five lumbar cord sections are cultured on each Millipore membrane (A). A one-week-old organotypic spinal cord culture is shown as seen with phase contrast microscopy (B). A motor neuron and its processes in a three-week-old culture are identified in a 2-μm semithin plastic-embedded section by light microscopy (C).

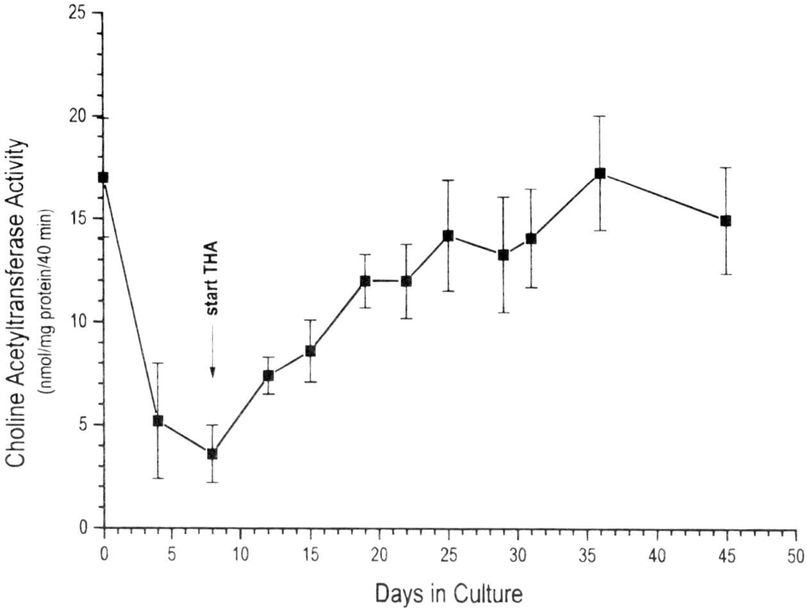

FIGURE 8. The time course of ChAT activity in organotypic spinal cord cultures shows a precipitous drop in activity for the first several days then a gradual return to baseline levels comparable to age-matched uncultured lumbar spinal cord tissue. (Adapted from Rothstein, J. D. et al., *Proc. Natl. Acad. Sci. U.S.A.*, 90, 6592, 1993.)

the laminar flow hood. Attach a pipette to vacuum and remove the medium from each well. Carefully refill each well with 1 ml of the incubation medium. **Keep the medium off the Millipore membrane surface.** Additional pharmacologic agents can be added conveniently to the medium in the wells at this time. Return the culture plates to the incubator.

IV. CULTURE ANALYSIS

The gross appearance of the cultures can be followed routinely using phase-contrast microscopy during the life of the culture (Figure 7B). Additional techniques utilized in our laboratory for culture analysis include: microscopic morphologic analysis of whole cultures using cresyl violet stains and 2-μm semithin plastic-embedded sections stained with toluidine blue (Figure 7C); immunohistochemical analysis of whole cultures for choline acetyltransferase (ChAT) and glutamate receptor proteins; western blot analysis for glutamate receptor and glutamate transporter proteins; and biochemical analysis for ChAT activity and somatostatin levels.

ChAT activity initially drops precipitously for the first several days after culturing but gradually returns to baseline levels comparable to uncultured lumbar spinal cord tissue (Figure 8). For this reason the cultures are routinely allowed an 8-d recovery period after preparation before experiments are begun. The initial drop in enzyme activity is a typical response of motor neurons to axotomy and is reflected in mild reversible chromatolytic changes morphologically during the first three weeks in culture. The return to baseline of the ChAT activity reflects the recovery from this initial injury. Counts of ChAT-positive neurons demonstrate a stable population of neurons throughout the period of the culture (Figure 2).

ACKNOWLEDGMENTS

Excellent technical assistance in the preparation of the organotypic spinal cord cultures and this manuscript was provided by C. Coccia, M. Dykes-Hoberg, L. Jin, and M. Lehar.

REFERENCES

1. Rothstein, J. D., Jin, L., Dykes-Hoberg, M., and Kuncl, R. W. (1993) *Proc. Natl. Acad. Sci. U.S.A.* 90, 6591–6595.
2. Delfs, J., Friend, S., Ishimoto, S., and Saroff, D. (1989) *Brain Res.* 488, 31–42.
3. Stoppini, L., Buchs, P. A., and Muller, D. (1991) *J. Neurosci. Methods* 37, 173–182.
4. Rothstein, J. D., Tsai, G., Kuncl, R. W., Clawson, L., Cornblath, D. R., Drachman, D. B., Pestronk, A., Stauch, B. L., and Coyle, J. T. (1990) *Ann. Neurol.* 28, 18–25.
5. Rothstein, J. D., Martin, L. J., and Kuncl, R. W. (1992) *N. Engl. J. Med.* 236, 1464–1468.
6. Corse, A. M., Kuncl, R. W., and Rothstein, J. D. (1993) *Soc. Neurosci. Abstr.* 19, 659.

Chapter 26

TISSUE CULTURE STUDIES OF NEURONAL RESPONSES TO PHYSICAL LESIONS OF THE PLASMA MEMBRANE: ENVIRONMENTAL INFLUENCES AND EXPERIMENTAL INTERVENTIONS

Jen Hill Lucas, Dennis G. Emery, and Guenter W. Gross

TABLE OF CONTENTS

0-8493-8094-4/95/$0.00+$.50

I. INTRODUCTION

Interest in central nervous system (CNS) injury and regeneration has inspired development of a variety of experimental models (Allen, 1911, 1914; Tarlov and Klinger, 1954; Khan and Griebel, 1983; Gennarelli and Thibault, 1985; Black et al., 1986; Bresnahan et al., 1987; Black et al., 1988). These models have involved delivery of a physical (mechanical) trauma to intact, three-dimensional CNS tissues. Investigations performed with these mostly *in vivo* models have contributed substantially to our understanding of CNS injury. However, a limitation of whole-animal preparations is that it is difficult to distinguish cellular responses to the separate components of the injury, which include free radical formation, release of neurotransmitters and neuromodulators, delayed tissue ischemia, acidosis, edema, and the direct effects of the mechanical violence. Thus, the relative contributions of the individual factors to cellular—particularly neuronal—deterioration and death are largely unknown. Likewise, proposals that some of these processes may, in fact, be more protective than injurious are still only intriguing speculations (Young, 1987).

Because of the complex layering of responses on the *in vivo* tissue level, a number of researchers have elected to use greatly simplified monolayer culture or isolated nerve/slice preparations to study the direct effects of mechanical injury on CNS cells (Chambers and Fell, 1931; Levi and Meyer, 1945; Easton, 1960, 1965; Hamburger and Rinder, 1966; Mire et al., 1970; Schlaepfer and Bunge, 1973; Shaw and Bray, 1977; Bird, 1978; Sole, 1980; Meiri et al., 1981; Yawo and Kuno, 1983, 1985; Balentine et al., 1988; Tecoma et al., 1989). In most of these studies individual neurons were subjected to membrane perforation or neurite transection injuries using various types of "microknives". Although useful, these mechanical methods of cell surgery do not offer the control necessary for creating experimental lesions without significant deformation of other areas of an operated cell. This limitation is particularly disadvantageous in studies of the responses of relatively small mammalian neurons. Cell distortion also precludes concurrent intracellular recording close to the lesion site during cell surgery. More important, without the capability of creating highly localized injuries, the physical parameters of the membrane lesions cannot be determined. These lesion parameters provide the only ready means of quantitatively assessing injury severity. Finally, the lack of precision can also result in damage to significant numbers of neighboring neurons and glia. This may increase variability by altering the chemical microenvironment around a target cell.

To address these difficulties, this laboratory developed techniques of laser microbeam cell surgery that permit the creation of lesions as small as 1 μm in diameter at precise locations on individual neurons growing in monolayer cultures of CNS cells (Gross et al., 1983). The precision of these methods (±1 μm positioning error) allows surgical manipulation of individual neurons with minimal disturbance of other portions of the target cells, surrounding structures, or intracellular electrodes. For the past 10 years, we have used this approach to investigate the cellular and subcellular responses of mammalian spinal cord neurons to a defined physical injury: dendrite amputation close (10 to 400 μm) to the perikaryon. These dendritic lesions have been characterized in terms of: (a) lesion and cellular physical parameters that govern injury severity (lesion diameter, distance of the lesion from the perikaryon, extent of proximal segment retraction after transection, and average diameter of the proximal segment) (Lucas et al., 1985; Lucas, 1987; Shi et al., 1989); (b) electrophysiological, morphological, and ultrastructural consequences of such injuries (Lucas et al., 1985; Kirkpatrick et al., 1985; Emery et al., 1987; Gross and Higgins, 1987); (c) some of the basic mechanisms of cellular injury and repair (ibid., Lucas, 1987; Lucas et al., 1990a; Emery et al., 1991); and (d) the probability of survival of neurons subjected to transection injuries of known severity (i.e., combination of lesion parameters; Lucas et al., 1985; Lucas, 1987). Also, because the effects of experimental interventions can be expressed in terms of absolute numbers of surviving cells, the transection injury model has proven particularly useful for initial testing of pharmacological and environmental manipulations designed to protect CNS neurons that have sustained membrane trauma (Shi et al., 1989; Lucas et al., 1990a,b,c; Wang et al., 1990, 1991).

In this paper, we review the foundational research on neuronal responses to dendrotomy under conditions that mimic the normal extracellular environment. The success of various strategies to increase the probability of survival by limiting ion-mediated intracellular damage spread is also described. Last, we report the results of recent investigations of metabolic inhibition as a tactic to protect cells after physical disruption of the plasma membrane.

II. METHODS

A. CELL CULTURE

Standard techniques utilized for preparation of monolayer cultures from dissociated embryonic (13 to 14 d) mouse spinal cords have been described elsewhere (Lucas et al., 1985). Most cultures were grown on glass coverslips, which were glued to the bottoms of 60-mm plastic culture dishes (Corning). For electron microscopy, cultures were grown directly on Lux permanox plastic culture dishes. Cultures were utilized for experiments 21 to 28 d postseeding when they were considered to be mature. At no time were the cultures exposed to antibiotic or antimycotic agents.

B. LASER CELL SURGERY

Figure 1 shows a schematic representation of the laser microbeam system, which was manufactured by BTG Biotechnik of Munich., Germany. Laser cell surgery was performed with a pulsed (8-ns pulse) nitrogen laser operating at 337 nm with a maximum output of 14 kW. A helium-neon (HeNe) laser (630 nm, 1 mW) was colinearly aligned with the nitrogen work laser and provided a visible target beam. The HeNe and UV laser beams were reflected into a Leitz Orthoplan microscope by a beam aligner and focused to a small diameter through Zeiss Ultrafluar quartz objectives. The minimum diameters of the laser beam with the 10× nonphase objective and 32× phase objective were 4.5 and 2.3 μm, respectively. All surgery was performed using the multiple-shot, low-energy method developed by Gross and coworkers (1983). In most experiments the firing frequency was 20 Hz and the energy density per pulse was 1 to 3 $\mu J/\mu m^2$. Complete transection of target neurites generally required between 10 to 20 s.

C. ARRANGEMENTS FOR LASER CELL SURGERY

Figure 2 shows two arrangements for laser cell surgery. The 10× nonphase quartz objective (working distance = 7.4 mm) was utilized for experiments requiring concurrent intracellular recording (Figure 2A). The 32× phase quartz objective was used for studies of neuronal survival after dendrite amputation (Figure 2B). In both arrangements the medium within the 60-mm culture dish was maintained at 37°C by a heater plate on the microscope stage. The pH of the culture medium was maintained at 7.3 to 7.4 by a 10% CO_2 in air atmosphere.

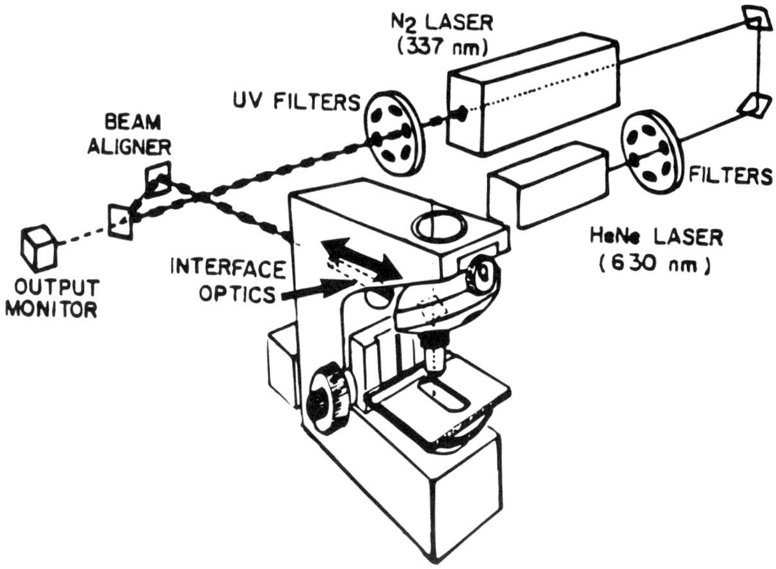

FIGURE 1. The laser microbeam system. The nitrogen (337 nm) and HeNe (630 nm) laser beams are aligned colinearly, reflected into a Leitz Orthoplan microscope, and focused onto a target on the microscope stage. (Reprinted with permission from Lucas et al., *J. Neurotrauma,* 2, 1985.)

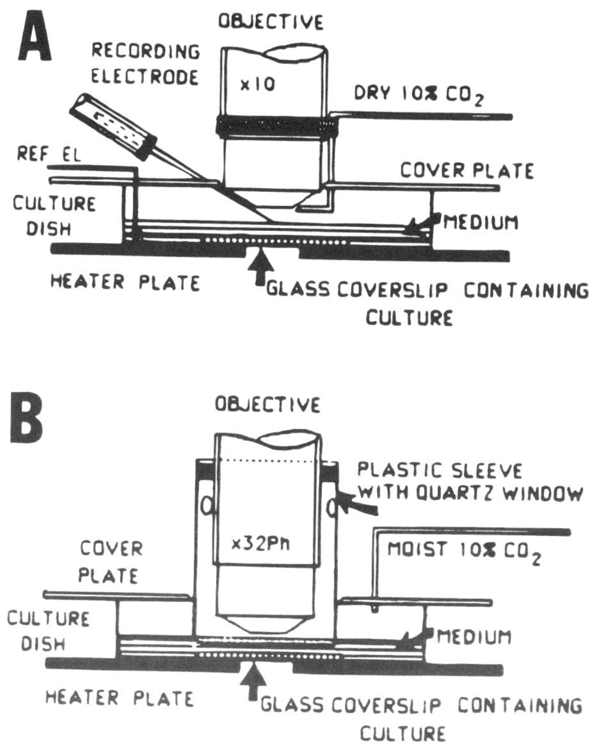

FIGURE 2. A. Arrangement for laser cell surgery and concurrent electrophysiologic recording using a 10× nonphase quartz objective (Zeiss Ultrafluar). A dry stream of 10% CO_2 in air controls medium pH and prevents condensation on the objective lens. B. Laser cell surgery arrangement used for neuronal survival experiments. The 32× phase Zeiss quartz Ultrafluar objective is protected from the warm culture medium by a plastic sleeve with a matched quartz coverslip at the end. (Reprinted with permission from Lucas et al., *J. Neurotrauma,* 2, 1985.)

Osmolarity was maintained by addition of 200 μl of distilled water to the 3.0 ml reservoir of medium at 30-min intervals. More detailed information about these arrangements and the precautions for prevention of bacterial or fungal contamination during surgery can be obtained from an earlier publication (Lucas et al., 1985).

D. SELECTION OF NEURONS AND RELOCATION AFTER SURGERY

Criteria for neuron selection and procedures for relocation and identification of selected neurons are described in the paper by Lucas and co-workers (1985).

E. PROTOCOL FOR CELL SURVIVAL STUDIES

For most studies of survival, control and experimental cultures constituted matched pairs or sets from the same seeding date. In each control or experimental culture, ten neurons were selected for surgery and ten others were selected to be unoperated controls. Twenty-four hours after neuron selection, a primary dendrite was transected from the ten neurons chosen for surgery (Figure 3). Unless otherwise stated, lesion distance in the survival studies was always 100 μm from the edge of the perikaryon. Cell surgery generally required 20 to

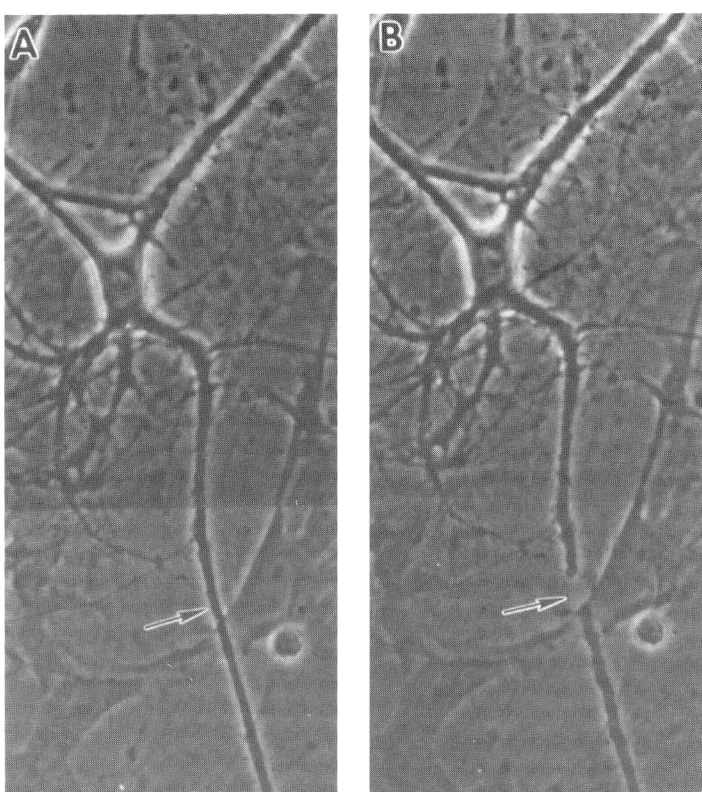

FIGURE 3. Amputation of a primary dendrite from a neuron growing in a monolayer culture of dissociated embryonic mouse spinal cord (25 DIV). A. Neuron before surgery. The target area (arrow) was 100 μm from the edge of the perikaryon. Lesion diameter (diameter of the process at the target point was 1.5 μm). Note that there was no major branching of the dendrite between the perikaryon and the target point. B. Same neuron 5 min after dendrite amputation. The dendrite was completely severed and the proximal and distal segments have retracted. Surgery was performed with the multiple shot method (Gross et al., 1983): firing frequency = 20 Hz, energy density per pulse = 1.7 $\mu J/\mu m^2$, firing interval = 17 s.

30 min. Survival of the ten lesioned and ten control (unlesioned) neurons in each culture was evaluated at 2 h after surgery by light microscopic examination (Lucas et al., 1985) and at 24 h by erythrosin B dye exclusion. The procedure for viability testing with erythrosin B is described in the paper by Emery and co-workers (1987).

F. ELECTRON MICROSCOPY STUDIES

The procedures for fixation, embedding, and staining of neurons for electron microscope (EM) examination can be found in the paper by Emery and co-workers (1987).

G. ION-SUBSTITUTION STUDIES

Procedures for cleaning of glassware to remove surface ions, for preparation of a low-calcium culture medium, and for measurement of calcium by atomic absorption spectrophotometry are described in the paper by Lucas and co-workers (1990a). For EM studies of the role of sodium in injury, choline chloride (GIBCO) was substituted for sodium chloride in the low-calcium medium. For EM studies of the role of chloride in injury, sodium isethionate or sodium propionate (Sigma) was substituted for sodium chloride in the low-calcium medium. Isoosmotic sucrose was used as a nonionic substitute in those studies in which all three major ions were reduced. Hepes buffer replaced sodium bicarbonate in these media.

H. HYPOTHERMIA STUDIES

The arrangements for maintaining neurons at reduced temperatures following dendrotomy are described in the papers by Lucas and co-workers (1990b,c).

I. PHARMACOLOGICAL INTERVENTIONS

Pharmacological agents were obtained from Sigma Chemical Company (St. Louis, MO). The toxicity of all compounds for uninjured spinal cord neurons was evaluated prior to testing on lesioned neurons. The maximum concentration at which there was no death of uninjured neurons and no visible signs of stress (maximum nontoxic concentration) was utilized in studies of the influence of these agents on neuronal survival after dendrotomy.

III. ION-MEDIATED DAMAGE AND DEATH OF NEURONS AFTER MEMBRANE TRAUMA

A. MEMBRANE RESEALING AFTER DENDROTOMY

At lesion distances greater than 50 μm, intraperikaryal recordings from over 100 neurons have consistently revealed a two-phase injury potential profile (Figure 4). The first (peak) phase occurs within the first 5 min after dendrotomy and consists of a rapid decrease in the original resting potential followed by a partial repolarization. This phase probably represents an initial incomplete lesion resealing. In the second (plateau) phase a new equilibrium potential is established which is always lower than the original resting potential. Complete recovery of the original resting potential was not observed within the 30 to 60-min intervals after dendrotomy (Figure 4; Lucas et al., 1985). Twenty-four hours after dendrotomy average resting potentials (RP) of surviving neurons were still lower than the average RP of unlesioned neurons in the same cultures (ibid.). Studies of transected invertebrate (Meiri et al., 1981) and nonmammalian vertebrate (Borgens et al., 1980) axons have also shown that normal membrane selectivity and electrophysiological properties were not restored for hours or even days. These RP changes were most likely the result of the ionic injury currents (particularly influxes of sodium and calcium), which were found to penetrate neurons through transection lesions (Borgens et al., 1980; Meiri et al., 1981).

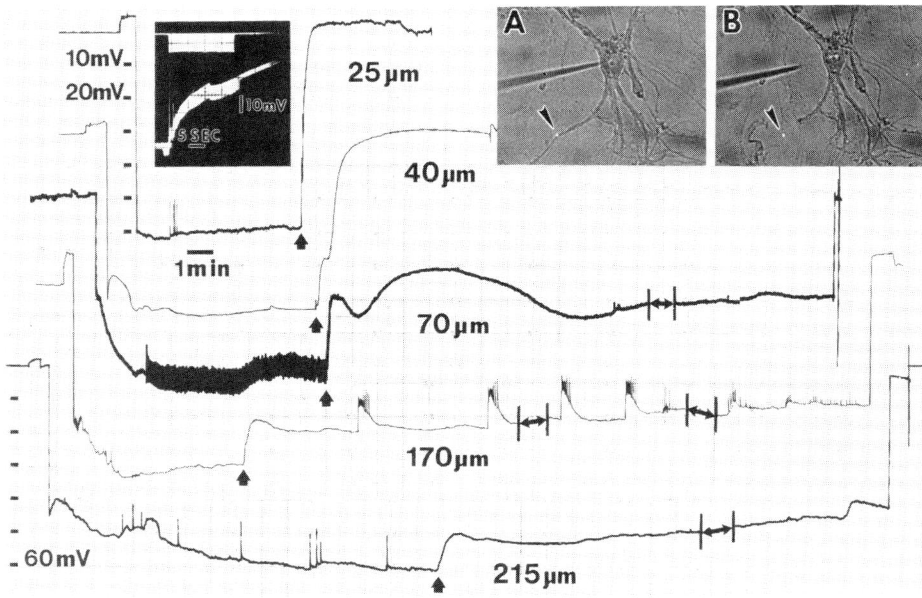

FIGURE 4. Injury potential profiles recorded from neurons subjected to dendrotomy at lesion distances ranging from 25–215 µm. Cells were held stably for at least 5 min prior to surgery. In each record the arrow indicates the time at which the dendrite was severed. There was generally a total loss of potential in neurons operated at distances of 50 µm or less (top two records). The oscilloscope record (top left) shows the loss of potential (lower trace) during and just after surgery on the cell operated at 25 µm. The period of laser firing is represented by the oscilloscope trace at the top. At longer lesion distances (three lower records) a two-phase injury potential profile was characteristic (see text). These cells were held for 30 to 60 min after surgery. Double arrows indicate where the postsurgery records have been truncated. Note that there was no recovery of the presurgery resting potential in any of these cells. The light micrographs on the right show (A) a neuron prior to dendrotomy with the electrode in the soma and (B) 35 min after dendrotomy. Transection lesion distance for this cell was 60 µm. Arrowheads indicate the target area and HeNe laser beam (A) before and (B) after surgery.

B. NEURONAL SURVIVAL AFTER DENDROTOMY: THE INFLUENCE OF LESION AND CELLULAR PHYSICAL PARAMETERS

Systematic investigations (over 800 neurons) of the influence of transection lesion physical parameters have revealed that the probability of survival: (a) increases as a function of distance of the lesion from the perikaryon (Figure 5A), (b) decreases as a function of lesion diameter (diameter of the dendrite at the point of transection) (Figure 5B), and (c) increases as a function of the extent of proximal segment retraction (Figure 5C) (Lucas et al., 1985; Lucas, 1987). It has been proposed that an influx of calcium perpetrates much of the cellular damage after a physical (or other) injury (for discussion, see Balentine, 1988). Therefore, because the lesion parameters probably govern the magnitude of the calcium current that penetrates neurons after neurite transection, they may be used for quantitative assessment of injury severity.

The influence of cellular physical parameters (somal volume, volume of the proximal segment, somal surface area, surface area of the proximal segment, numbers of other primary dendrites, and average diameter of the target process between the soma and the point of transection) on neuronal survival after dendrotomy was also evaluated (384 neurons). A first analysis revealed no obvious correlation between any of these cellular parameters and survival. However, in a group of 64 neurons with the same lesion physical parameters (distance, diameter, and retraction), it was found that as the average diameter of the section of dendrite between the soma and the target area increased, the probability of survival declined. Thus,

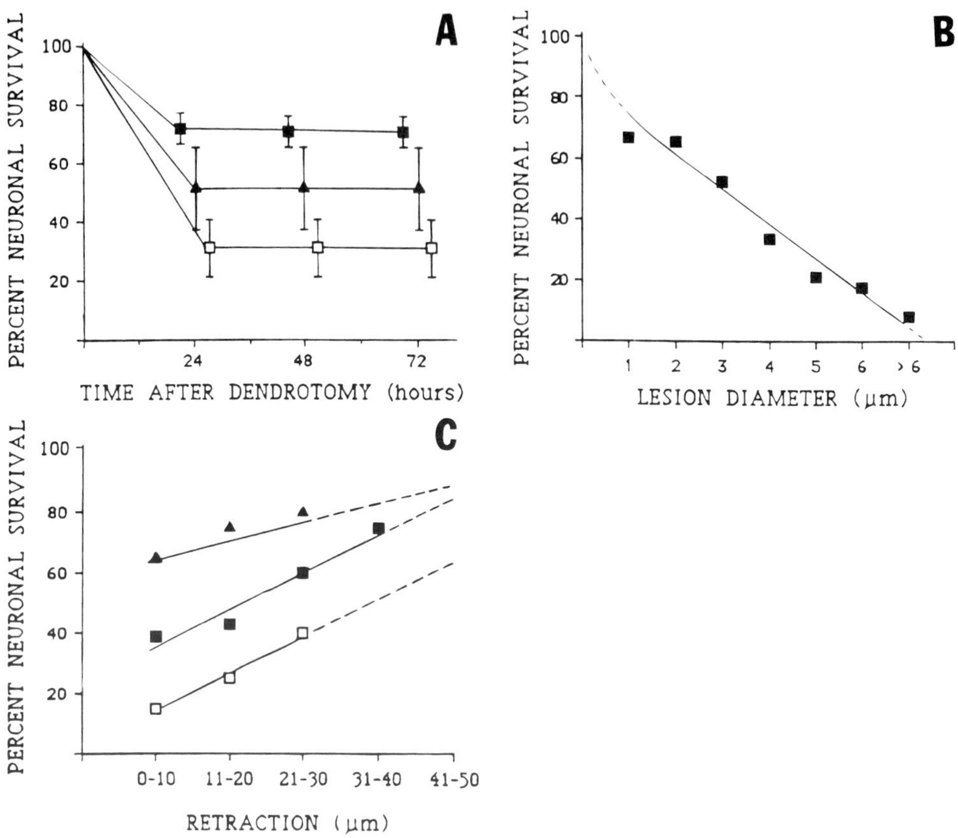

FIGURE 5. The influence of lesion physical parameters on neuronal survival after dendrotomy. A. The percentage of neurons that survive dendrotomy increases as a function of the distance of the transection lesion from the perikaryon. Survival was 30% in a group of neurons subjected to dendrite amputation at a lesion distance of 50 μm (closed squares), 53% in a group of neurons operated at 100 μm (triangles), and 71% in a group operated at 150 μm (open squares). Each function represents the mean percent survival in 10 experiments (60 neurons). In most cases, neurons which were alive 24 h after dendrotomy were still viable at 48 and 72 h. Survival among unlesioned control neurons (60 in each group) was close to 100%. (Reprinted with permission from Lucas et al. *J. Neurotrauma*, 2, 1985.) B. The percentage of neurons that survive dendrotomy decreases as the lesion diameter increases. A group of 384 neurons were subjected to dendrite amputation at a lesion distance of 100 μm. The diameter of the dendrite at the target point was measured for each cell prior to surgery and rounded to the nearest whole number. Each point represents between 6 and 74 evaluations of survival 24 h after surgery. C. The percentage of neurons that survive dendrotomy increases as the extent of proximal segment retraction (measured 5 min posttransection) increases. Lesion distance was 100 μm for all cells. The three functions represent survival data for 206 neurons with three ranges of lesion diameter: 2.0 to 2.9 μm (triangles), 3.0 to 3.9 μm (closed squares), and 4.0 to 4.9 μm (open squares). (Reprinted with permission from Lucas, J. H., *Brain Res.*, 425, 1987.)

when lesion parameters are constant, the simple physical resistance of the proximal segment to the influx of external calcium (and other ions) appears to determine the fate of neurons in the acute phase after dendrotomy.

C. THE ROLE OF CALCIUM IN NEURONAL DEATH AFTER DENDROTOMY

Replacement of normal (1.8 m*M*) calcium with low- (<30 μ*M*) calcium medium significantly delayed but did not prevent cell death after dendrite amputation 100 μm from the perikaryon (N = 14 experiments or 140 lesioned neurons per group) (Lucas et al., 1990a). Two-hour survival in low calcium was 63 ± 16% (SD) compared to 35 ± 8% in normal calcium ($p < 0.003$, independent Student's *t*-test). However, 6-h survival was 30 ± 7% and 23 ± 8% in the low- and normal calcium groups.

D. THE ROLES OF CALCIUM, SODIUM, AND CHLORIDE IN ULTRASTRUCTURAL DAMAGE AFTER DENDROTOMY

A drastic reduction of extracellular calcium delayed but ultimately did not prevent neuronal death after dendrite amputation (see above). The delayed death may have been the result of slow penetration by residual calcium in the medium which was still at a higher concentration than is normally found free in the cytoplasm (100 to 200 n*M*). Also, EM examinations of lesioned neurons which were dead 2 h after dendrotomy in this medium indicated that intracellular calcium eventually rose to micromolar levels (Lucas et al., 1990a). However, EM studies have also provided evidence that other ions may have played a role in neuronal deterioration in the low-calcium environment.

1. Dynamics of Ultrastructural Damage in a Normal Ionic Environment

In medium containing normal levels of all ions, neurons fixed within 30 min after dendrotomy displayed a gradient of damage that spread from the lesion toward the perikaryon (Figures 6 to 9) and dilation of the perinuclear Golgi (Figures 6 and 10) (Lucas et al., 1985; Emery et al., 1987; Lucas et al., 1990a). This damage gradient constituted a spatial representation of the temporal sequence of organelle and cytoskeletal alterations with early changes at the leading edge and more advanced deterioration near the lesion. Characteristic changes at the gradient front included the formation of vesicles (presumably derived from the smooth endoplasmic reticulum, SER) and the development of electron-lucent foci in mitochondria. Behind the front, a transition region could often be discerned where the mitochondria had lost their longitudinal orientation and had begun to dilate (Figure 8). In some cells the loss of mitochondrial orientation clearly coincided with an abrupt loss of cytoskeletal elements (Emery et al., 1987; Lucas et al., 1990a). Near the lesion the cytoplasm had a homogenous appearance with much open space and mitochondria were either grossly dilated or collapsed and electron dense (Figure 7).

2. Dynamics of Ultrastructural Damage in a Low-Calcium Environment

A very different pattern of ultrastructural change was seen when neurons were fixed within 30 min after dendrotomy in low- (<30 μ*M*) calcium medium (Figures 11 to 14; Lucas et al., 1990a). Except very near the lesion, mitochondria along the proximal segment were longitudinally oriented and had normal morphologies, and cytoskeletal elements were abundant (Figure 11). In the low-calcium milieu, mitochondria did appear to become more electron dense (Figures 11 and 13), and some mitochondria near the lesions had dilated intracristal spaces (Figure 13). The most striking feature of cells lesioned in low calcium was the formation of an extensive vesicular gradient within the proximal segment (Figures 11 and 12). At the trailing edge of the gradients (Figure 12, top) the vesicles generally appeared smaller as if the SER were, in fact, breaking apart. Golgi dilation still occurred within the perikarya of neurons lesioned in low-calcium medium (Figure 14).

3. Dynamics of Ultrastructural Damage in a Low-Calcium, Low-Sodium Environment

Neurons fixed within 30 min after dendrotomy in medium with reduced sodium and reduced calcium (choline chloride substituted for sodium chloride) displayed few of the ultrastructural changes seen in medium where only calcium was reduced (compare Figures 11 through 14 to Figures 15 through 17; Emery et al., 1991). No dramatic vesicular gradient developed in these cells (Figure 15) and there was no dilation of the perinuclear Golgi (Figures 15 and 16). In addition, there were numerous cytoskeletal elements and mitochondria

FIGURES 6 to 22. Ultrastructure of neurons

FIGURES 6 to 10. Ultastructure of a spinal neuron fixed 13 min after dendrite transection in medium with normal concentrations of calcium, sodium, and chloride (lesion distance = 100 μm). FIGURE 6. Section showing entire cell with gradient of damage in the proximal segment of the transected dendrite. The perikaryon contained clusters of dilated vesicles derived from the Golgi aparatus (see Figure 10 also). The top, middle, and bottom bracketed portions of the dendrite correspond to Figures 7, 8, and 9, respectively. Note that there was a region of virtually undisturbed (see Figure 9 also) cytoplasm between the dendritic gradient and the soma. Scale bar = 5 μm. FIGURE 7. Tip of the proximal segment (upper bracket of Figure 6). The cytoplasm in this region contained clear vesicles and numerous swollen mitochondria (M) but cytoskeletal elements were absent. Scale bar = 1 μm. FIGURE 8. More proximal portion of the dendrite stump (middle bracket of Figure 6). The SER has dilated and fragmented to form large clear vesicles (V). The more distal mitochondria have swollen (MS) but those closer to the soma (M) have not. Such a transition region along the dendritic damage gradient has frequently been observed in neurons fixed within 30 min of transection. Scale bar = 1 μm. FIGURE 9. Region of the dendrite close to the perikaryon (lower bracket of Figure 6). The cytoplasm in this area showed few changes as a result of the injury. Scale bar = 1 μm. FIGURE 10. Section of somal cytoplasm showing clusters of Golgi-derived vesicles (G) close to the nucleus (N). Scale bar = 1 μm.

FIGURES 11 to 14. Ultrastructure of a neuron fixed 11 min after dendrite amputation in medium with low (<30 μ*M*) calcium. Lesion distance = 100 μm. FIGURE 11. Section showing rows of dilated vesicles (V) within the dendritic stump. The vesicles were presumably derived from the swollen, fragmented SER. The presence of numerous, intact cytoskeletal elements (arrowheads) was probably responsible for the serried arrangement of the vesicles. Mitochondria (M) appeared more electron dense but had a normal, elongated morphology and were oriented parallel to the dendritic axis. Scale bar = 1 μm. FIGURE 12. Section showing almost all of the proximal segment of the severed dendrite (D). Compare the extensive, regularly arranged gradient of vesicles in this cell to the dendritic gradient in Figure 6. Scale bar = 5 μm. FIGURE 13. Section of proximal segment very close to the tip. Mitochondria (M) were very electron dense and several had dilated intracristal spaces (arrowheads) suggestive of the condensed configuration. Scale bar = 1 μm. FIGURE 14. Section of somal cytoplasm containing clusters of Golgi-derived vesicles (G) near the nucleus (N). The mitochondria (M) were not swollen but did appear darker than those in uninjured neurons. Scale bar = 1 μm.

FIGURES 15 to 17. Ultrastructure of a neuron fixed 12 min after dendrotomy in medium with reduced calcium and reduced sodium (choline chloride substituted for sodium chloride). Lesion distance = 100 μm. FIGURE 15. Section of the neuron from the perikaryon almost to the lesion. The cytoplasm showed few signs of reaction to injury other than darkening of the mitochondria. There are no dilated vesicles in either the dendrite or the perikaryon. Scale bar = 5 μm. FIGURE 16. Section of cytoplasm from the perikaryon showing that neither the mitochondria (M), nor the Golgi (G), nor the granular endoplasmic reticulum (ER) were swollen, although the mitochondria were dark. Some microtubules were visible (arrowheads) near the nucleus (N). Scale bar = 1 μm. FIGURE 17. Section of cytoplasm from the proximal segment showing intact microtubules (arrowheads) and mitochondria with dark matrices. Scale bar = 1 μm.

FIGURES 18 to 20. Ultrastructure of neuron fixed 10 min after dendrotomy in medium with reduced calcium and reduced chloride (sodium isethionate substituted for sodium chloride). Lesion distance = 100 μm. FIGURE 18. Section of perikaryal cytoplasm. At this magnification no pathological effects of injury were apparent. Mitochondria (M) were not dilated and there were only a few vesicles (arrowheads). Scale bar = 1 μm. FIGURE 19. A section of the proximal segment showing a nearly normal ultrastructure. There were no vesicles, cytoskeletal elements were abundant, and mitochondria (arrowheads) were elongated and longitudinally oriented. Scale bar = 1 μm. FIGURE 20. Section of perikaryal cytoplasm showing dilated Golgi cisternae. Scale bar = 0.1 μm.

FIGURES 21 to 22. Ultrastructure of an unlesioned spinal neuron fixed after exposure to the sodium ionophore monensin for a period of 105 min. FIGURE 21. Section of somal cytoplasm showing clusters of small vesicles (V). The clustering and perinuclear location indicated that they were derived from the Golgi. Scale bar = 1 μm. FIGURE 22. A higher magnification of one vesicle cluster in the perikaryon. Monensin treatment most closely mimicked the changes seen after dendrotomy in medium with reduced calcium and reduced chloride causing dilation of the Golgi cisternae but no swelling of the SER or mitochondria. The mitochondria showed some darkening. Scale bar = 1 μm.

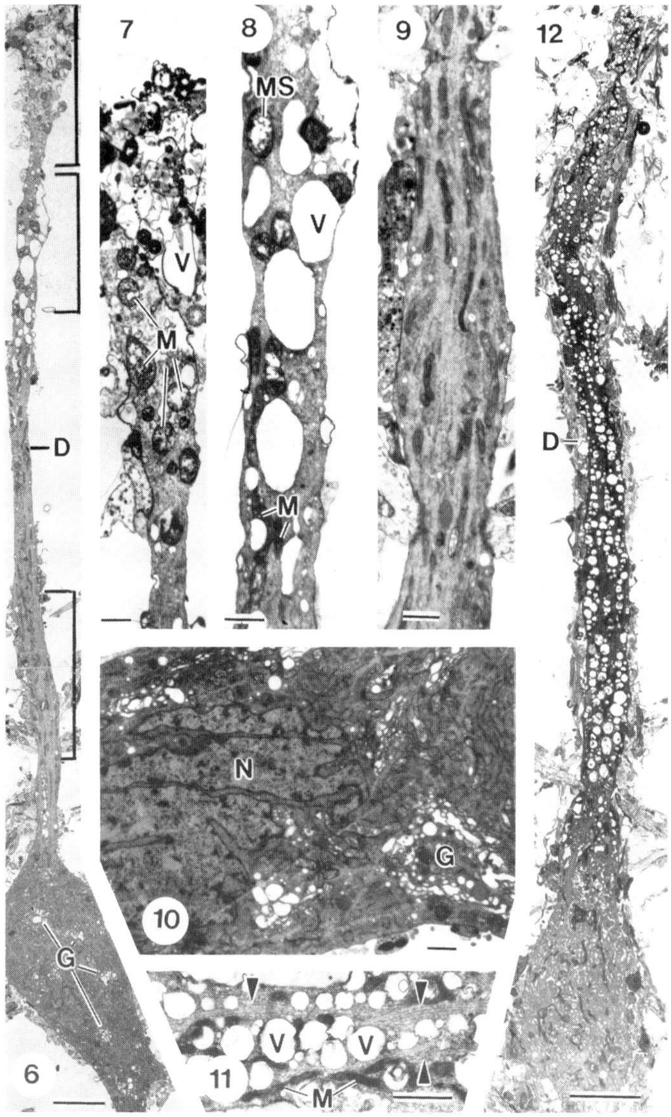

FIGURES 6 to 12.

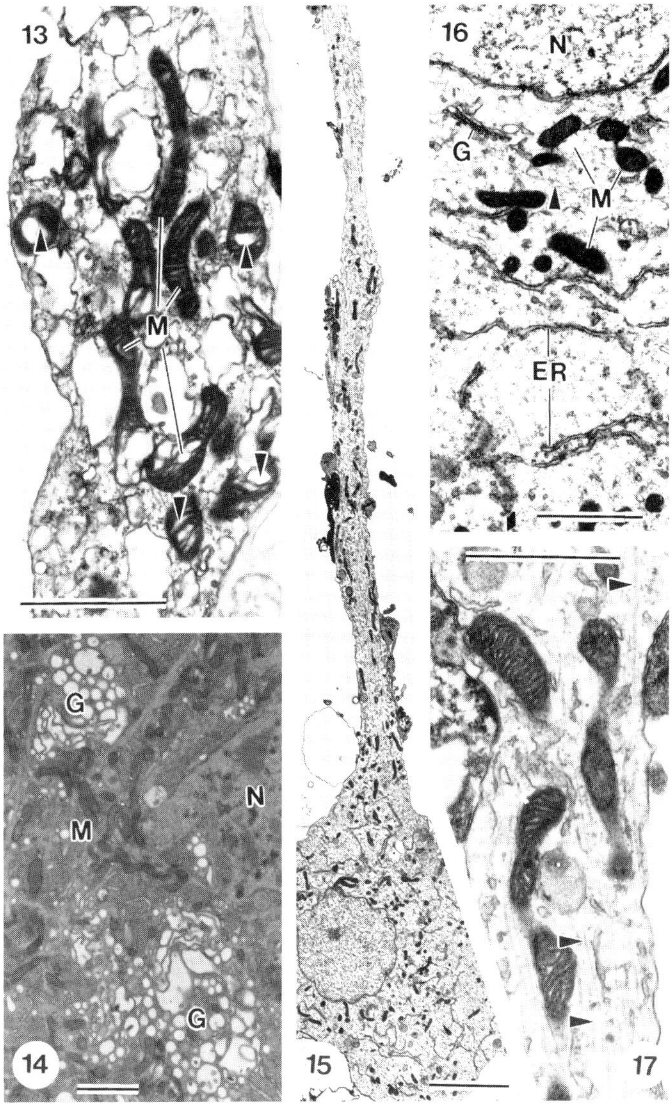

FIGURES 13 to 17.

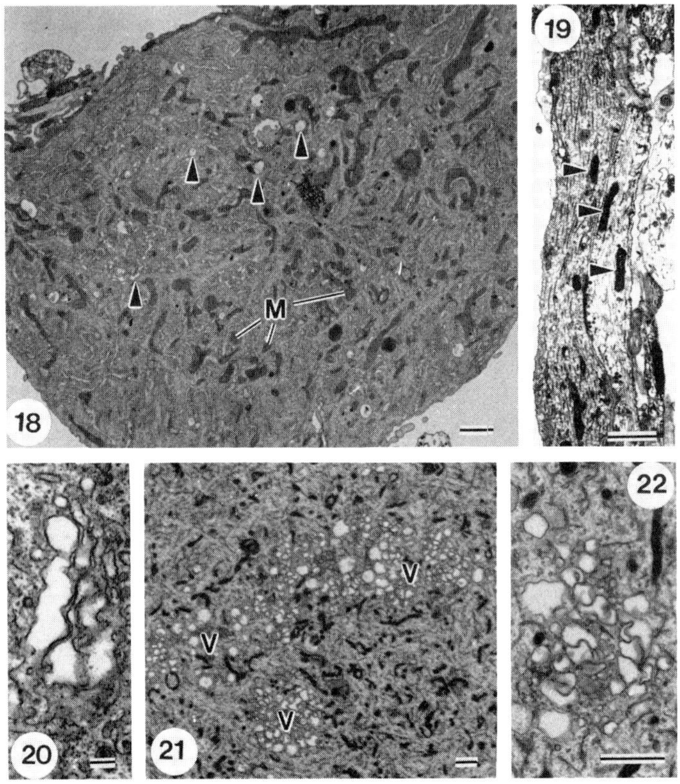

FIGURES 18 to 22.

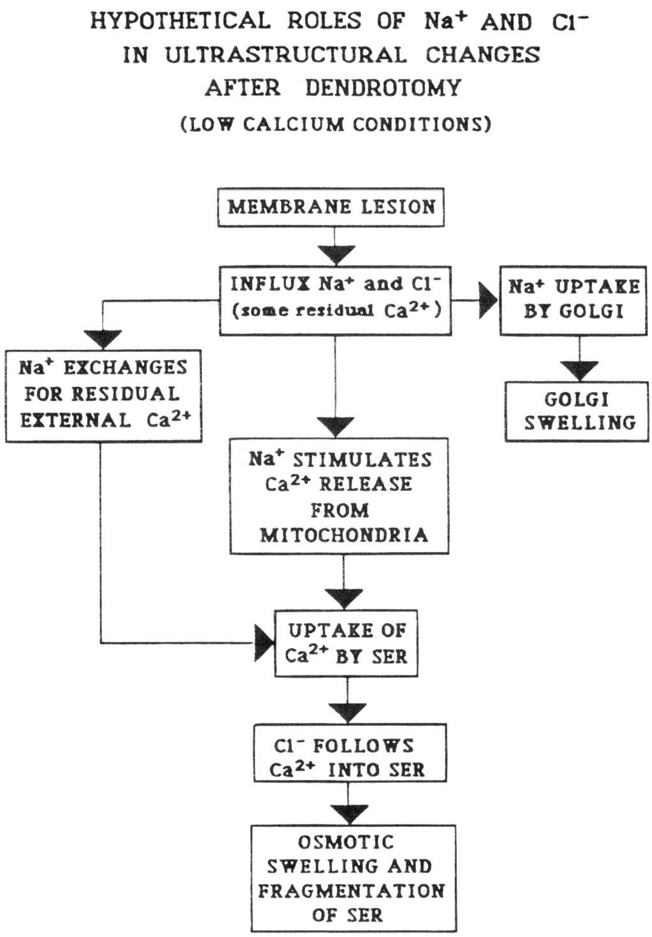

FIGURE 23. Sequence of ionic movements that are hypothesized to mediate the ultrastructural changes observed during the first 30 min after dendrotomy in medium with low (<30 μ*M*) calcium. (Adapted from Emery, D. G. et al., *Exp. Brain Res.*, 86, 20, 1991. With permission.)

appeared normal and were longitudinally oriented in the proximal segments (Figure 17). Mitochondrial matrices of some of the lesioned neurons appeared more electron dense (Figures 15 to 17).

4. Dynamics of Ultrastructural Damage in a Low-Calcium, Low-Chloride Environment

Neurons fixed within 30 min after dendrotomy in medium with reduced chloride and reduced calcium (sodium isethionate substituted for sodium chloride) showed a preservation of ultrastructure, which was very similar to that observed in neurons operated in medium where calcium and sodium were reduced (compare Figures 18 through 20 to Figures 15 through 17; Emery et al. 1991). The cytoskeleton was generally intact, mitochondria appeared normal and were oriented normally along the proximal segment, and there was no extensive gradient of SER vesicles (Figure 19). However, unlike neurons lesioned in the low-calcium, low-sodium milieu, neurons lesioned in the low-calcium, low-chloride medium (normal sodium) did display dilation of the Golgi (Figure 20). Similar Golgi dilation was also observed in uninjured spinal neurons exposed to the sodium ionophore monensin (2 μg/ml, Figures 21 and 22).

5. Dynamics of Ultrastructural Damage in a Low-Calcium, Low-Sodium, Low-Chloride Environment

In one set of experiments sucrose was used as a nonionic substitute for all three of the major components of the injury current. Neurons fixed within 30 min after dendrotomy in sucrose-substituted medium showed ultrastructural preservation comparable to that seen when calcium and sodium were reduced (data not shown).

6. Summary of Ultrastructural Damage Studies in Modified Ionic Environments

Table 1 summarizes the very early ultrastructural changes observed in the various media. The mechanisms that may mediate SER, Golgi, and mitochondrial changes after dendrotomy in low calcium are discussed later. Figure 23 presents the hypothesized sequence of events that underlie the SER swelling and Golgi dilation.

a. Swelling of the SER

Both chloride and sodium seem to be required to induce vesiculation of the SER after dendrotomy. However, although the SER is exquisitely sensitive to cytoplasmic calcium fluctuations, it does not, to our knowledge, transport sodium. Therefore, we have hypothesized that the sodium injury current affects the SER indirectly by causing an elevation of cytoplasmic calcium, and that this calcium is in turn moved into the SER. Mechanisms by which the sodium injury current could cause an increase in intracellular free calcium include reverse operation of the Na^+-Ca^{2+} antiporter of the plasma membrane (Naschen et al., 1986) and release of mitochondrial calcium stores via the Na^+-Ca^{2+} exchanger of the inner membrane (Nedergaard, 1984). By itself, SER sequestration of calcium probably does not cause dilation because of binding by proteins. We hypothesize, however, that when chloride is elevated, it will follow calcium into the lumen of the SER and cause this organelle to swell. Interestingly, Fishman and Tewari (1990) have shown that while elevating cytosolic calcium in squid axons induces vesiculation of the cytoplasm, sodium, and chloride enhance this response.

b. Golgi Dilation

In lesioned neurons this effect was observed only when levels of extracellular sodium were normal. It has also been observed in unlesioned neurons (Figures 21 and 22) and other

TABLE 1

		Dendritic Damage Gradient				Soma
				Mitochondria		
Medium	N (cells)	SER vesicles	Loss of MT	Swell/Condense		Swollen golgi
Normal ions	18	++	++	+++	–	++
Low Ca^{2+} $(<30\mu M)$	16	++++	–	–	++	++
Low Ca^{2+}, low Na^+	16	–	–	–	+	–
Low Ca^{2+}, low C^-	11	–	–	–	+	++
Sucrose for Na^+, Ca^{2+}, and Cl^-	5	–	–	–	–	–

Explanation of ion substitutions:

1. Low-calcium, low-sodium medium: choline chloride substituted for NaCl.
2. Low-calcium, low-chloride medium: sodium isethionate or propionate for NaCl.
3. Sucrose-substituted medium: NaCl and $CaCl_2$ replaced with sucrose.

(Adapted from Emery, D. G. et al., *Exp. Brain Res.*, 86, 69, 1991. With permission.)

types of cells exposed to monensin (Tartakoff and Vasalli, 1978). Therefore, it is our belief that it is probably an osmotic effect following the movement of sodium into the Golgi lumen. The early appearance of the Golgi change after dendrotomy suggests that the sodium component of the injury current spreads into the injured cells ahead of calcium. This would be consistent with the higher coefficient of diffusion for sodium and the relative paucity of binding, pumping, and sequestering mechanisms for this ion compared to calcium.

c. Mitochondrial Condensation

Mitochondrial condensation can be observed in many conditions where cellular energy requirements are increased and is indicative of increased respiration (Hackenbrock, 1966, 1968). The increased matrix density observed after dendrotomy in low calcium suggests a transition to the condensed configuration. We have speculated that this phenomenon was caused by elevated sodium as it was observed in neurons exposed to the ionophore monensin and in neurons lesioned in low-calcium medium with a normal sodium concentration. This change has also been observed in other cell types treated with monensin (Tartakoff and Vasalli, 1978). However, some cells lesioned in low-sodium medium also seem to have more dense mitochondria. Additional observations will be performed to determine the relationship of this more subtle change to the sodium and chloride injury currents.

E. NEURONAL SURVIVAL AFTER DENDROTOMY IN A LOW-CALCIUM, LOW-SODIUM, LOW-CHLORIDE ENVIRONMENT

Neurons were lesioned in low-calcium medium where sucrose +/−50 m*M* potassium chloride were substituted for sodium chloride. Two hours after cell surgery, viability of the lesioned neurons was evaluated by light microscopic examination and by application of erythrosin B. Using erythrosin B survival was found to be 62 ± 10% in the control (normal ions) group, 96 ± 5% in the low-calcium group, 89 ± 4% in the group where sucrose replaced sodium chloride, and 96 ± 5% in the group where sucrose and 50 m*M* potassium chloride replaced sodium chloride (four groups of five experiments, 50 lesioned neurons per group). The survival in the normal ion and low-calcium groups was higher than previously reported (see above). This was probably because replacement of sodium bicarbonate by Hepes buffer significantly reduced the sodium content of these media. When those neurons identified by light microscopy as moribund (dying but not yet dead, Lucas et al., 1985) were subtracted, the viable population in the control group, low-calcium group, and the low-calcium, low-sodium chloride groups with and without 50 m*M* potassium chloride was 50 ± 10, 80 ± 16, 86 ± 5, and 92 ± 8%, respectively. Although preliminary, this initial study indicates that other ions, especially sodium, also play a role in neuronal death after physical injury. Experiments are in progress to determine whether exposure to the modified ionic environments merely delay death or, in fact, improve long-term (24 h) survival.

F. THE EFFECT OF RECEPTOR/ION CHANNEL BLOCKADE ON NEURONAL SURVIVAL AFTER DENDROTOMY

Depolarization of lesioned neurons may open voltage-sensitive calcium channels (VSCC) or may remove the magnesium block from the *N*-methyl-D-aspartate (NMDA) receptor-governed ionophores. To determine whether channel-mediated currents contributed to neuronal death after dendrotomy, we lesioned neurons in the presence of antagonists to the VSCC or the NMDA complexes. In the study of VSCC blockers (Shi et al., 1989), survival was 49 ± 16% compared to 59 ± 7% with and without verapamil (80 μ*M*), respectively, 56 ± 11% compared to 50 ± 13% with and without nifedipine (10 μ*M*), respectively, and 42 ± 4% compared to 56 ± 6% with and without chlorpromazine (a VSCC-calmodulin antagonist, 10 μ*M*), respectively. Survival in the presence of the competitive NMDA blocker D-2-amino-5-phosphonovalerate (D-APV, 100 μ*M*) was 48 ± 5% compared to 43 ± 11%

in the control (no D-APV) group (Wang et al., 1990). VSCC blocker and D-APV dosages were the maximum nontoxic concentrations of these agents for uninjured spinal cord neurons. Each percentage represents the mean survival from five to ten experiments (50 to 100 lesioned neurons). These findings suggest that when the plasma membrane is mechanically disrupted, most of the calcium will enter through the physical breach and not via specific channels. A preliminary study of tetrodotoxin (2 μM, four pairs of experiments) also found no protection when sodium channels were blocked.

IV. METABOLIC INHIBITION INCREASES SURVIVAL AFTER DENDROTOMY

Much of the ion-linked deterioration observed in lesioned neurons may have been caused by calcium activation of degradative enzymes: phospholipases (Van den Bosch, 1980), proteases (Melloni and Pontremoli, 1989), and endonucleases (Nicotera et al., 1989). Reports that hypothermia and barbiturates are effective treatments for CNS trauma have suggested that inhibition of general cellular metabolism may be protective (see Hansebout, 1982 for review). We systematically evaluated whether cooling or application of the barbiturate thiopental could increase neuronal survival after dendrotomy.

A. HYPOTHERMIA

Survival of neurons 2 h after dendrotomy was increased when the cultures were cooled immediately after cell surgery (Lucas et al., 1990b). Survival was 70 ± 9% at 27°C compared to 60 ± 9% in the normothermic (37°C) control group, 86 ± 13% at 17°C compared to 68 ± 8% in the control group ($p < 0.05$), and 100% at 7°C compared to 56 ± 6% in the control group ($p < 0.001$). Lesioned neurons cooled to 17°C for 2 h and then rewarmed for 22 h at 37°C still had significantly higher survival than lesioned neurons in the control (24 h at 37°C) group (66 ± 9% vs. 46 ± 9%, $p < 0.01$) (Lucas et al., 1990b). However, both lesioned and unlesioned neurons exposed to temperatures below 17°C developed pronounced dendrosomatic swelling, and most of the swollen neurons died when normothermic conditions were restored (Lucas et al., 1990b,c). The low-temperature swelling and death after rewarming were prevented by competitive or noncompetitive NMDA antagonists (ibid.).

B. THIOPENTAL

Preliminary experiments have shown that neuronal survival was also significantly increased by addition of the maximum nontoxic concentration (200 μM) of the barbiturate thiopental (TP) to cultures immediately after cell surgery (Wang et al., 1991). Survival in the TP-treated group was 76 ± 5% compared to 50 ± 10% in the control (no TP) group. These data support the hypothesis that hypothermic protection of injured neurons is the result of metabolic inhibition.

V. CONCLUDING REMARKS

Tissue necrosis after CNS trauma is thought to involve a combination of cellular, local vascular and systemic processes. The relative importance of the various individual injury factors to the total damage is still unclear. However, there is evidence to suggest that physical disruption of cell membranes by the mechanical insult may make a persistent and, ultimately, significant contribution. Axonal fragmentation has been observed as early as 30 min after impact in experimental spinal cord injury (Balentine 1978a,b). Electrophysiological evidence suggests that membranes do not "heal" completely after traumatic disruption and that ionic currents may continue to penetrate surviving, injured cells through these breaches for hours or even days (Borgens et al., 1980; Meiri et al., 1981; Lucas et al., 1985). Measurements of extracellular free (Young et al., 1982; Stokes et al., 1983) and total tissue calcium and sodium (Happel et al., 1981; Young and Koreh, 1986; Kwo et al., 1989) during the first few hours indicate a continuous accumulation of these minerals in the vicinity of the zone of trauma with progressive tissue calcification (Balentine, 1978b). Other mechanisms that may contribute to this process include the leakage of ions into injured surviving cells because of free radical-mediated changes in membrane integrity (Chan et al., 1982; Chan et al., 1984) and depressed ion pump function (Clenendon et al., 1978), and precipitation of calcium by phosphate liberated from disrupted membranes (Young, 1987).

While there is little doubt that a massive increase in the levels of intracellular free calcium is sufficient to kill cells, the question of whether sustained calcium elevation is, indeed, the "final common pathway" in most forms of lethal neuronal (or other cellular) injury remains unanswered. The results of efforts to correlate changes in free calcium with the subsequent death of stressed cells have been contradictory (Lemasters et al, 1987; Ogura et al., 1988; Mattson et al., 1989; Michaels and Rothman, 1990). The investigations reviewed here and studies by others (Goldberg et al., 1986; Gallant, 1988; Fishman and Tewari, 1990) have indicated that other ions can contribute substantially to neuronal injury, and that modifications of the ionic environment to protect damaged cells may require more than a simple reduction of calcium.

The mechanism of protection of lesioned neurons by inhibitors of cellular metabolism is unknown. One possibility is that the rates of degradative and repair processes are slowed unequally—"buying time", in effect, for recovery. An intriguing alternative suggested by studies of axotomy and neuronal death after NGF withdrawal is that inhibition of anabolic processes (transcription, translation, etc.) prevents initiation of a cell death program (Martin et al., 1988; Johnson et al., 1989). Future investigations will explore both of these possibilities.

ACKNOWLEDGMENTS

The authors wish to acknowledge the consistently high level of culture support provided by Ms. Janet Douglas and Ms. Lynne Gordon. The authors also wish to thank Ms. Lisa Schaffer for the measurements of cellular physical parameters. This work was made possible by grants from the National Institutes of Health, the Hillcrest Foundation of Dallas, TX founded by Mrs. W. W. Caruth, and the International Foundation for Ethical Research (for the initial studies of lesioned neuron survival in a low ion environment).

REFERENCES

Allen, A. R. (1911) Surgery of experimental lesions of spinal cord equivalent to crush injury of fracture dislocation of spinal column: Preliminary report. *JAMA* 57, 878–880.

Allen, A. R. (1914) Remarks on the histopathological changes in the spinal cord due to impact: An experimental study. *J. Nerv. Ment. Dis.* 41, 141–147.

Balentine, J. D. (1978) Pathology of experimental spinal cord trauma. I. The necrotic lesion as a function of vascular injury. *Lab. Invest.* 39, 236–253.

Balentine, J. D. (1978) Pathology of experimental spinal cord trauma. II. Ultrastructure of axons and myelin. *Lab. Invest.* 39, 254–266.

Balentine, J. D. (1988) Spinal cord trauma: in search of the meaning of granular axoplasm and vesicular myelin. *J. Neuropathol. Exp. Neurol.* 47, 77–92.

Balentine, J. D., Greene, W. B., and Bornstein, M. (1988) *In vitro* spinal cord trauma. *Lab. Invest.* 58, 93–99.

Bird, M. M. (1978) Microsurgical transection of small nerve fiber bundles *in vitro*. *Cell Tissue Res.* 190, 539–553.

Black, P., Markowitz, R. S., Cooper, V., Mechanic, A., Kushner, H., Damjanov, I., Finkelstein, S. D., and Wachs, K. C. (1986) Models of spinal cord injury: Part 1. Static load technique. *Neurosurgery* 19, 752–762.

Black, P., Markowitz, R. S., Damjanov, I., Finkelstein, S. D., Kushner, H., Gillespie, J., and Feldman, M. (1988) Models of spinal cord injury: Part 3. Dynamic load technique. *Neurosurgery* 22, 51–60.

Borgens, R. B., Joffe, L. F., and Cohen M. J. (1980). Large and persistent electrical currents enter the transected lamprey spinal cord. *Proc. Natl. Acad. Sci. U.S.A.* 77, 1209–1213.

Bresnahan, J. C., Beattie, M. S., Todd, F. D., and Noyes, D. H. (1987) A behavioral and anatomical analysis of spinal cord injury produced by a feedback-controlled impaction device. *Exp. Neurol.* 95, 548–570.

Chambers, R. and Fell, H. B. (1931) Micro-operation on cells in tissue cultures. *Proc. R. Soc. Lond.* 109, 380–403.

Chan, P. H., Fishman, R. A., Schmidley, J. W., and Chen, S. F. (1984) Release of polyunsaturated fatty acids from phospholipids and alteration of brain membrane integrity by oxygen-derived free radicals. *J. Neurosci. Res.* 12, 595–605.

Chan, P. H., Yurko, M., and Fishman, R. A. (1982) Phospholipid degradation and cellular edema induced by free radicals in brain cortical slices. *J. Neurochem.* 38, 525–531.

Clenendon, N. R., Allen, N., Gordon, W. A., and Bingham, W. G. (1978) Inhibition of Na^+-K^+-activated ATPase activity following experimental spinal cord trauma. *J. Neurosurg.* 49, 563–568.

Easton, D. M. (1960) Nerve-end recording in a conducting volume. *Science* 132, 1312–1313.

Easton, D. M. (1965) Impulses at the artifactual nerve end. *Cold Spring Harbor Symp. Quant. Biol.* 30, 15–28.

Emery, D. G., Lucas, J. H., and Gross, G. W. (1987). The sequence of ultrastructural changes in cultured neurons after dendrite transection. *Exp. Brain Res.* 67, 41–51.

Emery, D. G., Lucas, J. H., and Gross, G. W. (1991) Contributions of sodium and chloride to ultrastructural damage after dendrotomy. *Exp. Brain Res.* In press.

Fishman, H. W. and Tewari, K. P. (1990) Ca^{2+}-induced vesiculation in dialyzed squid giant axon. *Soc. Neurosci. Abstr.* 16, 820.

Gallant, P. E. (1988) Effects of the external ions and metabolic poisoning on the constriction of the squid giant axon after axotomy. *J. Neurosci.* 8, 1479–1484.

Gennarelli, T. A. and Thibault, L. E. (1985) Biological models of head injury. In: *Central Nervous System Trauma Status Report* (Becker, D. P. and Povlishock, J. T., eds.) Bethesda, MD: National Institute of Neurological and Communicative Disorders and Stroke, National Institutes of Health, pp. 391–404.

Goldberg, W. J., Madingo., R. M., and Barnett, J. N. (1986) Effects of ischemia-like conditions on cultured neurons: protection by low Na^+ and low Ca^{+2} solutions. *J. Neurosci.* 6, 3144–3151.

Gross, G. W., Lucas, J. H., and Higgins, M. L. (1983). Laser microbeam surgery: ultrastructural changes associated with neurite transection in culture. *J. Neurosci.* 3, 1979–1993.

Gross, G. W. and Higgins, M. L. (1987). Cytoplasmic damage gradients in dendrites after transection lesions. *Exp. Brain Res.* 67, 52–60.

Hackenbrock, C. R. (1966). Ultrastructural basis for metabolically linked mechanical activity in mitochondria. I. Reversible ultrastructural changes with changes in metabolic steady state in isolated liver mitochondria. *J. Cell Biol.* 30, 269–297.

Hackenbrock, C. R. (1968). Chemical and physical fixation of isolated mitochondria in low-energy and high-energy states. *Proc. Natl. Acad. Sci. U.S.A.* 61, 598–605.

Hamburger, A. and Rinder, L. (1966) Experimental brain concussion: the early effect of sudden increase in intracranial pressure on the succinoxidase activity of isolated neurons and glial cells from the lateral vestibular nucleus of the rabbit. *J. Neuropathol. Exp. Neurol.* 25, 68–75.

Hansebout, R. R. (1982) A comprehensive review of methods of improving cord recovery after acute spinal cord injury. In: *Early Management of Acute Spinal Cord Injury*, (C. H. Tator, ed.) Raven Press: New York, pp 181–195.

Happel, R. D., Smith, K. P., Banik, N. L., Powers, J. M., Hogan, E. L., and Balentine, J. D. (1981) Ca^{+2}-accumulation in experimental spinal cord trauma. *Brain Res.* 211, 476–479.

Johnson, E. M., Chang, J. Y., Koike, T., and Martin, D. P. (1989) Why do neurons die when deprived of trophic factor? *Neurobiol. Aging* 10, 549–552.

Khan, M. and Griebel, R. (1983) Acute spinal cord injury in the rat: comparison of the three experimental techniques. *Can. J. Neurol. Sci.* 10, 161–165.

Kirkpatrick, J. B., Higgins, M. L., Lucas, J. H., and Gross, G. W. (1985). *In vitro* simulation of neural trauma by laser. *J. Neuropath. Exp. Neurol.* 44, 268–284.

Kwo, S., Young, W., and Decrescito, V. (1989) Spinal cord sodium, potassium, calcium, and water concentration changes in rats after graded contusion injury. *J. Neurotrauma* 6, 13–24.

Lemasters, J. J., DiGuiseppi, J., Nieminen, A., and Herman, B. (1987) Blebbing, free Ca^{+2} and mitochondrial membrane potential preceding cell death in hepatocytes. *Nature* 325, 78–81.

Levi, G. and Meyer, H. (1945) Reactive, regressive and regenerative processes of neurons, cultivated *in vitro* and injured with the micromanipulator. *J. Exp. Zool.* 99, 141–181.

Lucas, J. H., Gross, G. W., Emery, D. G., and Gardner, C. R. (1985). Neuronal survival or death after dendrite transection close to the perikaryon: correlation with electrophysiologic, morphologic, and ultrastructural changes. *J. Neurotrauma* 2, 231–255.

Lucas, J. H. (1987). Proximal segment retraction increases the probability of nerve cell survival after dendrite transection. *Brain Res.* 425, 384–387.

Lucas, J. H., Emery, D. G., Higgins, M. L., and Gross, G. W. (1990a). Neuronal survival and dynamics of damage after dendrotomy in low calcium. *J. Neurotrauma* 7, 169–192.

Lucas, J. H., Wang, G. F., and Gross, G. W. (1990b) Paradoxical effect of hypothermia on survival of lesioned and uninjured mammalian spinal neurons. *Brain Res.* 517, 354–357.

Lucas, J. H., Wang, G. F., and Gross, G. W. (1990c) NMDA antagonists prevent hypothermic injury and death of mammalian spinal neurons. *J. Neurotrauma* 7, 229–236.

Martin, D. P., Schmidt, R. E., DiStefano, P. S., Lowry, O. H., Carter, J. G., and Johnson, E. M. (1988) Inhibitors of protein synthesis and RNA synthesis prevent neuronal death caused by nerve growth factor deprivation. *J. Cell Biol.* 106, 829–844.

Mattson, M. P., Guthrie, P. B., Hayes, B. C., and Kater, S. B. (1989) Roles for mitotic history in the generation and degeneration of hippocampal neuroarchitecture. *J. Neurosci.* 9, 1223–1232.

Melloni, E. and Pontremoli, S. (1989) The calpains. *Trends Neurosci.* 12, 438–444.

Meiri, H., Spira, M. E., and Parnas, I. (1981) Membrane conductance and action potentials of a regenerating axonal tip. *Science* 211, 709–712.

Michaels, R. L. and Rothman, S. M. (1990) Glutamate neurotoxicity *in vitro:* antagonist pharmacology and intracellular calcium concentrations. *J. Neurosci.* 10, 283–292.

Mire, J. J., Hendelman, W. J., and Bunge, R. P. (1970) Observation on a transient phase of focal swelling in degenerating unmyelinated nerve fibers. *J. Cell Biol.* 45, 9–22.

Naschen, D. A., Sanchez-Armass, S., and Weinstein, A. M. (1986). The regulation of cytosolic calcium in rat brain synaptosomes by sodium-dependent calcium efflux. *J. Physiol.* 381, 17–28.

Nedergaard, J. (1984). Na^+-dependent regulation of extramitochondrial Ca^{2+} by rat-liver mitochondria. *Eur. J. Biochem.* 144, 159–168.

Nicotera, P., McConkey, D. J., Dypbukt, J. M., Jones, D. P., and Orrenius, S. (1989) Ca^{2+}-activated mechanisms in cell killing. *Drug Metab. Rev.* 20, 193–201.

Ogura, A., Miyamoto, M., and Kudo, Y. (1988) Neuronal death *in vitro:* parallelism between survivability of hippocampal neurons and sustained elevation of cytosolic Ca^{2+} after exposure to glutamate receptor agonist. *Exp. Brain Res.*, 73, 447–458.

Schlaepfer, W. W. and Bunge, R. P. (1973) Effects of calcium ion concentration on the degeneration of amputated axons in tissue culture. *J. Cell Biol.* 59, 456–470.

Shaw, G. and Bray, D. (1977) Movement and extension of isolated growth cones. *Exp. Cell Res.* 104, 55–62.

Shi, R-Y., Lucas, J. H., Wolf, A., and Gross, G. W. (1989) Calcium antagonists fail to protect mammalian spinal neurons after physical injury. *J. Neurotrauma* 6, 261–276.

Sole, G. M. (1980) The effects of microsurgical transection of neurite bundles *in vitro* before and after exposure to colchicine. *J. Anat.* 130, 777–788.

Stokes, B. T., Fox, P., and Hollinden, G. (1983) Extracellular calcium activity in the injured spinal cord. *Exp. Neurol.* 80, 561–572.

Tarlov, I. M. and Klinger, H. (1954) Spinal cord compression studies. II. Time limits for recovery after acute compression in dogs. *Arch. Neurol.* 71, 271–290.

Tartakoff, A. and Vassalli, P. (1978) Comparative studies of intracellular transport of secretory proteins. *J. Cell Biol.* 79, 694–707.

Tecoma, E. S., Monyer, H., Goldberg, M. P., and Choi, D. W. (1989) Traumatic neuronal injury *in vitro* is attentuated by NMDA antagonists. *Neuron* 2, 1541–1545.

Van den Bosch, H. (1980) Intracellular phospholipase A. *Biochim. Biophys. Acta* 604, 191–246.

Wang, G. F., Lucas, J. H., and Gross, G. W. (1990) Combined N-methyl-D-aspartate blockade and hypothermia as a treatment for neurite transection injury. *Soc. Neurosci. Abst.* 16, 1339.

Wang, G. F., Lucas, J. H., and Gross, G. W. (1991) Thiopental increases survival of spinal cord neurons after physical injury. *Soc. Neurosci. Abst.* 17, 161.

Yawo, H. and Kuno, M. (1983) How a nerve fiber repairs its cut end. *Science* 222, 1351–1353.

Yawo, H. and Kuno, M. (1985) Calcium dependence of membrane sealing at the cut end of the cockroach giant axon. *J. Neurosci.* 5, 1626–1632.

Young, W. (1987) The post-injury responses in trauma and ischemia: secondary injury or protective mechanisms? *J. Neurotrauma* 4, 27–52.

Young, W. and Koreh, I. (1986) Potassium and calcium changes in injured spinal cords. *Brain Res.* 253, 189–193.

Young, W., Yen, V., and Blight, A. (1982) Extracellular calcium ionic activity in experimental spinal cord contusion. *Brain Res.* 253, 105–113.

VI. Micro Techniques

Chapter 27

APPLICATION OF MICRODIALYSIS FOR CNS INJURY STUDY

Yoichi Katayama

TABLE OF CONTENTS

I. INTRODUCTION

There are three major advantages in using the microdialysis technique[1,2] for investigating neurochemical processes in the CNS. Firsty, this technique enables various neurochemical changes to be monitored simultaneously. Second, this technique provides a means of examining the mechanisms of the neurochemical changes by modifying various neurochemical environments *in situ* via the dialysis probe. Finally, and most importantly, the changes detected by this technique represent those occurring in the extracellular space (ECS), which permits neurochemical changes in the ECS to be distinguished from those in the intracellular space (ICS). These advantages can provide valuable data for elucidating the neurochemical processes occurring in the diseased CNS.

There are, however, several possibilities for misinterpretation of the data obtained from the diseased CNS. In order to avoid such misinterpretation, it is probably most important to note that the information provided by the microdialysis technique sensitively reflects the

0-8493-8094-4/95/$0.00+$.50

biochemical properties of the cellular membrane separating the ECS and ICS, which are not always the same. The concentration of a given substance within the ECS is dependent on the equilibrium between the ECS, ICS, and blood vessels. Accordingly, possible changes in function of the cellular membrane should always be taken into consideration before any conclusions are drawn from the data obtained by this technique.

Such properties of microdialysis can, on the other hand, be viewed as another advantage for detecting alterations in cellular membrane function in the diseased CNS. This chapter will discuss the data obtained from experimental models of ischemic and traumatic brain injury, focusing on the changes in cellular membrane function detected by the microdialysis technique.

During cerebral ischemia or anoxia, an abrupt increase in extracellular concentration of K^+ ($[K^+]_e$) to 50 to 60 m*M* is observed, following a slow increase.[3–7] A similar abrupt increase in $[K^+]_e$ can be induced by various forms of traumatic brain injury.[8–11] Such a rapid ionic flux may represent sudden changes in ion permeability of the cellular membrane. Since the ionic fluxes cause the osmotic pressure of intracellular impermeable anions to be no longer counterbalanced, water moves through the cellular membrane into the cells, and cellular swelling develops concomitantly with ionic shifts.[12–15] This chapter therefore also reviews how these changes in cellular membrane function influence the data obtained by the microdialysis technique. This, in turn, helps to delineate its unique value for investigating the mechanisms of the changes in cellular membrane function.

II. METHODOLOGICAL PROBLEMS

Methodological problems associated with the microdialysis technique have been the subject of numerous publications.[16–20] These difficulties are not repeated here, but a discussion is given of the methodological aspects that are important to note for maximizing the technique's unique capability of detecting changes in cellular membrane function. As mentioned earlier, these aspects must of course be borne in mind also in any experiments utilizing the microdialysis technique to study the diseased CNS.

A. DEPLETION OF SUBSTANCES FROM THE EXTRACELLULAR SPACE

The microdialysis technique involves insertion of a dialysis probe into the cerebral tissue, which causes a kind of focal traumatic brain injury. To exclude such initial effects of probe insertion, a certain period of time must be allowed to elapse before conducting the actual experiment. The duration of the period required for stabilization of the baseline data is highly variable depending on the substance of interest. Broadly speaking, the longer the duration for stabilization of the baseline data, the better the experimental data obtained appear to be. When the duration of dialysis is too long, however, there may be an *unphysiological stabilization*, which is sometimes critical for the data analysis. Extensive studies have not been made of how much of the period required for stabilization of the baseline data is due to the initial effects of probe insertion and how much is due to unphysiological stabilization.

The dialysis process is basically a removal of substances located within the ECS. With the microdialysis technique, the brain tissue is dialyzed for a certain period of time by the perfusate. This will inevitably lead to a slow but continuous removal of the substance of interest from the ECS, and could potentially produce an unphysiologically lowered level of many substances.

The best example of such a problem is provided by the outcome of Ca^{2+}-free dialysis (Figure 1).[20] Following initiation of the Ca^{2+}-free dialysis, the dialysate concentration of Ca^{2+} ($[Ca^{2+}]_d$) decreases rapidly from the higher values due to the previous Ca^{2+}-containing perfusate, and a further decrease in $[Ca^{2+}]_d$ continues. The baseline $[Ca^{2+}]_d$ is often decreased to less than half within 40 min following initiation of the Ca^{2+} free dialysis. Employing[45]

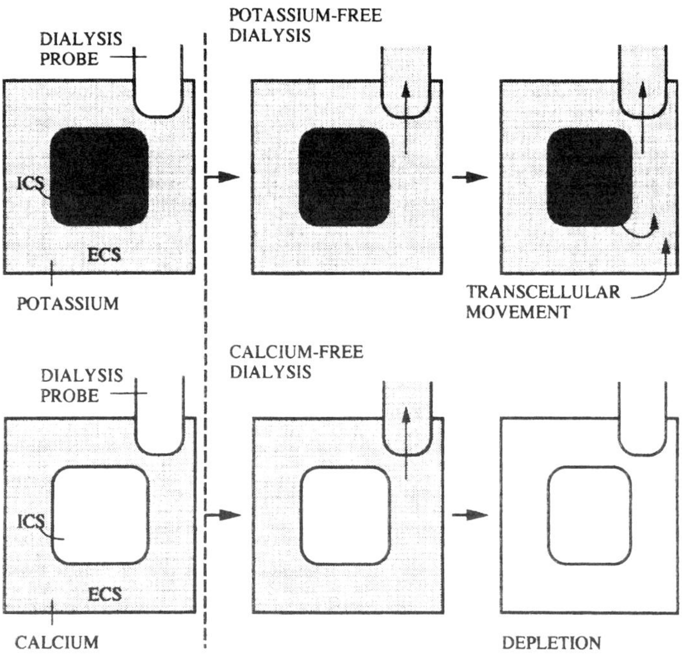

FIGURE 1. Effects of K^+-free (upper) and Ca^{2+}-free dialysis (lower). Following initiation of the K^+-free dialysis (dashed line), $[K^+]_d$ remains at approximately the same level for a considerable period of time, since K^+ is replenished continuously by the *transcellular movement* of K^+. In contrast, $[Ca^{2+}]_d$ decreases rapidly following initiation of the Ca^{2+}-free dialysis (dashed line), and Ca^{2+} can be depleted by Ca^{2+}-free dialysis. ECS, extracellular space; ICS, intracellular space. Darker tone indicates higher concentrations of K^+ (upper) or Ca^{2+} (lower).

Ca autoradiography, we have confirmed extensive Ca^{2+} depletion within a brain area of approximately 2 to 3 mm in diameter when Ca^{2+}-free dialysis is continued for more than 30 min (unpublished observations).

In contrast to Ca^{2+}, $[K^+]_e$ is quite resistant to removal by microdialysis (Figure 1).[21] Following the initiation of K^+-free dialysis, $[K^+]_d$ decreases rapidly from the higher values due to the previous K^+-containing perfusate, but a stationary level of $[K^+]_d$ is obtained within 5 min. Although a very slow, progressive decrease does occur thereafter, the $[K^+]_d$ remains at approximately the same level for a considerable period of time. The presence of a slight but significant decrease in the baseline $[K^+]_d$ becomes apparent only when the K^+-free dialysis is continued for 60 min.

If K^+-free dialysis removes K^+ only from the ECS, the removal of 0.2 m*M* at 5 μl/min as dialysate would result in a loss of 1.0 nmol/min from the ECS, which could deplete all of the K^+ in 3.3 μl ECS after 10 min. The lack of remarkable changes in baseline $[K^+]_d$ actually observed indicates that K^+ is replenished continuously by *transcellular movement* of K^+ during K^+-free dialysis.[22] It would appear therefore that K^+-free dialysis can provide reasonable information regarding alterations in $[K^+]_e$ and associated neurochemical changes, such as those observed during cerebral ischemia, in short-term experiments.

The above two examples clearly demonstrate the critical role of the transcellular movement of substances in determining the period required for stabilization of the baseline data. The diffusion parameters of each substance,[22] which may be altered by the volume and tortuosity of the ECS, must also be taken into account. In addition, passive movements of substances across the cellular membrane may be changed under certain pathological conditions. Some of these problems will be discussed later. Active uptake from and release into the ECS obviously represent other determinants of the data obtained with this technique. Although most experiments are designed to measure or modify these functions,[1,2,16–18] and no detailed analysis may be called for, it should be remembered that these are always determinants of the stabilization of the baseline data.

B. CELLULAR SWELLING AND SHRINKAGE OF THE EXTRACELLULAR SPACE

When rapid cellular swelling occurs, water moves from the ECS into the cells and causes shrinkage of the ECS, i.e., a decrease in water volume in the ECS. If the substance of interest does not move from the ECS into the cells and is left behind, the concentration of the substance in the ECS will increase and the data yielded by the microdialysis technique will thereby be influenced.

The occurrence of cellular swelling and resultant shrinkage of the ECS during cerebral ischemia, anoxia or spreading depression has been repeatedly demonstrated in *in vitro*[21,22] as well as *in vivo*[12,14,15,25] studies by various methods. The cellular swelling and ECS shrinkage can be detected as an increase in concentration of ECS markers, such as tetraethylammonium, tetraethyltris-methylammonium and choline, which do not move from the ECS into the cells.[14,25] These markers are introduced into the ECS by a superfusion technique, and changes in their ECS concentration are monitored by employing electrodes sensitive to ammonium ions.[14,25]

Cellular swelling during cerebral ischemia can be detected by microdialysis based on similar principles.[26,27] ^{14}C-sucrose is preperfused into the ECS through the dialysis probe. This substance has been widely employed in *in vitro* studies as an ECS marker. Changes in the ECS concentration of ^{14}C-sucrose ($[^{14}C\text{-sucrose}]_e$) are then determined from the dialysate with perfusate without ^{14}C-sucrose. Following termination of the ^{14}C-sucrose perfusion, $[^{14}C\text{-sucrose}]_d$ decreases rapidly during an initial period of a few minutes and decreases slowly thereafter. Autoradiograms have demonstrated that an area approximately 1.5 mm distant from the probe is perfused by the ^{14}C-sucrose.

A sudden and marked increase in $[^{14}C\text{-sucrose}]_d$ is observed at 1 to 3 min after the induction of cerebral ischemia.[26] The value of $[^{14}C\text{-sucrose}]_d$ usually remains elevated for 2 to 4 min and decreases thereafter (vide infra). Since sucrose is not taken up by either the cells or capillaries, the absolute amount of ^{14}C-sucrose in the ECS must be unchanged. The observed increase therefore represents a relative decrease in water volume in the ECS resulting from movement of water into the cells, i.e., cellular swelling (Figure 2). The increase in $[^{14}C\text{-sucrose}]_e$ is concomitant with an abrupt increase in $[K^+]_e$, indicating that the increase in $[^{14}C\text{-sucrose}]_e$ represents cellular swelling induced by massive ionic shifts across the cellular membrane.

The increase in $[^{14}C\text{-sucrose}]_d$ (approximately 1.4-fold) is generally smaller than the increase in ECS marker concentration detected by other methods.[14] This difference may be due partly to the continuous background decrease in $[^{14}C\text{-sucrose}]_d$ following termination of the ^{14}C-sucrose perfusion.[26,27] Another reason for the smaller increase in $[^{14}C\text{-sucrose}]_d$ could be a decreased recovery rate of the dialysis system *in vivo* during cerebral ischemia,[21] (vide infra). In addition, it is possible that the smaller response of $[^{14}C\text{-sucrose}]_d$ is caused by the sucrose administration into the ECS, which may prevent cellular swelling to a certain degree.

In any experiment, therefore, possible changes in ECS volume should be taken into consideration when interpreting the data obtained by the microdialysis technique. This is clearly the case in ischemic and traumatic brain injury.[26,27]

C. CHANGES IN EFFECTIVE SURFACE AREA AND RECOVERY EFFICIENCY

It is a routine procedure to determine the recovery rate of the dialysis system for the substances of interest before the experiment. It should be noted, however, that the recovery rate *in vivo* is smaller than the recovery rate *in vitro*.[20,21] For example, the recovery rate of

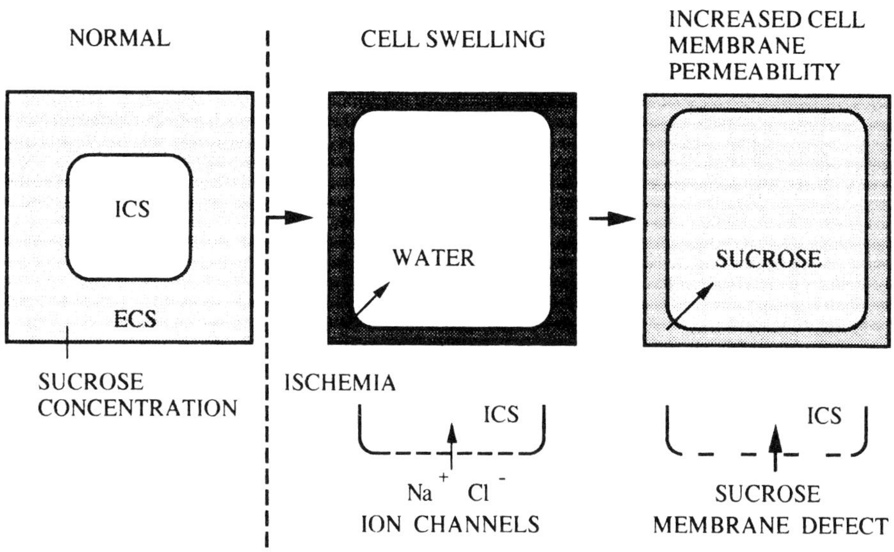

FIGURE 2. Detection of cellular swelling (middle) and an increase in cellular membrane permeability (right) after ischemia induction (dashed line). ^{14}C-sucrose was preperfused into the ECS through the dialysis probe as an ECS marker. A sudden and marked increase in [^{14}C-sucrose]$_d$ is observed at 1 to 3 min after the induction of cerebral ischemia. [^{14}C-sucrose]$_d$ usually remains elevated for 2 to 4 min. Since sucrose is not taken up by either the cells or capillaries, the absolute amount of ^{14}C-sucrose in the ECS must remain unchanged. The increase therefore represents a relative decrease in water volume in the ECS resulting from movement of water into the cells, i.e., cellular swelling. The increase in [^{14}C-sucrose]$_e$ is concomitant with an abrupt increase in $[K^+]_e$, indicating that the increase in [^{14}C-sucrose]$_e$ represents cellular swelling induced by massive *ionic shifts* across the cellular membrane. The value of [^{14}C-sucrose]$_d$, however, declines rapidly after this initial increase. This fall in [^{14}C-sucrose]$_d$ could result from penetration of the ^{14}C-sucrose into the cells. *Membrane defects* may be produced at this period of ischemia so that ^{14}C-sucrose can enter the cells. ECS, extracellular space; ICS, intracellular space. Darker tone indicates higher concentrations of sucrose.

the dialysis system for K^+ *in vivo* as determined from $[K^+]_d$ and $[K^+]_e$ each measured by microdialysis and a K^+-sensitive electrode ($[K^+]_d/[K^+]_e$) is always far smaller than the recovery rate *in vitro.*[21] This may reflect the lower effective surface area of the dialysis system *in vivo.*[20]

Furthermore, the ratio of the recovery rate *in vivo* to that *in vitro* is not constant in the diseased CNS. Following ischemia induction, this ratio drops dramatically in association with elevation of $[K^+]_e$ during ischemia.[21] This may be accounted for by a decrease in the effective surface area of the dialysis system *in vivo,* due presumably to cellular swelling and shrinkage of the ECS volume (Figure 3).[26,27] The decrease in the recovery rate *in vivo* can also be viewed as reflecting changes in the diffusion parameters[25] within the ECS. The effect of the shrunken ECS after the onset of cellular swelling would be to retard diffusion in general. Substances cannot therefore be replenished from areas adjacent to the site of dialysis, and the substances available for dialysis will become rapidly depleted. The smaller response of the ECS marker concentration as detected by microdialysis (vide supra) may be due partly to the decreased recovery efficiency following the onset of cellular swelling.

Without knowing the dynamic changes in effective surface area and recovery efficiency of the dialysis system *in vivo,* microdialysis can provide only a qualitative measure of the changes in concentration of any substance in the ECS in the diseased CNS.

D. INCREASE IN PERMEABILITY AND BREAKDOWN OF THE CELLULAR MEMBRANE

The ECS concentration of a given substance is dependent on the equilibrium across the ECS, ICS, and blood vessels. The permeability of the cellular membrane separating the ECS and ICS to a given substance is therefore a critical determinant of its ECS concentration.

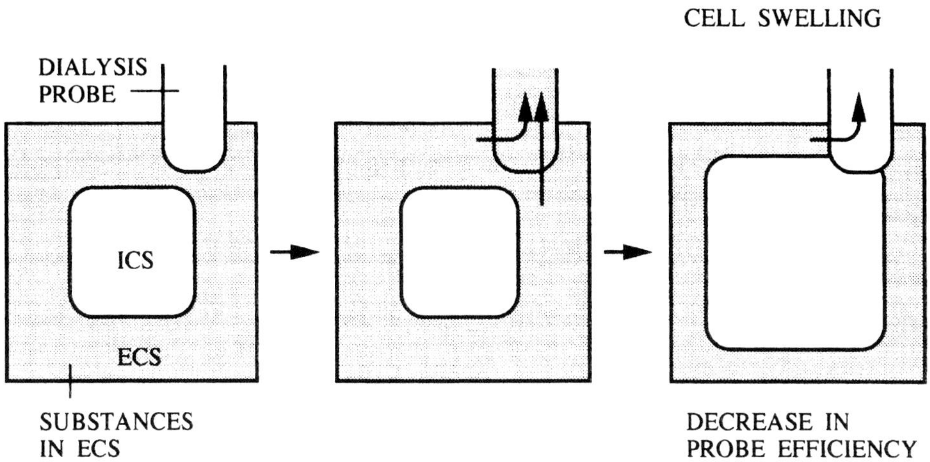

FIGURE 3. Decrease in recovery efficiency during ischemia. Following the ischemia induction (middle to right), the recovery rate of the dialysis system *in vivo* drops dramatically. This may be accounted for by a decrease in the effective surface area of the dialysis system *in vivo* due presumably to cellular swelling and shrinkage of the ECS volume (right). ECS, extracellular space; ICS, intracellular space.

During cerebral ischemia, the ECS concentration of ^{14}C-sucrose that has been preperfused through the dialysis probe increases suddenly, as mentioned earlier. The value of [^{14}C-sucrose]$_d$, however, declines rapidly after this initial increase.[28] Similar findings have been reported with trimethylammonium and choline employed as an ECS marker.[14] This fall in ECS marker concentration could result from either reexpansion of the ECS or penetration of the ECS marker into the cells. The latter possibility seems the more likely (Figure 2), since no reexpansion has been detected by other methods.[12]

Such a rapid decrease in ECS marker concentration is not clearly observed when ECS markers of much larger molecular size, such as ^{3}H-inulin, are used.[28] While the data regarding this phenomenon are not yet conclusive, it can be speculated that defects are starting to be produced in the cellular membrane at this period of ischemia so that ECS markers of smaller molecular size can enter the cells through the membrane defects (Figure 2). Similar observations, although over a different time course, have been reported in *in vitro* studies.[23,24] In support of this idea, liberation of free fatty acid, an indicator of membrane breakdown, has been demonstrated to commence at such an early period of ischemia *in vivo*.[29,30] Our preliminary data from microdialysis support this earlier finding.

In addition to passive movements of substances across the cellular membrane, active uptake from and release into the ECS are also determinants of the ECS concentration. Since the ECS concentration of any substance sensitively reflects the biochemical properties of the cellular membrane separating the ECS and ICS, their possible contribution to the data obtained by microdialysis should always be borne in mind.

E. CHANGES IN CEREBROVASCULAR PERMEABILITY

Another determinant of the ECS concentration of any substance is the capability of this substance to cross the capillary wall. Since the microdialysis technique involves insertion of a dialysis probe into the cerebral tissue, there is concern that this procedure might disrupt the normal cerebrovascular permeability through a kind of focal traumatic brain injury.[31,32] Changes in cerebrovascular permeability following experimental traumatic brain injury have been described previously employing various intravascular tracers.[33]

An altered cerebrovascular permeability can be detected by the microdialysis technique from the appearance of an intravascular tracer in the ECS.[31,32,34] For such a purpose, it is desirable that the intravascular tracer remains in the ECS after its entrance from the capillary into the ECS. If the intravascular tracer is taken up by the cells, detection of the tracer entering the ECS will also be dependent on the rate of uptake by the cells. As mentioned earlier, sucrose in the ECS is not taken up by the cells or capillaries for the most part. An increased cerebrovascular permeability can therefore be detected as the appearance of ^{14}C-sucrose in the ECS after intravascular administration of ^{14}C-sucrose.[34]

Only a very low radioactivity is evident in the dialysate as baseline data, suggesting that insertion of the dialysis probe into the cerebral tissue does not cause any marked disruption of the normal cerebrovascular permeability.[31,32,34] Immediately after traumatic brain injury induced by fluid-percussion (vide infra),[35,36] however, a clear increase in $[^{14}C\text{-sucrose}]_d$ is observed. Since the dialysis probe was temporarily withdrawn at the moment of injury and repositioned in the same place immediately after the injury, destruction of the moving brain tissue with the dialysis probe at the moment of injury was not the cause of this increase. In addition, these observations indicate that repositioning of the dialysis probe causes minimal changes in the cerebrovascular permeability.

Since the fluid percussion produces a transient cellular swelling (vide infra),[34,37,38] the increase in $[^{14}C\text{-sucrose}]_d$ immediately after fluid percussion could have been amplified by a decrease in ECS volume. $[^{14}C\text{-sucrose}]_d$ does not return to the baseline level, however, after the period of cellular swelling.[34] This indicates that a certain amount of sucrose has entered the ECS after the fluid percussion. In contrast, no change in radioactivity is detected in the dialysate after irreversible ischemia.

It should be remembered that substances originating in the blood can appear in the dialysate. This is dependent on passive as well as active movements of the substances across the capillary wall, and also across the cellular membrane separating the ECS and ICS.

III. EXPERIMENTAL PROCEDURES

A. DETECTION OF NEUROCHEMICAL CHANGES

The microdialysis technique can measure various neurochemical changes occurring within the ECS simultaneously. For example, microdialysis can identify the timing of the abrupt K^+ flux during cerebral ischemia.[21] It is then possible to evaluate various neurochemical changes, such as the excitatory amino acids (EAA) release, lactate accumulation or free fatty acid liberation, in relation to the abrupt K^+ flux, employing the same dialysate fractions. However, as mentioned above, the data yielded by microdialysis are influenced directly and/or indirectly by various changes in cellular membrane function. For this reason, the results should be interpreted in conjunction with possible changes in cellular membrane function. Moreover, the data obtained by this technique themselves frequently reveal changes in cellular membrane function in the diseased CNS. For example, massive changes in $[K^+]_d$ are likely to be a result of changes in the ion permeability of the cellular membrane. Cellular swelling detected as an increase in the ECS concentration of ^{14}C-sucrose preperfused into the ECS represents another example of the consequence of an altered cellular membrane function. Thus, the microdialysis technique can provide a unique opportunity to detect functional changes of the cellular membrane in the diseased CNS.

We usually place a pair of microdialysis probes at symmetrical sites of the brain bilaterally. The probes are initially perfused with Ringer's solution. The temperature of the perfusate is adjusted to 37°C, and the temperature is maintained using a chamber filled with water at 37°C into which the whole length of the inlet tubing is placed. Dialysate fractions are usually collected at 1-min intervals, except in some experiments where dialysate fractions are collected at 2-min intervals. The total volume of the outlet tube and the probe is adjusted to 5.0 μl. The second 1-min dialysate fraction collected after the induction of ischemia or trauma thus corresponds to the initial postischemic fraction.

B. MODIFICATION OF NEUROCHEMICAL ENVIRONMENTS

In addition to measurement of neurochemical changes in the ECS and functional changes in the cellular membrane, various modifications of the neurochemical environments can be made when employing dialysis probes. Modifications can be produced by the removal of certain bioactive substances from the ECS by dialysis. For example, we have demonstrated that Ca^{2+}-free dialysis can efficiently remove Ca^{2+} from the ECS (vide supra). Furthermore, the neurochemical processes can be manipulated experimentally through the administration of various agents *in situ* via the dialysis probes.

The effects of such modification can be evaluated not only by measurement with microdialysis but also by applying various morphological techniques. For example, we have demonstrated that perfusion of EAA antagonist through the dialysis probe can attenuate the hypermetabolism induced by traumatic brain injury as demonstrated by ^{14}C-deoxyglucose autoradiography and selective death of hippocampal pyramidal cells induced by transient forebrain ischemia (vide infra).

We perform experimental modification of the neurochemical environment through one of the two probes placed bilaterally (test probe). Dialysis with the other probe serves as the control (control probe). The effects of such experimental modification can then be evaluated by making comparisons between the control and test probes and examining the statistics for paired data.

IV. EXPERIMENTAL APPLICATIONS

A. CALCIUM-DEPENDENT COMPONENT OF POTASSIUM FLUXES

The abrupt increase in $[K^+]_e$ occurring during cerebral ischemia can be detected by microdialysis with K^+-free perfusate.[21] Induction of global ischemia produces a slight increase in $[K^+]_d$ in one or more fractions, in addition to the fraction for the dead space, and a marked increase in subsequent fractions. The onset of the sudden, large increase corresponds to an abrupt increase in $[K^+]_e$ as measured with a K^+-sensitive electrode.[21]

Similarly, the increase in $[K^+]_e$ following traumatic brain injury can be demonstrated by microdialysis.[39] Traumatic brain injury is induced by fluid percussion applied to the parietal epidural space.[35,36] This model of traumatic brain injury can produce widespread deformation of the brain without causing focal traumatic brain injury, so that the physiological and morphological consequences of the fluid percussion mimic diffuse traumatic brain injury. The probes were temporarily removed at the moment of injury induction. A large increase in $[K^+]_d$ occurred for 2 to 5 min immediately after the fluid percussion in agreement with the results of earlier studies employing ion-sensitive electrodes.[8,11]

The abrupt and large increase in $[K^+]_e$ during cerebral ischemia begins when $[K^+]_e$ reaches a level of 6 to 10 m*M*.[3–7] Neurotransmitter release and a sudden K^+ flux through ligand-gated ion channels of neuronal cells have been postulated to be responsible for this phenomenon.[40–42] Thus, when $[K^+]_e$ reaches this level, the nerve terminals may be depolarized, causing Ca^{2+} entry into the presynaptic nerve terminals, and neurotransmitters are thereby released.

Ca^{2+}-dependent exocytotic release of various neurotransmitters in response to an elevated $[K^+]_e$ has been demonstrated by *in vitro* studies, employing synaptosome preparations or brain slices.[43–47] Such release is abolished when Ca^{2+}-free perfusate containing cobalt ions (Co^{2+}), which blocks Ca^{2+} entry into the nerve terminals, is used.[47–49]

If the Ca^{2+}-dependent exocytotic release of neurotransmitters is responsible for the abrupt increase in $[K^+]_e$, inhibition of Ca^{2+} entry into the nerve terminals would be expected to alter the time course of the changes in $[K^+]_e$ during cerebral ischemia. In support of this hypothesis, the abrupt, large increase in $[K^+]_d$ following the induction of ischemia is significantly delayed by dialysis with Ca^{2+}-free perfusate containing Co^{2+} or Mg^{2+}, which blocks Ca^{2+} entry.[21] The effects of Co^{2+} or Mg^{2+} are dose-dependent, appearing as a prolonged latency of onset, an additional delay in reaching the maximum level, and a prolonged duration from the onset to the maximum level.

The above observations are consistent with the hypothesis that a major component of the initial increase in $[K^+]_e$ during cerebral ischemia is dependent upon the Ca^{2+}-dependent exocytotic release of neurotransmitters. The increase in $[K^+]_e$ after traumatic brain injury is also partially Ca^{2+}-dependent,[39] suggesting involvement of Ca^{2+}-dependent exocytotic release of neurotransmitters.

B. CALCIUM-DEPENDENT EXOCYTOTIC RELEASE OF EXCITATORY AMINO ACIDS

Employing brain microdialysis, the ECS concentration of glutamate ($[Glu]_e$) has been demonstrated to increase during cerebral ischemia.[31,50–52] It is well documented that glutamate at a sufficient concentration is neurotoxic.[53–56] The increase in $[Glu]_e$ has therefore been proposed to be a causative factor in the development of ischemic neuronal death. In addition, glutamate induces massive ionic fluxes across the plasma membrane.[54,55] For this reason, the elevation of $[Glu]_e$ has been suggested to play an important role in producing rapid ionic fluxes during cerebral ischemia or anoxia.

The elevation of $[Glu]_e$ during cerebral ischemia begins concomitantly with the large increase in $[K^+]_d$.[52,57] No increase in $[Glu]_d$ was noted in any fractions before the large increase in $[K^+]_d$. The increase in $[Glu]_e$ concomitant with the abrupt increase in $[K^+]_e$ appears therefore to be related to the elevated $[K^+]_e$, which would depolarize the nerve terminals. Following the initial rapid increase, a transient decline in $[Glu]_d$ is sometimes observed, and $[Glu]_d$ then reveals a slow increase. The aspartate in the dialysate is elevated similarly to $[Glu]_d$, although the increase is less marked.[57] It appeared that serine was also increased slightly following the onset of a large increase in $[K^+]_d$.

The increase in $[Glu]_d$ during cerebral ischemia is markedly attenuated by Ca^{2+}-free perfusate containing Co^{2+}.[57] This effect is due mainly to complete inhibition of the earlier rapid increase immediately after the onset of the large increase in $[K^+]_d$. The late increase in $[Glu]_d$ can be blocked only partially.[57] The rate of increase in $[Glu]_d$ is, however, significantly decelerated by Co^{2+}, suggesting the presence of a Ca^{2+}-dependent component also in the late increase.

These findings are consistent with *in vitro* studies that have demonstrated that elevation of $[K^+]_e$ induces both a rapid Ca^{2+}-dependent release of glutamate from the transmitter pool of the nerve terminals and a slow Ca^{2+}-independent release from the cytoplasmic pool.[46,47] The Ca^{2+}-dependency of the earlier rapid increase in $[Glu]_d$ during cerebral ischemia suggests that this increase represents a Ca^{2+}-dependent exocytotic release of glutamate from the transmitter pool. The Ca^{2+}-dependent exocytotic release of glutamate is energy dependent.[47] Since the massive ionic fluxes across the plasma membrane during cerebral ischemia could cause a precipitous fall in energy store, this type of glutamate release may be terminated shortly after the onset of the large increase in $[K^+]_e$. The transient decline in $[Glu]_d$ sometimes observed after the initial increase may be related to the fact that the data obtained by microdialysis during this period of ischemia can vary greatly depending on various parameters and conditions of the dialysis system used. Nevertheless, termination of Ca^{2+}-dependent exocytotic release and sustaining the function of the glutamate uptake system might represent another possible cause of the transient decline that needs to be tested.

In contrast to the rapid increase in $[Glu]_d$ occurring during cerebral ischemia, the slow increase appears to be Ca^{2+}-independent, at least partially. Ionic shifts across the plasma membrane, which eventually take place due probably to energy depletion, may cause a loss of the electrogenic component maintaining the glutamate gradient across the plasma membrane, and continuous Ca^{2+}-independent flux of glutamate from the cytoplasmic pool may thereby be induced. Loss of function of the glutamate uptake system and breakdown of the membrane function resulting from numerous other causes could also be responsible for the later slow increase in $[Glu]_d$.

$[Glu]_d$ increases simultaneously with the increase in $[K^+]_e$ following fluid percussion traumatic brain injury.[39] Since this sudden increase in $[K^+]_e$ is also partially Ca^{2+}-dependent,[39] exocytotic release of glutamate from the nerve terminals depolarized by the elevated $[K^+]_e$ may account for part of the $[Glu]_d$ increase, although Ca^{2+}-independent processes are also involved.

C. IONIC FLUXES THROUGH EXCITATORY AMINO ACID RECEPTOR-COUPLED ION CHANNELS

The ionic fluxes observed after ischemic and traumatic brain insults are too rapid to be accounted for merely by failure of the energy-dependent ion pump activity alone. Sudden changes in the ion permeability of the cellular membranes have therefore been postulated to occur presumably through indiscriminate release of neurotransmitters. EAAs are the most likely neurotransmitter that could produce such marked ionic shifts.[54,55] The increases in $[K^+]_d$ and $[Glu]_d$ are both inhibited by Ca^{2+}-free dialysis with Co^{2+}. On this basis, it is tempting to speculate that Ca^{2+}-dependent exocytotic release of EAAs could be the cause of the rapid ionic shifts and Ca^{2+}-dependent component of the increase in $[K^+]_e$ observed during cerebral ischemia[21] and after fluid percussion.[39]

Systemically administered EAA antagonists have been reported to inhibit spreading cortical depression,[58–62] but they fail to delay[59,60] and may even accelerate[62] sudden ionic shifts during cerebral ischemia or anoxia. Denervation of glutaminergic afferents has, however, been reported to delay ionic changes in the hippocampus during cerebral ischemia.[63] Furthermore, the abrupt increase in $[K^+]_d$ following ischemia induction is significantly delayed by the broad-spectrum EAA antagonist, kynurenic acid (KYN),[64–65] administered *in situ* through the dialysis probe.[67,68] Thus, a prolonged latency for the onset of the increase in $[K^+]_d$ and an additional delay in reaching the maximum level of increased $[K^+]_e$ are demonstrated.

The duration from onset to the maximum level was also prolonged. However, no effect is exerted by KYN on the maximum level of $[K^+]_d$ produced by ischemia. Dialysis with KYN causes no significant change in the baseline level of $[K^+]_d$. These findings indicate that the slope of the sudden increase in $[K^+]_d$ is less steep at the site of test dialysis perfused with KYN. The delay in latency induced by KYN was comparable to the above-mentioned delay induced by Ca^{2+}-free dialysis with Co^{2+} or Mg^{2+}.

The disparity in results from previous studies that failed to reveal inhibition of the sudden ionic shifts during cerebral ischemia with systemically administered EAA antagonists, may reflect the difference in methods of drug administration employed. It appears possible that higher doses of EAA antagonists are required in ischemia compared to those that can inhibit cortical spreading depression. Although KYN is a broad-spectrum antagonist of EAAs, different concentrations of KYN are needed to block the neuronal depolarization mediated by various EAA receptor subtypes.[66] Dose-response curves for the effects of KYN on the increase in $[K^+]_d$ indicate that the ionic fluxes mediated by EAAs may involve many, if not all, of the EAA receptor subtypes.[68]

Elevation of $[K^+]_e$ to a certain level is probably the cause of the Ca^{2+}-dependent exocytotic release of EAAs. It appears therefore that the sudden and rapid increase in $[K^+]_e$ occurring during cerebral ischemia represents a result of a malignant cycle beween the K^+ flux and EAA release (Figure 4).

The increase in $[K^+]_d$ after fluid percussion is also inhibited by KYN administered *in situ* through the dialysis probe.[39] This finding suggests that indiscriminate release of EAAs and the EAA-receptor-coupled ion channels may also play a vital role in the production of the massive ionic fluxes induced by traumatic brain injury.

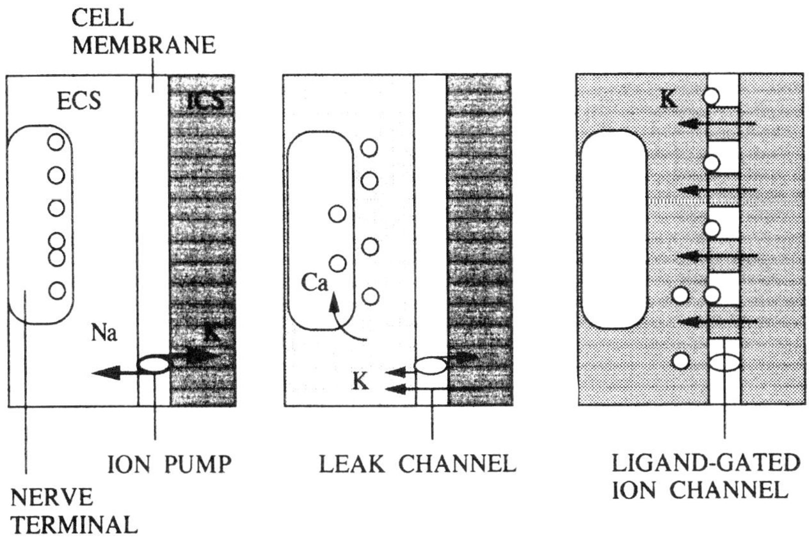

FIGURE 4. Possible mechanisms of massive ionic (e.g., K^+) shifts during ischemia. After ischemia induction, the energy-dependent ion pump fails to restore leaked K^+ completely so that $[K^+]_e$ slowly increases. Elevation of $[K^+]_e$ to a certain level causes depolarization of nerve terminals. This depolarization of the nerve terminals results in Ca^{2+}-dependent exocytotic release of EAAs (middle). Further flux of K^+ together with other ions occurs through EAA-coupled ion channels (right). It appears therefore that the sudden and rapid increase in $[K^+]_e$ observed during cerebral ischemia represents the result of a malignant cycle between the K^+ flux and EAA release. ECS, extracellular space; ICS, intracellular space. Open circles, EAA neurotransmitter. Darker tone indicates a higher concentration of K^+.

D. EXCITATORY AMINO ACID-MEDIATED CELLULAR SWELLING

As mentioned earlier, the microdialysis technique can identify the timing of development of cellular swelling *in vivo.* EAAs cause large fluxes of Na^+, K^+, Ca^{2+}, and Cl^- across the cellular membrane *in vitro,* which have been shown to generate cellular swelling as expected.[53–56] Among the various EAA antagonists, KYN attenuates such cellular swelling *in vitro* most effectively.[69] It would not be surprising therefore if ionic fluxes through EAA-coupled ion channels are also responsible for the cellular swelling during cerebral ischemia *in vivo.* As expected, KYN administered *in situ* through the dialysis probe inhibits both the cellular swelling and the massive ionic fluxes occurring simultaneously.[26,27] This implies that pathological processes quite similar to the EAA-mediated ionic fluxes and cellular swelling, which are seen *in vitro,* may underlie the early cellular swelling during cerebral ischemia *in vivo.*

Transient cellular swelling is also observed after fluid percussion,[34,37] as an increase in the ECS concentration of preperfused ^{14}C-sucrose concomitantly with the increase in $[K^+]_d$. Such cellular swelling is again inhibited by KYN administered *in situ* through the dialysis probe.[37] While ionic shifts are not the sole cause of the cellular swelling occuring during cerebral ischemia,[13,15,23,24] these findings indicate that Ca^{2+}-dependent exocytotic release of EAAs may play a major role in producing cellular swelling during the early period after ischemia induction and traumatic brain injury. This conclusion apparently accounts for the reported effect of KYN in reducing the brain edema induced by ischemia or anoxia.[70–73]

Following fluid percussion, an increase in cerebrovascular permeability occurs at approximately the same time.[34] The simultaneous occurrence of water movement from the ECS into the cells and an increase in the cerebrovascular permeability may contribute to the elevated intracranial and interstitial pressures observed during this period with this model.[74]

E. EXCITATORY AMINO ACID-MEDIATED HYPERMETABOLISM

The energy metabolism of the brain increases during the period soon after traumatic brain injury.[75–79] A diffuse increase in glucose utilization is observed by ^{14}C-deoxyglucose autoradiography following fluid percussion (Figure 5).[80] Sudden ionic fluxes across the cellular membrane, such as those observed in spreading depression, strongly activate the energy-dependent ion pump for restoration of ionic homeostasis.[81–88] The observed hypermetabolism after traumatic brain injury may thus be caused by the ionic perturbation across the cell membrane discussed above.

Since the major component of the ionic fluxes after fluid percussion appears to be mediated by EAA-coupled ion channels, KYN would attenuate the increase in glucose utilization rate if the ionic fluxes were indeed responsible for the observed hypermetabolism. Consistent with this hypotheis, the increase in glucose utilization rate is clearly inhibited in areas perfused with KYN through the dialysis probe (Figure 5).[89] No significant effect of KYN on the control levels of the glucose utilization rate is induced.[89] Removal of EAA-mediated afferents also prevents hypermetabolism following fluid percussion.[90]

F. EXCITATORY AMINO ACID-MEDIATED LACTATE ACCUMULATION

Lactate accumulation has been suggested to be one of the most important factors affecting cell viability under pathological conditions such as ischemia.[91–96] Although the lactate accumulation occurring during ischemia is generally taken as evidence of an impaired oxidative metabolism, lactate is accumulated in all metabolic states in which ATP hydrolysis exceeds ATP synthesis, for example, in response to a sudden increase in energy demand.[97,98]

The above data suggest that the sudden ionic shifts during cerebral ischemia may be mediated at their commencement by EAA-coupled ion channels, rather than be merely a result of depletion of energy, which terminates ion-pump activity. Such a sudden change in ionic gradients across the plasma membrane strongly activates the energy-dependent ion pumps and increases ATP hydrolysis for activating energy-dependent ion pumps, thereby stimulating lactate formation.[81–88] During a single passage of cortical spreading depression, which causes quite similar ionic events, almost parallel decreases in glucose content and pH, and elevation of the lactate concentration have been reported.[81–88,99–102]

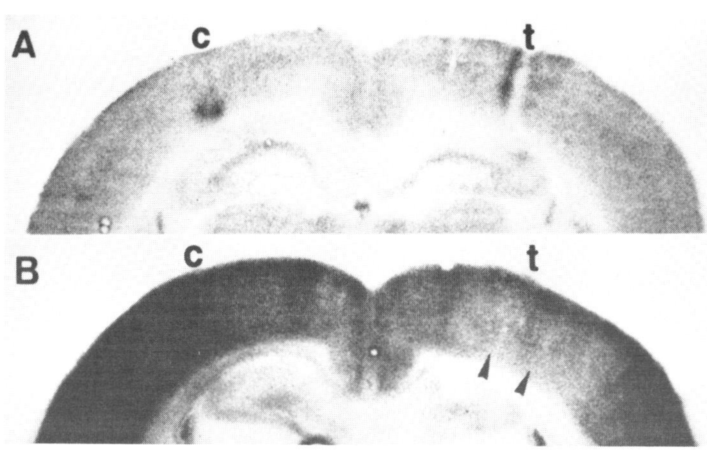

FIGURE 5. Representative examples of changes in local glucose utilization rate as evaluated by ^{14}C-deoxyglucose autoradiography. A, sham injury; B, fluid percussion injury. The control probe (c) perfused without kynurenic acid (KYN) and test (t) probe perfused with KYN were placed bilaterally in the cortex. The increase in ^{14}C-deoxyglucose uptake in B is inhibited in the area perfused by KYN (t), but not in the area perfused without KYN (c).

It appears possible, therefore, that a sudden increase in energy demand in response to the ionic shifts, in addition to a decreased capability of oxidative metabolism for ATP synthesis, may significantly contribute to lactate accumulation during ischemia. If this is in fact the case, inhibition of the ionic fluxes through EAA-coupled ion channels may attenuate any sudden increase in energy consumption and thereby decrease the lactate accumulation during the early period of ischemia.

Changes in the ECS concentration of lactate can be monitored by microdialysis.[103,104] The dialysate concentration of lactate ($[\text{lactate}]_d$) increases dramatically, beginning at a few minutes after ischemia induction.[105] This increase in $[\text{lactate}]_d$ is clearly delayed by KYN administered *in situ* through the dialysis probe.[105] The maximum increase is also inhibited. No change in baseline level of $[\text{lactate}]_d$ is seen. It appears most likely that the observed effect of KYN on the increase in $[\text{lactate}]_d$ is produced by inhibition of the early ionic shifts through EAA-coupled ion channels and the resultant increase in energy demand.

In complete ischemia, the amount of available glucose is predetermined. The decrease in lactate accumulation may therefore be accounted for by an increase in the fraction of glucose degraded through decarboxylation of pyruvate to acetyl CoA. It seems that lactate production is stimulated by a sudden increase in energy demand at the site of control dialysis, whereas more glucose was degraded through decarboxylation of pyruvate to acetyl CoA at the site of test dialysis perfused with KYN.

These inferences imply that glucose degradation through decarboxylation of pyruvate to acetyl CoA still comprises a significant part of the metabolism for energy production at the time of commencement of the massive ionic shifts. This is consistent with the hypothesis that the rapid ionic fluxes occurring during ischemia are a cause of early depletion of energy, rather than simply the result of energy failure.

It has been reported recently that systemically administered KYN does not affect the decrease in ATP level as measured by *in vivo* ^{32}P-nuclear magnetic resonance spectroscopy during cerebral ischemia.[106] It has also been demonstrated, however, that KYN significantly attenuates the decrease in phosphocreatine and the accumulation of inorganic phosphorus.[106] It is possible that KYN inhibits the rate of energy consumption, which may be more sensitively reflected by the phosphocreatine level during incomplete ischemia.

There is abundant evidence to indicate that uncontrolled release of EAAs may play a crucial role in producing ischemic neuronal damage. Transient forebrain ischemia in rodents leads to delayed death of hippocampal pyramidal cells.[107–109] Such cell death is selectively inhibited in the area perfused by KYN through the dialysis probe.[106] While the beneficial effects of EAA antagonists, such as KYN,[70–73] appear to be related mainly to inhibition of Ca^{2+} fluxes[110] through EAA-receptor coupled ion channels, inhibition of lactate accumulation may also contribute to the mechanisms whereby KYN is able to attenuate ischemic neuronal damage.

Lactate undergoes an increase in the brain tissue during the initial 60 min following fluid percussion.[111] The value of $[\text{lactate}]_d$ also increases after fluid percussion for 15 to 20 min following injury induction.[112,113] This may again be due to strong activation of the energy-dependent ion pump for the restoration of ionic homeostasis.[81–88,99–102] In support of this hypothesis, inhibition of the energy-dependent ion pump by oubain administered *in situ* through the dialysis probe can clearly attenuate the increase in $[\text{lactate}]_d$.[113] Furthermore, the increase in $[\text{lactate}]_d$ can be inhibited by KYN administered *in situ* through the dialysis probe.[113]

Based on these data, we believe that causal and associated relationships exist between EAA release, massive ionic fluxes and subsequent hypermetabolism with lactate accumulation occurring after ischemia induction as well as traumatic brain injury.

V. CONCLUSION

Changes in cellular membrane function in response to neurotransmitter release are characteristic of neurons as excitable cells. Such phenomena may thus represent the pathological process that is most unique to neuronal cells as compared to any other kinds of cells within the body. The microdialysis technique is useful for revealing such changes. Our recent studies further suggest that many other neurochemical changes occurring within the ECS in association with ischemic and traumatic brain injury are directly or indirectly related to changes in membrane function. Although the mechanisms for the observed neurochemical alterations are more complicated than those for neurochemical changes within the physiological range, application of this technique for investigating the diseased CNS can provide a deeper insight into the pathological processes involved.

ACKNOWLEDGMENTS

The data reviewed in this chapter are from experiments conducted in collaboration with Drs. T. Tamura, T. Kawamata, A. Yoshino, T. Kano, and T. Tsubokawa (Department of Neurological Surgery, Nihon University, Tokyo), and Drs. D.A. Hovda and D.P. Becker (Department of Neurosurgery, University of California at Los Angeles, Los Angeles). These investigations were supported by grants from the U.S. National Institutes of Health (NS27544), Lind Lawrence Foundation, Annie Laurie Aitken Charitable Trust, and World Boxing Council, and a research grant for Cardiovascular Disease (2A-2) from the Ministry of Health and Welfare of Japan.

REFERENCES

1. Delgado, J. M. R., Defeudis, F. V., Roth, R. H., Ryugo, D. K., and Mitruka, B. M. (1971) Dialytrode for long term intercerebral perfusion in awake monkeys. *Arch. Int. Pharmacodyn. Ther.* 198, 9–12.
2. Ungerstedt, U., Herrera-Manschitz, M., Jungnelius, U., Stahle, L., Tossman, U., and Zetterstrom, T. (1982) Dopamine synaptic mechanisms reflected in studies combining behavioral recordings and brain dialysis. *Adv. Dopamine Res.* 17, 219–231.
3. Vyscocil, F., Kriz, N., and Bures, J. (1973) Potassium selective microelectrodes used for measuring the extracellular brain potassium during spreading depression and anoxic depolarization in the rat. *Brain Res.* 39, 255–259.
4. Astrup, J., Symon, L., Branston, N. M., and Lassen, N. A. (1977) Cortical evoked potential and extracellular K^+ and H^+ at critical levels of brain ischemia. *Stroke* 8, 51–57.
5. Blank, Jr., W. F. and Kirshner, H. S. (1977) The kinetics of extracellular potassium changes during hypoxia and anoxia in the cat cerebral cortex. *Brain Res.* 123, 113–124.
6. Hansen, A. J. (1977) Extracellular potassium concentration in juvenile and adult rat brain cortex during anoxia. *Acta Physiol. Scand.* 99, 412–420.
7. Hossmann, K.-A., Sakaki, S., and Zimmermann, V. (1977) Cation activities in reversible ischemia of the cat brain. *Stroke* 8, 77–81.
8. Takahashi, H., Manaka, S., and Sano, K. (1981) Changes in extracellular potassium concentration in cortex and brain stem during the acute phase of experimental closed head injury. *J. Neurosurg.* 55, 708–717.
9. Tsubokawa, T. (1983) Cerebral circulation and metabolism in concussion. *Neurol. Surg. (Tokyo)* 11, 563–573.
10. Hubschmann, O. R. and Kornhauser, D. (1983) Effects of intraparenchymal hemorrhage on extracellular cortical potassium in experimental head trauma. *J. Neurosurg.* 59, 289–293.
11. DeSalles, A. A. E., Jenkins, L. W., Anderson, R. L. et al. (1986) Extracellular potassium activity following concussion. A microelectrode study in the cat. *Soc. Neurosci. Abstr.* 12, 967.
12. Van Harreveld, A. and Ochs, S. (1956) Cerebral impedance changes after circulatory arrest. *Am. J. Physiol.* 187, 180–192.
13. Maknight, A. D. and Leaf, A. (1977) Regulation of cellular volume. *Physiol. Rev.* 57, 510–573.
14. Hansen, A. J. and Olsen, C. E. (1980) Brain extracellular space during spreading depression and ischemia. *Acta Physiol. Scand.* 108, 355–365.

15. Hansen, A. J. (1985) Effect of anoxia on ion distribution in the brain. *Physiol. Rev.* 165, 101–148.
16. Benveniste, H., Drejer, J., Schousboe, A., and Diemer, N. H. (1987) Regional cerebral glucose phosphorylation and blood flow after insertion of a microdialysis fiber through the dorsal hippocampus in the rat. *J. Neurochem.* 49, 729–734.
17. Imperat, A. and DiChiara, G., (1984) Trans-striatal dialysis coupled to reverse phase high performance liquid chromatography with electrochemical detection: a new method for the study of the *in vivo* release of endogenous dopamine and metabolites. *J. Neurosci.* 4, 966–977.
18. Damsma, G., Westermink, B. H. X., Imperato, A., Rollema, H., DeVrics, J. B., and Horn, A. S. (1987) Automated brain dialysis of acetylcholine in freely moving rats: detection of basal acetylcholine. *Life Sci.* 41, 873–876.
19. Westerink, B. H. C., Damsma, G., Rollema, H., De Vries, J. B., and Horn, A. S. (1987) Scope and limitation of *in vivo* brain dialysis: a comparison of its application to various neurotransmitter systems. *Life Sci.* 41, 1763–1776.
20. Alexander, G. M., Grothusen, J. R., and Schwartzman, R. J. (1988) Flow dependent changes in the effective surface area of microdialysis probes. *Life Sci.* 43, 595–601.
21. Katayama, Y., Tamura, T., Becker, D. P., and Tsubokawa, T. (1991) Calcium-dependent component of massive increase in extracellular potassium during cerebral ischemia as demonstrated by microdialysis *in vivo. Brain Res.* 567, 57–63.
22. Nicholson, C., Phillips, J. M., and Gardner-Medwin, A. R. (1979) Diffusion from an iontophoretic point source in the rat brain. Role of torutuosity and volume fraction. *Brain Res.* 169, 580–584.
23. Parks, J. M., Shay, J., and Ames, A., III (1976) Cell volume and permeability of oxygen- and glucose-deprived retina *in vitro. Arch. Neurol.* 33, 709–714.
24. Ames, A., III and Nesbett, F. B. (1983) Pathophysiology of ischemic cell death: II. Changes in plasma membrane permeability and cell volume. *Stroke* 14, 227–233.
25. Phillips, J. M. and Nicholson, C. (1979) Anion permeability in spreading depression investigated with ion-sensitive microelectrodes. *Brain Res.* 173, 567–571.
26. Katayama, Y., Becker, D. P., Tamura, T., and Tsubokawa, T. (1990) Cellular swelling during cerebral ischemia demonstrated by microdialysis *in vivo:* preliminary data indicating the role of excitatory amino acids. *Acta Neurochir. Suppl.* 51, 183–185.
27. Katayama, Y., Tamura, T., Becker, D. P., and Tsubokawa, T. (1992) Early cellular swelling during cerebral ischemia *in vivo* is mediated by excitatory amino acids released from nerve terminal. *Brain Res.* 577, 121–126.
28. Katayama, Y., Tamura, T., Becker, D. P., and Tsubpkawa, T. (1991) Detection of cellular swelling and subsequent increase in plasma membrane permeability during cerebral ischemia *in vivo* using microdialysis. *J. Cereb. Blood Flow Metab.* 11, s479.
29. Aveldano, M. I. and Bazan, N. G. (1975) Rapid production of diacylglycerols enriched in arachidonate and stearate during early brain ischemia. *J. Neurochem.* 25, 919–920.
30. Yasuda, H., Kishiro, K., Izumi, N., and Nakanishi, M. (1985) Biphasic liberation of arachidonic and stearic acids during cerebral ischemia. *J. Neurochem.* 45, 168–172.
31. Benveniste, H., Drejer, J., Schousboe, A., and Diemer, N. H. (1984) Elevation of the extracellular concentrations of glutamate and aspartate in rat hippocampus during transient cerebral ischemia monitored by intracerebral microdialysis. *J. Neurochem.* 43, 1369–1374.
32. Tossman, U. and Ungerstedt, U. (1986) Microdialysis in the study of extracellular levels of amino acids in the rat brain. *Acta Physiol. Scand.* 128, 9–14.
33. Povlishock, J. T., Becker, D. P., Sullivan, H. G., and Miller, J. D. (1978) Vascular permeability alterations to horseradish peroxidase in experimental brain injury. *Brain Res.* 153, 223–239.
34. Katayama, Y., Becker, D. P., and Tamura, T. (1991) Changes in cerebrovascular permeability and excitatory amino acid-mediated cellular swelling in experimental concussive brain injury. In: *Intracranial Pressure,* Vol. 8 (Eds. C. J. J. Averzaat et al.), Springer, Berlin, pp. 484–487.
35. Katayama, Y., Glisson, J. D., Becker, D. P., and Hayes, R. L. (1985) Concussive head injury producing suppression of sensory transmission within the spinal cord. *J. Neurosurg.* 63, 97–105.
36. Dixon, C. E., Lyeth, B. G., Povlishock, J. T., et al. A fluid percussion model of experimental brain injury in the rat. *J. Neurosurg.* 67, 110–119.
37. Katayama, Y., Cheung, M. K., Alves, A., and Becker, D. P. (1989) Ion fluxes and cell swelling after experimental traumatic brain injury: the role of excitatory amino acids. In: *Intracranial Pressure,* Vol. 7 (Eds. J. T. Hoff and A. L. Betz), Springer, Berlin, pp. 584–588.
38. Katayama, Y., Becker, D. P., Tamura, T., and Ikezaki, K. (1990) Early cellular swelling in experimental traumatic brain injury: a phenomenon mediated by excitatory amino acids. *Acta Neurochi. Suppl.* 51, 271–273.
39. Katayama, Y., Becker, D. P., Tamura, T., and Hovda, D. (1990) Massive increase in extracellular potassium and indiscriminative glutamate release after concussive brain injury. *J. Neurosurg.* 73, 889–900.
40. Van Harreveld, A. (1987) Two mechanisms for spreading depression in chicken retina. *J. Neurobiol.* 9, 419–431.

41. Nicholson, C. and Kraig, R. P. (1981) The behavior of extracellular ions during spreading depression. In: *The Application of Ion Electrodes* (Ed. T. Zeuthen), Elsevier, North-Holland, Amsterdam, pp. 217–238.
42. Moghaddam, B., Schenk, J. O., Stewart, W. B., and Hansen, A. J. (1987) Temporal relationship between neurotransmitter release and ion flux during spreading depression and anoxia. *Can. J. Physiol. Pharmacol.* 54, 1105–1110.
43. Arnfred, T. and Hertz, L. (1971) Effects of potassium and glutamate on brain cortex slices: uptake and release of glutamic and other amino acids. *J. Neurochem.* 18, 259–265.
44. Nadler, J. V., Vaca, K. W., White, W. F., Lynch, G. S., and Cotman, C. W. (1976) Aspartate and glutamate as possible transmitters of excitatory hippocampal afferents. *Nature* 260, 538–540.
45. Bosley, T. M., Wookhams, P. L., Gordon, R. D., and Balazs, R. (1983) Effects of anoxia on the stimulated release of amino acid neurotransmitters in the cerebellum *in vitro. J. Neurochem.* 40, 188–201.
46. Nicholls, D. G. and Sihra, T. S. (1986) Synaptosomes process an exocytotic pool of glutamate. *Nature,* 321, 772–773.
47. Sanchez-Prieto, J., Sihra, T. S., and Nicholls, D. G. (1987) Characterization of the exocytotic release of glutamate from guinea-pig cerebral cortical synaptosomes. *J. Neurochem.* 49, 58–64.
48. Drejer, J., Benveniste, H., Dlemer, N. H., and Schousboe, A. (1985) Cellular origin of ischemia-induced glutamate release from brain tissue *in vivo* and *in vitro. J. Neurochem.* 45, 145–151.
49. Martins-Ferreira, H., DeOliveira Castro, G., Struchiner, C. J., and Rodrigues, P. S. (1974) Circulating spreading depression in isolated chick retina. *J. Neurophysiol.* 37, 778–784.
50. Hagberg, H., Lehmenn, A., Sandberg, M., Nystrom, B., Jacobson, I., and Hamberger, A. (1985) Ischemia-induced shift of inhibitory and excitatory amino acids from intra- to extracellular compartments. *J. Cereb. Blood Flow Metab.* 5, 413–419.
51. Globus, M. Y.-T., Busto, R., Dietrich, W. D., Martinez, E., Valdes, I., and Ginsberg, M. D. (1988) Effect of ischemia on the *in vivo* release of striatal dopamine, glutamate and g-aminobutyric acid studied by intracerebral microdialysis. *J. Neurochem.* 51, 1455–1464.
52. Scheller, D., Heister, U., Peters, U., and Hoeller, M. (1989) Glutamate and aspartate are released concomitantly with the terminal DC-negativation after global cerebral ischemia. *J. Cereb. Blood Flow Metab.* 9, s372.
53. Olney, J. W., Price, M. T., Samson, L., and Lambuyere, J. (1986) The role of specific ions in glutamate neurotoxicity. *Neurosci. Lett.* 65, 65–71.
54. Choi, D. W. (1987) Ionic dependence of glutamate neurotoxicity. *J. Neurosci.* 7, 369–379.
55. Mayer, M. L. and Westbrook, G. L. (1987) Cellular mechanisms underlying excitotoxicity. *TINS* 10, 59–61.
56. Rothman, S. M. and Olney, J. W. (1987) Excitotoxicity and the NMDA receptor. *TINS* 10, 299–302.
57. Katayama, Y., Kawamata, T., Tamura, T., Becker, D. P., and Tsubokawa, T. (1991) Calcium-dependent glutamate release concomitant with massive potassium flux during cerebral ischemia *in vivo. Brain Res.* 558, 136–140.
58. Gorelova, N. A., Koroleva, V. I., Amemori, T., Pavlik, V., and Bures, J. (1987) Ketamine blockade of cortical spreading depression in rats. *Electroenceph. Clin. Neurophysiol.* 66, 440–447.
59. Hernandez-Caceres, J., Macias-Gonzalez, R., Brozek, G., and Bures, J. (1987) Systemic ketamine blocks spreading depression but does not delay the onset of terminal anoxic depolarization in rats. *Brain Res.* 437, 360–364.
60. Hansen, J. A., Lauritzen, M., and Wieloch, T. (1988) MK-801 inhibits spreading depression but not anoxic depolarization. In: *Recent Advances in Excitatory Amino Acid Research* (Eds. J. Lehman and L. Turski), Allan Liss, New York.
61. Marrannes, R., Willems, R., De Prins, E., and Wauquier, A. (1988) Evidence for a role of the N-methyl-D-aspartate (NMDA) receptor in cortical spreading depression in the rat. *Brain Res.* 457, 226–240.
62. Marrannes, R., De Prins, E., Willems, R., and Wauquier, A. (1988) NMDA antagonists inhibit cortical spreading depression but accelerate the onset of neuronal depolarization induced by asphyxia. In: *Mechanisms of Cerebral Hypoxia and Stroke* (Ed. G. Sanjen), Plenum, New York, pp. 303–304.
63. Benveniste, H., Jorgensen, M. B., Lundbaek, J. A., and Hansen, A. J. (1989) Ionic changes in the normal and denervated hippocampus during ischemia. *J. Cereb. Blood Flow Metab.* 9, s46.
64. Perkins, M. N. and Stone, T. W. (1982) An iontophoretic investigation of the actions of convulsant kynurenines and their interaction with the endogenous excitant quinolinic acid. *Brain Res.* 247, 184–187.
65. Ganong, A. H., Lanthorn, T. H., and Cotman, C. W. (1983) Kynurenic acid inhibits synaptic and acidic amino acid-induced responses in the rat hippocampus and spinal cord. *Brain Res.* 273, 170–174.
66. Ganong, A. H. and Cotman, C. W. (1986) Kynurenic acid and quinolinic acid act as N-methyl-D-aspartate receptors in the rat hippocampus. *J. Pharmacol. Exp. Ther.* 236, 293–299.
67. Katayama, Y., Becker, D. P., Tamura, T., Martin, N., and Tsubokawa, T. (1989) Inhibition of massive ionic fluxes during cerebral ischemia (terminal depolarization) with excitatory amino acid antagonist as demonstrated by microdialysis. *J. Cereb. Blood Flow Metabol.* 9, s57.
68. Katayama, Y., Tamura, T., Becker, D. P., and Tsubokawa, T. (1992) Inhibition of rapid potassium flux during cerebral ischemia *in vivo* with excitatory amino acid antagonist. *Brain Res.* 568, 294–298.

69. Choi, D. W., Koh, J.-Y., and Peters, S. (1988) Pharmacology of glutamate neurotoxicity in cortical cell culture. Attenuation by NMDA antagonist. *J. Neurosci.* 8, 185–196.
70. Simon, R. P., Young, R. S. K., Stout, S., and Cheung, J. (1986) Inhibition of excitatory neurotransmission with kynurenate reduces brain edema in neonatal anoxia. *Neurosci. Lett.* 71, 361–364.
71. Simon, R. P., Swan, J. H., Griffith, T., and Meldrum, B. S. (1984) Blockade of methyl-D-aspartate receptors may protect against ischemic damage in the brain. *Science* 226, 850–852.
72. Germano, I. M., Pitts, L. H., Meldrum, B. S., Bartkowski, H. M., and Simon, R. P. (1987) Kynurenate inhibition of cell excitation decreases stroke size and deficits. *Ann. Neurol.* 22, 730–734.
73. Gill, R. and Woodruff, G. N. (1990) The neuroprotective actions of kynurenic acid and MK-801 in gerbils are synergistic and not related to hypothermia. *Eur. J. Pharmacol.* 176, 143–149.
74. Wako, N., Shima, K., and Marmarou, A. (1989) Time course of brain tissue pressure in temporal fluid percussion injury. In: *Intracranial Pressure,* Vol. 7 (Eds. J. T. Hoff and A. L. Betz), Springer, Berlin, pp. 593–597.
75. Nelson, S. R., Lowry, O. H., and Passonneau, J. V. (1966) Changes in energy reserves in mouse brain associated with compressive head injury. In: *Head Injury* (Eds. Careness, W. F. and Walker, A. W.), Lippincott, Philadelphia, pp. 444–447.
76. Nilsson, B. and Nordstrom, C. H. (1977) Experimental head injury in the rat. Part 3: Cerebral blood flow and oxygen consumption after concussive impact acceleration. *J. Neurosurg.* 47, 262–273.
77. Nilsson, B. and Nordstrom, C. H. (1977) Rate of cerebral energy consumption in concussive head injury in the rat. *J. Neurosurg.* 47, 274–281.
78. Nilsson, B. and Ponten, U. (1977) Experimental head injury in the rat. Part 2: Regional brain energy metabolism in concussive trauma. *J. Neurosurg.* 47, 252–261.
79. Duckrow, R. B., LaManna, J. C., Rosenthal, M., Levasseur, J. E., and Patterson, J. L., Jr. (1981) Oxidative metabolism activity of cerebral cortex after fluid-percussion head injury in the cat. *J. Neurosurg.* 54, 607–614.
80. Yoshino, A., Hovda, D. A., Kawamata, T., Katayama, Y., and Becker, D. P. (1991) Dynamic changes in local cerebral glucose utilization following fluid-percussion injury: evidence of a hyper- and subsequent hypometabolic state. *Brain Res.* 561, 106–119.
81. Tschirgi, R. D., Kazutoyo, I., Taylor, J. L., Walker, R. M., and Sonnenschein, R. R. (1957) Changes in cortical pH and blood flow accompanying spreading cortical depression and convulsion. *Am. J. Physiol.* 190, 557–562.
82. Krivanek, J. (1961) Some metabolic changes accompanying Leao's spreading cortical depression in the rat. *J. Neurochem.* 6, 183–189.
83. Rosenthal, M. and Somjen, G. (1973) Spreading depression sustained potential shifts and metabolic activity of cerebral cortex in cats. *J. Neurophysiol.* 36, 739–745.
84. Mayevsky, A. and Chance, B. (1974) Repetitive patterns of metabolic changes during cortical spreading depression of the awake rat. *Brain Res.* 65, 529–533.
85. LaManna, J. C. and Rosenthal, M. (1975) Effect of ouabain and phenobarbital on oxidative metabolic activity associated with spreading cortical depression in cats. *Brain Res.* 88, 145–149.
86. Shinohara, M., Dollinger, B., Brown, G., Rapoport, S., and Sokoloff, L. (1979) Changes in local cerebral glucose utilization during and following recovery from spreading depression. *Science* 203, 188–190.
88. Gjedde, A., Hansen, A. J., and Quistorff, B. (1981) Blood brain glucose transfer in spreading depression. *J. Neurochem.* 37, 807–812.
89. Kawamata, T., Katayama, Y., Hovda, D. A., Yoshino, A. and Becker, D. P. (1992) Administration of excitatory amino acid antagonists via microdialysis attenuates the increase in glucose utilization seen following concussive brain injury. *J. Cereb. Blood Flow Metab.* 12, 12–24.
90. Yoshino, A., Katayama, Y., Kawamata, T., and Becker, D. P. (1992) Hippocampal CA3 lesion prevents the post-concussive metabolic derangement in CA1. *J. Cereb. Blood Flow Metab.* 12, 996–1006.
91. Myers, R. E. (1979) A unitary theory of causation of anoxic and hypoxic brain pathology. *Adv. Neurol.* 26, 195–217.
92. Ginsberg, M. D., Welsh, F. A., and Budd, W. W. (1980) Deleterious effect of glucose pretreatment on recovery from diffuse cerebral ischemia in the cat. *Stroke* 11, 347–354.
93. Welsh, F. A., Ginsberg, M. D., Rieder, W., and Budd, W. W. (1980) Deleterious effect of glucose pretreatment on recovery from diffuse cerebral ischemia in the cat. *Stroke* 11, 355–363.
94. Kalimo, H., Rehncrona, S., Soderfelt, B., Olsson, Y., and Siesjo, B. K. (1981) Brain lactic acidosis and ischemic cell damage. 2. Histopathology. *J. Cereb. Blood Flow Metab.* 1, 313–327.
95. Pulsinelli, W. A., Waldman, S., Rawlinson, D., and Plum, F. (1982) Moderate hyperglycemia augments ischemic brain damage. A neuropathologic study in the rat. *Neurology* 32, 1239–1246.
96. Rehncrona, S., Rosen, I., and Siesjo, B. K. (1981) Brain lactic acidosis and ischemic cell damage. 1. Biochemistry and neurophysiology. *J. Cereb. Blood Flow Metab.* 1, 2297–311.
97. Howse, D. C. and Duffy, T. E. (1975) Control of the redox state of the pyridine nucleotides in the rat cerebral cortex. Effect of electroshock-induced seizures. *J. Neurochem.* 24, 935–940.

98. Paschen, W., Djuricic, B., Mies, G., Schmidt-Kastner, R., and Linn, F. (1987) Lactate and pH in the brain. Association and dissociation in different pathophysiological states. *J. Neurochem.* 48, 154–159.
99. Somjen, G. G. (1984) Acidification of interstitial fluid in hippocampal formation caused by seizures and by spreading depression. *Brain Res.* 311, 186–188.
100. Mutch, W. A. C. and Hansen, A. J. (1984) Extracellular pH changes during spreading depression and cerebral ischemia. Mechanisms of brain pH regulation. *J. Cereb. Blood Flow Metab.* 4, 17–27.
101. Csiba, L., Paschen, W., and Mies, G. (1985) Regional changes in tissue pH and glucose content during cortical spreading depression in rat brain. *Brain Res.* 336, 167–170.
102. Harris, R. J., Richards, P. G., Symon, L., Habib, A.-H. A., and Rosenstein, J. (1987) pH, K^+, and PO_2 of the extracellular space during ischemia of primate cerebral cortex. *J. Cereb. Blood Flow Metab.* 7, 599–604.
103. Kuhr, W. G. and Korf, J. (1988) Extracellular lactic acid as an indicator of brain metabolism: continuous on-line measurement in conscious, freely moving rats with intrastriatal dialysis. *J. Cereb. Blood Flow Metab.* 8, 130–137.
104. Hallstroem, A., Carlsson, A., Hillered, L., and Ungerstedt, U. (1989) Simultaneous determination of lactate, pyruvate, and ascorbate in microdialysis samples from rat brain, blood, fat, and muscle using high-performance liquid chromatography. *J. Pharmacol. Methods* 21, 113–124.
105. Katayama, Y., Kawamata, T., Kano, T., and Tsubokawa, T. (1991) Excitatory amino acid antagonist administered via microdialysis attenuates lactate accumulation during cerebral ischemia and subsequent hippocampal damage. *Brain Res.* 584, 329–333.
106. Roucher, P., Meric, P., Correze, J. L., Mispelter, J., Tiffon, B., Lhoste, J. M., and Seylaz, J. (1991) Metabolic effects of kynurenate during reversible forebrain ischemia studied by *in vivo* ^{31}P-nuclear magnetic resonance spectroscopy. *Brain Res.* 550, 54–60.
107. Kirino, T. (1982) Delayed neuronal death in the gerbil hippocampus following ischemia. *Brain Res.* 239, 57–69.
108. Pulsinelli, W. A., Brierley, J. B., and Plum, F. (1982) Temporal profile of neuronal damage in a model of transient forebrain ischemia. *Ann. Neurol.* 11, 491–498.
109. Katayama, Y., Tsubokawa, T., Koshinaga, M., and Miyazaki, S. (1991) Temporal pattern of survival and dendritic growth of fetal hippocampal cells transplanted into ischemic lesions of the adult rat hippocampus. *Brain Res.* 562, 352–355.
110. Siesjo, B. K. and Wieloch, T. (1985) Molecular mechanisms of ischemic brain damage: calcium related events. In: *Cerebrovascular Diseases* (Eds. F. Plum and W. Pulsinelli), Raven Press, New York, pp. 187–197.
111. Yang, M. S., DeWitt, D. S., Becker, D. P., and Hayes, R. L. (1985) Regional brain metabolite levels following mild experimental head injury in the cat. *J. Neurosurg.* 63, 617–621.
112. Hovda, D. A., Becker, D. P., and Katayama, Y. (1992) Central nervous system trauma: secondary injury and acidosis. *J. Neurotrauma*, 9 suppl., 47–60.
113. Kawamata, T., Katayama, Y., Hovda, D. A., Yoshino, A., and Becker, D. P. (1995) Lactate accumulation following concussive brain injury. The role of ionic fluxes induced by excitatory amino acids. *Brain Res.* in press.

Chapter 28

MEASUREMENT OF GLUTAMATE AND OTHER NEUROTRANSMITTERS IN ISCHEMIC BRAIN BY MICRODIALYSIS: EFFECTS OF HYPOTHERMIA AND HYPERTHERMIA

Myron D. Ginsberg, Raul Busto, and Mordecai Y.-T. Globus

TABLE OF CONTENTS

I. INTRODUCTION

The technique of intracerebral microdialysis is uniquely useful in studying the pathophysiology of cerebral ischemia in that it permits local sampling of the extracellular fluid environment of the brain in serial fashion; samples may then be assayed by high-performance liquid chromatography (HPLC) or other methods for a great variety of neurotransmitters, neuromodulators, and other chemical intermediates of interest.

II. BASIC METHODS

Microdialysis probes are double-lumen steel cannulae fitted with ports for perfusion (via the inner cannula) and perfusate collection (via the outer cannula). At the distal end of the probe is a common space ensheathed by a semipermeable membrane that permits the perfusate to dialyse the extracellular environment in which it is placed. Reliable probes, used in all of our studies, are manufactured by Carnegie Medicin, Sweden. Probes may be acquired

0-8493-8094-4/95/$0.00+$.50

with membrane-tip diameters ranging from 380 to 640 μm; the exposed membrane length is typically 4 mm. Smaller probes are desirable for sampling from small structures (e.g., hippocampus); however, the smaller the membrane length, the lower the recovery fraction for substances being sampled. Microdialysis probes are inserted into brain regions of interest (e.g., striatum, hippocampus, neocortex, or thalamus) via stereotaxic electrode holders. A microinfusion pump (Carnegie Medicin, Sweden) is used to perfuse the probe at a rate of approximately 2 μl/min with either mock cerebrospinal fluid or with a modified Ringer's solution, whose composition (m*M*) is NaCl, 147; KCl, 3.0; and $CaCl_2$, 1.3. The perfusate effluent from the probe may be collected manually or via an automated sample collector (Bioanalytical Systems, Inc.).

Probes may be selected with pore sizes to permit upper molecular weight cutoffs of up to 20 kD. Thus, classic neurotransmitters and intermediary metabolites are readily sampled whereas larger peptides are not. Fractional recovery for a given substance via a microdialysis probe may be measured by means of a standard solution *in vitro*. Recovery fraction generally ranges from 5 to 20% and diminishes with increasing flow rate of the perfusate, and with decreasing membrane size. Data from microdialysis experiments may then be corrected for recovery fraction and expressed in units of concentration per milliliter of extracellular fluid. Microdialysis probe recovery decreases somewhat as temperature is lowered.[2] Therefore, when microdialysis is used in studies of brain hypothermia, *in vitro* determination of probe recovery fraction should be performed at the same temperatures as those of the animal study itself. Benveniste[2] has considered the quantitative aspects of microdialysis probe recovery in detail.

III. ASSAY METHODS

Amino acid neurotransmitters such as glutamate, glycine, and GABA in microdialysis perfusate samples are assayed by HPLC with electrochemical detection[18] following pre-column derivatization.[1] A two-step derivatization process is used. In the first step, a combination of *o*-phthalaldehyde and *t*-butylthiol is used for the formation of isoindole. The substitution of *t*-butylthiol for the more commonly used 2-mercaptoethanol improves the derivative's stability. To scavenge leftover *t*-butylthiol, iodoacetamide is added in a second step. Both the derivatization and injection of the samples may be carried out under automatic control using a refrigerated microsampler system (BAS/CMA 200, Bioanalytical Systems). We are currently using a BAS 200 liquid chromatograph equipped for gradient operation and amperometric detection (Bioanalytical Systems). Amino acids are separated on a 0.32 × 10-cm Biophase II C-8 cartridge (particle size, 3 μm) (Bioanalytical Systems). The glassy carbon electrode is maintained at 0.7 V vs. Ag/AgCl. The mobile phases consist of pH 6.8 acetate/acetonitrile/tetrahydrofuran mixtures (solvent A, 67.8:27.4:4.8 by volume, respectively; solvent B, 15:45:40, respectively). The gradient profile is based on simple linear changes in both solvents. This method provides good resolution of the common amino acids with an elution time of approximately 30 min. A sharp solvent ramp is used to strip the column after the leucine peak. From peak heights, dialysate concentrations of glutamate, GABA, and glycine may be computed by the use of external standards.[18]

To assay **catecholamine neurotransmitters** such as dopamine, the HPLC system consists of a solvent delivery module (BAS PM-30A) and a coulometric detector (ESA Model 5100A) with multielectrode capability, and a laboratory computing integrator (Spectraphysics Model SP 4270) for data acquisition. The multielectrode combination consists of a conditioning cell (Model 5021) and a high-sensitivity analytical cell (Model 5011). The separation of catecholamines is accomplished by a reverse-phase analytical column (Beckman Ultrasphere ODS 5 μm; 4.6 mm ID × 25 cm length) maintained at a temperature of 40.1°C, with a mobile phase consisting of 0.09 *M* sodium acetate, 0.04 *M* citric acid (pH 4.5), 16.2 mg octane sulfonate, and 8% methanol. The electrode potentials selected for our determinations are as follows: conditioning cell, +0.3 V; analytical cell detector 1, 0.3 V; detector 2, 0.43 V. A current-voltage curve was previously analyzed to arrive at these potentials. The identification of peaks is accomplished by running parallel external standards. The sensitivity of the system allows us to detect dopamine and its major metabolites (DOPAC and HVA) at the level of picomoles.

IV. DATA REDUCTION AND ANALYSIS

Commercial software is available to facilitate the computation of dialysate concentrations from HPLC peak heights of samples and external standards. The current software is capable of performing these functions on-line. The recovery fraction determined *in vitro* with standard solutions may then be used to convert microdialysate concentrations from micro- or nanomoles per milliliter of perfusate to micro- or nanomoles per milliliter of extracellular fluid. The constraints of the recovery fraction and the assay sensitivity of HPLC for the substance being measured lead to the requirement that each microdialysis sample be collected typically over at least a 5 to 10 min epoch, to assure sufficient quantities to permit reliable assay. The multiple data points resulting from the assay of sequential samples in a given study are appropriately analyzed by a repeated-measures analysis of variance statistical design.

V. APPLICATION OF MICRODIALYSIS TO CEREBRAL ISCHEMIA: AMINO ACID NEUROTRANSMITTERS

Glutamate is the major excitatory neurotransmitter in the central nervous system. Under pathologic conditions, there is now abundant evidence that glutamate-mediated excitotoxicity contributes to the pathophysiology of ischemic, traumatic, and epileptic brain damage.[3,10,26] Studies from many laboratories have documented the massive release of glutamate from cells into the brain extracellular space during both global and focal ischemic insults (summarized in References 3 and 14). Below, we review our own studies carried out at the Cerebral Vascular Disease Research Center at the University of Miami.

In rats subjected to 20 min of forebrain ischemia by four-vessel occlusion combined with systemic hypotension, extracellular neurotransmitter levels measured in the striatum by microdialysis showed an approximately sevenfold increase in glutamate and a fivefold increase in GABA contents during ischemia; these levels fell toward baseline over the first 30 min of the reperfusion period.[15]

We next compared whether differences in neurotransmitter release among various brain structures could explain the relative vulnerability of these regions to ischemia.[18] Extracellular levels of glutamate, GABA, and glycine were measured in rats with 10 or 20 min of global cerebral ischemia produced by bilateral carotid occlusions plus hypotension. Two microdialysis probes were placed in the brain of each rat: either in the dorsal hippocampus and contralateral somatosensory cortex; or in the dorsolateral striatum and contralateral anterior thalamus. In a matched series, neuropathologic evaluation established that hippocampal CA_1 neurons were consistently injured by both 10 and 20 min of ischemia; that the striatum was injured by 20 but not 10 min of ischemia; and that the thalamus and cortex were largely spared following both durations. Despite these differences in regional vulnerability, the temporal profiles of ischemia-induced increases in glutamate and GABA were the same for all forebrain regions assayed. Glycine increases were also similar among regions during ischemia, but glycine levels remained higher during the recirculation period in the hippocampus and striatum (after the 20-min insult) than in the thalamus or cortex.[18] These data thus revealed a similar pattern of glutamate release in both vulnerable and nonvulnerable brain structures and suggested that glutamate release of itself is not sufficient to engender ischemic injury.

As glycine is a coneurotransmitter necessary for activation of NMDA receptors by glutamate,[30] and as GABA is a ubiquitous inhibitory neurotransmitter, we posited that the interplay of *multiple* neurotransmitters might be a factor influencing selective vulnerability. We defined an "excitatory index"[20] as

$$\text{Excitotoxic index} = [\text{glutamate}] \times [\text{glycine}]/[\text{GABA}]$$

During 20 min of ischemia in rats produced by two-vessel occlusion, the excitotoxic index increased to a similar degree in striatum and thalamus, while during postischemic recirculation, a marked increase in the excitotoxic index was observed in the striatum, a structure vulnerable to this degree of ischemia, but not in the thalamus—a structure spared from ischemic injury in this setting (Figure 1).[20] When the effects of 10- and 20-min ischemic insults were compared, the marked increase in the excitotoxic index in the striatum seen with the 20-min insult was not observed following only 10 min of ischemia (Figure 1).[18] In the hippocampus, the excitotoxic index increased following both durations of ischemia, but to a greater degree following the 20-min insult (Figure 1).[18] These observations parallel closely the relative vulnerability of these structures to the two durations of ischemia. These data thus support the view that multiple neurotransmitters are important in selective ischemic injury.

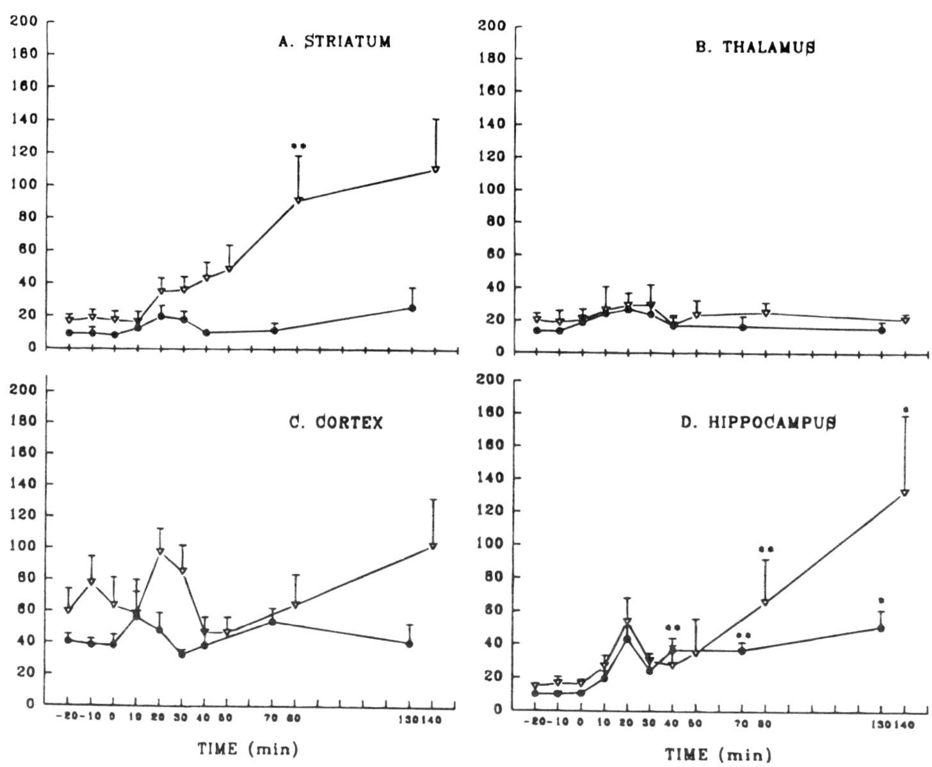

FIGURE 1. Time course of changes in the excitotoxic index computed from extracellular levels of glutamate, glycine, and GABA in (A) dorsolateral striatum, (B) anterior thalamus, (C) somatosensory cortex, and (D) dorsal hippocampus in rats subjected to 10 or 20 min of forebrain ischemia. The significance of increases from baseline values was assessed by two-way analysis of variance. $*p < 0.01$; $**p < 0.05$. (From Globus, M. Y.-T. et al., *J Neurochem*, 57, 470–478, 1991. With permission.)

Interestingly, *multiple* ischemic insults (e.g., three 5-min periods of global ischemia separated by 1 h) lead to an accelerated evolution of neuropathological changes and to damage to structures (e.g., thalamus) not ordinarily injured by a single insult of similar total duration (e.g., 15 min).[24] We have shown that there is a delayed secondary release of glutamate in the striatum following the third multiple insult, not seen following single insults; and a corresponding marked increase (7- to 12-fold) in the excitotoxic index at 1–3 h following the third multiple insult.[24]

In *focal ischemia*, we have shown that during middle cerebral artery (MCA) occlusion in rats, extracellular glutamate measured from the penumbral cortex (i.e., a zone of moderate ischemia adjacent to the severely ischemic core region) undergoes a massive increase (26-fold) over the first 20 to 30 min of ischemia but subsequently falls to baseline levels even before the MCA occlusion is reversed at 2 h.[28] This finding suggests that glutamate reuptake mechanisms still remain functional in the penumbral cortex in this setting. Correlative laser Doppler CBF measurements from the same cortical site showed that the CBF threshold for glutamate release was 48% of control CBF, and *massive* glutamate release was observed at CBF values of 20% of control or below. Infarct volume correlated with the magnitude of glutamate release.[28] Contrasting with our findings in the ischemic penumbra, other laboratories using microdialysis in the ischemic core of focal occlusion models have shown that increases in extracellular glutamate are sustained following permanent MCA occlusion.[9]

VI. EFFECT OF ALTERED BRAIN TEMPERATURE

Our group has demonstrated,[6] and many other groups have confirmed, that mild-to-moderate reductions of brain temperature (e.g., from 36–37°C to 33 or 30°C) confer marked protection from ischemic injury produced by 10 or 20 min of global ischemia (see Reference 13 for review). In studies utilizing 20 min of global ischemia by four-vessel occlusion, in which microdialysis was used to sample extracellular neurotransmitters, glutamate levels increased sevenfold and extracellular dopamine increased 500-fold above baseline during the ischemic and early recirculation periods in rats with normothermic brain temperature.[7] By contrast, if brain temperature were reduced to 33 or 30°C during ischemia, glutamate release was completely inhibited (Figure 2A).

In the case of dopamine, intraischemic brain hypothermia caused peak levels during ischemia to fall by about 60% compared to normothermic ischemia (Figure 2B). These effects may help to explain why mild intraischemic hypothermia protects the brain.

In contrast to the protective effects of mild hypothermia, elevations of brain temperature of even mild degree (e.g., 39°C), if present during ischemia, markedly exacerbate outcome by extending neuropathologic changes to brain regions that are unaffected by normothermic insults,[12] and by accentuating blood-brain barrier disruption.[11] In global ischemia carried out at a brain temperature of 39°C, the magnitude of glutamate release was 1.8-fold greater than during ischemia at 37°C, and glutamate levels tended to remain higher during the postischemic recirculation period.[27] Extracellular GABA levels were also higher with hyperthermic than with normothermic ischemia (50-fold vs. 41-fold), as were glycine levels. The computed excitotoxic index differed markedly, increasing only twofold above baseline following normothermic ischemia but 20-fold following hyperthermic ischemia.[27]

In *focal* ischemia produced by a 2-h period of MCA occlusion in rats, a brain temperature of 39°C was associated with a 31-fold increase in extracellular glutamate in the ischemic penumbra, compared to only a sevenfold increase in normothermic rats.[29] The ischemic CBF threshold (expressed as a percentage of control CBF) for glutamate release in this study was 33% in normothermic rats and 61% in the hyperthermic group, while the actual mean level of CBF during ischemia (24% of control) did not differ between the two groups. Thus, mild

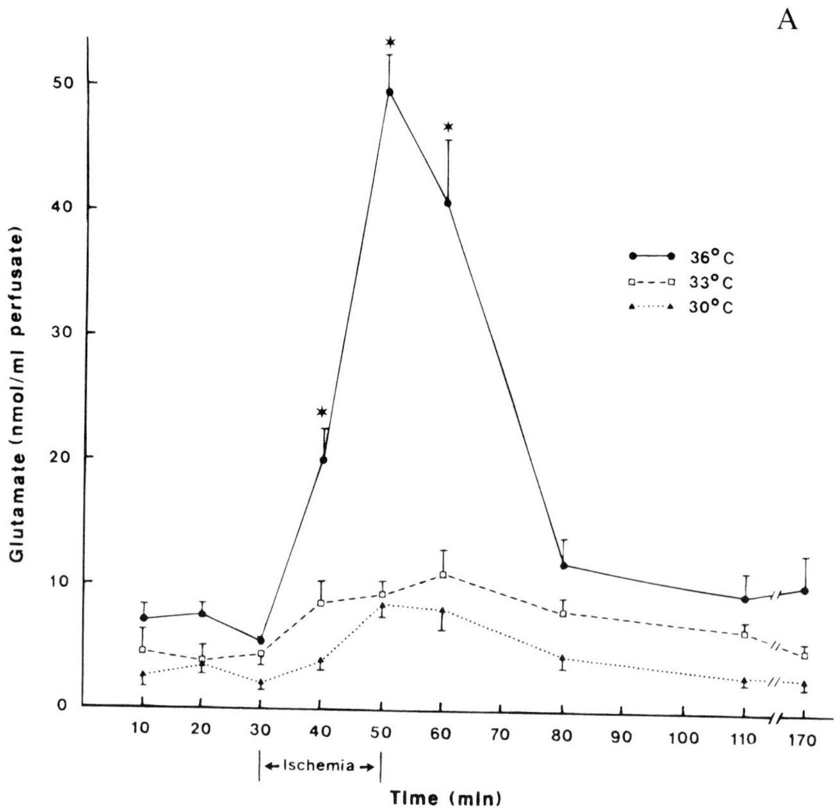

FIGURE 2. Plots of the time-course of changes in striatal microdialysis perfusate levels of glutamate (nmol/ml) (A) and dopamine (pmol/ml) (B) in animals whose intraischemic brain temperature was maintained at either 36°C (n = 10, filled circles), 33°C (n = 4, open squares), or 30°C (n = 8, filled triangles). The data are shown as mean values ± SEM. Statistical significance was assessed by two-way ANOVA. In animals whose intraischemic brain temperature was maintained at 36°C, massive increases in extracellular glutamate (A) and dopamine levels (B) were observed (* = significantly higher than control values, $p < 0.01$). By contrast, in rats with intraischemic brain temperatures of 33 or 30°C, glutamate levels failed to rise (A), and ischemia-induced increases in striatal dopamine levels were significantly lower than in the 36°C group (a = significantly lower than the corresponding level in the 36°C group, $p < 0.01$ in Figure 2B). (From Busto, R. et al., *Stroke* 20, 904–910, 1989. With permission.)

brain hyperthermia both accentuates glutamate release and alters the CBF threshold at which this release first occurs.

VII. OTHER NEUROTRANSMITTERS IN ISCHEMIA

The striatum is known to be vulnerable to 20 to 30 min of global ischemia. We suspected that *dopamine*, a major striatal neurotransmitter, might play a role in this process. We were able to demonstrate that a unilateral lesion of the substantia nigra (SN) was capable of protecting the ipsilateral striatum from neuronal pathologic changes following a bilateral global ischemic insult of 20 min carried out two weeks after the SN lesion,[21] and diminished the extent of postischemic metabolism/blood flow uncoupling.[22] Microdialysis in normal rats with 20 min ischemia revealed 500-fold increases in extracellular dopamine, peaking within the first 10 min of ischemia and falling to baseline by 30 min of recirculation. By contrast, in rats with prior SN lesion, there was a complete blunting of dopamine release from the

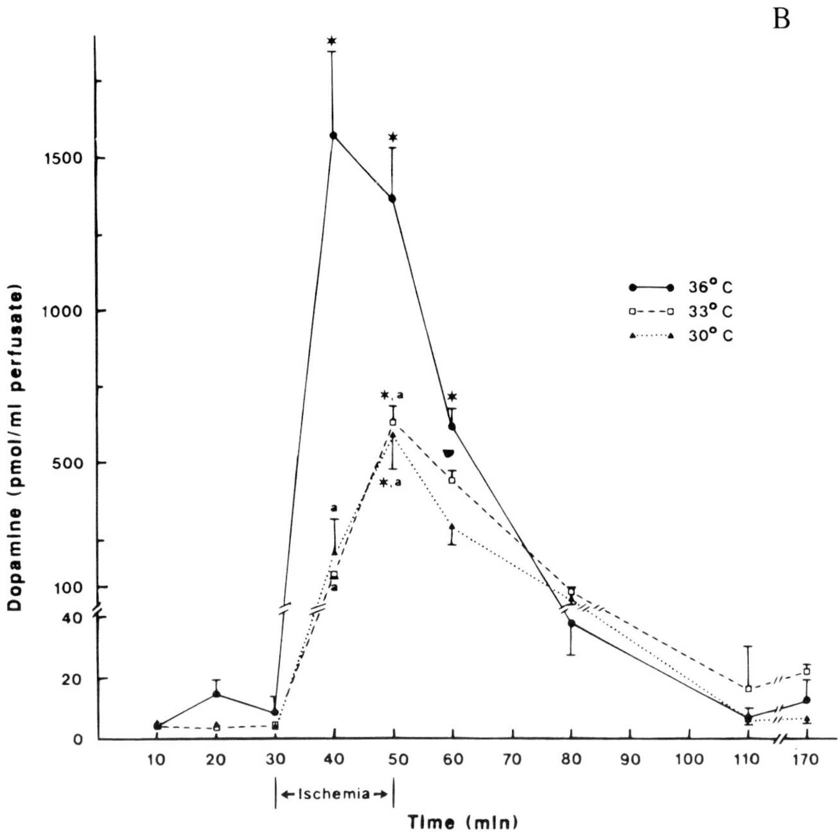

FIGURE 2. Continued.

striatum on the SN-lesioned side.[15,16] Prior SN lesion also attenuated striatal glutamate release during ischemia by approximately 40%.[15]

Many other neurotransmitters have been implicated in the process leading to (or protecting from) ischemic injury. These include norepinephrine, serotonin, and adenosine. Studies from our laboratory have explored the release of these neurotransmitters during ischemia by intracerebral microdialysis.[8,17,23] The role of nonglutamatergic neurotransmission in ischemic injury has been reviewed by our group.[19]

VIII. OTHER APPLICATIONS OF INTRACEREBRAL MICRODIALYSIS

Microdialysis can be used to sample the extracellular space for alterations in second messengers. In a recent microdialysis study from our laboratory, 20 min of global ischemia was shown to induce a threefold rise in extracellular levels of cyclic AMP (cAMP) in the striatum. Treatment with a dopamine D1 antagonist significantly attenuated this rise, indicating that activation of D1 receptors is involved in the generation of cAMP during ischemia.[25]

The microdialysis technique can also be used for local administration of drugs. The advantage of this approach is that the drug can be continuously infused locally into the brain without increasing the volume of extracellular space. Furthermore, this enables one to evaluate the effects of drugs on biochemical changes occurring in the vicinity of the probe. A recent study using microdialysis combined with magnetic resonance imaging has demonstrated that infusion of glutamate or NMDA affects a zone of approximately 1.5 mm around the probe.[5] This indicates that the effects of drugs administered via the probe extend to a significant number of neurons around the probe site.

IX. TECHNICAL CONSIDERATIONS

A disadvantage of the brain dialysis technique is the possibility of tissue trauma produced by implantation of the probe. We and others have performed histological studies and have found that the probe cannula produces only a narrow track through the tissue, leaving cells surrounding the dialysis fiber relatively intact. We routinely allow a 2-h period to elapse following probe placement to achieve tissue stabilization prior to collection of microdialysis perfusate. We then begin to perfuse the microdialysis probe and to collect samples over a 30-min baseline period (typically, three baseline samples of 10 min each). The expected finding in these samples is of low and stable levels of neurotransmitters throughout this baseline. The ischemic insult is then initiated.

Studies of local cerebral glucose utilization and blood flow have revealed consistent changes within 3 h following implantation of microdialysis probes.[4] Thus, some investigators prefer to insert the probe up to 24 h prior to the actual experiment to achieve thorough stabilization *in situ.*[2] Microdialysis probes may also be implanted chronically and used to record neurotransmitters in awake rats over a several-day period by means of a swivel assembly that permits free movement of the animal within a bowl-shaped enclosure (Bioanalytical Systems). Microdialysis membranes tend to deteriorate after several days in tissue owing to protein deposition and clogging of the membrane.

X. CONCLUSION

Microdialysis is a powerful means of obtaining serial regional information as to the neurochemistry of the brain's extracellular space under normal and abnormal conditions. Its potential has yet to be fully exploited.

ACKNOWLEDGMENTS

Our research is supported by USPHS Grant NS 05820. We are indebted to Ms. Elena Martinez for superb technical assistance in our microdialysis studies, and to Ms. Helen Valkowitz for preparing the typescript.

REFERENCES

1. Allison, L. A., Mayer, G. S., and Shoup, R. E., *o*-Phthalaldehyde derivatives of amines for high-speed liquid chromatography/electrochemistry, *Anal Chem,* 56, 1089–1096, 1984.
2. Benveniste, H., Short review. Brain microdialysis, *J Neurochem,* 52, 1667–1679, 1989.
3. Benveniste, H., The excitotoxin hypothesis in relation to cerebral ischemia, *Cerebrovasc Brain Metab Rev,* 3, 213–245, 1991.
4. Benveniste, H., Drejer, J., Schousboe, A., and Diemer, N. H., Regional cerebral glucose phosphorylation and blood flow after insertion of a microdialysis fiber through the dorsal hippocampus in the rat, *J Neurochem,* 49, 729–734, 1987.
5. Benveniste, H., Hedlund, L. W., and Johnson, G. A., Mechanism of detection of acute cerebral ischemia in rats by diffusion-weighted magnetic resonance microscopy, *Stroke,* 23, 746–754, 1992.
6. Busto, R., Dietrich, W. D., Globus, M. Y.-T., Valdes, I., Scheinberg, P., and Ginsberg, M. D., Small differences in intraischemic brain temperature critically determine the extent of ischemic neuronal injury, *J Cereb Blood Flow Metab,* 7, 729–738, 1987.
7. Busto, R., Globus, M. Y.-T., Dietrich, W. D., Martinez, E., Valdes, I., and Ginsberg, M. D., Effect of mild hypothermia on ischemia-induced release of neurotransmitters and free fatty acids in rat brain, *Stroke,* 20, 904–910, 1989.

8. Busto, R., Globus, M. Y.-.T., Martinez, E., Valdes, I., and Ginsberg, M. D., Comparative effect of transient global ischemia on extracellular levels of adenosine in vulnerable and non-vulnerable brain regions in the rat, *J Cereb Blood Flow Metab,* 11 Suppl 2, S210, 1991.
9. Butcher, S. P., Bullock, R., Graham, D. I., and McCulloch, J., Correlation between amino acid release and neuropathologic outcome in rat brain following middle cerebral artery occlusion, *Stroke,* 21, 1727–1733, 1990.
10. Choi, D. W., Excitotoxic cell death, *J Neurobiol,* 23, 1261–1276, 1992.
11. Dietrich, W. D., Busto, R., Halley, M., and Valdes, I., The importance of brain temperature in alterations of the blood-brain barrier following cerebral ischemia, *J Neuropathol Exp Neurol,* 49, 486–497, 1990.
12. Dietrich, W. D., Busto, R., Valdes, I., and Loor, Y., Effects of normothermic versus mild hyperthermic forebrain ischemia in rats, *Stroke,* 21, 1318–1325, 1990.
13. Ginsberg, M. D., Sternau, L. L., Globus, M. Y.-.T., Dietrich, W. D., and Busto, R., Therapeutic modulation of brain temperature: relevance to ischemic brain injury, *Cerebrovasc Brain Metab Rev,* 4, 189–225, 1992.
14. Ginsberg, M. D., Takagi, K., and Globus, M. Y.-.T., Release of neurotransmitters in cerebral ischemia: relevance to ischemic injury, in *Pharmacology of Cerebral Ischemia,* Krieglstein, J. and Oberpichler-Schwenk, H., Eds., Wissenschaftl Verlagsgesellschaft, Stuttgart, 1992, 177–189.
15. Globus, M. Y.-.T., Busto, R., Dietrich, W. D., Martinez, E., Valdes, I., and Ginsberg, M. D., Effect of ischemia on the *in vivo* release of striatal dopamine, glutamate, and gamma-aminobutyric acid studied by intracerebral microdialysis, *J Neurochem,* 51, 1455–1464, 1988.
16. Globus, M. Y.-.T., Busto, R., Dietrich, W. D., Martinez, E., Valdes, I., and Ginsberg, M. D., Intra-ischemic extracellular release of dopamine and glutamate is associated with striatal vulnerability to ischemia, *Neurosci Lett,* 91, 36–40, 1988.
17. Globus, M. Y.-.T., Busto, R., Dietrich, W. D., Martinez, E., Valdes, I., and Ginsberg, M. D., Direct evidence for acute and massive norepinephrine release in the hippocampus during transient ischemia, *J Cereb Blood Flow Metab,* 9, 892–896, 1989.
18. Globus, M. Y.-.T., Busto, R., Martinez, E., Valdes, I., Dietrich, W. D., and Ginsberg, M. D., Comparative effect of transient global ischemia on extracellular levels of glutamate, glycine, and gamma-aminobutyric acid in vulnerable and nonvulnerable brain regions in the rat, *J Neurochem,* 57, 470–478, 1991.
19. Globus, M. Y.-.T. and Ginsberg, M. D., Non-glutamatergic neurotransmission: role in ischemic brain injury, in *Neurotransmission and Cerebrovascular Function II,* Seylaz, J. and Sercombe, R., Eds., Elsevier Science Publishers, Amsterdam, 1989, 467–480.
20. Globus, M. Y.-.T., Ginsberg, M. D., and Busto, R., Excitotoxic index—a biochemical marker of selective vulnerability, *Neurosci Lett,* 127, 39–42, 1991.
21. Globus, M. Y.-.T., Ginsberg, M. D., Dietrich, W. D., Busto, R., and Scheinberg, P., Substantia nigra lesion protects against ischemic damage in the striatum, *Neurosci Lett,* 80, 251–256, 1987.
22. Globus, M. Y.-.T., Ginsberg, M. D., Harik, S. I., Busto, R., and Dietrich, W. D., Role of dopamine in ischemic striatal injury: metabolic evidence, *Neurology,* 37, 1712–1719, 1987.
23. Globus, M. Y.-.T., Wester, P., Busto, R., and Dietrich, W. D., Ischemia-induced extracellular release of serotonin plays a role in CA1 neuronal cell death in rats, *Stroke,* 23, 1595–1601, 1992.
24. Lin, B., Globus, M. Y.-.T., Dietrich, W. D., Busto, R., Martinez, E., and Ginsberg, M. D., Differing neurochemical and morphological sequelae of global ischemia: comparison of single- and multiple-insult paradigms, *J Neurochem,* 59, 2213–2223, 1992.
25. Prado, R., Busto, R., and Globus, M. Y.-.T., Ischemia-induced changes in extracellular levels of striatal cyclic AMP: role of dopamine neurotransmission, *J Neurochem,* 59, 1581–1584, 1992.
26. Rothman, S. M. and Olney, J. W., Glutamate and the pathophysiology of hypoxic-ischemic brain damage, *Ann Neurol,* 19, 105–111, 1986.
27. Sternau, L. L., Globus, M. Y.-.T., Dietrich, W. D., Martinez, E., Busto, R., and Ginsberg, M. D., Ischemia-induced neurotransmitter release: effects of mild intraischemic hyperthermia, in *The Role of Neurotransmitters in Brain Injury,* Globus, M. Y.-.T. and Dietrich, W. D., Eds., Plenum Press, New York, 1992, 33–38.
28. Takagi, K., Ginsberg, M. D., Globus, M. Y.-.T., Dietrich, W. D., Martinez, E., Kraydieh, S., and Busto, R., Changes in amino acid neurotransmitters and cerebral blood flow in the ischemic penumbral region following middle cerebral artery occlusion in the rat: correlation with histopathology, *J Cereb Blood Flow Metab,* 13, 575–585, 1993.
29. Takagi, K., Ginsberg, M. D., Globus, M. Y.-.T., Martinez, E., and Busto, R., The effect of hyperthermia on the release of glutamate in the ischemic penumbral region following middle cerebral artery occlusion in rats, *Am J Physiol.,* 35, H1770–1776, 1994.
30. Wong, E. H. F. and Kemp, J. A., Sites for antagonism on the *N*-methyl-D-aspartate receptor channel complex, *Annu Rev Pharmacol Toxicol,* 31, 401–425, 1991.

Chapter 29

A MICRODIALYSIS MODEL FOR IDENTIFYING AGENTS OF SECONDARY DAMAGE IN SPINAL CORD INJURY

Danxia Liu and David J. McAdoo

TABLE OF CONTENTS

ABSTRACT

We present an *in vivo* model for characterizing agents of secondary damage upon spinal cord injury. Applications to glutamate, potassium ions, and the hydroxyl radical are discussed.

I. INTRODUCTION

Damage continues to worsen for hours and days after primary injuries to the central nervous system.[1] Some of this secondary damage is attributed to the release of substances such as amines, excitatory amino acids, and eicosanoids, and the generation of reactive oxygen species such as the hydroxyl radical. Blocking the damage caused by these substances seems the only possible short-term way to reduce long-term impairments resulting from CNS injury.

In vivo experiments are "absolutely necessary to fully characterize the multifactorial secondary injury and to test pharmacological agents."[2] To establish that an agent causes secondary damage, it must be demonstrated that the substance is released upon injury, and that concentrations comparable to those released damage neurons. We accomplish this by using microdialysis to sample substances released upon experimental spinal cord injury[3,4] and to administer candidate released agents into the spinal cord to characterize the damage

0-8493-8094-4/95/$0.00+$.50

they produce.[5–8] Evoked potentials are recorded to determine the effects of the candidate on electrical conduction. Histological examination of the exposed tissue reveals whether neurons are lost. Analysis of microdialysates shows whether other damaging substances are released. Combining release measurements with determining the effects of administering substances into tissue is a powerful approach to identifying agents of secondary damage. We illustrate the usefulness of this approach in studies of the excitatory amino acid (EAA) glutamate, the depolarizing potassium ion, the hydroxyl free radical, and the effect of methylprednisolone (MP), a drug that ameliorates long-term impairments from spinal cord injury,[9] on glutamate release.[10]

II. SYSTEMS STUDIED

A. THE EFFECTS OF POTASSIUM IONS

We investigated K^+ as a candidate damaging agent because the concentration of extracellular K^+ ($[K^+]_e$) rises substantially upon injury to the spinal cord,[11,12] and elevated K^+ can kill neurons.[13,14] Loss of electrical conduction accompanies the rise in $[K^+]_e$ and a transient recovery in conduction accompanies a subsequent decline in $[K^+]_e$.[15] A role of K^+ in injury-induced damage has been suggested.[16] However, the possibility that elevated $[K^+]_e$ contributes to secondary damage upon spinal cord injury has not been emphasized.

Our experimental setup is shown in Figure 1. K^+ administration very rapidly blocked electrical signals in the spinal cord rostral to the dialysis zone (Figure 2) that were elicited by stimulation of the sciatic nerve. Responses to the same stimulus caudal to the dialysis zone were little affected. Pre- and postsynaptic potentials were rapidly blocked during exposure to K^+. Return to artificial cerebrospinal fluid (ACSF) after 2 h of exposure of axons to 100 m*M* K^+ was followed by about 18% recovery. Postmortem histological examination revealed

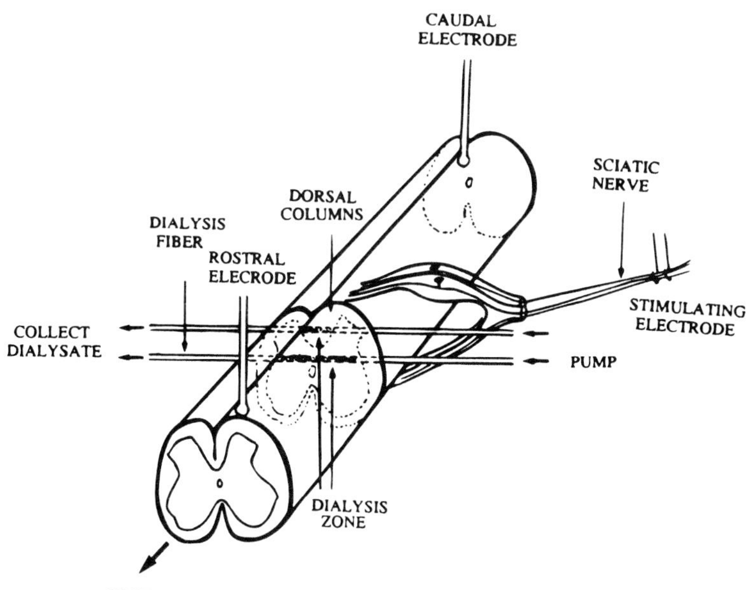

FIGURE 1. Experimental setup for collecting and administering substances in the spinal cord by microdialysis while recording effects on responses to electrical stimulation. Microdialysis fibers with zones in the white (upper fiber) and gray (lower fiber) matter are shown. The rostral caudal electrode is used to record responses to sciatic nerve stimulation before they pass the dialysis zone and the rostral electrode is used to record responses to sciatic nerve stimulation after they pass the dialysis zone. Experimental procedures are described in more detail in the text. (From Liu, D. and McAdoo, D., *J. Neurotrauma*, 10, 351, 1993. With permission.)

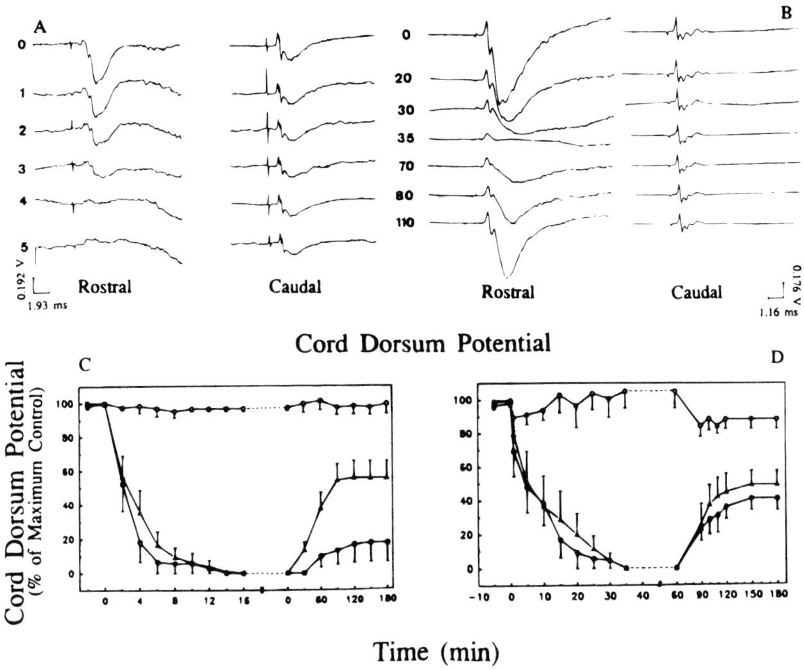

FIGURE 2. Cord dorsum potentials caused by sciatic nerve stimulation while administering K^+ through a microdialysis fiber. Recording arrangements are illustrated in Figure 1. The left records were to administration in the white matter and the right responses to administration in the gray matter. The upper panels are actual records from individual experiments and the lower panels are average responses (N = 6 in C and 5 in D; error bars are SD). The numbers on the left of the traces in the upper panel are the number of minutes after the onset of K^+ administration. In C and D, open circles are caudal recordings, filled circles are rostral recordings of presynaptic responses, and triangles are rostral records of postsynaptic responses. At time 0, the fluid passing through the fiber was switched from one containing ACSF to one containing 1000 m*M* K^+. This gave a concentration of about 100 m*M* K^+ in the tissue near the fiber. The perfusion fluid was returned to ACSF after 2 h for the last 3 h of the experiment. The decreasing responses in C and D are blockage by K^+ and the increasing responses on the right sides are recovery upon return to ACSF. Zero time for the recovery trace in D is not shown. (From Liu, D. and McAdoo, D., *J. Neurotrauma*, 10, 355, 1993. With permission.)

loss of most neuronal cell bodies around the dialysis zone of the fiber (Figure 3). K^+ also caused release of amino acids, particularly neurotransmitters. The substantial release of K^+ that accompanies spinal cord injury may contribute to secondary damage. Of particular interest is that K^+ blocks axonal conduction, because damage to axons produces most of the crippling that follows spinal cord injury.

B. THE EFFECTS OF THE NEUROTRANSMITTER AND EXCITOTOXIN GLUTAMATE ON SPINAL CORD NEURONS

Elevated glutamate and aspartate are highly toxic to neurons,[17] and they are released upon spinal cord injury.[4,18] We used microdialysis to estimate the concentrations of amino acids released upon injury[4] and then to characterize the damage glutamate causes by administering it into the spinal cord.[6]

We estimated that glutamate concentrations in the extracellular space following injury are in the range 0.14 to 1 m*M*. Based on an estimated tenfold concentration gradient across the wall of the administering fiber and a decreasing concentration with distance from the fiber, we passed 10- and 100-m*M* solutions of glutamate in ACSF through the fiber. This gave estimated 1- and 10-m*M* concentrations in the tissue around the fiber. Histological examination revealed a significant loss of neuronal cell bodies in both treatments with an estimated

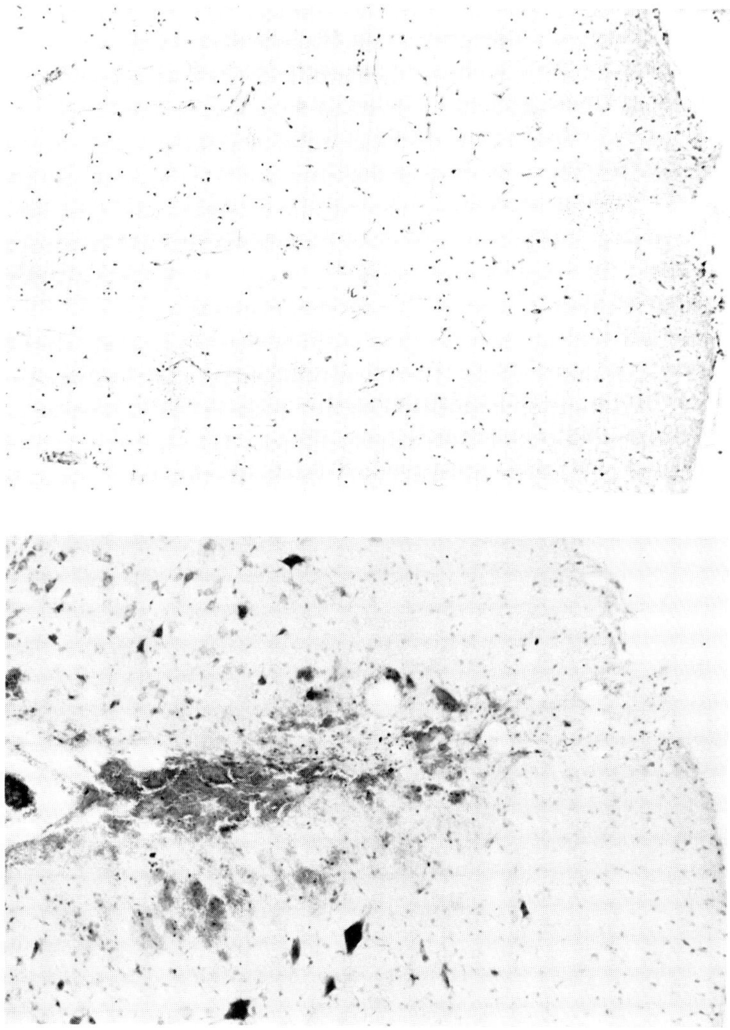

FIGURE 3. Destruction of neurons upon administration of KCl into the rat spinal cord as described in Figure 2. Sections were prepared as described in the text. All sections are at the same magnification. The larger dark bodies in each section are neuronal cell bodies and the smaller dark bodies are predominantly glia. (A) After exposure to KCl. The fiber track passes horizontally across the middle of the section. (B) Following administration of ACSF. (From Liu, D. and McAdoo, D., *J. Neurotrauma*, 10, 457, 1993. With permission.)

threshold of 1 m*M* glutamate, the highest level of glutamate released upon experimental injury. The real threshold is probably lower following injury because: the loss of neurons near threshold may take longer than we waited to become detectable; the depolarizing conditions that exist following injury would increase the sensitivity of the NMDA receptor to glutamate; removal mechanisms probably reduced the concentration below that which we estimated at short distances from the fiber; and the effects of other released damaging substances probably add to those of glutamate.

C. THE EFFECT OF METHYLPREDNISOLONE, A PROTECTIVE DRUG, ON GLUTAMATE RELEASE

It is important to understand the protective mechanisms of drugs. MP reduces long-term impairment in humans.[9] This is the first treatment that reduces crippling from spinal cord injury in humans; it raises hope for further developing ways to alleviate permanent impairments caused by such injury. EAA release probably contributes to secondary damage following trauma to the spinal cord, so decreasing the release of EAAs following injury to it should reduce damage to the spinal cord. MP is thought to block free radical attack.[19] Free radicals cause release of EAAs.[20] Therefore, we investigated whether the clinically effective level of MP reduces release of amino acids following experimental spinal cord injury. We injected 30 mg/kg of MP sodium succinate into the jugular vein of rats immediately after injuring the cord. A repeated-measures analysis of variance indicated a significant reduction in glutamate release in MP-treated animals (Figure 4).

D. *IN VIVO* STUDIES OF THE EFFECT OF THE HYDROXYL RADICAL ON SPINAL CORD NEURONS

Reactive oxygen species seem to contribute to tissue damage in many traumas and degenerative diseases.[21] The highly destructive hydroxyl radical is thought to be released by the sequence

$$O_2 \rightarrow O_2^{-\bullet} \rightarrow H_2O_2 \xrightarrow{Fe^{2+}} {}^{\bullet}OH + {}^{-}OH$$

The Fe^{2+}-catalyzed conversion of H_2O_2 to $^{\bullet}OH$ is known as Fenton's reaction. $^{\bullet}OH$ is capable of attacking nearly all molecules in biological systems. $^{\bullet}OH$ could be especially important following injury due to the release of iron ions into the extracellular space by rupture of cells. Most of the evidence for such roles in indirect, based on either effects of drugs, detection of products expected of free radical reactions, or *in vitro* experiments.

We wished to determine the consequences of generating $^{\bullet}OH$ inside tissue *in vivo*. We administered Fe^{2+} and H_2O_2 into the spinal cord through parallel microdialysis fibers such

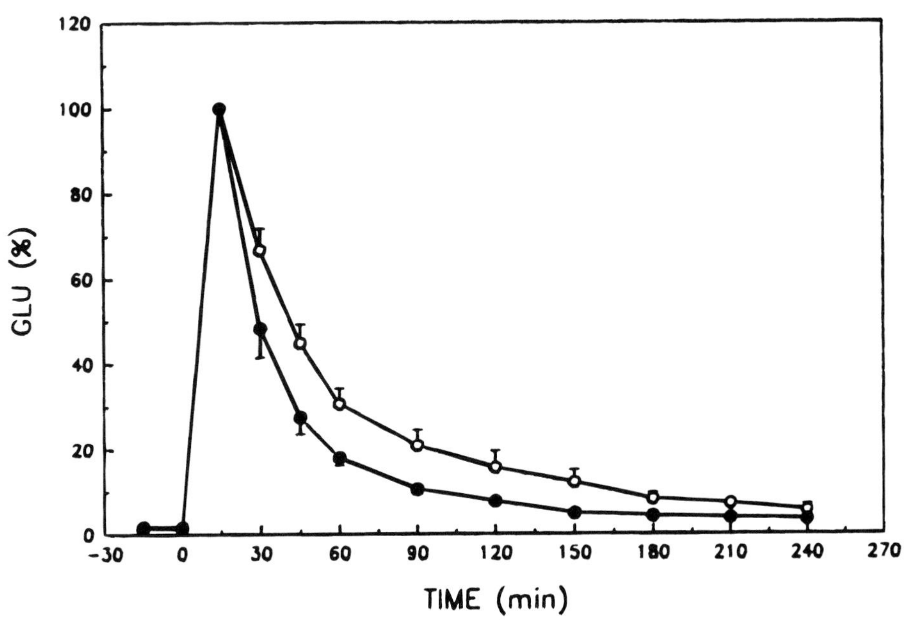

FIGURE 4. Effect of administering MP on glutamate concentrations in microdialysates collected following experimental injury to the rat spinal cord. Open circles, control animals; closed circles, treated animals. MP was injected into the jugular vein immediately after injury to the cord. Results were normalized such that the first point after injury = 100%. MP caused glutamate concentrations to decline significantly faster following injury.

that these Fenton's reagents mixed to produce ˙OH in the surrounding tissue.[8] The combined agents caused significantly more loss of neurons than did either agent alone. Electrical responses rostral to the dialysis zone disappeared in 3 to 5 h. $FeCl_2$ caused a lesser decline, and H_2O_2 was without a detectable effect on electrical conduction. Generation of ˙OH killed neurons and caused loss of axonal conduction, so this species could contribute to secondary damage. All of the treatments increased release of amino acids.

III. EXPERIMENTAL PROCEDURES

A. PROCEDURES FOR IMPACT INJURING AND MICRODIALYSIS SAMPLING

Male Sprague-Dawley rats (300 to 400 g) are anesthetized first with pentobarbital (35 mg/kg IP) and then maintained with urethane (780 mg/kg IP). Animals are considered to be adequately anesthetized when there is no withdrawal upon noxious foot pinch. Upon full anesthesia, a laminectomy is performed from vertebra T12 to vertebra L3. The sciatic nerve is also exposed when it is to be stimulated. The animal is clamped in a frame by its dorsal vertebral processes. Body temperature is maintained at 37 to 38°C throughout an experiment by feedback from a rectal probe to a heating blanket.

Following exposure of the cord, a microdialysis fiber is inserted laterally through the lumbar enlargement. Slack is left in the fiber external to the cord to prevent pull on the fiber from exacerbating damage when the cord is injured. Our methods for preparing the fibers are described elsewhere.[4–8,22] ACSF (composition in m*M:* Na^+, 151.1; K^+, 2.6; Mg^{2+}, 0.9; Ca^{2+}, 1.3; Cl^-, 22.7; HCO_3^-, 21.0; HPO_4^{2-}, 2.5) is pumped through the fiber at a rate of 5 μl/min. The ACSF is bubbled with 95% O_2 − 5% CO_2 prior to each experiment. After release due to insertion injury declines (75 min), fluid is collected from the outlet end of the fiber in plastic tubes in ice. Three 15-min samples are collected to establish a baseline level. The cord is then injured over the fiber by dropping a 10-g weight 5 cm down a guide tube, and samples are collected for several hours thereafter.

During sample collection at flow rates of 5 μl/min there are large concentration gradients across the fiber membrane, and substantial depletion of substances from the fluid around the fiber occurs.[23] At very slow flow rates, concentrations in the dialysate become close to those in the extracellular fluid, and there is little perturbation of concentrations.[24] To calibrate the collection efficiency at 5 μl/min, samples were collected from fibers inserted into rat spinal cords at flow rates of 0.15 and 5 μl/min. Averaging results for glutamate and aspartate gave a value of 35 for the ratio of the concentration in the dialysate at 0.15 μl/min to that at 5 μl/min. Therefore under control conditions at a 5-μl/min flow rate *in vivo* extracellular concentrations were at least 35 times those in the collected fluids.

B. PROCEDURE FOR CHEMICAL INJURY AND MICRODIALYSIS ADMINISTRATION

Our standard experimental setup is shown in Figure 1. The microdialysis fiber here was used to administer the candidate damaging agent and to sample the substances released by the administered agent. The microdialysis fiber is inserted laterally through the cord at L3–L4 just rostral to the main entry zone of the sciatic nerve. The fiber can be inserted in white or gray matter. After the control samples are collected, the perfusion solution is switched to the ACSF containing the candidate harmful agent.

When microdialysis is used to administer the candidate substance, the concentration achieved outside the fiber is needed to implement our approach. This requires determining the relationship between the concentrations inside the dialysis fiber and the concentrations outside the fiber. To find this, two fibers are glued together in parallel with their dialysis zones aligned and inserted into the spinal cord. A 1000-m*M* KCl solution in ACSF was passed through fiber A and ACSF through fiber B. $[K^+]_e$ was computed using the formula

$$[K^+]_e = ([K^+]_A[K^+]_B)^{1/2}$$

where $[K^+]_A$ = concentration administered in fiber A and $[K^+]_B$ = concentration recovered in fiber B. A tenfold K^+ gradient[5] and an 11-fold gradient for kainic acid[6] across the fiber walls were calculated. Thus, a factor of about ten seems fairly general for small molecules at a 5-μl/min flow rate in our fibers.

C. ELECTROPHYSIOLOGICAL RECORDING

Cord dorsum potentials are recorded in conjunction with administering a substance into the spinal cord by microdialysis (Figure 1). Single platinum ball electrodes are placed in contact with the dorsal surface of the cord rostral and caudal to the fiber on the side of the exposed sciatic nerve. This arrangement allows recording of activity in sensory neurons evoked by stimulation of the sciatic nerve. The caudal electrode is placed as far as possible from the microdialysis fiber. Signals recorded at the rostral electrode pass the dialysis zone, while those at the caudal electrode do not. Caudal recordings served to distinguish decreases caused by the administered substances from general decline of the preparation. ACSF or Ringer's solution is pumped through the fibers while control recordings are made. The perfusion fluid is then switched to one containing the administered substance, and the response to sciatic nerve stimulation recorded for the remainder of the experiment. A pair of hook electrodes is placed under the sciatic nerve. The sciatic nerve is stimulated by applying 1-mA, 0.1-ms pulses between the hook electrodes. The signals are amplified and displayed on an oscilloscope and simultaneously digitized and stored in a computer.

D. HISTOLOGICAL EXAMINATION

At the end of an experiment, the animal is fixed by perfusion through the heart with a 10% formalin solution in sodium phosphate buffer. The section of the cord containing the dialysis zone is removed and placed in formalin solution. This piece of cord is embedded in paraffin, cut into 15-μm sections, stained with a Kluver-Barrera stain (Luxol Fast Blue to stain white matter and Neutral Red to stain cell bodies), and examined for loss of neurons. Damage is assessed by counting the numbers of neurons in 200 × 500-μm areas near and parallel to the dialysis track. The densities of neurons in treated tissue is compared to the densities in the corresponding area in other animals following exposure to ACSF and to similar areas anterior or posterior to the dialysis region in the same spinal cord. Significance of differences between counts is assessed using the Wilcoxon-Mann-Whitney rank sum test.

E. ANALYSIS OF AMINO ACIDS

We analyze amino acids as naphthalene-2,3-dicarboxaldehyde derivatives[25] using high-pressure liquid chromatography (HPLC). The derivatives are analyzed on an HPLC equipped with a fluorescence detector (excitation wavelength, 250 nm emission cutoff filter, 470 nm). A C_{18} cartridge column (150 × 4.6 mm, 5 μm particle diameter) is used. The components of the elution system are buffer A: 10% tetrahydrofuran + 90% potassium phosphate buffer (0.05 *M*), pH 6.05, buffer B: 55% acetonitrile + 10% methanol + 35% potassium phosphate buffer. The elution gradient starts at 12% B for 1 min, increases to 35% B over 25 min, to 50% B in 13 min, and finally to 85% B in 10 min.

F. *IN VIVO* GENERATION OF •OH

We use the double-fiber technique to administer H_2O_2 through one microdialysis fiber and $FeCl_2$-EDTA through a parallel fiber to generate •OH.[7,8] H_2O_2 and $FeCl_2$ are administered alone as controls and simultaneously from parallel fibers. The ˙OH radical is generated in the overlapping dialysis zones in the tissue around the two fibers. To prepare a $FeCl_2$-EDTA solution, NaCl, KCl, $CaCl_2$, and EDTA are dissolved in a solution of NaH_2PO_4 (10 m*M*) and NaOH (20 m*M*). This solution is degassed by sonication for 15 min, and then bubbled with argon for 15 min and sealed under argon. $FeCl_2$ is added at the start of each experiment. Unless the solution is degassed and addition of $FeCl_2$ delayed until the start of an experiment, a brown precipitate forms. The final pH of the $FeCl_2$ solution is 7.2 to 7.4, and the final Na^+, K^+, and Ca^{2+} concentrations are those in regular Ringer's solution. Inside the fibers the EDTA concentration is 8.2 m*M;* Fe^{2+}, 5 m*M;* and the H_2O_2, 50 m*M*. ˙OH generation was demonstrated by administering phenylalanine and analyzing the hydroxyphenylalanines produced by ˙OH attack on phenylalanine.[7]

ACKNOWLEDGMENTS

The authors thank the Spinal Cord Research Foundation (#1025), the American Paralysis Association (LB1-9104-1) and NIH (NS11255) for financial support, and Debbie Pavlu for typing.

REFERENCES

1. Kao, C. C. (1980) Spinal cord reconstruction after traumatic injury, in *The Spinal Cord and Its Reaction to Traumatic Injury.* W. F. Windle (ed). Marcel Dekker, NY, 18: 249–270.
2. Anderson, D. K. (1992) Chemical and cellular mediation in spinal cord injury. *J. Neurotrauma* 9: 143–146.
3. Liu, D., Valadez, V., Sorkin, L. S., and McAdoo, D. J. (1990) Norepinephrine and serotonin release upon impact injury to rat spinal cord. *J. Neurotrauma* 7: 219–227.
4. Liu, D., Thangnipon, W., and McAdoo, D. J. (1991) Excitatory amino acids rise to toxic levels upon impact injury to the rat spinal cord. *Brain Res.* 547: 344–348.
5. Liu, D. and McAdoo, D. J. (1993) An experimental model combining microdialysis with electrophysiology, histology, and neurochemistry to establish mechanisms of secondary damage in spinal cord injury: the effect of potassium. *J. Neurotrauma* 10: 349–362.
6. Liu, D., McAdoo, D. J., and Pan, E. Glutamate neurotoxicity in the spinal cord: a neurochemical and histological study, submitted.
7. Liu, D. (1993) Generation and detection of hydroxyl radical *in vivo* in rat spinal cord by microdialysis administration of Fenton's reagents and microdialysis sampling. *J. Biochem. Biophys. Methods* 27: 281–291.
8. Liu, D., Yang, R., Yan, X., and McAdoo, D. J. (1994) Hydroxyl radical generated *in vivo* kills neurons in the rat spinal cord: electrophysiological, histological and neurochemical results. *J. Neurochem.* 62: 37–44.
9. Bracken, M. B., Shepard, M. J., Collins, W. F., Holford, T. R., Young, W., Baskin, D. S., Eisenberg, H. M., Flamm, E., Leo-Summers, L., Maroon, J., Marshall, L. F., Perot, P. L., Piepmeier, J., Sonntag, V. K. H., Wagner, F. C., Wilberger, J. E., and Winn, H. R. (1990) A randomized, controlled trial of methylprednisolone or naloxone in the treatment of acute spinal-cord injury. *N. Engl. J. Med.* 322: 1405–1411.
10. Liu, D. and McAdoo, D. J. (1993) Methylprednisolone reduces excitatory amino acid release following experimental spinal cord injury. *Brain Res.* 609: 293–297.
11. Young, W. and Koreh, I. (1986) Potassium and calcium changes in injured spinal cords. *Brain Res.* 365: 42–53.
12. Chesler, M., Sakatani, K., and Hassan, A. Z. (1991) Elevation and clearance of extracellular K^+ following contusion of the rat spinal cord. *Brain Res.* 556: 71–77.
13. Rothman, S. M. (1985) The neurotoxicity of excitatory amino acids is produced by passive chloride influx. *J. Neurosci.* 5: 1483–1489.
14. Kawasaki, K., Czeh, G., and Somjen, G. G. (1988) Prolonged exposure to high potassium concentration results in irreversible loss of synaptic transmission in hippocampal tissue slices. *Brain Res.* 457: 322–329.
15. Young, W., Koreh, I., Yen, V., and Lindsay, A. (1982) Effect of sympathectomy on extracellular potassium ionic activity and blood flow in experimental spinal cord contusion. *Brain Res.* 253: 115–124.
16. Eidelberg, E., Sullivan, J., and Brigham, A. (1975) Immediate consequences of spinal cord injury: possible role of potassium in axonal conduction block. *Surg. Neurol.* 3: 317–321.

17. Coyle, J. T. and Puttfarcken, P. (1993) Oxidative stress, glutamate, and neurodegenerative disorders. *Science* 262: 689–695.
18. Panter, S. C., Yum, S. W., and Faden, A. I. (1990) Alteration in extracellular amino acids after traumatic spinal cord injury. *Ann. Neurol.* 27: 96–99.
19. Hall, E. D. and Braughler, J. M. (1989) Central nervous system trauma and stroke II. Physiological and pharmacological evidence for involvement of oxygen radicals and lipid peroxidation. *Free Rad. Biol. Med.* 6: 303–313.
20. Pellegrini-Giampietro, D. E., Cherici, G., Alesiani, M., Carla, V., and Moroni, F. (1988) Excitatory amino acid release from rat hippocampal slices as a consequence of free radical formation. *J. Neurochem.* 51: 1960–1963.
21. Halliwell, B. and Gutteridge, J. M. C. (1986) Oxygen free radicals and iron in relation to biology and medicine: some problems and concepts. *Arch. Biochem. Biophys.* 246: 501–514.
22. Sorkin, L. S., Steinman, J. L., Hughes, M. G., Willis, W. D., and McAdoo, D. J. Microdialysis recovery of serotonin released in spinal cord dorsal horn. *J. Neurosci. Methods* 23: 131–138.
23. Bungay, P. M., Morrison, P. F., and Dedrick, R. L. (1990) Steady-state theory for quantitative microdialysis of solutes and water *in vivo* and *in vitro*. *Life Sci.* 46: 105–119.
24. Johnson, R. D. and Justice, J. B. (1983) Model studies for brain dialysis. *Brain Res.* Bull. 10: 567–571.
25. Matuszewski, B. K., Givens, R. S., Srinivasachar, K., Carlson, R. G., and Higuchi, T. (1987) N-substituted 1-cyanobenz[f]isoindole: evaluation of fluorescence efficiencies of a new fluorogenic label for primary amines and amino acids. *Anal. Chem.* 59: 1102–1105.

VII. Molecular Biological Approach to CNS Injury

Chapter 30

POSTISCHEMIC GENE REGULATION

C. Y. Hsu, Z. Y. Hu, A. Salminen, B. J. Gwag, and P. K. Liu

TABLE OF CONTENTS

I. INTRODUCTION

Excitotoxins, especially glutamate, appear to play a primary role in hypoxic-ischemic neuronal degeneration (Choi and Rothman, 1990). Neurons may survive at least temporarily after exposure to glutamate that is released into the extracellular space after cerebral ischemia (Benveniste et al., 1984). A major pathophysiological consequence after ischemia is a massive influx of calcium through glutamate receptor-gated calcium channels (Choi and Rothman, 1990). For the surviving neurons, the ensuing elevation of intracellular Ca^{2+} triggers a multitude of signal events. Second messengers (e.g., inositol triphosphate, diacylglycerol, cyclic AMP) appear to mediate some of these processes linking the receptor activation to changes in nuclear regulatory mechanisms that affect gene expression (Figure 1). Recent observations of the alteration of gene expression in ischemic brain (Sagar, 1993) have broadened our insight into postischemic gene regulation as a possible molecular mechanism to affect functional recovery after stroke. These postischemic events may play an important role in initiating repairing processes in the injured brain (Figure 2). This chapter summarizes recent studies of postischemic gene regulation from our and other laboratories using a cortical infarct model. Studies employing other stroke models (Jacewicz et al., 1986; Gunn et al., 1990; Dempsey et al., 1991; Nowak, 1991; Sharp et al., 1991; Uemura et al., 1991; Abe et al., 1991; Lindvall et al., 1992; Welsh et al., 1992) will not be reviewed except for those data that are cited for comparison with our stroke model. The focal cerebral ischemia-reperfusion model in rats employed in these studies has been detailed in another chapter in this volume.

II. IMMEDIATE EARLY GENES

Alteration of gene expression and synthesis of new gene products have been initially demonstrated in a stroke model in which the appearance of new gene products (proteins)

0-8493-8094-4/95/$0.00+$.50

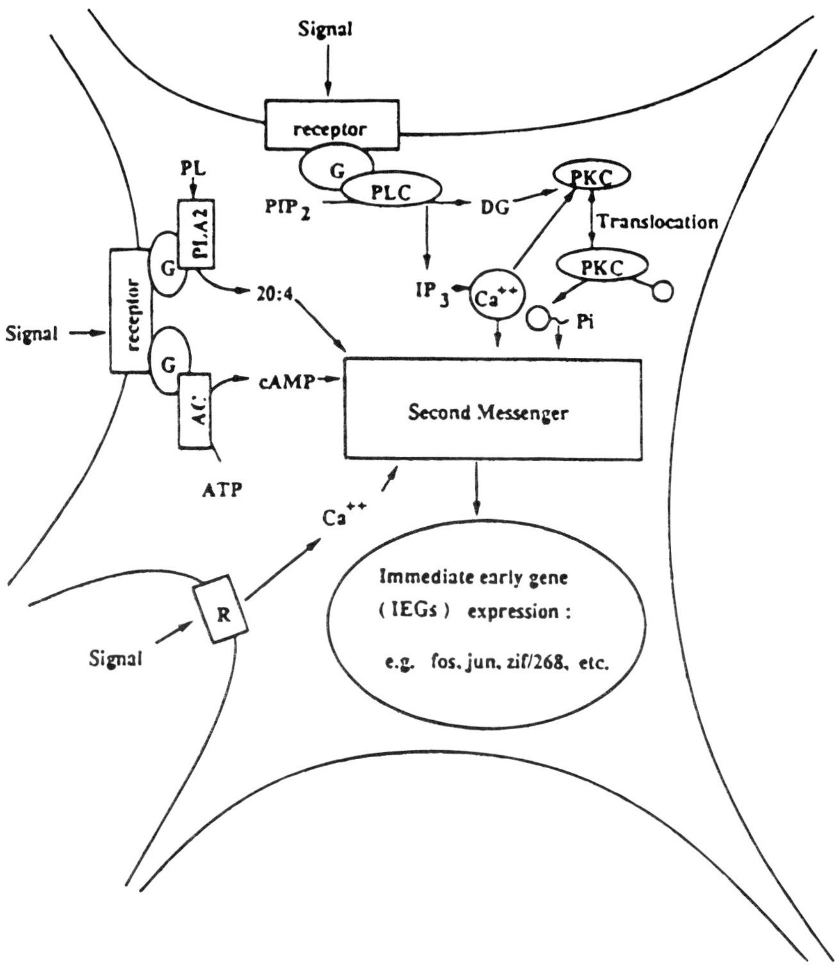

FIGURE 1. Biochemical cascades of postischemic gene regulation. An alteration of gene expression and subsequent phenotypic changes may be caused by receptor-induced signal transduction and second messenger-mediated activation of the expression of selected genes. (Hsu et al., 1993, reproduced with permission from *Stroke*.)

were noted in the ischemic brain using two-dimensional gel electrophoresis (Jacewicz et al., 1986). Subsequently, the increased expression of several immediate early genes and neurotrophin genes have been reported in a number of stroke models (Gunn et al., 1990; Dempsey et al., 1991; Nowak et al., 1991; Sharp et al., 1991; Uemura et al., 1991; Lindvall et al., 1992; Welsh et al., 1992). The protooncogenes, including c-*fos* and c-*jun* gene families, are immediate early genes that are expressed rapidly and transiently in response to ischemia-reperfusion. An essential part of regulation in gene expression is the induction or activation of sequence specific transcription factors (Kruijer et al., 1984; Chiu et al., 1988; Ransone et al., 1989; Sonenberg et al., 1989; Kornhauser et al., 1992). The gene products of c-*fos* and c-*jun* families are transcription factors (see Section III). The activation of these transcription factors may induce the expression of downstream genes, some of which encode other transcription factors, which in turn regulate the expression of many long-term responsive genes (Barber and Verma, 1987; Karin, 1992). Studies of the expression and regulation of immediate early genes and subsequent events will help us understand the molecular mecha-

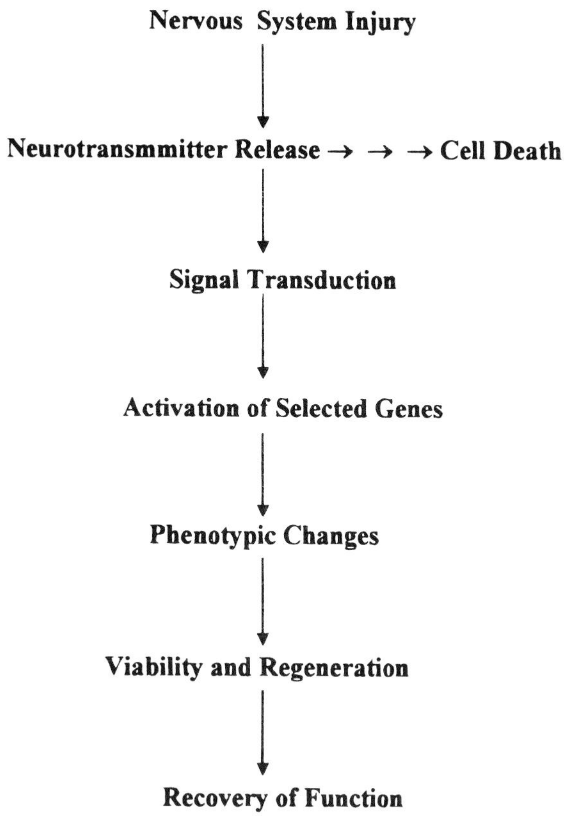

FIGURE 2. Schematic depiction of the possible pathophysiological consequences of neuronal exposure to excitatory neurotransmitter(s) following ischemic insult.

nisms that link the extracellular signals to the delayed neuronal response to ischemic insult (Figure 3).

In addition to the c-*fos* and c-*jun* families, other immediate early genes have also been implicated in transcription regulation. *Krox*-20 is a member of the zinc finger transcription factor family and an inducible immediate early responsive gene (Chavrier et al., 1988; 1989). Similar to other immediate early genes such as c-*fos,* c-*jun,* NGFI-A, NGFI-B, and NGFI-C, the expression of *Krox*-20 was activated within minutes by serum and growth factors and was induced during G_0/G_1 transition in cell cultures (Chavrier et al., 1988). Recent studies showed that *Krox*-20 was present in both embryonic and adult CNS, suggesting that it may play a role in the normal function of neurons (Wilkinson et al., 1989; Mack et al., 1992). The expression of *Krox*-20 was investigated in the focal ischemia model by Northern blot analysis and *in situ* hybridization. Northern blot analysis showed that ischemia for 30 min, which caused little cortical infarction, induced a 29-fold increase in *Krox*-20 mRNA signal exclusively in the ischemic cortex. Ischemia for 90 min, which led to large cortical infarction, induced *Krox*-20 mRNA not only in the ischemic cortex but also in the ipsilateral hippocampus. The induction of *Krox*-20 mRNA was rapid, transient, and appeared to be controlled at the transcriptional level, as indicated by nuclear run-on assays. The regional induction of *Krox*-20 mRNA was further confirmed by *in situ* hybridization. These results suggest that the expression of transcription factor genes may play a role in the postischemic changes of the injured brain (An et al., 1992).

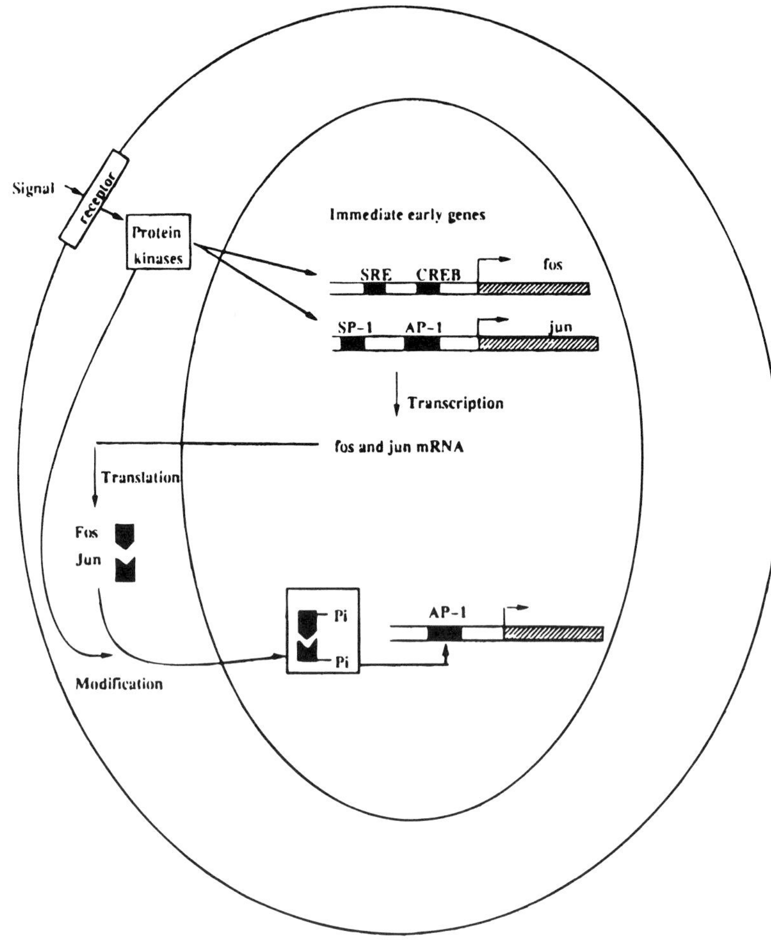

FIGURE 3. A diagrammatic scheme of postischemic transcription regulation. A possible mechanism to regulate late effector genes by an increase in AP-1 binding activity following c-*fos* and c-*jun* expression. (Hsu et al., 1993, reproduced with permission from *Stroke.*)

III. TRANSCRIPTION FACTORS

The gene products of the c-*fos* and c-*jun* families can interact via the leucine-zipper to form heterodimers (Turner and Tjian, 1989) that bind specifically to the AP-1 binding sites or 12-*o*-tetradecanoylphorbol-13-acetate (TPA) responsive elements (TRE) of selected genes (Barber and Verma, 1987; Karin, 1992). The AP-1 factor has been shown to transcriptionally activate a number of important genes (Kruijer et al., 1984; Chiu et al., 1988; Ransone et al., 1989; Sonenberg et al., 1989; Morgan and Curran, 1991; Kornhauser et al., 1992). An increase in AP-1 binding activity may in turn induce or repress the expression of other so-called late effector genes, which are important for basic cellular functions and cell survival (Figure 3). The exact roles of increased AP-1 binding in the induction or repression of selected late genes remain to be determined. Among these late genes, the neurotrophic factor genes are possible targets for AP-1 activity. An AP-1 binding site has been found at the junction of first exon and first intron of the nerve growth factor (NGF) gene (Zheng and Heinrich, 1988). Recent experimental evidence suggests that the expression of NGF is regulated by c-*fos* (Hengerer et al, 1990) and is induced by TPA (D'Mello et al., 1990). The coinduction of c-*fos* and *jun* B, and the enhanced expression of c-*jun* mRNAs by ischemia-reperfusion (An et al., 1993) implies the possibility of an increase in DNA binding activity of transcription factor AP-1 (*fos-jun* heterodimer). The AP-1 binding activity was determined by mobility shift assays. Nuclear extract from the right cerebral cortex of sham-operated rats showed a basal binding activity to the AP-1 DNA sequence. Ischemia for 30 min followed by reperfusion for 30 min showed only a slight increase in AP-1 binding activity. Reperfusion for 4 h resulted in a four to sixfold increase of AP-1 binding activity. The DNA binding activity persisted for as long as 24 h after reperfusion. Qualitatively similar results were also obtained from rats subjected to ischemia for 90 min (An et al., 1993).

IV. NEUROTROPHIN GENES

Focal cerebral ischemia-reperfusion caused a transient increase in the expression of two neurotrophin genes: NGF and brain-derived neurotrophic factor (BDNF) mRNAs after reperfusion for 30 min to 4 h in the ischemic right MCA cortex and the ipsilateral hippocampus. There was an apparent suppression of the expression of these neurotrophin genes 1 to 3 d after reperfusion. A second wave of neurotrophin gene expression appeared between two to four weeks after ischemia. The biphasic pattern of expression was also observed in the contralateral hemisphere, but to a lesser extent. *In situ* hybridization study using ^{32}P-NGF and BDNF probes showed increased signal in the ischemic core as well as the adjacent cortex and hippocampus. Finklestein and colleagues (1990) also showed a time-dependent expression of beta fibroblast growth factor in the ischemic right MCA cortex.

V. ANTISENSE STRATEGY

The realization that the naturally occurring antisense RNA is involved in the regulation of plasmid copy number in bacteria has been followed by the discovery of the antisense-nucleic acid technology (Inouye, 1988; Simons and Kleckner, 1988). This technology has a broad application to the regulation of gene expression from single cells to transgenic mice in medicine, plants in agriculture, and the biotechnical industry. In addition to homologous recombination (Field et al., 1992), the plasmid clones that express DNA or RNA, and ribozymes ODN have been used in the antisense strategy to transiently suppress the expression of c-*fos* gene (Colotta et al., 1992; Chiasson et al., 1992; Holt et al., 1986; Nishikura and Murray, 1987; Mercola et al., 1987; Edwards et al., 1988; Kerr et al., 1988; Levi and Ozato, 1988; Schonthal et al., 1988). The oligodeoxynucleotides that are used in the antisense strategies generally are single-stranded DNA that are 15 to 30 nucleotides in length. The oligodeoxynucleotide specifically blocks transcription, splicing, or translation of the target mRNA; therefore, it inhibits the expression of the target protein. The route of cellular uptake is thought to be receptor-mediated endocytosis. Moreover, genes cloned into a plasmid can be delivered to the neuron of the brain with the aid of lipofectin by *in vivo* gene transfer (Holt et al., 1990; Ono et al., 1990).

To study the postischemic regulation of immediate early genes and the subsequent events, we sought to suppress the post-ischemic expression of c-*fos* by an antisense oligodeoxynucleotide of c-fos mRNA. The effectiveness of the antisense oligodeoxynucleotide was first confirmed by its capability to block *in vitro* c-*fos* mRNA translation into Fos. In a subsequent study of the distribution of this oligodeoxynucleotide following intraventricular administration, we noted that the ^{32}P-labeled antisense oligodeoxynucleotide was internalized within 6 h and detectable in high-molecular weight species up to 41 h in the cytosol fraction. Treatment with RNase H released the oligodeoxynucleotide from the RNA fraction, suggesting an association of the labelled oligodeoxynucleotide with cellular RNA after uptake. When focal cerebral ischemia was induced 6 to 16 h after the infusion of the antisense oligodeoxynucleotide, the postischemic increase in Fos expression and AP-1 binding activity were suppressed. A lack of suppression of the binding activity of a cyclic AMP response element by the same oligodeoxynucleotide suggests the specific effect of this antisense strategy on c-*fos*-related events. Results from this study support the contention that postischemic expression of c-*fos* can be suppressed by an antisense strategy *in vivo* and that Fos expression is at least partially responsible for an increase in AP-1 binding activity after focal cerebral ischemia (Liu et al., 1994).

The ability of the antisense oligodeoxynucleotide of c-*fos* mRNA to block the postischemic expression of Fos and suppress AP-1 binding activity raises the possibility that the subsequent expression of AP-1 that regulates late effector genes may also be affected by antisense oligodeoxynucleotide of c-*fos* mRNA. In a preliminary study using *in situ* hybridization, we noted that expression of BDNF, but not NGF, mRNA after ischemia was inhibited by antisense oligodeoxynucleotide (Liu et al., unpublished observation). This preliminary observation is consistent with the contention that regulation of c-*fos* expression may be linked to the expression of late effector genes such as BDNF. However, the lack of effect of the same antisense oligodeoxynucleotide of c-*fos* mRNA on NGF expression also suggests the complexity of gene regulation under different pathological conditions. Further studies using different antisense strategies are needed to delineate the complex but delicate control of late effector genes after focal cerebral ischemia.

VI. CONCLUSION

In a focal cerebral ischemia model in the rat with well-defined ischemic zone, we demonstrated the expression of a number of immediate early genes. Severe focal ischemia also caused expression of these immediate early genes in distant structures such as ipsilateral and contralateral hippocampus. However, we have never noted the expression of these immediate early genes in the contralateral (left) MCA cortex. These observations suggest that the expression of immediate early genes, such as c-*fos, jun* B, and *Krox*-20, in nonischemic regions removed from the ischemic cortex may be related to ischemia-induced stimulation of selected neuronal pathways at a distance (Welsh et al., 1992; An et al., 1993). The molecular events leading to the recovery of brain function after stroke remains largely undefined. Recent studies showing an alteration of neurotrophin and receptor gene expression in ischemic brain (Lindvall et al., 1992; Merlio et al., 1993) open a new avenue for exploring the biochemical cascades that may lead to the restoration of brain function after stroke. These postischemic events entail an ischemia-induced increase in the extracellular concentration of excitatory amino acid neurotransmitters that may either lead to cell death (Choi and Rothman, 1990) or trigger signal transduction (Morgan and Curran, 1991). In the latter, the glutamate receptor-mediated activation of phospholipases and protein kinases resulted in the alteration of nuclear regulatory processes, including the expression of immediate early genes such as c-*fos, jun* B, and *Krox*-20. Our preliminary studies demonstrate the feasibility of an *in vivo* antisense strategy to suppress the postischemic expression of an immediate early gene. The subsequent reduction of the binding activity of a transcription factor relevant to Fos supports the contention that Fos is a major component of AP-1 binding activity in the ischemic brain. Results from this study also raise the possibility that the antisense oligodeoxynucleotide thus designed is capable of modulating the functional consequences related to c-*fos* expression. This mode of gene suppression may be useful in dissecting the molecular mechanism and in understanding the pathophysiological significance of postischemic c-*fos* expression in relation to late effector gene function that may be regulated by AP1.

ACKNOWLEDGMENT

Studies cited in this review are supported by NIH grants NS25545 and NS 28995. The authors thank Drs. G. An, M. H. Jiang, T. N. Lin, J. S. Liu, and J. J. Xue for their contribution to works presented here.

REFERENCES

Abe K, Tanzi RE, and Kogure K. (1991) Selective induction of Kunitz-type protease inhibitor domain-containing amyloid precursor protein mRNA after persistent focal ischemia in rat cerebral cortex. *Neurosci. Lett.* 125, 172–174.

An G, Lin TN, Liu JS, and Hsu CY. (1992) Induction of Krox-20 expression after focal cerebral ischemia. *Biochem. Biophys. Res. Commun.* 188, 1104–1110.

An G, Lin TN, Liu JS, Xue JJ, He YY, and Hsu CY. (1993) Expression of c-fos and c-jun family genes after focal cerebral ischemia. *Ann. Neurol.* 33, 457–464.

Barber J and Verma IM. (1987) Modification of *fos* proteins: phosphorylation of c-*fos*, but not v-*fos*, is stimulated by 12-tetradecanoyl-phorbol-13-acetate and serum. *Mol. Cell Biol.* 7, 2201–2211.

Benveniste H, Drejer J, Schousboe A, and Diemer NH. (1984) Elevation of the extracellular concentrations of glutamate and aspartate in rat hippocampus during transient cerebral ischemia monitored by intracerebral microdialysis. *J. Neurochem.* 43, 1369–1374.

Chavrier P, Janssen-Timmen U, Mattei MG, Zerial M, Bravo R, and Charnay P. (1989) Structure, chromosome location, and expression of the mouse zinc finger gene Krox-20: multiple gene products and coregulation with the proto-oncogene c-fos. *Mol. Cell. Biol.* 9, 787–797.

Chavrier P, Zerial M, Lemaire P, Almendral J, Bravo R, and Charnay P. (1988) A gene encoding a protein with zinc fingers is activated during G0/G1 transition in cultured cells. *EMBO J.* 7, 29–35.

Chiasson BJ, Hooper ML, Murphy PR, and Robertson HA. (1992) Antisense oligonucleotide eliminates *in vivo* expression of c-fos in mammalian brain. *Eur. J. Pharmacol. Mol. Pharmacol. Sec.* 227, 451–453.

Chiu R, Boyle WJ, Meek J, Smeal T, Hunter T, Karin M. (1988) The c-Fos protein interacts with c-Jun/AP ~ 1 to stimulate transcription of AP ~ 1 responsive genes. *Cell* 54, 541–552.

Choi DW and Rothman SM. (1990) The role of glutamate neurotoxicity in hypoxic-ischemic neuronal death. *Annu. Rev. Neurosci.* 13, 171–182.

Colotta F, Polentarutti N, Sironi M, and Mantovani A. (1992) Expression and involvement of c-fos and c-jun protooncogenes in programmed cell death induced by growth factor deprivation in lymphoid cell lines. *J. Biol. Chem.* 267, 18278–18283.

Dempsey RJ, Carney JM, and Kindy MS. (1991) Modulation of ornithine decarboxylase mRNA following transient ischemia in the Gerbil brain. *J. Cereb. Blood Flow Metab.* 11, 979–985.

D'Mello SR and Heinrich G. (1990) Induction of nerve growth factor gene expression by 12-O-tetradecanoyl phorbol 13-acetate. *J. Neurochem.* 55, 718–721.

Edwards SA, Rundell AYK, and Adamson ED. (1988) Expression of c-fos antisense RNA inhibits the differentiation of F9 cells to parietal endoderm. *Dev. Biol.* 129, 91–102.

Field SJ, Johnson RS, Mortensen RM, Papaioannou VE, Spiegelman BM, and Greenberg ME. (1992) Growth and differentiation of embryonic stem cells that lack an intact c-fos gene. *Proc. Natl. Acad. Sci. U.S.A.* 89, 9306–9310.

Finklestein SP, Caday CG, Kano M, Berlove DJ, Hsu CY, Moskowitz MA, and Klagsbrun M. (1990) Growth factor expression after stroke. *Stroke* 21, III122–124.

Gunn AJ, Dragunow M, Faull RL, and Gluckman PD. (1990) Effects of hypoxia-ischemia and seizures on neuronal and glial-like c-*fos* protein levels in the infant rat. *Brain Res.* 531, 105–116.

Hengerer B, Lindholm D, Heumann R, et al. (1990) Lesion-induced increase in nerve growth factor mRNA is mediated by c-*fos*. *Proc. Natl. Acad. Sci. U.S.A.* 87, 3899–3903.

Holt JT, Gopal TV, Moulton AD, and Nienhuis AW. (1986) Inducible production of c-fos antisense RNA inhibits 3T3 cell proliferation. *Proc. Natl. Acad. Sci. U.S.A.* 83, 4794–4798.

Inouye M. (1988) Antisense RNA: its functions and applications in gene regulation—a review. *Gene* 72, 25–34.

Jacewicz M, Kiessling M, and Pulsinelli WA. (1986) Selective gene expression in focal cerebral ischemia. *J. Cereb. Blood Flow Metab.* 6, 263–272.

Karin M. (1990) The AP-1 complex and its role in transcriptional control by protein kinase C. In *Molecular Aspects of Cellular Regulation.* Cohen P, Foulkes G, Eds. Amsterdam, Elsevier/North Holland Biomedical Press. 6:143–161.

Kerr LD, Holt JT, and Matrisian LM. (1988) Growth factors regulate transient gene expression by c-fos-dependent and c-jun-dependent pathways. *Science* 242, 1424–1427.

Kornhauser JM, Nelson DE, Mayo KE, and Takahashi JS. (1992) Regulation of *jun* B messenger RNA and AP-1 activity by light and a circadian clock. *Science* 255, 1581–1584.

Kruijer W, Cooper JA, Hunter T, and Verma IM. (1984) Platelet-derived growth factor induces rapid but transient expression of the c-*fos* gene and protein. *Nature (London)* 312, 711–716.

Levi BZ and Ozato K. (1988) Constitutive expression of c-fos antisense RNA blocks c-fos gene induction by interferon and by phorbol ester and reduces c-myc expression in F9 embryonal carcinoma cells. *Genes Dev.* 2, 554–566.

Lindvall O, Ernfors P, Bengzon J, Kokaia Z, Smith M-L, Siesjo BK, and Persson H. (1992) Differential regulation of mRNAs for nerve growth factor, brain-derived neurotrophic factor, and neurotrophin 3 in the adult rat brain following cerebral ischemia and hypoglycemic coma. *Proc. Natl. Acad. Sci. U.S.A.* 89, 648–652.

Liu K, Salminen A, He YY, Jiang MH, Xue JJ, Liu JS, and Hsu CY. (1994) Suppression of ischemia-induced Fos expression and AP-1 activity by an antisense oligodeoxynucleotide to c-*fos* mRNA. *Ann. Neurol.*, in press.

Mack KJ, Cortner J, Mack P, and Farnham PJ. (1992) Krox 20 messenger RNA and protein expression in the adult central nervous system. *Mol. Brain Res.* 14, 117–123.

Malone RW, Felgner PL, and Verma IM. (1989) Cationic liposome-mediated RNA transfection. *Proc. Natl. Acad. Sci. U.S.A.* 86, 6077–6081.

Mercola D, Rundell A, Westwick J, and Edwards SA. (1987) Antisense RNA to the c-fos gene: restoration of density-dependent growth arrest in a transformed cell line. *Biochem. Biophys. Res. Commun.* 147, 288–294.

Merlio J-P, Ernfors P, Kokaia Z, Middlemas DS, Bengzon J, Kokaia M, Smith M, Siesjo BK, Hunter T, Lindvall O, and Persson H. (1993) Increased production of the TrkB protein tyrosine kinase receptor after brain insults. *Neuron* 10, 151–164.

Morgan JI and Curran T. (1991) Stimulus-transcription coupling in the nervous system: involvement of the inducible proto-oncogenes fos and jun. *Annu. Rev. Neurosci.* 14, 421–451.

Nishikura K and Murray HM. (1987) Antisense RNA of protooncogene c-fos blocks renewed growth of quiescent 3T3 cells. *Mol. Cell Biol.* 7, 637–649.

Nowak TS, Jr. (1991) Localization of 70 kDa stress protein mRNA induction in gerbil brain after ischemia. *J. Cereb. Blood Flow Metab.* 11, 432–439.

Ono T, Fujino Y, Tsuchiya T, and Tsuda M. (1990) Plasmid DNAs directly injected into mouse brain with lipofectin can be incorporated and expressed by brain cells. *Neurosci. Lett.* 117, 259–263.

Ransone RJ, Visvader J, Sassone-Corsi P, and Verma IM. (1989) *fos-jun* Interaction: mutational analysis of the leucine zipper domain of both proteins. *Genes Dev.* 3, 770–781.

Sagar SM. (1993) The molecular biology of brain ischemia: trendy genes in sick neurons. *Ann. Neurol.* 33, 437–438.

Schonthal A, Herrlich P, Rahmsdorf HJ, and Ponta H. (1988) Requirement for fos gene expression in the transcriptional activation of collagenase by other oncogenes and phorbol esters. *Cell* 54, 325–334.

Sharp FR, Lowenstein D, Simon R, and Hisanaga K. (1991) Heat shock protein hsp72 induction in cortical striatal astrocytes and neurons following infarction. *J. Cereb Blood Flow Metab.* 11, 621–627.

Simons RW and Kleckner N. (1988) Biological regulation by antisense RNA in prokaryotes. *Annu. Rev. Genet.* 22, 567–600.

Sonenberg JL, Rauscher FJ, Morgan JI, and Curran T. (1989) Regulation of proenkephalin by *fos* and *jun. Science* 246, 1622–1624.

Turner R and Tjian R. (1989) Leucine repeats and adjacent DNA bind domain mediate the formation of functional cFos-cJun heterodimers. *Science* 243, 1689–1694.

Uemura Y, Kowall NW, and Moskowitz MA. (1991) Focal ischemia in rats causes time-dependent expression of c-*fos* protein immunoreactivity in widespread regions of ipsilateral cortex. *Brain Res.* 552, 99–105.

Welsh FA, Moyer DJ, and Harris VA. (1992) Regional expression of heat shock protein 70 mRNA and c-*fos* mRNA following focal ischemia in rat brain. *J. Cereb. Blood Flow Metab.* 12, 204–212.

Wilkinson DG, Bhatt S, Chavrier P, Bravo R, and Charnay P. (1989) Segment-specific expression of a zinc-finger gene in the developing nervous system of the mouse. *Nature* 337, 461–464.

Zheng M, and Heinrich G. (1988) Structural and functional analysis of the promoter region of the nerve growth factor gene. *Mol. Brain Res.* 3, 133–140.

Chapter 31

SOD-1 TRANSGENIC MICE: AN APPLICATION TO THE STUDY OF ISCHEMIC BRAIN INJURY

Hideyuki Kamii, Hiroyuki Kinouchi, Sylvia F. Chen, Frank R. Sharp, Charles J. Epstein, and Pak H. Chan

TABLE OF CONTENTS

I. INTRODUCTION

A rapid increase in the need to explore the molecular basis of cellular function and injury in the brain has led neuroscientists to employ state-of-the-art molecular biological techniques, including transgenic (Tg) mouse technology. With the Tg technology, it is now becoming routine and feasible to increase the level of (overexpress) particular foreign proteins or enzymes in a certain species of animal, or to mutate or delete a particular gene using a "knock out" approach in certain animal species. One of the more powerful aspects of Tg mouse techniques is the ability to manipulate gene expression and its particular translational product in the mouse. This is useful in studying cell physiology and pathology and in developing animal models of particular human diseases. Over the past 10 years, the success in gene cloning and the subsequent production of Tg mice has led to new insights into the regulation of gene expression, the process of tumorigenesis and sexual alteration, the functioning of the immune system, the molecular basis of cystic fibrosis, and aging. Tg mice have been developed in the areas of neuroscience and neurobiology (Koretsky, 1992; Popko et al., 1990; Rosenfeld et al., 1988).

Recently, Tg mice overexpressing Cu, Zn-superoxide dismutase (SOD-1) have been successfully developed (Epstein et al., 1987) and applied to the study of Down's syndrome, Parkinson's disease, brain ischemia, brain trauma, and aging (Chan et al., 1993; Chan et al., 1991; Epstein et al., 1992; Kinouchi et al., 1991; Yang et al., 1994). This chapter describes an application of SOD-1 Tg mice to the study of oxidative mechanisms in ischemic brain injury.

II. SOD-1 TRANSGENIC MICE

Heterozygous Tg mice of strain TgHS/SF-218-3 carrying human SOD-1 genes were derived from the founder stock described by Epstein et al. (1987) and were bred on a

0-8493-8094-4/95/$0.00+$.50

CD-1 mouse background. Tg mice were identified by qualitative demonstration of human SOD-1 using nondenaturing gel electrophoresis followed by nitroblue tetrazolium staining (Chan et al., 1991; Epstein et al., 1987). There were no observable phenotypic differences between Tg mice and nontransgenic (nTg) normal littermates.

Figure 1 shows the presence of human SOD-1 enzymatic activity in the cerebral cortex, cerebellum, and spinal cord of Tg mice. In nTg mice, only mouse SOD activity was expressed, while human SOD homodimer activity as well as the human/mouse SOD heterodimer activity were found in Tg mice. The actual amount of total SOD-1 activity in the cerebral cortex, cerebellum, and spinal cord is shown in Table 1. The cortex and cerebellum of Tg mice had a near threefold increase in SOD-1 activity as compared to that of nTg mice.

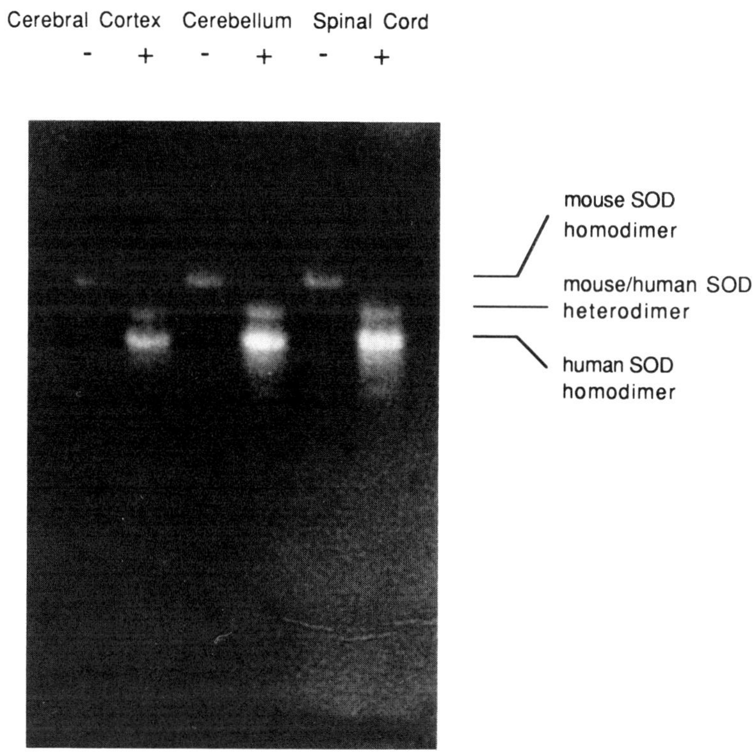

FIGURE 1. Expression of human SOD-1 enzymatic activity in cerebral cortex, cerebellum, and spinal cord of transgenic mice. Both mouse and human SOD enzymatic activity are detected using nondenaturing gel electrophoresis. This is followed by nitroblue tetrazolium staining. +, transgenic mice; −, nontransgenic mice. (Data obtained from Chan et al., 1991.)

TABLE 1
SOD-1 Levels (units/mg protein) in Neural Tissues of Nontransgenic and Transgenic Mice

	Nontransgenic (nTg)	Transgenic (Tg)	Tg/nTg
Cerebral cortex	7.9 ± 0.5	22.7 ± 1.4*	2.9
Cerebellum	9.0 ± 2.0	25.8 ± 1.3*	2.9
Spinal cord	15.4 ± 2.0	34.9 ± 1.5*	2.3

Note: Values are mean ± SEM (n = 3). *$p < 0.001$ (Tg vs. nTg mice, by Student's *t*-test). Data obtained from Chan et al. (1991).

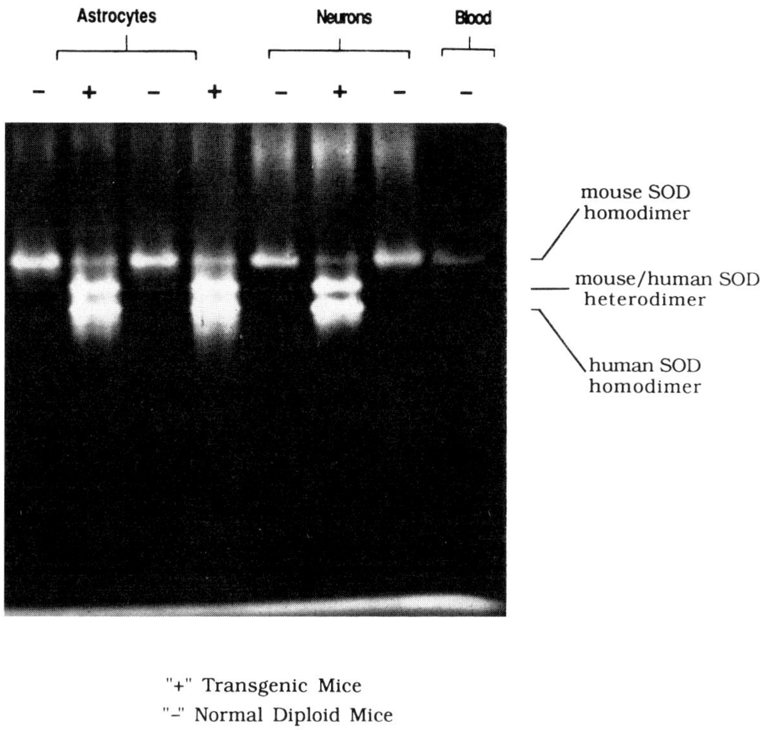

FIGURE 2. Expression of human SOD-1 enzymatic activity in primary culture of cortical neurons and astrocytes of transgenic mice. Both mouse and human SOD enzymatic activity are detected using nondenaturing gel electrophoresis followed by nitroblue tetrazolium staining. +, transgenic mice; −, nontransgenic mice.

The expression of human SOD-1 in primary culture of neurons and astrocytes of Tg mice is demonstrated in Figure 2. The expression of human SOD homodimer, as well as human/mouse SOD heterodimer, were readily identified in both the neurons and astrocytes of Tg mice, while the neurons and astrocytes obtained from nTg mice exhibit only mouse SOD activity.

III. EXPERIMENTAL

A. FOCAL CEREBRAL ISCHEMIA

Male Tg mice and nTg littermates (30 to 35 g) were subjected to focal cerebral ischemia according to the method of Yang et al. (1994) with minor modifications. Mice were anesthetized with chloral hydrate (350 mg/kg IP) and xylazine (4 mg/kg IP). The rectal temperature of the animals was maintained at 37 ± 0.5°C with a heating pad. The left femoral artery was cannulated for measurement of mean arterial blood pressure, PO_2, PCO_2, and pH during surgical procedures. The left common carotid artery was exposed, and the external carotid artery and its branches were isolated and coagulated. A 5-0 monofilament nylon suture, blunted at the tip, was introduced into the internal carotid artery through the external carotid artery stump, up to the anterior cerebral artery. At this point, the origin of the middle cerebral artery (MCA) was occluded. When reperfusion to the MCA distribution was needed, the suture was pulled back into the external carotid artery. Sham-operated control mice underwent surgery but did not have the suture inserted.

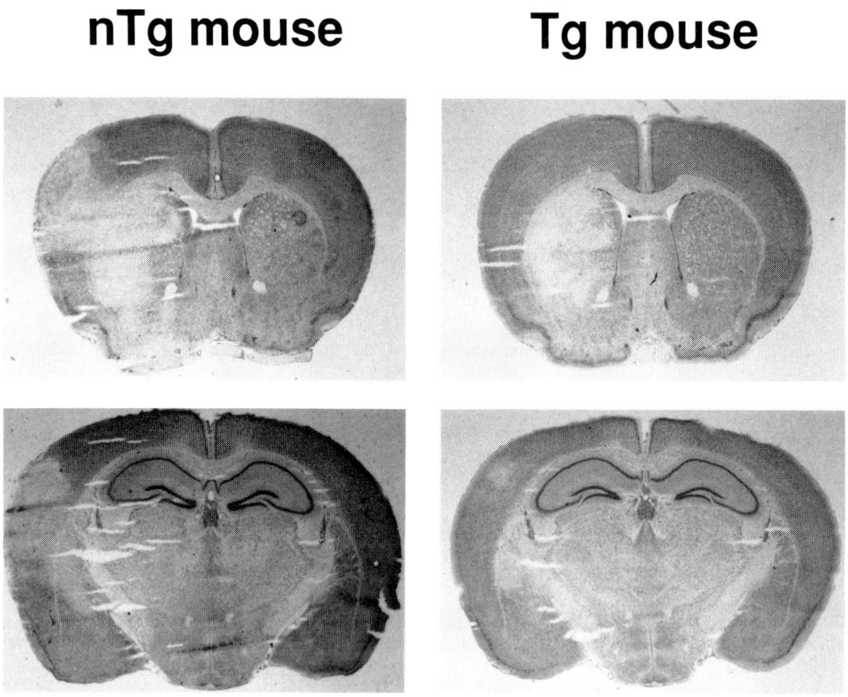

FIGURE 3. Representative photographs of brain sections at 24 h after 1 h of MCA occlusion in nontransgenic (nTg) and transgenic (Tg) mice. Infarcted areas were seen in the lateral part of the caudate putamen and the dorsolateral cortex in nTg mice, while Tg mice demonstrated them only in the caudate putamen with a small portion of infarcted cortex.

B. INFARCT SIZE AND VOLUME FOLLOWING FOCAL CEREBRAL ISCHEMIA

To compare the infarct size and volume following focal cerebral ischemia between nTg and Tg mice, the brains were histopathologically examined. At 24 h after 1 h of MCA occlusion, or following 24 h of permanent MCA occlusion, the mice were decapitated and the brains were frozen in powdered dry ice. Brain sections (20 μm) were cut at 500-μm intervals on a cryostat and stained with cresyl violet. The infarct size of the brain sections were quantitated by an image-analysis system (Swanson et al., 1990), and the infarct volume was calculated based on these measurements.

Figure 3 shows representative photographs of brain sections at 24 h after 1 h of MCA occlusion in nTg and Tg mice. Infarcted areas were seen in the lateral part of the caudate putamen and the dorsolateral cortex in nTg mice, while Tg mice demonstrated them only in the caudate putamen with a small portion of infarcted cortex. The infarct size and volume of nTg and Tg mice are shown in Figure 4. Tg mice demonstrated reduced infarct size and volume compared to those of nTg mice. On the other hand, a 24-h permanent MCA occlusion produced similar infarcted areas in nTg and Tg mice (Figure 5). There were also no significant differences in the infarct size and volume between nTg and Tg mice following a 24-h permanent MCA occlusion (Figure 6).

These findings suggest that an increased level of antioxidant in SOD-1 Tg mice is more beneficial following transient focal cerebral ischemia than following permanent ischemia. Therefore, with respect to clinical treatment of cerebral ischemia, our results suggest that antioxidant pharmacotherapy is essential when the blood flow is likely to be restored to ischemic regions. In the treatment of thrombotic stroke with thrombolytic agents (e.g.,

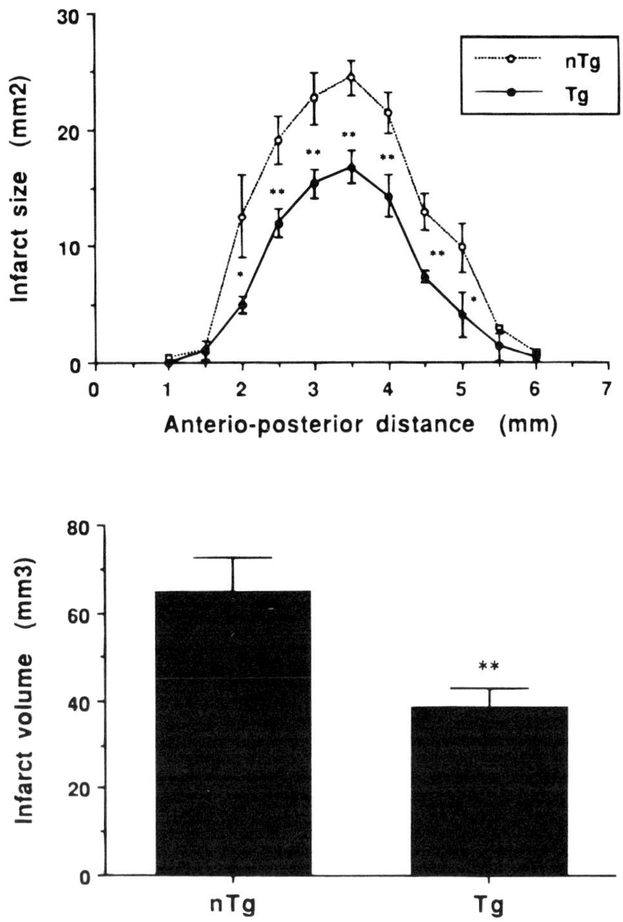

FIGURE 4. Infarct size (upper) and volume (lower) of nontransgenic (nTg) and transgenic (Tg) mice at 24 h after 1 h of MCA occlusion. The infarct size of the brain sections were quantitated by a image-analysis system and the infarct volume was calculated using the values. Tg mice demonstrated significantly reduced infarct size and volume compared to those of nTg mice. Values are mean ±SD (n = 6). $*p < 0.05$, $**p < 0.01$ (Tg vs. nTg mice, by Student's *t*-test).

t-PA), antioxidant therapy should be employed to reduce reperfusion injury by oxygen free radicals.

C. EXPRESSION OF c-fos mRNA AFTER A MILD FOCAL BRAIN ISCHEMIA

The protooncogene c-fos is one of the immediate early genes that are expressed in response to a variety of physiological and pathological stimuli (Morgan and Curran, 1991; Sagar et al., 1988; Sharp et al., 1993). Recent reports suggest that ischemic induction of c-fos may play an important role in the pathogenesis of ischemic brain injury (Combs et al., 1992; Nowak et al., 1993). To clarify the role of oxygen free radicals in expression of the c-fos protooncogene, the distribution of c-fos mRNA was investigated in SOD-1 Tg mice compared to control nTg littermates after a mild (i.e., 10 min) focal cerebral ischemia.

Following 10 min of MCA occlusion, the mice were allowed to recover for up to 7 d. At each time point, the mice were decapitated and their brains were frozen in powdered dry ice. Brain sections (20 μm) were cut on a cryostat and processed for *in situ* hybridization using the method of Schalling et al. (1988) with minor modifications. The brain sections

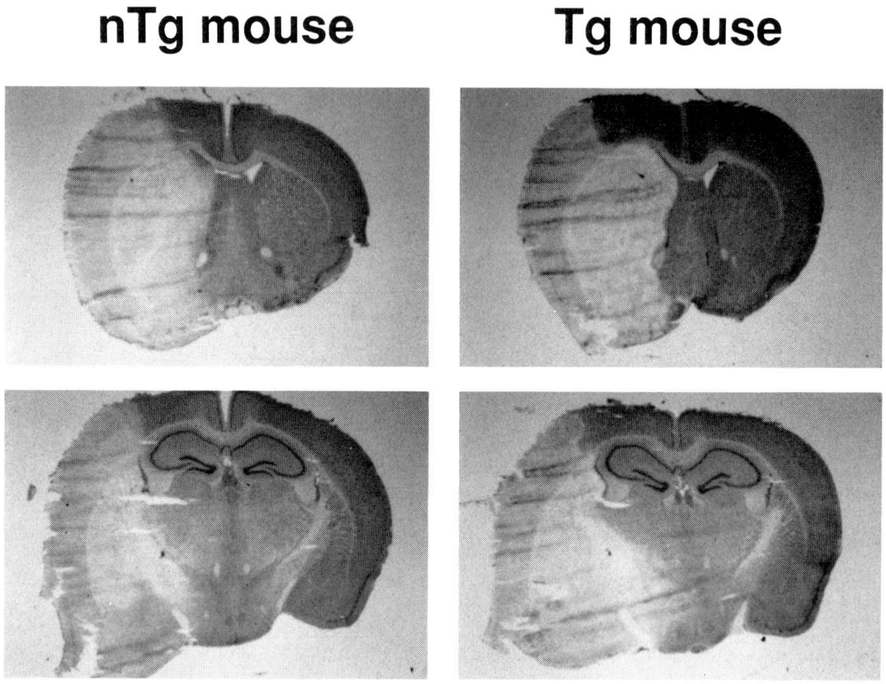

FIGURE 5. Representative photographs of brain sections following 24 h of permanent MCA occlusion in non-transgenic (nTg) and transgenic (Tg) mice. Infarcted areas were similar between nTg and Tg mice.

were hybridized at 37°C for 18 h with 1 × 10^6 cpm of a ^{35}S-labeled synthetic oligonucleotide probe. After the hybridization, the sections were rinsed in 1 × SSC (150 m*M* sodium chloride, 15 m*M* sodium citrate, pH 7.4) at 52°C for 60 min with several changes of 1 × SSC, dehydrated, and exposed to Kodak SB-5 film for a week.

The distribution of c-fos mRNA in the postischemic brains of nTg and Tg mice was assessed at caudate and dorsal hippocampal levels (Figure 7). Expression of c-fos mRNA occurred at 1 to 6 h after reperfusion in the ipsilateral hippocampus and thalamus in Tg mice, while it did so only at 1 h in the same regions in nTg mice. In the ipsilateral cortex, there were no significant differences in the pattern of the expression between nTg and Tg mice. Figure 8 shows the total areas of c-fos mRNA expression at dorsal hippocampal levels of nTg and Tg mice following 0 to 6 h reperfusion. The areas were significantly larger in Tg mice than in nTg mice at 2 and 6 h.

These findings suggest that oxygen free radicals may play an important role in expression of c-fos following cerebral ischemia. In addition, the differences in c-fos expression in the hippocampus and thalamus, but not in the cortex, of Tg compared to nTg mice suggest c-fos gene may be activated by a number of mechanisms in ischemia, including oxidative stress.

IV. CONCLUDING REMARKS

We described a feature of SOD-1 Tg mice and the use of such mice in the study of ischemic brain injury. It is believed that oxidative stress is associated with development of a number of neurological disorders including cerebral ischemia, traumatic brain injury, and some forms of neurodegenerative diseases. The SOD-1 Tg mice are a unique tool to elucidate oxidative mechanisms in the pathogenesis of these neurological disorders.

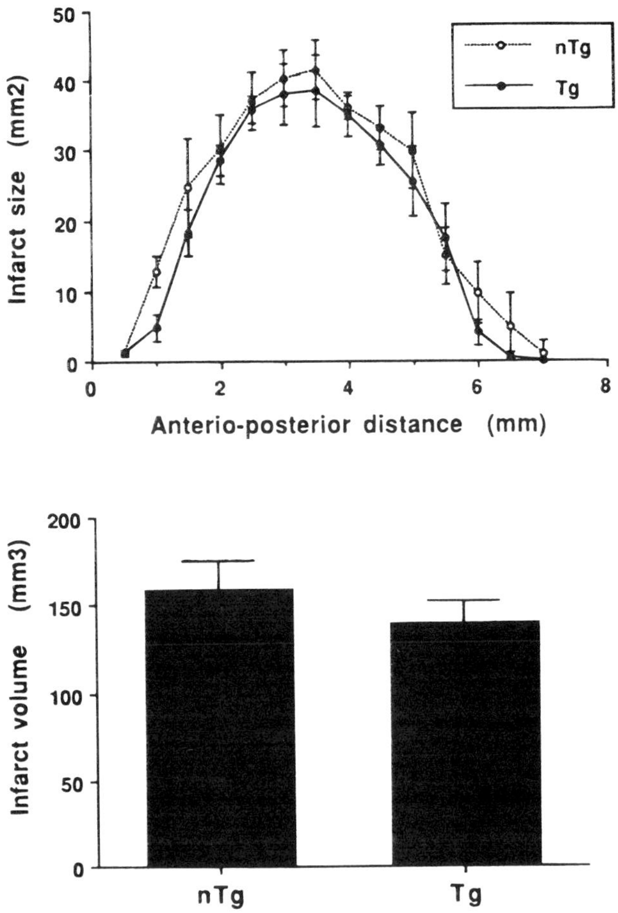

FIGURE 6. Infarct size (upper) and volume (lower) of nontransgenic (nTg) and transgenic (Tg) mice following 24 h of permanent MCA occlusion. There were no significant differences in the infarct size and volume between nTg and Tg mice. Values are mean ±SD (n = 6). (Data obtained from Chan et al., 1993.)

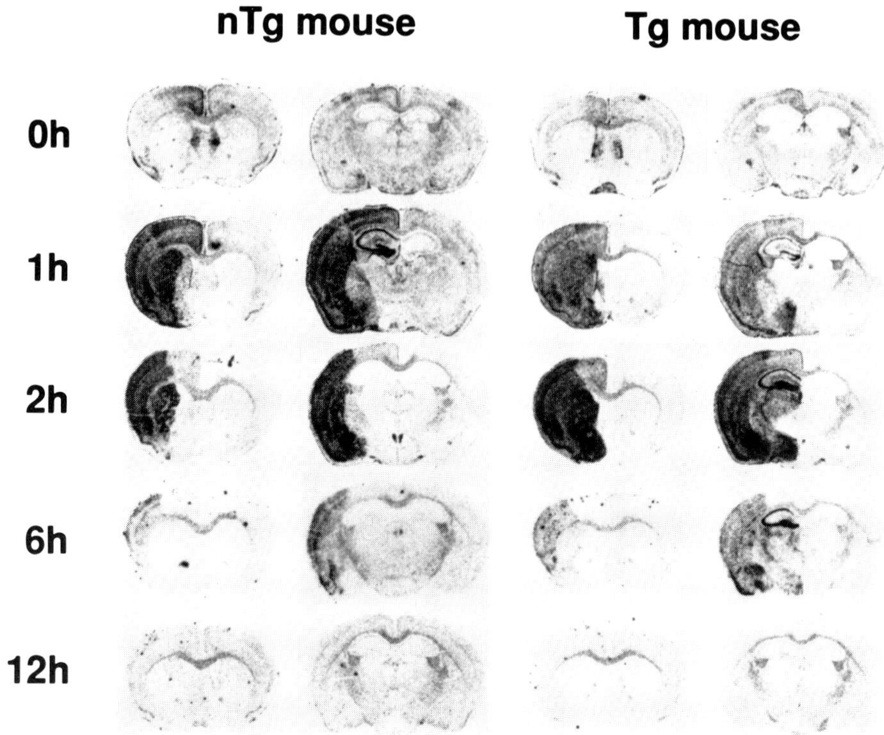

FIGURE 7. Distribution of c-fos mRNA expression at caudate and dorsal hippocampal levels in nTg (two left columns) and Tg mice (two right columns) following a 10-min MCA occlusion. 0 h in the figure indicates a time just before reperfusion and 1, 2, 6, and 12 h indicate periods after reperfusion. Note that the c-fos expression in the ipsilateral hippocampus and thalamus, but not in the cortex, was more sustained in Tg mice than in nTg mice.

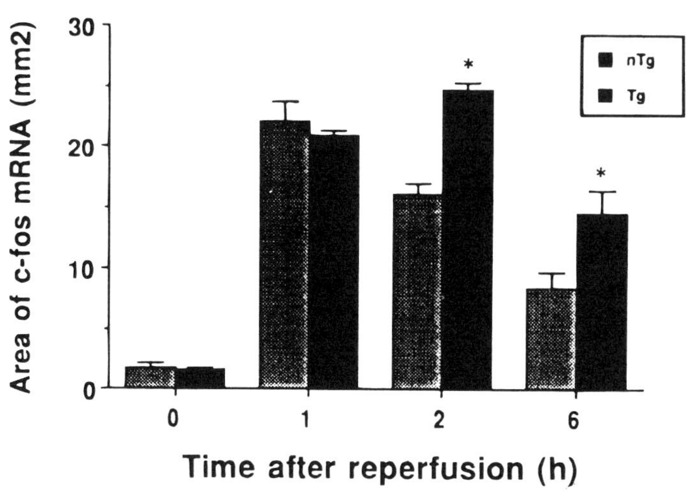

FIGURE 8. Total areas of c-fos mRNA expression at dorsal hippocampal levels in nTg and Tg mice. The areas were significantly larger in Tg mice than in nTg mice at 2 and 6 h after reperfusion. Values are mean ±SD (n = 3). $^*p < 0.01$ (Tg vs. nTg mice, by Student's *t*-test).

ACKNOWLEDGMENTS

The authors thank Jari Koistinaho, Rhodora Gacayan, and Liza Reola for their technical assistance. This work was supported by NIH grants AG-08938, NS-25372, and NS-14543.

REFERENCES

Chan PH, Kamii H, Yang G, Gafni J, Epstein CJ, Carlson E, Reola L (1993) Brain infarction is not reduced in SOD-1 transgenic mice after a permanent focal cerebral ischemia. *Neuroreport* 5, 293–296.

Chan PH, Yang GY, Chen SF, Carlson E, Epstein CJ (1991) Cold-induced brain edema and infarction are reduced in transgenic mice overexpressing CuZn-superoxide dismutase. *Ann Neurol* 29, 482–486.

Combs DJ, Dempsey RJ, Donaldson D, Kindy MS (1992) Hyperglycemia suppresses c-fos mRNA expression following transient cerebral ischemia in gerbils. *J Cereb Blood Flow Metab* 12, 169–172.

Epstein CJ, Avraham KB, Lovett M, Smith S, Elory-Stein O, Bry C, Groner Y (1987) Transgenic mice with increased CuZn-superoxide dismutase activity: animal model of dosage effects in Down's syndrome. *Proc Natl Acad Sci USA* 83, 8044–8048.

Epstein CJ, Chan PH, Cadet JL, Carlson E, Chen S, Chu L, Fahn S, Jackson-Lewis V, Kinouch H, Kostic V, Kujirai K, Mizui A, Naini A, Przedborski S, Yang G (1992) Resistance of SOD-transgenic mice to oxidative stress. In: *Gene Transfer and Therapy in the Nervous System* (Gage FH, Christen Y, eds.), Berlin, Springer-Verlag, pp. 106–117.

Kinouchi H, Epstein CJ, Mizui T, Carlson E, Chen SF, Chan PH (1991) Attenuation of focal cerebral ischemic injury in transgenic mice overexpressing CuZn superoxide dismutase. *Proc Natl Acad Sci USA* 88, 11158–11162.

Koretsky AP (1992) Investigation of cell physiology in the animal using transgenic technology. *Am J Physiol* 262, C261–C275.

Morgan JI, Curran T (1991) Stimulus-tanscription coupling in the nervous system: involvement of the inducible proto-oncogenes fos and jun. *Annu Rev Neurosci* 14, 421–451.

Nowak TS Jr., Osborne OC, Suga S (1993) Stress protein and proto-oncogene expression as indicators of neuronal pathophysiology after ischemia. In: *Progress in Brain Research, Vol. 96* (Kogure K, Hossmann K-A, Siesjö BK, eds.), Amsterdam, Elsevier Science Publishers, pp. 195–208.

Popko B, Readhead C, Dausman J, Hood L (1990) Transgenic mice in neurobiological research. In: *Neuromethods, Vol. 16: Molecular Neurobiological Techniques.*

Rosenfeld MG, Crenshaw EBI, Lira SA (1988) Transgenic mice: applications to the study of the nervous system. *Annu Rev Neurosci* 11, 353–372.

Sagar SM, Sharp FR, Curran T (1988) Expression of c-fos protein in brain: metabolic mapping at the cellular level. *Science* 240, 1328–1331.

Schalling M, Dagerlind A, Brene S, Hallma H, Djurfeldt M, Persson H, Terenius L, Goldstein M, Schlesinger D, Hokfelt T (1988) Coexistence and gene expression of phenylethanolamine N-methyltransferase, and neuropeptide tyrosine in the rat and bovine adrenal gland: effects of reserpine. *Proc Natl Acad Sci USA* 85, 8306–8310.

Sharp FR, Sagar SM, Swanson RA (1993) Metabolic mapping with cellular resolution: c-fos vs. 2-deoxyglucose. *Crit Rev Neurobiol* 7, 205–228.

Swanson RA, Morton MT, Wu GT, Davidson C, Sharp FR (1990) A semiautomated method for measuring brain infarct volume. *J Cereb Blood Flow Metab* 10, 290–293.

Yang G, Chan PH, Chen J, Carlson E, Chen SF, Weinstein P, Epstein CJ, Kamii H (1994) Human copper-zinc superoxide dismutase transgenic mice are highly resistant to reperfusion injury after focal cerebral ischemia. *Stroke* 25, 165–170.

VIII. Free Radicals and Lipid Peroxidation

Chapter 32

IMPROVEMENTS IN THE SALICYLATE TRAPPING METHOD FOR MEASUREMENT OF HYDROXYL RADICAL LEVELS IN THE BRAIN

John S. Althaus, Paula K. Andrus, Edward D. Hall, and Philip F. VonVoigtlander

TABLE OF CONTENTS

ABSTRACT

We present a clear, concise method for measuring hydroxyl radical ˙OH production based on hydroxylation of salicylate. The procedure involves *in vivo* administration of salicylate followed by *ex vivo* measurement. Salicylate, after systemic administration, enters brain tissue and rapidly reacts with ˙OH to form dihydroxybenzoic acid (DHBA) species. The two principal benzoic acids formed are the 2,3- and 2,5-dihydroxy isomers. These can be separated using high-performance liquid chromatography and quantified electrochemically. We found that exogenous and/or artificial sources of DHBA in gerbil and mouse brain tissues can distort the hydroxyl radical measurement. Therefore, techniques were used to minimize experimental errors from these and other factors. These included the addition of antioxidants in the extraction media to eliminate artificial sources of DHBA formation. Also, an internal standard was added to the extraction media to reduce the variability of measurement by correcting for chemical losses and dilution factors. Finally the ratio of DHBA brain concentration to salicylate brain concentration is remarkably constant, allowing the data to be normalized. The constant DHBA/salicylate ratio likely results because salicylate itself does not appear to affect the steady-state concentration of ˙OH in the brain over the dose range tested. We found that D-penicillamine (which does not penetrate the brain), a known chelator of copper and iron, reduced DHBA formation in mouse plasma, but not in the mouse brain. Thus, it appears that systemic concentrations of DHBA do not contaminate the brain. In a separate experiment, IP administrations of SKF-525A (an inhibitor of cytochrome-P450-induced oxidation) in gerbils had no effect on brain levels of 2,5-DHBA. This argues against a significant amount of 2,5-DHBA production via cytochrome-P450 mechanisms in the central nervous system. In conclusion we present a method of measuring ˙OH that is simple, concise, and reliable.

0-8493-8094-4/95/$0.00+$.50

I. INTRODUCTION

Reactive oxygen species (ROS) have been implicated as a causal factor in posttraumatic cell loss in tissues subjected to ischemia, followed by reperfusion.[1] The hydroxyl radical ($^{\bullet}OH$) is one of several ROSs thought to play a predominant role.[2] There are a number of possible pathways giving rise to $^{\bullet}OH$ following ischemia/reperfusion. These pathways begin with the enzymatic or nonenzymatic generation of superoxide (O_2^-). Thus, O_2^- can react with nitric oxide (NO) to produce peroxynitrite (^-OONO), which rapidly degenerates to $NO_2^{\bullet}$ and $^{\bullet}OH$.[3] Alternatively, O_2^- is converted to hydrogen peroxide (H_2O_2) by spontaneous reaction or more rapidly by catalase.[4] When H_2O_2, from these chemical sources or from enzymatic sources (e.g., monoamine oxidase), interacts with ferrous ion (Fe^{2+}) the reaction yields $^{\bullet}OH$. Fe^{2+} can also catalyze the conversion of peroxynitrite to $^{\bullet}OH$.

Once $^{\bullet}OH$ has been formed, its subsequent reactivity is diffusion rate limited based on a large reaction rate constant (10^9 to 10^{10} M^{-1} s^{-1}).[3] Therefore, Fe^{2+} becomes a critical component in directing the site of $^{\bullet}OH$ reactivity.[5] In this regard, the phospholipid head groups of cell membranes are known to chelate Fe^{2+}.[6] Thus, $^{\bullet}OH$, catalyzed by Fe^{2+} bound to lipid membranes, is well positioned to initiate a chemical process known as lipid peroxidation.[4] Lipid peroxidation is a chain reaction involving the propagation of lipid radicals, which can occur within a lipid bilayer, such as a cell membrane. The chain reaction can be initiated by $^{\bullet}OH$ with the abstraction of an allelic hydrogen of a phospholipid. Therefore, $^{\bullet}OH$ increases can be diagnostic of lipid peroxidation and cell damage.

Aromatic hydroxylation is a potentially useful measure of $^{\bullet}OH$ radical formation *in vitro*[7] and *in vivo*.[8] Floyd et al. first reported the use of salicylate (SAL) to trap hydroxyl radical while employing HPLC with electrochemical detection to measure the resulting formation of hydroxylated products.[9] This assay is based on the known reaction involving the addition of H_2O_2 to Fe^{2+} and SAL. In this reaction three major products form: catechol, 2,3-DHBA, and 2,5-DHBA. Although these products may be formed in low yield, because they are electrochemically active, detection is sensitive and measurement is possible in a variety of biological matrixes.

In recent years, there have been over 50 published reports of SAL hydroxylation used to detect hydroxyl radical generation *in vivo* and *in vitro*. In cases where *in vivo* brain $^{\bullet}OH$ radical generation under conditions of oxidative stress was investigated, different methods of SAL administration were utilized. These include systemic administration (e.g., IP),[10] intracranial administration (e.g., ICV),[11] and intracranial perfusion (e.g., microdialysis).[12] Each method has its own set of advantages and disadvantages. We find IP administration of SAL adequate, as well as not requiring specialized, technical skills.

The purpose of this report is to present an improved technique that is both simple and reliable for measuring $^{\bullet}OH$ in brain.

II. METHODS

A. ANIMAL DOSING

Animals (CF-1 mice, 18 to 20 g, Charles River, Kalamazoo, MI or Mongolian gerbils, 50 to 60 g, Tumblebrook Farms, West Brookfield, MA) were injected (SC) with test drug or saline (vehicle). Fifteen minutes later animals received IP injections of SAL (Aldrich, Milwaukee, WI; dose specified in text) or saline. After 15 min, animals were killed by decapitation, the right hemisphere of the forebrain was dissected, immediately frozen in liquid nitrogen, and stored at −80°C.

B. TISSUE ANALYSIS

Each frozen brain hemisphere was weighed and placed in a 1.5-ml microcentrifuge tube containing extraction media consisting of 0.5 ml mobile phase (see below), an internal standard of 3,4-dihydroxy cinnamic acid (DHCA) (1 μg/ml; Aldrich, Milwaukee, WI), 0.1% desferal (deferoxamine, CIBA, NJ), 0.5% mannitol, and 1.0% phenylalanine. The tissue was sonicated in the mobile phase mixture for 20 to 30 s with a Virtis Virsonic 300 sonicator at a power setting of 4. The milky solution was frozen on dry ice and then centrifuged twice for 5 min each at 14,000 rpm. The supernatant (50 μl) was injected onto a BAS ODS 5 μm column with a Perkin-Elmer SS-100 autosampler. Each liter of mobile phase consisted of 20% acetonitrile, 10% tetrahydrofuran, and 14.15 g of monochloroacetic acid. An external standard consisting of DHCA (1 μg/ml), 2,5-DHBA, 2,3-DHBA (0.1 μg/ml; Aldrich, Milwaukee, WI), and salicylic acid (10 μl/ml) was initially injected onto the HPLC for later use to determine individual response factors used in the calculations. The column effluent flowed at 1 ml/min into a BAS electrochemical detector set at 650 mV and then into a UV detector set a 295 nm. The 2,5-DHBA and 2,3-DHBA were detected electrochemically, and SAL was detected spectrophotometrically (destruction of SAL by the electrochemical detector prior to entry into the spectrophotometric detector was negligible). The data are expressed as the ratio of DHBA to SAL in order to normalize DHBA levels at differing and widely varying brain concentrations of SAL. Chromatographic peaks were integrated with Waters Maxima 820 software. On-column masses of DHBAs and SAL in samples were calculated by convention using injection volume, concentration of standards, peak area of standards, and peak area of samples. Unless otherwise stated, the data are expressed as the ratio of 2,5-DHBA to SAL. Statistical differences were determined using ANOVA at 95% confidence level.

C. PLASMA ANALYSIS

In one experiment blood was drawn for plasma analysis. Animals were anesthetized with metofane (methoxyflurane, Pitman-Moore, Mundeteir, IL) 1 min before decapitation. Blood samples were obtained via cardiac puncture with a 1-ml syringe fitted with a 22-gauge needle rinsed with heparin; samples were centrifuged at 1000 rpm for 5 min. A 20-μl sample of plasma was added to 80 μl of extraction media and the mixture was centrifuged at 14,000 rpm for 5 min. The supernatant was assayed and analyzed as described earlier.

III. RESULTS AND DISCUSSION

Before attempting to use the SAL ˙OH trapping method it was important to identify exogenous and/or artificial sources of DHBA. We were concerned that the commercially available SAL used to dose the animals was contaminated with DHBA. If so, what was the amount and was this relevant? Using our procedure, the sodium SAL used contributed less than 0.005% of the DHBA measured in brain.

Another source of DHBA may have originated at the extraction step. Given the nature of the extraction, SAL oxidation may have occurred. To test this possibility, mice were dosed with SAL at 1000 mg/kg; the brains were removed and processed as previously described. In a separate experiment, SAL was added exogenously to untreated mouse brain samples at a concentration that was roughly equivalent to the concentration found when mice were dosed at 1000 mg/kg. Analysis of the tissue indicated that roughly 30% of the DHBA measured was formed during the extraction step of the assay. This fact alone would not necessarily invalidate conclusions regarding increases or decreases in ˙OH radical formation if appropriate controls were incorporated. Nevertheless, this artificial source would dilute the ˙OH measurement, making statistical comparisons more difficult. The exact nature of this oxidation is unknown. However, it is likely that iron and oxygen from internal and/or possibly external sources (e.g., sonicator probe) may provide the proper condition for *ex vivo* hydroxyl radical formation and SAL oxidation. To help eliminate these artificial sources of DHBA formation, three antioxidants (phenylalanine 1%, mannitol 0.5%, and desferyoxamine 0.1%) were added to the extraction solution. Under these conditions, less than 10% of the DHBA measured appeared to come from the extraction step.

Another important technique employed to reduce the variability of measurement was the inclusion of an internal standard to the extraction media. In this case, DHCA was selected as the compound of choice because it mimicked the basic structural and antioxidant characteristics of the DHBAs. As an internal standard, DHCA corrects for chemical losses and dilution factors. Calculations were conveniently related to the amount of DHCA in the extraction media (usually 1 μg/ml). In this regard, an essential part of the calculation was the determination of the unit response factors (RF) for DHBAs and SAL. This calculation was based on detection of a standard solution of DHBAs, DHCA, and SAL (described in Section II) and is illustrated in the equation:

$$RF_i = (A/\mu g)_i/(A/\mu g)_{DHCA} \tag{1}$$

where i denotes 2,3-DHBA, 2,5-DHBA, or SAL; A is the standard peak area. In this case, peak areas for DHBAs and DHCA were determined electrochemically and peak area for SAL was determined spectrophotometrically.

Once the RF for each component is determined, calculation of unknown sample tissue concentration (TC) can be found as follows:

$$TC_i = ((A_i/A_{DHCA}) \times [DHCA] \times V \times RF_i)/m \tag{2}$$

where i denotes 2,3-DHBA, 2,5-DHBA, or SAL; A is the sample peak area; [DHCA] is the concentration of DHCA; V is the volume of the extract; and m is the mass of the tissue.

Finally, the ratio (R) of 2,3-DHBA or 2,5-DHBA to SAL can be easily found as follows:

$$R_n = TC_n/TC_{SAL} \tag{3}$$

where n is 2,3-DHBA or 2,5-DHBA.

Previously, in reports from our laboratory describing measurement of $^{\bullet}OH$ formation, data was expressed as the ratio of DHBA to salicylate.[10,13,14] Figure 1 shows that in gerbils dosed with SAL (10 to 1000 mg/kg), measurement of brain DHBAs (μg/g) vary 100-fold. However, the ratio of DHBA brain concentrations to SAL brain concentrations is remarkably constant over this same dose range. The reason for this observation, which was also demonstrated in mice and discussed extensively,[10] may be as follows: SAL does not appear to affect the steady-state concentration of $^{\bullet}OH$ in the brain over the dose range tested. There are at least two reasons to explain this phenomenon. First, the turnover rate for $^{\bullet}OH$ (rate of formation over rate of destruction) is not affected by even large concentrations of SAL because of diffusion rate limitations. In other words, given a constant formation of $^{\bullet}OH$ formation (e.g., based on basal levels of Fe^{2+}, H_2O_2, and $ONOO^-$), no amount of SAL will change the rate of $^{\bullet}OH$ destruction because this is diffusion rate limited (10^9 to 10^{10} M^{-1} s^{-1}). Thus, the steady-state concentration of $^{\bullet}OH$ remains constant. Only changes in the formation of $^{\bullet}OH$ (e.g., changes in levels of Fe^{2+}, H_2O_2, and $ONOO^-$) will cause an increase or decrease in DHBA formation. Second, it may be possible that SAL only acts as a probe (i.e., trapping, but also generating a secondary radical) and not as a scavenger (i.e., trapping, but not generating a secondary radical). For example, if the SAL radical ($^{\bullet}SAL$) once formed ($SAL + {}^{\bullet}OH \rightarrow {}^{\bullet}SAL + H_2O$) reacts with water, generating a hydrogen atom (${}^{\bullet}SAL + H_2O \rightarrow SAL\text{-}OH + {}^{\bullet}OH$), which in turn reacts with water to yield $^{\bullet}OH$ and hydrogen gas (${}^{\bullet}H + H_2O \rightarrow H_2 + {}^{\bullet}OH$), then there would be no net change in the $^{\bullet}OH$ concentration.

The fact that the DHBA/SAL ratio is constant over SAL dose allows the data to be normalized. This is an important consideration because not all changes in DHBA are related

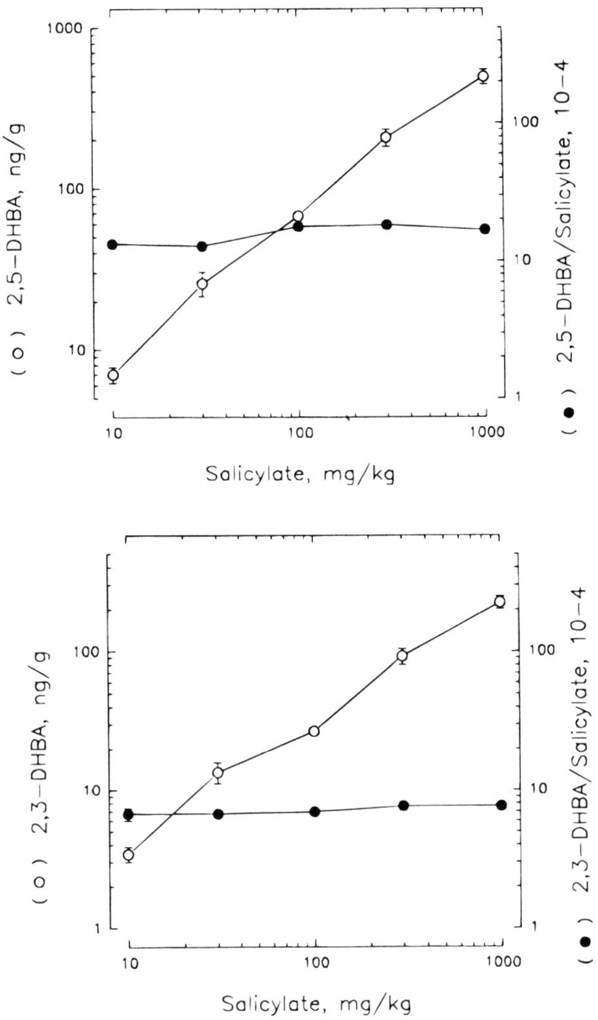

FIGURE 1. Ratio of 2,5-DHBA (top graph) or 2,3-DHBA (bottom graph) to salicylate in whole gerbil brain as a function of the salicylate dose. Salicylate was administered iP to gerbils 15 min before sacrifice. Values equal mean ± SEM for ten animals per dose. The DHBA/salicylate ratio remained constant, regardless of the dose given.

exclusively to increased or decreased oxidative stress. Simple changes in DHBA might also result from effects of blood flow and/or SAL pharmacokinetics. As stated previously, the ratio DHBA/SAL will change if the rate of formation of ˙OH changed but not if the brain concentration of SAL changed.

In the study of ˙OH formation in brain, a variety of SAL administration techniques have been used. These include systemic administration (iP),[10] intracranial administration (ICV),[11] and intracranial perfusion (microdialysis).[12] The latter two methods have the advantage of being devoid of systemic complications related to metabolism of SAL leading to blood-borne sources of DHBAs. However, the disadvantages of these latter techniques are related to suboptimal delivery of SAL to brain tissue and requirement of specialized skills and equipment (i.e., stereotaxis and dialysis probes, etc.).

We were concerned about blood-borne sources of DHBAs with IP administration of SAL used in our experiments. In the evaluation of this potential problem, we discovered that D-penicillamine, a chelator of copper and iron[15] and a scavenger of peroxynitrite (one source of ·OH),[16] effectively reduced DHBA formation in the plasma (Figure 2). More importantly, the same dose of D-penicillamine did not effect DHBA formation in the brain. These results can be explained based on the pharmacokinetics of D-penicillamine. Penicillamine, after systemic administration, is readily absorbed into the blood stream.[17] However, it is also known that neutral D-amino acids do not penetrate the blood-brain barrier.[18] Perhaps penicillamine does not affect brain levels of DHBA because it is not likely to penetrate into the brain parenchyma. The data, therefore, suggests that DHBA formation systemically does not predominate and does not influence the measurement of DHBA formation in the brain.

The use of 2,5-DHBA as a measure of ·OH has been criticized due to its possible production via cytochrome-P450-dependent metabolism, at least in the case of *in vitro* studies with liver

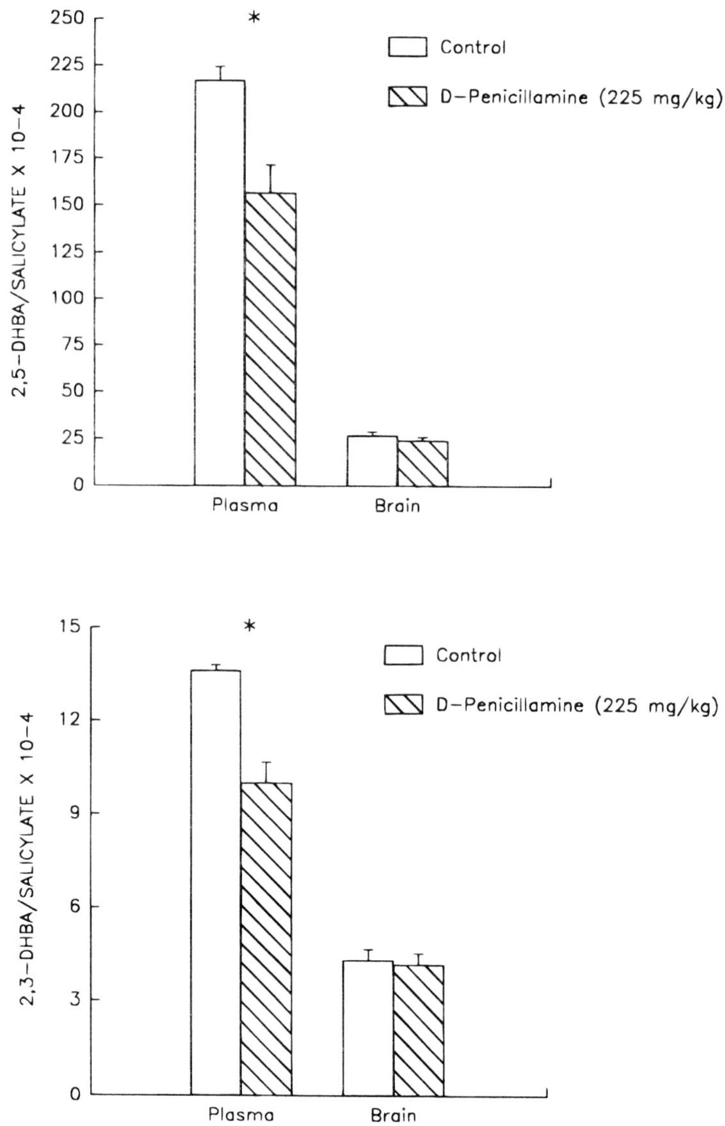

FIGURE 2. Ratio of 2,5-DHBA (top graph) or 2,3-DHBA (bottom graph) to salicylate in right hemisphere mouse brain or in mouse plasma. Penicillamine (225 mg/kg) was administered SC 30 min before sacrifice. Salicylate (300 mg/kg) was administered iP 15 min before sacrifice. Values equal mean ± SEM for six animals. Penicillamine effectively reduced DHBA formation in the plasma, but not in the brain ($p < 0.05$ by ANOVA).

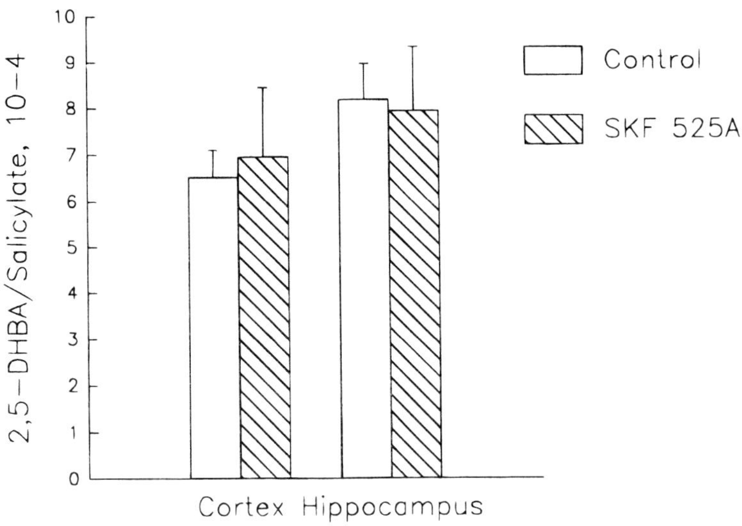

FIGURE 3. Effects of cytochrome-P450 inhibition by SKF-525A (50 mg/kg iP, 30 min before sacrifice; 300 mg/kg salicylate iP administration 15 min before sacrifice) on levels of 2,5-DHBA formation in cortex and hippocampus in sham-operated nonischemic gerbils. Values equal mean ± SEM for six or seven animals.

microsomes.[19] Therefore, SKF-525A (an inhibitor of cytochrome-P450 drug oxidations) was administered at 50 mg/kg IP to examine the effect in nonischemic gerbils. As shown in Figure 3, SKF-525A had no effect on the brain levels of 2,5-DHBA, which argues against a significant amount of its production via cytochrome-P450 mechanisms in the central nervous system. It appears that systemically, 2,5-DHBA may come from metabolic sources. The 2,5-DHBA/SAL ratio is much larger in plasma (214×10^{-4}) than in the brain (27×10^{-4}). However, we have already demonstrated that DHBA formation systemically does not appear to influence the measurement of DHBA in the brain. In addition, previously we showed that the time course of ˙OH formation in brain following injury was qualitatively the same, whether measured using 2,5-DHBA or 2,3-DHBA.[13]

IV. CONCLUSION

We described in detail a method for measuring ˙OH formation based on the hydroxylation of SAL. By analyzing the appearance of 2,5-DHBA and/or 2,3-DHBA following SAL administration in tissue subjected to ischemia/reperfusion, changes in ˙OH formation can be determined. We have pointed out several caveats related to the analysis and we described techniques that can be used to minimize experimental errors. Addition of antioxidants to the extraction media can help reduce artificial sources of SAL oxidation. The addition of an internal standard can account for chemical losses and dilution factors to give a more accurate measurement. Data expressed as a ratio between DHBAs and SAL can correct for changes resulting from variance in blood flow and pharmacokinetics. We showed that enzymatic and blood-borne sources of DHBAs do not appear to influence nonenzymatic (i.e., ˙OH reactivity) sources of SAL hydroxylation in the brain. In conclusion, we presented a method of measuring ˙OH that is simple, concise, and reliable.

REFERENCES

1. Hall, E. D., Yonkers, P. A., Andrus P. K., Cox J. W., Anderson D. K. (1992) Biochemical and pharmacology of lipid antioxidants in acute brain and spinal cord injury. *J. Neurotrauma* 9, S424–S442.

2. Halliwell, B. (1992) Reactive oxygen species and the central nervous system. *J. Neurochem.* 59, 1609–1623.
3. Radi, R., Beckman, J. S., Bush, K. M., Freeman, B. A. (1991) Peroxynitrite-induced membrane lipid peroxidation: the cytotoxic potential of superoxide and nitric oxide. *Arch. Biochem. Biophys.* 288, 481–487.
4. Braughler J. M., Hall, E. D. (1989) Central nervous system trauma and stroke I. Biochemical considerations for oxygen radical formation and lipid peroxidation. *J. Free Radic. Biol. Med.* 6, 289–301.
5. Fujii, T., Hiramoto, Y., Terao, J., Fukuzawa, K. (1991) Site-specific mechanism of initiation by α-tocopherol of lipid peroxide-dependent lipid peroxidation in charged micelles. *Arch. Biochem. Biophys.* 284, 120–126.
6. Yoshida, K., Terao, J., Suzuki, T., Takama, K. (1991) Inhibitory effects of phosphatidylserine on iron-dependent lipid peroxidation. *Biochem. Biophys. Res. Commun.* 179, 1077–1081.
7. Gelvan, D., Saltman, P. (1991) Small chelators and the nature of the metal ion in site-specificity of ˙OH damage, in *Oxidative Damage & Repair: Chemical, Biological and Medical Aspects* (Davies K. J. A., ed.). pp. 825–830, Pergamon, New York.
8. Grootveld, M., Halliwell, B. (1986) Aromatic hydroxylation as a potential measure of hydroxyl-radical formation *in vivo*. *Biochem. J.* 237, 499–504.
9. Floyd, R. A., Watson, J. J., Wong, P. K. (1984) Sensitive assay of hydroxyl free radical formation utilizing high pressure liquid chromatography with electrochemical detection of phenol and salicylate hydroxylation products. *J. Biochem. Biophys. Methods* 10, 221–235.
10. Althaus, J. S., Andrus, P. K., Williams, C. M., VonVoigtlander, P. F., Cazers, R. C., Hall, E. D. (1993a) The use of salicylate hydroxylation to detect hydroxyl radical generation in ischemic and traumatic brain injury: reversal by tirilazad mesylate (U-74006F). *Mol. Chem. Neuropath.* 20, 147–162.
11. Giovanni, A., Hastings, T. G., Liang, L. P., Zigmond, M. J. (1992) Methamphetamine increases hydroxyl radicals in rat striatum: role of dopamine. *Soc. Neurosci. Abstr.* 18, 1444.
12. Chiueh, C. C., Huang, S. J., Murphy, D. L. (1992) Enhanced hydroxyl radical generation by 2′ methyl analog of MPTP: suppression by clorgyline and deprenyl. *Synapse* 116, 183–191.
13. Hall, E. D., Andrus, P. K., Yonkers, P. A. (1993a) Brain hydroxyl radical generation in acute experimental head injury. *J. Neurochem.* 60, 588–594.
14. Hall, E. D., Andrus, P. K., Althaus, J. S., VonVoigtlander, P. V. (1993b) Hydroxyl radical production and lipid peroxidation parallels selective post-ischemic vulnerability in gerbil brain. *J. Neurosci. Res.* 34, 107–112.
15. Ciuffi, M., Gentilini, G., Franchi-Micheli, S., Zilletti, L. (1992) D-penicillamine affects lipid peroxidation and iron content in the rat brain cortex. *Neurochem. Res.* 17, 1241–1246.
16. Althaus, J. S., Oien, T. T., Fici, G. J., Scherch, H. M., Strautman, A. F., Sethy, V. H., VonVoigtlander, P. F. (1993b) Peroxynitrite (POX) scavengers: possible neuroprotectants against nitric oxide (NO)-dependent toxicity. *Soc. Neurosci. Abstr.* 19, 1482.
17. Michael, B., Carruthers, G., Manfred, H., Freeman, D., Percy, J., Pabenstein, D. (1982) Pharmacokinetics of reduced D-penicillamine in patients with rheumatoid arthritis. *Arthritis Rheum.* 25, 111–116.
18. Oldendorf, W. H. (1971) Brain uptake of radiolabelled amino acids, amines, and hexoses after arterial injection. *Am. J. Physiol.* 221, 1629–1639.
19. Axelrod, J., Reichenthal, J., Brodie, B. B. (1954) Mechanism of the potentiating action of beta-diethylaminoethyl diphenylpropylacetate. *J. Pharmacol. Exp. Ther.* 112, 49–54.

Chapter 33

TRAPPING OF OXYGEN FREE RADICALS DURING BRAIN ISCHEMIA AND REPERFUSION

Atsuhiro Sakamoto, Ryo Ogawa, Tomoko Ohnishi, and S. Tsuyoshi Ohnishi

TABLE OF CONTENTS

I. INTRODUCTION

It has been hypothesized that reactive oxygen free radicals are formed during ischemia-reperfusion injury of the brain.[1–4] These radicals have been considered to attack the major classes of cellular components, such as lipids,[5,6] proteins,[7] and nucleic acids.[8] Therefore, they may play an important role in causing extensive secondary damage in brain tissue and possibly death. However, the evidence for the participation of radicals in ischemia-reperfusion injury has been indirect. The central roles of free radicals have been proposed through the activation of lipid peroxidation, the alteration of membrane phospholipid components, the changes of the levels of endogenous radical scavengers, and the protection provided by free radical scavengers.[9–12] Thus, direct detection of free radicals during ischemia or reperfusion has long been desired.

Recently, it has been reported that the inhibition of free radicals or lipid peroxidation with PEG-SOD, PEG-catalase,[13] allopurinol, dimethylthiourea,[14] or deferoxamine[15] has the potential to attenuate cerebral ischemia-reperfusion injury. Thus, a drug that can prevent the formation of free radicals may have therapeutic value in the treatment of CNS injury. Ohnishi et al. synthesized an acetoxy-methyl ester of prostaglandin E_1 oligomer. It demonstrated reproducible protective effects on the brain and the heart against ischemia.[16,17] Evidence suggests that prostaglandin oligomers have antioxidant activity.[18] Ohnishi et al. have attempted to increase the antioxidant activity of these compounds and synthesized a new oligomer from PGB_2 that has ascorbic acids as ester groups (termed OC-5186).[19,20] In this chapter, we report on some *in vivo* actions of this new compound.

0-8493-8094-4/95/$0.00+$.50

To determine the effect of antioxidants, it is also necessary to detect free radicals directly. By using the spin-trapping technique, we attempted to directly detect oxygen-derived radicals in the rat brain exposed to ischemia and reperfusion. We also measured the effect of OC-5186 on the formation of oxygen-derived free radicals.

II. METHODS[21,22]

A. EXPERIMENTAL PROTOCOL

We used the forebrain ischemia that was produced by the bilateral occlusion of the common carotid arteries combined with hemorrhagic hypotension (30 mmHg), because it was easy to prepare, to reperfuse the brain, and to maintain normal systemic conditions.

Adult male Sprague-Dawley rats weighing 325 to 375 g were used. They were allowed free access to food and water prior to experiments. Anesthesia was induced with isoflurane. Following tracheal intubation, rats were ventilated with 1.4% isoflurane in 70% N_2O − 30% O_2 using a Harvard Rodent Ventilator Model 683. The tracheal intubation is better than tracheotomy when the morbidity after ischemia-reperfusion injury is to be estimated, because the respiratory complication is one of the important factors for the mortality and the morbidity. So, we used the tracheal intubation for ventilatory support. Vecuronium bromide (0.1 to 0.15 mg/kg) was administered as needed to maintain paralysis. A catheter was inserted into the tail artery for blood withdrawal and blood pressure monitoring, and another catheter was inserted into the tail vein for drug and blood infusion. We used the tail artery and the tail vein, because we do not want to injure the cervical and the femoral circulation for the purpose of estimating the mortality and the paralysis after brain damage. If the catheterization from above the tail skin is difficult, the skin can be cut to expose blood vessels for cannulation. The bilateral common carotid arteries were isolated for inducing ischemia later when all other procedures were completed. Two needle electrodes were placed bilaterally into the temporal area to monitor the interhemispheric electroencephalogram during the ischemic period. An electrocardiogram was used to obtain the electrocardiogram and the heart rate. Following these preparatory procedures, isoflurane was discontinued while mechanical ventilation with 70% N_2O− 30% O_2 was continued throughout. Then, the rats were allowed a 30-min period to reach steady state.

Cerebral ischemia was produced by the occlusion of both common carotid arteries followed by spontaneously bleeding to a reservoir from the femoral artery to decrease the mean arterial pressure. The pressure of 30±3 mmHg was achieved by adjusting the height of the reservoir. This mean arterial pressure was maintained for 10 or 20 min. Rectal temperature was maintained at 37°C using a heat lamp. PCO_2 was kept between 35 and 45 mmHg by adjusting the ventilator throughout the study. Heparin (150 IU/kg) was administered intravenously to prevent coagulation. Arterial pH was maintained at normal levels by bicarbonate infusion. During the experiment, mechanical ventilation, correction of acid-base balance, and stable body temperature were necessary for obtaining the stable base cerebral blood flow. At the end of the ischemic period, the clamps on the carotid arteries were released and the blood (which was withdrawn previously and kept at 37°C) was reinfused into the rat. Then animal was maintained for various reperfusion periods (0, 2, 5, 10, 20, and 40 min). Blood samples (0.2 ml) were taken for arterial gas measurement at the end of each period.

B. PREPARATION OF SOLUTIONS

N-tert-butyl-α-phenylnitrone (PBN), chloroform (catalog no. 31,998-8) and methyl alcohol (catalog no. 27,047-4) were obtained from Aldrich Chemicals Co. (Milwaukee, WI). PBN was dissolved to a 130 m*M* concentration in argon purged deionized water (Milli Q cartridges, Bedford, MA) and was treated with activated charcoal according to Buettner et al.[23] until electron paramagnetic resonance spectroscopy (EPR) signals from impurities were eliminated.

Final concentration of PBN was determined using the molar extinction coefficient of 16,700 at 294 nm in ethanol. To a Krebs-Ringer phosphate (KRP) buffer solution (120 m*M* NaCl, 5 m*M* KCl, 1.3 m*M* $CaCl_2$, 1.3 m*M* $MgSO_4$, and 10 m*M* Na_2HPO_4) containing 80 m*M* PBN, 0.1 m*M* diethylenetriamine-pentaacetic acid (DTPA, Aldrich) was added. To the mixture of chloroform and methanol (2:1), 1 m*M* DTPA was added and kept overnight. Undissolved DTPA was removed by filtration, and the solution was deoxygenated by bubbling with nitrogen. This solution was used throughout these experiments.

C. DETECTION OF PBN SPIN ADDUCTS

After reperfusion, the rats were decapitated and the cerebral cortex was removed and homogenized in a 2.5-ml ice-cold, deoxygenated KRP solution within 15 seconds after decapitation. This produced a concentration of about 45 mg tissue protein per milliliter (protein as determined by the biuret method). Lipids were extracted by 6 ml of the chloroform-methanol mixture using the procedure of Folch et al.[24] Each lipid extract was evaporated and redissolved by adding 0.1 ml of the chloroform-methanol mixture. Using a microsyringe with a fine, long needle, the extract solution was transferred into a glass capillary pipette (i.d. 1.55 mm, length 90 mm; one side had been flame-sealed) and EPR measurements were performed using a Varian E-109 spectrometer. The spectrometer settings were as follow: gain, 1.25×10^5; microwave power, 20 m*W;* modulation amplitude, 0.2 mT; time constant, 0.5 s; scan range, 10 mT; scan time, 4 min; the mean temperature was 10 ± 1°C. Two scans were made and the signals were averaged.

D. DETECTION OF PBN SPIN ADDUCTS *IN VIVO*

For detection of *in vivo* production of free radicals, a 3-ml solution of PBN (80 m*M* in 0.9% NaCl) was administered IV 30 min before ischemia. After ischemia or ischemia-reperfusion, the brain was removed, the cortex was extracted with the chloroform-methanol solution as described above, and EPR spectra were taken.

III. RESULTS

A. PHYSIOLOGICAL PARAMETERS

We attempted to compensate for the changes by adjusting ventilation and by infusing sodium bicarbonate, so that the changes in blood pressure, heart rate, blood gases and pH, were not significantly different among groups. Flat EEGs were observed in all animals during the ischemic period. This indicated that the blood flow to the brain was reduced enough to make ischemic damage.

B. *IN VIVO* SPIN TRAPPING

EPR spectra of the cortex from the rats, to which PBN was injected IP and were subsequently exposed to ischemia-reperfusion, demonstrated three doublet signals shown in Figure 1. In this *in vivo* model, the signal was not large enough to allow us to perform further analysis. Therefore, we used an *in vitro* trapping method for quantitative analysis.

C. *IN VITRO* SPIN TRAPPING

Typical EPR spectra of spin adducts in cortex extracts are shown in Figure 2. No signal was detected in the solution without the cortex homogenate. As shown in the figure, all signals obtained demonstrated three doublets. The spectra had parameters of $a_N = 14.3$ G and $a_H{}^{\beta} = 2.8$ G. We assumed that the signal height (the average of both peaks of the doublet) of a low-field doublet is proportional to the amount of free radicals in the samples, and plotted the time course curve of the spin-adduct formation (Figure 3A). Upon reperfusion, there occurred a burst of spin-adduct formation. The formation reached a peak at 5 min. A

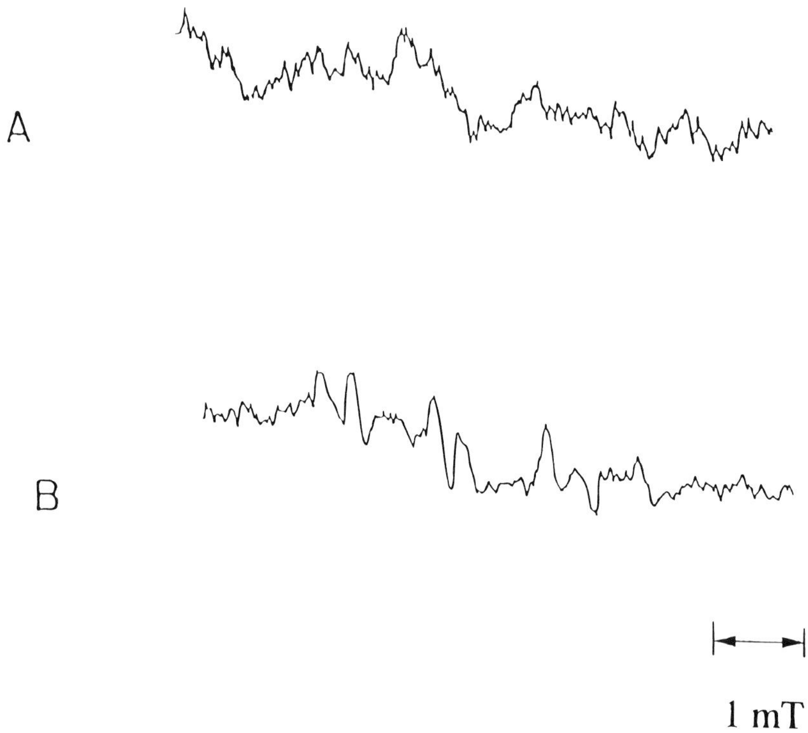

FIGURE 1. *In vivo* spin-trapping of free radicals. PBN (80 m*M* × 3 ml) was injected (IP) 30 min before the onset of ischemia. A: no ischemia; B: 20 min ischemia—5 min reperfusion. (From Sakamoto, A. et al., *Brain Research*, 554, 186–192, 1991. With permission.)

similar but less marked burst was observed when the ischemia time was shortened to 10 min (Figure 3B).

In order to examine whether or not superoxide anions were produced during ischemia-reperfusion injury, we tested the effect of SOD added to the homogenizing solution. Without SOD, the intensity of spectra from the rat brain exposed to 20 min of ischemia-reperfusion increased by 239±41% (n = 6) from the control. When the cortex was homogenated with a high concentration of SOD (200 μg/ml), the intensity of spectra increased by 167±11% (n = 5). The spin-adduct formation was decreased by 52%.

D. INHIBITION OF SPIN-ADDUCT FORMATION BY PROSTAGLANDIN OLIGOMERS

Figure 4 shows that upon reperfusion, there occurred a burst of spin-adduct formation in nontreated rats, and the formation reached a peak at 5 min. With pretreatment of the new antioxidant OC-5186[19,20] (9 mg/kg IP 30 min before ischemia), the spin-adduct formation at the peak point was inhibited by 98%. The dose-response curve of the effect of OC-5186 is shown in Figure 5.

IV. DISCUSSION

It has been well recognized that ischemia-reperfusion induces severe injury in many organs, and that a mechanism of injury may be linked to the production of oxygen free radicals.[25,26] However, the evidence for the production of such radicals in ischemia-reperfusion

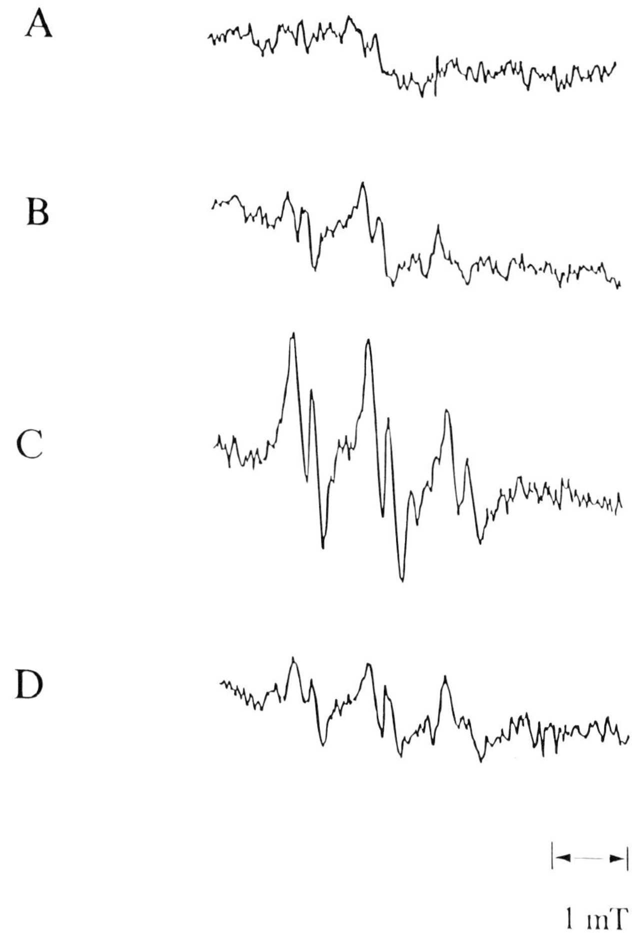

FIGURE 2. Electron spin resonance spectra of spin adducts extracted from the cortex of the rat exposed to: (A) no ischemia; (B) 20 min ischemia—no reperfusion; (C) 20 min ischemia—5 min reperfusion; (D) 20 min ischemia—40 min reperfusion. (From Sakamoto, A. et al., *Brain Research*, 554, 186–192, 1991. With permission.)

injury has been indirect. Because of the unstable nature of free radicals, direct demonstration of their production has been difficult.

In order to demonstrate the production of free radicals, we employed the spin-trapping technique, which could provide evidence for free radical formation. Recently, using the spin-trapping technique and EPR, the formation of oxyradicals during ischemia-reperfusion injury was directly demonstrated in several organs *in vivo*.[21,22,27–29]

As a spin trap, PBN has been used frequently because it forms relatively long-lived radical adducts and because it is not effected by photoirradiation.[30] Oliver et al.[31] reported that free radical signals could be detected during reperfusion of the gerbil brain *in vivo* by using injected PBN. However, samples were only collected at a single time point during reperfusion and the time course of free radical formation was not demonstrated. We chose a method to detect free radicals by homogenizing the cortex with a PBN solution after decapitation, because the metabolic degradation of PBN was shown to be very rapid.[32] This method enabled us to measure the effect of the ischemia time as well as the reperfusion time on the free radical formation. If PBN is injected into animals, the *in vivo* concentration of PBN

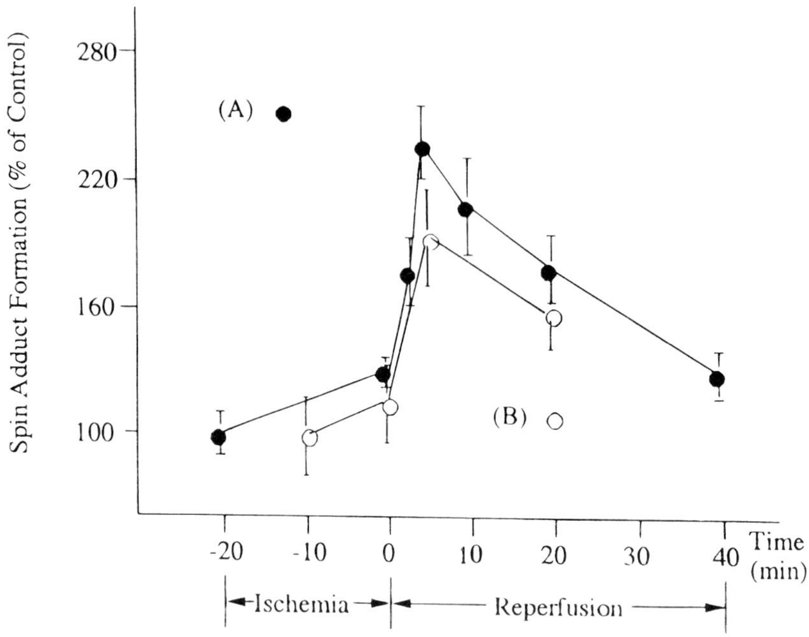

FIGURE 3. (A) Time course of the formation of PBN spin adducts during a 20-min ischemia period and during the subsequent reperfusion of the rat brain. (B) Time course of the formation of PBN spin adducts during a 10-min ischemia period and during the subsequent reperfusion of the rat brain.

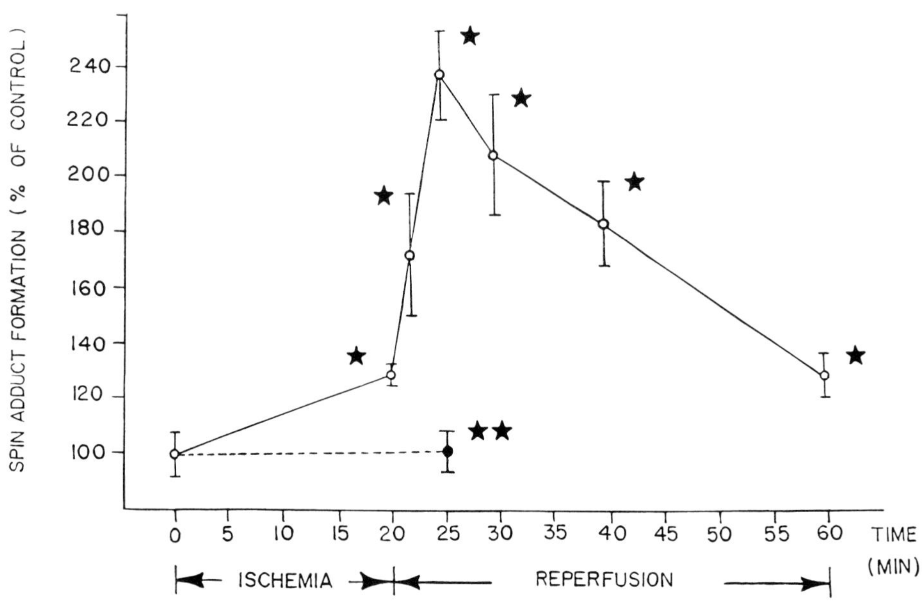

FIGURE 4. Time course of the formation of PBN spin adducts during a 20-min ischemia period and the subsequent reperfusion of the rat brain: (○) nontreated group; (●) OC-5186 (9 mg/kg) pretreated group. (From Sakamoto, A. et al., *Brain Res.*, 554, 186–192, 1991. With permission.)

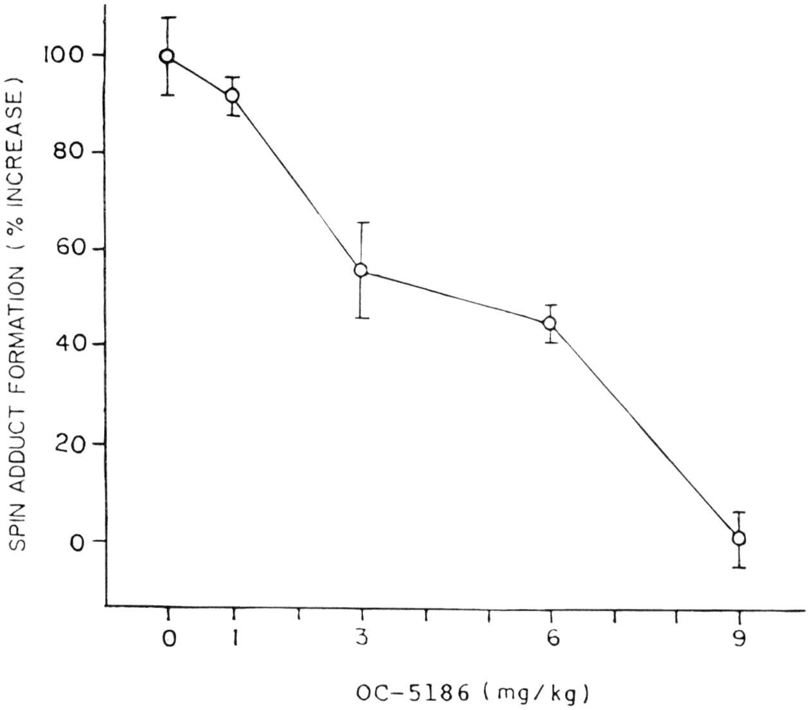

FIGURE 5. Dose-response curve of the effect of OC-5186 (N = 5, each) on the spin-adduct formation at 5-min after reperfusion. (From Sakamoto, A. et al., *Free Radical Biol. Med.*, 11, 385–391, 1991. With permission.)

may change rapidly so that we cannot measure the time course of free radical formation under the same conditions.

We added DTPA to both the buffer solution and the chloroform-methanol mixture since the contamination of trace metals is known to interfere with spin trapping experiments.[33] Since we had to use a high gain setting of the EPR spectrometer, we had to purify PBN with a charcoal treatment to eliminate trace contaminants. We also found that some organic solvents could produce impurity signals. The solvents we used in these experiments were found to be satisfactory.

At this point, the nature of the free radicals observed in our studies is not known. Although the shape of the EPR signals is similar to those of adducts of oxygen-centered or carbon-centered species,[34] we could not identify the type from our results. Comparing the values of superfine splitting constant in the previous studies[33,35,36] with those of our results, it is possible that our adducts may consist of a mixture of oxygen- and carbon-centered adducts. Kramer et al. used toluene for spin-adduct extraction and found PBN/alkoxyl adducts.[37,38] Since we used chloroform-methanol for extraction, we might have formed PBN/methyl radical adducts.

In order to examine the production of superoxide anions during ischemia-reperfusion injury, we tested the effect of the addition of SOD to the PBN solution before the brain cortex was homogenized. The signal decreased by 52% in the presence of a high concentration of SOD. This indicates that at least a half of the free radicals trapped by PBN may be derived from superoxide anions. How the superoxide anions could be transformed to the observed oxygen- or carbon-centered adducts is not clear at this point. It is possible that the Harber-Weiss reaction in the medium may be involved, or that the superoxide radicals might have reacted with nitric oxide to form peroxynitrite, a highly reactive compound.[39]

OC-5186, a prostaglandin oligomer with a high antioxidant activity was shown to inhibit the production of free radicals upon reperfusion. The compound also inhibited the ischemia-reperfusion-induced lipid peroxidation as measured by the level of thiobarbituric acid reactive substances (TBAR) and conjugated diene formation.[22]

In summary, applying the spin-trapping technique and EPR to a cerebral ischemia-reperfusion model, we demonstrated an abrupt burst of free radical formation upon reperfusion. The production of radicals peaked at 5 min. The formation of free radicals was sensitive to the inhibition of both SOD and an antioxidant prostaglandin oligomer.

NOTE ADDED IN PROOF

PBN manufactured by OMRF Spin Trap Source (825 N.E. 13th St., Oklahoma City, Oklahoma 73104) may be used without further purification.

REFERENCES

1. Demppoulos, H. D., Flamm, E. S., Pitroniigro, D. D., Seligman, M. L.: The free radical pathology and the microcirculation in the major central nervous system disorders. (1980) *Acta. Physiol. Scand.* 492, 91–119.
2. Raiche, M. E.: The pathophysiology of brain ischemia. (1983) *Ann. Neurol.* 13, 2.
3. Shaller, C. A., Jaques, S., Shelden, C. H.: The pathophysiology of stroke: a review with molecular considerations. (1981) *Surg. Neurol.* 14, 433–443.
4. Siesjo, B. K.: Cell damage in the brain: a speculative synthesis. (1981) *J. Cereb. Blood Flow Metab.* 1, 155–162.
5. Girotti, A. W.: Mechanism of lipid peroxidation. (1985) *J. Free Radic. Biol. Med.* 1, 87–95.
6. Floyd, R. A.: Role of oxygen free radicals in carcinogenesis and brain ischemia. (1990) *FASEB J.* 4, 2587–2597.
7. Wolff, S. P., Garner, A., Dean, R. T.: Free radicals, lipids and protein degradation. (1986) *Trends Biol. Sci.* 11, 27–31.
8. Floyd, R. A., Schneider, J. E.: Hydroxy free radical damage to DNA. In: *Membrane Lipid Oxidation,* Vol. 1. Vigo-Pelfrey, C., ed., CRC Press, Florida (1990).
9. Chan, P. H., Schmidley, J. W., Fishman, R. A., Longer, S. M.: Brain injury edema, and vascular permeability changes induced by oxygen-derived free radicals. (1984) *Neurology* 34, 315–320.
10. MacMillan, V.: Cerebral Na^+,K^+-ATPase activity during exposure to and recovery from acute ischemia. (1982) *J. Cereb. Blood Flow Metab.* 2, 457–465.
11. Watson, B. D., Busto, R., Goldberg, W. J., Santiso, M., Yoshida, S., Ginsberg, M. D.: Lipid peroxidation *in vivo* induced by reversible global ischemia in rat brain. (1984) *J. Neurochem.* 42, 268–274.
12. Wieloch, T., Siesjo, B. K.: Ischemic brain injury: the importance of calcium, lipolytic activities, and free fatty acids. (1982) *Pathol. Biol.* 30, 269–277.
13. Liu, T. H., Beckman, J. S., Freeman, B. A., Hogan, E. L., Hsu, C. Y.: Polyethylene glycol-conjugated superoxide dismutase and catalase reduce ischemic brain injury. (1989) *Am. J. Physiol.* 256, H589–H593.
14. Martz, D., Rayos, G., Schielke, G. P., Betz, Å. L.: Allopurinol and deimethylthiourea reduce brain infarction following middle cerebral artery occlusion in rats. (1989) *Stroke* 20, 488–494.
15. Cerchiari, E. I., Hoel, T. M., Safer, P., Sclabassi, R. J.: Protective effects of combined superoxide dismutase and deferoxamine on recovery of cerebral blood flow and function after cardiac arrest in dogs. (1987) *Stroke* 18, 869–878.
16. Onhini, S. T., Tominaga, T., Katsuoka, M.: Inhibition of ischemic brain edema formation by post-ischemic administration of a prostaglandin oligomer (1989) *Prosta. Leuko. Essentl. Fat. Acids* 37, 107–111.
17. Ohnishi, S. T., Katsuoka, M.: Protection of the heart from ischemia by a new oligomeric derivative of prostaglandin E1. (1989) *Prosta. Leuko. Essentl. Fat. Acids* 36, 159–163.
18. Katsuoka, M., Ohnishi, S. T.: Pharmacologic protection of perfused rat heart against global ischemia. (1989) *Prosta. Leuko. Essentl. Fat. Acids* 38, 151–156.
19. Ohnishi, S. T., Ohnishi, T.: Biologically significant psychochemical properties of antioxidative prostaglandin derivatives. (1991) *Arzneim. Forsch.* 41, 1201–1205.
20. Ohnishi, S. T., Sakamoto, A., Ohnishi, T., Ogawa, R.: Inhibition of lipid peroxidation by prostaglandin oligomeric derivatives. (1992) *Prosta. Leuko. Essentl. Fat. Acids* 45, 217–221.
21. Sakamoto, A., Ohnishi, S. T., Ohnishi, T., Ogawa, R.: Relationship between free radical production and lipid peroxidation during ischemia-reperfusion injury in the rat brain. (1991) *Brain Res.* 554, 186–192.

22. Sakamoto, A., Ohnishi, S. T., Ohnishi, T., Ogawa, R.: Protective effect of a new anti-oxidant on the rat brain exposed to ischemia-reperfusion injury. (1991) *Free Radic. Biol. Med.* 11, 385–391.
23. Buettner, G. B., Oberley, L. W.: Considerations in the spin trapping of superoxide and hydroxyl radical in aqueous systems using 5,5-dimethyl-1-pyrroline-1-oxide. (1978) *Biochem. Biophys. Res. Commun.* 83, 69–74.
24. Folch, J., Lees, M., Sloane-Stanley, G. H.: A simple method for the isolation and purification of total lipids from animal tissues. (1957) *J. Biol. Chem.* 226, 497–509.
25. Chanbers, D. E., Parks, D. A., Patterson, G.: Xanthine oxidase as a source of free radical damage in myocardial ischemia. (1985) *J. Mol. Cell Cardiol.* 17, 145–152.
26. Parks, D. A., Bulkley, G. B., Granger, D. N.: Ischemic injury in the cat small intestine: role of superoxide radicals. (1982) *Gastroenterology* 82, 9–15.
27. Garlick, P. B., Davies, M. J., Hearse, D. J.: Direct detection of free radicals in the reperfused rat heart using electron spin resonance spectroscopy. (1987) *Circ. Res.* 61, 757–760.
28. Kuzuya, T., Hoshida, S., Kim, Y.: Detection of oxygen-derived free radical generation in the canine postischemic heart during late phase of reperfusion. (1990) *Circ. Res.* 66, 1160–1165.
29. Nillson, U. A., Lundgren, E., Haglind, A.: Radical production during intestinal ischemia and reperfusion *in vivo* in cat—an ESR study. *Proc. 4th Int. Cong. Oxygen Radicals* (1987) pp. 150–152.
30. Lai, E. K., Crossley, C., Sridhar, R., Misra, H. P., Janzen, E. G., McCay, P. B.: *In vivo* spin trapping of free radicals generated in brain, spleen, and liver during radiation of mice. (1986) *Arch. Biochem. Biophys.* 244, 156–160.
31. Oliver, C. N., Starke-Reed, P. E., Stadtman, E. R., Liu, G. L., Carney, J. M., Floyd, R. A.: Oxidative damage to brain proteins, loss of glutamine synthetase activity, and production of free radicals during ischemia/reperfusion-induced injury to gerbil brain. (1990) *Proc. Natl. Acad. Sci. U.S.A.* 87, 5144–5147.
32. Chen, G., Griffin, M., Poyer, J. L., McCay, P. B.: HPLC procedure for the pharmacokinetics study of the spin-trapping agent, α-phenyl-N-tert-butyl nitrone (PBN). (1990) *Free Radic. Biol. Med.* 8, 93–98.
33. Buettner, G. R.: Spin trapping of hydroxyl radical. In: *CRC Handbook of Methods for Oxygen Radical Research,* Greenwald, R. A., ed., CRC Press, Florida (1985), pp. 151–155.
34. Bolton, J. R., Borg, D. C., Schwart, H. M.: *Biological Applications of Electron Spin Resonance.* Wiley Interscience, New York (1972).
35. Thornalley, P. J., Banister, J. V.: The spin trapping of superoxide radicals. In: *CRC Handbook of Methods for Oxygen Radical Research,* Greenwald, R. A., ed., CRC Press, Florida (1985), pp. 133–136.
36. Weglicki, W. B., Arroyo, J. H., Kramer, J. H., Mak, I. T., Leiboff, R. H., Mergner, G. W., Dickens, B. F.: Application of spin trapping/ESR techniques in models of cardiovascular injury. In: *Oxy-Radicals in Molecular Biology and Pathology.* Cerutti, P. A., Fridrich, I., McCord, S. M., eds., Alan R. Liss, New York (1988), pp. 357–364.
37. Tortolani, A. J., Powell, S. R., Misik, V., Weglicki, W. B., Pogo, G. J. and Kramer, J. H. Detection of alkoxyl and carbon-centered free radicals in coronary sinus blood from patients undergoing elective cardioplegia. (1993) *Free Radic. Biol. Med.* 14, 421–426.
38. Kramer, J. H., Misik, V. and Weglicki, W. B. Magnesium-deficiency potentiates free radical production associated with postischemic injury to rat hearts: Vitamin E affords protection. (1994) *Free Radic. Biol. Med.* 16, 713–723.
39. Beckman, J. S. The double-edged role of nitric oxide in brain function and superoxide-mediated injury. (1991) *J. Dev. Physiol.* 15, 53–59.

Chapter 34

TRAPPING OF NITRIC OXIDE RADICALS IN BRAIN ISCHEMIA

Shinya Sato, Teiji Tominaga, Tomoko Ohnishi, and S. Tsuyoshi Ohnishi

TABLE OF CONTENTS

I. INTRODUCTION

Recently, the biological roles of nitric oxide (NO) have attracted considerable attention. NO plays a wide variety of roles in the central nervous system, the cardiovascular system, and in the immune system.[3,4,6,15,17] NO has been known as an endothelium-derived relaxing factor (EDRF), which is associated with the maintenance and regulation of the vascular tone.[18–20] It has also been speculated that NO may form a harmful radical, peroxynitrite, by interacting with superoxide anions.[1,21] Therefore, possible roles of NO in ischemic tissue damage have been a subject of concern for many investigators.

NO is a free radical, and its short half-life (on the order of seconds) has rendered direct measurement difficult. So far, the production of NO was suggested indirectly by chemiluminescence, conversion of oxyhemoglobin to metohemoglobin or by an increase of its stable metabolites such as nitrites and nitrates.[15]

Since NO is paramagnetic, it can be detected by electron paramagnetic resonance (EPR). A great number of EPR studies have been conducted on NO using two methods. (a) The interaction of NO with heme produces specific EPR signals.[7,26] Hemoglobin or CO-bound

0-8493-8094-4/95/$0.00+$.50

hemoglobin were used as a NO-trapping agent.[8] (b) NO and Fe were used for analyzing the structure of proteins because they form a complex with protein thiol groups.[22] This is a very specific method to detect NO, because NO, Fe, and thiol produce a clear triplet signal. Using this property, Vanin's group[9,29,31] developed a NO-trapping technique using diethyldithiocarbamate (DETC) and iron (Fe^{2+}). They measured NO production in the brain of gamma-irradiated mice.[31] Using this technique, we have previously reported that NO is produced during rat brain ischemia.[23–25,27,28]

In this chapter, we present the application of this NO-trapping technique to brain ischemia in the rat.

II. GENERAL PROCEDURES

The NO-trapping procedure is shown in a flowchart (Figure 1).

A. INSTRUMENTATION AND MATERIALS

1. EPR spectrometer with a liquid nitrogen flow system.
2. EPR tube. We used a 4-mm (i.d.) EPR tube (Wilmad Glass Co., Buena, NJ).
3. Plastic syringes (3 ml and 5 ml) and infusion tubing. We used 8-in. long tubing (we cut off the needle portion from the 21-gauge Bufferfly infusion set manufactured by Abbott Hospital, North Chicago, IL). An IV catheter can also be used.
4. DETC and Fe-citrate (the weight ratio of $FeSO_4 \cdot 7H_2O$:sodium citrate = 1:5) were prepared in saline solutions, distributed into small test tube, and frozen at −20°C until use.

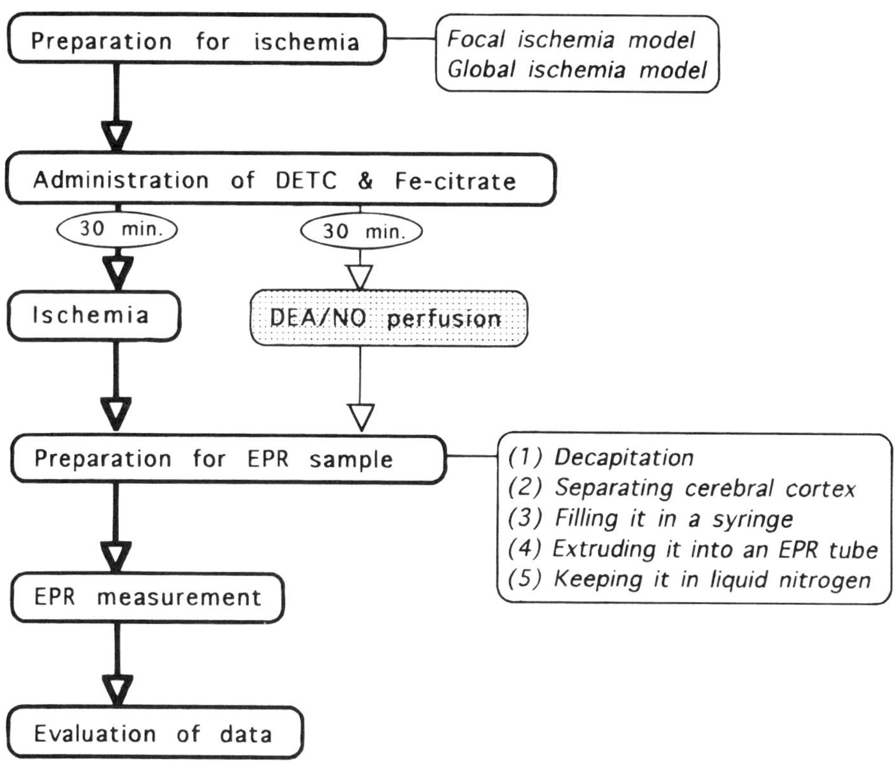

FIGURE 1. Flowchart of our NO-trapping procedure. Thick arrows show the direction of the procedure, while thin arrows indicate the procedure for the DEA/NO perfusion, which is performed for the quantitation of NO production. (From Sato, S. et al., *Biochim. Biophys. Acta*, 1181, 195–197, 1993. With permission.)

B. PREPARATION OF EPR SAMPLE

The rat is sacrificed by decapitation under isoflurane anesthesia. The cerebral cortex is placed in a 3-ml plastic syringe and extruded into a 5-ml plastic syringe mounted with the infusion tubing. It is again extruded into an EPR tube. With this procedure, the tissue is homogenized. The specimen is immediately frozen in an ethanol solution cooled by dry ice, and is kept at liquid nitrogen temperature until its EPR measurement.

C. EPR CONDITIONS

EPR spectra are recorded using a Varian E-109 spectrometer with a liquid nitrogen flow system at 110 to 130 K. The conditions for EPR measurements are 9.22 GHz microwave frequency, 20 mW microwave power, 100 kHz modulation frequency, 0.32 mT modulation, and 0.25 s time constant.

III. NO SIGNALS IN EPR SPECTRA

Originally, Kubrina et al.[9] administered DETC (500 mg/kg) intraperitoneally and $FeSO_4$ (40 mg/kg)/Na-citrate (200 mg/kg) subcutaneously 30 min before decapitation. Figure 2 shows the structure of a possible NO-trapping compound. When the reagents were administered to nonischemic rat by the same manner, two types of signals were detected; a quartet signal from Cu-DETC and a signal derived from NO-Fe-DETC characterized by a small peak and a trough appeared at g = 2.047 and at g = 2.025, respectively (Figure 3a). By administration of N^G-nitro-L-arginine methyl ester (L-NAME, 45 mg/kg), a NO synthase inhibitor, the small peak and the trough disappeared (Figure 3b). Therefore, it was confirmed that these signals were related to the production of NO from L-arginine. When only DETC was administered, the EPR spectrum showed only the signal of Cu-DETC (Figure 3c). The result indicates that Fe-citrate should be simultaneously administered with DETC in order to obtain the NO-Fe-DETC signal from the rat brain. We must inject DETC and Fe-citrate from separate sites because Fe-DETC complex is not water soluble (the solubility is $<10^{-5}$). When Fe-citrate was administered simultaneously with DETC, and the rat was exposed to 5 min of severe global ischemia produced by the removal of the heart, the typical NO-Fe-DETC signal, which is a clear triplet signal centered at g = 2.039 with a hyperfine coupling constant a_N of 13 G, was observed (Figure 3d).

IV. APPLICATION OF THE NO-TRAPPING METHOD TO BRAIN ISCHEMIA

The EPR study of NO metal complexes has several decades of history. However, NO trapping in brain ischemia is a novel approach and, therefore, requires methodological characterization.

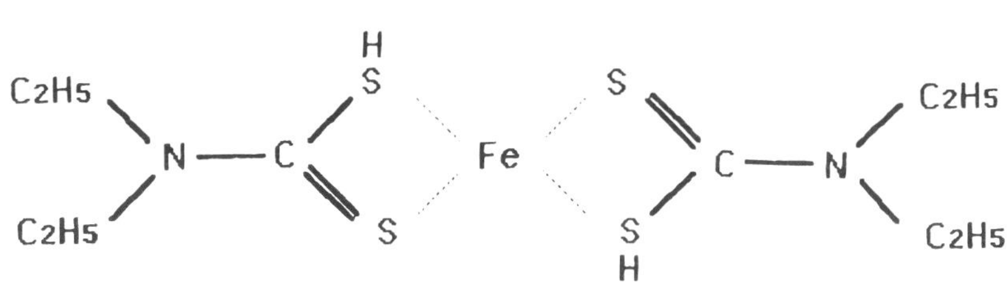

FIGURE 2. The structure of a **possible** NO-trapping complex. (From Sato, S. et al., *Biochim. Biophys. Acta*, 1181, 195–197, 1993. With permission.)

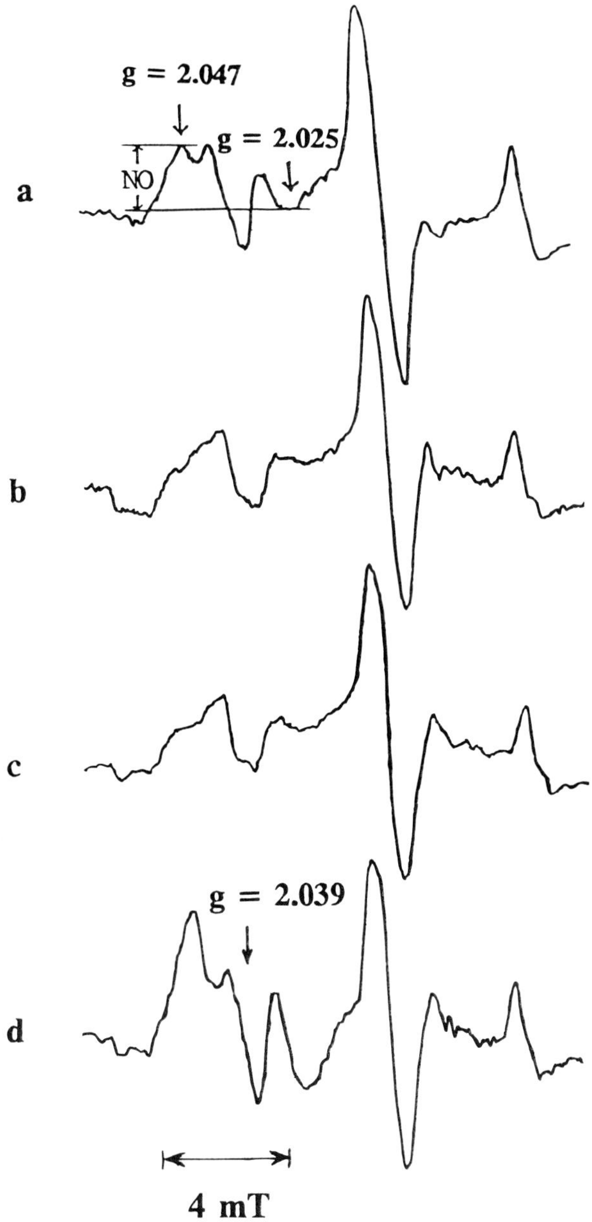

FIGURE 3. EPR spectra of NO-Fe-DETC and Cu-DETC obtained from the rat cerebral cortices. (a) DETC (500 mg/kg) + $FeSO_4$ (40 mg/kg)/Na-citrate (200 mg/kg); a small peak (g = 2.047) and a trough (g = 2.025) are portions of NO-Fe-DETC. (b) DETC + Fe + L-NAME (45 mg/kg); note the disappearance of the NO-Fe-DETC found in the signal (a). (c) DETC alone; the arrows indicate a quartet spectrum of Cu-DETC. (d) DETC + Fe; severe ischemia (removal of the heart, 5 min). EPR conditions: 20 mW microwave power, 1.0×10^4 gain, 0.32 mT modulation, 9.22 GHz microwave frequency, and 100 kHz modulation frequency. Temperature was 110 ± 1 K. (Adapted from Tominaga, T. et al., *J. Cereb. Blood Flow Metab.*, 14, 715–722, 1994.)

A. CHARACTERIZATION OF EPR CONDITIONS

We examined the microwave power dependence of the NO-Fe-DETC and Cu-DETC signals. We found that an optimal signal-to-noise ratio of the NO-Fe-DETC signal was obtained at a microwave power of 20 mW.[28] We also found that the lower the sample temperature was, the easier we could distinguish the NO-Fe-DETC signal from the Cu-DETC signal, because the Cu-DETC signal became small at low temperatures (Figure 4). Therefore, we used 110 to 130 K for EPR measurements. The power saturation profile of NO-Fe-DETC and Cu-DETC at 110 K are shown in Figure 5. At approximately 2 mW, both signals started to deviate from the nonsaturated line, which has a slope of 0.5. At 20 mW, the Cu-DETC signal was already quite saturated, but the level of saturation of the NO-Fe-DETC signal was acceptable for the measurement.[28]

B. AMOUNTS OF REAGENTS AND ROUTES OF ADMINISTRATION[28]

The amount of DETC—Since it was noticed that DETC has a hypotensive effect, we examined the effects of the route of administration and the dosage of DETC on mean arterial blood pressure (MABP) in the rat. Originally, Kubrina et al.[1] used 500 mg/kg of DETC for NO trapping in mice. However, in our experiment, 500 mg/kg of DETC evoked significant hypotension in the rat (below 70 mmHg, regardless of the route of its administration), which is under the lower limit of autoregulation of cerebral circulation. The values of MABP at 30 min after the administration of DETC are shown in Table 1. When 400 mg/kg of DETC was administered subcutaneously, instead of intraperitoneally, the MABP was kept in an acceptable range. In this case, the signal of Cu-DETC reached a maximum at 30 min after the administration of DETC, and then declined (Figure 6). Therefore, the concentration of DETC in brain tissue seems to reach a maximum at 30 min after the administration of DETC.

The amount of Fe—Although Fe did not have a significant effect on blood pressure, a generation of extremely large NO-Fe-DETC signals occurred with higher dosages of $FeSO_4$ (over 40 mg/kg). This suggests that an excessive amount of Fe induces NO production by itself. A question would be raised as to whether a small amount of Fe (less than 40 mg/kg) might exacerbate ischemic damage so that NO was then produced. Regarding this point, we believe that a small amount of Fe does not by itself aggravate the endothelial cells to produce NO.[28]

Taking these results into consideration, we decided to administer both 400 mg/kg DETC (200 mg/ml $\times$ 2 ml/kg) and 20 mg/kg $FeSO_4$-100 mg/kg Na-citrate (a mixture of 40 mg/ml $FeSO_4$ + 200 mg/ml Na-citrate $\times$ 0.5 ml/kg) subcutaneously for the study of ischemia.

C. EXAMPLE OF EPR SPECTRA IN BRAIN ISCHEMIA[23]

Figure 7a shows a spectrum for a nonischemic condition. When the bilateral common carotid arteries (CCAs) were occluded, the production of NO was increased (Figure 7b). When CCAs were not occluded and the MABP was reduced to 50 mmHg by hemorrhagic hypotension, we obtained a spectrum that is almost the same as that of CCAs occlusion (Figure 7c). When the occlusion of CCAs was combined with hemorrhagic hypotension, a clear triplet signal of the NO-Fe-DETC appeared (Figure 7d). These results clearly demonstrate that the NO production increases in response to the extent of ischemic insult, and that it occurs not only during severe ischemia, but also during mild ischemia and oligemia that do not produce brain infarction. Since it is well accepted that NO regulates cerebral blood flow under physiological conditions,[18–20] the NO production during mild ischemia and oligemia may be a protective response for increasing cerebral blood flow.

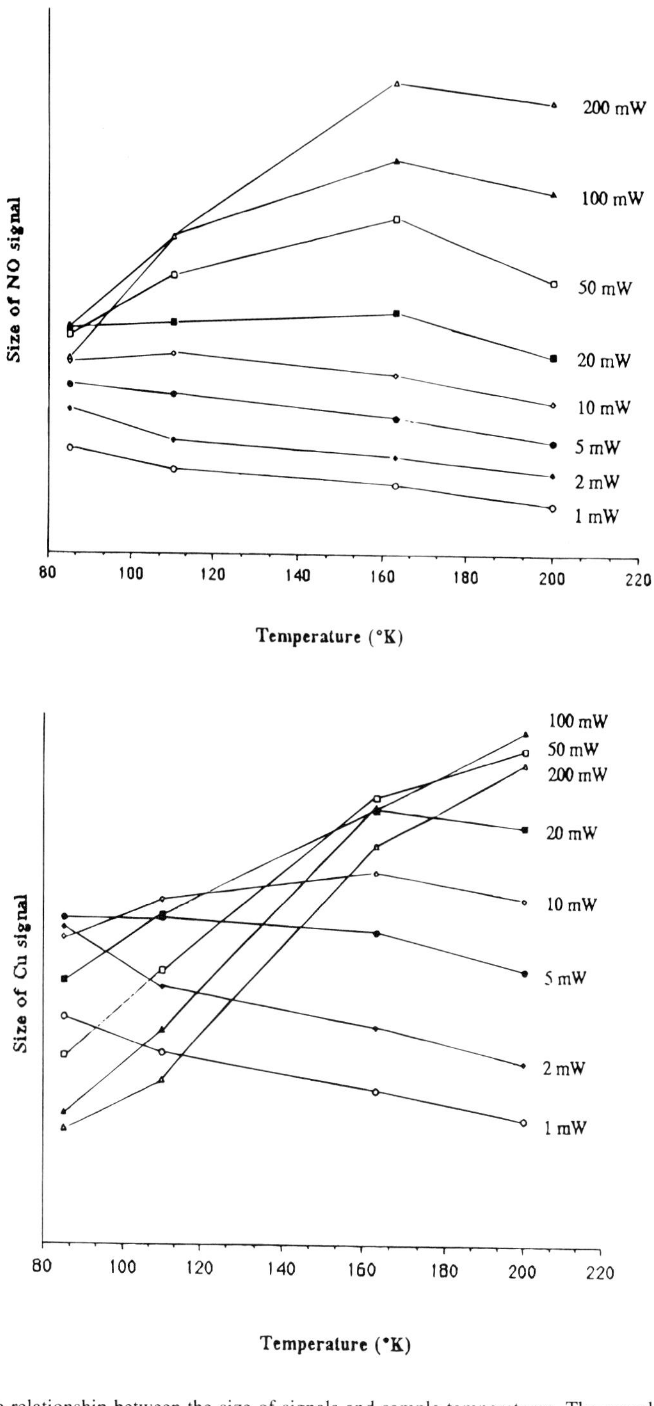

FIGURE 4. The relationship between the size of signals and sample temperatures. The sample shown in Figure 3d was used. EPR spectra obtained at 1, 2, 5, 10, 20, 50, 100, and 200 mW microwave power. Other EPR conditions: 1.0×10^4 gain, 0.32 mT modulation, 9.22 GHz microwave frequency, and 100 kHz modulation frequency.

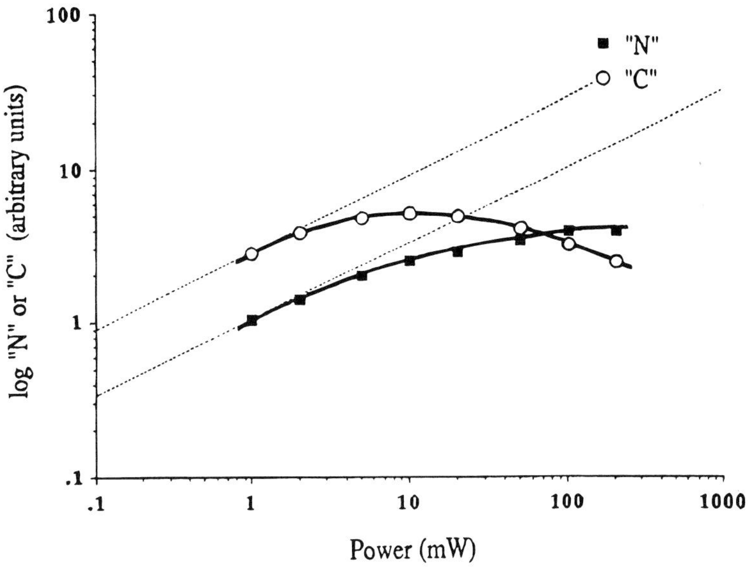

FIGURE 5. The saturation profile of the NO-Fe-DETC and the Cu-DETC signals at 110 ± K. The sample shown in Figure 3d was used. The NO-Fe-DETC signal and the Cu-DETC signal are indicated "N" and "C", respectively. The nonsaturated lines (slope of 0.5) are shown with the dashed lines. Other EPR conditions: 2.5×10^4 gain, 0.32 mT modulation amplitude, 9.22 GHz microwave frequency, and 100 kHz modulation frequency. (From Tominaga et al., *J Cereb Blood Flow Metab.*, 14, 715–722, 1994. With permission.)

TABLE 1
Relationships Among Amounts of Reagents, Routes of Administration, and Blood Pressure

	DETC 500 mg/kg, intraperitoneally	DETC 500 mg/kg, subcutaneously	DETC 400 mg/kg, subcutaneously
Mean arterial pressure (mean ± SD)	63.5 ± 12.9 (n = 10)	68.5 ± 5.4 (n = 5)	81.3 ± 7.8 (n = 10)

Comparisons: intraperitoneally 500 mg/kg vs. subcutaneously 500 mg/kg: N.S.*; subcutaneously 500 mg/kg vs. subcutaneously 400 mg/kg: $p \leq 0.01$*; intraperitoneally 500 mg/kg vs. subcutaneously 400 mg/kg: $p \leq 0.01$*

Note: Mean arterial pressure was evaluated at 30 min after the administration of DETC.
* Student's *t*-test.

V. AN ATTEMPT TO QUANTITATE THE PRODUCTION OF NO *IN VIVO*[28]

In order to quantitate the production of NO, we devised a method to measure the relationship between the concentration of NO in blood vessels produced by the perfusion of 1,1-diethyl-2-hydroxy-2-nitroso-hydrazine (DEA/NO, Midwest Research Institute, Kansas City, MO)[11] and the size of NO-Fe-DETC signals. DEA/NO is stable in alkaline solutions, but the molecules of DEA/NO releases three NO molecules under neutral condition within a few minutes (Figure 8).[11] Thus, DEA/NO was used as a NO generator to calibrate the NO concentrations. In performing this examination, we made the following assumptions: (a) the Fe-DETC complex is absorbed into the surface membrane of the endothelial cells,[30] (b) under normal or ischemic condition, NO radicals produced in the brain are trapped by this absorbed complex; (c) in the experiments with DEA/NO, NO in the DEA/NO perfusate is also similarly trapped by this absorbed complex; and (d) the NO trapping method does not

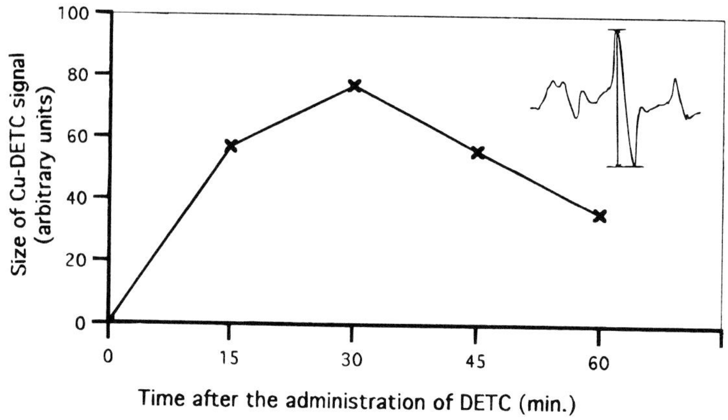

FIGURE 6. The time course of the size of Cu-DETC signal after the administration of DETC. DETC (400 mg/kg) and $FeSO_4$ (20 mg/kg)/Na-citrate (100 mg/kg) were administered subcutaneously. The Cu-DETC signal was estimated as indicated in the inset. EPR conditions: 20 mW microwave power, 2.5×10^4 gain, 0.32 mT modulation, 9.22 GHz microwave frequency, and 100 kHz modulation frequency. Temperature was 110 ± 1 K.

provide a straightforward measure of NO production or concentration. Detected signals are the result of the rate of production of the trapped NO and its rate of destruction. Therefore, this method also needs the assumption that the production of NO becomes steady state rapidly after the onset of ischemia.

To make a neutral DEA/NO solution, an alkaline DEA/NO stock solution was diluted into a N_2-equilibrated, 3 m*M* phosphate buffered saline solution (pH 7.4) at room temperature. All solutions used in this experiment were degassed under vacuum and equilibrated with N_2 by bubbling. All containers were also filled with N_2. The concentrations of NO generated in the perfusates were 0, 10, 20, and 50 μM. Thirty minutes after the administration of DETC (400 mg/kg) and $FeSO_4$ (20 mg/kg)/Na-citrate (100 mg/kg), the rat was perfused under anesthesia with the neutral DEA/NO solution via the ascending aorta at the pressure of 100 mmHg for 15 min. After the perfusion, the brain was removed quickly and the EPR sample was prepared in the same manner as mentioned earlier.

Along with the increase of the NO concentration in the perfusate up to 50 μM, the NO-Fe-DETC signal exhibited proportional growth. The signal height "N" was well correlated with the NO concentration in the perfusate (Figure 9). For example, the asterisk in Figure 9 indicates the size of NO signal in global forebrain ischemia produced by the occlusion of CCAs combined with hemorrhagic hypotension. In this ischemic condition, the local concentration of NO around blood vessels was estimated to be 10.9 μM with this method, and concentrations of NO in nonischemic rat (control) and hypotensive rat (MABP = 50 mmHg) were determined to be 1.6 and 5.8 μM, respectively.[28]

VI. STABILITIES OF Fe-DETC AND NO-Fe-DETC COMPLEXES

In severe ischemic region, Fe-DETC may be metabolized and decrease during ischemia. On the other hand, Fe-DETC may continue to be supplied to mild ischemic or oligemic regions. Therefore, the amounts of Fe-DETC are not homogeneous in the brain tissue after prolonged ischemia. We cannot compare signals taken from short-ischemic animals with those taken from long-ischemic animals quantitatively. In addition, the exact stability of the NO-Fe-DETC complex is not known at this point and it may be relatively low (its lifetime may be less than 10 min). Therefore, we may underestimate the production of NO in prolonged ischemia.

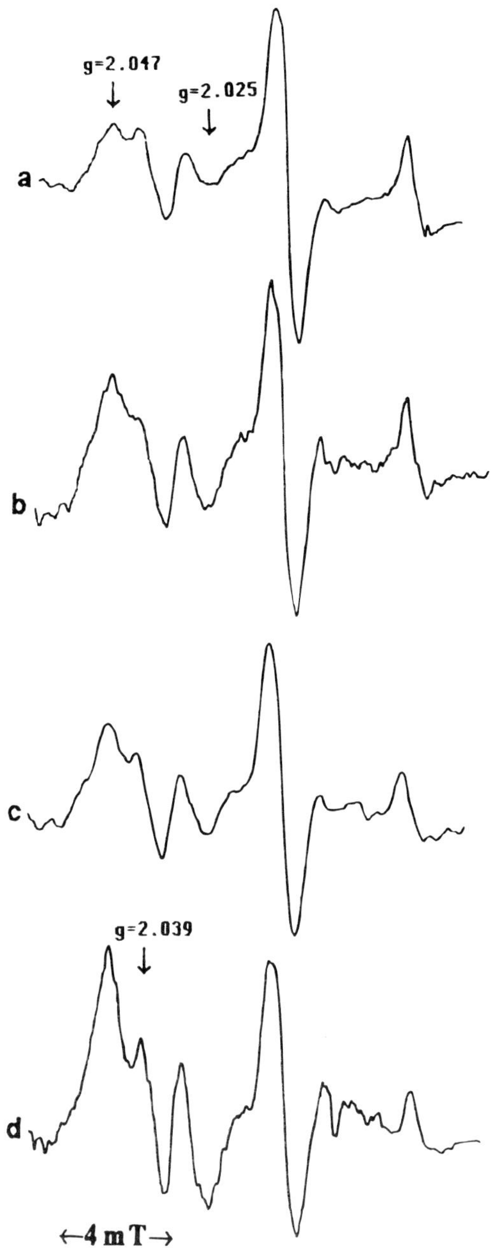

FIGURE 7. EPR signals of the brain cortices of rats: (a) no ischemia; (b) only CCAs occlusion; (c) only hemorrhagic hypotension (MABP is 50 mmHg); (d) CCAs occlusion with hemorrhagic hypotension (MABP is 60 mmHg). EPR conditions: 20 mW microwave power, 1.0×10^4 gain, 0.32 mT modulation, 9.22 GHz microwave frequency, and 100 kHz modulation frequency. Temperature was 110 ± 1 K. (From Sato et al., *Biochim Biophys Acta*, 1181, 195–197, 1993. With permission.)

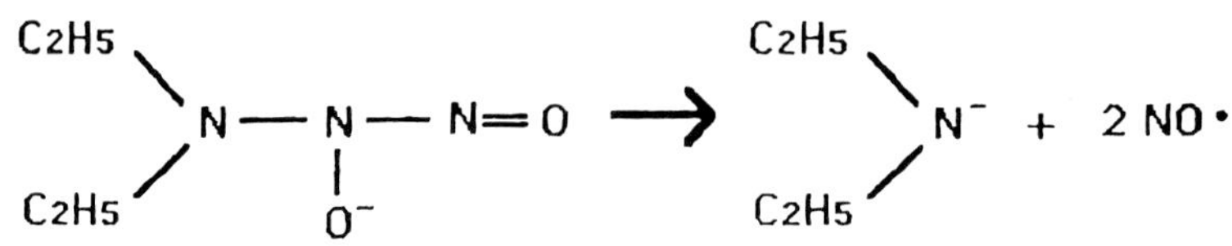

FIGURE 8. NO radical generating compound: 1,1-diethyl-2-hydroxy-2-nitroso-hydrazine (DEA/NO).

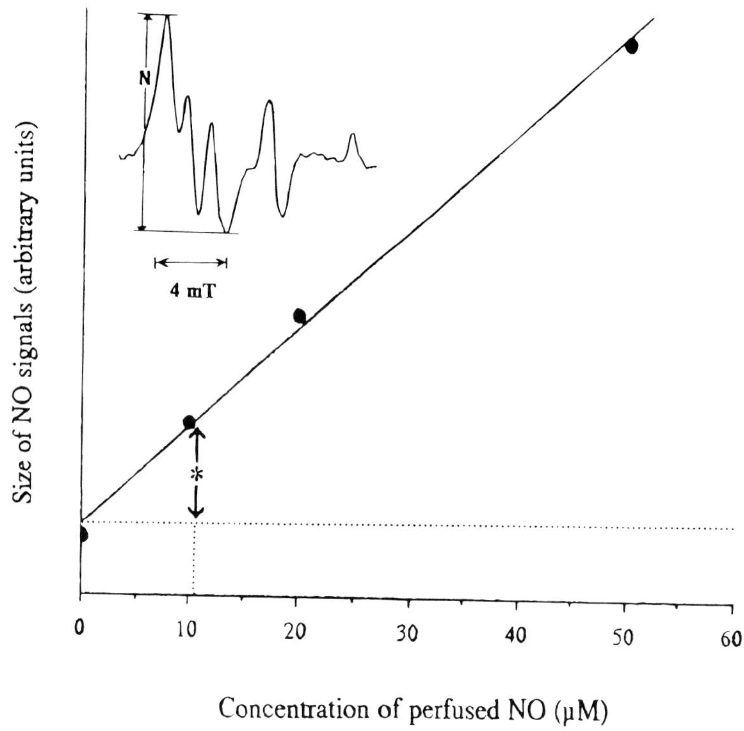

FIGURE 9. The relationship between the signal heights of NO-Fe-DETC complex and NO concentrations in the perfusate. EPR signals obtained from the rat cerebral cortices perfused with DEA/NO solutions. The inset shows an example of EPR spectra. The size of the NO signal indicated as "N" was well correlated with the concentration of NO in the perfusate (the correlation coefficient r = 0.998, $p < 0.01$). The asterisk indicates the size of NO signal in global forebrain ischemia. (From Tominaga et al., *J Cereb Blood Flow Metabol,* 14, 715–722, 1994. With permission.)

VII. ORIGIN OF Cu-DETC SIGNAL

Figure 10a indicates the EPR signal obtained from the rat brain cortex. The spectrum shows a typical quartet spectrum of Cu-DETC as reported previously.[23–26,30] The NO-Fe-DETC signal overlaps the Cu-DETC signal. In severe ischemic condition, the Cu-DETC signal decreased rapidly. This decrease may be caused by the metabolic decomposition of the Cu-DETC complex.

To date, the exact source of Cu ions is not clear. We attempted to identify possible sources of Cu ions. When 5 μM $CuSO_4$ and 2 mM DETC in 10 mM sodium phosphate buffer (pH 7.4) were incubated *in vitro* for 5 min, we could not reproduce this *in vivo* Cu-DETC signal (Figure 10b). When this mixture was incubated for 5 min in the presence of 20 mM phosphatidyl choline liposomes, the typical Cu-DETC signal was reproduced *in vitro* (Figure 10c). Phosphatidyl choline liposomes were prepared by extensive sonication using a Branson

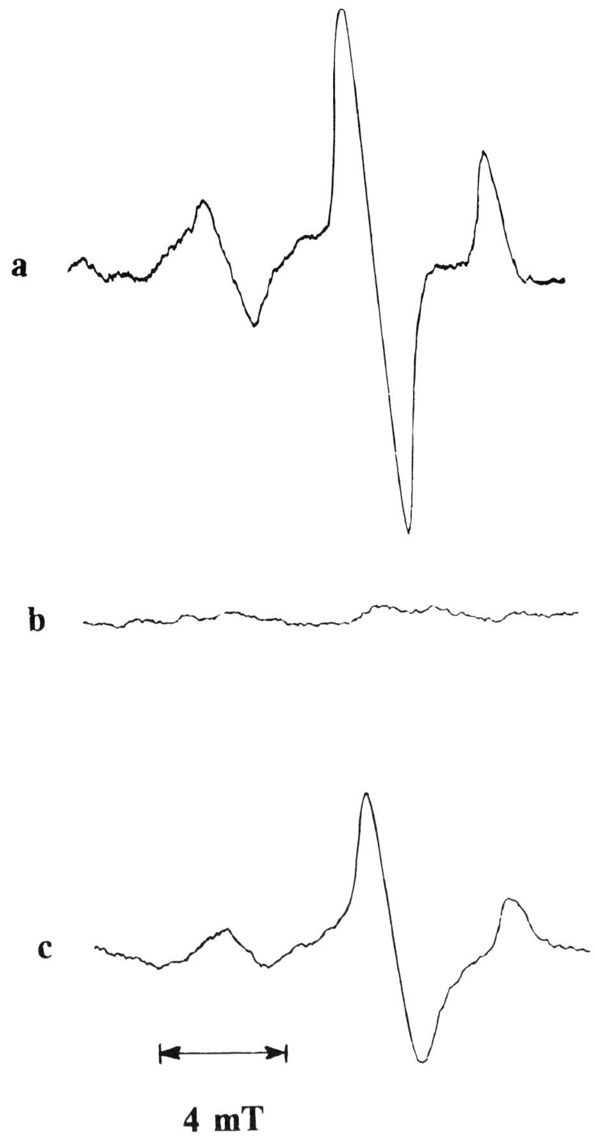

FIGURE 10. Cu-DETC signals *in vivo* and its reconstitution *in vitro.* (a) An EPR signal obtained from rat brain cortex; 500 mg/kg DETC was injected intraperitoneally 30 min before decapitation. (b) 5 μ*M* $CuSO_4$ and 2 mM DETC in 10 m*M* sodium phosphate buffer (pH 7.4) were incubated for 5 min at 25°C. (c) 20 m*M* phosphatidyl choline liposomes made in 10 m*M* sodium phosphate buffer (pH 7.4), 5 μ*M* $CuSO_4$, and 2 m*M* DETC were incubated for 5 min at 25°C. EPR conditions: 20 mW microwave power, 0.32 mT modulation, 9.22 GHz microwave frequency, and 100 kHz modulation frequency. Temperature was 220 K. The receiver gain was 5.0×10^3 for (a) and 2.5×10^3 for (b) and (c).

sonifier model 185 with a tip converter (Branson Sonic Power Co., Danbury, CT) in 10 m*M* sodium phosphate buffer (pH 7.4) for 45 min. This suggests that the Cu-DETC complex should be predominantly located in the membrane system *in vivo.*

The Fe-DETC complex is also located in membranes. Vanin's group[16] found that yeast cells that absorbed the Fe-DETC complex can be used for the NO assay. We found that liposomes that absorbed Fe-DETC complex can also be used for the NO assay as shown by the clear triplet signal in Figure 11.

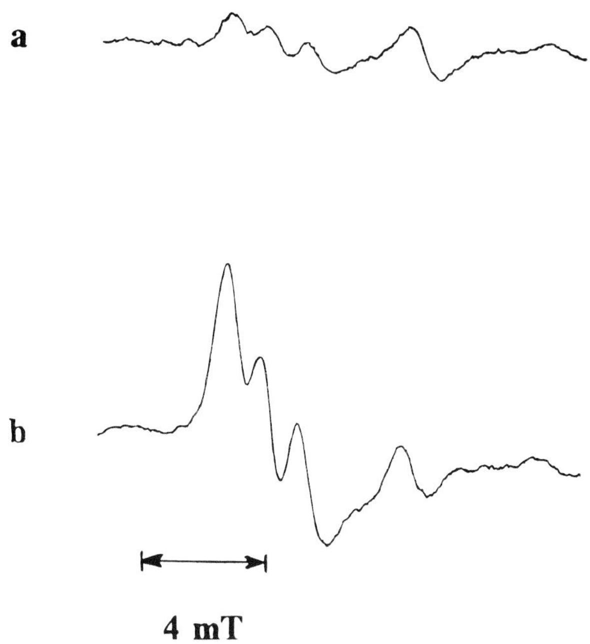

FIGURE 11. Reconstruction of NO-Fe-DETC signals *in vitro*. (a) 20 m*M* phosphatidyl choline liposomes prepared in 10 m*M* sodium phosphate buffer (pH 7.4), 0.2 m*M* $FeSO_4$/1 m*M* Na-citrate, 2 m*M* DETC, and 7 μM DEA/NO (=10 μM NO) were incubated for 5 min at 37°C. (b) 20 m*M* phosphatidyl choline liposomes prepared in 10 m*M* sodium phosphate buffer (pH 7.4), 0.2 m*M* $FeSO_4$/1 m*M* Na-citrate, 2 m*M* DETC, and 35 μM DEA/NO (=50 μM NO) were incubated for 5 min at 37°C. EPR conditions: 20 mW microwave power, 0.32 mT modulation, 9.22 GHz microwave frequency, and 100 kHz modulation frequency. Temperature was 220 K. The receiver gain was 2.0×10^3.

We evaluated whether SOD could be a source of Cu ions, because SOD is a major Cu-containing protein in brain tissues.[5] When 5 μM SOD and 2 m*M* DETC in 10 m*M* sodium phosphate buffer (pH 7.4) were mixed and incubated together for 15 min at 25°C, we could not reproduce the Cu-DETC signal (Figure 12a). However, when the mixture was incubated for 15 min in the presence of 20 m*M* phosphatidyl choline liposomes, the Cu-DETC signal appeared (Figure 12b). We made the interesting observation that this Cu-DETC signal became larger as time passed (Figure 12c). It is possible that DETC could gradually remove Cu ions from SOD. In order to confirm that SOD is the source of Cu ions in the brain, the use of transgenic mice in which greater activity of SOD is expressed[2] would be a good animal model.

We also observed Cu-DETC signals from blood samples that were taken from the DETC-injected rat (Fig. 13a-e). In this case, the Cu-DETC complex may be localized in the cell membrane of red blood cells. DETC was reported to react slowly with red blood cell SOD at this concentration.[14] Therefore, the signals observed within several minutes after the injection of DETC may be derived from Cu ions in plasma. In plasma, there are two kinds of Cu-containing proteins, namely, ceruloplasmin and Cu-albumin complex. It was suggested that the Cu-albumin complex can react with DETC readily while cereloplasmin cannot.[12] Therefore, DETC may react with the Cu-albumin complex in plasma, and then dissolves into cell membranes of not only red blood cells but also endothelium as a Cu-DETC complex.

All these data suggest that the EPR signal in this NO-trapping method was derived from both the Cu-DETC complex and the NO-Fe-DETC complex, which are localized in membranes. As for Cu ions, in addition to free intracellular Cu ions as speculated in the past, intracellular Cu proteins including SOD and the Cu-albumin complex in plasma seem to be the most possible sources.

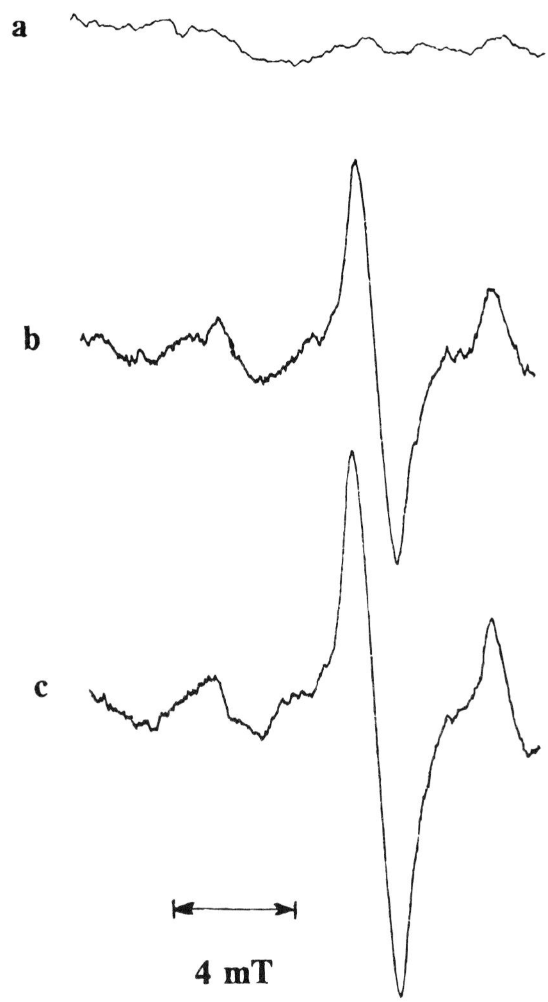

FIGURE 12. Reconstitution of the Cu-DETC signal *in vitro* from SOD. (a) 5 μ*M* SOD and 2 m*M* DETC in 10 m*M* sodium phosphate buffer (pH 7.4) were incubated for 15 min at 25°C. (b) 20 m*M* phosphatidyl choline liposomes made in 10 m*M* sodium phosphate buffer (pH 7.4), 5 μ*M* SOD, and 2 m*M* DETC were incubated for 15 min at 25°C. (c) A mixture of (b) was incubated for 60 min at 25°C. EPR conditions: 20 mW microwave power, 5.0×10^3 gain, 0.32 mT modulation, 9.22 GHz microwave frequency, and 100 kHz modulation frequency. Temperature was 220 K.

VIII. CONCLUSION

The EPR NO-trapping technique is useful for detecting NO production *in vivo.* The greatest feature of this method is that EPR can specifically and directly detect NO radicals. In addition, it is possible to estimate the amount of NO production quantitatively. This method has been successfully applied to detect the production of NO radicals in the lung and the liver of rats exposed to endotoxin and ozone (Laskin, D. and Ohnishi, S.T., to be published).

Recently, a porphyrinic microsensor has been used for the direct measurement of NO.[10,13] Compared to the EPR method, the electrode method has the advantage that it allows continuous recording. However, it has the disadvantage that it measures NO production only around the

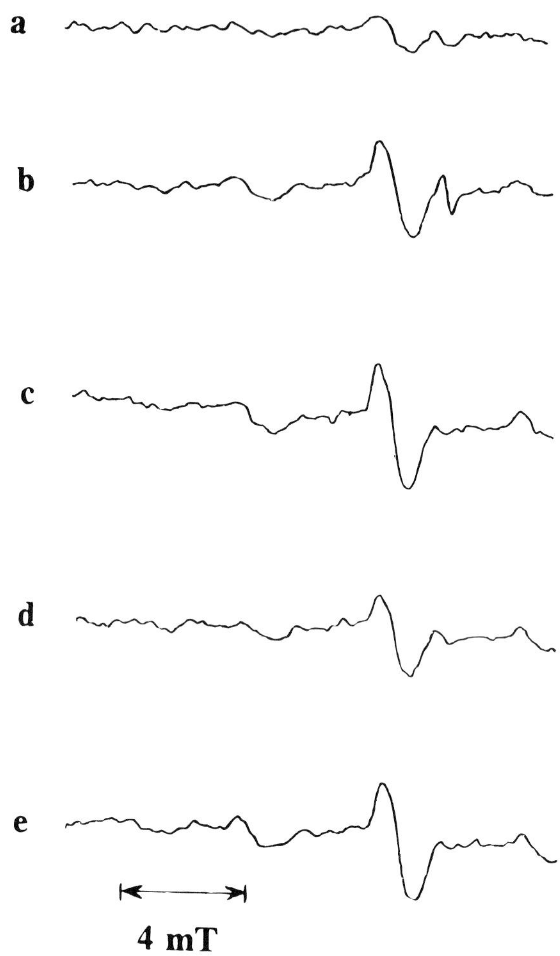

FIGURE 13. EPR signals obtained from blood samples of the rat that was administered with 500 mg/kg of DETC intraperitoneally. The time after the administration: (a) 1 min, (b) 5 min, (c) 15 min, (d) 30 min, and (e) 60 min. EPR conditions: 20 mW microwave power, 2.5×10^4 gain, 0.32 mT modulation, 9.22 GHz microwave frequency, and 100 kHz modulation frequency. Temperature was 110 K.

microelectrode tip. The electrode method also has a disadvantage in that the interpretation of the signals is not easy because the microsensor may be influenced by temperature, ion compositions, pH, and other radicals. For this reason, these two methods are considered to be complementary, not competitive.

To date, the production of NO in brain ischemia has been speculated using indirect methods. Now, we can detect NO production directly. This EPR NO-trapping method would make a significant contribution to the progress of research on the production of NO in brain ischemia.

ACKNOWLEDGMENTS

The authors are grateful to Prof. Takashi Yoshimoto, Chairman of the Department of Neurosurgery, Tohoku University School of Medicine for his encouragement. Thanks are also due to Dr. Larry Keefer of the Chemistry Section at the National Cancer Institute for his valuable suggestion. The work was supported by NIH grants NS30186 to STO and GM30736 to TO.

REFERENCES

1. Beckman, J. S., The double-edged role of nitric oxide in brain function and superoxide-mediated injury, *J Dev Physiol,* 15, 53–59, 1991.
2. Chan, P. H., Yang, G. Y., Chen, S. F., Carlson, E., and Epstein, C. J., Cold-induced brain edema and infarction are reduced in transgenic mice overexpressing CuZn-superoxide dismutase, *Ann Neurol,* 29, 482–486, 1991.
3. Dawson, T. M., Dawson, V. L., and Snyder, S. H., A novel neuronal messenger molecule in brain: the free radical, nitric oxide, *Ann Neurol,* 32, 297–311, 1992.
4. Garthwaite, J., Glutamate, nitric oxide and cell-cell signalling in the nervous system, *Trends Neurosci,* 14, 60–67, 1991.
5. Heikkila, R. E., Cabbat, F. S., and Cohen, G., *In vivo* inhibition of superoxide dismutase in mice by diethyldithiocarbamate, *J Biol Chem,* 251, 2182–2185, 1976.
6. Knowles, R. G., Palacios, M., Palmer, R. M. J., and Moncade, S., Formation of nitric oxide from L-arginine in the central nervous system: a transduction mechanism for stimulation of the soluble guanylate cyclase, *Proc Natl Acad Sci USA,* 86, 5159–5162, 1989.
7. Kon, H. and Kataoka, N., *Biochemistry,* 8, 4757, 1969.
8. Kosaka, H., Watanabe, M., Yoshihara, H., Harada, N., and Shiga, T., Detection of nitric oxide production in lipopolysaccharide-treated rats by ESR using carbon monoxide hemoglobin, *Biochem Biophys Res Commun,* 184, 1119–1124, 1992.
9. Kubrina, L. N., Caldwell, W. S., Mordvintcev, P. I., Malenkova, I. V., and Vanin, A. F., EPR evidence for nitric oxide production from guanidino nitrogens of L-arginine in animal tissues *in vivo, Biochim Biophys Acta,* 1099, 233–237, 1992.
10. Malinski, T., Bailey, F., Zhang, Z. G., and Chopp, M., Nitric oxide measured by a porphyrinic microsensor in rat brain after transient middle cerebral artery occlusion, *J Cereb Blood Flow Metab,* 13, 355–358, 1993.
11. Maragos, C. M., Morley, D., Wink, D. A., Dunams, T. M., Saavedra, J. E., Hoffman, A., Bove, A. A., Isaac, L., Hrabie, J. A., and Keefer, L. K., Complexes of NO with nucleophiles as agents for the controlled biological release of nitric oxide. Vasorelaxant effect, *J Med Chem,* 34, 3242–3247, 1991.
12. Martin, D. W., in *Harper's Review of Biochemistry* 18th edition (Martin, D. W., Mayes, P. A., and Rodwell, V. W. Eds.), 553–564, Lange Medical Publications, Los Altos, 1981.
13. Matsui, T., Nagafuji, T., Sugiyama, S., Asano, T., and Naito, S., Enhanced activity of calcium/calmodulin-independent nitric oxide synthase in the early stage of cerebral ischemia, *Soc Neurosci Abstr,* 19, 849, 1993.
14. Misra, H. P., Reaction of copper-zinc superoxide dismutase with diethyldithiocarbamate, *J Biol Chem,* 254, 11623–11628, 1979.
15. Moncada, S., Palmer, R. M. J., and Higgs, E. A., Nitric oxide: physiology, pathophysiology, and pharmacology, *Pharmacol Rev,* 43, 109–142, 1991.
16. Mordvintcev, P., Mülsch, A., Busse, R., and Vanin, A., On-line detection of nitric oxide formation in liquid aqueous phase by electron paramagnetic spectroscopy, *Anal Biochem,* 199, 142–146, 1991.
17. Nozaki, K., Moskowitz, M. A., Maynard, K. I., Koketsu, N., Dawson, T. M., Bredt, D. S., and Snyder, S. H., Possible origins and distribution of immunoreactive nitric oxide synthase-containing nerve fibers in cerebral arteries, *J Cereb Blood Flow Metab,* 13, 70–79, 1993.
18. Palmer, R. M. J., Ashton, D. S., and Moncada, S., Vascular endothelial cells synthesize nitric oxide from L-arginine, *Nature* 333, 664–666, 1988.
19. Pelligrino, D. A., Koenig, H. M., and Albrecht, R. F., Nitric oxide synthesis and regional cerebral blood flow response to hypercapnia and hypoxia in the rat, *J Cereb Blood Flow Metab,* 13, 80–87, 1993.
20. Prado, R., Watson, B. D., Kuluz, J., and Dietrich, W. D., Endothelium-derived nitric oxide synthase inhibition. Effects on cerebral blood flow, pial artery diameter, and vascular morphology in rats, *Stroke,* 23, 1118–23, 1992.
21. Rubanyi, G. M. and Vanhoutte, P. M., Superoxide anions and hyperoxia inactivate endothelium-derived relaxing factor, *Am J Physiol,* 250, H822–H827, 1986.
22. Salerno, J. C. and Ohnishi, T., Tetranuclear and binuclear iron-sulfur clusters in succinate dehydrogenase: a method of iron quantitation by formation of paramagnetic complexes, *Biochem Biophys Res Commun,* 73, 833–839, 1976.
23. Sato, S., Tominaga, T., Ohnishi, T., and Ohnishi, S. T., EPR spin-trapping study of nitric oxide formation during bilateral carotid occlusion of rat, *Biochim Biophys Acta,* 1181, 195–197, 1993.

24. Sato, S., Tominaga, T., Ohnishi, T., and Ohnishi, S. T., Electron paramagnetic resonance study on the time course of nitric oxide production during rat brain ischemia and reperfusion, *Brain Res,* 647, 91–96, 1994.
25. Sato, S., Tominaga, T., Ohnishi, T., and Ohnishi, S. T., Role of nitric oxide in brain ischemia, *Ann NY Acad Sci,* in press.
26. Shiga, T., Hwang, K. J., and Tyuma, I., *Biochemistry,* 8, 378, 1969.
27. Tominaga, T., Sato, S., Ohnishi, T., and Ohnishi, S. T., Potentiation of nitric oxide formation following bilateral carotid occlusion and focal cerebral ischemia in the rat: *in vivo* detection of the nitric oxide radical by electron paramagnetic resonance spin trapping, *Brain Res,* 614, 342–346, 1993.
28. Tominaga, T., Sato, S., Ohnishi, T., and Ohnishi, S. T., Electron paramagnetic resonance (EPR) detection of nitric oxide produced during forebrain ischemia of the rat, *J Cereb Blood Flow Metab,* 14, 715–722, 1994.
29. Vanin, A. F., Hypothesis: Endothelium-derived relaxing factor is a nitrosyl iron complex with thiol ligands, *FEBS,* 289, 1–3, 1991.
30. Vedernikov, Y. P., Mordvintcev, P. I., Malenkova, I. V., and Vanin, A. F., Effect of diethyldithiocarbamate on the activity of nitric oxide-releasing vasodilators, *Eur J Pharmacol,* 212, 125–128, 1992.
31. Voevodskaya, N. V. and Vanin, A. F., Gamma-irradiation potentiates L-arginine-dependent nitric oxide formation in mice, *Biochem Biophys Res Commun,* 186, 1423–1428, 1992.

Chapter 35

DIRECT MEASUREMENT OF NITRIC OXIDE IN THE RAT BRAIN SUBJECTED TO OCCLUSION OF THE MIDDLE CEREBRAL ARTERY

Toru Matsui

TABLE OF CONTENTS

I. INTRODUCTION

The pathogenesis underlying ischemic cerebral damage remains to be clarified. Recent evidence has suggested that nitric oxide (NO) may mediate the excitotoxic effects of excessive *N*-methyl-D-aspartate (NMDA) receptor activation, through generation of peroxynitrite anion and hydroxyl radical.[1–3] Blockade of nitric oxide formation has been reported to decrease hypoxic-ischemic brain damages. However, some investigators reported that inhibition of nitric oxide synthase (NOS) activity did not prevent either glutamate or NMDA-induced neurotoxicity in hipocampal neurons nor in whole-brain culture.[4] In addition, nitroso compounds can protect neurons from NMDA receptor-mediated neurotoxicity. Very recently, Lei and Lipton provided a good explanation to this controversial point, stating that neurocytotoxicity or neuroprotection of NO may be dependent upon its redox state.[5,6]

Our recent study demonstrated that the intraperitoneal administration of N^{ω}-nitro-L-arginine (LNA) resulted in a great deal of amelioration of brain edema following both permanent and temporary ischemia.[7,8]

0-8493-8094-4/95/$0.00+$.50

Since hitherto demonstrated experimental results showed indirect evidence that NO may play a crucial role in ischemic brain damage, much attention has been paid to examining whether NO is yielded *in vivo* in the brain subjected to focal cerebral ischemia.

Malinski et al.[9] have reported that NO concentration in the brain increased during and after transient middle cerebral artery occlusion (MCAO) in rats. They measured NO during 2 h of MCAO, or 2 h of MCAO followed by 1 h of recirculation by using a porphyrinic NO microsensor. The main purpose of the present study is to investigate whether or not NO is dramatically produced *in vivo* in the rat brain subjected to MCAO, using a newly developed NO measurement system.

In this manuscript, it is shown how the present microsensor calibrated by *NO* donor and NO itself can be used to measure NO in *in vivo* models such as MCAO in rats.

II. MATERIALS AND METHODS

A. MEASUREMENT OF NO

1. Introduction of Electrode and Its Calibration with SNAP (*S*-nitroso-*N*-acetyl-DL-penicillamine)

We used a working electrode (microsensor) of Pt/Ir alloy and coated it with polymer film (Figure 1), by which the electrode is made stronger and easier to manipulate than previous electrodes.[9–14]

The principle behind the NO meter is measurement of the picoampere-order redox current between the following two electrodes.[15] A working electrode was made of 0.2-mm-diameter Pt/Ir alloy coated with a three-layered membrane, which consists of KCl, NO-selective resin, and normal silicone membranes. A counterelectrode was made of carbon fiber. A KCl membrane was deposited electrochemically on the Pt/Ir wire to suppress overvoltages in the discharge of NO. NO-selective resin was made of a sort of nitrocellulose, pyroxyline lacquer. The outermost membrane was introduced to avoid nonspecific ionic effect and electrochemical reactions. This polarographic current was detected with a current-voltage converter circuit.

Electrocurrent (Figure 2) was monitored in Krebs-Henseleit solutions (KHS) containing 10^{-6} *M*, 3×10^{-6} *M*, 10^{-5} *M*, 3×10^{-5} *M*, and 10^{-4} *M* SNAP, respectively, which were kept at 37°C in the presence of superoxide dismutase (SOD, 100 U/ml). Although the sensitivity of electrodes varied one by one because the thickness of membranes could not be constant, the representative correlation line is shown in Figure 3. Under controlled temperature (37°C), the electrocurrent increased linearly from 2 to 1000 pA when SNAP concentration increased from 1 *M* to 0.1 m*M*. In addition, effects of hemoglobin solution on electrocurrent was examined in the 10^{-4} *M* SNAP-containing KHS. As shown in Figure 4, repeated addition to hemoglobin (Hb, 2.17 μ*M*) resulted in significant diminution of NO concentration.

2. Calibration of Microsensor by NO

1. KHS was bubbled with nitrogen gas for replacement of the air component except nitrogen overnight.
2. Two kinds of KHS with pH = 6.4 and 7.4 were prepared using 1-*N* NaOH and 0.3 NHCl.
3. A standard NO-containing solution (e.g., concentration = 0.2 m*M*) was prepared, and various concentrations of NO were made up (Figures 5 and 6).

Since the maximum concentration of NO in the KHS is 1.73 m*M* at 1 atm and 37°C, it is possible to make up the NO-containing solution less than 1.73 m*M* depending upon the combination of NO and nitrogen gas. For example, mixture of NO (0.5 l/min) and nitrogen gas (3.8 l/min) provides 0.2 m*M* NO solution at 1 atm and 37°C. Then, addition of 10, 30, 70, 130, 210, 310, 435, 580, 745, and 955 μl of the NO solution into the N_2-replaced KHS (20 ml) provides 0.1, 0.3, 0.7, 1, 2, 3, 4, 6, 7, and 9 μ*M* NO solution, respectively. Calibration

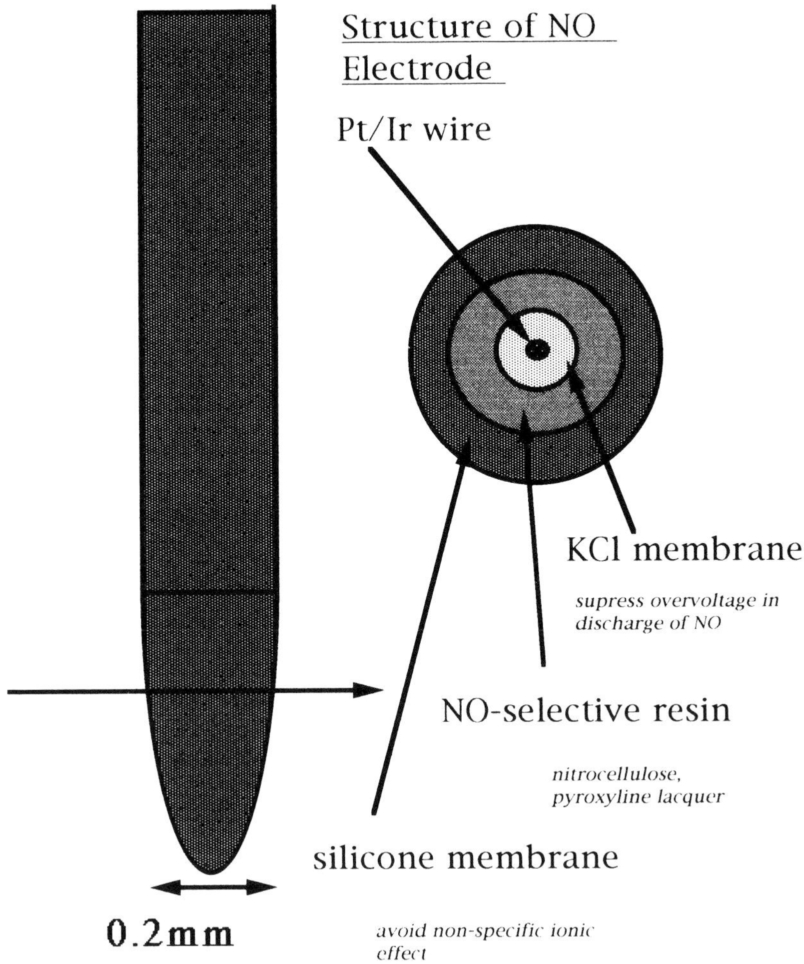

FIGURE 1. Structure of NO microsensor.

of the NO microsensor was performed in each solution. The same procedure was repeated in the other KHS (pH = 6.4).

4. The correlation line of each electrode was obtained by linear regression analysis.

The correlation of NO concentration and amplitude obtained was examined with linear regression analysis. The linear equation was different among electrodes. Electrodes with a correlation coefficient less than 0.95 were not used in the *in vivo* study. Using electrodes with coefficients greater than 0.95, the following *in vivo* study was carried out (Figures 7 and 8).

3. Measurement of NO *In Vivo*

a. *General Procedures*

Sprague-Dawley rats were anesthetized with 2% halothane for an induction and halothane (0.3 to 1%) for maintenance in 2:1 N_2O/O_2 mixture. Animals were tracheostomized and connected to the artificial respirator (Model 683, Harvard Instrument Co., Cambridge, MA).

The femoral artery was cannulated to monitor arterial blood pressure and to sample blood gases. The femoral vein was also catheterized to monitor blood sugar and administer

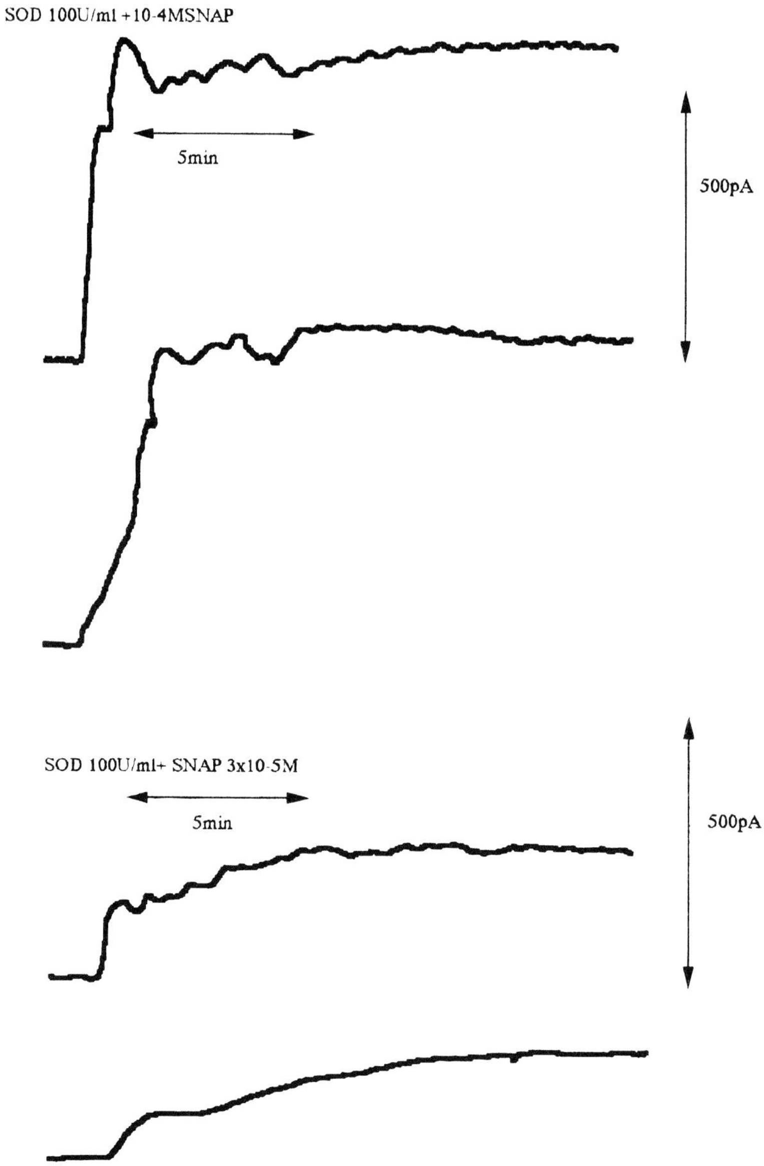

FIGURE 2. Detection of nitric oxide released from SNAP under the presence of SOD (100 U/ml).

bicarbonate according to necessity. The rectal and temporal muscle temperatures were maintained at 37.5 ± 0.5 and 36.5 ± 0.5°C, respectively, using heating pads and lamp. According to the modified method of Tamura, the proximal trunk of the MCA was first exposed. Then, animals were fixed to the stereotaxic apparatus for implantation of the NO microsensor, CBF probe, and temperature lead, as described in Section II.A.3.b. After the stabilization of the NO microsensor, which was allowed for 2 h, saline (Group 1) or LNA (1 mg/kg body weight, IP, Group 2) was given to animals (Figure 9). Since there was no major change in the electrocurrent, the left MCA was occluded by miniclip. Three hours after induction of MCAO, administration of saline (Group 1) or LNA (Group 2) was repeated. NO was monitored until animals were killed at 4 h after MCAO.

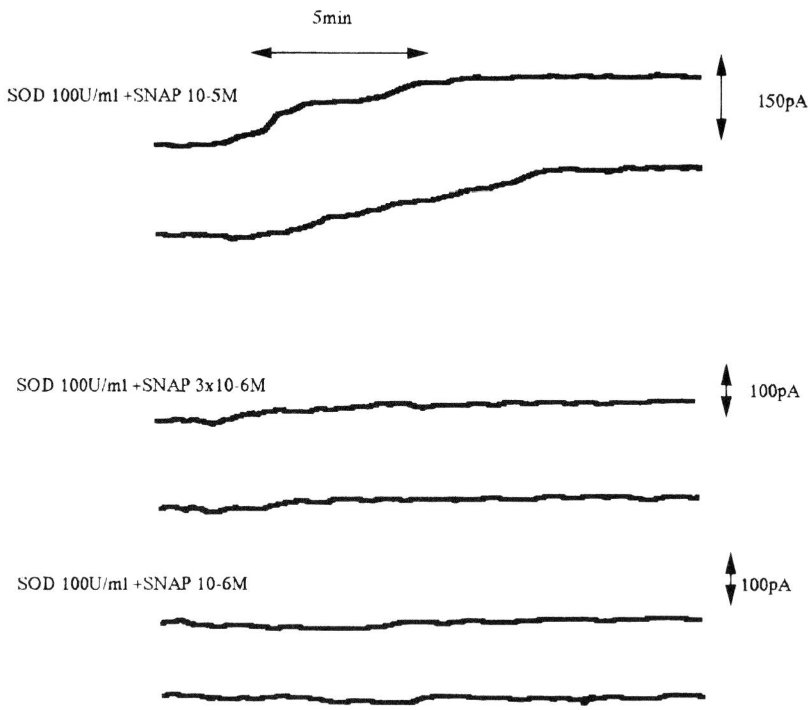

FIGURE 2. Continued.

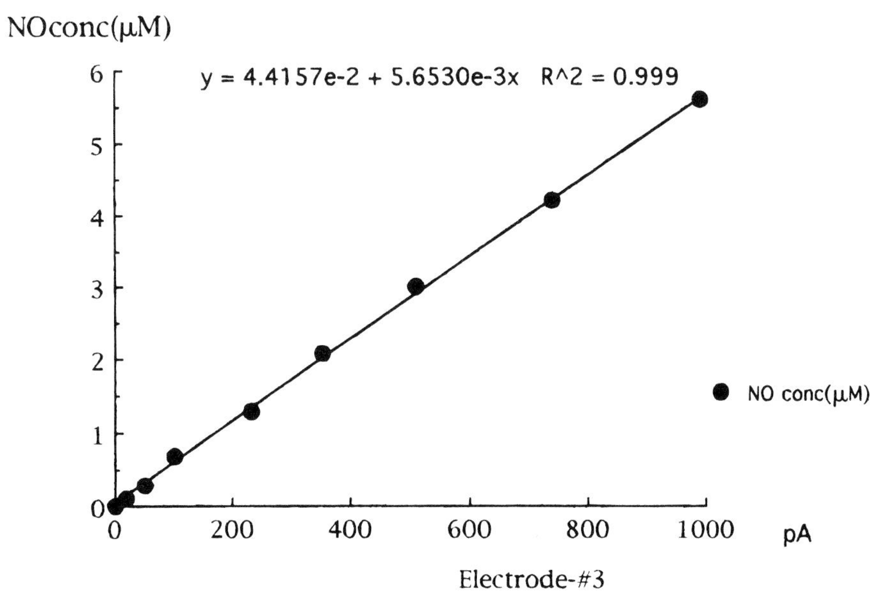

FIGURE 3. Regression line obtained.

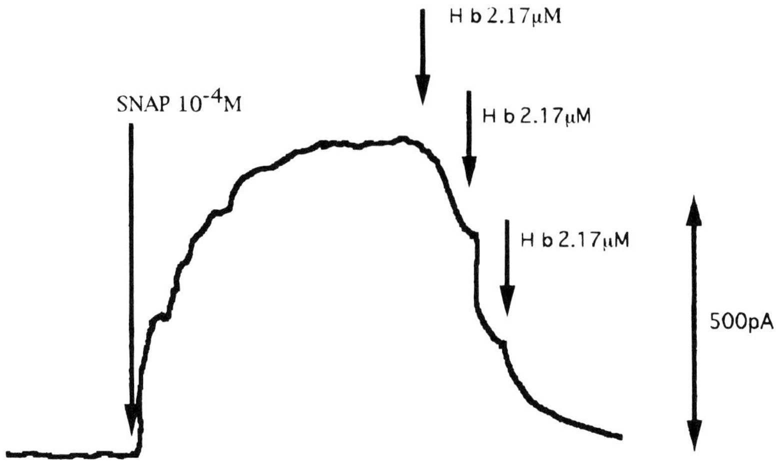

FIGURE 4. Effect of hemoglobin on NO detection.

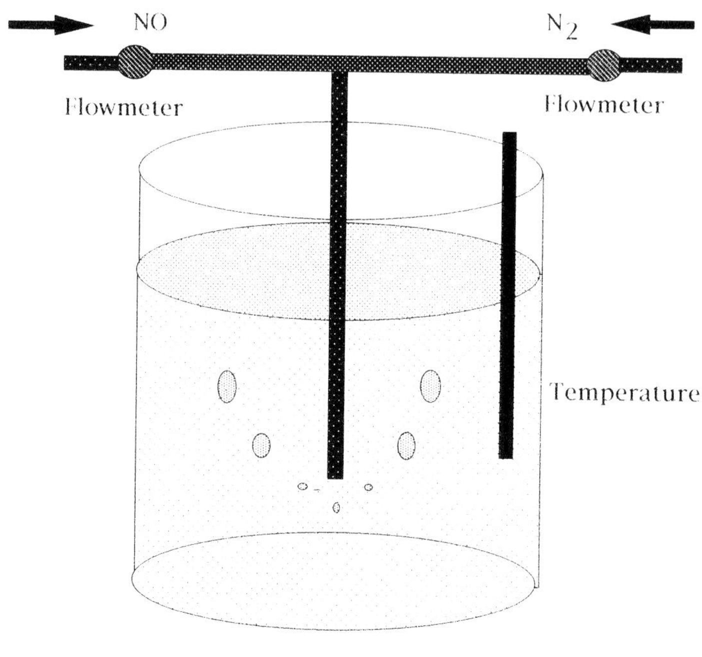

FIGURE 5. Composing NO solution.

At the end of each experiment, animals were transcardially perfused with heparinized saline and 10% neutral buffered formalin. The head was immersed in formalin solution for 24 h, after which the brain was removed and stained with hematoxylin and eosin for evaluation of ischemic cell damage and brain injury by an implanted electrode.

b. Implantation of NO Microsensor, CBF Probe, and Temperature Lead

All surgical procedures should be performed under an operating microscope, and therefore microneurosurgical technique is required. As shown in the drawing, a small craniotomy was bilaterally made. In this step of craniotomy, the saline (room temperature) was poured over

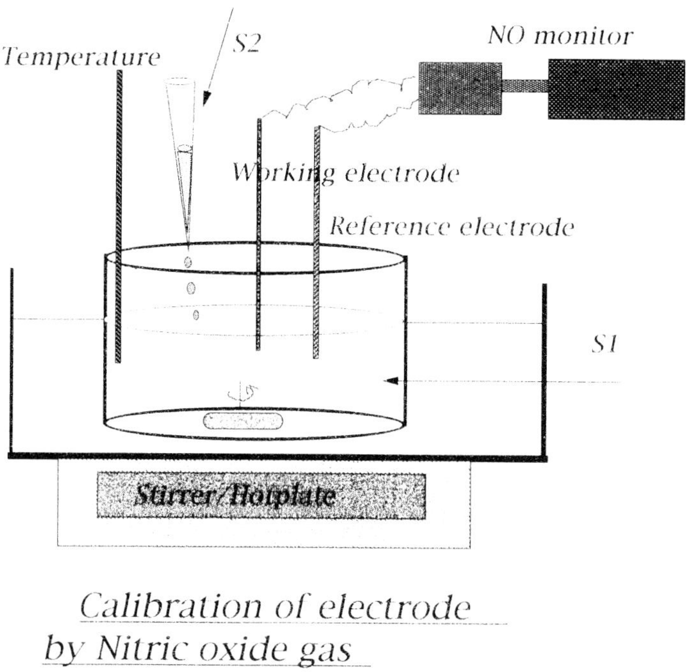

FIGURE 6. Calibration of the microsensor by nitric oxide.

the operative field for preventing heat injury to brain tissue. Care was taken to prevent traumatic brain injury and to not totally remove the cranium (thin bone should remain over the surface of the dura matter). The remaining thin bone was dissected from the dura mater and removed (Figure 7). In the second step, the tip of the NO microsensor was carefully inserted into the brain tissue (3 mm deep at 30 to 40° to the brain surface) through the cranial window. The electrode was inserted into the ischemic core region in the affected side and parietal region in the contralateral side.

CBF was continuously measured with a laser Doppler flowmeter. The region of ischemic core and penumbra was constant in the present MCAO model, as previously reported.[7] If even minimal bleeding was observed with the naked eye, the results obtained from the study were omitted. Appropriate insertion of the electrode was necessary, and so it was determined whether any blood was attached to the electrode and whether there was any evidence of bleeding around the insertion area (Figure 9).

4. Parameters Monitored

During the course of the experimental procedures, oxygen tension in the brain tissue (oxygen monitor, Intermedical Co., Ltd., Tokyo, Japan), temporal muscle temperature (digital thermometer, Unique Medical Co., Ltd., Tokyo, Japan), regional cerebral blood flow (laser Doppler flowmeter, Advance Co., Ltd., Tokyo, Japan), and systemic arterial blood pressure (7754 system, Hewlett-Packard Co., Tokyo, Japan) were monitored. Arterial and venous blood samples were periodically collected for blood gas analysis (Model 170 pH/blood gas analyzer, Corning, Tokyo, Japan) and blood sugar measurement (Toh-Echo-Super, Kyoto Dai Ichi Kagaku, Kyoto, Japan).

5. Statistical Analysis

The correlation between electrocurrent and NO concentration was analyzed with linear regression analysis, using a Macintosh program (Cricket graph).

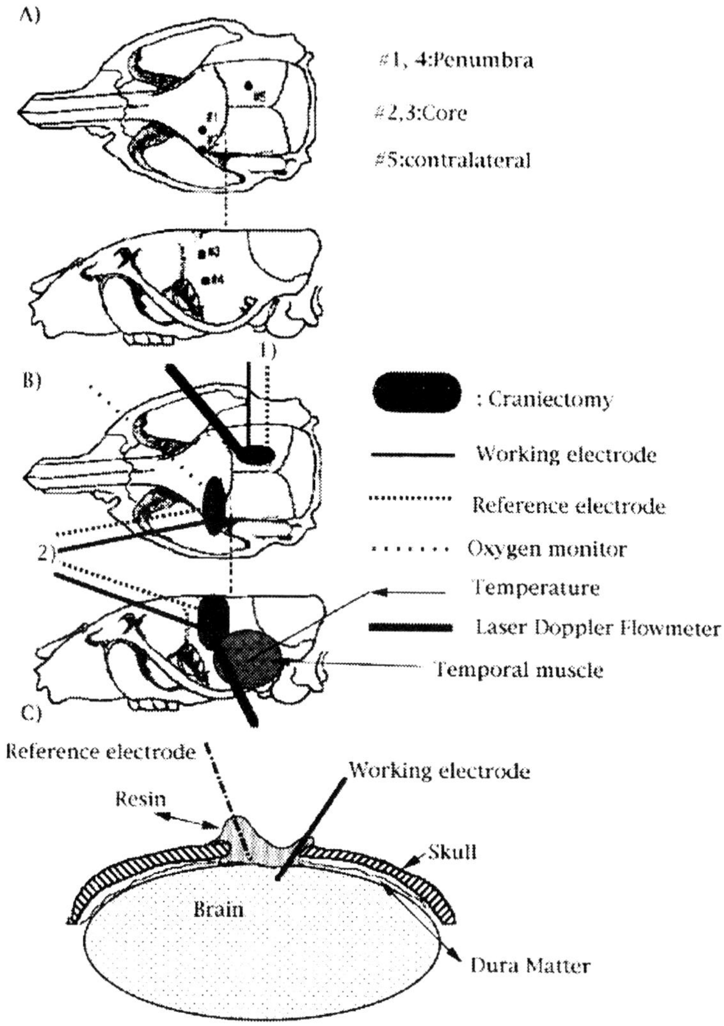

FIGURE 7. Implantation of NO microsensor, reference electrode, CBF proble, oxygen monitor, and temperature probe.

III. RESULTS

A. CALIBRATION OF ELECTRODE BY NO OR SNAP

The study using NO gas revealed that the present electrode detected NO and the amplitude increase was found to be linearly correlated with concentration of NO as shown in Figures 7 and 8. The measurement was not influenced by pH, if pH was maintained 6.4 to 7.6. In the same way, the electrode was calibrated by SNAP (Figures 2 and 3).

B. DIRECT MEASUREMENT OF NO *IN VIVO*

The NO-sensor was allowed to stabilize for 2 h and then the present study was started (Figure 10).

Immediately after induction of MCAO (Group 1), NO was detected and the level of electrocurrent reached 500 to 800 pA at 15 to 30 min of MCAO (the first peak). Thereafter, it returned to the baseline value. Reproduction of NO, ranged around 200 pA, was monitored

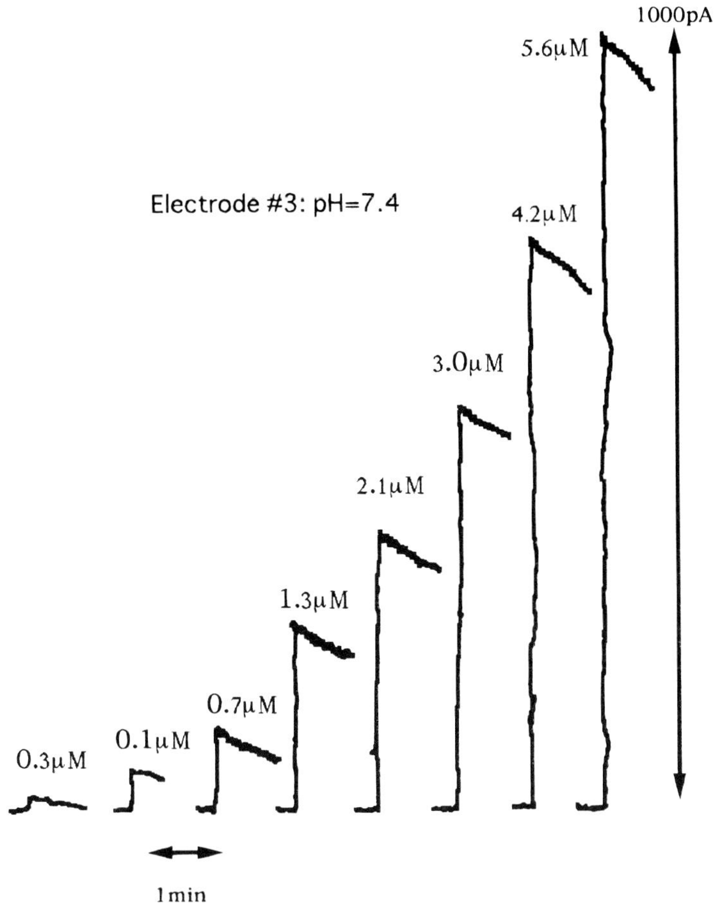

FIGURE 8. Actual tracing of the calibration of the microsensor by nitric oxide-containing solution in the presence of SOD.

3 h after MCAO (the second peak). Preischemic administration of LNA (1 mg/kg, IP) significantly inhibited both NO production peaks (Group 2). In particular, the second peak of NO production was almost completely suppressed (Figure 10). According to calibration by the NO-containing solution, the first and second peaks corresponded to 1.5 to 2 μM and 0.5 to 1 μM of NO concentration, respectively.

Pathological examination showed no evidence of intracerebral hemorrhage due to insertion of the NO microsensor.

IV. DISCUSSION

A. METHODOLOGY

Since NO is gaseous, there are many difficulties in the calibration of the present microsensor. As explained in Section II, the composition of the standard NO-containing solution is based upon the known value of maximum NO solubility in KHS (1.73 mM at 37°C and 1 atm). The present calibration was performed in the acidotic (pH = 6.4) and neutral conditions, because cerebral ischemia induces metabolic failure, leading to the acidotic environment. All microsensors used in the present study showed identical response to NO in both pH solutions.

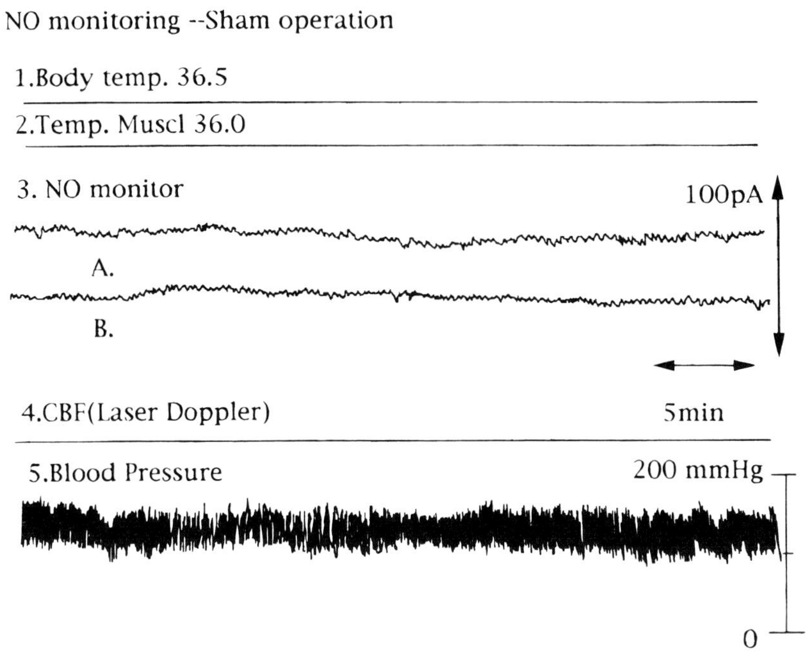

FIGURE 9. Stabilization of the microsensor.

The electrocurrent was plotted against the coresponding NO concentration and the regression line of each sensor was obtained (Figure 8). Based upon the linear regression equation, NO concentration corresponding to each peak of the electrocurrent was calculated. Although a very small number of investigators reported the methods by which NO could be directly measured in the *in vitro* conditions,[9–14] an accurate method of NO measurement in the *in vivo* model has not been established. Malinski et al. were the first investigators to report direct measurement of NO in stroked rats, employing the porphyrin-coated microsensor. However, their observation lacked some of the key phenomena of permanent focal cerebral ischemia, as mentioned below.

Immediately after permanent MCAO, the electrocurrent, up to about 500 to 800 pA, due to NO production began to be detected in the present study. According to the calibration of the relevant electrode, that value was corresponded to about 1.5 to 2 μM NO. However, over 60 min MCAO, the electrocurrent declined to the baseline. The phenomenon might be explained by the following two reasons. First, NO production itself was decreased. Second, although NO was still produced, it was coupled with oxygen radical to form peroxynitrite anion.

The second peak, ascribable to NO overproduction, was ranged about 100 to 200 pA, which corresponds to about 0.5 to 0.7 μM NO. Surprisingly, preischemic administration of 1 mg/kg LNA (IP) suppressed both peaks of NO production and thereby protected the brain from ischemic injury.[7,8,16,17]

From the viewpoint of interpreting the data obtained, the following should be considered:

1. Is it possible to estimate NO concentration in the *in vivo* from the calibration line obtained in the *in vitro* condition? Some differences between the actual NO value in the stroked brain and the value obtained, which was calculated on a basis of the calibration performed in the absence of oxygen (replaced by nitrogen), might be brought about. However, there have been no accurate methods for NO calibration except the present one.

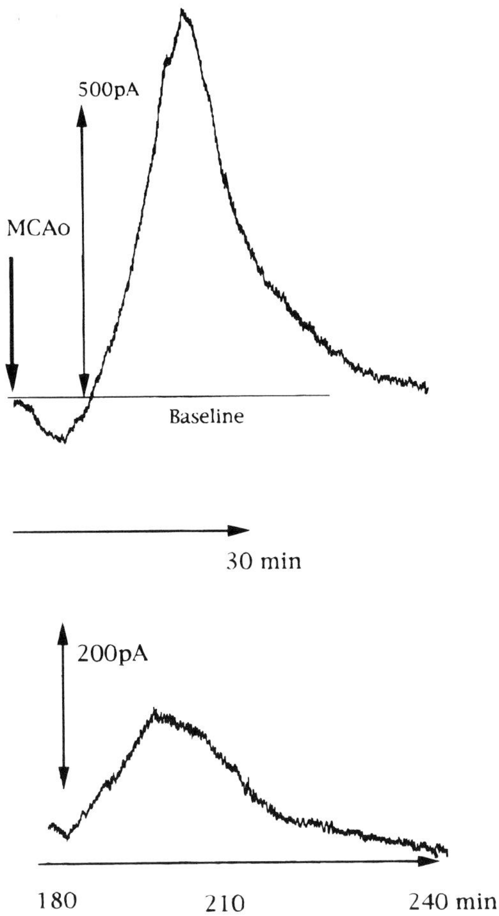

FIGURE 10. The first and second peaks of NO production after MCAO.

2. Is this microsensor specifically sensitive to NO? This might be explained by the fact that this microsensor is sensitive to various NO donors such as SNAP and SIN-1 (Figures 11 and 12).
3. Does the decline of the electrocurrent express a decrease of NO production, although the electrocurrent rises according to an increase of NO production? In another words, the possibility should be kept in mind that degradation of NO exceeding its production (e.g., forming ONOO) may result in an apparent decline of the electrocurrent, which is mistaken as a decrease in NO production.
4. Since the NO monitor was set to zero after initial stabilization of the microsensor under physiological status and the electrocurrent continued to be monitored, relative but not absolute value of NO concentration was obtainable. Along this line of thought, the interpretation of the data obtained focused only on the increase of electrocurrent. Thus far, the decline of the electrocurrent remains to be clarified.
5. The present method is based upon measuring the electrocurrent and is thus likely to be influenced by external electrical atmosphere. Thus, the reproducibility of experimental results require keeping the experimental circumstances as electrically stable as possible.

B. ROLE OF NO IN CEREBRAL ISCHEMIA

Although there are several issues to be resolved, the *in vitro* study indicated that the present microsensor is likely to be sensitive to NO and that there is a linear correlation

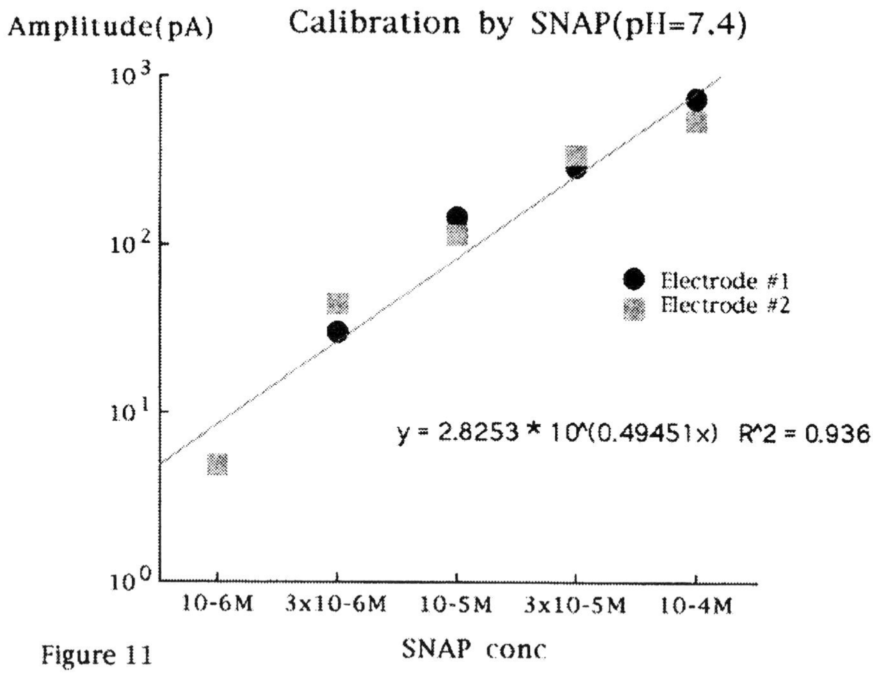

FIGURE 11. Calibration of the microsensor by SNAP.

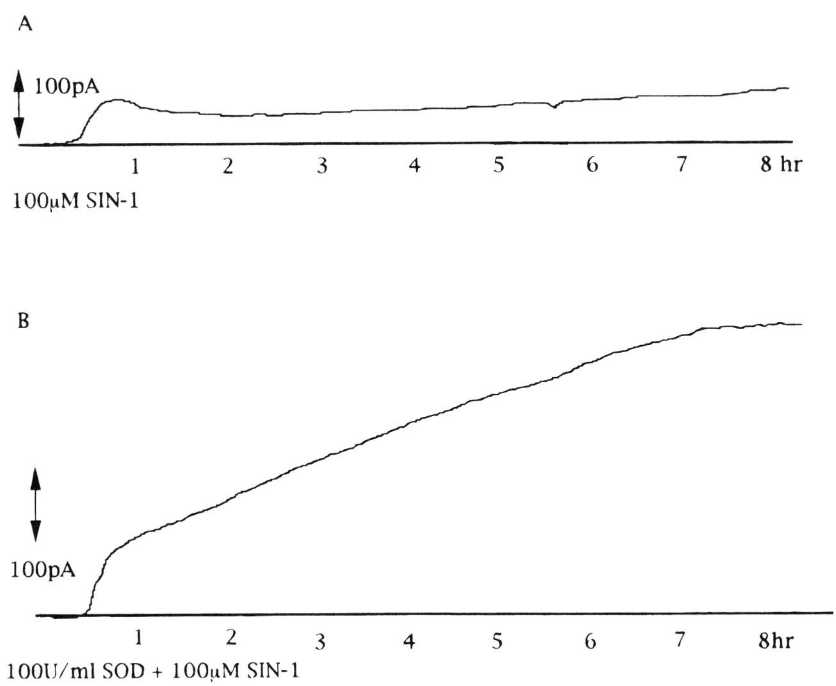

FIGURE 12. Calibration of the microsensor by SIN-1 in the absense and presence of SOD.

between electrocurrent increase and NO concentration. The *in vivo* study elucidated that NO production in the ischemic brain was remarkably enhanced immediately and 3 h after induction of MCAO. Administration of LNA, at the dosage of 1 mg/kg (IP, 5 min before and 3 h after MCAO), inhibited NO production induced by MCAO.

To date, it remains controversial whether NO has a major role in pathogenesis of ischemic brain damage.[1,2,7,18–20] We have already reported that 1 mg/kg LNA inhibited ischemic cerebral damage.[8] Although it remains to be clarified whether LNA acted on vascular or neuronal sites, this study led us to believe that total abolition of NOS activity in the ischemic brain might worsen cerebral microcirculation of the ischemic brain.[7] In concert with this thinking, the effective dosage of LNA against MCAO in rats was found to be very low (0.01 to 1 mg/kg, IP).[21]

Brain microvessels (MVs) have been known to play crucial roles as the functional sites of the blood-brain barrier, which regulates the permeability of intrinsic/extrinsic substances. Furthermore, experimental evidence indicates that significant increases in the water and sodium contents within the brain develop as early as 3 h after MCAO.[22] Thus, we have already reported the temporal profiles of constitutive and inducible NO synthase (cNOS and iNOS) activity in the brain MV following MCAO, suggesting that cNOS and iNOS were activated 15 min and 4 h after MCAO, respectively. This phenomenon seems to be consistent with NO measurement in the present study.

The present results might be interpreted as follows. The prominent influx of calcium following onset of ischemia induced activation of cNOS, leading to exaggerated production of NO. Overproduction of NO potentially deteriorates the mitochondrial electron transfer function, causing a reduced redox state of NO: NO•.[1–3] In focal cerebral ischemia, NO binds existing oxygen radical to form peroxynitrite anion, which is thought to be extremely cytotoxic. According to direct NO measurement, the first prominent peak of NO production was detected within 30 min after induction of MCAO and the second moderate phase was found 3 h after the induction. It is reasonably speculated that the former might originate from activation of the cNOS and the later from iNOS, although no direct evidence of iNOS induction was obtained in this model of ischemia. Thus, initial increases of cNOS activity observed in the present study seems to correspond strongly with the development of ischemic brain edema, namely, increased iNOS activity in cerebral MVs precedes the maximum formation of brain edema. Further supporting evidence for our idea comes from the reports that dexamethasone, which inhibits the expression of iNOS but not the activity of cNOS in vascular endothelial cells,[23] ameliorates ischemic brain edema and infarction.[24]

It seems very likely that activation of iNOS, particularly in the initial stage of ischemia, plays a pathological role in the development of ischemic brain damage, in combination with explosive activation of cNOS immediately after onset of brain ischemia. However, in order to confirm the above hypothesis, it will be necessary to estimate the temporal profile of NOS activity in synatosome and glial tissues following cerebral ischemia.

V. CONCLUSION

The present study presents intriguing evidence that direct monitoring of NO in the ischemic brain demonstrated the interesting aspects characteristic of permanent ischemia. Taking into account our previous report that a NOS inhibitor reduced ischemic brain damage,[7,10] the present findings render it very likely that blockade of i- and cNOS expression/activity in brain MVs at the early stage of ischemia becomes a novel pharmacotherapeutic tactic for the treatment of cerebral stroke. Further, we have almost completed NO measurement in temporary MCAO in rats, which will be reported in the near future.

REFERENCES

1. Dawson, T. M., Dawson, V. L., London, E. D., Bredt, D. S., and Snyder, S. H., *Proc. Natl. Acad. Sci. U.S.A.* 88, 6368–6371 (1991).
2. Dawson, T. M., Dawson, V. L., and Snyder, S. H., *Ann. Neurol.*, 32, 297–331 (1992).
3. Beckman, J. S., *J. Dev. Physiol.* 15, 53–59 (1991).
4. Demerle-Pallardy C., Lonchampt, M. O., Chabrier, P. E., and Braquet, P., *Biochem, Biophys, Res. Commun.* 181, 456–464 (1991).
5. Lipton, S. A., Choi, Y-B., Pan, Z-H., Lei, S. Z., Chen, H. S. V., Sucher, N. J., Loscalzo, J., Singel, D. J., and Stamler, J. S., *Nature* 364, 626–632 (1993).
6. Lei, S. Z., Pan, Z-H., Aggarwal, S. K., Chen, H-S., V., Hartman, J., Sucher, N., and Lipton, S. A., *Neuron* 8, 1087–1099, (1992).
7. Matsui, T., Nagafuji, T., Auer, R. N., Koide, T., Tsutsumi, K., and Asano, T., *J. Cereb. Blood Flow Metab.* 13 (suppl. 1), S150, (1993).
8. Nagafuji, T., Matsui, T., Koide, T., and Asano, T., *Neurosci. Lett.* 147, 159–162 (1992).
9. Malinski, T., Bailey, F., Zhang, G., and Chopp, M., *J. Cereb. Blood Flow Metab.* 13, 355–358 (1993).
10. Palmer, R. M. et al., *Nature* 327, 524–526 (1987).
11. Bredt's et al., *Pro. Natl. Acad. Sci. U.S.A.* 86, 9030–9033 (1989).
12. Kelm, M. et al., *J. Circ. Res.* 66, 1561–1575 (1990).
13. Palmer, R. M., Ashton, D. S., and Moncada, S., *Nature* 333, 664–666 (1988).
14. Kikuchi, K., Hayakawa, H., Nagano, T., Hirata, Y., Sugimoto, T., and Hirobe, M., *Chem. Pharm. Bull.* 40, 2233–2235 (1992).
15. Ichimori, K., Ishida, H., Yamashita, T., Nakazawa, H., and Murakami, E., Abstract, *The Third International Meeting on Biology of Nitric Oxide,* Cologne, Germany (1993). In press.
16. Bredt, D. S. and Snyder, S. H., *Proc. Natl. Acad. Sci. U.S.A.* 87, 682–685 (1990).
17. Nagafuji, T., Sugiyama, M., Matsui, T., and Koide, T., *Eur. J. Pharmacol.* (1993). In press.
18. Hewett, S. I., Corbett, J. A., McDaniel, M. I., and Choi, D. W., *Brain Res.* 625, 337–341 (1993).
19. Nishikawa, T., Kirsch, J. R., Koehler, R. C., Bredt, D. S., Snyder, S., and Traystman, R. J., *Stroke* 24, 1717–1724 (1993).
20. Kader, A., Frazzini, V. I., Solomon, R. A., and Trifiletti, R. R., *Stroke* 241, 1709–1716 (1993).
21. Nagafuji, T., Sugiyama, M., and Matsui, T., Brain edema 93 proceeding, in press.
22. Gotoh, O., Asano, T., Koide, T., and Takakura, K., *Stroke* 16, 101–109 (1985).
23. Moncada, S., Palmer, R. M. J., and Higgs, E. A., *Pharmacol. Rev.* 43, 109–142 (1991).
24. Koide, T., Wieloch, T. W., and Siesjo, B. K., *J. Cereb. Blood Flow Metab.,* 6, 395–404 (1986).

Chapter 36

NITRIC OXIDE- AND SUPEROXIDE-MEDIATED TISSUE INJURY: DETECTION OF PEROXYNITRITE

Harry Ischiropoulos, John P. Crow, Yao Zu Ye, and Joseph S. Beckman

TABLE OF CONTENTS

I. INTRODUCTION

Experimental evidence has implicated oxidative stress as a major cause of tissue injury in ischemia-reperfusion, in glutamate neurotoxicity, in neurodegenerative diseases, and from activation of inflammatory cells.[1–3] Oxidative stress is the result of overproduction of reactive species that overwhelms the cellular antioxidant capacity leading to inactivation of key cellular functions and ultimately to cell death. Most of the existing data have concentrated on the contribution of oxygen-derived reactive species such as superoxide in tissue injury. However, recent reports have also implicated excessive production of nitric oxide in neurotoxicity associated with glutamate receptor stimulation,[4–6] ischemia-reperfusion,[7–10] and cold-induced brain edema.[11] Nitric oxide is released in response to activation of glutamate receptors and is involved in neuronal plasticity. Nitric oxide regulates vascular tone, inhibits the aggregation of platelets and adherence of neutrophils, and mediates the microbicidal and tumorstatic activities of macrophages. Nitric oxide and superoxide are weak one-electron oxidants and individually are unlikely to cause tissue injury. However, since both nitric oxide and superoxide are free radicals they react at a near-diffusion limited rate to form peroxynitrite, a relatively long-lived and highly reactive species. Peroxynitrite is a stronger oxidant than either nitric oxide or superoxide alone and may provide the common link that explains the protective effects of superoxide dismutase and nitric oxide synthesis inhibitors in neuronal injury resulting from activation of glutamate receptors. The major goal of this overview is

0-8493-8094-4/95/$0.00+$.50

to describe some of the oxidative chemistry of peroxynitrite focusing in the reactions that may be useful for its detection.

II. NITRIC OXIDE- AND SUPEROXIDE-MEDIATED TISSUE INJURY

Figure 1 is an outline of possible processes that can lead to the simultaneous generation of nitric oxide and superoxide and promote oxidative stress in the CNS. Stimuli such as ischemia can induce release of glutamate, which activates glutamate receptors.[12] Reduced blood flow can also activate flow-sensing channels in endothelial cells. Activation of glutamate receptors as well as flow-sensing channels in endothelial cells can induce cell depolarization, which permits the influx of calcium.[13] The increase in intracellular calcium concentration triggers a cascade of events that lead to simultaneous production of nitric oxide, superoxide, and peroxynitrite. Kader et al.[14] and Malinski et al.[15] found transient increases in nitric oxide synthase (NOS) activity and release of nitric oxide production following focal cerebral ischemia. Calcium efflux can activate phospholipase A_2 (PLA_2) to increase the levels of arachidonic acid (AA). Depolarization of neurons and endothelium can potentially increase ATP consumption by the Na^+-K^+-ATPase. The augmented demand for ATP increases oxidative phosphorylation and as a consequence the production of superoxide from mitochondria. The reintroduction of oxygen during reperfusion can further increase the generation of nitric oxide by NOS activity as well as the generation of superoxide by the metabolism of AA by lipoxygenase (LPO) and cyclooxygenase (CO) pathways[1] and by other sources such as cytochrome-P450. Recent evidence has also revealed that microglia are activated early in the brain following ischemia and trauma.[16] Activated microglia have been shown to release both superoxide[17] and nitric oxide.[18] Although we and others have been able to show production of peroxynitrite by activated rat alveolar macrophages[19] and Kupffer cells,[20] generation of peroxynitrite by activated microglia has not been reported.

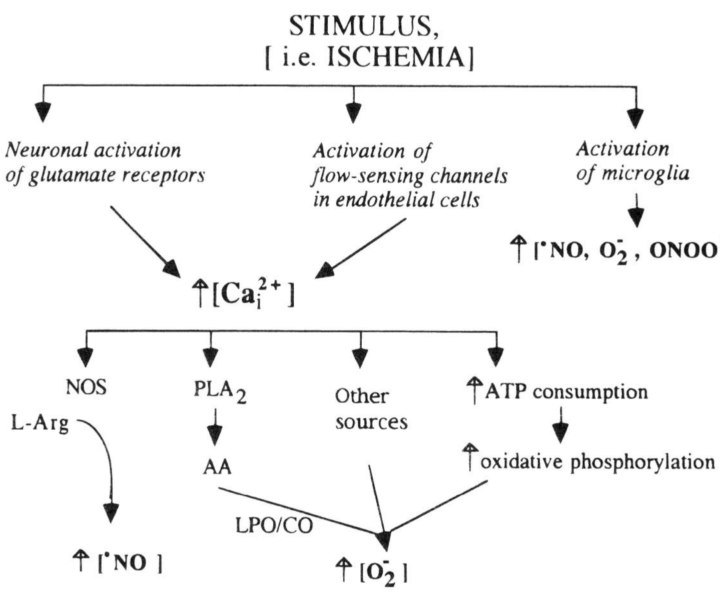

FIGURE 1. Schematic representation of potential interactions between cellular activation, release of nitric oxide, superoxide, and peroxynitrite in the CNS.

III. THE REACTION OF NITRIC OXIDE WITH SUPEROXIDE

Peroxynitrite is formed by the reaction between two free radicals, nitric oxide, and superoxide, as follows:[21]

$$^{\bullet}NO + O_2^- \longrightarrow ONOO^-$$

Peroxynitrite is not a free radical because the unpaired electrons of nitric oxide and superoxide are combined to form a new bond in peroxynitrite. The second-order rate constant of the reaction between nitric oxide and superoxide is $6.7 \times 10^9\ M^{-1}\ s^{-1}$,[22] approximately 30 times faster than the reaction of nitric oxide with oxyhemoglobin and three times faster than the reaction of superoxide with superoxide dismutase.

IV. OXIDATIVE CHEMISTRY OF PEROXYNITRITE

The pKa of peroxynitrite is 6.8 and at pH 7.4, approximately 25% will be protonated to form peroxynitrous acid (ONOOH), a reactive oxidant as follows:

$$ONOO^- + H^+ \longrightarrow ONOOH$$

Peroxynitrous acid isomerizes to nitrate with a pseudo-first-order rate of $0.6\ s^{-1}$ in phosphate buffer and 37°C.[23]

$$ONOOH \longrightarrow NO_3^-$$

Although peroxynitrous acid does not appear to physically separate to form hydroxyl radical and nitrogen dioxide ($^{\bullet}OH$ - - $^{\bullet}NO_2$),[23–26] it will oxidize organic molecules (i.e., DMSO to formaldehyde) and biological molecules to give the same oxidized products as hydroxyl radical.

$$ONOOH \longrightarrow \text{“}^{\bullet}OH\text{ - - }^{\bullet}NO_2\text{”} \xrightarrow{\text{DMSO}} H_2C{=}O + NO_2^-$$

The yield of the oxidized product can be competitively decreased by $^{\bullet}OH$ scavengers.[24] Because the oxidation products are similar to those derived solely from hydroxyl radical peroxynitrite cannot be assayed by detecting the $^{\bullet}OH$-like species reactivity. Furthermore, the hydroxyl radical-like reactivity may not be as important as other direct oxidative pathways such as the initiation of lipid peroxidation in the absence of metal catalysis, the oxidation of sulfhydryls (Sh), and the nitration of free and protein tyrosine (Tyr).

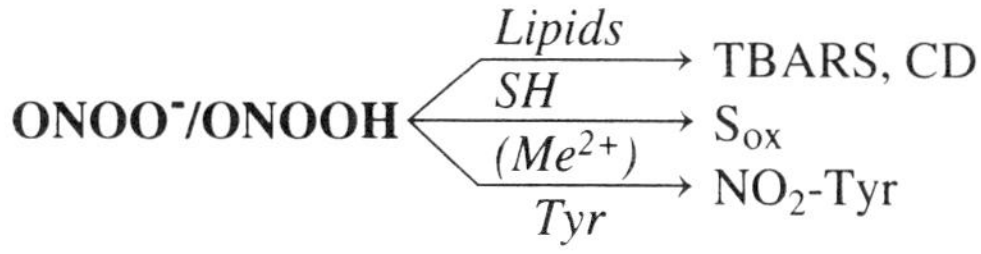

Peroxynitrite and/or peroxynitrous acid will initiate the oxidation of unsaturated fatty acids to TBARS and conjugated dienes (CD).[27] More importantly, the oxidation of lipids by peroxynitrite does not require a metal catalyst and may significantly contribute to tissue injury. Sulfhydryls are oxidized directly by peroxynitrite anion.[28] Cysteine is oxidized by peroxynitrite at a rate of $5900\ M^{-1}\ s^{-1}$, while the single cysteine residue of bovine serum albumin is oxidized at a rate $2600\ M^{-1}\ s^{-1}$. This pathway may significantly contribute to the *in vivo* toxicity of peroxynitrite by inactivating mitochondrial enzymes[29,30] and depleting cells of antioxidant defenses such as glutathione. Because superoxide and hydrogen peroxide oxidize sulfhydryls, although at lower rates compared to peroxynitrite, this reactive pathway cannot be used for peroxynitrite detection. However, cysteine can be used as a scavenger of peroxynitrite in competition assays.

V. CHEMILUMINESCENCE AND FLUORESCENCE DETECTION OF PEROXYNITRITE

Peroxynitrite will induce luminol chemiluminescence with a quantum yield of approximately 10^{-3} in bicarbonate buffer,[31] while superoxide or nitric oxide alone do not induce any significant luminol chemiluminescence. Luminol-enhanced chemiluminescence, indicative of peroxynitrite formation, has been detected from activated Kupffer cells[20] and from bovine aortic endothelial cells stimulated with either bradykin or a Ca^{2+} ionophore (A23187).[32] Recently, we found that luminol-enhanced chemiluminescence following depolarization of bovine aortic endothelial cells was inhibited by the nitric oxide inhibitor N^G-nitro-L-arginine methyl ester as well as by superoxide dismutase but not by hydroxyl radical scavengers DMSO and mannitol. Peroxynitrite also oxidized dihydrorhodamine 123 (DHR 123) to fluorescent rhodamine 123.[33] The yield of oxidation was approximately 50% of the initial concentration of peroxynitrite. Superoxide, hydrogen peroxide, or nitric oxide alone did not induce any significant oxidation of DHR 123. Hydroxyl radical, hypochlorous acid hydrogen peroxide plus horseradish peroxidase, and horseradish peroxidase alone also oxidize DHR 123. The oxidation of DHR 123 has been useful in documenting the generation of peroxynitrite from nitric oxide- and superoxide-generating systems such as 3-morpholinosydonimine *N*-ethylcarbamide (SIN-1). Since DHR 123 can be oxidized by other reactive species, indirect evidence for the existence of peroxynitrite can be obtained by studying the effects of superoxide dismutase and nitric oxide synthesis inhibitors. However, this approach may not be suitable because exogenously added superoxide dismutase may not be able to compete for superoxide and nitric oxide that are generated intracellularly or close to the cell membrane. Even in an *in vitro* assay superoxide dismutase was not able to completely inhibit the reaction of superoxide with nitric oxide. The oxidation of deoxyribose to malondialdehyde by 1 m*M* SIN-1 was inhibited only by 43% in the presence of 500 U/ml superoxide dismutase.[34]

Inhibition of nitric oxide synthesis also requires caution because NOS inhibitors have different IC_{50} values for the NOS isoforms[35–38] and the kinetics of inhibition can vary in different cell types containing different isoforms of NOS. When NOS activity inhibitors are administered in the whole animal or organ, changes in blood flow and hemodynamics of the tissue should be considered (i.e., vasoconstriction, increase adherence of neutrophils, platelet aggregation) that may alter the outcome of the experiment. Finally, the inhibition of nitric oxide synthesis should be confirmed by either measuring the nitric oxide concentration directly or by measuring both nitrate and nitrite, the stable decomposition products of nitric oxide and peroxynitrite.

VI. PEROXYNITRITE-MEDIATED NITRATION OF TYROSINE

The major product from the reaction of peroxynitrite with tyrosine is nitrotyrosine (Figure 2). Low-molecular weight metal catalysts such as Fe^{3+} EDTA or Cu^{2+} and metal-containing enzymes like Cu,Zn superoxide dismutase can catalyze the nitration of phenolic compounds and protein tyrosine residues to give nitrophenols and nitrotyrosine.[39,40] Nitrotyrosine was previously considered to be an amino acid modification that resulted from exposure to nitrosating or nitrating agents.[41–45] The role of specific tyrosine residues in the function of proteins has been commonly elucidated by nitrating tyrosine residues by tetranitromethane,

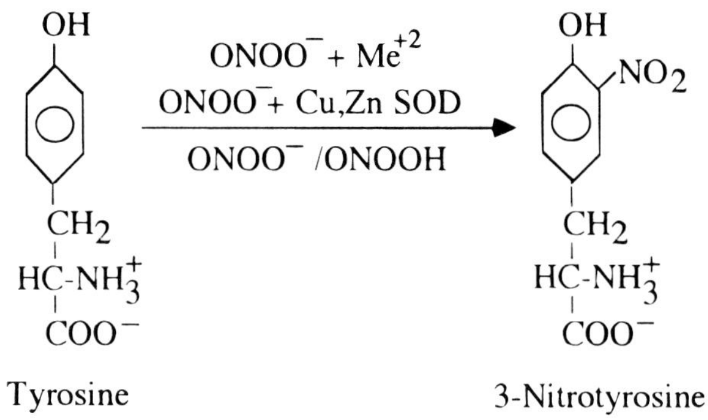

FIGURE 2. Nitration of the ortho position of tyrosine is a major product of peroxynitrite reaction with proteins.

a well-known nitrating agent.[43–45] Nitrotyrosine in proteins can be also formed by the reaction with nitrogen dioxide which can be generated from the reaction of nitric oxide with oxygen. However, the potential contribution of nitrogen dioxide in tyrosine nitration *in vivo* is greatly diminished by the slow rate of nitrogen dioxide formation, as compared to the rapid reactions of nitric oxide with superoxide and hemoglobin. Nitration of tyrosine by nitrogen dioxide proceeds via a free radical mechanism requiring two molecules of nitrogen dioxide to form nitrotyrosine. The first nitrogen dioxide molecule will oxidize tyrosine to form a tyrosyl radical intermediate. A second nitrogen dioxide then reacts with the tyrosyl radical intermediate to form nitrotyrosine. However, the tyrosyl radical intermediate may also react with another tyrosyl radical intermediate to form dityrosine[46] or react with other cellular components at physiologic pH. Therefore, the efficiency of nitration by nitrogen dioxide is much lower than peroxynitrite. For example, bolus addition of 1 m*M* peroxynitrite in protein solution results in the formation of nitrotyrosine, while neither nitrosotyrosine nor nitrotyrosine was formed by the addition of 1.5 m*M* nitric oxide under aerobic conditions.

VII. METHODS FOR NITROTYROSINE DETECTION

Nitration of tyrosine residues in purified nonheme proteins is easy to detect by UV-visible spectroscopy due to the characteristic yellow color. The pKa of nitrotyrosine is between 7.4 and 7.8 depending upon the protein environment. At alkaline pH the absorbance of nitrotyrosine is maximal at the 420- to 440-nm range ($\epsilon_{438\ nm} = 4200\ M^{-1}\ cm^{-1}$), while at acidic pH the absorbance maximum shifts to the 350- to 360-nm range ($\epsilon_{350\ nm} = 3400\ M^{-1}\ cm^{-1}$). There is no further modification of nitrotyrosine by additional peroxynitrite, hydroxyl radical, nitrogen dioxide, or reductants such as ascorbate, glutathione, and even borohydride. The simple spectrophotometric assay is not applicable for detection of nitrotyrosine in tissues and biological samples due to lack of sensitivity or the presence of other absorbing species such as hemeproteins. However, nitrotyrosine can be specifically detected by the following methods:

Gas chromatography-with thermal analyzer and by gas chromatography-mass spectroscopy—Nitrotyrosine can be detected by gas chromatography apparatus equipped with a thermal analyzer detector after derivatization with *N*-methyl-*N*-(*tert*-butyldimethylsilyl) trifluoroacetamide. To detect the major metabolite of nitrotyrosine in urine 3-nitro-*p*-hydroxyphenylacetic acid by gas chromatography and mass spectroscopy *tert*-butyldimethylsilyl derivatization was utilized.[47] Using these methods Oshima et al. found nitrotyrosine and 3-nitro-*p*-hydroxyphenylacetic acid in human urine and plasma.[47]

HPLC analysis of phenolic compounds—The nitration product(s) of phenolics such as *p*-hydroxyphenylacetic acid can be readily analyzed by HPLC using a reverse phase C18 column and a mobile phase consisting of an acetonitrile gradient at acidic pH.[19] Commercially available authentic nitrated compounds for some of these phenolics can be used to identify and quantify the nitrated compounds, as well as serve as internal standards. We used the Cu,Zn superoxide dismutase catalyzed nitration of *p*-hydroxyphenylacetic acid to form 3-nitro-hydroxyphenylacetate to detect peroxynitrite formation by activated rat alveolar macrophages.[19] However, this assay is limited by the relative low efficiency of the superoxide dismutase catalysis; only 8% of synthetic peroxynitrite will form 3-nitro-hydroxyphenylacetate in the presence of catalytic concentrations of superoxide dismutase. Another limitation is that peroxynitrite must be formed or diffuse into the extracellular environment in order to be detected.

Amino acid analysis—Amino acid analysis has been used to detect nitrotyrosine in purified proteins after treatment with a nitrating agent.[41–44] Figure 3 shows the elution profile of a clarified homogenate from an ischemic lung. The tissue was homogenized in cold saline containing 0.01% butylated hydroxytoluene under nitrogen. Lung homogenates were clarified by a 15-min centrifugation at 12,000 *g*. Membrane-bound proteins can be solubilized by the use of detergents, but detergents must be efficiently removed by gel filtration, dialysis, or ultrafiltration prior to chromatography since they interfere with column separation. The samples were then hydrolyzed in 6 *N* HCl containing 1% phenol for 24 h at 110°C. The hydrolyzed samples were dried under vacuum, neutralized with triethylamine, redried, and derivatized at room temperature with phenylisothiocyanate (PTC). The derivitized amino acids were separated on a reverse-phase Waters PicoTag column (3.9 × 150 mm) and detected at 254 nm. The PTC-amino acids in the hydrolysates were identified and quantified by comparing their retention times and peak areas with a freshly derivitized standard mixture of amino acids containing authentic nitrotyrosine (Aldrich). Under these conditions authentic

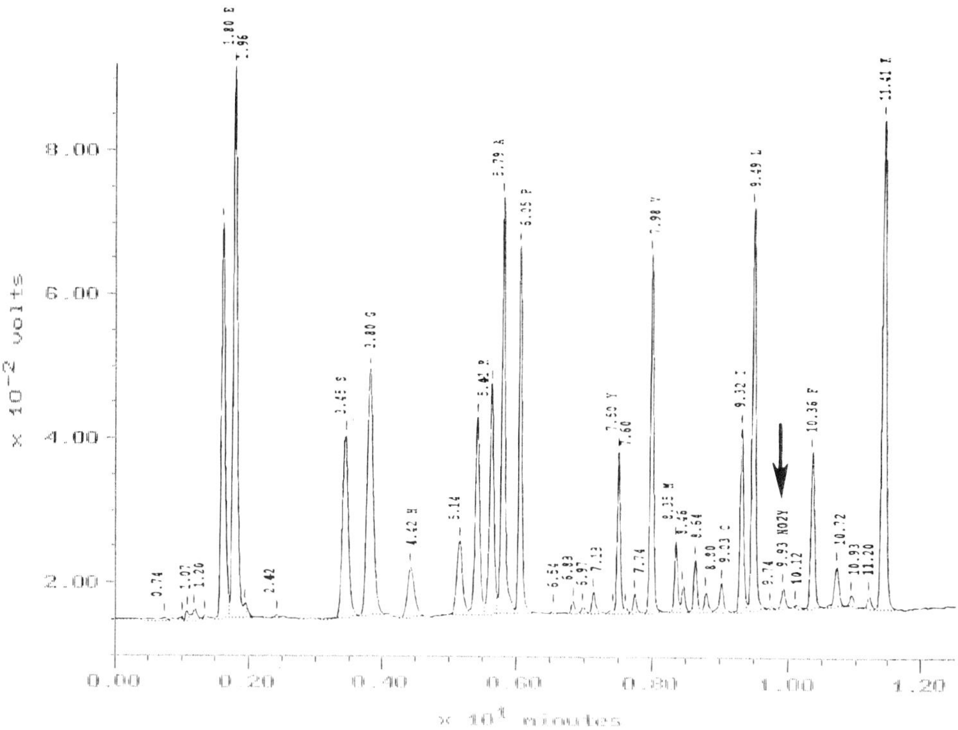

FIGURE 3. Amino acid analysis elution profile of clarified homogenate from an ischemic lung. The arrow indicates the nitrotyrosine (NO_2-Y) peak.

nitrotyrosine was eluted at 9.84 to 10.01 min and approximately 90% of nitrotyrosine standard was recovered. The retention time and recovery of nitrotyrosine was confirmed by using bovine serum albumin that was nitrated by chemically synthesized peroxynitrite. The same tissue sample can be used as a control by reducing the nitrotyrosine residues to aminotyrosine. This will result in the disappearance of the nitrotyrosine peak because the aminotyrosine formed is not stable under this hydrolysis procedure and will not be recovered quantitatively. Nitrotyrosine can be reduced to aminotyrosine with dithionite under alkaline conditions (pH = 9). The dithionite solution should be made fresh every time by placing the dithionite crystals in a plastic tube and degassing by nitrogen or argon. The desired volume of deaerated buffer is added with a syringe and purging is continued for few minutes after the dithionite is dissolved. The sample can be made alkaline by the addition of sodium hydroxide and dithionite as high as 1 *M* can then be added. The dithionite-reduced samples should be prepared fresh every time because aminotyrosine will slowly autooxidize back to nitrotyrosine. Heating and acid hydrolysis might result in diminished recovery of nitrotyrosine, especially if the biological sample contains small quantities of nitrotyrosine. To improve quantitation of small amounts of nitrotyrosine in complex biological samples such as tissue homogenates, an HPLC method has been developed recently.[48] The samples do not require derivatization; therefore, only aromatic amino acids are detected. The method utilizes the intrinsic absorbance of nitrotyrosine and aminotyrosine at 365 nm and tyrosine at 280 nm to quantify percent nitration. Because the 365-nm absorbance of nitrotyrosine-containing peptides is the same as free nitrotyrosine, protein samples can be hydrolyzed using proteases, thereby avoiding harsh acid hydrolysis, which may affect yields and does not allow detection of aminotyrosine.

Polyclonal and monoclonal antibodies specific for nitrotyrosine—Antibodies to nitrotyrosine can be raised using methods similar to that used for making phosphotyrosine antibodies. A significant advantage is the immunogenicity (ability to induce an adaptive immunity) of the nitrophenol entity in nitrotyrosine and the excellent ability to bind to the antibody. 2,4-Dinitrophenol has been commonly used to increase immunogenicity of antigens. Polyclonal antibodies against nitrotyrosine were made previously by using nitrotyrosine coupled to serum albumin.[49,50] We produced a polyclonal antibody in rabbits immunized with peroxynitrite-treated keyhole limpet hemocyanin. Using this antibody, nitrotyrosine was detected in human coronary artery atherosclerotic lessions.[51] For immunohistochemistry and other immunochemical techniques the following conditions should be tested: (a) Reactivity with the antibody; after positive results are obtained with an initial dilution of 1:100 to 1:1,000, the reactivity is repeated at higher dilution. (b) Reactivity with antigen-competed primary antibody. This is accomplished by diluting the primary antibody in 10 m*M* nitrotyrosine in 0.1 *M* PBS or 0.05 *M* Tris-HCl pH 7.4 for 10 min to 1 hr. The antigen-competed primary antibody is then used at the same dilution as the noncompeted primary. Nitrotyrosine at 10 m*M* can also be included in the blocking buffer used for dilution. (c) Reactivity with nonspecific IgG. (d) The samples are treated for approximately 1 min in a PBS pH 9 containing 1 *M* dithionite. Dithionite, a strong reducing agent, will reduce nitrotyrosine to aminotyrosine and should eliminate binding. Neither aminotyrosine nor phosphotyrosine at 10 m*M* should affect binding.

VIII. CONSEQUENCES OF PROTEIN NITRATION

The nitration of tyrosine residues may not result in an immediate toxicity but in a delayed multilevel cellular injury. For example, nitration of tyrosine residues will inactivate the mitochondrial Mn superoxide dismutase[37] diminishing the ability of cells to cope with additional oxidative stress. Nitration of tyrosine residues in surfactant proteins may be in part responsible for the inactivation of surfactant by peroxynitrite.[52] In addition, nitration of tyrosine residues can inactivate enzymes and receptors that depend on tyrosine residues for their activity.[43–45] More importantly, we recently found that nitrotyrosine markedly inhibits phosphorylation of tyrosine residues *in vitro* and in Type II alveolar cells and that nitrotyrosine formation is in part responsible for the ability of AP-1, a transcription factor, to bind to DNA. Therefore, nitrotyrosine formation may interfere with normal signal transduction pathways and further work is required to elucidate all the consequences of protein tyrosine nitration.

ACKNOWLEDGMENTS

This work was supported, in part, by grants from the Alabama Heart, Southeastern Pennsylvania Heart and Pennsylvania Lung Associations (HI) and from the National Institutes of Health HL 46407 and the Shannon Award (JSB). H. Ischiropoulos is a Parker B. Francis Fellow in Pulmonary Research. J.S. Beckman is an Established Investigator of the American Heart Association.

REFERENCES

1. Kontos, H. A. (1989) *Chem. Biol. Interact.* 72, 229–255.
2. Fahn, S. and Cohen, G. (1992) *Ann. Neurol.* 32, 804–812.
3. Coyle, J. T. and Puttfarcken, P. (1993) *Science* 262, 689–700.
4. Dawson, V., Dawson, T., London, E., Brent, D., and Snyder, S. (1991) *Proc. Natl. Acad. Sci. U.S.A.* 88, 6368–6371.
5. Cazevieille, C., Muller, A., Meynier, F., and Bonne, C. (1993) *Free Rad. Biol. Med.* 14, 389–395.
6. Lipton, S. A., Choi, Y-B., Pan, Z-H., Lei, S. Z., Chen, H-SV., Sucher, N. J., Loscalzo, J., Singel, D. J., and Stamler, J. S. (1993) *Nature (London)* 364, 626–632.
7. Nowicki, J. P., Duval, D., Poignet, H., and Scatton, B. (1991) *Eur. J. Pharmacol.* 204, 339–340.
8. Buisson, A., Plotkine, M., and Boulu, R. G. (1992) *Br. J. Pharmacol.* 106, 766–767.
9. Triffiletti, R. R. (1992) *Eur. J. Pharmacol.* 218, 197–198.
10. Beckman, J. S., Chen, J., Ischiropoulos, H., Zhu, L., and Conger, K. A. (1992) In: *Pharmacology of Cerebral Ischemia,* J. Krieglstein and H. Oberpichler, Eds., Wissenschaftliche Varlagsgesellschaft mbH, Stuttgart, 1–12.
11. Oury, T. D., Piantadosi, C. A., and Crapo, J. D. (1993) *J. Biol. Chem.* 268, 15394–15398.
12. Choi, D. W. and Rothman, S. M. (1990) *Annu. Rev. Neurosci.* 13, 171–182.
13. Choi, D. W. (1992) *J. Neurobiol.* 23, 1261–1276.
14. Kader, A., Frazzini, V. I., Solomon, R. A., and Trifiletti, R. R. (1993) *Stroke* 24, 1709–1716.
15. Malinski, T., Bailey, F., Zhang, Z. G., and Chopp, M. (1993) *J. Cereb. Blood Flow Metab.* 13, 355–358.
16. Gehrmann, J., Boonekoh, P., Miyazawa, T., Hossman, K. A., and Kreutzberg, G. W. (1992) *J. Cereb. Blood Flow Metab.* 12, 257–269.
17. Colton, A. C. and Gilbert, D. L. (1987) *FEBS Lett.* 223, 284–288.
18. Chao, C. C., Hu, S., Molitor, T. W., Shaskan, E. G., and Peterson, P. K. (1992) *J. Immunol.* 149, 2736–2741.
19. Ischiropoulos, H., Zhu, L., and Beckman, J. S. (1992) *Arch. Biochem. Biophys.* 298, 446–451.
20. Wang, J-F., Komarov, P., Sies, H., and DeGroot, H. (1991) *Biochem. J.* 279, 311–314.
21. Blough, N. V. and Zaferiou, O. C. (1985) *Inorg. Chem.* 24, 3502–3504.
22. Huie, R. E. and Padjama, S. (1993) *Free Rad. Res. Comm.* 18, 195–199.
23. Kopenol, W. H., Moreno, J. J., and Pryor, W. A., Ischiropoulos, H., and Beckman, J. S., (1992) *Chem. Res. Toxicol.* 5, 834–842.
24. Beckman, J. S., Beckman, T. W., Chen, J., Marshall, P. A., and Freeman, B. A. (1990) *Proc. Natl. Acad. Sci. U.S.A.* 87, 1620–1624.
25. Yang, G., Candy, T. E. G., Boaro, M., Wilkin, H. E., Jones, P., Nazhat, N. B., Saasalla-Nazhat, R. A., and Blake, D. R. (1992) *Free Rad. Biol. Med.* 12, 327–330.
26. Crow, J. P., Spruell, C., Chen, J., Gunn, C., Ischiropoulos, H., Tsai, J-H. M., Smith, C. D., Radi, R., Koppenol, W. H., and Beckman, J. S. (1994) *Free Rad. Biol. Med.* 16, 331–338.
27. Radi, R., Beckman, J. S., Bush, K. M., and Freeman, B. A. (1991) *Arch. Biochem. Biophys.* 288, 481–487.
28. Radi, R., Beckman, J. S., Bush, K. M., and Freeman, B. A. (1991) *J. Biol. Chem.* 266, 4244–4250.
29. Radi, R., Rodriguez, M., Castro, L., and Telleri, R. (1994) *Arch. Biochem. Biophys.* 308, 89–95.
30. Rubo, H., Denicola, A., and Radi, R. (1994) *Arch. Biochem. Biophys.* 308, 96–105.
31. Radi, R., Cosgrove, T. P., Beckman, J. S., and Freeman, B. A. (1993) *Biochem. J.* 290, 52–57.
32. Kooy N. W. and Royall J. A. (1993) *Arch. Biochem. Biophys.* 310, 352–359.

33. Kooy, N. W., Ischiropoulos, H., Beckman, J. S., and Royall, J. A. (1994) *Free Rad. Biol. Med.* 16, 149–156.
34. Hogg, N., Darley-Usmar, V. M., Wilson, M. T., and Moncada, S. (1992) *Biochem. J.* 281, 419–424.
35. Gross, S. S., Stuehr, D. J., Aisaka, K., Jaffe, E. A., Levi, R., and Griffith, O. W. (1990) *Biochem. Biophys. Res. Comm.* 170, 96–103.
36. Lambert, L. E., Whitten, J. P., Baron, B. M., Cheng, H. C., Doherty, N. S., and McDonald, I. A. (1991) *Life Sci.* 48, 69–75.
37. McCall, T. B., Feelisch, M., Palmer, R. M. J., and Moncada, S. (1991) *Br. J. Pharmacol.* 102, 234–238.
38. Olken, N. M., Rusche, K. M., Richards, M. K., and Marletta, M. A. (1991) *Biochem. Biophys. Res. Comm.* 177, 828–833.
39. Ischiropoulos, H., Zhu, L., Chen, J., Tsai, J-HM, Martin, J. C., Smith, C. D., and Beckman, J. S. (1992) *Arch. Biochem. Biophys.* 298, 431–437.
40. Beckman, J. S., Ischiropoulos, H., Zhu, L, van der Woerd, M., Smith, C. D., Harrison, J., Martin, J. C., and Tsai, J-HM. (1992) *Arch. Biochem. Biophys.* 298, 438–445.
41. Knowles, M. E., McWeeny, D. J., Couchman, L., and Thorogood, M. (1974) *Nature (London)* 247, 288–289.
42. Natake, M. and Ueda, M. (1986) *Nutr. Cancer.* 8, 41–45.
43. Janig, G. R., Rabe, K. H., Makower, A., and Ruckpaul, K. (1988) *Biophys. Biochim. Acta* 47, 565–579.
44. Deckers-Hebestreit, G., Schmid, R., Kiltz, H-H., and Altendorg, K. (1987) *Biochemistry* 26, 5486–5492.
45. Glazer, A. N., Delange, R. J., and Sigman, D. S. (1975) *Chemical Modification of Proteins,* Glazer, A. N., Ed., Elsevier North-Holland, Amsterdam, 95–99.
46. Prutz, W. A., Monig, H., Butler, J., and Land, E. J. (1985) *Arch. Biochem. Biophys.* 243, 125–134.
47. Ohshima, H., Friesen, M., Brouet, I., and Bartsch, H. (1990) *Fd. Chem. Toxic.* 28, 647–652.
48. Crow, J. P. and J. S. Beckman (1995) *Methods,* in press.
49. Helman, M. and Givol, P. (1971) *Biochem. J.* 125, 971–974.
50. Korner, M., Thiem, N. V., Cardinaud, R., and Lacombe, G. (1983) *Biochemistry* 22, 5843–5847.
51. Beckman, J. S., Ye, Y. Z., Anderson, P. G., Chen, J., Accavitti, M. A., Tarpey, M. M., and White, C. R. (1994) *Biol. Chem. Hoppe-Syler.* 375, 81–88.
52. Haddad, I. Y., Ischiropoulos, H., Holm, B. A., Beckman, J. S., Baker, J. R., Matalon, S. (1993) *Am. J. Physiol.* 265, L555–564.

IX. Histological Techniques

Chapter 37

HISTOPATHOLOGY AFTER TRAUMATIC BRAIN INJURY

Richard L. Sutton

TABLE OF CONTENTS

I. INTRODUCTION

Assessment of histopathology is a necessity for experimental studies of injury to the central nervous system (CNS) and the study of potentially neuroprotective pharmacotherapeutic agents. While the pathobiology of CNS damage is immensely varied and complicated, many sensitive and specific histologic methods developed by researchers in experimental neuroanatomy are available for assessing alterations and/or damage to neural components, the occurrence of neuronal cell death, and reactions of nonneuronal elements within the CNS.

A comprehensive presentation of the available techniques for histopathological studies is not practical in this space. The current chapter is intended to serve as a brief introduction to some histological procedures that can be utilized for large-scale screening of some aspects of neuronal and glial alterations following traumatic brain injury (TBI). The focus in the following discussion is on the use of relatively inexpensive staining procedures that have been useful in characterizing the time course and distribution of histopathology following unilateral cortical contusion in the rat. The procedures discussed can be used to: (1) identify

0-8493-8094-4/95/$0.00+$.50

regions with moribund neurons early after TBI; and (2) characterize the astroglial response to TBI.

II. TISSUE PREPARATION

Prior to processing of tissue for histological staining or immunohistochemistry, the tissue must be prepared to ensure adequate preservation of cellular membranes. General principles of perfusion and tissue preparation in the rat are discussed below. Some specific recipes used in this author's laboratory for preparing buffers, perfusates, and cryoprotectant solutions are detailed at the end of this section.

In preparation for perfusion the animal should be anesthetized (e.g., with sodium pentobarbital, 75 to 100 mg/kg, IP) and restrained in a supine position. After opening the thorax to expose the heart, inserting a blunt needle (e.g., 16- to 18-gauge) through the left ventricle to cannulate the aorta and cutting the right atrium, the blood is flushed from the vasculature by infusing a physiological saline solution. Subsequently, aldehyde fixatives are infused to harden the brain tissue and thus preserve tissue morphology. The duration of vascular rinsing and the type and concentration of aldehydes used can be varied, depending on the particular histological staining procedures that will be utilized. Vascular rinsing should generally be performed rapidly (e.g., 30 s to 2 min. duration) with small volumes (ranging from 10 to 250 ml) of saline solution. The choice of the type and amount (ranging from 100 to 500 ml) of fixative to be used during perfusion, an appropriate buffering agent, and whether or not post fixation is appropriate, should be guided by published methods relevant to the particular application being used or should be experimentally determined by each laboratory.

Following fixation and removal of the brain from the calvarium, tissue must be prepared for sectioning. If the tissue is adequately hardened by the fixative, a Vibratome may be used to obtain sections (20 to 100 μm thickness). If frozen sections are to be obtained (4 to 50 μm thickness), using a freezing microtome or cryostat, tissue should be cryoprotected. This can be performed *in situ* by perfusion of a sucrose solution immediately following fixation, by postmortem saturation of the tissue in sucrose solutions (e.g., placing tissue in sucrose solutions ranging from 10 to 30% until the tissue sinks), or a combination of the two methods. Frozen tissue sections may either be mounted onto glass slides and then processed for staining, or cut and placed into buffered solutions, stained in the free-floating condition, and subsequently mounted onto glass slides for microscopic examination. The microscope slides should be coated with gelatin to ensure that tissue adheres to the slide throughout the staining process.

III. PREPARATION OF BUFFERS AND PERFUSION SOLUTIONS

The chemicals and reagents needed to prepare buffers, solutions for perfusions, cyroprotection, and staining of brain tissue can be purchased from any number of commercial suppliers (e.g., Baxter or Fisher Scientific, Sigma, etc.). Standard laboratory glassware, staining boats, slide carriers, etc. can be purchased from similar sources.

A. SODIUM PHOSPHATE SOLUTIONS

In our laboratory stock solutions of 0.2 *M* sodium phosphate solutions are prepared, filtered, and stored at 4°C for future use. Solution 1 (sodium phosphate dibasic, anhydrous, FW = 142.0) is prepared by dissolving 28.4 g of Na_2HPO_4 in 1 l of distilled water. Solution 2 (sodium phosphate monobasic, anhydrous, FW = 120.0) consists of 24 g of NaH_2PO_4 per liter of distilled water.

A working solution of 0.1 *M* phosphate buffer (PB) is made by mixing these dibasic and monobasic stock solutions. While stirring 1 l of solution 1 and monitoring the pH, solution 2 is slowly added (approximately 200 ml) until the desired pH (e.g., 7.4) is reached. Distilled water is then added (1:1 ratio) to attain a 0.1 *M* PB solution and the pH is rechecked (minor adjustments in pH can be made with concentrated NaOH or HCl).

B. PHOSPHATE BUFFERED SALINE (PBS)

Dissolve 9 g of NaCl in 1 l of 0.1 *M* PB to make a 0.1 *M* PBS solution (adjust pH to 7.4). This solution is generally used for flushing of the vasculature during perfusions. In some cases sections taken with a sliding freezing microtome are cut into a 0.01 *M* PBS solution. This latter solution is made by mixing 9 g of NaCl in 900 ml of distilled water and 100 ml of 0.1 *M* PB (adjust pH to 7.4).

C. PHOSPHATE BUFFERED FORMALIN

Concentrated formaldehyde can be purchased from any number of commercial scientific suppliers. These are diluted with 0.1 *M* PB to create a 10% formalin solution (adjust pH to 7.4 to 7.6).

D. PARAFORMALDEHYDE (3%)

Under a fume hood, stir and heat (do not exceed 60°C) 200 ml of 0.1 *M* PB and add 30 g of paraformaldehyde; stir vigorously for 15 to 20 min. If the solution is cloudy in appearance, it can be cleared by adding a few drops of concentrated NaOH. Filter this solution through a paper filter and add 800 ml of 0.1 *M* PB; adjust the pH to 7.6 with concentrated NaOH or HCl.

E. BUFFERED SUCROSE SOLUTIONS

Prior to cutting frozen sections, brain tissue is moved through increasing concentrations (10, 20, and 30%) of sucrose dissolved in either 0.1 *M* PB or 10% buffered formalin. The tissue remains in each successive solution until it sinks to the bottom of the container (18 to 24 h per solution).

IV. PREPARATION OF SLIDES AND GLASSWARE

Microscope slides are loaded into glass staining racks and submerged in 5% nitric acid for at least 24 h. Following this acid wash, slides are placed in a plastic container and rinsed under running tap water for a minimum of 15 min followed by at least three rinses in distilled water. Slides are then dipped in a warm gelatin solution consisting of 10 g of gelatin (275 bloom), 0.25 g of chromium potassium sulphate, and 0.2 g of sodium azide in 1 l of distilled water (stored at 4°C). After excessive gelatin has dripped off the slides they are placed in a 57 to 60°C oven overnight to dry.

Glass staining boats, glass staining racks (carriers), and beakers/storage jars used for processing of tissue are generally washed well with warm water and laboratory detergent, rinsed with tap water, and then further cleaned by overnight submersion in 5% nitric acid. Glassware is rinsed well with tap water followed by distilled water and then dried prior to use. For some staining procedures (e.g., silver staining), the same storage, mixing, and staining glassware should be used for each subsequent batch of prepared solutions.

V. DETECTION OF ACUTE NEURONAL DAMAGE AFTER TBI

One of the simplest and least expensive methods for assessing neuronal loss is to examine Nissl-stained tissue sections utilizing light microscopy. This technique is well suited for histological assessments where TBI has led to somewhat large-scale neuronal damage days to weeks after injury. If damage is minor or diffusely located throughout the cerebrum, however, light microscopy to identify regions with neuronal loss in Nissl-stained tissue can be extremely time-consuming and some slight damage may not be detected. In addition, during acute phases after injury to the CNS, the lack of Nissl staining in neurons does not necessarily indicate their death and/or removal. Therefore, in the initial stages of work with an injury model it is desirable to use stains that can help identify or locate brain regions that exhibit moribund neurons.

In the author's experience, two staining methods have proven particularly useful in markering damaged neuronal elements in the acute phase after TBI in the rat. These are the acid fuchsin and silver staining methods. Both of these methods can be performed on brain tissue after fixation with a 10% formalin solution and acid fuchsin staining can also be performed on tissue fixed with 3% paraformaldehyde.

A. ACID FUCHSIN

The presence of acidophilic neurons in acid fuchsin-stained tissue has been used to identify regions with dead or dying neurons after brain injury produced by ischemia (Dienel, 1984), hypoglycemia (Auer et al., 1985), kianic acid (Lees, 1989), and fluid percussion (Cortez et al., 1989). In initial work to characterize neuronal damage that follows differing severity of unilateral cortical contusion induced with a pneumatically controlled impact device, acid fuchsin staining has also proved useful (Sutton et al., 1993; Sutton et al., 1994; Sutton et al., submitted). In these latter studies rats were sacrificed from 6 h to 30 d after contusion of the right sensorimotor cortex. Perfusions were carried out using 250 ml of PBS (pH = 7.4) followed by 500 ml of 3% paraformaldehyde (pH = 7.6; see recipes above) and brains were sectioned in the frontal plane (40 μm thickness) on a freezing microtome after sinking in a 30% sucrose solution. Acid fuchsin staining (with and without Nissl counterstaining; see Table 1) was found to be an excellent marker for identifying regions with moribund neurons at early stages (e.g., 6 and 24 h or 8 d) after unilateral TBI. In the microscopic analyses of the histological materials stained with acid fuchsin, low power (e.g., 4× to 10×) objectives are adequate for initial rapid scanning of many (e.g., 20 to 30) sections from each brain. Higher power (20× to 40×) objectives are only infrequently needed to verify the presence of sparse acidophilic neurons in some brain regions or when trying to discern if a neuron is acid fuchsin-positive but Nissl-negative (which is thought to more reliably indicate morbidity than is acid fuchsin-positive staining only).

In the ipsilateral contused cortex, acid fuchsin-positive neurons were most numerous and widespread at 6 and 24 h after TBI. Loss of Nissl staining was also apparent in contused cortical tissue at these acute time points. By 8 d postinjury, matching regions of cortex were essentially necrotic and only a few acidophilic neurons were observed in cortical tissue bordering regions of necrotic cavitation. In the dentate/hippocampal regions, peak numbers of acid fuchsin-positive neurons were observed at 6 and 24 h after TBI, although numerous hilar neurons and CA_3 pyramidal cells were also acidophilic in rats sacrificed at 8 d. As shown in Plate 3a* and 3b,* acid fuchsin-positive profiles were most frequently observed in the ipsilateral hilar region (polymorphic neurons) and in the CA_3 sector (pyramidal neurons) after injury velocities of either 2.25 or 3.22 m/s. Less severe and more variable labeling of pyramidal neurons in the CA_1 region occurred after either injury severity, and the 3.22 m/s severity of TBI resulted in far greater numbers of acid fuchsin-positive granule cells in the dentate gyrus ipsilateral to injury than did the 2.25 m/s impact. Variable and slight loss of

*See the frontispiece.

TABLE 1
The Stages and Recipes for Solutions Used for Staining Histological Sections with Thionin, Acid Fuchsin, or a Combination of Thionin and Acid Fuchsin

Process	Steps	Solution	Time
Delipidization and hydration	1	95% ethanol	5 min
	2	Chloroform	5 min
	3	100% ethanol	2 min
	4	Xylene	5 min
	5	95% ethanol	2 min
	6	70% ethanol	2 min
	7	Distilled water	2 min
Stain	8	Thionin	2–5 min
Rinsing	9,10,11	Distilled water	5 dips
Differentiation	12	Acid-formalin	2–3 min, agitate
Rinsing	13–14	Distilled water	5 dips
Stain	15	Acid fuchsin	1 min
Rinsing	16,17	Distilled water	5 dips
Dehydration and differentiation	18	70% ethanol	2 min
	19	95% ethanol	2 min
	20	100% ethanol	2 min
	21	Butanol	2 min
Clearing	22–24	Xylene	2 min

Coverslip using Permount.
For thionin staining only, use steps 1–14 and 18–24.
For acid fuchsin staining only, use steps 1–7 and 15–24.

Thionin
Stock solution:
Dissolve 10 g of thionin in 100 ml of distilled water (stir using gentle heat); filter.
Buffer solution:
Dissolve 7 gm of sodium acetate and 2 ml of glacial acetic acid in 1 l of distilled water; pH to 4.8.
Working solution:
Mix 27 ml of filtered stock with 273 ml of buffer.
Acid-formalin
Mix 300 ml of distilled water, 3 ml of glacial acetic acid and 3 ml of 40% formaldehyde.
Acid fuchsin
Mix 300 ml of distilled water with 1 ml of glacial acetic acid and 0.3 g of acid fuchsin; pH to 3.0 to 3.2 using a dilute solution of NaOH or HCl. Use no longer than one week.

Nissl staining within these dentate/hippocampal regions was observed by 6 or 24 h after TBI, but was readily apparent by 8 d or 30 d postinjury (see Figure 1a to c).

In the ipsilateral thalamic nuclei and the caudate-putamen, acidophilic cells were first observed at 24 h after TBI. However, in contrast to cortex and hippocampal regions, acidophilic neurons in these subcortical regions were more widespread and numerous at 8 d than at more acute time points after TBI. By 8 d after cortical contusion a reduction of Nissl-positive neurons and substantial tissue atrophy was apparent in the ipsilateral dorsal caudate-putamen and in the ipsilateral dorsal/anterior nuclei and ventral/posterior nuclei of the thalamus. Neuronal loss and atrophy in both regions had increased by 30 d after TBI (see Figure 1d). In all animals with unilateral cortical contusion injury examined to date, acidophilic neurons have not been observed in contralateral structures.

We are currently utilizing the acid fuchsin stain to examine regional neuronal damage from 2 h to 7 d following delivery of a fluid percussion injury (2.2 to 2.6 atm) to the right sensorimotor cortex. After fluid percussion injury, acidophilic polymorphic neurons are

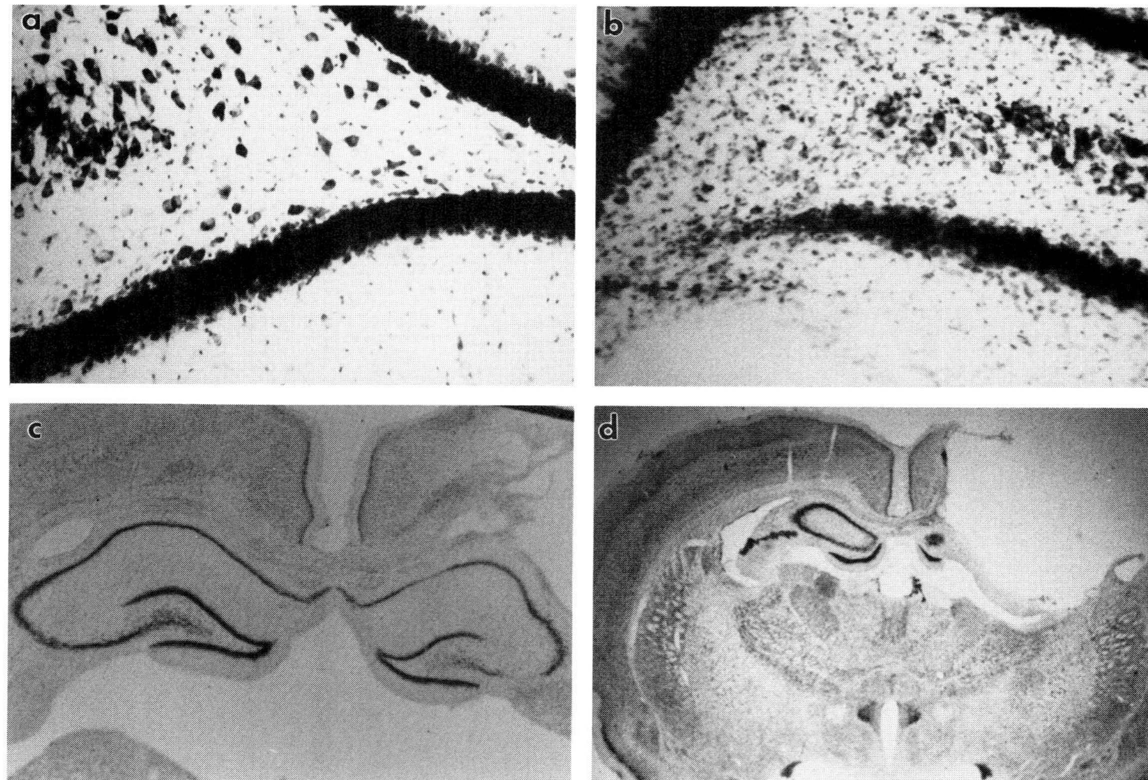

FIGURE 1. Thionin-stained brain tissue of rats sacrificed 30 d after contusion (3.22 m/s impact velocity; 2mm cortical compression) of the right sensorimotor cortex. Substantial loss of thionin-positive neurons occurs in regions where acidophilic and/or argyrophilic neurons were noted acutely (6 h to 8 d) after TBI (see Plate 3). While dentate granule cells, polymorphic hilar neurons, and CA_3 pyramidal neurons contralateral to injury appear intact (a), substantial numbers of these have degenerated in the dentate/hippocampal regions ipsilateral to contusion (b). These rats with TBI have substantial atrophy of the hippocampus ipsilateral to injury (c) and also show extensive cortical cavitation necrosis and shrinkage of the hemithalamus ipsilateral to injury (d).

frequently observed bilaterally (ipsilateral > contralateral) in the dentate hilar regions and more rarely in the CA_3 sectors of the hippocampus within the first week after injury. While this finding of bilateral cell damage after fluid percussion injury contrasts with the results found with cortical contusion models of TBI (Krobert et al., 1992; Sutton et al., 1993; Sutton et al., 1994; Sutton et al., submitted), these data are in agreement with those of other investigators who report bilateral loss of dentate hilar neurons following fluid percussion injury (Lowenstein et al., 1992).

B. SILVER STAINING

Experimental neuroanatomists have developed a number of histologic silver staining methods that can be used to label dead and dying neurons and degenerative neuronal elements (Fink and Heimer, 1967; Carlsen and de Olmos, 1981; Van den Pol and Gallyas, 1990). As with the acid fuchsin staining method, silver impregnation (argyrophilia) of neuronal cell bodies and dendrites can be used to identify CNS regions with neuronal damage at acute time points after brain injury (see Iizuka et al., 1989; Lowenstein et al., 1992).

In a preliminary study (Krobert et al., 1992), we examined the time course of neuronal damage in the dentate/hippocampus of rats with unilateral cortical contusion induced via a weight-drop method. In this study thionin, acid fuchsin, and a modified Fink-Heimer silver staining procedure were used to examine cell damage at 2 or 6 h and at 1, 2, 3, and 4 d following injury. Similar to findings after contusion induced with the pneumatically controlled contusion device (described earlier), acid fuchsin-positive neurons were primarily observed in the ipsilateral CA_3 and hilar regions after contusion induced using the weight-drop model of TBI. Peak numbers of acidophilic cells within the dentate/hippocampus were observed 2 to 3 d after injury and loss of Nissl staining in these same regions also became clearly evident from 2 to 3 d postinjury. Argyrophilic neurons in the CA_3 and hilar sectors of the hippocampus ipsilateral to contusion were first observed at 2 h postinjury and increased in numbers and distribution by 6 h. Peak numbers of argyrophilic neurons appeared at 24 h postinjury and the argyrophilic neurons decreased in numbers and appeared atrophic from 2 to 4 d after injury. The contralateral hippocampus of contused rats had no discernible cell damage or loss at any of the time points analyzed, and argyrophilic or acidophilic neurons were never observed in sham operations.

More recently, our lab has begun studies to compare and contrast neuropathology that develops from 2 h to 7 d following TBI induced via the fluid percussion vs. the cortical contusion models. Perfusions are performed using a small volume (40 to 50 ml) of PBS ($pH = 7.4$) followed by a 10% buffered formalin solution (100 ml; $pH = 7.4$) and the brains are post fixed *in situ* for one week prior to removing them from the calvarium. After sinking in a 20% sucrose/10% buffered formalin solution, frontal sections (10 to 40 μm thickness) are cut using a cryostat, and thaw-mounted onto gelatin-coated slides. In addition to other stains, the modified Fink-Heimer silver staining procedure (see Table 2) is being utilized to examine regional damage following unilateral injury to the cortex. Unilateral contusion results in numerous argyrophilic pyramidal cells within the ipsilateral, but not contralateral, hilar, and CA_3 (see Plate 3c) regions 6 h after TBI. At this time after unilateral fluid percussion injury, few argyrophilic neurons appear in the ipsilateral CA_3 region, but neurons are labeled bilaterally in the dentate hilar regions (ipsilateral > contralateral). Although data for other time points and for additional regions are only preliminary, it appears that peak numbers of argyrophilic neurons in hippocampal regions occur from 6 to 24 h postinjury in both injury models while contusion results in more widespread argyrophilia in subcortical neurons than does the fluid percussion injury.

TABLE 2
The Stages and Recipes for Solutions Used for Silver Staining (Modified Fink-Heimer Procedure) of Brain Tissue Sections Mounted on Glass Slides

Steps	Solutions and incubation durations
1–3	3 × 5-min rinses in distilled water.
4	Agitate tissue during 11–12 min incubation in 0.05% potassium permanganate (0.15 g in 300 ml of distilled water). Optimal procedure is to use this solution within 10 min after preparation.
5	Rinse in distilled water.
6	Into bleach solution (1:1 mixture of 1% oxalic acid and 1% hydroquinone) until sections are completely white (about 1 min). Repeat if needed in a second batch of bleach solution. The solution is made the day of staining.
7–9	3 × 2-min rinses in distilled water.
10	Agitate tissue during a 1-h incubation in solution A (0.3 g of uranyl nitrate and 1.5 g silver nitrate in 300 ml of distilled water). This solution can be prepared days prior to use if kept in a tightly sealed container and protected from light.
11	Agitate tissue during a 30-min incubation in solution B (0.51 g of uranyl nitrate and 4.88 g of silver nitrate in 300 ml of distilled water). This solution can be prepared days prior to use if kept in a tightly sealed container and protected from light.
12–14	3 × 2 min-rinses in distilled water.
15	Into ammoniacal silver for 2 min (agitate; less time may be optimal). Prepare a fresh batch of "base mix" (10 ml of concentrated ammonium hydroxide and 15 ml of 2.5% NaOH) and a 2.5% solution of silver nitrate. Prepare the ammoniacal silver solution by adding 25 ml of the base mix to 250 ml of 2.5% silver nitrate; use within 30 min. The solution should be clear—if not, remake and/or order new ammonium hydroxide.
16	Into the reducing solution (628 ml of distilled water, 25 ml of 95% ethanol, 5.8 ml of 10% formalin and 6.7 ml of 1% citric acid) until sections are brown (agitate). This solution can be made in advance if stored in a dark, sealed container.
17	Into second batch of reducing solution for 1 min (agitate).
18	Rinse in distilled water for 1 min.
19	Into 1.0% sodium thiosulphate (3 g in 300 ml of distilled water) for 1 min (agitate; longer time may be needed).
20–22	3 × 5-min rinses in distilled water.

Tissue can be dried overnight, dehydrated in ethanols (95%, 100%), cleared in three changes of xylenes and coverslipped.

VI. ASSESSMENT OF ASTROGLIOSIS FOLLOWING TBI

Alterations in glial cell populations within the CNS can be used as a potential, indirect marker of damage to neuronal elements. The increase of glial cells (reactive gliosis or astrogliosis) and/or an increase in the synthesis of glial fibrillary acidic protein (GFAP) are frequently used as indicators of damage to both neurons and glia after injury caused by neurotoxins, disease, aging, or physical damage to the CNS (Eng, 1985; Eng, 1988; Norton et al., 1992).

In a recent experiment, we (Sutton, Lescaudron, and Stein, submitted for publication) examined the astrogliosis response to TBI in rats sacrificed at 6 or 24 h and at 8 or 30 d after unilateral cortical contusion. Animals were perfused with 250 ml of PBS (pH = 7.4) followed by 500 ml of 3% paraformaldehyde (pH = 7.6). After sinking in a 30% sucrose solution brains were sectioned in the frontal plane (40 μm thickness) on a freezing microtome and tissue sections were collected in 0.01 *M* PBS (pH = 7.4). Peroxidase-antiperoxidase (PAP) immunohistochemical staining procedures (see Sternberger et al., 1970) were used to stain sections for GFAP. Equipment, materials, and recipes for incubation solutions used in the staining procedures are detailed at the end of this section.

Qualitative analyses of the distribution of GFAP-positive cells throughout the cerebrum of rats sacrificed at each time point were performed using light microscopy (4× to 10×

objectives offer adequate magnification) and noting changes in regional GFAP expression, intensity of stain, and morphological changes in GFAP-positive cells of TBI rats compared to sham-operated animals. By 6 h after TBI increased members of GFAP-positive astrocytes were readily apparent in the cortex and corpus callosum under the contusion site, and reactive astrocytes were also noted in the ipsilateral dorsal hippocampus. The GFAP-positive cells in these same regions were more darkly stained, slightly hypertrophied, and more widely distributed in anterior-posterior directions by 24 h after contusion. At the 24-h time point, GFAP-positive profiles were also noted in the ipsilateral striatum and thalamic nuclei as well as in the contralateral neocortex and dorsal hippocampus. Maximal GFAP expression (as judged by distribution, staining intensity, and cell hypertrophy) was seen at 8 d after contusion. Intensely GFAP-positive cells were present bilaterally throughout the neocortex (from +5.2 to −8.0 mm relative to bregma), although larger numbers were present in the ipsilateral cortex (see Figure 1d). Increased numbers of hypertrophied astrocytes were present in virtually all structures ipsilateral to TBI, most notably in the hippocampus, basal ganglia, thalamus, amygdala, and substantia nigra. A substantial increase in GFAP-positive cells was also observed within the contralateral dorsal hippocampus, caudate nucleus, and globus pallidus of TBI rats 8 d postinjury. As illustrated in Plate 3, the hypertrophy of the GFAP-positive cells from 24 h to 8 d after TBI was quite remarkable. One month after TBI the GFAP staining intensity, size of GFAP-positive cells, and the numbers of GFAP-positive cells appeared reduced compared to that observed 8 d postinjury, although the regional distribution of astrogliosis at this time roughly approximated that seen at 8 d.

While qualitative analyses indicated that the numbers of reactive astrocytes in many brain regions increased from 6 h to 8 d and then decreased by 30 d after TBI, it is possible that increased intensity of GFAP staining and astrocyte hypertrophy in these regions contributed to this conclusion. To verify this conclusion, the numbers of GFAP-positive cells were counted (400× magnification, 240 × 240 μm microscopic field) in various cortical and subcortical structures. Figure 2 illustrates the results of cell counts in the CA_3 sector of the dorsal hippocampus (three nonsequential sections, two microscopic fields per tissue section) in four sham-operated rats and four-rats sacrificed at each of four time points after unilateral TBI (3.22 m/s impact velocity). As for most regions examined (Sutton, Lescaudron, and

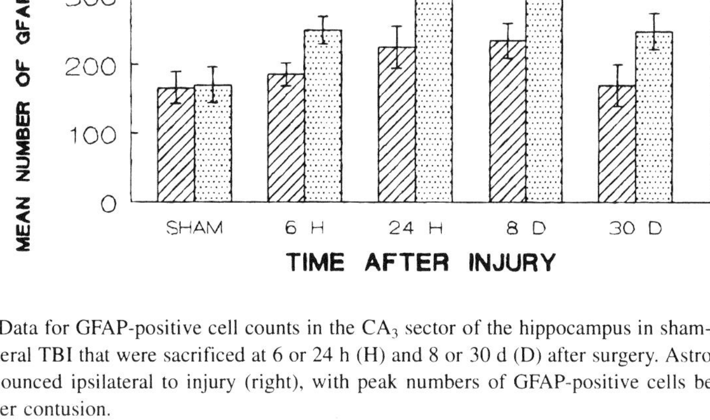

FIGURE 2. Data for GFAP-positive cell counts in the CA_3 sector of the hippocampus in sham-operated rats and rats with unilateral TBI that were sacrificed at 6 or 24 h (H) and 8 or 30 d (D) after surgery. Astrocyte proliferation was most pronounced ipsilateral to injury (right), with peak numbers of GFAP-positive cells being present from 24 h to 8 d after contusion.

Stein, submitted for publication), these data indicate that the numbers of GFAP-positive cells in the ipsilateral CA_3 region increased substantially following contusion of the right sensorimotor cortex. Peak numbers of astrocytes in ipsilateral structures are found from 1 to 8 d after TBI, with fewer numbers being observed contralateral to injury. These findings indicate that increased GFAP expression after unilateral cortical contusion reflects both hyperplasia (see Cavanagh, 1970) as well as hypertrophy of astrocytes.

VII. PROCEDURES FOR GFAP STAINING

In contrast to the thionin, acid fuchsin, and silver staining procedures where tissue is mounted onto slides prior to staining, we perform GFAP staining on free-floating sections. To reduce incubation volumes (and thus expenses) and ensure that tissue is gently agitated to suspend and separate sections during the various incubation stages, some specialized equipment is needed. Rotary shakers with variable speeds can be purchased from commercial scientific supply companies for use during the incubation/agitation steps of the staining procedures. Small carriers (2 to 3 cm diameter) for transferring tissue sections through incubation mediums can be made by cutting plastic cylinders (e.g., pill bottles) in half and securing nylon mesh over one end with epoxy. Incubation vessels (e.g., glass crystallizing dishes) that allow for a relatively tight fit of these tissue carriers should be used in order to reduce volumes of the diluted antibodies. The number of tissue sections per carrier, volume of incubation solutions, and speed settings of a shaker that will allow adequate suspension and separation of tissue sections during agitation should be empirically determined.

A. SOLUTIONS AND INCUBATION FORMULAS

Solutions and incubation mediums should be made just prior to use. The volumes and formulas given below have proven adequate for processing up to 100 tissue sections simultaneously.

0.4% Triton—Weigh 0.2 g of Triton X-100 into a small beaker. Add 50 ml of 0.01 *M* PBS and stir vigorously.

Diluent with 3% swine serum—Weigh 0.05 g of Triton X-100 into a beaker. Add 0.05 g of glycine, 0.5 g of bovine serum albumin, and 50 ml of 0.01 *M* PBS; stir well. To this diluent solution, add 1.5 ml of normal swine serum (Dako Corporation).

Primary antibody—Rabbit anti-GFAP (20 μl) is diluted (1:1000) in 20 ml of diluent with 3% swine serum. This primary antibody can be purchased from Chemicon International (for GFAP data presented in this manuscript, the primary antibody was generously donated by Dr. L. F. Eng of Stanford University).

Secondary antibody—Swine anti-rabbit immunoglobulins (200 μl; Dako Corporation) is diluted (1:100) in 20 ml of diluent with 3% swine serum.

Tris buffer—Dissolve 13.2 g of Trizma HCl (Sigma) and 2.78 gm of Trizma Base (Sigma) in 1 l of distilled water. Final pH should be 7.6.

PAP solution—Rabbit PAP (20 μl; Dako Corporation) is diluted (1:1000) in 20 ml of diluent with 3% swine serum.

Chromagen solution—Dissolve 50 mg of 3,3′-diaminobenzidine (Sigma) in 100 ml of Tris buffer. Filter this solution and add 30 μl of 30% hydrogen peroxide.

B. STAGES OF TISSUE PROCESSING

During all incubation and rinsing stages, tissue is gently agitated. After cutting of the tissue sections into 0.01 *M* PBS they are transferred into the plastic/nylon mesh carriers and given 3 × 10-min rinses in refrigerated 0.01 *M* PBS. Sections are then incubated in 0.4% Triton (room temperature) for 30 min, rinsed in 0.01 *M* PBS (3 × 15 min, room temperature) and incubated in cold diluent with 3% swine serum for 1 h (refrigerator). After a single

15-min rinse in cold 0.01 *M* PBS, sections are incubated in the primary antibody overnight (18 h) at 4°C. On the second day of tissue processing, all incubations/rinses are performed at room temperature. While still in the primary antibody, tissue is removed from the refrigerator and allowed to warm to room temperature for 30 min. Sections are then rinsed (3 × 10 min) in 0.01 *M* PBS, placed in the secondary antibody solution for 1 h, rinsed in 0.01 *M* PBS (3 × 10 min) and incubated in the PAP solution for 1 h. After final rinses in 0.01 *M* PBS (2 × 10 min) and Tris buffer (1 × 10 min), tissue is transferred into the chromagen solution. The duration of incubation in the chromagen solution required to obtain the desired staining intensity should be determined empirically (45 min is generally sufficient). When the stain intensity is satisfactory (determined by examining a section or two under the microscope), the reaction is stopped by rinsing the sections in Tris buffer (3 × 5 to 10 min). Tissue can then be mounted onto gelatin-coated slides, dried, dehydrated and cleared through alcohols and xylenes, and coverslipped.

ACKNOWLEDGMENTS

The author would like to thank his coauthors and collaborators at Rutgers University, the University of New Mexico, the University of California at Los Angeles, the University of Minnesota, and Hennepin County Medical Center who contributed greatly to the research findings described in this manuscript.

The research described herein and preparation of this manuscript was supported, in part, by the Minneapolis Medical Research Foundation, NIH Grant NS30322, and the U.S. Army Medical Research and Development Command under Grant No. DAMD17-91-Z-1006. Opinions, interpretations, conclusions, and recommendations are those of the author and are not necessarily endorsed by the funding sources. In conducting research using animals, the investigator adhered to the *Guide for the Care and Use of Laboratory Animals,* prepared by the Committee on Care and Use of Laboratory Animals of the Institute of Laboratory Animal Resources, National Research Council (NIH Publication No. 86-23, Revised 1985).

REFERENCES

Auer, R. N., Kalimo, H., Olsson, Y., and Siesjo, B. K. (1985) The temporal evolution of hypoglycemic brain damage. I. Light and electron microscopic findings in the rat cerebral cortex. *Acta Neuropathol. (Berl)* 67: 13–24.

Carlsen, J. and de Olmos, J. S. (1981) Silver impregnation of degenerating neurons and their processes: a modified cupric-silver technique. *Brain Res.,* 208: 426–431.

Cavanagh, J. B. (1970) The proliferation of astrocytes around a needle wound in the rat brain. *J. Anat.,* 106: 471–487.

Cortez, S. C., McIntosh, T. K., and Noble, L. J. (1989) Experimental fluid percussion brain injury: vascular disruption and neuronal and glial alterations. *Brain Res.,* 482: 271–282.

Eng, L. F. (1985) Glial fibrillary acidic protein (GFAP): the major protein of glial intermediate filaments in differentiated astrocytes. *J. Neuroimmunol.,* 8: 203–214.

Eng, L. F. (1988) Regulation of glial intermediate filaments in astrogliosis. In: *Biochemical Pathology of Astrocytes* (Norenberg, M. D., Hertz, L., and Schousboe, A., Eds.), pp. 79–90, Alan R. Liss, New York.

Dienel, G. (1984) Regional accumulation of calcium in post-ischemia rat brain. *J. Neurochem.,* 43: 913–925.

Fink, R. P. and Heimer, L. (1967) Two methods for selective silver impregnation of degeneration axons and their synaptic endings in the central nervous system. *Brain Res.,* 4: 369–374.

Iizuka, H., Sakatani, K., and Young, W. (1989) Selective cortical neuronal damage after middle cerebral artery occlusion in rats. *Stroke* 20: 1516–1523.

Krobert, K. A., Salazar, R. A., Sutton, R. L., and Feeney, D. M. (1992) Temporal evolution of histopathology and unit activity in rat hippocampal CA3 region after focal cortical contusion. *J. Neurotrauma,* 9: 64.

Lees, G. J. (1989) *In vivo* and *in vitro* staining of acidophilic neurons as indicative of cell death following kainic acid-induced lesions in rat brain. *Acta Neuropathol.,* 77: 519–524.

Lowenstein, D. H., Thomas, M. J., Smith, D. H., and McIntosh, T. K. (1992) Selective vulnerability of dentate hilar neurons following traumatic brain injury: a potential mechanistic link between head trauma and disorders of the hippocampus. *J. Neurosci.,* 12: 4846–4853.

Norton, W. T., Aquino, D. A., Hozumi, I., Chiu, F.-C., and Brosnan, C. F. (1992) Quantitative aspects of reactive gliosis: a review. *Neurochem. Res.,* 17: 877–885.

Sternberger, L. S., Hardy, P. H., and Cuculis, J. J. (1970) The unlabeled antibody enzyme method of immunohistochemistry: preparation and properties of soluble antigen-antibody complex (horseradish peroxidase-antiperoxidase) and its use in the identification of spirochetes. *J. Histochem. Cytochem.,* 18: 315–333.

Sutton, R. L., Hovda, D. A., Adelson, P. D., Benzel, E. C., and Becker, D. P. (1994) Metabolic changes following cortical contusion: relationships to edema and morphological changes. *Acta Neurochir.* (Suppl. 60), 446–448.

Sutton, R. L., Lescaudron, L., and Stein, D. G. (1993) Unilateral cortical contusion injury in the rat: vascular disruption and temporal development of cortical necrosis. *J. Neurotrauma,* 10: 135–149.

Sutton, R. L., Lescaudron, L., and Stein, D. G. (Submitted for publication) Unilateral cortical contusion injury in the rat: temporal development of histopathology in hippocampus and subcortical regions.

Van den Pol, A. N. and Gallyas, F. (1990) Trauma-induced Golgi-like staining of neurons: a new approach to neuronal organization and response to injury. *J. Comp. Neurol.,* 296: 654–673.

Chapter 38

BIOCHEMICAL IMMUNOHISTOLOGY

Parthena M. Martin and James P. O'Callaghan

TABLE OF CONTENTS

I. INTRODUCTION

Gliosis is a condition associated with neurodegenerative diseases, or neural damage resulting from trauma or exposure to toxic chemicals, and is characterized by the presence of reactive astrocytes. These astrocytes are differentiated from normal astrocytes by their large size and the increased production of glial filaments, one of which is glial fibrillary acidic protein (GFAP), the primary intermediate filament protein of astrocytes. Although gliosis is used to identify areas of neuropathology or neuronal damage in the central nervous system, there is no commonly accepted definition of gliosis at the cellular or molecular level. The identification of gliosis traditionally is based upon light microscopic evaluation of tissue that has been immunostained using an antibody to GFAP. The criteria most frequently cited as indicators of gliosis include increases in the size, number, and GFAP immunoreactivity of astrocytes, as well as changes in the length of processes, the presence of mitotic figures, and other changes in the metabolic activity of astrocytes (Norton et al., 1992; Malhotra et al., 1990; Malhotra et al., 1992). In a recent study, however, where three qualified observers were asked independently to identify gliosis in samples of human forebrain tissue, the diagnosis of gliosis was found to be correlated with increases in astrocyte size (hypertrophy), but not with increases in the density or staining of astrocytes (de Cunha et al., 1993). The concept of gliosis clearly needs to be better defined, and a less subjective method of identifying gliosis would be helpful.

0-8493-8094-4/95/$0.00+$.50

The primary advantage of immunocytochemistry as a method of studying gliosis, is that it is an anatomical technique that allows the direct visualization of substances within cells in their normal environment. Consequently, this method permits the examination of specific cells, and, with double-staining techniques, it is possible to examine the relationship between types of cells, for example, between reactive astrocytes and microglia, or between astrocytes and neurons, in specific anatomical locations. Immunocytochemistry, however, also has several disadvantages. First, a number of variables (e.g., source of antisera, method of fixation, etc.) contribute to variations in immunoreactivity. It also is a time-intensive method, requiring microscopic examination of multiple slides, and finally, the evaluation of the data is qualitative rather than quantitative.

GFAP can be quantified using radioimmunoassays or enzyme-linked immunosorbent assay (ELISA) (Eng, et al., 1976; Kretzschmar, et al., 1985; Albrechtsen and Bock, 1985; Stewart et al., 1986; Eng et al., 1986; Wang et al., 1990; O'Callaghan, 1991), but in many experiments, both qualitative and quantitative data are desirable. Separate sets of animals usually are used for the two types of procedures, however, because the aldehyde-based fixatives used to prepare brain tissue for immunocytochemistry make that tissue unsuitable for use in biochemical assays. Although it is possible to carry out the ELISA and the immunocytochemistry on fresh, unfixed tissue, the unfixed tissue is fragile, and difficult to handle for immunocytochemistry. The method described in this chapter, which is outlined in the flowchart shown in Figure 1, uses a 70% ethanol solution as a fixative, permitting both immunocytochemical and biochemical assays to be done on brain tissue from a single animal. (For a discussion of fixatives, see Larsson, 1988.) In this procedure, brains that are fixed in an ethanol solution

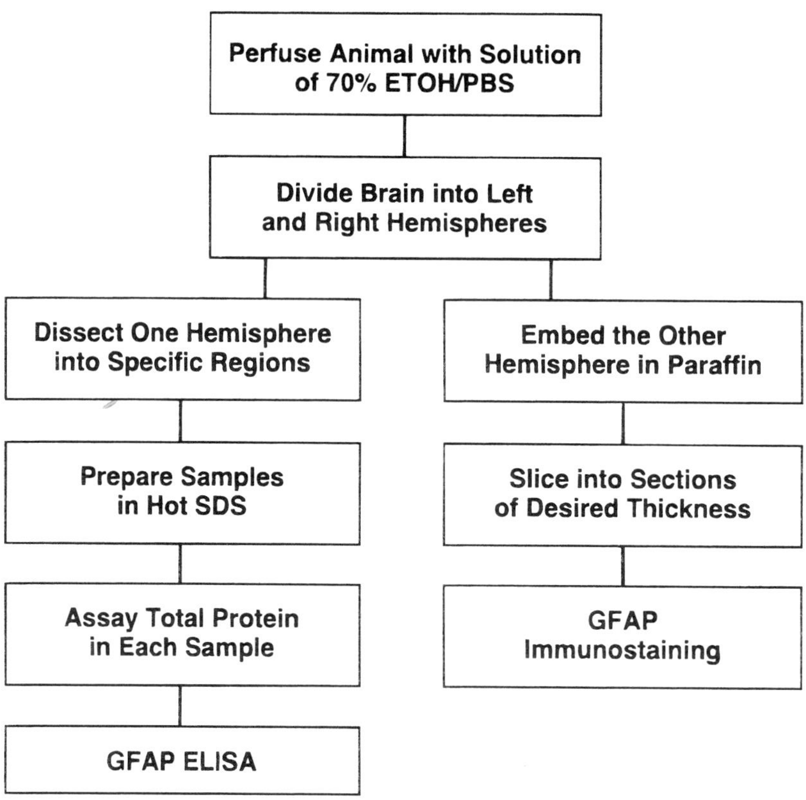

FIGURE 1. Flowchart illustrating the procedure used to prepare tissue for both GFAP immunocytochemistry and GFAP ELISA.

are hemisectioned. One hemisphere is dissected into the regions of interest, and a sandwich ELISA is used to quantify the GFAP in these regions. The other hemisphere is paraffin embedded, then sectioned so that it can be processed for GFAP immunocytochemistry. We have found this procedure to work well for both rat and mouse brain tissue.

II. PROTOCOL FOR SANDWICH ELISA AND IMMUNOSTAINING IN ETHANOL-FIXED TISSUE

A. PERFUSION AND TISSUE PREPARATION

After animals are deeply anesthetized with sodium pentobarbital, perfuse with 100 ml of cold 0.9% NaCl containing 10 iu/ml heparin and then with 100 ml of 70% ethanol containing 150 m*M* NaCl. Following perfusion-fixation, remove the brain, place it in 70% ethanol fixative, and allow it to sit on an oscillating table overnight. On the following day, place the brains in fresh fixative, and store them at 4°C until ready for processing. At that time the brain can be hemisectioned, so that one hemisphere is ready for use in the ELISA, while the other hemisphere can be processed for immunostaining.

B. ELISA

1. Preparation of Samples

Prepare a solution of 1% (w/v) sodium dodecyl sulfate (SDS; purchased from Bio Rad Laboratories, Richmond, CA) and place on stir plate/heater. Heat until the SDS reaches a temperature of 90 to 95°C. At this point the SDS will be steaming, and the container will be hot to the touch. Maintain the SDS at this temperature while preparing the samples, and discard the remaining SDS when done. When the SDS is ready, dissect the hemisphere of the brain into the desired regions. Place each region in a microcentrifuge tube, weigh, and add a tenfold volume of hot 1% SDS. Homogenize with a sonic probe (e.g., Kontes MicroUltrasonic Cell Disrupter, Model XL 2005, Heat Systems, Inc., Farmingdale, NY). At this point the homogenates can be stored at −80°C until further processing. One important feature to note is that tissue fixed in ethanol develops a rubbery texture, and does not homogenize easily compared to fresh tissue. Care must be taken to insure that the tissue is completely homogenized. In lieu of sonification, homogenization can be achieved with a motor-driven pestle and homogenizing vessel.

2. Protein Assay

Measure the total protein concentration of the detergent homogenates using a standard method where detergent interference is not a problem, such as the bicinchoninic acid assay (Smith et al., 1985), using bovine serum albumin to generate the standard curve.

3. Sandwich ELISA Assay

Prepare 1 to 2 l of phosphate buffered saline (PBS). We use Modified Dulbecco's Phosphate Buffered Saline Pack (Pierce, Rockford, IL). Each pack can be dissolved in a volume of 500 ml deionized water.

a. Preparation of Standard Curve

Standards are prepared using aliquots of pooled brain tissue, obtained from naive subjects, homogenized in hot 1% SDS as described above, and stored at −80°C until needed for the assay. The points of the standard curve for the ELISA are based upon the total protein content of the aliquot. Therefore, the aliquot is first diluted to a concentration of 10 μg of protein per 100 μl using PBS containing 0.5% Triton X-100, then further diluted to concentrations ranging from 0.05 μg/100 μl to 10 μg/100/μl (e.g., 0.05, 0.1, 0.25, 0.5, 1.0, 5.0, and 10 μg/100μl).

When this assay was developed (O'Callaghan, 1991), the concentration of GFAP in brain homogenates used as standards was determined by comparing the immunoreactivity of a single aliquot with that of purified GFAP isolated from rat spinal cord according to the method of Eng and DeArmond (1983) (see O'Callaghan, 1990, for additional details). It is not necessary, however, to quantify the concentration of GFAP in the standards, since data can be expressed as percent of control.

b. Preparation of Samples

Dilute the 1% SDS samples (control and experimental) in PBS + 0.5% Triton X-100 to a concentration of 10 μg of protein per milliliter. Samples high in GFAP (e.g., cerebellum) may have to be diluted to 5 μg/ml to avoid readings above the linear portion of the standard curve, while samples low in GFAP (e.g., striatum) may have to be diluted to 20 μg/ml to avoid readings below the linear portion of the standard curve. Alternatively, samples may be diluted by volume, as necessary, to be read on the linear portion of the curve, and the numerical value can be adjusted later for protein. For example, we dilute 10 μl of sample with 990 μl of PBS + Triton X-100, then go through additional dilutions as necessary, so that samples are read on the linear portion of the standard curve. Later, the numerical values are adjusted for protein.

c. The Sandwich ELISA

Coat Immulon-2 flat-bottom, 96-well microtiter plates (Dynatech Laboratories, Chantilly, VA) with an IgG fraction of a polyclonal anti-GFAP, (we use the DAKO polyclonal rabbit anti-cow GFAP antibody) using a concentration of 1.0 μg total immunoglobulin protein per 100 μl, per well, diluted in PBS. This can be done most easily using eight-channel pipetters. Incubate the plate 1 to 2 h at 37°C, and then overnight at 4°C. The overnight incubation may be omitted if necessary. After incubation, wash the wells four times with PBS, 200 μl per well. Again, this can be done most easily using an eight-channel micropipetter. After each rinse, discard the PBS, and make sure all the PBS is removed from the well between rinses by inverting the plate and tapping it firmly face down against some absorbent material such as paper towels, absorbent pads, etc. Block for 1 h with BLOTTO (5% w/v Carnation nonfat dry milk in PBS), 100 μl per well, at room temperature, then discard the BLOTTO, again making sure the BLOTTO is removed by inverting the plate and tapping it firmly against absorbent material. Load standards and samples in duplicates, 100 μl per well. Incubate for 1 h at room temperature. Wash the plate four times with PBS containing 0.5% Triton X-100, 200 μl per well, then incubate 1 h with monoclonal anti-GFAP (Boehringer Mannheim), 1:500, made up in BLOTTO containing 0.5% Triton X-100, 100 μl per well. Wash four times with PBS containing 0.5% Triton X-100, 200 μl per well, then incubate 30 min at room temperature with alkaline phosphatase conjugated anti-mouse IgG (Jackson ImmunoResearch), 1:3000, made up in BLOTTO containing 0.5% Triton X-100, 100 μl per well. Wash four times with PBS containing 0.5% Triton X-100, using a volume of 200 μl per well, then add 100 μl of P-nitrophenylphosphate substrate (Bio Rad) and stop the reaction 10 to 15 min later with 100 μl of 0.4 *N* NaOH. Read at 405 nm. We use a Molecular Devices UV Max Microplate Reader (Menlo Park, CA) coupled to a Macintosh computer running a Soft Max (Molecular Devices, Menlo Park, CA) program.

d. Special Equipment

In our laboratory, a liquid dispenser (Model EAD1/8; available from Tecan, Research Triangle Park, NC) is used for the PBS washes, while a robot (Tecan Model 5052; Tecan, Research Triangle Park, NC) is used to dilute and pipette the standards and samples. Although use of the liquid dispenser and the robot makes this assay easier and faster to perform, neither are necessary for the assay.

C. IMMUNOCYTOCHEMISTRY

1. Embedding

Trim the block to a final thickness of 0.5 to 1.0 cm. Place the tissue on an automatic paraffin-embedding machine and dehydrate through an ascending series of alcohols and clearing agents, prior to infiltration with paraffin. The dehydration sequence is 70% alcohol for 3 h; 80% alcohol for 1 h; 95% alcohol for 1 h; 95% alcohol for 2 h; 100% alcohol for 2 h; 100% alcohol for 1.5 h; 100% alcohol for 1.5 h; xylene for 1 h; xylene for 1 h; xylene for 1 h; paraffin for 1.5 h; paraffin for 1.5 h. All of the solvents should be used at room temperature and the paraffin at 60°C. Tissue should be embedded in paraffin molds, cooled, and stored for sectioning.

2. Staining

a. Preparation for Staining

Serial 6-μm-thick sections should be cut and mounted onto poly-L-lysine-coated slides and incubated overnight in a 39°C oven. When ready for staining (at least 24 h later), the sections should be deparaffinized in the following series of washes: two changes of xylene, 5 mins each; two changes of 100% ethanol, 1 min each; 95% ethanol, 1 min; 70% ethanol, 1 min, then place in PBS, preparatory to staining.

b. Immunostaining

Wash sections in PBS (pH 7.4) + 0.3% Triton X-100 for 30 min, then rinse in 10% solution of normal goat serum/PBS + 0.3% Triton X-100 for 10 min. Incubate slices for 48 h at 4°C in GFAP antibody (DAKO polyclonal; rabbit anti-cow) at a concentration of 1:500 in a 0.3% Triton-X/PBS solution. After incubation, rinse slices with PBS four to five times, 5 min each. Incubate slices for 90 min in an anti-rabbit secondary antibody, (Vector biotinylated anti-rabbit IgG) using a concentration of 1:500. Incubation should be done on a mixing plate at room temperature. Rinse with PBS three times, 5 min each, then incubate with avidin-biotin peroxidase using the Vector elite kit, as follows: 40 μl of Vial A, + 40 μl Vial B in 8 ml of PBS. Incubate slices in the mixture 90 min on a mixing plate at room temperature. Rinse slices in PBS two times, 5 min each, and visualize using 3,3′-diaminobenzidine tetrachloride monohydrate (DAB). Rinse slices in PBS twice rapidly to stop the reaction, then rinse three more times (5 min each) in PBS. Dehydrate through an ascending series of alcohols, then through two 5-min washes of xylene. Coverslip with Permount.

III. EXAMPLES OF DATA FROM ETHANOL-FIXED TISSUE

The regional distribution of GFAP, as assayed by an ELISA in normal, ethanol-perfused rat brains (n = 6) is shown in Figure 2. Figures 3, 4, and 5 illustrate GFAP immunostaining in the striatum, brainstem, and cerebellum, respectively, from brains used in the ELISA assay. As can be seen in Figure 2, there is a great deal of variation in the regional distribution of GFAP, and this variation corresponds to the density of GFAP immunostained-astrocytes observed in Figures 3, 4, and 5. One important point illustrated by these data, however, is that the ELISA seems to be more sensitive in showing clear differences in the relative amounts of GFAP in brain regions that, immunocytochemically, appear to be similar. For example, the hindbrain and cerebellum show striking patterns of GFAP-immunoreactive astrocytes, but it would be difficult to judge, on the basis of a visual impression, the relative amounts of GFAP in these two regions. Furthermore, the ELISA provides low, but readily measurable amounts of GFAP in the striatum, a region in which astrocytes are sparse.

Figure 6 shows the elevation of GFAP that we typically find following the administration of a well-characterized neurotoxicant, trimethyltin (TMT). Rats were given either vehicle or TMT (8 mg/kg calculated as the base; administered IV in the lateral tail vein) then singly

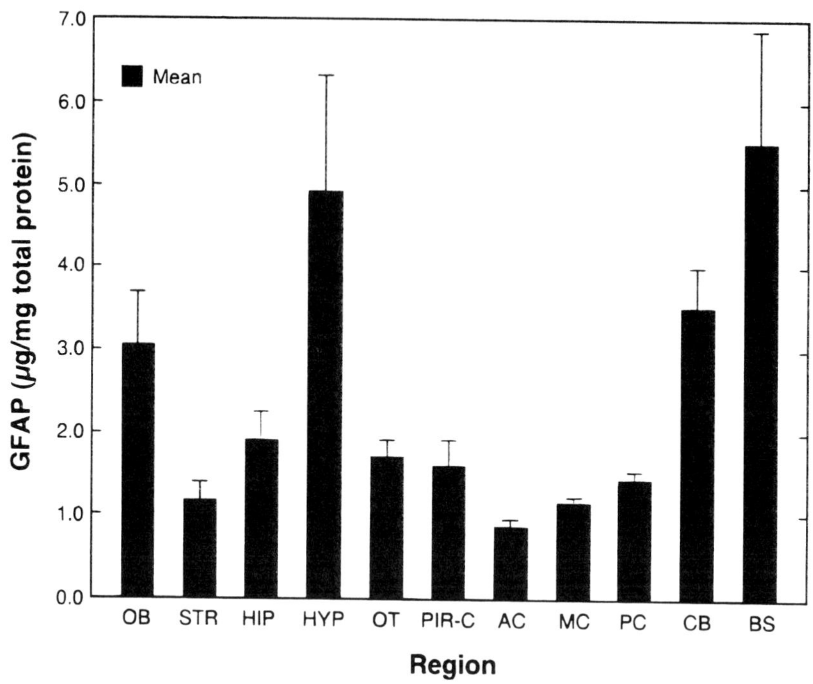

FIGURE 2. Regional distribution of GFAP, as assayed by a sandwich ELISA, in ethanol-fixed rat brain. Brain regions shown on the graph are as follows: OB, olfactory bulbs; STR, striatum; HIP, hippocampus; HYP, hypothalamus; OT, olfactory tracts; PIR-C, piriform cortex; AC, anterior cortex; MC, midcortex; PC, posterior cortex; CB, cerebellum; BS, brainstem. Values shown are the means ± standard errors. (Adapted from Martin and O'Callaghan, 1995.)

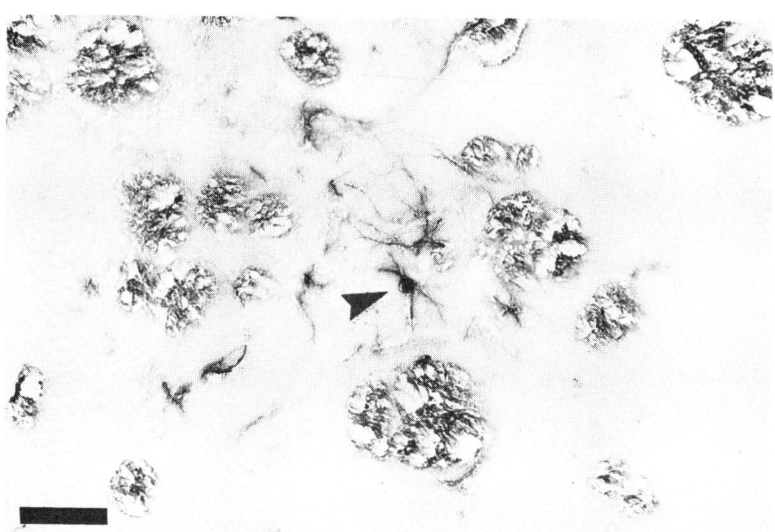

FIGURE 3. GFAP immunostained cell (indicated by arrow) in striatum of rat brain. Bar = 50 µm. (Adapted from Martin and O'Callaghan, 1995.)

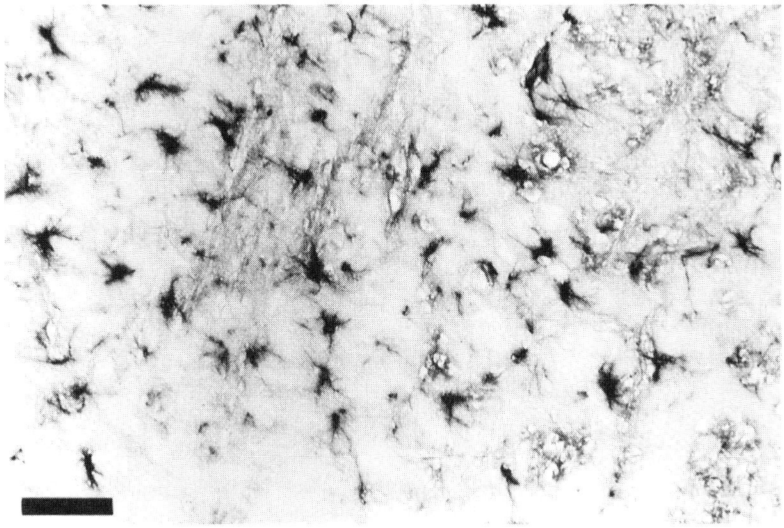

FIGURE 4. GFAP immunostained cells in brainstem of rat brain. Bar = 50 μm. (Adapted from Martin, P. and O'Callaghan, J. P., 1995.)

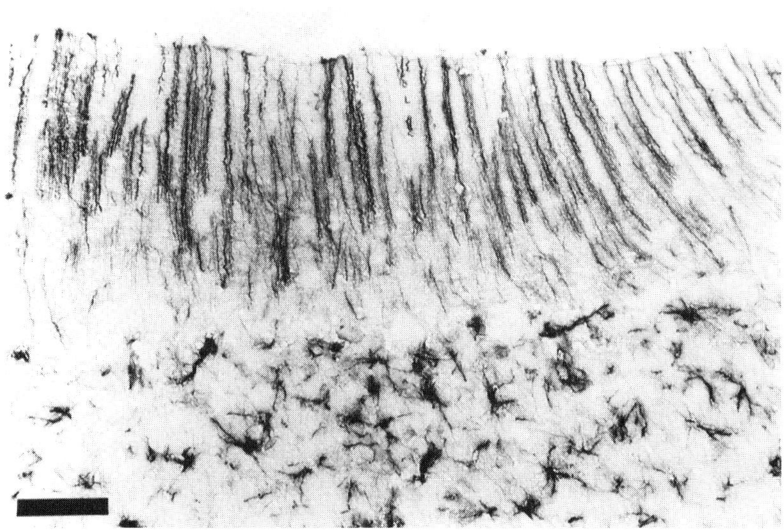

FIGURE 5. GFAP immunostained cells in cerebellum of rat brain. Bar = 50 μm. (Adapted from Martin, P. and O'Callaghan, J. P., 1995.)

housed for five weeks at the ambient room temperature. Brains were fixed in ethanol, and GFAP was measured in the hippocampus and several other brain regions as an indication of gliosis. As can be seen in Figure 6, TMT was elevated approximately eightfold in the hippocampus, one region that is specifically vulnerable to TMT-induced insult.

In summary, immunocytochemistry and the ELISA may be viewed as complementary methods, since immunocytochemistry provides anatomical information while the ELISA provides quantification. As the data presented above show, ethanol fixation allows GFAP-immunoreactive cells to be examined visually in the same set of tissue from which GFAP is quantified, without a sacrifice in the sensitivity of either method.

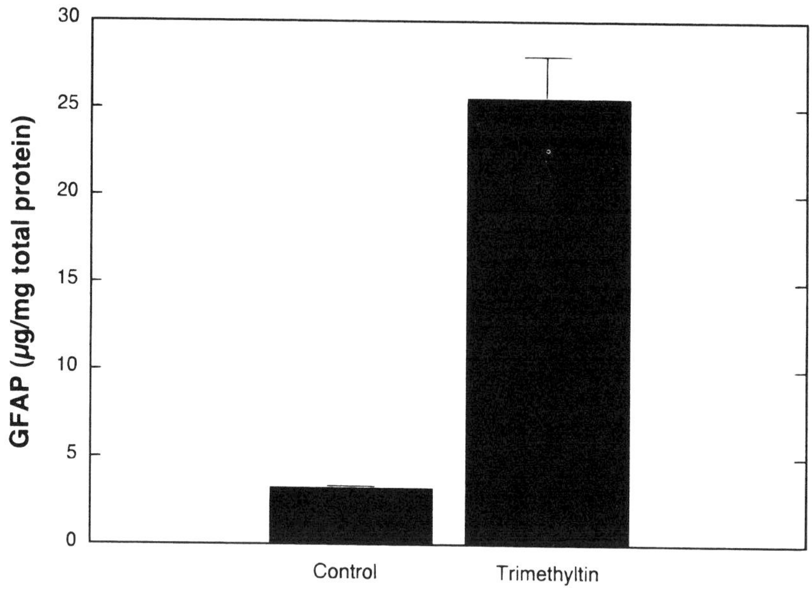

FIGURE 6. Concentration of GFAP in the hippocampus of control rats and rats treated with a neurotoxicant, trimethyltin. Values shown are the means ± standard errors.

REFERENCES

Albrechtsen, M. and Bock, E. (1985) Quantification of glial fibrillary acidic protein (GFAP) in human body fluids by means of ELISA employing a monoclonal antibody. *J. Neuroimmunol.* 8, 301–309.

de Cunha, A., Jefferson, J., Tyor, W., Glass, J., Jannotta, F., and Vitkovic, L. (1993) Gliosis in human brain: relationship to size but not other properties of astrocytes. *Brain Res.* 600, 161–165.

Eng, L. and DeArmond, S. (1983) Immunochemistry of the glial fibrillary acidic protein. In: Zimerman, H., ed. *Progress in Neuropathology*, Vol. 5 (Raven Press, New York) pp. 19–30.

Eng, L., Lee, Y., and Miles, L. (1976) Measurement of glial fibrillary acidic protein by a two-site immunoradiometric assay. *Anal. Biochem.* 71, 243–259.

Eng, L., Stocklin, E., Lee, Y.-L., Shiurba, R., Coria R., Halks-Miller, M., Mozsgai, C., Fukayama, G., and Gibbs, M. (1986) Astrocyte culture on nitrocellulose membranes and plastic: detection of cytoskeletal proteins and mRNAs by immunocytochemistry and *in situ* hybridization. *J. Neurosci. Res.* 16, 239–250.

Kretzschmar, H., DeArmond, S., and Forno, L. (1985) Measurement of GFAP in hepatic encephalopathy by ELISA and transblots. *J. Neuropathol Exp. Neurol.* 44, 459–471.

Larsson, L-I. (1988) *Immunocytochemistry: Theory and Practice*, (CRC Press, Boca Raton) pp. 41–76.

Malhotra, S., Shnitka, T., and Elbrink, J. (1990) Reactive astrocytes—a review. *Cytobios*, 61, 133–160.

Malhotra, S., Aldskogius, H., and Shnitka, T. (1992) Biochemical diversity among reactive astrocytes. *Dendron*, 1, 55–68.

Martin, P. and O'Callaghan, J. P. (1995) A direct comparison of GFAP immunocytochemistry and GFAP concentration in various regions of ethanol-fixed rat and mouse brain. *J. Neurosci. Methods*, in press.

Norton, W., Aquino, D., Hozumi, I., Chiu, F.-C., and Brosnan, C. (1992) Quantitative aspects of reactive gliosis: a review. *Neurochem. Res.*, 9, 877–885.

O'Callaghan, J. (1991) Quantification of glial fibrillary acidic protein: comparison of slot-immunobinding assays with a novel sandwich ELISA. *Neurotoxicol. Teratol.*, 13, 275–281.

O'Callaghan, J., Miller, D., and Reinhard, J. (1990) Characterization of the origins of astrocyte response to injury using the dopaminergic neurotoxicant, 1-methyl-4-phenyl-1,2,3,6-tetrahydropyridine. *Brain Res.*, 521, 73–80.

Smith, P., Krohn, R., Hermanson, G., Mallia, A., Gartner, F., Provenzano, M., Fujimoto, E., Goeke, N., Olson, B., and Klenk, D. (1985) Measurement of protein using bicinchoninic acid. *Anal. Biochem.*, 150, 76–85.

Stewart, M., Bourne, R., and Gabbott, P. (1986) Decreased levels of an astrocytic marker, glial fibrillary acidic protein, in the visual cortex of dark-reared rats: measurement by enzyme-linked immunosorbent assay. *Neurosci. Lett.*, 63, 147–152.

Wang, S., Rosengren, L., Karlsson, J., Stigbrand, T., and Haglid, K. (1990) A simple quantitative dot-immunobinding assay for glial and neuronal marker proteins in SDS-solubilized brain tissue extracts. *J. Neurosci. Methods*, 33, 219–227.

X. Topics with Growing Interest

Chapter 39

CHRONIC ALCOHOLISM AND BRAIN ISCHEMIA

Teiji Tominaga and S. Tsuyoshi Ohnishi

TABLE OF CONTENTS

I. INTRODUCTION

Chronic alcohol ingestion is known to cause damage at cellular and subcellular levels. In most of these cases, a measure of physiologic compensation occurs and the individual appears to be functionally normal. However, with additional trauma such as ischemia or injury in an automobile accident, the reduced tolerance and recuperative capacity become obvious. An alcoholic suffers more than a normal patient. Very often the accumulated damage from alcoholism combines with a normally survivable insult to become life-threatening. Through the study of renal function of alcoholic rats, we have already demonstrated this serious problem.[2] We have also demonstrated that the membranes of the alcoholics are abnormal with respect to enhanced calcium permeability.[5] We hypothesized that this may be a mechanism of the enhanced vulnerability of the alcoholic to additional insults. The purpose of this study is to examine whether alcoholic rat brain is vulnerable to ischemic injury. For this purpose, we used chronic alcoholic rat (ethanol ingestion for 35 d) and evaluated ischemic brain edema measured by water and ion contents.

II. MATERIALS AND METHODS

Male Sprague-Dawley rats were used for the experiments. The body weight of the rat ranged from 300 to 360 g. Rats were fed for five weeks with a liquid diet containing ethanol (ethanol providing 36% of total chlories). In nonalcoholic diet, the calories of ethanol was replaced by that of carbohydrates. The liquid diets were obtained from Bio-Serv Inc. (Frenchtown, NJ). The alcoholic rat is very aggressive. Therefore, it is particularly important to maintain the rat individually in a cage to avoid injury. The focal cerebral ischemia was produced by the procedures described previously.[7] In brief, the bilateral common carotid arteries were exposed under halothane anesthesia with care taken to avoid injury especially to the vagal nerve. Unilateral common carotid artery was ligated with 2-0 silk suture in two places. Under a surgical microscope, a small subtemporal craniotomy was performed in the same side of the carotid ligation, and the middle cerebral artery (MCA) was exposed. The dura matter was carefully opened using a 27-gauge needle. The MCA was ligated at the level just proximal to its bifurcation using 10-0 suture (Ethicon, mononylon filament). Immediately, the common carotid artery in the other side was occluded by an aneurysm clip. After the procedure, the bone window was covered with a small piece of gel-foam, and the wounds were sutured. The rat was allowed to recover from anesthesia. One hour after the carotid clipping, the rat was hand-held and the clip was released.

0-8493-8094-4/95/$0.00+$.50

Twenty-four hours after the onset of focal ischemia, the rat was sacrificed under halothane or overdose of barbiturate anesthesia. The ipsilateral and contralateral cerebral hemispheres were quickly removed and their wet weights (WW) are measured in a crucible. The samples were dried up in an oven at 105°C for 5 to 7 d, and then the dry weights (DW) were measured. The water content (percent wet weight) of the cerebral hemisphere was calculated as follows:

$$\text{Water content (\% WW)} = \frac{\text{WW} - \text{DW}}{\text{WW}} \times 100$$

The dried brain sample was digested in concentrated nitric acid, which was approximately diluted with distilled and deionized water. An atomic absorption spectrophotometer (Hitachi, model Z-8000) was used for measurement of ion contents. The concentrations of sodium, potassium, and calcium were measured at 589.6, 766.5, and 422.7 nm, respectively.[7]

III. RESULTS AND DISCUSSION

The effect of chronic alcoholic ingestion on ischemic brain edema including ion movements are shown in Figure 1. The brain edema caused by focal ischemia was more profound in the chronic alcoholic than the nonalcoholic rats at 24 h after ischemia. Water, sodium, and potassium accumulated to a greater degree in alcoholic than nonalcoholic rat brain. Potassium was also decreased to a greater degree in the alcoholic rat. Therefore, chronic ingestion of alcohol for 35 d makes the brain tissue more susceptible to ischemic insult. These findings are consistent with the increased vulnerability of the kidney of a chronic alcoholic rat to the insult of mercury chloride, a nephrotoxin, and ischemia.

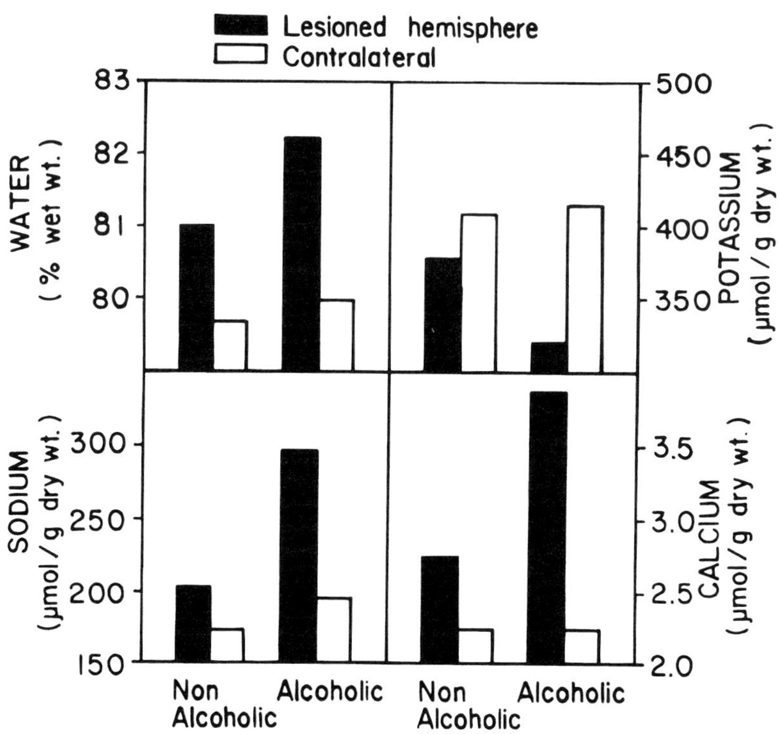

FIGURE 1. Ischemic brain edema caused by focal cerebral ischemia in alcoholic and nonalcoholic rats. The water and ion contents were determined 24 h after focal ischemia.

Characteristics of ionic movements, increase of sodium and calcium, and decrease of potassium, are commonly found in CNS injuries including ischemia and trauma. The excessive influx of sodium into ischemic tissue plays a major role in the development of brain edema. The increase of intracellular calcium level causes an activation of self-destructive enzymes. An activation of phospholipases leads to degradation of membrane phospholipids into free fatty acids, and evokes cascade reactions starting from arachidonic acid. Calcium-dependent protease catalyzes cytoskeletal proteins. Calcium-dependent endonucleases breaks chromosomal DNA into small fragments.[8] Therefore, excessive calcium influx has a crucial role in the ischemic cell death. Although the mechanism of the enhancement of ischemic brain edema in alcoholic rat is yet to be determined, one possible explanation may be the increased membrane permeability to calcium ion. Chronic alcohol ingestion alters the calcium permeability of sarcoplasmic reticulum obtained from skeletal muscle of rat.[3–5] Ethanol may affect the calcium pump, the regulatory subunit of the calcium channel, or membrane fluidity.[5] When membrane permeability increases, calcium influx into cell or release from intracellular store may be accelerated, thereby resulting in enhancement of ischemic cellular injury. Alternatively, impairment of mitochondrial function and Na-K-ATPase enzyme activity may be associated with the development of enhanced brain edema.[1,6] Further detailed study is necessary to elucidate the enhanced vulnerability of alcoholic brain to ischemic insult.

ACKNOWLEDGMENT

The authors thank Mr. S. R. Narayan for his excellent technical assistance.

REFERENCES

1. Hasumura, Y., Teschkike, R., and Lieber, C. S. Acetaldehyde oxidation by hepatic mitochondria; its decrease after chronic ethanol consumption. *Gastroenterology* 66: 415–422, 1974.
2. Ishigami, M., Ohnishi, S. T., Maeda, T., and Yabuki, S. The increased vulnerability of kidneys of chronic alcoholic rats to the insult of a Nephrotoxin. *Ann. N.Y. Acad. Sci.*, 492: 335–337, 1987.
3. Ohnishi, S. T., Waring, A. J., Fang, S.-R. G., Horiuchi, K., and Ohnishi, T. Sarcoplasmic reticulum membrane of rat skeletal muscle is disordered with chronic alcohol ingestion. *Membrane Biochem.*, 6: 49–63, 1985.
4. Ohnishi, S. T. Chronic alcohol ingestion alters the calcium permeability of sarcoplasmic reticulum of rat skeletal muscle. *Membrane Biochem.*, 6: 33–47, 1985.
5. Ohnihi, S. T., Flick, J. L., and Rubin, E. Ethanol increases calcium permeability of heavy sarcoplasmic reticulum of skeletal muscle. *Arch. Biochem. Biophys.*, 233: 588–594, 1984.
6. Pachinger, O. M., Tillmans, H., Mao, J. C., Fauvel, J. M., and Bing, R. J. The effect of prolonged administration of ethanol on cardiac metabolism and performance on the dog. *J. Clin. Invest.* 52: 2690–2696, 1973.
7. Tominaga, T. and Ohnishi, S. T. Interrelationship of brain edema, motor deficits and memory impairment in the rat exposed to focal ischemia. *Stroke,* 20: 513–518, 1989.
8. Tominaga, T., Kure, S., Narisawa, K., and Yoshimoto, T. Endonuclease activation following focal cerebral ischemic injury in the rat brain. *Brain Res.* 608: 21–26, 1993.

Chapter 40

GUNSHOT WOUNDS TO THE BRAIN

Dan Torbati

TABLE OF CONTENTS

0-8493-8094-4/95/$0.00+$.50

I. INTRODUCTION

Gunshot wound (GSW) to the brain causes multiple insults resembling all other types of neurotrauma combined together. When a high-speed object (missile) penetrates a tissue, besides its direct crushing effect, it also generates pressure waves from its kinetic energy transfer. This creates a temporary cavity, which is many times the cross-sectional area of the missile itself (Butler et al., 1945; Harvey et al., 1947; Harvey and McMillen, 1947). This temporary cavity subsides in a few seconds into a much smaller but permanent lesion that constitutes the missile path or trajectory, initially filled with extravasated blood and pulped cells. The displacement of the tissue, by momentary formation of the temporary cavity in the brain, greatly stretches and damages the cells beyond the direct crushing effect of the missile (Horsley, 1894; Butler et al., 1945; Harvey et al., 1947; Barnette and Meirowsky, 1955; Allen et al., 1982, 1983; Fackler et al., 1988; Shaffrey et al., 1992). In some cases, more tissue damage is expected because of generation of a high-amplitude, short-duration shock wave passing through the tissue with the velocity of sound in water (1460 m/s). These destructive events create a chaos, affecting many brain physiologic functions including hemodynamic, circulatory, metabolic, electrical, permeability, and neurotransmission. Consequently, the normal activity of the respiratory, cardiovascular, and endocrine systems that are controlled by the brain may become severely impaired (Horsley, 1894; Webster and Gurdjian, 1943; Crockard et al., 1977, 1979; Brown et al., 1982; Carey et al., 1989; Torbati et al., 1992; Soblosky et al., 1992). This review focuses on the early pathophysiologic responses to experimental GSW of the brain, comparing them to those in the civilian craniocerebral GSWs. The immediate attention required for field resuscitation and possible surgical therapy will be reiterated. Finally, the contribution of GSW-induced secondary intracranial and systemic insults to delayed neurological damage is discussed.

II. DESTRUCTIVE FORCES OF MISSILES (BALLISTICS) IN TISSUES

Horsley (1894), in his classical lecture and demonstration on the destructive effects of small projectiles at the Royal Institution in London, has provided us with an elegant analysis of physical and pathologic events of missile injury in general, and of GSW to the brain in particular. In his monograph, Horsley discussed the physical properties of a missile, such as speed, rotation, size, cross-sectional area, heating, and destructive effects in relation to the physical characteristics of the object pierced by a bullet. His observations, concepts, and analysis of both the physics and pathology of GSW have been repeatedly confirmed during the past century, as is described in the following discussion.

The extent of damage inflicted by a missile to tissues is directly related to the missile velocity and the amount of its energy, according to the formula:

$$E = \frac{1}{2} mV^2$$

where, E is the energy of the missile (in joules), m is its mass (kg), and V is its velocity (m/sec). The velocity of a missile is the major factor in producing irreversible damage rather than its shape or mass (Kirkpatrick and DiMaio, 1978; Fackler et al., 1988; Cheng et al., 1988; Phillips, 1992). Theoretically, this can be derived from the above formula, indicating that the energy of a missile is proportional to the square of its velocity, but only directly proportional to its mass. The GSW to the brain in humans requires about 60 m/s velocity in order to penetrate skin and bone, but once through, there is little resistance to the force of the missile in the brain (Crockard, 1979).

The kinetic energy of a missile is transmitted to the surrounding tissues along its path. This generates a crushing force because of creation of a temporary cavity that forcefully displaces the neighboring tissues and generates a short-lasting pressure wave of a few milliseconds (Butler et al., 1945; Harvey et al., 1947). An additional destructive force that is generated by the impact of a missile to any fluid-like object (tissues) is a very high-amplitude pressure wave (shock wave pressure), which is transmitted with the velocity of sound in water (about 1460 m/s). This pressure wave may last only for a few microseconds and is comparable to injury from an underwater blast where pressure changes cause damage to men and animals in the water. However, this type of tissue damage apparently does not occur during GSW to the brain, but only by GSW to air-containing organs, particularly, lungs and gastrointestinal tract (Horsley, 1894; Harvey and McMillen, 1947; Harvey 1948; Richmond et al., 1968; Huller and Bazini, 1970; Harmon and Haluszka, 1983). Nevertheless, studies by Berlin (1979) and Tikka et al. (1982) suggest that pressure waves from high-velocity missiles may give rise to injuries even distant from the target area. More recent studies, in fact, have demonstrated the incidence of microscopic CNS damage, as well as EEG depression, caused by high-velocity missiles aimed at extremities (Suneson et al., 1990; Goransson et al., 1988). However, the extent of CNS damage due to a peripheral GSW, and its mechanism of injury in both humans and experimental animals is not known.

III. LOW AND HIGH VELOCITY MISSILES

Military rifles are designed to generate a relatively high-velocity missile that can cause extensive damage, as compared to the civilian or the low-velocity missiles (DeMath, 1969). The latter is less destructive because of a smaller temporary cavity produced by small-caliber, low muzzle exit velocity missiles (Harvey, 1948; Raimondi and Samuelson, 1970). The dividing line between low- and high-velocity missiles is arbitrary. Kirkpatrick and DiMaio (1978), reporting on 42 fatal civilian GSW to the brain, found a marked increase in severity of wounds that occurred with muzzle velocity of about 610 m/s (2000 ft/s). They considered a missile with a velocity of about 457 m/s (1500 ft/s), which is above the range of most civilian handguns, to be a low-velocity missile. Crockard (1979), however, classified all handguns and submachine guns that fire missiles below the speed of sound of 300 m/s (about 1000 ft/s) as low-velocity missiles. Comparatively, rifles firing high-velocity bullets will be those exceeding the speed of sound. Clinically, the difference between a low- and high-velocity missile is the extent of tissue damage, and a higher mortality rate with high-velocity military rifles, compared to low-velocity civilian handguns. This major difference suggests that the basic principles of critical care for a low-velocity GSW to the brain could be different from a high-velocity missile wound to the brain (Hubschmann et al., 1979; Nagib et al., 1986; Taha et al., 1991; Levy et al., 1993). In addition, brain wounds created by low or high velocity missiles are not the same as the nonmissile penetrating brain injury, such as that caused by an electric drill or a stab wound. The latter generates very low energy for penetration (Crockard et al., 1982), and is not the subject of this review.

IV. GSW TO THE BRAIN AS A PUBLIC HEALTH PROBLEM

The civilian GSW to the brain is a major health problem in some large urban areas (Raimondi and Samuelson, 1970; Kirkpatrick and DiMaio, 1978; Lillard, 1978; Hubschmann et al., 1979; Nagib et al., 1986; Jaggar and Dietz, 1986; Selden et al., 1988; Miner et al., 1990; Benzel et al., 1991; Phillips, 1992; Kaufman, 1993). In addition to a continuous suffering of the brain-wounded victims and their families (Jagger and Dietz, 1986), there is also an overwhelming social and financial impact of GSW to the brain (McMordie and Barker, 1988). However, the bulk of clinical data on GSW to the brain has been derived from the neurosurgical reports of the military conflicts of the 20th century (Jefferson, 1919; Lewin and Gibson, 1956; Hammon, 1971; Carey et al., 1972; West, 1980; Rish et al., 1983; Carey, 1987; Aarabi, 1990; Brandvold et al., 1990). Thus, the systematic study of GSW to the brain in animals has usually been motivated by the needs of the military medicine (Butler et al., 1945; Djorjevic et al., 1976; Allen et al., 1982; Carey et al., 1989). Except for a relatively large number of clinical reports on both civilian and military types of GSWs to the brain (Hammon, 1971; Carey et al., 1972; Kirkpatrick and DiMaio, 1978; Hubschmann et al., 1979; West, 1980; Kaufmann et al., 1986; Clark et al., 1986; Brandvold et al., 1990; Graham et al., 1990; Shaffrey et al., 1992), only a few interrupted but systematic attempts have been made to investigate the pathogenesis of the cranial GSW and to devise a scientific base for therapeutic intervention (Crockard et al., 1977, 1979; Crockard, 1979; Brown et al., 1979; Allen et al., 1982, 1983; Carey et al., 1989, 1990; Torbati et al., 1989, 1992; Awasthi et al., 1991; Soblosky et al., 1992).

V. GSW TO THE BRAIN: A MULTIPLE INSULT

In military medicine, piercing brain wounds are usually known as missile wounds to the brain, while in civilian practice they are usually referred to as gunshot wounds (GSW) to the brain. The civilian craniocerebral GSW may often imply that the injury is caused by low-velocity missiles, compared to the high-velocity of a military type of missile (DeMath, 1969; Raimondi and Samuelson, 1970; Shaffrey et al., 1992). In this review, both terms, namely, gunshot wound and missile wound, will be used interchangeably. As mentioned earlier, direct damage to brain parenchyma is caused by the crushing action of the missile itself, or indirectly from the damage created by the temporary cavity. These injuries disrupt the delicate connections (white fiber tracts) among various neuroanatomical formations (Jefferson, 1919; Butler et al., 1945), producing primary damage (Harvey and McMillen, 1947; Harvey, 1948). The current concept of a secondary insult or delayed brain injury, consequent to GSW to the brain, is not the damage caused by high-amplitude, short-duration pressure waves, as described by Harvey et al., (1947) and Harvey and McMillan (1947), but it is attributed to the pathophysiologic and biochemical reactions that may occur after the brain has survived the initial mechanical injury at the time of impact (Torbati et al., 1992). This concept is compatible with Miller and Becker's definition (1982) for other models of traumatic brain injury, indicating that the secondary or delayed brain injury includes all events other than the mechanical injury sustained at the time of impact. Miller and Becker (1982) have also divided the secondary phenomena of brain trauma, in general, into systemic (hypoxemia and hypotension), and intracranial insults (hypertension, brain swelling, edema, and vasospasm).

GSW induces an immediate brainstem dysfunction, which may lead to apnea, reduced ventilation, and hypoxemia (Horsley, 1894; Crockard et al., 1979; Carey et al., 1989). Further complications include the formation of hematomas, and a sudden drastic increase in intracranial pressure (ICP), (Barnette and Meirowsky, 1955; Djordjevic et al., 1976). In addition, the breakdown of cerebral vessels and capillaries, in both the temporary cavity and the missile permanent trajectory, leads to extravasation of the blood thereby reducing, or even abolishing, capillary blood flow. In fact, studies have demonstrated significant GSW-induced reductions in cerebral blood flow (CBF), (Crockard et al., 1977a, b; Torbati et al., 1989, 1992). Concurrently, energy metabolism is reduced (Gerber and Moody, 1972; Crockard et al., 1976), and electrical activity is abnormally modified (Crockard et al., 1977a; Torbati et al., 1992). These events may be followed by the opening of the blood-brain barrier (BBB) and development of vasogenic edema (Allen et al., 1983; Carey et al., 1990). In the worst-case-scenario, cerebral ischemia may ensue, resulting in irreversible neurologic damage and death. Parallel to the CNS effects, GSW of the brain may produce significant alterations in

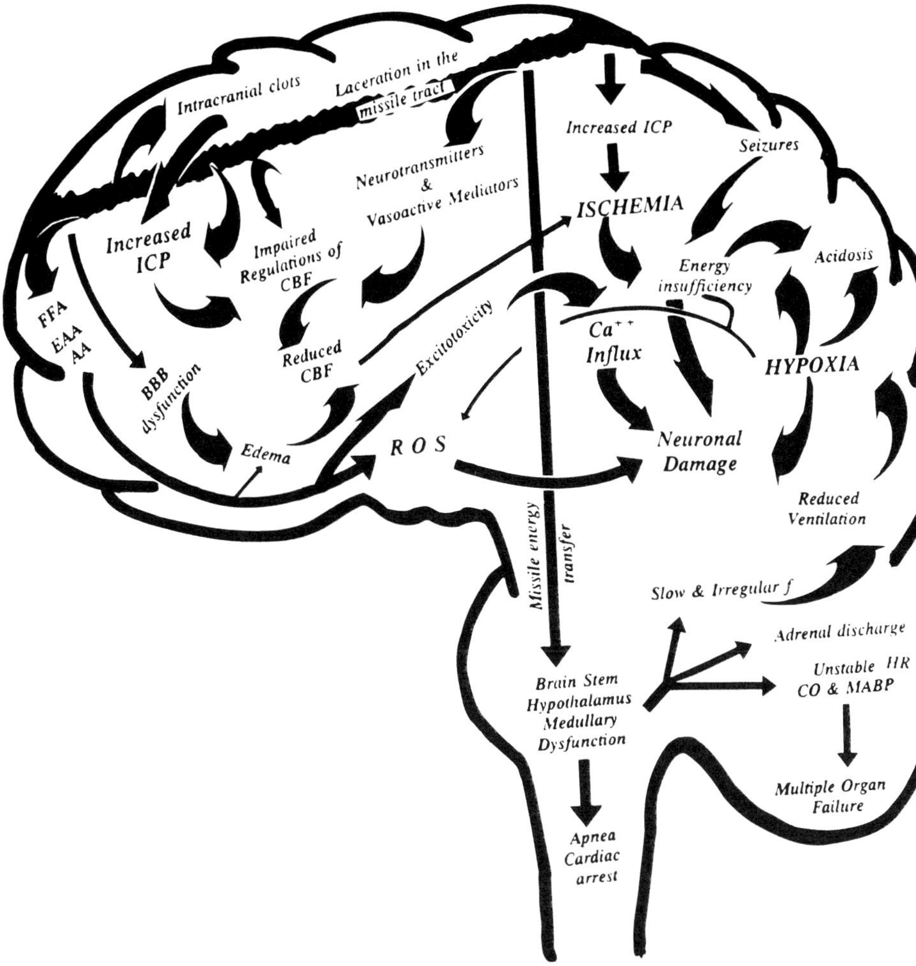

FIGURE 1. Gunshot wound to the brain creates primary (brainstem effects) and secondary pathophysiologic changes (circulatory, metabolic, neurochemical, oxidative stress) that may lead to an immediate death or delayed neurological deficiencies. This illustration demonstrates speculative pathways that are also common features of other experimental models of brain injury or neurodegenerative events, including hypoxic-ischemia, edema, and persisting intracranial hypertension. AA = Arachidonic acid; BBB = blood-brain barrier; CBF = cerebral blood flow; CO = cardiac output; EAA = excitatory amino acids; f = frequency of breathing; FFA = free fatty acids; HR = heart rate; ICP = intracranial pressure; MABP = mean arterial blood pressure; ROS = reactive oxygen species.

cardiovascular and respiratory functions. These include reduction in cardiac output (CO), primary systemic hypertension followed by secondary hypotension, alterations in blood flow to peripheral organs, and hematological changes (Crockard et al., 1977b; Brown et al., 1982; Carey et al., 1989; Torbati et al., 1988, 1988a, 1992; Awasthi et al., 1991). Figure 1 illustrates the possible relations among various GSW-induced pathophysiologic changes that are discussed in this review.

VI. PIONEERING STUDIES ON EXPERIMENTAL GSW TO THE BRAIN

Kramer and Horsley (1894) and Horsley (1894, 1915) were the first investigators who experimentally studied GSW to the brain in ether-anesthesized dogs, while recording ICP, mean arterial blood pressure (MABP), heart rate (HR), and respiratory frequency (f). They found that penetration of a low-velocity (about 180 m/s; 600 ft/s) bullet into the cranial cavity caused complete arrest of the respiration, with a remarkable rise in MABP at 5 to 10 s later. It was concluded that the cause of death was not, as previously believed, due to cardiac arrest, but due to complete cessation of breathing. This was attributed to the effect of the mechanical pressure of the missile on the medullary cardiorespiratory centers, or perhaps due to (the brainstem) anemia (ischemia). These investigators also demonstrated that a quick artificial respiration prevented fatal consequences, and they, therefore, strongly recommended that first aid providers should administer artificial respiration to GSW victims. Kramer and Horsley (1894), also identified a second cause of death due to hemorrhage into the missile trajectory. They reasoned that the lacerated blood vessels could severely raise ICP, which in turn, leads to paralysis of the respiratory center and a marked slowing of the HR. Supporting evidence for this concept was a violent increase in ICP at the moment of missile penetration. They also demonstrated that the initial GSW-induced intracranial hypertension recoils to a steady increase in ICP, creating the secondary cause of death. To prevent the secondary cause of death, Kramer and Horsley (1894) simply suggested an ordinary surgical measure.

VII. EXPERIMENTAL MODELS OF GSW TO THE BRAIN DURING WORLD WAR II

For almost half a century, since the pioneering studies of Horsley and Kramer, no serious attempt was made to further elucidate the pathophysiology of the GSW to the brain. During World War II, however, Webster and Gurdjian (1943) investigated alterations in ICP, MABP, and respiration in dogs under morphine anesthesia (large quantities of 20 mg/kg), and 1% procaine hydrochloride for local anesthesia. They used bullets of varying velocity and size to produce relatively minimal, moderate, and profound physiologic effects of GSW to the brain with intact cranium. The profound effect was uniformly characterized by respiratory paralysis, abolishment of lid and corneal reflexes, marked hypertension, and death. Hypertension, however, did not occur when the medulla was directly injured. It also failed to appear after adequate decompression before injury, by opening of the cistern and exposing of the medulla. This procedure also spared the respiratory function in the acute phase of the injury. The profound injury also created substantial pathological damage including petechial hemorrhages of varying size and number, which were adjacent to the site of penetration, as well as the pons, medulla, and upper cervical spinal cord. In the moderate injury, immediate mortality was low and respiratory arrest was temporary (5 to 30 s), and a rise in MABP with tachycardia was consistently observed. With the moderate injury the palpebral reflexes, which were entirely lost after the profound injury, reappeared 3 to 145 s after injury.

The gross pathologic damage in the brain was also less severe in the moderate group, compared to the profound group, as judged by less frequent distant lesions. In the minimally traumatized group the brain was penetrated by the bullet after making a hole in the bone with a mechanical drill. Dogs in this group survived the injury, and both respiration and reflexes were preserved. In fact, in some animals respiration was rather stimulated and consciousness was maintained. The MABP usually remained unchanged, and postmortem pathological examination showed mainly local damage with or without hemorrhage. Webster and Gurdjian's experiments provided visual evidence for the importance of an intense rise in the ICP that accompanies the GSW to the brain. They also demonstrated that by exposing the cisterna magna, at the moment of injury the cerebellum was expelled into the opening made. A suboccipital cranioctomy followed by a supratentorial GSW resulted in expulsion of the cerebellum with such a force as to lacerate that tissue. The sudden rise in ICP in some cases was as high as 406 mmH_2O (about 30 mmHg). Thus, it was concluded that the acute pathological response following GSW to the brain is due to an intense and transitory increase in ICP at the moment of injury. The increased ICP is dispersed throughout the closed system and acts upon all the brain and it is specifically manifested by the response of the vital medullary centers. In addition, Webster and Gurdjian (1943) projected that further GSW-induced neural cell injury may result from the indirect effects of intrinsic vascular damage.

A similar effort to study GSW to the brain during World War II was made by Butler et al. (1945). Their investigations were carried out in pentobarbital (nembutal)-anesthetized cats and dogs, using small steel spheres. The spheres were shot with an especially constructed experimental gun, capable of generating high-velocity wounds. For the first time, these investigators were able to take microsecond roentgenogram during and immediately after the passage of a sphere through a part of the animal body. They demonstrated the formation of a large temporary cavity that pulsates several times before a relatively small, permanent wound trajectory subsides. This temporary cavity was created in both animal tissues and gelatin blocks, and existed for only a few microseconds after the passage of the high-velocity missile (sphere). The displacement of the tissue about the temporary cavity occurred with such a destructive force that it could fracture bones that were several centimeters away from the missile trajectory. Butler et al. (1945), thus emphasized the importance of the high-velocity missiles-induced temporary cavity in stretching nerves and breaking small vessels with consequent extravasation of blood.

The widespread skull destruction in a study by Butler et al. (1945) was actually caused by a relatively small sphere of 3.1 mm diameter with a velocity of approximately 1220 m/s. However, the amount of energy deposited in the brain was sufficient to virtually transect the brainstem, while the cerebral hemisphere and cerebellum were in the condition of a "macerated pulp." They also used a relatively large sphere of about 6.2 mm diameter in a cat with approximate velocity of 1160 m/s, showing a similar destructive effect as in the dog. In contrast, when an intact cranial cavity of a cat (open foramen magnum) was impacted with the same velocity, the skull damage was relatively minor and restricted to the points of entrance and exit of the missile. Thus, Butler et al. (1945) concluded that the high-velocity missiles, in passing through the brain, create a high internal pressure to surrounding tissues within the cranial cavity with extensive disarticulation of bones and maceration of the brain tissue.

VIII. EXPERIMENTAL GSW TO THE BRAIN AFTER WORLD WAR II

Many years after the Webster and Gurdjian's study, Djordjevic et al. (1975) investigated the time course and nature of ICP changes consequent to GSW to the brain in pentobarbital-anesthetized, mechanically ventilated dogs. They used bullets of different velocities (122 to 785 m/s), correlating alterations in ICP with changes in CBF (xenon technique). In theses experiments, a GSW to the brain with a heavy bullet of 0.87 g (5.6 mm diameter) produced variable increases in ICP to as high as 300 mmHg followed by a secondary decline, after a few minutes, to a steady-state value of about 30 to 70 mmHg. These increases in ICP corresponded with rises in MABP, which sometimes reached values exceeding 300 mmHg. The MABP ultimately stabilized on a slightly hypotensive level. The CBF measurements with low-velocity GSW showed a consistent decrease to about 50% of the control, starting a "few minutes" after the wounding with some tendency toward later restitution. The decreases in the CBF could not be completely explained on the basis of a reduced cerebral perfusion pressure (CPP), calculated as the difference between MABP-ICP. Although a progressive increase in cerebrovascular resistance (CVR) occurred after wounding, particularly with high-velocity injuries, autoregulation was abolished as indicated by a linear pressure-flow relationship (Djordjevic et al., 1976).

IX. PRIMATE MODELS OF GSW TO THE BRAIN

One of the important factors in choosing a suitable model for head trauma studies is the selection of animal models that closely resemble human physiology and anatomy. Thus, the primate has remained the animal of choice because of brain structure, angulation of brain-brainstem-spinal cord, and head to brain mass ratio, which are similar to man (Gennarelli et al., 1981, 1982; King and Yarbrough, 1985; King et al., 1988). Gerber and Moody (1972) were the first investigators to study experimental GSW to the brain in 11 pentobarbital-anesthetized rhesus monkeys. They tried to standardize GSW to the brain by monitoring a variety of physiologic parameters (ICP, MABP, blood gases, ventilation and f, EEG, and carotid blood flow), correlating them with survival and other pathophysiologic evidence of the injury. They used a low-velocity missile of approximately 170 m/s, but the bullets were shot through a burr hole in the skull, thereby creating an open cranium. Of the 11 animals, only 6 survived the GSW until elective sacrifice 5 to 7 d later. Five other monkeys died within 46 min of the GSW to the brain. The survivors were soon able to get about their cages easily and feed themselves. All animals manifested some alteration in respiration at the moment of impact, and two of the nonsurvivors did not recover spontaneous breathing. Gerber and Moody's (1972) data analysis revealed that the single parameter that correlated most closely with survival or death was the carotid blood flow, which was measured 3 min after the GSW. At this time, the lowest flow for the survivors was at least 60% of the baseline value, while the highest, for nonsurvivors, was 23% of the baseline. Other physiologic variables, and pathologic examinations, failed to yield a positive correlation with survival or death. Gerber and Moody (1972) concluded that the carotid flow provides a parameter by which GSW-induced death of the brain could be consistently predicted. This conclusion, however, was based only on a small number of animals in each category. In addition, the effect of an open cranium on survival rate in these monkeys remains unknown.

Perhaps the most comprehensive studies on experimental GSW of the brain were those conducted by a group of investigators from the University of Chicago who used rhesus monkeys (5 to 8 kg) as models (Crockard et al., 1976, 1977, 1977a, 1979; Crockard, 1979; Brown et al., 1979, 1982; Levett et al., 1980). They produced a standard occipitofrontal missile injury using an air rifle and a 3-mm diameter (0.31 g) BB pellet. The monkeys were first tranquilized with 1 to 2 mg/kg phencyclidine hydrochloride for insertion of an endotracheal tube, and then supplemented with 2 mg/kg pentobarbital IV for the surgical preparation about 1 h before injury. If required, after injury, the animals were further supplemented hourly with 1 mg/kg phencyclidine hydrochloride and barbiturate. They produced GSWs with velocities of 56, 90, and 180 m/s with trajectories 5° divergent from the sagittal plane. The missiles passed through an occipital burr hole that was made toward the upper outer quadrant of the right orbit. The burr hole was closed immediately after GSW with the missile trajectory producing a clean wound in the brain tissue, away from the vital structures and blood vessels (nonsignificant intracranial clots or subdural/epidural hematomas). The missile passed through the white matter lateral to the lateral ventricle, across the top of the thalamus. Using this standard model, they recorded various central and peripheral physiological parameters before and after wounding. The main variables studied were ICP, CBF (xenon technique), MABP, HR, ECG, respiration, arterial and sagittal blood gases, cerebral metabolic rate for oxygen (CMR-O_2), and CMR-lactate. The CVR and CPP were calculated and EEG and somatosensory evoked potentials (SEP) from the sensory cortex were recorded. The energy levels chosen for these studies were 1.26 J (standard energy), 5.02 J (high energy), and 0.49 J (low energy). Since the missile was fired through the exposed skull and dura, all the missile energy was considered to have been absorbed by the brain and its covering. These investigators then tried to correlate the extent of alterations in the above parameters to the wounding energy.

X. GSW-INDUCED INTRACRANIAL AND SYSTEMIC PRESSURE CHANGES IN THE PRIMATE MODEL

The low-energy injury of 0.49 J in this model was not lethal and produced no significant changes in respiration, blood gases, and MABP. Exceptionally, a moderate increase in ICP to about 25 mmHg was developed. The standard energy, on the other hand, produced 45% mortality in about 6 h after injury, with most of the measured parameters exhibiting various degrees of abnormality. In animals that survived less than 1 h, the MABP fell during the first few minutes after wounding. This was then followed by a rise above control values at 5 to 7 min after injury, and a secondary drop to severe hypotensive levels. The survivors, on the other hand, exhibited a slight increase in MABP within few minutes of wounding, followed by a mild hypotension about 10 min after wounding (Crockard et al., 1976, 1977). In contrast, the ICP rose up to 60 to 90 mmHg within a few seconds of impact, subsiding to still moderately elevated levels of 20 to 35 mmHg for the rest of the animal's life. Consequently, the CPPs were reduced by 50%. The high-energy injury produced persisting apnea in four of seven monkeys, so they required immediate ventilatory support. The ICP in these animals rose to peak values of 80 to 200 mmHg, leading to CPPs of only 30 to 35 mmHg, which were significantly lower than those in the standard energy group, resulting in a 100% mortality within 1 h of injury (Crockard et al., 1976, 1977).

XI. GSW-INDUCED CIRCULATORY, METABOLIC, AND ELECTRICAL CHANGES IN THE PRIMATE MODEL

The standard energy injury reduced the CBF by 50% (Crockard et al., 1979b). Concurrently, CVR fell initially and then rose to very high values at 30 min postwounding. No relationship was found between CPP and CBF, indicating autoregulatory breakdown. However, while the CPP is very important for maintaining a normal CBF, following GSW to the brain the CPP as conventionally calculated (MABP−ICP) bears little relationship to the events at the capillary level (Crockard et al., 1977a). The CMR-O_2 also fell after injury and remained low for the rest of the animal's life. Conversely, CMR-lactate rose immediately after injury, persisted for 5 h, and then fell again, indicating progressive reduction in oxidative metabolism. Brain functional activity as determined by EEG showed that during the first minute after the wounding there were increasing EEG changes, consisting of flattening and slowing of the baseline activity. With high energy, the EEG showed periods of isoelectric recording, but there was never an immediate extinction of the EEG after wounding. In the standard injury group, 20% of monkeys had epileptic seizures with generalized or focal convulsions. Recording of the SEP, at the same time periods when CBF and CMR-O_2 were measured, showed a significant reduction in the amplitude with the standard injury. Although, at the same time, CPP was greatly reduced, no correlation between these two variables was found. In contrast, a close relationship existed between SEP and the hemispheric CBF as early as 5 min after injury. The relationship between CBF and SEP also existed in individual animals with serial measurements. The lowest flows were observed at the time when SEP was absent (CBF of 11 and 18 ml/100g/min) in two cases that required ventilatory support. On the other hand, SEP reappeared in several animals subsequent to increased flow.

The clinical significance of these flow-function relationships are obvious, indicating the need for adequate capillary perfusion and tissue oxygen and nutrition (Crockard et al., 1977a). In practical terms, however, the therapeutic strategies to achieve such goals are less clear. Nevertheless, the investigators recommended that the flow must be maintained by preserving a good CO, which was found to be a major factor in the control of CBF following a GSW to the brain (Crockard, 1979; Crockard et al., 1979). In addition, tissue oxygen should be raised by high inspired oxygen level, and CPP should be increased by the reduction of GSW-induced increase in ICP. In fact, they tried to achieve part of these goals by administration of mannitol 15 and 30 min after a standard injury (Brown et al., 1979). In this study, control CBF fell dramatically to as low as 16 ml/100g/min at 30 min postwounding. Administration of mannitol vastly improved the CBF to 32.5 ml/100g/min ($p < 0.001$) at 30 min, and to virtually baseline of 39.4 ml/100g/min at 6 h postwounding. The net effect of mannitol was to stabilize the CVR. Monkeys that started to receive mannitol at 1 h following GSW to the brain also showed similar improvement in MABP, CPP, and CVR, but the ICP did not differ significantly from that of the untreated animals. Moreover, early administration of mannitol (15 min) significantly improved CMR-O_2 at 6 h, as compared to the group that received mannitol at 1 h after GSW (Brown et al., 1979).

XII. THE FELINE MODEL OF GSW TO THE BRAIN

More recently, Carey et al. (1989) developed a cat model of GSW to the brain in order to study pathophysiology and to evaluate treatments to lower mortality and morbidity. This promising project, however, was prematurely curtailed because of political pressure imposed by animal-rights activists (Carey, 1990). At the time of this review, only a small number of full articles and abstracts on the cat model had been published, and more data remain to be evaluated for publication in the future. The following discussion, therefore, concerns only a brief review of the already published articles or abstracts (Torbati et al., 1988, 1988a, 1988b, 1989, 1990, 1990a, 1990b, 1991, 1992; Torbati and Carey, 1989, 1989a; Carey et al., 1989, 1990; McKowen et al., 1989; Awasthi et al., 1991; Soblosky et al., 1992). Scientists who are interested in further details of the feline model of GSW to the brain may refer to several annual reports to the U.S. Army Medical Research and Development Command, which have been approved for public release (Carey, 1983–1985, 1986–1991; Carey et al., 1986–1987, 1987–1988, 1988–1989, 1989–1990).

These investigators used unselected mongrel cats of 3 to 4 kg to study a standard GSW to the brain under pentobarbital anesthesia. In this model, a small (31 mg, 2 mm diameter) steel sphere was used to create a relatively nonlethal GSW. To achieve a precise trajectory, the anterior wall of the right frontal sinus was removed, while the posterior sinus wall remained intact, thus maintaining a closed cranium before wounding. Another characteristic of the cat model was a well-defined anteroposterior trajectory, passing through the right cerebral hemisphere with occiputs angled 20° to the left. This deviation from the midsagittal line prevented intracerebral bleeding into the ventricles avoiding blocking of the cerebrospinal fluid (CSF) passages, and thus created a reproducible trajectory with frontal entry towards the lateral-occipital lobe (Carey et al., 1989). In this model, a 100% mortality (LD_{100}) occurred with missile energies of 2.4 J, mostly because of an immediate respiratory arrest, and the LD_{50} was obtained at about 1.4 J. These observations in cats once again confirmed the classic observations of Horsley (1894) and others (Webster and Gurdjian, 1943; Crockard et al., 1979), indicating that the primary cause of death is an immediate cessation of breathing, in which case prompt respiratory support, in many cases, can restore spontaneous respiratory activity. This also implies that humans who receive a brain wound may die from missile-induced apnea instead of brain damage per se (Carey et al., 1989).

XIII. GSW-INDUCED INTRACRANIAL AND SYSTEMIC PRESSURE CHANGES IN THE FELINE MODEL

Carey et al. (1989), using 0.9, 1.4, and 2.5 J energy for injury, usually observed a sharp rise in MABP, starting a few seconds after woundings. This GSW-induced systemic hypertension peaked at approximately 1 min and at 4 min returned to control levels, where it stabilized. The transient postwounding hypertension was significant for all three energy groups, but it did not correspond with missile energy. All wounded cats developed an abrupt rise in ICP, which then fell slowly to a still-elevated, but lower, level than its peak. Unlike MABP, both the immediate and long-lasting intracranial hypertensions were proportional to wound energy. The CPP became uniformly elevated only at 1 min after wounding due to markedly elevated MABP. However, because the elevated MABP declined to normotensive levels within 5 min of wounding, while the increase in ICP was relatively maintained, the corresponding CPPs fell significantly below the prewounding baseline. The decrease in CPP was energy dependent, showing 5, 23, and 46% reduction with 0.9, 1.4, and 2.5 J, respectively (Carey et al., 1989). The pattern of GSW-induced alterations in MABP in the feline model was not similar to that observed in the primate model, where a standard energy of 1.26 J was used for wounding (Crockard et al., 1976, 1977). The difference in MABP between the two models was only evident when comparing wounded cats with nonsurviving monkeys, which died within 1 h of injury. Surviving monkeys, like in the cat model, initially exhibited an increase in MABP. However, unlike the cat model, the MABP in the surviving monkeys declined significantly, at 30 min postwounding (Crockard et al., 1976, 1977). On the other hand, alterations in ICP were more similar in both models, where they led to a significant decrease in the CPP (50% in monkeys and 23 to 46% in cats). Since the brainstem effects of GSW in the feline model resemble those of the primate model (see below), despite the species differences in MABP response, one may infer that perhaps the neuroanatomical differences between them do not substantially affect brain responses to GSW. In that case, the economic advantages of the feline model vs. primate model, at least for drug screening, is obvious.

XIV. GSW-INDUCED CBF REGULATORY IMPAIRMENT IN THE FELINE MODEL

Regulation of CBF is a complex process involving mechanical (systemic and intracranial pressures), chemical (O_2, CO_2), metabolic (adenosine, NO), and neurogenic factors (Miller et al., 1978; Kontos et al., 1978; Wagner and Traystman, 1986; Bevan and Bevan, 1993; Pelligrino et al., 1993; Buchanan and Phillis, 1993). By definition, autoregulation of CBF is the ability to maintain CBF unchanged despite a decrease in MABP, CPP, or an increase in ICP (Miller et al., 1978). CBF autoregulation may be impaired following brain injury, and by a variety of cerebral and extracerebral insults, specifically intracranial hypertension or hemorrhage, cerebral hypoxemia, ischemia, and hypercapnia, as well as systemic hemorrhagic hypotension (Crockard et al., 1979; Saunders et al., 1979; Lewelt et al., 1980, 1982; Feustel and Nelson, 1987; DeWitt et al., 1986, 1992; Bouma and Muizelaar, 1992). While most of these pathologic events are a common by-product of GSW to the brain, high concentrations of oxygen (hyperoxia), and hyperventilation-induced hypocapnia (to reduce ICP) may be used for emergency resuscitation. Thus, regulatory mechanisms of CBF are confronted with both pathological insults and therapeutic measures.

To examine how a wounded brain may react to mechanical and chemical regulatory factors, the cerebrovascular responses to changes in systemic and intracranial pressures, hypoxia, hypercapnia, hypocapnia (hyperventilation), hyperoxia, and a combination of hyperoxia-hypercapnia were studied in the feline model of GSW to the brain (Torbati and Carey, 1989, 1989a; McKowen et al., 1989; Torbati et al., 1988, 1988a, 1988b, 1989, 1990, 1990a). These investigators measured regional CBF (rCBF) by the microspheres technique before and after a GSW to the brain in artificially ventilated, anesthetized, paralyzed cats. The published results and conclusions indicate that both CBF autoregulation, reactivity to oxygen from hypoxia to hyperoxia, and reactivity to CO_2 from hypocapnia to hypercapnia, are severely impaired following a GSW to the brain. A clinically important observation was that brain-wounded cats, which were made simultaneously hypotensive by arterial bleeding, did not autoregulate CBF, and their brain became ischemic at relatively high levels of MABP. Blood reinfusion, while not completely restoring MABP, further decreased CBF. The CVR was significantly reduced only during hemorrhagic hypotension without injury (indicating intact autoregulatory ability), whereas it was significantly increased during hemorrhagic hypotension accompanied by GSW to the brain or even with GSW alone (Torbati et al., 1988, 1989). These results are in agreement with other studies in cats subjected to fluid percussion brain injury (Lewelt et al., 1980; DeWitt et al., 1992). In fact, mild hemorrhagic hypotension in cats subjected to percussion injury was associated with significant reductions in rCBF, O_2 delivery, and neurologic deficiency (DeWitt et al., 1992).

Torbati et al., (1990) also examined whether GSW-induced autoregulatory impairment is due to intracranial hypertension, the wound itself, or the associated hemorrhagic hypotension. In this study, they measured changes in rCBF at MABP of 131 (controls), 111, 96, and 70 mmHg, while the corresponding ICPs (artificially increased by infusion of mock CSF through cisterna magna) were 7, 82, 59, and 42 mmHg, respectively. The resultant CPPs were on the average 123 (controls), 30, 35, and 28 mmHg, respectively. Despite these severe reductions in CPP, the CVR in all hypotensive challenges was significantly reduced, indicating a fair autoregulatory ability in the face of high levels of ICPs and blood loss without injury. In fact, combination of severe hemorrhagic hypotension with a mild increase in ICP (42 mmHg) did not produce significant changes in rCBF even in the face of CPPs as low as 28 mmHg. In addition, blood reinfusion in the presence of intracranial hypertension (about 60 mmHg), restored any reduction in rCBF caused by the combination of hemorrhagic hypotension and intracranial hypertension. These data indicate that GSW-induced, or other brain injury-induced, CBF autoregulatory impairment is not entirely related to intracranial hypertension, hemorrhagic hypotension, or even their combination (Torbati et al., 1990).

The vasodilatory effects of hypoxia and hypercapnia (Kontos et al., 1978; Pelligrino et al., 1993) after GSW to the brain were absent, and these challenges did not produce increases in rCBF. In some neuroanatomical structures, hypoxia and hypercapnia were even able to reverse the expected vasodilatory effect into a vasoconstrictive effect (Torbati et al., 1988a). These observations are in agreement with impaired hypoxic and hypercapnic reactivity after injury in other models of brain trauma (Saunders et al., 1979; Lewelt et al., 1982; Feustel and Nelson, 1987; Ishige et al., 1987; Bouma and Muizelaar, 1992; Takeuchi et al., 1992). Thus, GSW-induced hypoventilation, or transient apnea, may be detrimental to postwounding recovery because of the potential to create both hypoxemia and hypercapnia (Torbati et al., 1988a).

Torbati and Carey (1989, 1989a), and Torbati et al. (1990a) also briefly studied the effect of hyperventilation-induced hypocapnia, hyperoxia (100% O_2), and hyperoxia-hypercapnia (95% O_2 plus 5% CO_2) on cerebrovascular reactivity in cats subjected to GSW of the brain. The data from five cats demonstrated that postwounding hyperventilation (PCO_2 about 21 mmHg), unlike the prewounding response, was able to significantly reduce rCBF only about the wound track and in the white matter. While the responsiveness to hypocapnia was attenuated elsewhere in the brain, it had no effect on ICP reduction (Torbati and Carey, 1989a). In addition, following a GSW to the brain in six cats, the well-known vasoconstrictive effect of hyperoxia in the normal brain (see Chapter 2 in this publication) was significantly enhanced. This effect of hyperoxia may counteract the expected beneficial effect of hyperoxygenation to reduce cerebral hypoxia after wounding (Torbati and Carey, 1989). Interestingly, hyperoxia (95% O_2) in the presence of hypercapnia (5% CO_2), in three wounded cats, seemed to attenuate the previously observed postwounding vasoconstrictive effect of the hyperoxia alone, except in the damaged tissue around the wound track (Torbati et al., 1990a). Whether hyperoxygenation after GSW to the brain could be improved by addition of CO_2 remains to be reevaluated.

XV. GSW-INDUCED BRAINSTEM EFFECTS IN THE FELINE MODEL

It appears that brainstem dysfunction is a common pathway for both primary and secondary damage to the brain in several models of traumatic brain injury, including GSW to the brain (Crockard et al., 1979; Levett et al., 1980; Gennarelli et al., 1982a; Carey et al., 1989; Shima and Marmarou, 1991). Exceptionally, there are now also brain trauma models that are purposely designed to prevent extensive brainstem damage (Marmarou et al., 1994). Torbati et al. (1992) evaluated the cardiovascular and respiratory variables in the cat model of the GSW to the brain. In this study, various cerebral, cardiovascular, and respiratory variables were simultaneously recorded in pentobarbital-anesthetized cats up to 90 min after wounding by 1.4 J. Four wounded cats, who survived a 90-min postwounding period, showed only a transient brainstem effect, including a 50% increase in MABP, concurrently with 50% reductions in the breathing frequency (f) and HR. Eleven nonsurviving cats lived from 1 to 41 min after wounding. These cats initially demonstrated similar changes in MABP, f, and HR to survivors, but these variables became unstable, possibly indicating persisting brainstem damage. Although apnea accounted for death in 10 of 11 nonsurvivors, only one cat died from reduced respiration alone. Others exhibited one or several postwounding secondary complications, including abrupt increased ICP with negative CPP, and extreme reduction in CO or CBF and ventilation. Interestingly, one cat died from cardiac arrest (Torbati et al., 1992). Thus, the immediate cardiorespiratory brainstem effects of GSW to the brain in the feline model was similar to that noted in many other species sustaining brain injury. While both missile and nonmissile injuries to the brain may cause an immediate fatal apnea, the later intracranial and systemic events may be species-specific. In the feline model of GSW to the brain, mortality was not consistently related to a single physiological variable (Torbati et al., 1992), as experienced in some cases in the primate model of missile wounding, or angular acceleration of the head (Gerber and Moody, 1972; Crockard et al., 1977b; Gennarelli, 1982a).

Neurochemical alterations may mediate important components of brain pathophysiology associated with brain injury (Hayes et al, 1992). In this regard, Soblosky et al. (1992) investigated the acute effects of GSW on brainstem and hypothalamic biogenic amines in the feline model. They also tried to separate the effects of brain wounding from its associated increase in ICP (infusion of mock CSF into the cisterna magna) without injury. This study demonstrated that changes in hypothalamic and brainstem biogenic amines, whether resulting from GSW to the brain or increased ICP without wounding, were reflective of a stress response. GSW-induced brainstem distortion did not appear to be a factor and monoaminergic systems appeared to remain intact despite a severe and eventually lethal brain injury (Soblosky et al., 1992).

XVI. NEUROPATHOLOGY OF GSW TO THE BRAIN

Allen et al. (1982, 1983), presented a short historical review of GSW neuropathology in humans, and investigated the neuropathology of a high-velocity GSW to the brain in rhesus monkeys. The animals in that study were anesthesized with chloral urethane, and injured with a steel sphere (3.2 mm) with a mass that was proportional in volume to the rhesus monkey's cranial cavity as a modern small-caliber, military rifle bullet would be to that of the human. The impact velocity (1000 m/s) was also similar to that of a military rifle bullet at short range. A left-to-right transfrontal trajectory at the temporal region just above the orbit was produced. Light microscopic examination showed pathologic lesions not only in the missile trajectory, but also at sites more remote from the track, including hypothalamus, brainstem, and cerebellum. Diffuse subarachnoid hemorrhage was commonly found, and intraventricular hemorrhage was a constant feature. Perivascular hemorrhage was frequently observed, and was suggested to be the site of early edema. In the ultrastructural study of high-velocity GSW to the brain the monkeys were subjected to the same procedures for injury (Allen et al., 1983). The investigators found widespread swelling of perivascular astrocytes within 30 min of the injury, attributing it to a direct effect of force, chemical mediation secondary to extravasation of blood, alterations in the permeability of the BBB, and, finally, ischemia.

The above findings were somewhat different from those observed in the feline model by Carey et al. (1991). The latter investigators studied the type of edema and its magnitude and time course following an uncomplicated GSW to the brain with a velocity of 240 to 300 m/s (0.9 to 1.4 J) and surveyed brain edema in the entire brain, by the wet weight/dry weight technique, 6 h to 7 d postwounding. Using their standard frontooccipital right cerebral hemispheric injury (Carey et al., 1989), they found brain edema only in the white matter about the missile trajectory in the injured hemisphere. Corresponding significant increases in white matter Na also occurred, while K levels remained unchanged. This vasogenic edema around the wound track was relatively mild, peaked at 24 to 48 h after wounding, and resolved spontaneously within a week. Carey et al. (1990) inferred that brain edema following an uncomplicated GSW to the brain in humans is also mild, and may resolve spontaneously. This speculation, however, does not account for nonsurviving cats, in 1.4 J injury, who might have died because of extensive development of cerebral edema and perhaps ischemia.

XVII. COMMON FEATURES OF GSW TO THE BRAIN AND OTHER MODELS OF NEUROTRAUMA

The question is whether the pathophysiology of the experimental GSW of the brain shares similar mechanisms with other models of neurotrauma. By definition, the latter category may include not only closed head injuries produced by impact acceleration to the whole head, angular acceleration models, or cerebral concussion produced by the fluid percussion method, but also models simulating the consequences of a physical impact to the brain, namely, intracranial hypertension, ischemia, hypoxia, and edema. Theoretically, a positive answer to the above question may imply that one can learn and extrapolate from studying various models of neurotrauma (either *in vivo* or *in vitro*) to understand the pathogenesis of the missile wound to the brain. To clarify this question, Brown et al. (1982) designed specific experiments in rhesus monkeys comparing cerebral and cardiac changes between GSW to the brain and the closed head injury by the impact-acceleration technique. They found a comparable reduction in CO in the impact-acceleration model to that occurring in the GSW to the brain. They suspected that perhaps the reduced CO is a common problem in the early stages of both closed head injury and GSW to the brain. In both models, the CO remained low even after the HR had returned to normal. In contrast to GSW to the brain, MABP was not reduced significantly in the acceleration model in which CO was partially compensated for by an increase in peripheral vascular resistance. Since ICP increases in the acceleration injury were relatively modest, less reduction in CPP occurred, as compared to GSW of the brain. Similarly, CBF did not fall as dramatically in the acceleration injury as in the GSW. Interestingly, however, the CMR-O_2 did fall dramatically in the acceleration injury. In the GSW to the brain, the CMR-O_2 changes paralleled those of CBF, while in the acceleration injury the CMR-O_2 fell much more than CBF. Accordingly, Brown et al. (1982) suggested that the reduced oxygen extraction from the microcirculation, which is out of proportion to the flow changes, indicates that brain cells may be more diffusely impaired in the acceleration injury than in the GSW to the brain. These investigators also suggested that changes in CO after both types of CNS injury could be reversed through autonomic intervention (cholinergic blockade).

Another neuropathologic effect of GSW to the brain could be an early or delayed development of cerebral hypoxic-ischemia. One reason GSW to the brain can lead to an early cerebral ischemia is that the sudden increase in ICP may be severe, or MABP may fail to maintain an adequate pressure differential (CPP) between itself and the increased ICP. Another possibility for the onset of focal or extensive cerebral ischemia is the gradual development of cerebral edema. Although in the feline model of GSW to the brain mortality from cerebral ischemia alone was a rare event, the total CBF was significantly reduced at 90 min postwounding (Torbati et al., 1989, 1992). It is therefore possible that with a delayed development of cerebral edema, systemic hypotension, inadequate oxygenation, or cerebral regulatory impairment, these cats too may become prone to hypoxic ischemia.

Clinical studies measuring CBF and other physiological variables repeatedly up to 10 d after severe injury have demonstrated that patients with reduced CBF had the highest mortality and the poorest outcome compared to patients with a normal or raised CBF (Robertson et al., 1992). In this study 10% of the 102 patients had GSW to the brain, but there were no differences in the type of injury, initial score on the Glasgow Coma Scale (GCS), mean ICP, or the highest ICP among groups with low, normal, and high CBF. In fact, patients with GSW of the brain were evenly distributed among all CBF groups. Among variables such as age, initial GCS, hemoglobin concentration, CPP, and CMR-O_2, a reduced CBF remained significantly associated with an unfavorable neurologic outcome. Robertson et al. (1992), therefore, reasoned that transient periods of hypoxia, hypotension, or intracranial hypertension are likely to result in ischemia in patients with reduced CBF reservoirs. The author's opinion on delayed neurodegenerative consequences of GSW to the brain (cellular, neurochemical, and molecular mechanisms) is that they may follow a similar pathologic pathway as those present in cerebral edema and hypoxic-ischemia. The development of the GSW-induced secondary insults, as speculatively illustrated in Figure 1, particularly, the increased level of oxidative stress following brain injury (Torbati et al., 1990b, 1991), may indicate that the outcome of GSW to the brain could benefit from other studies concerning various models of neural trauma, particularly, closed head injury, hypoxia, edema, and ischemia. In other words, if the proposed pathologic events are indeed responsible for delayed neurological deficiencies following GSW to the brain, then relatively rich knowledge of the underlying mechanisms of local or global ischemia, hypoxia, and/or edema are useful for understanding, at least in part, the complex pathology of GSW to the brain.

XVIII. PROGNOSIS AND MANAGEMENT OF GSW TO THE BRAIN

In the primate model of GSW to the brain, using conventional forms of data analysis, the length of survival after injury correlated well with alterations in MABP, ICP, and CBF. However, each of these parameters, separately, were not satisfactory for predicting an individual's prognosis (Crockard et al., 1977b). Using the technique of multiple linear regression analysis, on the other hand, the life expectancy of the individual animals could be predicted with great accuracy, about 30 min postwounding (Crockard et al., 1977b). Whether prognosis in the primate model of GSW to the brain would be exactly applicable to civilian GSWs to the brain is not yet clear. However, some of the early brainstem effects of experimental GSW to the brain may be useful prognostic indicators. These include, for example, unstable MABP, f, and HR that developed in the nonsurviving spontaneously breathing cats subjected to GSW to the brain with 1.4 J (Torbati et al., 1992). Clinical studies that retrospectively have reviewed the factors that may be predictive of survival following GSW to the brain indicate that the GCS scores on admission can serve as the most important factor in predicting survival. Age, the presence of extruded brain, and use of a shotgun could not be shown to be factors independent of admision GCS score (Selden et al., 1988; Kennedy et al., 1993).

West (1980) and more recently Carey (1987) and Kaufman (1993a) have reviewed the history of the management of GSW from ancient times to the modern era. Accordingly, it appears that the Horsley's (1894) recommendations for the immediate treatment of GSW victims have not been basically changed over the last 100 years. During World War II, Ascroft (1943) laid down the principles and priorities for the first aid of the GSW to the brain inflicted by military missiles, starting with resuscitation, followed by preparation for operation, and finally operation of different craniocerebral injuries. Generally, these therapeutic concepts are also valid for the civilian GSW to the brain and had not been radically altered at the time of this review (Crockard, 1979; Hubschmann et al., 1979; Nagib et al., 1986; Helling et al., 1992). In his comprehensive review article on penetrating craniocerebral missile injuries, Crockard (1979) laid the foundation of the critical care and prognosis of GSW patients. He stated that the recognition and treatment of the secondary pathophysiologic effects of GSW, both intracranially and systemically, are vital in the management and eventual outcome of patients. In his opinion, it is only after vigorous attention to the epiphenomena of the injury that there is a place for surgery, and the timing of surgery is comparatively less important than the immediate institution of resuscitative measures, especially care of the airways.

The current emphasis in the treatment of the GSW to the brain is again on the prompt instigation of resuscitative measures, including an immediate care of the airways, replacement of fluid loss by intravenous transfusion, and the control of the ICP. Prevention of cerebral swelling and administration of antibiotics are also among the first aid priorities (Ascroft, 1943; Hubschmann et al., 1979; West, 1980). Once the vital functions have stabilized, the extent of injury can be precisely assessed using modern radiologic techniques (Selden et al., 1988; Phillips, 1992; Levy et al., 1993; Kennedy et al., 1993). At this stage, surgical intervention aimed at arresting hemorrhage, evacuating intracranial hematoma, debridment, and removal of bone, missile, and metal fragments (if readily accessible), as well as repairing of the dura matter and closing of the scalp should take place (Byrnes et al., 1974; West, 1980; Helling et al., 1992). Surgical management of GSW to the brain has remained a controversial issue in the care of civilian patients. In an attempt to determine who might benefit from aggressive surgical intervention, Helling et al. (1992) examined 89 patients who had sustained GSW to the brain. Their survey suggests that early surgical intervention seemed to result in better survival but was frequently undertaken in those patients who were less neurologically impaired and who manifested evidence of mass lesions on scanning. In addition, there was evidence to suggest that even pessimistic situations (bihemispheric injuries) might benefit from aggressive management (Helling et al., 1992). On the other hand, Brandvold et al. (1990) suggest a less aggressive surgical approach. This conclusion was based on their analysis of the clinical outcome of brain-wounded subjects who were followed up during an average period of about six years after GSW to the brain. These investigators suggest that the initial debridement could be tempered in an effort to preserve more cerebral tissue and it is unnecessary to reoperate for retained bone fragments (Brandvold et al., 1990).

In conclusion, Graham et al. (1990) provide a general guideline using the results of their prospective study of the civilian GSW to the head in 100 consecutive patients. They have presented an algorithm for treatment priorities, which is based on GCS score. It is recommended that all cranial GSW patients should initially receive aggressive resuscitation. Afterward, patients with stable vital signs should be examined by computerized tomographic scan. Providing that the patient's GCS score after resuscitation is three to five, and no operable hematomas are present, then no further therapy should be offered. However, all patients with a GCS score greater than five should receive aggressive surgical therapy (Graham et al., 1990).

ACKNOWLEDGMENT

The GSW studies in the cat model reported in this review were supported by a U.S. Army research fund to Michael M. Carey, M.D. The author was an investigator in a part of the feline model studies on GSW to the brain from 1987 to 1991. Data reviewed here on the feline model are only those that have already been published, either as full articles or as abstracts. Nevertheless, because of previous involvement in the feline project, the author wishes to declare that the views and conclusions expressed in this review article are those of the author only, and should not be construed as an official Department of the Army position, or as the opinions of any other investigator in that project.

REFERENCES

Aarabi, B. (1990) Surgical outcome in 435 patients who sustained missile head wounds during the Iran-Iraq war. *Neurosurgery* 27, 692–695.

Allen, I. A., Scott, R., and Tanner, J. A. (1982) Experimental high-velocity missile head injury. *Injury* 14, 183–193.

Allen, I. A., Kirk, J., Maynard, R. L., Cooper, G. K., Scott, R., and Crockard, H. A. (1983) An ultrastructural study of experimental high velocity penetrating head injury. *Acta Neuropathol. (Berl)* 59, 277–282.

Ascroft, P. B. (1943) Treatment of head wounds due to missiles. *Lancet* 2, 211–218.

Awasthi, D., Rock, W. A., Carey, M. E., and Farrell, J. B. (1991) Coagulation changes after an experimental missile wound to the brain in the cat. *Surg. Neurol.* 36, 441–446.

Barnette, J. C. and Meirowsky, A. M. (1955) Intracranial hematomas associated with penetrating wounds of the brain. *J. Neurosurg.* 12, 34–38.

Benzel, E. C., Day, W. T., Kesterson, L., Willis, B. K., Kessler, C. W., et al. (1991) Civilian craniocerebral gunshot wounds. *Neurosurgery* 29, 67–72.

Berlin, R. (1979) Energy transfer and regional blood flow changes following missile trauma. *J. Trauma* 19, 170–176.

Bevan, J. A. and Bevan, R. D. (1993) Is innervation a prime regulator of cerebral blood flow? *NIPS* 8, 149–153.

Bouma, G. J. and Muizelaar, J. P. (1992) Cerebral blood flow, cerebral blood volume, and cerebrovascular reactivity after severe head injury. *J. Neurotrauma* 9(Suppl 1), S333–S348.

Brandvold, B., Levi, L., Feinsod, M., and George, E. D. (1990) Penetrating craniocerebral injuries in the Israeli involvement in the Lebanese conflict, 1982–1985. *J. Neurosurg.* 72, 15–21.

Brown, F. D., Johns, L. M., Crockard, H. A., and Mullan, S. (1979) Response of mannitol following experimental cerebral missile injury. In: *Neural Trauma,* Popp, A. J. ed., Raven Press, New York, pp. 281–287.

Brown, F. D., Jafar, J. J., Krieger, K., Johns, L., Leipzig, T. J., and Mullan, S. (1982) Cardiac and cerebral changes following experimental head injury. In: *Head injury: Basic and Clinical Aspects.* Grossman, R. G. and Gillenberg, P. L. eds., Raven Press, New York, pp. 151–157.

Buchanan, J. E. and Phillis, J. W. (1993) The role of nitric oxide in the regulation of cerebral blood flow. *Brain Res.* 610, 248–255.

Butler, E. G., Puckett, W. O., Harvey E. N., and McMillan, J. H. (1945) Experiments on head wounding by high velocity missiles. *Neurosurgery,* 2, 358–363.

Byrnes, D. P., Crockard, H. A., Gordon, D. S., and Gleadhill, C. A. (1974) Penetrating craniocerebral missile injuries in the civil disturbances in Northen Ireland. *Br. J. Surg.* 61, 169–176.

Carey, M. E., Young, H. F., and Mathis, J. L. (1972) The neurosurgical treatment of craniocerebral missile wounds in Vietnam. *Surg. Gynecol. Obstet.* 135, 386–390.

Carey, M. E. (1983–1985, 1986–1991) The effect of experimental missile wound to the brain on brain electrolytes, regional cerebral blood flow and blood brain barrier permeability. Final summary reports to U.S. Army Med. Res. Dev. Command., Fort Detrick, Frederick, MD.

Carey, M. E. (1987) Learning from traditional combat mortality and morbidity data used in the evaluation of combat medical care. *Mil. Med.* 152, 6–13.

Carey, M. E. (1990) Neurosurgeons must fight animal activists. *Surg. Neurol.* 33, 373–377.

Carey, M. E., Torbati, D., Soblosky, J., McKown, H. C., Farrell, J. B., and Davidson, J. (1986–1987) An experimental brain missile wound: ascertaining pathophysiology and evaluating treatments to lower mortality and morbidity. Annual report to U.S. Army Med. Res. Dev. Command., Fort Detrick, Frederick, MD.

Carey, M. E., Torbati, D., Soblosky, J., Farrell, J. B., and Davidson, J. (1987–1988; 1988–1989; 1989–1990) An experimental brain missile wound: ascertaining pathophysiology and evaluating treatments to lower mortality and morbidity. Annual reports to U.S. Army Med. Res. Dev. Command., Fort Detrick, Frederick, MD.

Carey, M. E., Sarna, G. S., Farrell, J. B., and Happel, L. T., (1989) Experimental missile wound to the brain, *J. Neurosurg.* 71, 754–764.

Carey, M. E., Sarna, G. S., and Farrell, J. B. (1990) Brain edema following an experimental missile wound to the brain. *J. Neurotrauma,* 7, 13–20.

Cheng, X. Y., Feng, T. S., Liu, Y. Q., Ma, Y. Y., Wu, B. J. et al. (1988) Wounding properties of steel pellets with different velocity and quality on soft tissues of dogs. *J. Trauma* 28, S33–S36.

Clark, W. C., Muhlbaner, M. S., Watridge, C. B., and Ray, M. W. (1986) Analysis of 76 civilian craniocerebral gunshot wounds. *J. Neurosurg.* 65, 9–14.

Crockard, H. A., Brown, F. D., Calica, A. B., and Mullans, S. (1976) ICP, CVR and cerebral metabolism following experimental cerebral missile injury. In: *Intracranial Pressure III,* Beks J. W. F., Bosch D. A., and Brock M., eds., Springer-Verlag, New York, pp. 73–78.

Crockard, H. A., Brown, F. D., Johns, L. M., and Mullan, S. (1977) An experimental cerebral missile injury model in primates. *J. Neurosurg.* 46, 776–783.

Crockard, H. A., Brown, F. D., Trimble, J., and Mullan, J. F. (1977a) Somatosensory evoked potentials, cerebral blood flow and metabolism following cerebral missile trauma in monkeys. *Surg. Neurol.* 7, 281–287.

Crockard, H. A., Brown, F. D., Calica, A. B., Johns, L. M., and Mullan, S. (1977b). Physiological consequences of experimental cerebral missile injury and use of data analysis to predict survival. *J. Neurosurg.* 46, 784–794.

Crockard, H. A. (1979) Penetrating craniocerebral missile injuries. *Int. Anesth. Clin.* 17, 307–326.

Crockard, H. A., Johns, L., Levett, J., Brown, F., and Mullan, S. (1979) "Brainstem" effects of experimental cerebral missile injury. In: *Neural Trauma.* Popp, A. J. ed., Raven Press, New York, pp. 19–25.

Crockard, A., Iannotti, F., and Kang, J. (1982) Posttraumatic edema in the gerbil. In: *Head Injury: Basic and Clinical Aspects.* Grossman, R. G. and Gildenberg, P. L. eds., Raven Press, New York, pp. 159–168.

DeMath, W. E., Jr. (1969) Bullet velocity as applied to military rifle wounding capacity. *J. Trauma* 9, 27–38.

DeWitt, D. S., Jenkins, L. W., Wei, E. P., Lutz, H., Becker, D. P., and Kontos, H. A. (1986) Effects of fluid-percussion brain injury on regional cerebral blood flow and pial arteriolar diameter. *J. Neurosurg.* 64, 787–794.

DeWitt, D. S., Proughm, D. S., Taylor, C. L., and Whitley, J. M. (1992) Reduced cerebral blood flow, oxygen delivery, and electroencephalographic activity after traumatic brain injury and mild hemorrhage in cats. *J. Neurosurg.* 76, 812–821.

Djordjevic, M., Lofgren, J., Steiner, L., and Zwetnow, N. N. (1976) Intracranial pressure effects of missile wounds. In: *Intracranial Pressure III.* Becks J. W. F., Bosch D. A., and Brock M. eds., New York, Springer-Verlag, pp. 79–84.

Fackler, M. L., Bellamy, R. F., and Malinowski, J. A. (1988) The wound profile: illustration of the missile-tissue interaction. *J. Trauma* 28, S-21-S29.

Feustel, P. J. and Nelson, L. R. (1987) Cortical blood flow regulation during hypoxemia in experimental head injury. *J. Surg. Res.* 43, 86–93.

Gennarelli, T. A., Adams, J. H., and Graham, D. I. (1981) Acceleration induced head injury in the monkey. I. The model, its mechanical and physiological correlates. *Acta Neuropathol. (Suppl.)* VII, 23–25.

Gennarelli, T. A., Thibault, L. E., Adams, J. H., Graham, D. I., Thompson, C. J., and Marcincin, R. P. (1982) Diffuse axonal injury and traumatic coma in the primate. *Ann. Neurol.* 12, 564–574.

Gennarelli, T. A., Segawa, H., Wald, U., Czernicki, Z., Marsh, K., and Thompson, C. (1982a) Physiologic response to angular acceleration of the head. In: *Head Injury: Basic and Clinical Aspects.* Grossman, R. G. and Gildenberg, P. L., Raven Press, New York, pp. 129–140.

Gerber A. M. and Moody R. A. (1972) Craniocerebral missile injuries in the monkey: an experimental physiological model. *J. Neurosurg.* 36, 43–49.

Goransson, A. M., Ingvar, D. H., and Kutyna, F. (1988) Remote cerebral effects on EEG in high-energy missile trauma. *J. Trauma* 28, S204–S205.

Graham, T. W., Williams, F. C., Jr., Harrington, T., and Spetzler, R. F. (1990) Civilian gunshot wound to the head: a prospective study. *Neurosurgery* 27, 696–700.

Hammon W. M. (1971) Analysis of 2187 consecutive penetrating wounds of the brain from Vietnam. *J. Neurosurg.* 34, 127–131.

Harmon, J. W. and Haluszka, M. (1983) Care of blast-injured casualties with gastrointestinal injuries. *Mil. Med.* 148, 586–588.

Harvey, E. N., Korr, I. M., Oster, G., et al. (1947) Secondary damage wounding due to pressure changes accompanying the passage of high-velocity missile. *Surgery* 21, 218–239.

Harvey, E. N. and McMillen, J. H. (1947) An experimental study of shock waves resulting from the impact of high velocity missiles on animal tissues. *J. Exp. Med.* 85, 321–328.

Harvey, E. N. (1948) Studies on wound ballistics. In: *Advances in Military Medicine.* Andres E. C., Bronk D. W., Carden G. A. Jr., et al. eds., Atlantic Little Brown, Boston. Vol. 1, pp. 191–205.

Hayes, R. L., Jenkins, L. W., and Lyeth, B. G. (1992) Neurotransmitter-mediated mechanisms of traumatic brain injury: acetylcholine and excitatory amino acids. *J. Neurotrauma,* 9(Suppl 1), pp. S-173–S-187.

Helling, T. S., McNabney, W. K., Whittaker, C. K., Schultz, C. C., and Watkins, M. (1992) The role of early surgical intervention in civilian gunshot wound to the head. *J. Trauma* 32, 398–400.

Horsley V. (1894) The destructive effects of small projectiles. *Nature* 50, 104–108.

Horsley, V. (1915) Remarks on gunshot wounds of the head. *Br. Med. J.* 1, 321–323.

Hubschmann O., Shapiro K., Baden M., and Shulman K. (1979) Craniocerebral gunshot injuries in civilian practice—prognostic criteria and surgical management: experience with 82 cases. *J. Trauma* 19, 6–12.

Huller, T. and Bazini, Y. (1970) Blast injuries of the chest and abdomen. *Arch. Surg.* 100, 24–30.

Ishige, N., Pitts, L. H., Berry, I., Carlson, S. G., Nishimura, M. C., Moseley, M. E., and Weinstein, P. R. (1987) The effect of hypoxia on traumatic head injury in rats: alterations in neurologic function, brain edema, and cerebral blood flow. *J. Cereb. Blood Flow Metab.* 7, 759–767.

Jefferson, J. (1919) The physiological pathology of gunshot wounds of the head. *Br. J. Surg.*, 7, 262–289.

Jagger, J. and Dietz, P. E. (1986) Death and injury by fire arms: who cares? *JAMA* 255, 3143–3144.

Kaufmann H. H., Makela M. G., Lee K. F., Haid R. W., and Gildenberg P. L. (1986) Gunshot wounds to the head: a perspective. *Neurosurgery* 18, 689–695.

Kaufman, H. H. (1993) Civilian gunshot wounds to the head. *Neurosurgery* 32, 962–963.

Kaufman, H. H. (1993a) Historical Vignette. Treatment of head injuries in the American Civil War. *J. Neurosurg.* 78, 838–845.

Kennedy, F., Gonzalez, P., Dang, C., Fleming, A., and Sterling-Scott, R. (1993) The Glasgow Coma Scale and prognosis in gunshot wounds to the brain. *J. Trauma,* 35, 75–77.

King, F. A. and Yarbrough, C. J. (1985) Medical and behavioral benefits from primate research. *Physiologist,* 28, 75–87.

King, F. A., Yarbrough, C. J., Anderson, D. C., Gordon, T. P., and Gould, K. G. (1988) *Primate Sci.* 240, 1475–1482.

Kirkpatrick J. B. and DiMaio V. (1978) Civilian gunshot wounds of the brain. *J. Neurosurg.* 49, 185–198.

Kontos, H. A., Wei, E. P., Raper, A. J., Rosenblum, W. I., Navari, R. M., and Patterson, J. L., Jr. (1978) Role of tissue hypoxia in local regulation of cerebral microcirculation. *Am. J. Physiol.* 234, H582–H591.

Kramer, S. P. and Horsley, V. (1894) On the effects produced on the circulation and respiration by gunshot injuries of the cerebral hemispheres. (Abstr.) *Proc. R. Soc.*, 55, 278.

Levett, J. M., Johns, L. M., Replogle, R. L., and Mullan, S. (1980) Cardiovascular effects of experimental cerebral missile injury in primates. *Surg. Neurol.* 13, 59–64.

Levy M. L., Rezai A., Masri L. S., Litofsky S. N., Giannotta S. L. et al. (1993) The significance of subarachnoid hemorrhage after penetrating craniocerebral injury: correlations with angiography and outcome in a civilian population. *Neurosurgery* 32, 532–540.

Lewelt, W., Jenkins, L. W., and Miller, J. D. (1980) Autoregulation of CBF after experimental fluid percussion injury of the brain. *J. Neurosurg.* 53, 500–511.

Lewelt, W., Jenkins, L. W., and Miller, J. D. (1982) Effects of experimental fluid percussion injury of the brain on cerebrovascular reactivity to hypoxia and hypercapnia. *J. Neurosurg.* 56, 332–338.

Lewin W. and Gibson R. M. (1956) Missile head wounds in the Korean campaign. A survey of British casualties. *Br. J. Surg.* 43, 628–632.

Lillard P. L. (1978) Five years experience with penetrating craniocerebral gunshot wounds. *Surg. Neurol.* 9, 79–83.

Marmarou, A., Abd-Elfattah Foda, M. A., Van Den Brink, W., Kita, J. C., and Demetriadou, K. (1994) A new model of diffuse brain injury in rats. Part I: Pathophysiology and biomechanics. *J. Neurosurg.* 80, 291–300.

McKowen, H. C., Farrell, J. B., and Carey, M. E. (1989) rCBF following an experimental missile wound to the brain. In: *Intracranial Pressure VII,* Hoff, J. T. and Betz, A. L. eds., Springer-Verlag, Berlin, pp. 620–622.

McMordie, W. R. and Barker, S. L. (1988) The financial trauma of head injury. *Brain Injury,* 21, 357–364.

Miller, J. D., Sweet, R. C., Narayan, R., and Becker, D. P. (1978) Early insults to the injured brain. *JAMA* 240, 439–442.

Miller, J. D. and Becker, D. P. (1982) Secondary insults to the injured brain. *J. R. Coll. Surg. Edinburgh* 27, 292–298.

Miner M. E., Ewing-Cobbs L., Kopaniky D. R., Cabrera J., and Kaufmann P. (1990) The results of treatment of gunshot wounds to the brain in children. *Neurosurgery* 26, 20–25.

Nagib, M. G., Rockswold, G. L., Sherman, R. S., and Lagaard, M. W. (1986) Civilian gunshot wound to the brain: prognosis and management. *Neurosurgery* 18, 533–537.

Pelligrino, D. A., Koenig, H. M., and Albrecht, R. F. (1993) Nitric oxide synthesis and regional cerebral blood flow responses to hypercapnia and hypoxia in the rat. *J. Cereb. Blood Flow Metab.* 13, 80–87.

Phillips C. D. (1992) Emergent radiologic evaluation of the gunshot wound victim. *Radiol. Clin. N. Am.* 30.2, 307–324.

Raimondi A. J. and Samuelson G. H. (1970) Craniocerebral gunshot wounds in civilian practice. *J. Neurosurg.* 32, 647–653.

Richmond, D. R., Damon, E. G., Fletcher, E. R., Bowen, I. G., and White, C. S. (1968) The relationship between selected blastwave parameters and the response of mammals exposed to air blast. *Ann. N.Y. Acad. Sci.* 152, 103–120.

Rish B. L., Dillon J. D., and Weiss G. H. (1983) Mortality following penetrating craniocerebral injuries. *J. Neurosurg.* 59, 775–780.

Robertson, C. S., Contant, C. F., Gokaslan, Z. L., Narayan, R. K., and Grossman, R. G. (1992). Cerebral blood flow, arteriovenous oxygen difference, and outcome in head injured patients. *J. Neurol. Neurosurg. Psych.* 55, 594–603.

Saunders, M. L., Miller, J. D., Stablein, D., and Allen, G. (1979) The effects of graded experimental trauma on cerebral blood flow and responsiveness to carbon dioxide. *J. Neurosurg.* 51, 18–26.

Selden, B., Goodman, J. M., Cordell, W., and Rodman, G. H., Jr. (1988) Outcome of self-inflicted gunshot wounds of the brain. *Ann. Emerg. Med.* 17, 247–253.

Shaffrey, M. E., Polin, R. S., Phillips, C. D., Germanson, T., Shaffrey, C. I., and Jane, J. A. (1992) Classification of civilian craniocerebral gunshot wounds: a multivariable analysis predictive of mortality. *J. Neurotrauma* 9(Suppl), S279–S285.

Shima, K. and Marmarou, A. (1991) Evaluation of brain-stem dysfunction following severe fluid-percusion head injury to the cat. *J. Neurosurg.* 74, 270–277.

Soblosky, J. S., Rogers, N. L., Adams, J. A., Farrell, J. B., Davidson, J. F., and Carey, M. E. (1992) Central and peripheral biogenic amine effects of brain missile wounding and increased intracranial pressure. *J. Neurosurg.* 76, 119–126.

Suneson, A., Hansson, A.-A., and Seeman, T. (1990) Pressure wave injuries to the nervous system caused by high-energy missiles extremity impact. Part II. Distant effects on the CNS—a light and electron microscopic study on pig. *J. Trauma* 30, 295–306.

Taha, J. M., Saba, M. I., and Brown, J. A. (1991) Missile injuries to the brain treated by simple wound closure: results of a protocol during the Lebanese conflict. *Neurosurgery* 29, 380–384.

Takeuchi, T., Horiuchi, J., Terada, N., Nagao, M., and Terajima, H. (1992) Effects of hypoxia, hyperoxia and hypercapnia on graded cerebral ischemic responses in rabbits. *Am. J. Physiol.* 263, H1839–H1846.

Tikka, S., Cederberg, A., and Rokkanen, P. (1982) Remote effects of pressure waves in missile trauma. The intra abdominal pressure changes in anesthetized pigs wounded in one thigh. *Acta Chir. Scand. Suppl.* 508, 167–173.

Torbati, D., Carey, M. E., Davidson, J. F., and Farrell, J. B. (1988) CBF autoregulation after penetrating brain injury in anesthetized cats. *FASEB J.* 2, A-1291.

Torbati, D., Carey, M. E., Davidson, J. F., and Farrell, J. B. (1988a). Regional cerebral blood flow responsiveness to hypoxia and hypercapnia before and after brain missile wounding in anesthetized paralyized cats. (Abstract), *Neurosci. Abs. Book,* 14, 621.

Torbati, D., Carey, M. E., Davidson, J. F. (1988b) Peripheral and circulatory correlates of brain missile wounding in anesthetized paralyized cats. (Abstract), *Physiologist* 31, A-149.

Torbati, D., Farrell, J. B., Davidson, J. F., and Carey, M. E. (1989) Effect of hemorrhagic hypotension on CBF and ICP after brain missile wounding in anesthetized paralyized cats. In: *Intracranial Pressure VII,* Hoff, J. T. and Betz, A. L. eds., Springer-Verlag, Berlin, pp. 657–659.

Torbati, D. and Carey, M. E. (1989) Effect of normobaric hyperoxia on regional cerebral blood flow before and after brain missile wounding in anesthetized cats. (Abstract), *Undersea Biomed. Res.* (Suppl.), 16, 78.

Torbati, D. and Carey, M. E. (1989a) Effect of hyperventilation on regional CBF following brain missile wounding in anesthetized cats. (Abstract), *J. Neurotrauma* 6, 205.

Torbati, D., Davidson, J. F., and Carey, M. E. (1990) Autoregulation of regional cerebral blood flow during increased intracranial hypertension associated with hemorrhagic hypotension in anesthetized cats. (Abstract), *FASEB J.* 4, A-1097.

Torbati, D., Carey, M. E., and Davidson, J. F. (1990a) Cerebrovascular responsiveness to hyperoxic-hypercapnia after brain missile wounding in pentobarbital anesthetized cats. (Abstract), *Neurosci. Abs. Book* 16, Part 1, p. 741.

Torbati, D., Church, D. F., Pryor, W. A., and Carey, M. E. (1990b) Ascorbyl radical and oxidative stress in the central nervous system trauma. (Abstract), *Free Radic. Biol. Med.* 9(Suppl), 43.

Torbati, D., Church, D. F., Carey, M. E., and Pryor, W. A. (1991) Generation of free radicals immediately after closed-head injury in rat brain. (Abstract), *FASEB J.* 5, A-891.

Torbati, D., Jacks A. F., Carey M. E., Davidson J. F., and Farrell J. B. (1992) Cerebral cardiovascular and respiratory variables after an experimental brain missile wound. *J. Neurotrauma* 9 (Suppl-I), S143–S155.

Wagner, E. M. and Traystman, R. J. (1986) Hydrostatic determination of cerebral perfusion. *Crit. Care Med.* 14, 484–490.

Webster, J. E. Gurdjian, E. S. (1943) Acute physiological effects of gunshot and other penetrating wounds of the brain. *J. Neurophysiol.* 6, 255–262.

West, C. G. (1980) A short history of management of penetrating missile injuries of the head. *Surg. Neurol.* 16, 145–149.

Index

INDEX

0-8493-8094-4/95/$0.00+$.50

D

I

N

O

P

R

S

T

U

V

W

X

Z